Cellulose and Other Natural Polymer Systems

Biogenesis, Structure, and Degradation

Cellulose and Other Natural Polymer Systems

Biogenesis, Structure, and Degradation

EDITED BY

R. MALCOLM BROWN, JR.

University of North Carolina
Chapel Hill, North Carolina

PLENUM PRESS • NEW YORK AND LONDON

Library of Congress Cataloging in Publication Data

Main entry under title:

Cellulose and other natural polymer systems

Includes bibliographical references and index.
1. Cellulose. 2. Polymers and polymerization. 3. Botanical chemistry. I. Brown, R. Malcolm (Richard Malcolm), 1939-

QK898.C42C44 581.19'2482 82-3796
ISBN-13: 978-1-4684-1118-8 e-ISBN-13: 978-1-4684-1116-4 AACR2
DOI: 10.1007/978-1-4684-1116-4

Softcover reprint of the hardcover 1st edition 1982
A Division of Plenum Publishing Corporation
233 Spring Street, New York, N.Y. 10013

Dr. Hans-Henning Heunert 1920–1980

The volume is dedicated to Dr. Hans-Henning Heunert who died on 24 November, 1980 in Göttingen, Federal Republic of Germany. Dr. Heunert was a member of the staff of the Institut fur den Wissenschaftlichen Film.

I became acquainted with Dr. Heunert in 1968–69 when I was in Germany on an NSF postdoctoral fellowship studying in the laboratory of Professor Peter Sitte at the University of Freiburg. I was introduced to the Institut fur den Wissenschaftlichen Film (IWF) through Dr. Heinz Reichenbach of Freiburg who had made several scientific films with the IWF. During the winter and spring months of 1969, I made several trips to Göttingen to produce scientific films on the role of the golgi apparatus in cell wall formation, algal viruses, and phototaxis in selected algae. I rapidly became aware of the tremendous talents of Dr. Heunert and his colleagues at the IWF. During this period, I learned many "tricks of the trade." One major "take-home lesson" was that scientific filming made it possible to record phenomena which often cannot be presented by other analytical means, such as the case with cellulose synthesis. No one had ever observed microfibrils being formed *in vivo*. This observation was made possible through the time-lapse cinematographic techniques developed and taught by Dr. Heunert.

In 1978, we achieved success in first filming cellulose microfibril assembly from the surface of the bacterium, *Acetobacter xylinum*. To Dr. Heunert, I owe a debt of gratitude for being able to accomplish this task. My colleagues and I adapted many of the excellent filming techniques devised by Dr. Heunert. Without his early help and encouragement, our attempts at filming cellulose microfibril assembly might never have materialized. Today, the modern era of video microscopy is with us. The future developments and achievements in the area will owe much to the expertise and wisdom of Dr. Heunert.

Dr. Heunert began his career in scientific photography when he and his wife Hedwig initiated a workshop for scientific photography. During these sessions, they taught many students the principles of the subject, and this led to a number of microcinematography and photographic projects for industry and university personnel. In 1950, Dr. Heunert joined the Leitz Company where he also served as an adviser for many difficult optical problems submitted by scientists from many fields.

In 1953, Dr. Heunert joined the IWF in Göttingen where he continued to provide his expertise in microscopy and scientific cinematography. Since 1956, Dr. Heunert was the head of the microcinematographic team at the IWF. During his association with the IWF, he demonstrated prolific advances in filmmaking and publications. He has published four books on microcinematography and microphotography and has published more than 40 scientific papers. He is the author of more then 260 scientific films. Of these, 146 are research films, and 114 films have been produced for education.

In memory of Dr. Heunert I express my sincere appreciation for his expertise, advice, friendliness, and above all, his remarkable impact upon the scientific community. He will be sorely missed. It is my hope that this dedication will serve as a reminder that science makes tremendous progress when it is possible to integrate diversified techniques and approaches for the solving of scientific problems.

R. Malcolm Brown, Jr.

Contributors

YEHOSHUA ALONI, Department of Biological Chemistry, Institute of Life Sciences, The Hebrew University of Jerusalem, 91904 Jerusalem, Israel

MOSHE BENZIMAN, Department of Biological Chemistry, Institute of Life Sciences, The Hebrew University of Jerusalem, 91904 Jerusalem, Israel

JOHN BLACKWELL, Department of Macromolecular Science, Case Western Reserve University, Cleveland, Ohio 44106

NICHOLAS C. CARPITA, Department of Botany and Plant Pathology, Purdue University, West Lafayette, Indiana 47907

KAY M. COOPER, Department of Botany, University of North Carolina at Chapel Hill, Chapel Hill, North Carolina 27514

H. MAELOR DAVIES, MSU-DOE Plant Research Laboratory, Michigan State University, East Lansing, Michigan 48824

DEBORAH P. DELMER, MSU-DOE Plant Research Laboratory, Michigan State University, East Lansing, Michigan 48824

W. GEORGE, North Carolina State University, Raleigh, North Carolina 27650

TOSHIYUKI GOTO, Department of Microbiology, Osaka Medical College, Takatsuki, Osaka 569, Japan

CANDACE H. HAIGLER, Department of Botany, University of North Carolina at Chapel Hill, Chapel Hill, North Carolina 27514

HIROSHI HARADA, Department of Wood Science and Technology, Faculty of Agriculture, Kyoto University, Kyoto 606, Japan

WERNER HERTH, Zellenlehre, Universität Heidelberg, D-6900 Heidelberg, Federal Republic of Germany

A. S. KLEIN, MSU-DOE Plant Research Laboratory, Michigan State University, East Lansing, Michigan 48824

VISHNU KUMAR, Department of Biology, McGill University, Montreal, Quebec, H3A 1B1 Canada

GORDON A. MACLACHLAN, Department of Biology, McGill University, Montreal, Quebec, H3A 1B1 Canada

MARIANNE MARX-FIGINI, Division Macromoleculas, Instituto de Investigaciones Fiscoquimicas Teoricas y Aplicadas, Universidad Nacional de la Plata, Conicet y CIC, La Plata, Argentina

YVES MEYER, Faculté des Sciences Exactes et Naturelles, Université de Perpignan, 66025 Perpignan, France

DAVID MONTEZINOS, Department of Biological Sciences, Wayne State University, Detroit, Michigan 48202

SUSETTE C. MUELLER, The Biological Laboratories, Harvard University, Cambridge, Massachusetts 02138

RALPH S. QUATRANO, Department of Botany and Plant Pathology, Oregon State University, Corvallis, Oregon 97331

GERALD J. F. RING, The Institute of Paper Chemistry, Appleton, Wisconsin, 54912

DWIGHT K. ROMANOVICZ, Department of Biology, West Georgia College, Carrollton, Georgia 30118

JOSÉ RUIZ-HERRERA, Instituto de Investigacion en Biologia Experimental, Facultad de Quimica, Universidad de Guanajuato, Apartado Postal N'187, Guanajuato, Gto. Mexico

EBERHARD SCHNEPF, Zellenlehre, Universität Heidelberg, D-6900 Heidelberg, Federal Republic of Germany

L. H. SAWYER, North Carolina State University, Raleigh, North Carolina 27650

DESH PAL S. VERMA, Department of Biology, McGill University, Montreal, Quebec, H3A 1B1 Canada

BRIGITTE VIAN, Laboratoire de Cytologie Expérimentale, Ecole Normale Supérieure, Université Pierre et Marie Curie, 75 231 Paris Cedex 05, France

ALAN R. WHITE, Department of Botany, University of North Carolina at Chapel Hill, Chapel Hill, North Carolina 27514

J. H. M. WILLISON, Biology Department, Dalhousie University, Halifax, Nova Scotia, Canada B3H 4J1

Introduction

The concept for a treatise covering selected natural polymer systems was initiated during a national meeting in cell biology in 1978. The challenge to the editor was to organize a book dealing principally but not exclusively with cellulose.

A brief background may help to provide the reader with information to better understand the reasons for the specific selections within this volume. During the past decade, we have witnessed significant changes in the sciences as well as the day-to-day life styles of our citizens. It will not be forgotten that during the early seventies, a significant change was to take many Americans by surprise. The oil embargo on The United States caused unexpected shortages of fuels. The long gasoline lines impressed in the minds of Americans that our energy-rich future with non-renewable resources is limited. The modelling of ecosystems, population growth, urban development, etc., have continued to raise our awareness that life on earth, including renewable resources, is indeed fragile. Contrary to popular belief, even wood and wood products are not limitless.

What is less understood to the general public and many scientists is that we know very little about how cellulose, the most abundant macromolecule on earth, is synthesized. In spite of the tremendous economic potential for wood and cellulose products, and in spite of the vast accumulation of scientific data on the composition and structure of wood and textiles, the mechanisms of cellulose biogenesis are much less understood than one might have suspected. Why is this? An old adage, "If it were an easy task, one would expect to have already read about it", is certainly true with respect to the field of cellulose biogenesis. Repeated attempts to assemble cellulose microfibrils *in vitro* have been made without success, yet a close relative, chitin, has been synthesized in microfibrillar form *in vitro*.

In spite of the relative paucity of information on cellulose microfibril biogenesis, we have begun, at last, to make progress in the area. The time is now ripe for a treatise dealing with the subject. New technological discoveries and approaches have provided us with a better understanding of how the cellulose microfibril is assembled. In this volume, some of the most recent and innovative approaches are presented dealing with the biochemistry, macromolecular structure, and cytology of cellulose and its assembly and degradation.

Cellulose, chitin, lignin, and other natural polymer systems are so essential to our daily lives that it is of great importance that we strive to better understand how these products are assembled by living systems. Because these are complex, interdependent systems, the skills and efforts of biochemists, cell biologists, molecular geneticists, and physicists are needed to help solve the many extremely complex problems in the field.

We must not forget the exciting new perspectives in the biotechnology field. With the expectations of gene splicing, it is conceivable that we will come much closer to an understanding of the genetic control for natural polymer synthesis. This means that many opportunities for technological exploitation of natural polymer synthesis and/or degradation will become available. In order to be prepared to meet these challenges, we must be in a position to understand the basics of the processes. This book was not intended to be "all inclusive" with respect to natural polymer systems, but is an account of the latest developments in this exciting field. The editor has omitted many important natural polymers, largely because it would have been impossible to have included all of these topics in a volume of the planned size.

What the reader hopefully will find in this book is a new approach. . .one which will serve as a stimulus to those in basic research as well as to those in industrial research and development. The strength of this volume depends primarily on the imagination and expertise of the contributors, all of whom have made and are continuing to make strides in their respective fields of study.

R. Malcolm Brown, Jr.
Chapel Hill, North Carolina

Contents

I. BIOGENESIS

Chapter 1

A Cytological Model of Cellulose Biogenesis in the Alga *Oocystis apiculata*

David Montezinos

1. Introduction 3
2. Microfibril Assembly 4
3. Microfibril Orientation 13
4. Discussion 16
5. Concluding Remarks 19
References 20

Chapter 2

Organized Microfibril Assembly in Higher Plant Cells

Brigitte Vian

1. Introduction 23
2. Evolution of Ideas about the Organization of Cell Walls 24
3. Positioning and Ordering of Wall Subunits 32
4. Concluding Remarks 40
References 41

Chapter 3

Cell-Wall Formation in *Fucus* Zygotes: A Model System to Study the Assembly and Localization of Wall Polymers

Ralph S. Quatrano

1. Introduction 45
2. Experimental System 46
3. Experimental Framework and Questions 50
References 57

Chapter 4

Cell-Wall Regeneration by Protoplasts Isolated from Higher Plants

J. H. M. WILLISON and A. S. KLEIN

1. Introduction 61
2. Studies of the Regenerated Wall: Perspective 63
3. Cellulose Microfibrils 64
4. Cell-Wall Constituents Other Than Cellulose 78
References 82

Chapter 5

Cellulose-Microfibril Assembly and Orientation in Higher Plant Cells with Particular Reference to Seedlings of *Zea mays*

SUSETTE C. MUELLER

1. Introduction 87
2. Factors That Affect the Visualization of Terminal Complexes 90
3. Oriented Microfibril Deposition 91
4. Unidirectional Microfibril Deposition 91
5. Experimental Procedures 95
6. Influence of Cytoplasmic Structures on Microfibril Deposition 97
7. Conclusions 102
References 102

Chapter 6

Microfibril-Tip Growth and the Development of Pattern in Cell Walls

J. H. M. WILLISON

1. Introduction 105
2. *Glaucocystis nostochinearum* 106
3. Cotton Fibers 116
4. Conclusions 123
References 123

Chapter 7

The Role of the Golgi Apparatus in the Biosynthesis of Natural Polymer Systems with Particular Reference to Cellulose

DWIGHT K. ROMANOVICZ

1. Introduction 127
2. *Pleurochrysis*: A Model System for Golgi-Derived Cellulose 129
3. Relationship of *Pleurochrysis* to Other Systems 140
4. Concluding Remarks 143
References 143

Chapter 8

Interaction of Cell-Wall Formation and Cell Division in Higher Plant Cells

Yves Meyer and Werner Herth

1. Introduction 149
2. Experimental Data with Tobacco Mesophyll Protoplasts 151
3. Discussion 159
4. Conclusions 162
References 163

Chapter 9

Callose-Deposit Formation in Radish Root Hairs

Kay M. Cooper

1. Introduction 167
2. UDP-Glucose-Incubated Root Hairs 169
3. Discussion 177
4. Summary 181
References 182

Chapter 10

Chitin-Fibril Formation in Algae

Werner Herth and Eberhard Schnepf

1. Introduction 185
2. β-Chitin-Fibril Formation of the Centric Diatoms *Cyclotella* and *Thalassiosira* 186
3. Lorica Microfibrils of *Poterioochromonas* 194
4. Conclusions 204
References 205

Chapter 11

Synthesis of Chitin Microfibrils *in Vitro*

José Ruiz-Herrera

1. Introduction 207
2. General Properties of Chitin Synthetase 208
3. Effect of Nucleosides and Nucleotides 209
4. Polyoxins 210
5. Activation and Inactivation of Chitin Synthetase 210
6. Autogenous Inhibitors of Chitin Synthetase 211
7. Biosynthesis of Chitin Microfibrils 212
8. Chitosomes and Chitin Biosynthesis 213
9. Dissociated Forms of Chitin Synthetase 217

10. On the Mechanism of Chitin-Chain Formation 218
11. Perspectives and Future Developments............................ 219
References.. 220

Chapter 12
Cellulose Synthesis in Detached Cotton Fibers
Nicholas C. Carpita

1. Introduction.. 225
2. Criteria for Cell-Free Synthesis of Cellulose 227
3. Why Is Cellulose Synthesis Lost on Homogenization of Plant Tissues?.. 227
4. Protection of Cellulose Synthesis in Detached Cotton Fibers.......... 228
5. Role of Polyethylene Glycol in the Protection of Cellulose Synthesis.. 233
6. Does Cellulose Synthesis Require a Membrane Potential? 236
7. Where Do We Go from Here?.. 238
References.. 238

Chapter 13
The Control of Molecular Weight and Molecular-Weight Distribution in the Biogenesis of Cellulose
Marianne Marx-Figini

1. Introduction to the Concepts of Molecular Weight and Molecular-Weight Distribution and Their Relationships to the Polymerization Mechanism.. 243
2. Molecular Weight and Molecular-Weight Distribution in the Biogenesis of Cellulose 252
3. Discussion 265
References.. 268

Chapter 14
Biogenesis of Cellulose I Microfibrils Occurs by Cell-Directed Self-Assembly in *Acetobacter xylinum*
Candace H. Haigler and Moshe Benziman

1. Introduction.. 273
2. *Acetobacter xylinum* as a Model System for Cellulose Biogenesis...... 275
3. Effect of Calcofluor White ST on Cellulose Biogenesis............... 275
4. Effect of Carboxymethylcellulose on Cellulose Biogenesis............ 285
5. Cellulose Microfibrils Are Synthesized by Cell-Directed Self-Assembly in *Acetobacter xylinum* 287
6. Does Cell-Directed Crystallization Occur in Other Organisms? 293
7. Theoretical Implications of Cell-Directed Self-Assembly 294
References.. 295

Chapter 15

A Study of the Polymerization Kinetics of Bacterial Cellulose through Gel-Permeation Chromatography

GERARD J. F. RING

1. Introduction 299
2. Bacterial-Cellulose Production 301
3. Theoretical and Experimental Objectives 304
4. Poisson Polymerization Function 305
5. Experimental Procedures 308
6. Results and Discussion 311
7. Conclusions 322
8. Nomenclature 322

References 324

Chapter 16

Does β-Glucan Synthesis Need a Primer?

GORDON A. MACLACHLAN

1. Introduction 327
2. Results 328
3. Discussion and Conclusions 335

References 337

Chapter 17

Intermediates of Cellulose Synthesis in *Acetobacter*

YEHOSHUA ALONI and MOSHE BENZIMAN

1. Introduction 341
2. Studies with Whole Cells 342
3. Studies with Cell-Free Preparations 349
4. Inhibitors of Cellulose Synthesis 354
5. General Assessment 356

References 359

Chapter 18

Protein Glycosylation in Higher Plants: Recent Developments

H. MAELOR DAVIES and DEBORAH P. DELMER

1. Introduction 363
2. *In Vitro* Glycosylation Systems from Higher Plants 364
3. The Evidence for Core Glycosylation *in Vitro* 366
4. Glycosylation by Pathways Other Than Oligosaccharide Transfer from Lipid-Oligosaccharide 373
5. Glycoprotein Products of the *in Vitro* Systems 373
6. Subcellular Localization 374

7. Other Aspects 376
References 377

II. STRUCTURE

Chapter 19

The Structure of Cellulose Microfibrils in *Valonia*

HIROSHI HARADA and TOSHIYUKI GOTO

1. Introduction 383
2. Dimensions of Cellulose Microfibrils 383
3. Cross-Sectional View of Cellulose Microfibrils 385
4. Internal Structure of Cellulose Microfibrils 394
References 399

Chapter 20

The Macromolecular Organization of Cellulose and Chitin

JOHN BLACKWELL

1. Introduction 403
2. X-Ray Diffraction of Fibrous Polymers 405
3. Polymorphic Structures of Cellulose 408
4. X-Ray Studies of the Structures of Cellulose 408
5. Chitin 417
References 427

Chapter 21

Comparisons between Synthetic and Natural Microfiber Systems

L. H. SAWYER and W. GEORGE

1. Introduction: Fibers, Fibrils, and the Ultimate Fibril 429
2. Synthetic Fiber-Forming Technologies: Common Characteristics 431
3. Specific Fiber-Forming Processes 435
4. Direct Fibril-Forming Processes 439
5. Fibrils in Natural Fibers—Comparison and Integration with Synthetic Systems 442
6. Fibril Formation: Anisotropic Fluids and Solidification 447
7. Discussion: Nonequilibrium Thermodynamics 451
References 453

III. DEGRADATION

Chapter 22

β-Glucanases in Higher Plants: Localization, Potential Functions, and Regulation

DESH PAL S. VERMA, VISHNU KUMAR, and GORDON A. MACLACHLAN

1. Introduction 459

2. β1,3-Glucanases 461
3. β1,4-Glucanase (Cellulase) 465
4. Regulation of β-Glucanases 477
References 480

Chapter 23
Visualization of Cellulases and Cellulose Degradation
ALAN R. WHITE

1. Introduction 489
2. High-Resolution Electron Microscopy of Cellulase Components 497
3. Cellulose Degradation by the Complete Cellulase Enzyme System 498
4. Cellulose Degradation by Purified and Reconstituted Cellulase Components 502
5. Conclusions and Future Perspectives 505
References 506

Index 511

COLOR PLATES

Plate I

Figure 1. *Oocystis apiculata* observed with polarization microscopy (first-order red compensator inserted). Note wall birefringence. Photograph by Alan White and Candace Haigler.

Figure 2. *Oocystis apiculata*. Same as Fig. 1, but cell rotated 90°. Photograph by Alan White and Candace Haigler.

Figure 3. *Pleurochrysis scherfellii* observed with polarization microscopy. Note birefringence of cell-wall and parietal chloroplasts. Photograph by R. Malcolm Brown, Jr.

Figure 4. Germinating pollen tube of *Nicotiana* incubated in UDP-glusoe and stained with aniline blue. Note callose deposit at tip of pollen tube that has ruptured. Photograph by Kay Cooper.

Figure 5. Germinating pollen tube of *Nicotiana* incubated in the absence of UDP-glucose and stained with aniline blue. Note the absence of callose staining. Photograph by Kay Cooper.

Plate II

Figure 6. *Acetobacter xylinum* and ribbons of cellulose microfibrils detected with Calcofluor White after synthesis. Photography by Candace Haigler and R. Malcolm Brown, Jr. Reproduced by permission of the American Association for the Advancement of Science (from *Science* **210**:4472, November, 1980).

Figure 7. *Acetobacter xylinum* and altered cellulose made in the presence of Calcofluor White. Note distinct band material emerging from the cell surface. This cellulose has no detectable crystallinity in the wet state. Photograph by Candace Haigler and R. Malcolm Brown, Jr.

Plate III

Figure 8. *Oocystis apiculata* grown in the presence of Calcofluor White. Note intensive staining of daughter-cell walls that were assembled in the presence of Calcofluor, while the mother walls synthesized before incubation in Calcofluor fluoresce less. Photograph by Candace Haigler, Alan White, Kay Cooper, and R. Malcolm Brown, Jr.

Figure 9. *Oocystis apiculata*. Same view as Fig. 8, but observed with polarization microscopy. Photograph by Candace Haigler, Alan White, Kay Cooper, and R. Malcolm Brown, Jr.

Figure 10. Young ovary of cotton (*Gossypium*) approximately 15 days postanthesis. Photograph by R. Malcolm Brown, Jr.

Figure 11. *Glaucocystis nostocinearum*. Two young daughter cells. Note nucleus and prominent chloroplasts. Photograph by R. Malcolm Brown, Jr.

Plate IV

Figure 12. *Poterioochromonas*, a chrysophycean flagellate. These cells were stained with Calcofluor White after they synthesized chitin. Note the prominent stalks or loricae. Photograph by Werner Herth.

Figure 13. *Poterioochromonas* cells that synthesized chitin in the presence of Congo Red (0.02 mg/ml), and then stained with Calcofluor. Note amorphous chitin walls with altered crystallinity. Photograph by Werner Herth.

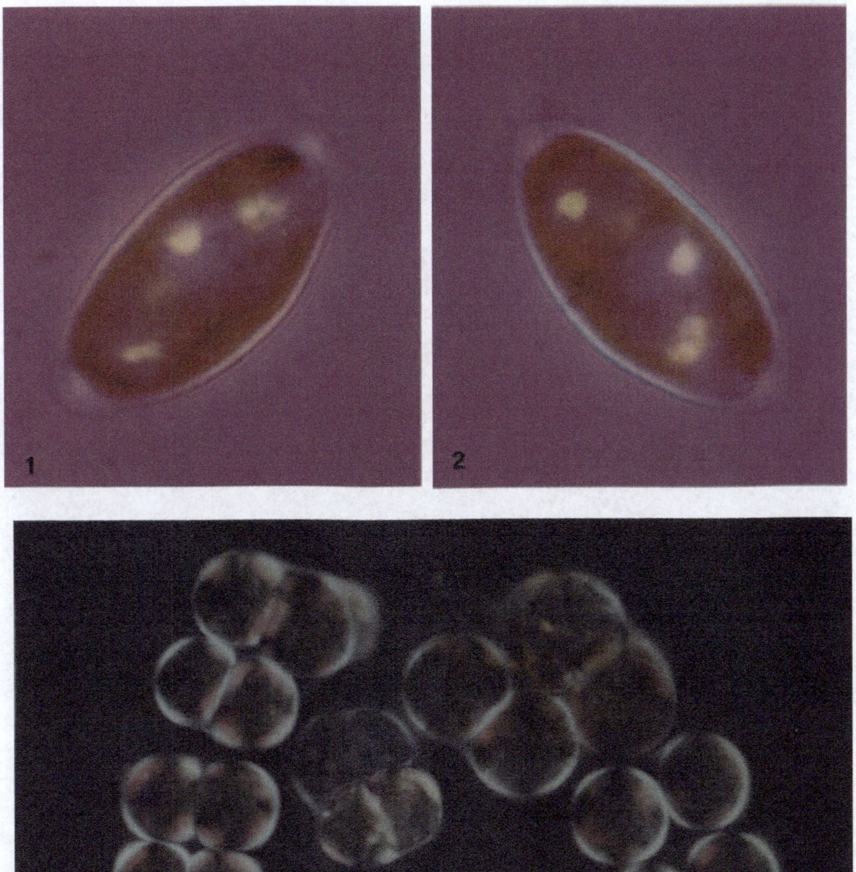

PLATE I

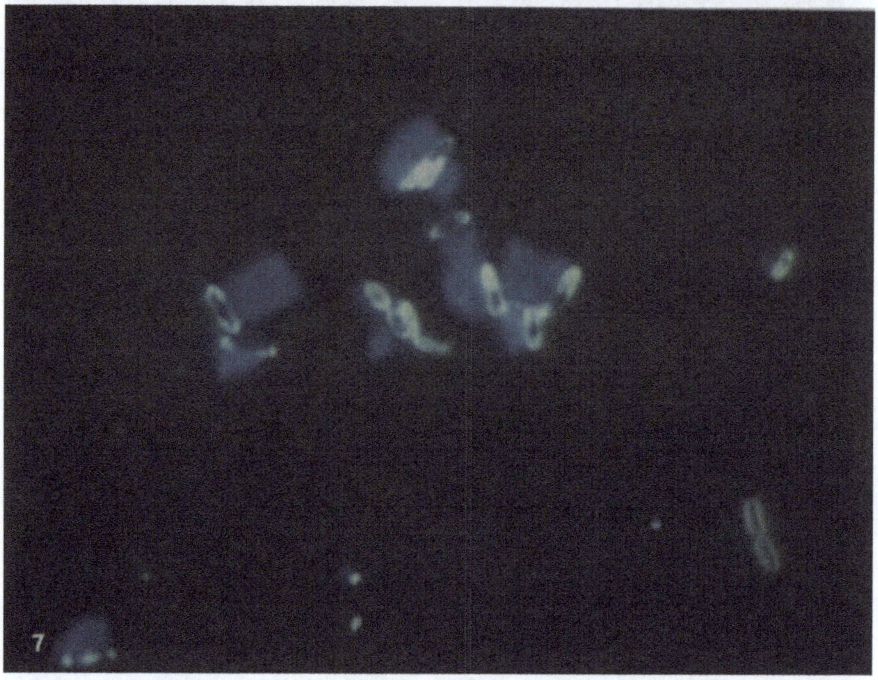

PLATE II

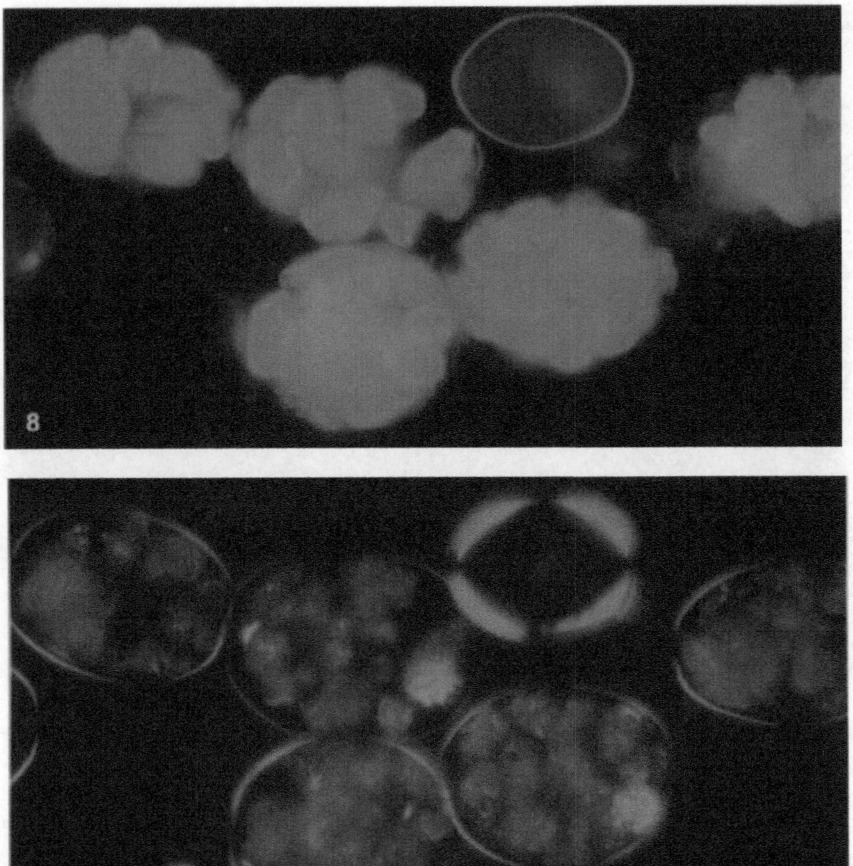

Plate III

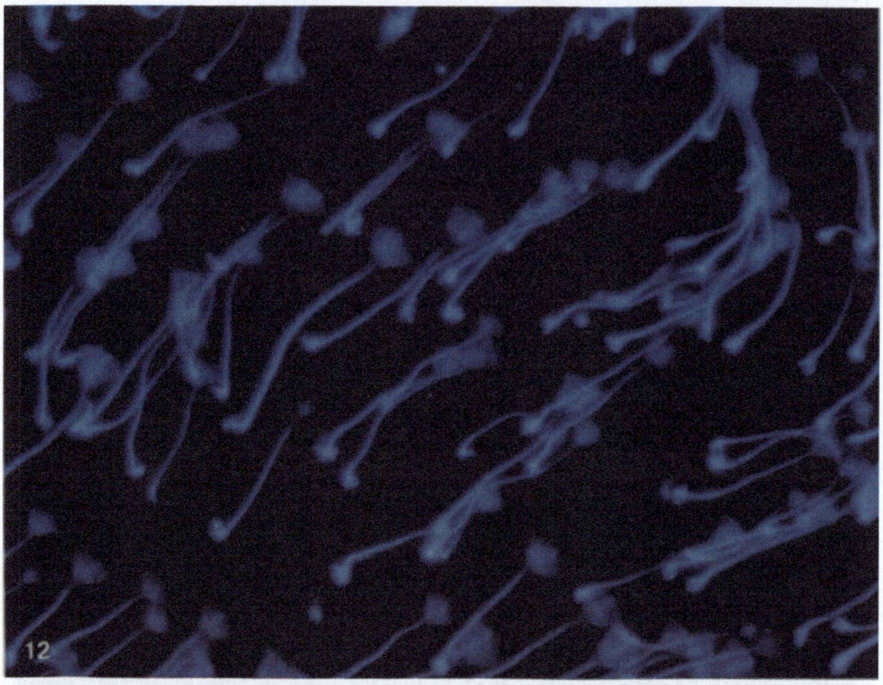

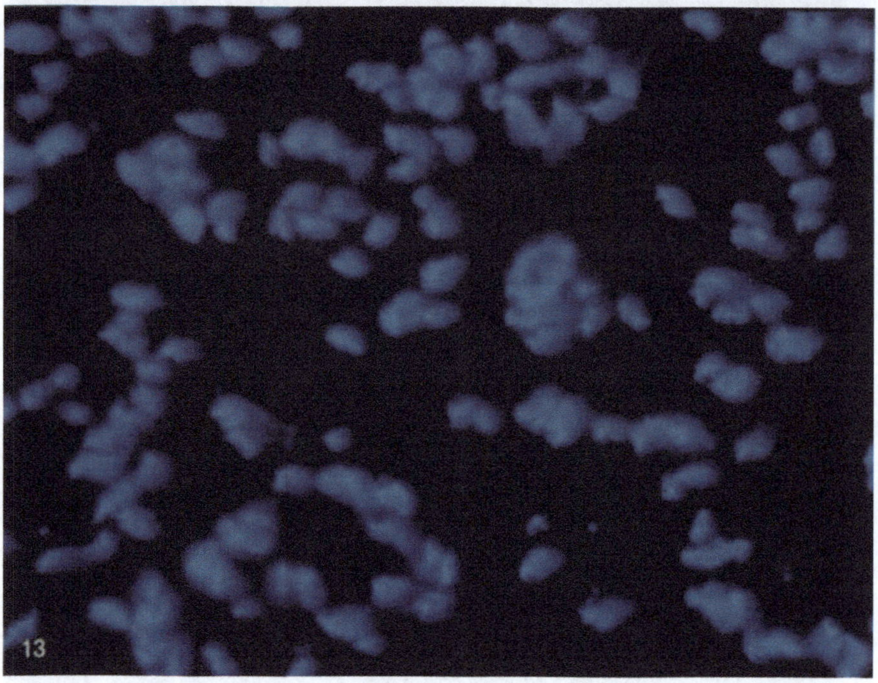

Plate IV

I
BIOGENESIS

1

A Cytological Model of Cellulose Biogenesis in the Alga *Oocystis apiculata*

DAVID MONTEZINOS

1. Introduction

Cytological observations on plant-cell walls began with the earliest use of light microscopes, and the study of cell-wall structure was one of the earliest biological applications of electron microscopy. Nonetheless, progress in understanding the cytology of cell walls and especially the biogenesis of cellulosic microfibrils is slow at best. Where are microfibrils assembled, and what structures are involved in assembly? How are microfibrils oriented, what structures are involved in orientation, and how do orienting and assembling structures interact? How does the cell control an extracellular assembly process? These are preeminent questions concerning the cytology of cell-wall formation.

Beginning in the 1930s, and especially since the 1950s, work in the laboratory of R. D. Preston was directed toward the use as model systems of algae with large cellulosic microfibrils that are arranged in distinctive patterns to form the framework of the cell wall. Investigating *Valonia, Cladophora, Chaetomorpha, Glaucocystis,* and *Oocystis* with polarization microscopy, X-ray diffraction analysis, chemical extraction, and electron microscopy, Preston and co-workers confirmed both the validity of using these organisms as experimental models for growth and development and value of their cell walls for testing hypotheses concerning wall biogenesis (see Preston, 1974).

During the past two decades, the model-systems approach to cellulose biogenesis was embraced by R. M. Brown, Jr., and co-workers in studies on the alga

DAVID MONTEZINOS • Department of Biological Sciences, Wayne State University, Detroit, Michigan, 48202.

Pleurochrysis (see Brown and Romanovicz, 1976). Within the last ten years, this approach was expanded in his laboratory to include studies on *Acetobacter*, *Glaucocystis*, cotton fibers, and corn hypocotyls, (Brown and Willison, 1977; Brown, 1978) as well as my own efforts using *Oocystis apiculata*.

Using *Oocystis* as a model system provides a unique set of advantageous conditions. Unlike higher plant systems, the alga is unicellular. The cells can be mass cultured on defined media, and the cell cycles within a culture can be synchronized to provide numerous examples of each stage in the biogenesis of the cell wall. At the same time, *Oocystis*, as opposed to other algae that have been used as model systems, is a Chlorophyte, taxonomically and morphologically more closely related to higher plants.

2. *Microfibril Assembly*

2.1. *Life Cycle of Oocystis*

Freshly inoculated cultures of *Oocystis* on solidified modified Kantz medium (Kantz and Bold, 1969) (Table I) held in the dark for 2 days and then exposed to continuous illumination at 21°C will undergo one complete cell cycle in 8–10 days. The synchrony of cells cultured according to this regime is sufficient to allow the preparation of cells for electron microscopy with nearly all the cells in approximately the same stage of cell-wall biogenesis.

Cell division typically occurs during the 4th day following inoculation. Two to four mitotic events occur, followed by cytokinesis, resulting in 4–16 autospores within each mother-cell wall. The autospores have no cell walls initially but form a thin fibrous initial wall layer within 1 day of division. By the 7th day after inoculation, large cellulosic microfibrils are produced to form the

TABLE I
Modified Kantz Medium, Ingredients and Procedure

5 ml of each of the following macrostocks is added per liter of medium:		
$NaNO_3$	75g/l	
$MgSO_4 \cdot 7H_2O$	7.5g/l	
NaCl	2.5g/l	
$CaCl_2 \cdot 2H_2O$	3.32g/l	
K_2HPO_4	7.5g/l	refrigerate
Tris buffer (pH 7.6)	50g/l	refrigerate
1 ml of each of the following microstocks is added per liter of medium:		
H_3BO_3	11.42g/l double distilled water	
$FeSO_4 \cdot 7H_2O$	4.98g/l acidified water (0.1% H_2SO_4)	
EDTA	50g EDTA + 31g KOH/l distilled water	
$H\text{-}H_5$	8.82 $ZnSO_4 \cdot 7H_2O$ + 1.44g $MnCl_2 \cdot 4H_2O$ + 0.71g MoO_3 + 1.57g $CuSO_4 \cdot 5H_2O$ + 0.49g $Co(NO_3)_2 \cdot 6H_2O$/l acidified water	
Vitamins	0.7mg vitamin B_{12} + 8.6mg Thiamine HCl + 6.1mg Biotin/100ml distilled water	
To solidify, use 15g Difco Bacto Agar per liter of medium.		

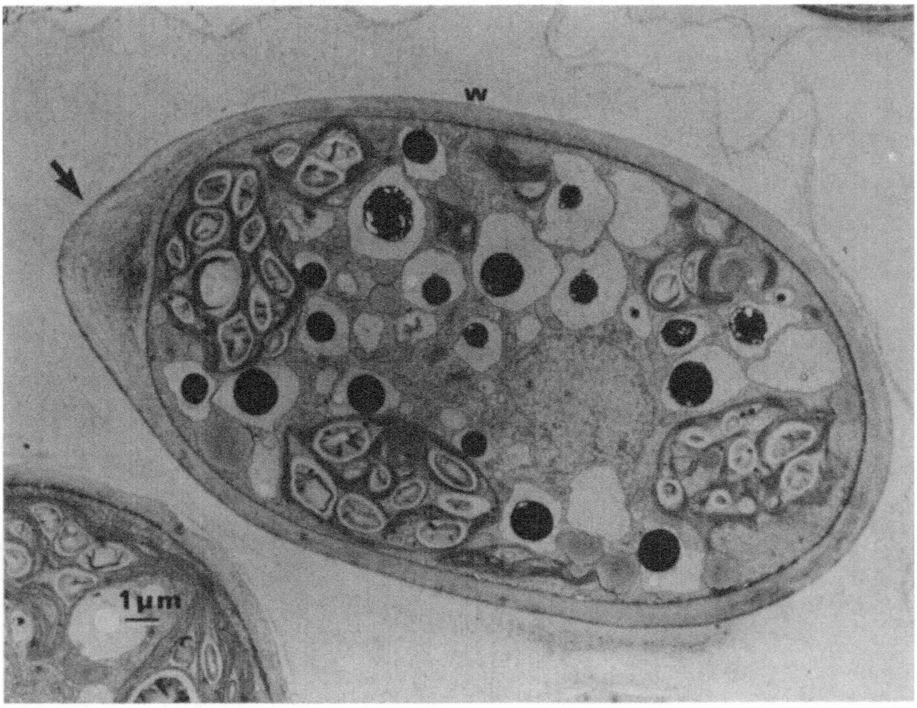

FIGURE 1. Section through a mature cell of *Oocystis* revealing the thick cellulosic cell wall (w) and one of the two polar thickenings (→) in the wall of this alga.

thick well-organized cell wall characteristic of this alga (Fig. 1). Subsequently, the cells enlarge to 2–4 times their original volume, rupturing the old mother-cell wall (Montezinos and Brown, 1976).

2.2. *Structure of the Cell Wall of Oocystis*

The well-organized cell wall of *Oocystis* is composed of cellulosic microfibrils with rectangular cross sections 5.5 × 7.1 nm in area. The wall is arranged in layers composed of single rows of parallel microfibrils (Fig. 2). The cell is ellipsoidal and the microfibrils spiral from pole to pole, with those of each layer perpendicular to microfibrils in adjacent layers at the equator of the cell. The helices tighten at the poles, but the layers separate slightly to form polar thickenings. All the microfibrils are coated and cross-linked by noncellulosic wall components.

This complex organization of the cell wall of *Oocystis* is valuable in terms of the use of this system as a model for two reasons: First, it provides a recognizable pattern with which to compare the organization of structures suspected of being involved in cellulose biogenesis. Second, it appears to be the product of an elegant orientation mechanism.

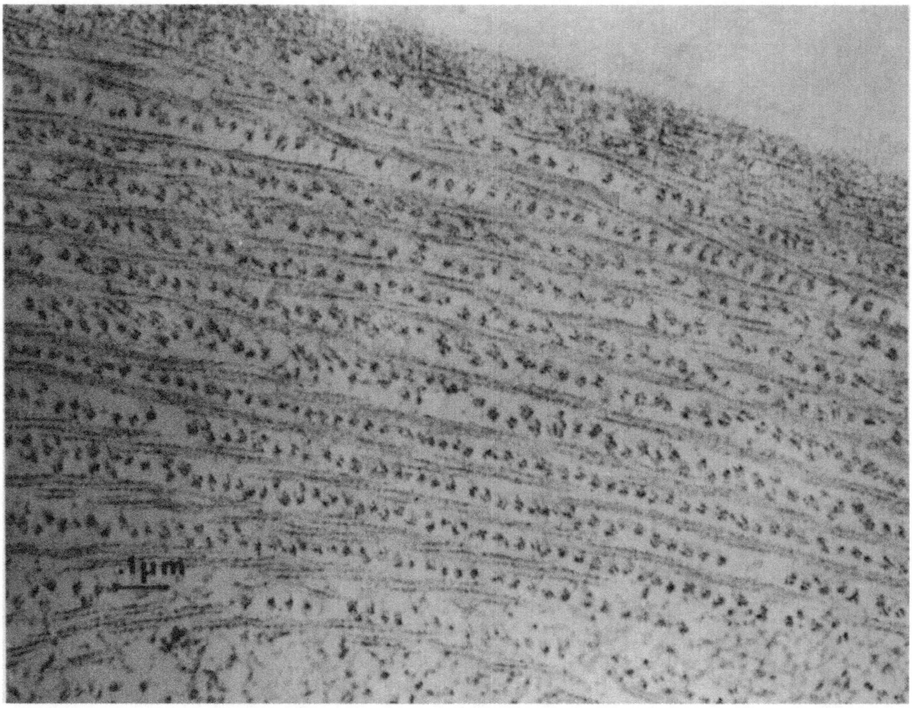

FIGURE 2. Section through the cell wall of *Oocystis*. Note the rows of parallel microfibrils forming the layers that constitute the framework of the cell wall.

2.3. *Plasma Membrane of Oocystis as Revealed by Freeze-Etching Studies*

Robinson and Preston (1972) conducted a freeze-etching study of the plasma membrane of *Oocystis*, and they considered this membrane a likely site for the synthesis and assembly of microfibrils. However, this investigation preceded the widespread acceptance of two major developments in the fields of freeze-etching and membrane research. First, the standardization of freeze-etching interpretation formalized by Branton *et al.* (1975) stipulates that the fracturing process reveals the hydrophobic inner surfaces of the lipid monolayers of membranes (for a description of the surfaces of membranes revealed by freeze-etching, see Fig. 3). On this basis, structures formerly believed to be on the outside of the plasma membrane would actually be exposed on the hydrophobic side of the inner monolayer of the membrane (the PF face). Second, an accumulation of evidence led Singer and Nicolson (1972) to the formulation of the fluid mosaic model of membrane structure, in which proteins are not rigidly affixed to the surfaces of lipid bilayers, but rather embedded in, and free to move within, the plane of the lipid bilayers of membranes. With these concepts as background, it has been possible to identify the EF and PF faces of the plasma membrane and to postulate interpretations of the particle patterns found in each of them in cells of *Oocystis* undergoing cell-wall biogenesis.

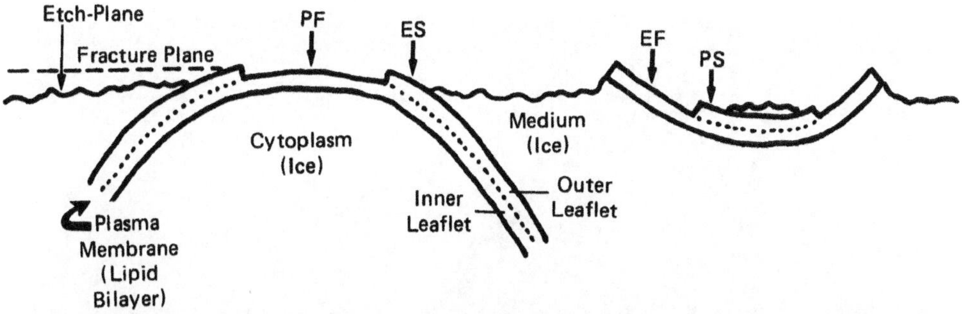

FIGURE 3. Freeze-etch terminology. The process of freeze-etching involves the fracturing of membrane lipid bilayers to expose the interior of the membrane (fracture faces) and etching of the ice surface also exposed by fracturing to expose the surfaces of membranes. Thus, the views of the plasma membrane possibly exposed by freeze-etching are the cytoplasmic surface (PS), the fracture face of the lipid monolayer adjacent to the cytoplasm (PF), the fracture face of the lipid monolayer adjacent to the exterior of the cell (EF), and the exterior surface of the membrane (ES). Based on Branton *et al.* (1975).

When cells engaged in secondary wall biosynthesis are freeze-etched, and the EF face of the plasma membrane is exposed, the most notable features of this fracture face are ridges that are the impressions of the innermost microfibrils of the secondary wall. These ridges allow as many as three layers of microfibrils to be discerned, although the layers are not actually exposed. Occasionally, microfibrils are seen to have been torn back through the membrane during the fracturing process, leaving a rip in the outer leaflet, often with a piece of microfibril protruding from the surface at the end of the rip (Fig. 4). These tears in the membrane are always continuous with ridges, indicating that microfibrils probably underlie all such ridges.

Also notable on the EF face of the plasma membrane is the presence of linear arrays of particles that are often associated with the termini of ridges (Fig. 5). Such terminal complexes have three rows of 30 or more particles, each about 7.0 nm in diameter. These complexes are all oriented parallel to one another on any given membrane; when they are associated with short ridges, they are seen to be associated with only one end of each ridge; and when they are found not associated with ridges, they are associated with each other in pairs (Fig. 6).

The morphology of these terminal complexes and their orientation and association with microfibril ridges provide compelling evidence that the structures being visualized are indeed those that assemble microfibrils in *Oocystis*. Furthermore, an assembly sequence can be reconstructed in which paired complexes assemble one microfibril each, moving apart in opposite directions in the process, and leaving the two new parallel microfibrils behind. By studying polar views of the EF face of the plasma membrane, it is possible to conclude that each complex moves in a spiral to the pole of the cell, then reverses direction and produces another microfibril parallel to the one it just produced, so that after several passes up and down the cell, the complexes have assembled an entire layer composed of nonoverlapping parallel microfibrils. This is followed by a reorientation of the complex pairs to produce the opposing helix of the next wall layer and a repetition of the assembly process.

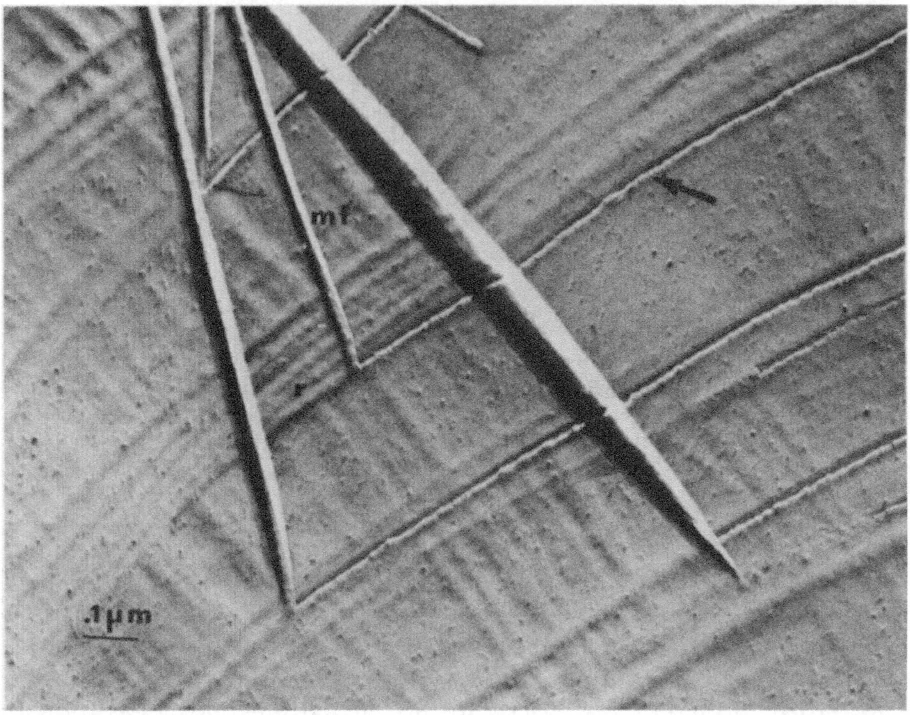

FIGURE 4. Rips (←) in the EF face of the plasma membrane. These rips are always continuous with ridges (r) and are produced when microfibrils (mf) are pulled up through the membrane during the fracturing process.

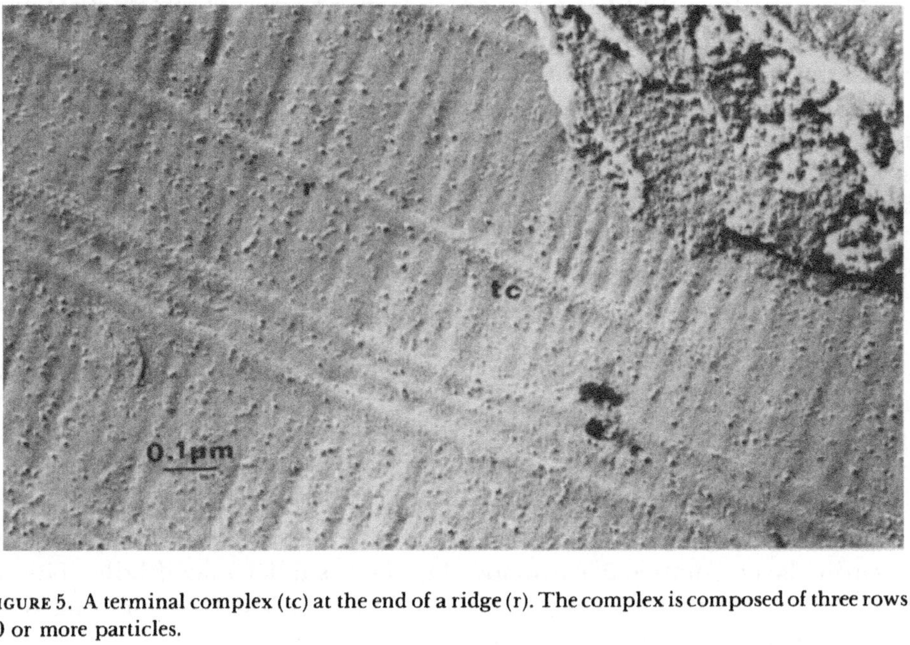

FIGURE 5. A terminal complex (tc) at the end of a ridge (r). The complex is composed of three rows of 30 or more particles.

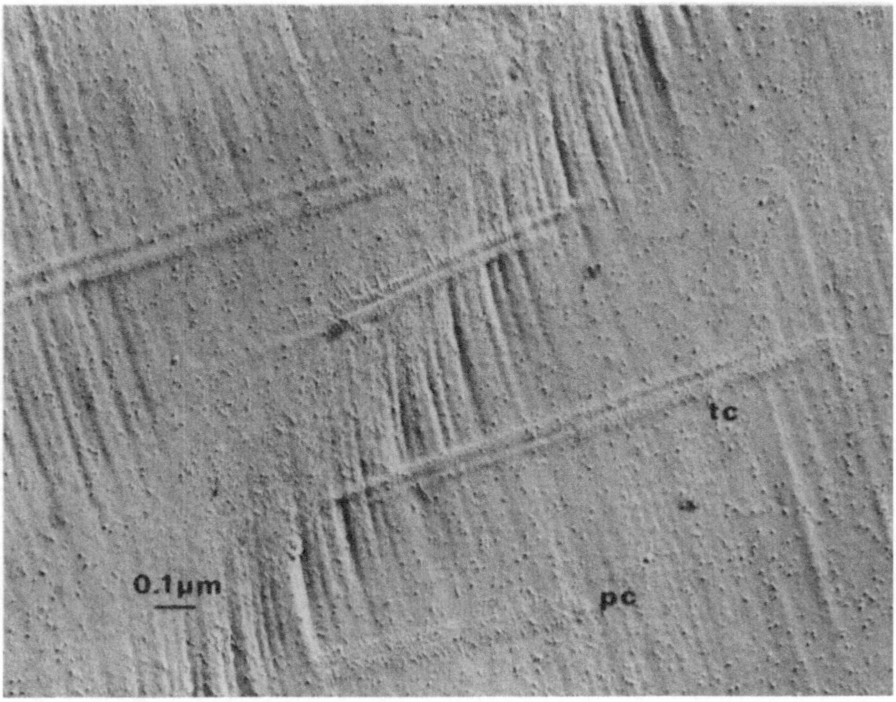

FIGURE 6. Terminal complexes (tc). These complexes associate with opposite ends of short parallel ridges or, when not associated with ridges, occur in pairs (pc).

2.4. *Biogenesis of the Cell Wall of Oocystis as Observed in Fixed, Sectioned Specimens*

Results from fixed and sectioned preparations of *Oocystis* provide some additional support for the hypothesis of microfibril assembly derived from freeze-etching data. The innermost microfibrils in cells undergoing cell-wall formation can be seen within the periplasm that underlies the bulk of the cell wall, and in fortuitous sections, microfibrils appressed to the plasma membrane can be found to end in structures that appear to be terminal complexes. (Fig. 7).

Perhaps the most important data from studying fixed material are those that provide evidence supporting the role of the Golgi apparatus in producing the periplasm and the role of the periplasm in coating and cross-linking microfibrils (Montezinos and Brown, 1976).

It has in fact been generally accepted that noncellulosic cell-wall components are synthesized intracellularly, packaged in the Golgi, and then transported to the cell surface via exocytosis of Golgi vesicles (Morré and Mollenhauer, 1974). It is significant that the morphology of the Golgi apparatus and its associated vesicles in *Oocystis* forms just the type of secretory mechanism required for exocytosis. The existence of such a pathway is supported by the staining pattern of membranes treated with phosphotungstic acid–chromic acid (Montezinos and Brown, 1976). Only the secretory face of the Golgi apparatus,

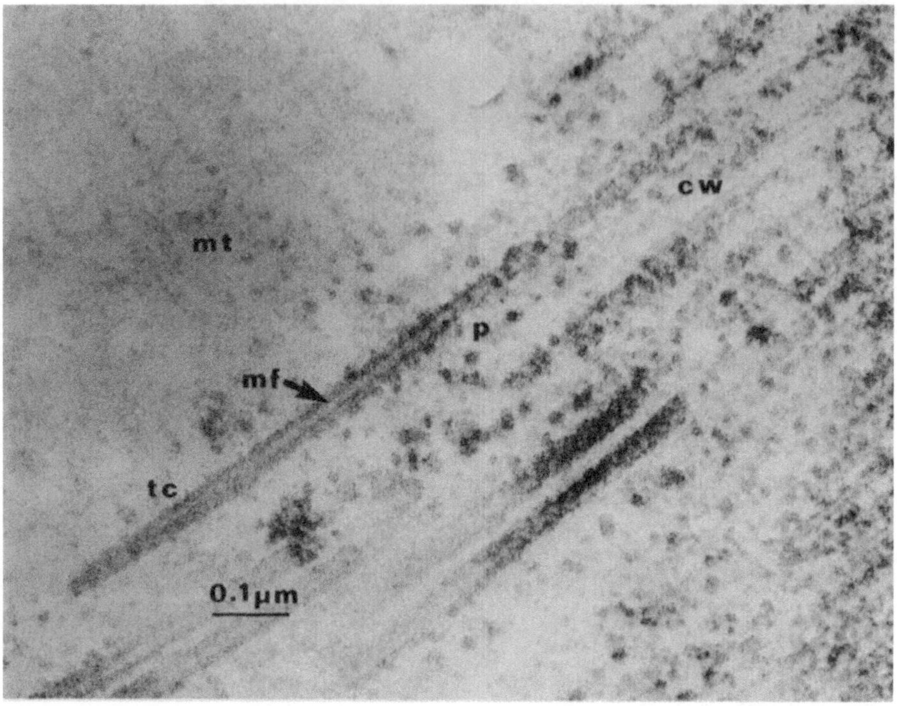

FIGURE 7. A terminal complex (tc) as seen in a tangential section of the plasma membrane, with a microfibril (mf) emerging from it, passing through the periplasm (p), and merging with other microfibrils in the cell wall (cw). Note that a microtubule (mt) subtending the plasma membrane is oriented perpendicular to the terminal complex.

the Golgi-derived vesicles, and the plasma membrane are stained by this procedure. Even more convincing is the cytochemical demonstration of carbohydrates in the Golgi apparatus, vesicles, and the periplasm (Fig. 8).

The postulation of a role for the periplasm in cell-wall biogenesis is not a new idea (Roland, 1973). In the case of *Oocystis*, it is graphically clear that new cellulosic microfibrils must traverse the periplasm to become part of the complete wall, and in the process they become coated and cross-linked with noncellulosic wall components. In addition to this, the presence in the periplasm of carbohydrates, and the suggestion of protein as demonstrated by the precipitation of the periplasm by tannic acid, establishes a strong circumstantial case for the periplasm as the site of assembly of the final three-dimensional structure of the cell wall.

2.5. *Use of Freeze-Etching to Investigate the Cytophysiology of Oocystis*

2.5.1. *Experiments with the Chelating Agent EDTA*

To study the forces of interaction that maintain subunit particles within terminal complexes, a series of experiments were performed involving EDTA, a chelator of divalent cations. Pretreatment of cells in 1% sodium EDTA for 15 min prior to freeze-etching results in complete dissociation of the particles from the

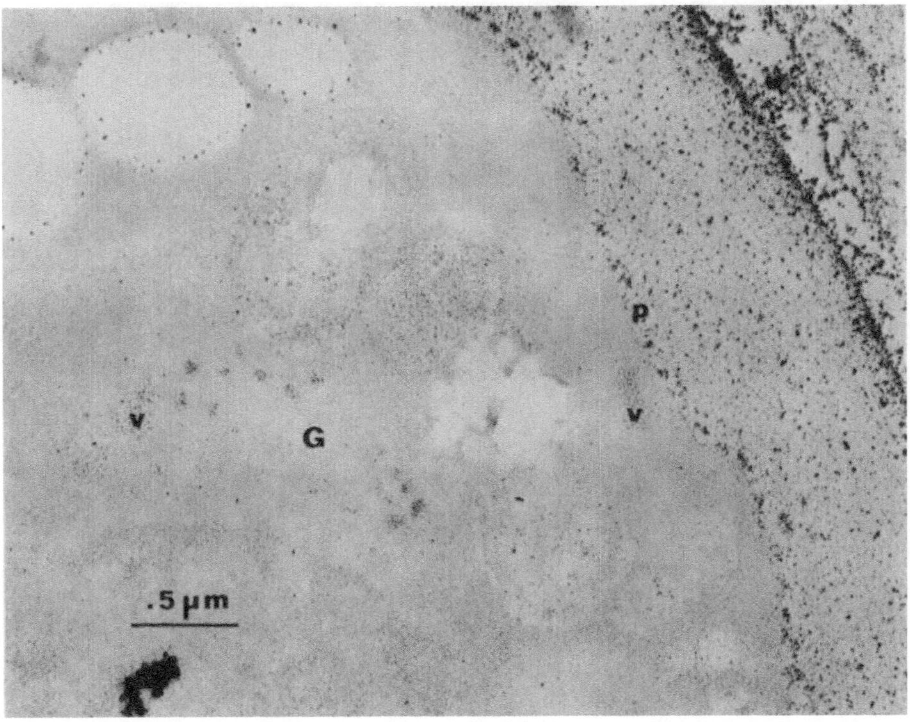

FIGURE 8. Section of cells of *Oocystis* active in cell-wall formation treated with periodic acid followed by silver methenamine (see Montezinos and Brown, 1976). This treatment results in the staining of carbohydrates in the Golgi apparatus (G), Golgi-derived vesicles (v), and the periplasm (p).

ends of ridges (Fig. 9). This same pretreatment followed by a treatment in 1% magnesium sulfate or magnesium chloride results in reaggregation of particles into paired complexes, usually at right angles to the microfibrils from which they dissociated (Fig. 10). Complexes are not seen reassociated at the ends of microfibrils. Further, if the cells are held for 1.5 hr following magnesium treatment, new ridges are formed, indicating that the reassociated complexes are functional in the assembly of microfibrils (Montezinos and Brown, 1978).

The EDTA effect clearly indicates that the terminal complexes of *Oocystis* are held together by ionic interactions, as would be expected for proteinaceous structures. This is in accord with evidence from studies on artificial membrane systems that indicates that particles observed in freeze–etched membrane fracture faces are globular proteins embedded in the lipid bilayer (Segrest *et al.*, 1974; Vail *et al.*, 1974). Thus, the terminal complexes appear to be striking examples of well-defined aggregates of membrane-bound protein.

2.5.2. Effect of the Cellulose Synthesis Inhibitor Coumarin Compared with the Effect of Glutaraldehyde–Tannic Acid Fixation

Coumarin is known to be a powerful inhibitor of cellulose synthesis. Nonetheless, its effect on the morphology of the plasma membrane during cell-

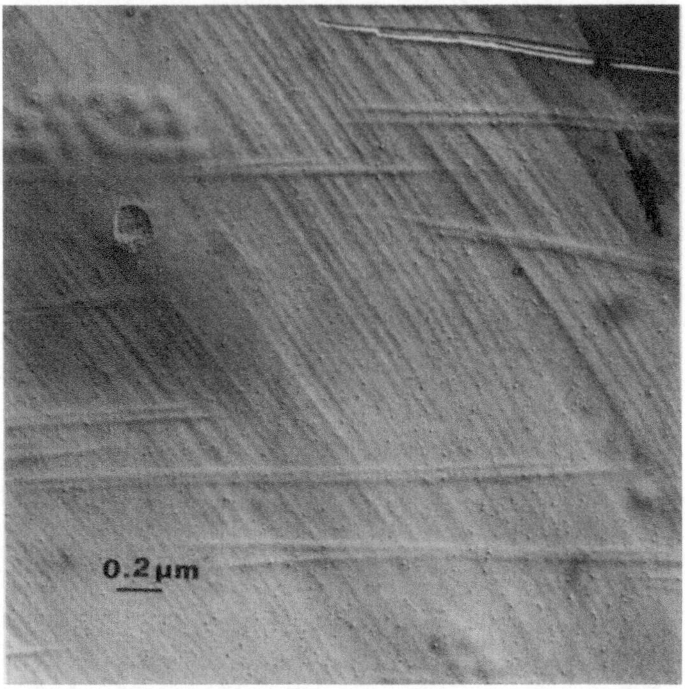

FIGURE 9. Short paired ridges on the EF face of the plasma membrane with no associated terminal complexes following a 15-min treatment with 1% EDTA.

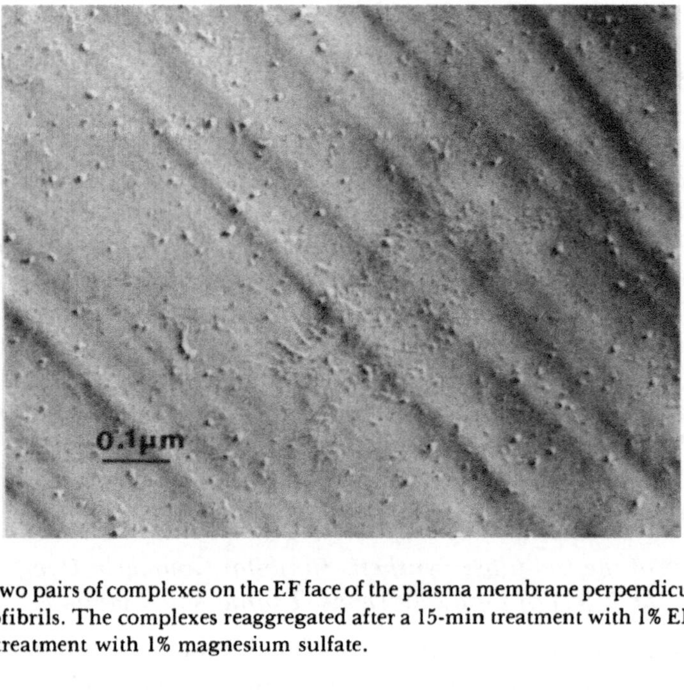

FIGURE 10. Two pairs of complexes on the EF face of the plasma membrane perpendicular to the last-formed microfibrils. The complexes reaggregated after a 15-min treatment with 1% EDTA followed by a 15-min treatment with 1% magnesium sulfate.

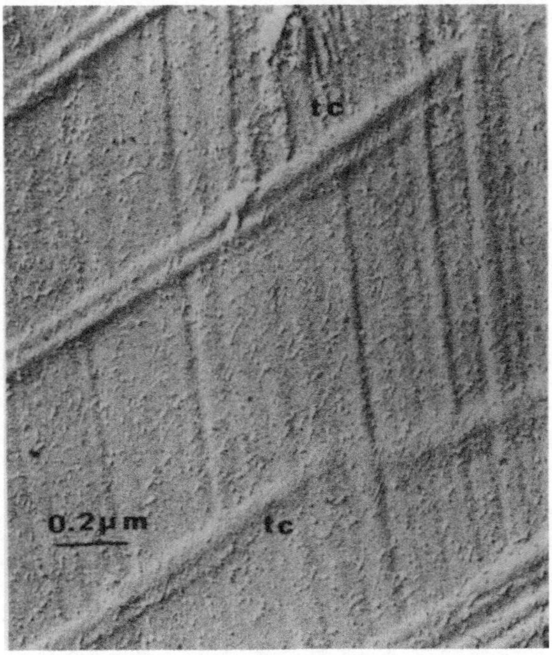

FIGURE 11. Following treatment for 2 hr with 10^{-3} M coumarin, terminal complexes (tc) on the EF face of the plasma membrane can be seen at the ends of ridges, but the particulate substructure of the complexes is no longer visible.

wall formation is surprisingly subtle. After 2 hr exposure to 10^{-3} M coumarin, freeze-etching reveals apparently normal ridge patterns. The outlines of terminal complexes are also observed; however, the particles in the complexes are not visible (Fig. 11).

In comparison, treatment for 2 hr in 1% glutaraldehyde with 1% tannic acid results in a loss of ridge patterns on the EF face [an effect after fixation in other systems as well (i.e., Willison, 1975, 1976)]. As with coumarin, the outlines of terminal complexes are visible, but the individual particles are not (Fig. 12).

If, as appears to be the case, the effect of fixation is to dislocate the proteinaceous complexes from the lipids of the membrane to the outer membrane surface, then it is possible to suggest that an analogous effect of coumarin is responsible for its inhibition of cellulose synthesis.

3. Microfibril Orientation

3.1. Granule Bands and Their Patterns and Orientation

In their early freeze-etching study of *Oocystis*, Robinson and Preston (1972) ascribed the roles of both synthesis and orientation of microfibrils to particulate arrays that they termed "granule bands." These bands are comprised of strings of approximately eight particles, which occur in pairs. The pairs are aligned side by side to form extensive rows (Fig. 13).

Although originally interpreted as attached to the outer surface of the

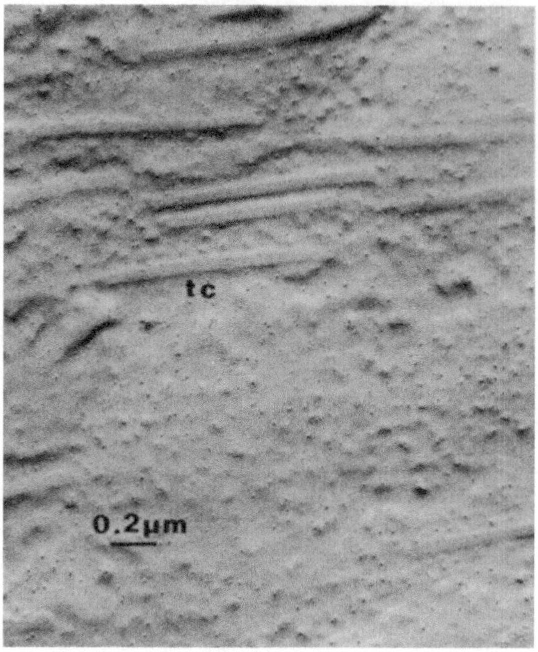

FIGURE 12. Following treatment for 2 hr with 1% glutaraldehyde and 1% tannic acid, terminal complexes (tc) are visible as featureless rectangles on the EF face of the plasma membrane.

FIGURE 13. Granule bands (gb) forming rows on the PF face of the plasma membrane underlying ridges that are the impressions of microfibrils.

plasma membrane, granule bands are actually embedded within the membrane, remaining with the PF face during fracturing. The apparent impressions left by granule bands in the EF face of the plasma membrane form crosshatches on the microfibril-impression ridges.

Through examination of granule-band rows and their impressions, it can be seen that granule bands associate with the most recently assembled microfibrils, cutting across rows formed by older microfibrils, and granule bands appear to organize into rows directly behind terminal complexes. Thus, circumstantial evidence exists that granule bands have some role in microfibril orientation. This hypothesis is further supported by experiments with colchicine, which causes disruption of granule-band rows and changes in microfibril orientation (Montezinos and Brown, 1978).

One serious problem with the postulation of a function for granule bands is that they are unique to *Oocystis*. Nonetheless, a role for granule bands in the complex orientation mechanism for aligning microfibrils in this alga remains a distinct possibility (see Brown, 1978).

3.2. *Microtubule Orientation in Oocystis*

The idea that cortical microtubules are involved in microfibril orientation began with the earliest observations of microtubules (Ledbetter and Porter, 1963). Unfortunately, this hypothesis has attained general acceptance merely on the basis of repeated observations of the parallel orientation of microtubules adjacent to the plasma membrane and microfibrils in the cell wall (see Hepler and Palevitz, 1974). The only other substantial observation brought to bear on the problem concerns the effect of colchicine, an inhibitor of microtubule polymerization that disrupts the normal orientation of microfibrils (see Newcomb, 1969; Hepler and Palevitz, 1974).

Cortical microtubules are found in *Oocystis* directly subtending the plasma membrane and oriented parallel to one of the orientations of microfibrils in the wall. Furthermore, as Robinson *et al.* (1976) have reported, colchicine does disrupt the mechanism of orientation that produces the unusual pattern of microfibrils in the wall of *Oocystis*. In particular, the effect of colchicine appears to be a disruption of the mechanism that changes the direction of microfibril formation from layer to layer.

In addition to these observations, the following data are significant. First, evidence from both sectioned and freeze-etched specimens indicates that microtubules do not necessarily parallel the most recently synthesized microfibrils (see Fig. 7). Second, concentrations of colchicine and vinblastine that clearly cause loss of microfibril orientational control in *Oocystis* do not cause the disappearance of microtubules (Montezinos and Brown, 1978).

What these data indicate is that only a weak case exists for a direct causal role for microtubules in microfibril orientation. In fact, studies on microtubule structure have established the variable effectiveness of the so-called mitotic poisons in destabilizing microtubules *in vivo* (Raff, 1979). More important, other studies have revealed the existence of microtubule accessory proteins

(MAPs) that may be of vital importance in the functioning of microtubules (Petzelt, 1979; Sloboda, 1980). Clearly, until these structural components have been visualized and investigated in plant cell-wall synthesizing systems, the exact role of microtubules in wall biogenesis remains an open question.

4. Discussion

4.1. Oocystis Model of Microfibril Assembly and Orientation

With evidence available at present, an overall hypothetical model for microfibril assembly and cell-wall organization for the alga *Oocystis* can be constructed with the following components (Fig. 14):

1. Terminal complexes form in pairs parallel to each other across the

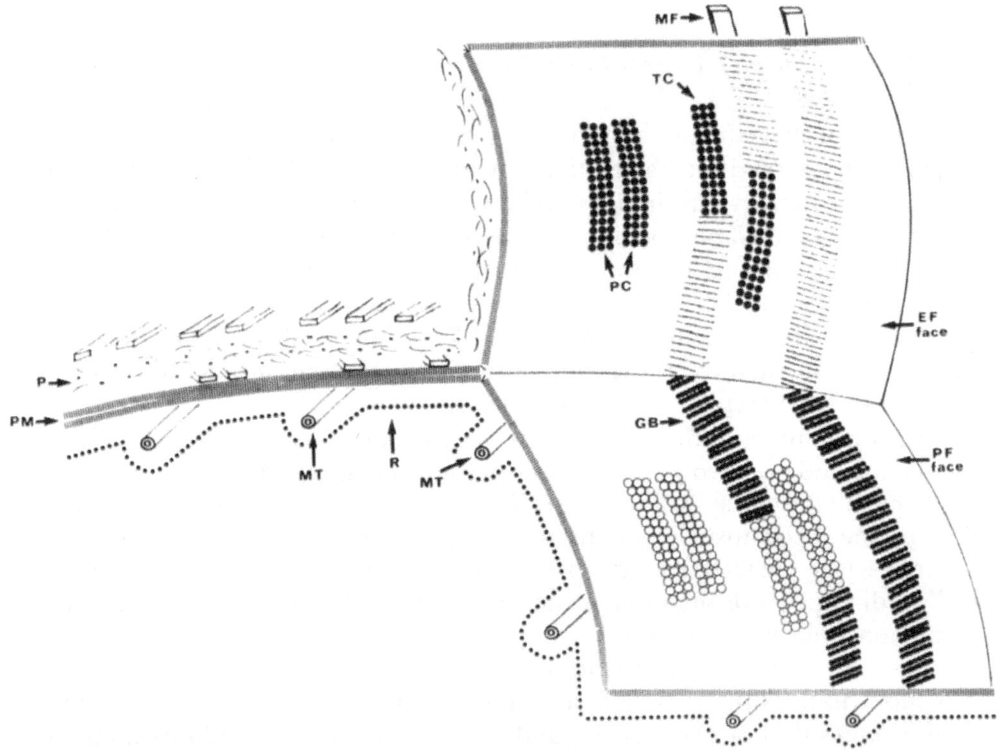

FIGURE 14. Diagrammatic portrayal of the hypothetical model of cell-wall biogenesis in *Oocystis*. The components of the model include: (1) Periplasm (P). (2) Plasma membrane (PM). Terminal complexes (TC) associate with the ends of ridges that are the impressions of microfibrils (MF) or with each other in pairs (PC) on the EF face of the plasma membrane. The ridges are crosshatched by the impressions of granule bands (GB) that associate with the PF face of the plasma membrane. The granule bands form rows underlying the ridges and end at the impressions in the PF face produced by terminal complexes. (3) Microtubules (MT). Also note a region (R) of possible transmembrane control that is yet to be elucidated, underlying the plasma membrane and surrounding the microtubules.

plasma membrane. Each complex assembles a microfibril, with pairs producing microfibrils in opposite directions, the complexes moving across the membrane along spiral paths from pole to pole, leaving the microfibrils in place as they go. By reversing directions at the poles, the complexes make several traverses up and down the cell until a single row of microfibrils covers nearly the entire cell surface. Then the complexes dissociate from the microfibrils and reassemble in pairs reoriented appropriately to begin assembly of microfibrils along spiral paths oppositely handed from those of the layer just produced. Thus, a new layer is produced crossing under the last-formed microfibrils and resulting in the final pattern typical of the cell wall of *Oocystis*.

2. Granule bands within the plasma membrane aggregrate in rows beneath newly assembled microfibrils. Unique to *Oocystis*, these particulate structures may have a role in the complex orientation mechanism that results in the well-organized cell wall of this alga.

3. Microtubules subtending the plasma membrane may also have a role in orienting microfibrils. In particular, they may be involved in the mechanism that reorients the direction of assembly of microfibrils from layer to layer.

4. A periplasm, derived at least in part from Golgi products via exocytosis, coats and cross-links microfibrils with noncellulosic wall components. The periplasm would be the site of final three-dimensional assembly of the cell wall.

It should be recognized immediately that the observed interactions of these components do not entirely account for the biogenesis of microfibrils in *Oocystis*. There may be other components of the biogenetic mechanism as yet unobserved or other observed structures that contribute but have been overlooked. In this regard, it is notable that endoplasmic reticulum (ER) underlies much of the plasma membrane, separated from it during cell-wall formation only by microtubules. Although there is no direct evidence to suggest a role for the ER in wall biogenesis, in view of the data involving EDTA–magnesium treatments, it can be postulated that the ER may serve a regulatory role. By sequestering or releasing divalent cations via a membrane-bound ion pump, the ER could cause disaggregation and reaggregation of terminal complexes to accomplish reorientation from assembly of one layer to assembly of the next.

4.2. Microfibril Assembly and Synthesis

The plasma membrane of plant cells has long been considered the likely site for the synthesis and assembly of cellulosic microfibrils (Preston, 1974). The few studies that provide direct support for this concept include the work of Northcote and co-workers (Northcote and Pickett-Heaps, 1966; Harris and Northcote, 1971; Bowles and Northcote; 1972, 1974) and one study by Wooding (1968). These indicate that [^{14}C]glucose is incorporated into cellulose at the cell surface rather than passing through the Golgi–exocytotic pathway. Therefore, the discovery of terminal complexes in *Oocystis* provides much-needed cytological support for this hypothesis.

There is as yet no direct evidence of a biosynthetic role for terminal complexes. However, considering the widespread acceptance of the concept that microfibrils are simultaneously synthesized and assembled, first postulated on theoretical grounds by Roelofson (1958), it is reasonable to argue that the proteinaceous terminal complex is in fact an enzyme complex. In addition, the terminal complex fits Roelofsen's suggestion that the synthesizing complex should be at the end of the growing microfibril, and being at only one end of the microfibril, it is in agreement with the contention of Gardner and Blackwell (1974) that the glucan chains within the microfibril are assembled in parallel orientation rather than antiparallel.

4.3. Complexes on the Plasma Membrane

The concept of particulate synthesizing complexes actually originated with the earliest freeze-etching studies of yeast by Moor and Mühlethaler (1963). Despite the lack of evidence linking plasma-membrane particles with microfibrils [with the possible exception of a study on plasmolyzed cells by Roland (1967)], Mühlethaler (1969) proposed a model for microfibril biosynthesis that consisted of a complex of particles working together to produce a single microfibril. This marked similarity to terminal complexes is offset, however, by Mühlethaler's contention that the complex is fixed on the plasma membrane.

During the same period, Preston (1974) was formulating his ordered-granule hypothesis, in which a cuboidal arrangement of membrane-bound particles simultaneously synthesizes and orients microfibrils. Although this bears less similarity to the *Oocystis* model, the concept of a directed synthesis implicit in Preston's hypothesis does fit well the directional emergence of microfibrils from the ends of terminal complexes.

Strikingly similar to the terminal complexes of *Oocystis* are analogous structures actually observed in other systems by R. M. Brown, Jr., and co-workers. The terminal complexes found in *Glaucocystis* (Brown and Willison, 1977) are linear, like those of *Oocystis*, and also assemble large microfibrils in a complicated but well-organized pattern. Terminal complexes observed in corn root stele (Mueller *et al.*, 1976) and probably in cotton fibers (Willison and Brown, 1977) are smaller and globular, producing smaller microfibrils in less highly oriented arrays. Investigations revealing terminal complexes associated with growing microfibrils have also been carried out in other laboratories. Recent work by Giddings *et al.* (1980) on cell-wall biogenesis in *Micrasterias* also include evidence in support of this concept. Clearly, the results of all these investigations lend credence both to the observations underlying the *Oocystis* model and to the value of the model as such.

4.4. Microfibril-Orientation Mechanisms

Although a role for microtubules in microfibril orientation remains a distinct probability, the sort of causal relationship postulated by Newcomb (1969), Heath (1974), and Hepler and Palevitz (1974) is not supported by the *Oocystis* model. Microtubules in *Oocystis* would appear more likely to serve as cytoskeletal supports than as guiding elements. Direct movement of terminal complexes along microtubules does not seem to occur. Terminal complexes reorient at the poles of the cell during the formation of wall layers and dissociate and reorganize in new orientations between the formation of wall layers, while no concurrent changes in microtubules are noted. Thus, it is worthwhile to suggest that less attention is due to the orientation of microtubules and the effects of colchicine, and more is due to the structures as yet undefined that connect microtubules to the plasma membrane and may have more active roles in microfibril orientation.

More definitive is the fact that the *Oocystis* model fails to support the concept of passive wall orientation espoused by Green *et al.* (1970). On the contrary, the orientation of microfibrils in *Oocystis* constitutes a clear example of the kind of cytoplasmic control postulated for all cell walls by Roland and Vian (1979). In fact, when cells of *Oocystis* expand following cell-wall formation, the orientation of microfibrils remains the same.

4.5. Structure of the Plasma Membrane

Essential to an understanding of the *Oocystis* model is acceptance of the concept of membrane fluidity formalized by Singer and Nicolson (1972). In fact, the plasma membrane of *Oocystis* may be regarded as a striking example where movement of membrane particles (of granule bands and of aggregated and disaggregated terminal complexes) does occur in the plane of the plasma membrane.

More important, these movements of particles in the plasma membrane are highly organized. Nicolson (1976) and Edelman (1974) have documented specific cases in which microtubules act as skeletal elements with microfilaments and other proteins associated with, but not part of, the plasma membrane to provide transmembrane control of particle movements. This is already being regarded as a general rule (see Shinitsky and Henkart, 1979), and it could easily be true in the case of *Oocystis*. There is, perhaps, a more pressing need for investigation of structures lying between the plasma membrane and the microtubules than for the study of any other aspect of cell-wall biogenesis in *Oocystis*.

5. Concluding Remarks

It has been approximately a decade since investigations of cell-wall biogenesis in *Oocystis* began, and during that time, *Oocystis* has served profitably as a model system. Our knowledge of this system remains remarkably incomplete. Moreover, in many respects, the study of this alga cannot provide information directly transferable to other systems. Nonetheless, as long as the model continues to provide useful insights into the general mechanisms of how plants produce cell walls, it will be worthwhile to continue investigations on *Oocystis apiculata*.

References

Bowles, D. J., and Northcote, D. H., 1972, The sites of synthesis and transport of extracellular polysaccharides in the root tissues of maize, *Biochem. J.* **130**:1133.

Bowles, D. J., and Northcote, D. H., 1974, The amounts and rates of export of polysaccharides found within the membrane system of maize root cells, *Biochem. J.* **142**:139.

Branton, D., Bullivant, S., Gilula, N. B., Karnovsky, M. J., Moor, H., Mühlethaler, K., Northcote, D. H., Packer, L., Satir, B., Satir, P., Speth, V., Staehelin, L. A., Steere, R. L. and Weinstein, R. S., 1975, Freeze-etching nomenclature, *Science* **190**:54.

Brown, R. M., Jr., 1978, Biogenesis of natural polymer systems with special reference to cellulose assembly and deposition, The Third Phillip Morris Science Symposium.

Brown, R. M., Jr., and Romanovicz, D. K., 1976, Biogenesis and structure of Golgi-derived cellulosic scales in *Pleurochrysis*. I. Role of the endomembrane system in scale assembly and exocytosis, *Appl. Polym. Symp.* **28**:537.

Brown, R. M., Jr., and Willison, J. H. M., 1977, Golgi apparatus and plasma membrane involvement in secretion and cell surface deposition, with special emphasis on cellulose biogenesis, *in*: *International Cell Biology 1976–77* (B. R. Brinkley and K. R. Porter, eds.), pp. 267–283, The Rockefeller University Press, New York.

Edelman, G. M., 1974, Surface alterations and mitogenesis in lymphocytes, *in*: *Control of Proliferation in Animal Cells* (B. Clarkson and R. Baserga, eds.), pp. 357–377, Cold Spring Harbor Laboratory, New York.

Gardner, K. H., and Blackwell, J., 1974, The structure of native cellulose, *Biopolymers* **13**:1975.

Giddings, T. H., Jr., Brower, D. L., and Staehelin, L. A., 1980, Visualization of particle complexes in the plasma membrane of *Micrasterias denticulata* associated with the formation of cellulose fibrils in primary and secondary cell walls, *J. Cell Biol.* **84**:327.

Green, P. B., Erickson, R. O. and Richmond, P. A., 1970, On the physical basis of cell morphogenesis, *Annu. Rev. Plant Physiol.* **20**:365.

Harris, P. J., and Northcote, D. H., 1971, Polysaccharide formation in plant and Golgi bodies, *Biochim. Biophys. Acta* **237**:56.

Heath, I. B., 1974, A unified hypothesis for the role of membrane bound enzyme complexes, and microtubules in plant cell wall synthesis, *J. Theoret. Biol.* **48**:445.

Hepler, P. K., and Palevitz, B. A., 1974, Microtubules and microfilaments, *Annu. Rev. Plant Physiol.* **25**:309.

Kantz, T., and Bold, H. C., 1969, Phycological studies. IX. Morphological and taxonomic investigations of *Nostoc* and *Anabaena* in culture, pp. 1–67, Publication 6924, University of Texas Press, Austin.

Ledbetter, M. C., and Porter, K. R., 1963, A "microtubule" in plant cell fine structure, *J. Cell. Biol.* **19**:239.

Montezinos, D., and Brown, R. M., Jr., 1976, Surface architecture of the plant cell: Biogenesis of the cell wall, with special emphasis on the role of plasma membrane in cellulose biogenesis, *J. Supramol. Struct.* **5**:277.

Montezinos, D., and Brown, R. M., Jr., 1978, Cell wall biogenesis in *Oocystis*: Experimental alteration of microfibril assembly and orientation, *Cytobios* **23**:119.

Moor, H., and Mühlethaler, K., 1963, Fine structure in frozen-etched yeast cells, *J. Cell. Biol.* **17**:609.

Morré, D. J., and Mollenhauer, H. H., 1974, The endomembrane concept: A functional integration of endoplasmic reticulum and Golgi apparatus, in: *Dynamic Aspects of Plant Ultrastructure* (A. W. Robards, ed.), pp. 84–137, McGraw-Hill, New York.

Mueller, S. C., Brown, R. M., Jr., and Scott, T. K., 1976, Cellulosic microfibrils: Nascent stages of synthesis in a higher plant cell, *Science* **194**:949.

Mühlethaler, K., 1969, Fine structure of natural polysaccharide systems, *J. Polymer Sci. Part C* **28**:305.

Newcomb, E. H., 1969, Plant microtubules, *Annu. Rev. Plant Physiol.* **20**:253.

Nicolson, G. L., 1976, Transmembrane control of the receptors on normal and tumor cells. I. Cytoplasmic influence over cell surface components, *Biochem. Biophys. Acta* **457**:57.

Northcote, D. H., and Pickett-Heaps, J. D., 1966, A function of the Golgi apparatus in polysaccharide transport in the root cap cells of wheat, *Biochem. J.* **98**:159.

Petzelt, C., 1979, Biochemistry of the mitotic spindle, *Int. Rev. Cytol.* **60**:53.

Preston, R. D., 1974, *The Physical Biology of Plant Cell Walls*, Chapman and Hall, London.

Raff, E. C., 1979, The control of microtubule assembly *in vivo*, *Int. Rev. Cytol.* **59**:1.

Robinson, D. G., and Preston, R. D., 1972, Plasmalemma structure in relation to microfibril biosynthesis in *Oocystis*, *Planta* **104**:234.

Robinson, D. G., Grimm, I., and Sachs, H., 1976, Colchicine and microfibril orientation, *Protoplasma* **89**:375.

Roelofsen, P. A., 1958, Cell wall structure as related to surface growth: Some supplemental remarks on multinet growth, *Acta Bot. Neerl.* **7**:77.

Roland, J. C., 1967, Sur la sécrétion de la cellulose dans les parois des cellules végétales, *C. R. Acad. Sci. Ser. D* **264**:2757.

Roland, J. C., 1973, The relationship between the plasmalemma and plant cell wall, *Int. Rev. Cytol.* **36**:45.

Roland, J. C., and Vian, B., 1979, The wall of the growing plant cell: Its three-dimensional organization, *Int. Rev. Cytol.* **61**:129.

Segrest, J. P., Gulik-Krzywicki, T., and Sardet, C., 1974, Association of the membrane penetrating polypeptide segment of the human erythrocyte MN-glycoprotein with phospholipid bilayers. I. Formation of freeze-etch intramembranous particles, *Proc. Natl. Acad. Sci. U.S.A.* **71**:3294.

Shinitsky, M., and Henkart, P., 1979, Fluidity of cell membranes—current concepts and trends, *Int. Rev. Cytol.* **60**:121.

Singer, S. J., and Nicolson, G. L., 1972, The fluid mosaic model of the structure of cell membranes, *Science* **175**:720.

Sloboda, R. D., 1980, The role of microtubules in cell structure and cell division, *Am. Sci.* **68**:290.

Vail, W. J., Papahadjopoulos, D., and Moscarello, M. A., 1974, Interaction of a hydrophobic protein with liposomes: Evidence for particles seen in freeze-fracture as being proteins, *Biochim. Biophys. Acta* **345**:463.

Willison, J. H. M., 1975, Plant cell wall microfibril deposition revealed by freeze-fractured plasmalemma not treated with glycerol, *Planta* **126**:93.

Willison, J. H. M., 1976, An examination of the relationship between freeze-fractured plasmalemma and cell-wall microfibrils, *Protoplasma* **88**:187.

Willison, J. H. M., and Brown, R. M., Jr., 1977, An examination of developing cotton fibers: Wall and plasmalemma, *Protoplasma* **92**:21.

Wooding, F. P. B., 1968, Radioautographic and chemical studies of incorporation into sycamore vascular tissue walls, *J. Cell Sci.* **3**:71.

2
Organized Microfibril Assembly in Higher Plant Cells

BRIGITTE VIAN

1. Introduction

The cell wall represents one of the most characteristic features of plant cells. Its occurrence around the protoplasts accounts for a great difference between plant cells and animal cells, the latter being considered as "naked cells" limited only by the plasmalemma and its sugar-rich coat, the glycocalyx. In plants, the thick wall enclosing the cells can be seen in two different ways that, at first sight, may appear antinomic. On the one hand, the wall is an essentially rigid formation, a skeleton responsible for the shape of the cell and for the strength of the plant. On the other hand, the wall must be plastic and extensible so that a surface increase during extension may occur.

In higher plants, the cell wall has long been considered mainly as an outer envelope, inert and inactive, the behavior of which during extension is essentially passive. During the last decade, due to technical refinements and diversification, the knowledge of wall structure and functions has improved greatly. In particular, the passive behavior of the wall during extension has been questioned, and new proposals have appeared. The role of the cell wall as a barrier compartment between external medium and protoplasm, across which the substances necessary for cell metabolism move and are exchanged, is also important. Except for naked cells of lower groups and for gametes, which represent the only plant cells deprived of walls, cell recognition and interactions must negotiate through the cell wall. Recent data suggest that polysaccharide polymers that constitute cell walls could even act as "hormones" in recognition systems between hosts and pathogens (Albersheim *et al.*, 1980).

Numerous books and reviews have been devoted to the study of the cell wall

BRIGITTE VIAN • Laboratoire de Cytologie Expérimentale, Ecole Normale Superieure, Université Pierre et Marie Curie, 24 rue Lhomond, 75 231 Paris Cedex 05, France.

of higher plants (Roelofsen, 1959; Muhlethaler, 1959; Frey-Wyssling, 1953, 1976; Northcote, 1972; O'Brien, 1972; Roland, 1973; Preston, 1974; Albersheim, 1975; Roland and Vian, 1979)* and the reader is referred to them for a complete description of the structural components and of main characteristics of the cell wall. The classic areas defined on the basis of light-microscopic observations have been confirmed at the electron-microscopic level. The middle lamella is the outer area common to contiguous cells. It is essentially composed of gel-like polysaccharides of pectic types that form a cementing medium. Internal areas are typically biphasic: a discontinuous microfibrillar phase made of crystalline polysaccharides—mainly cellulose, but also other linear polysaccharides—and a continuous matrix phase consisting of a wide variety of polysaccharides.

At present, an important part of wall research is at the stage of identification of macromolecules and determination of their primary structures (McNeil *et al.*, 1979). What appears most significant is that the wall subunits are actually interconnected—by means of covalent or noncovalent bondings or both—in a defined network. In fact, a gap still exists between the precise and sophisticated data concerning the molecular structure of wall subunits and those concerning the three-dimensional organization of the wall. The fine texture of the wall is still an object of controversy. The purpose of this chapter is to focus on results that indicate that the three-dimensional organization of the wall is basically ordered. Most of the discussion will concern the question of the architecture of elongating walls, which is at present the most debated. The morphogenetic capacities of cells for building up such defined arrangements will also be analyzed.

2. *Evolution of Ideas about the Organization of Cell Walls*

2.1. *Classic Opposition: Dispersed Wall vs. Ordered Wall*

In higher plants, it is traditional to distinguish two types of walls according to whether they are laid down during or after elongation.

The primary wall, i.e., the first envelope built up very early by growing cells, is typically defined by its dispersed texture. This is mainly based on observations in both light microscopy (polarizing microscope) and electron microscopy (shadow-casting). According to the classic descriptions, the recently deposited microfibrils of the inner part of the wall are mostly parallel and transversely oriented as related to the growth axis. On the external face, they are more dispersed and generally longitudinally oriented. These observations led Roelofsen and Houwink to propose the multinet growth hypothesis in 1953 (see Roelofsen, 1959). In this hypothesis, the microfibrils are first deposited according to the major stress, which is transverse, and are passively shifted during expansion. The result is that the microfibrils become more or less dispersed and reoriented. The behavior of the wall during expansion is passive.

**Cell Walls '81* was also recently published following the 2nd Cell Wall meeting held in Göttingen in April 1981 (see Roland, 1981, for exact reference).

The secondary wall, i.e., the wall built up when the cell has stopped growth and almost reached a differentiated state, qualifies as an ordered wall. It corresponds to a thick and nonextensible wall that is typically crossed polylamellate. In the concentric layers, the microfibrils are parallel and precisely oriented according to the cellular axis. Successive zones have been described (the S_1, S_2, and S_3 zones) that have become standard. This organization is constant and specific.

2.2. *Possible Causes of Controversy Concerning the Dispersed Texture of Expanding Walls*

Despite a general agreement about these definitions of walls and of their corresponding textures, the universality of the dispersed texture applied to elongating walls has been questioned. It has been suggested that elongating walls could be ordered and polylamellate. Since recent literature indicates that this is still a controversial subject, it seems worthwhile to recall some possible causes of such conflicting data.

2.2.1. *Problem of Techniques*

Investigations of wall texture at the electron-microscopic level have long been performed almost exclusively with the shadow-casting technique. This method utilizes macerations with dilute alkali and hot acids to separate the cells and to eliminate the amorphous matrix. When specimens are embedded in plastics for further sectioning, the plastic embedding must also be removed prior

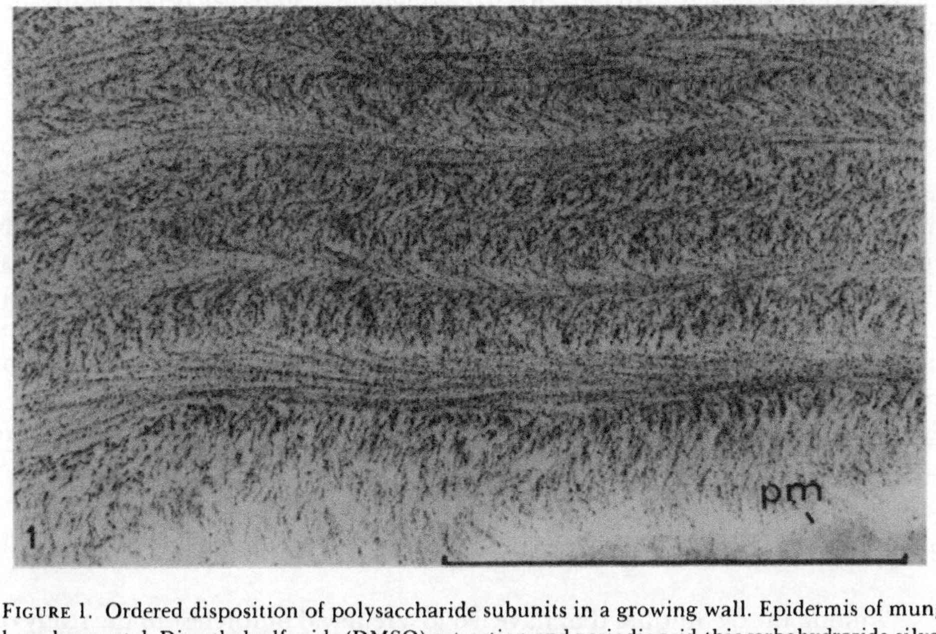

FIGURE 1. Ordered disposition of polysaccharide subunits in a growing wall. Epidermis of mung bean hypocotyl. Dimethylsulfoxide (DMSO) extraction and periodic acid-thiocarbohydrazide-silver proteinate (PATAg) reaction. (pm) Plasmalemma in oblique view. Scale bar: 500 nm.

to shadow-casting. All these treatments necessary to unmask the microfibrils are very drastic and destroy the cohesion of the assembly. In the case of the thin elongating walls, which are highly hydrated and in which the fibrillar framework is loose, the result is an artifactual shifting of wall subunits leading to a disorder in texture.

The polarizing microscope has also been widely used as a tool to analyze wall texture. Interpretation of the results also requires caution, since the method can provide only a mean indication of the microfibril orientation in the wall analyzed (Preston, 1974).

2.2.2. *Problem of Level*

Another source of difficulties is that the actual state of growth is not always accurately measured at the organ or even at the cellular level.

The physiological methods of growth measurement have greatly improved, and it appears that even in an organ highly elongating, only a narrow zone is actually growing (for a discussion, see Prat, 1978). Studies of growth kinematics indicate that growing tissues are simultaneously increasing in size and changing their position in relation to other parts of the plant (Erickson and Silk, 1980). Therefore, it is necessary to take into account measurements for both spatial and temporal variation in growth rates within tissues. It is likely that in experiments using specimens that are too large, information is provided for cells having different growth characteristics, so that cells recorded as elongating may actually have ceased growth.

Even at the cellular level, experiments involving labeling of the cell-surface sites by means of cytochemical markers and following the spacing of these markers during cell growth indicate that elongation could even occur locally and arrhythmically in minute zones within one wall (Vian and Roland, 1980).

2.2.3. *Problem of Cell Specificity*

Different extension processes may also occur simultaneously within the same organ. Classic experiments using partially split organs have already shown antagonisms and distinct potentialities and sensitivities among tissues. By means of split-tests and measurements of segments with displacement transducers after removal of pith or cortical tissues of mung bean hypocotyl, recent experiments have indicated that internal (pith) and external (cortical parenchyma, epidermis) tissues of the same level in the elongating zone have different potentialities for growth (Prat and Roland, 1980).

It will be shown later (see Section 2.5) that a duality of cell-wall architecture corresponds to such duality of extensibility.

2.3. *Ordered Architecture of Elongating Walls*

Among the techniques that have led to a renewal of interest in the study of wall architecture at the electron-microscopic level, ultrastructural cytochemistry is especially suitable for visualizing the wall subunits in ultrathin sections. In particular, the periodic acid-thiocarbohydrazide-silver proteinate (PATAg) test for polysaccharides provides sharp and reproducible results (for details, see Roland, 1978). The reactivity is massive due to the compactness of polysaccharide components within the wall. When mild chemical or enzymatic extractions are performed, microfibrillar subunits are sharply unmasked and the fine texture appears (see Figs. 1, 5, 6, and 10).

In parallel, the methods using cryofixation have been widely and successfully applied to investigate the organization of the walls. Since these methods eliminate embedding in plastic, the risks of chemical artifacts are greatly

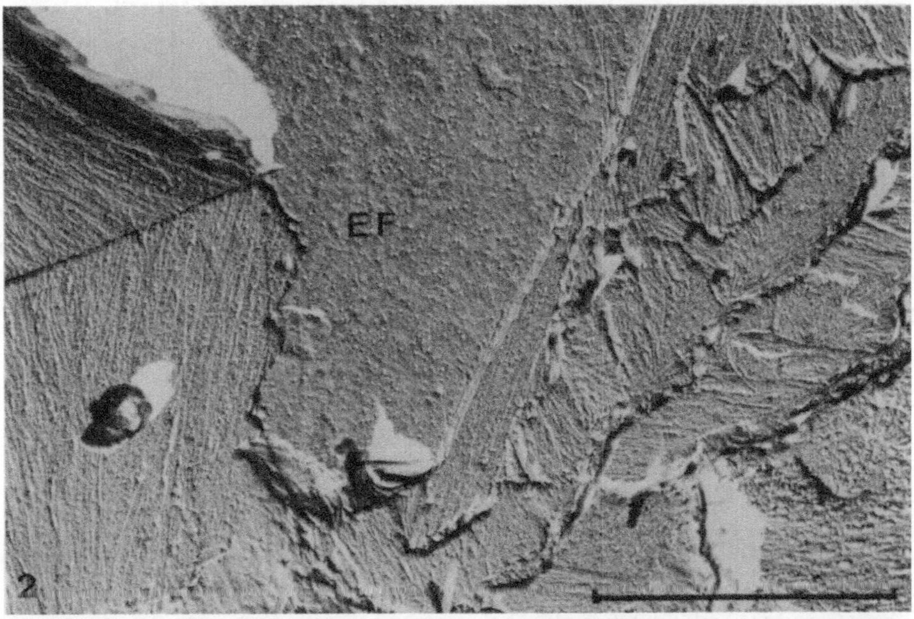

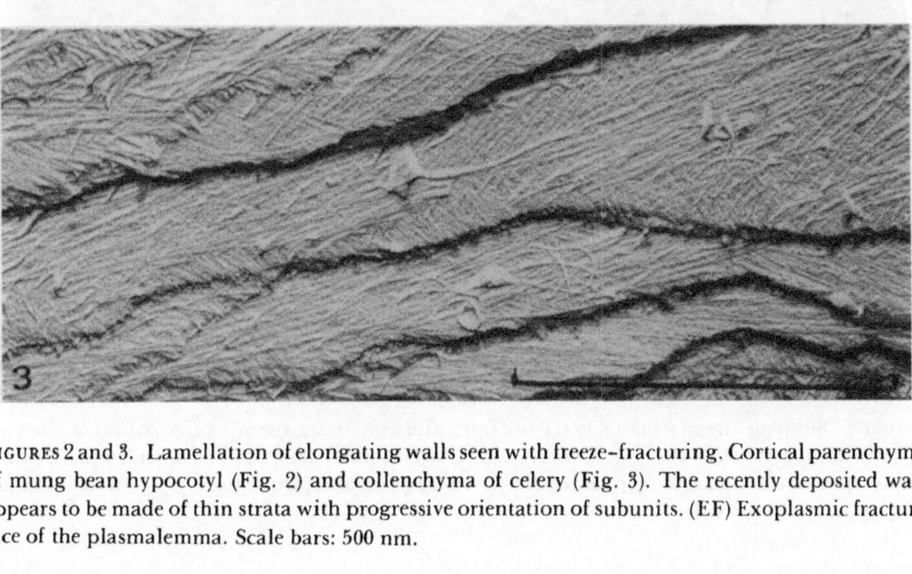

FIGURES 2 and 3. Lamellation of elongating walls seen with freeze-fracturing. Cortical parenchyma of mung bean hypocotyl (Fig. 2) and collenchyma of celery (Fig. 3). The recently deposited wall appears to be made of thin strata with progressive orientation of subunits. (EF) Exoplasmic fracture face of the plasmalemma. Scale bars: 500 nm.

reduced. In particular, the freeze-fracturing technique is favorable to observe replicas of the cell surface (wall-plasmalemma interface). The results have appeared greatly improved when both fixative and cryoprotectant were omitted prior to freezing (Figs. 2 and 3) (Willison, 1976; Brown and Montezinos, 1976; Vian *et al.*, 1978; Mueller and Brown, 1980).

Ultracryotomy, which allows negative staining to be performed directly on ultrathin frozen sections, provides a sharp visualization of the *in situ* microfibrillar arrangement within the wall. Mild extractions can be performed directly on sections, which eliminates problems of penetration and allows following and controlling the progressive decrustation of the wall subunits (Roland *et al.*, 1977; Vian, 1978, 1981). The dimensions and the substructure of the microfibrils are directly revealed and may be compared in various specimens (Fig. 4).

All repeated observations performed with different tissues have led to the modification of the classic view of the wall in elongating cells as being characterized by a scattered texture (Itoh, 1975) (for a discussion, see Roland and Vian, 1979). When mild techniques are used and when actual sites of growth are accurately checked, elongating walls appear typically multilayered and ordered. Therefore, a distinction between primary and secondary walls on the basis of structural arguments no longer seems possible (Figs. 5–8). The differences in texture observed in both types of walls following classic shadow-casting are easily explained by the fact that the thin primary wall, more fragile than the

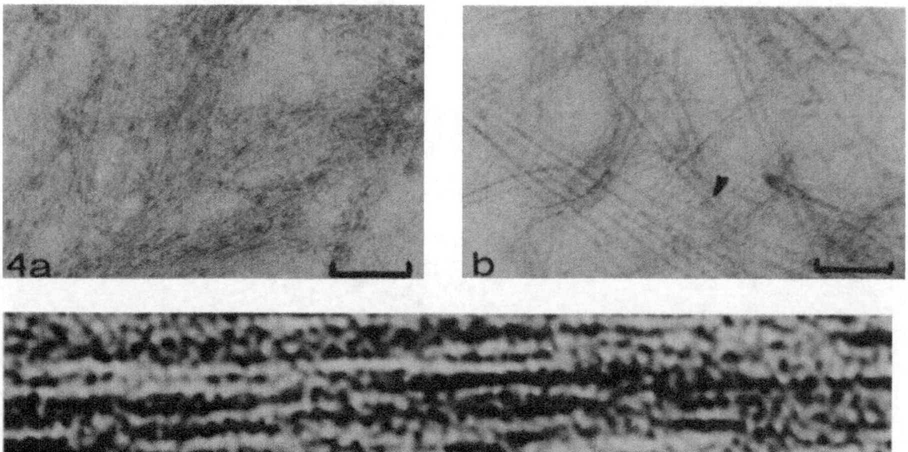

FIGURE 4. Substructure of negatively stained microfibrils. Ultracryotomy. Comparison, at the same magnification (scale bar: 100 nm), between microfibrils of small diameter from a higher plant [mung bean (a)] and larger and more rigid microfibrils from a green alga [Oocystis (b)]. (c) Detail at a higher magnification (pea root).

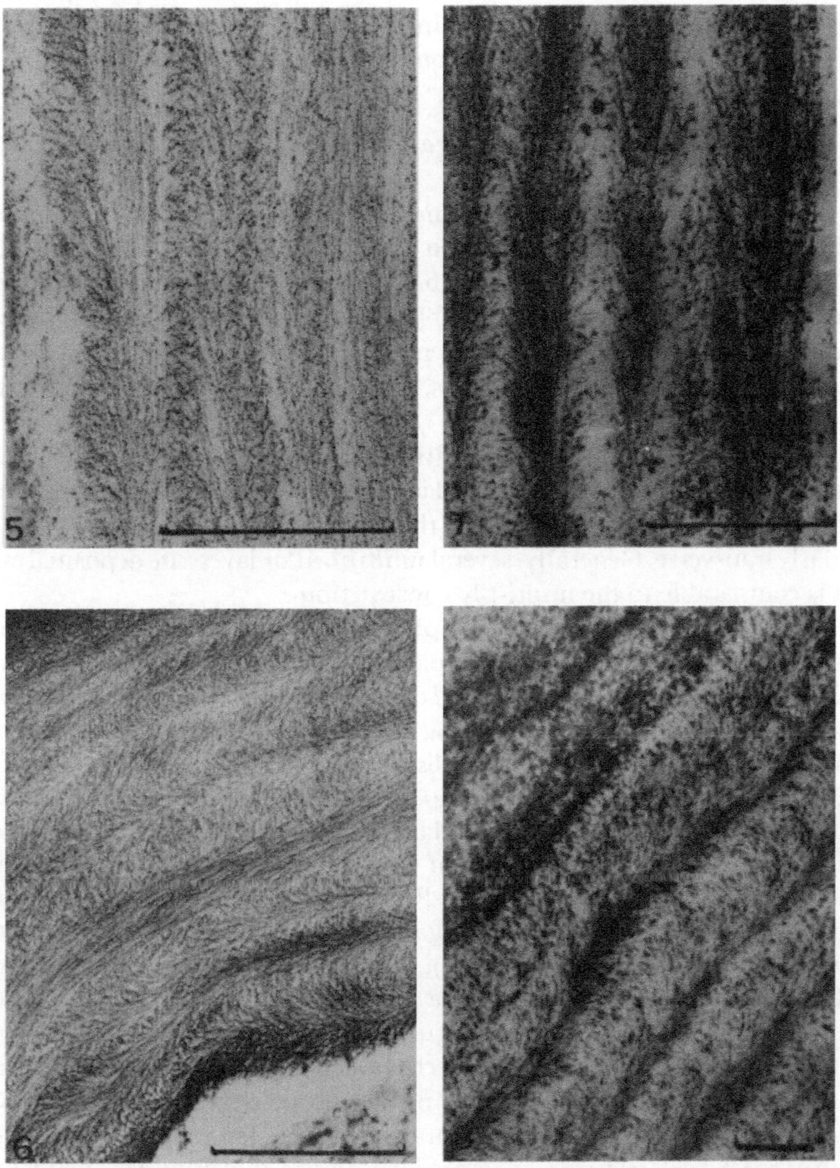

FIGURES 5–8. Various examples of multilayered walls. Cortical parenchyma of elongating zone of mung bean hypocotyl (Fig. 5); stellate pith cell of *Juncus* [Fig. 6. (Courtesy of J. C. Roland)]; delignified sclereids of beech bark [Figs. 7 and 8 (courtesy of N. Parameswaran)]. Whatever the type of wall (growing or nongrowing) and the method of decrustation used to unmask the microfibrils (DMSO for Figs. 5 and 6, xylanase and avicelase for Fig. 7, 1.3% H_2SO_4 for Fig. 8), a typical polylamellate construction with bow-shaped arrangements is seen. Scale bars: 500 nm.

thicker wall laid down in mature cells, is more susceptible to disorganization by the successive macerations. The fragility is likely due to the lability of chemical bonds and to the relative importance of alkali-soluble polysaccharides. Morphogenetic properties of elongating and nonelongating walls are probably more homogeneous than has been generally claimed (Roland, 1981).

2.4. Plywood Models and Bow-Shaped Arrangements

The regularity observed in the construction of walls has led to a comparison of them with a plywood construction model (Roland and Vian, 1979). In the wall, according to this model, multifibrillar layers are deposited on both sides of the middle lamella, which is comparable to the core of the plywood. The fibrillar subunits are parallel within the layers, and their orientation is criss-crossed. Even if an apparent variability exists, the plane of construction is always the same.

In the simplest case, only one multifibrillar layer is deposited on each side of the middle lamella; therefore, the wall is comparable to a three-ply construction. In the case of "primordial walls," the orientation of subunits seems to be regularly transverse. Generally, several multifibrillar layers are deposited and the wall is comparable to the multi-ply construction.

In fact, the interpretation of such plywoodlike structures can be complicated by the occurrence of bow-shaped patterns, generally observed on oblique sections. There has been a great deal of discussion concerning the significance of twisted arrangements; yet many biological materials display a twisted plywood architecture, such as cuticles of crabs and other crustacea, chromosomes of dinoflagellates, and connective tissues. On the basis of the work of Bouligand and his group (Bouligand, 1972; Livolant *et al.*, 1978), it has been shown that the bow-shaped patterns are not relevant to an actual curvature of subunits (for discussions, see Neville *et al.*, 1976; Roland and Vian, 1979). By combining several complementary ways of investigation (cytochemistry, stereo-observations, freeze-fracturing, and ultracryotomy), it appears that in the case of plant-cell walls, the twisted patterns indicate the existence of intermediate strata in which the polysaccharide orientation progressively rotates through a small angle intermediate between orthogonal directions (Fig. 9). The twisting is an illusion and corresponds to an oblique sectioning of the successive layers. Similar aspects of twisted patterns have recently been described by Parameswaran and Sinner (1979) in beech bark sclereids (compare Figs. 5 and 7, and 6 and 8). Their work led to the same interpretation.

The bow-shaped patterns are regularly seen with most of the extracting agents, chemical or enzymatic, provided that the extraction is mild. The use of specific enzymes, for example, purified endopolygalacturonase, provides information on their chemical nature (Roland and Vian, 1981). Endopolygalacturonase is an enzyme specific for polysaccharides of pectic type, i.e., polymers in covalent association with galacturonosyl-containing polysaccharides free of

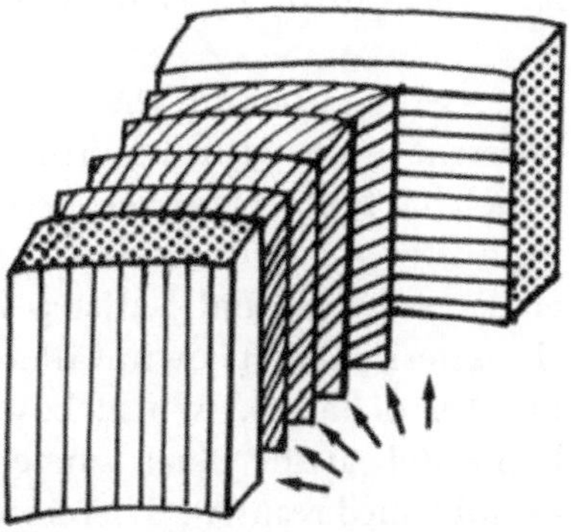

FIGURE 9. Model of a wall built up with successive layers. The microfibril orientation changes progressively through small angles intermediate between 0 and 90° (arrows). An oblique sectioning gives rise to bow-shaped patterns.

methyl ester groups. When the enzyme-treated wall is subjected to the PATAg test, the bow-shaped arcs are particularly well defined (Fig. 10). The rotating subunits themselves are resistant to the extraction. The intense decrustation around them indicates the occurrence of a massive embedding of unesterified polysaccharides extractable by endopolygalacturonase. The importance of this pectic embedding in the morphogenesis of the three-dimensionally ordered wall will be considered in Section 3.4.

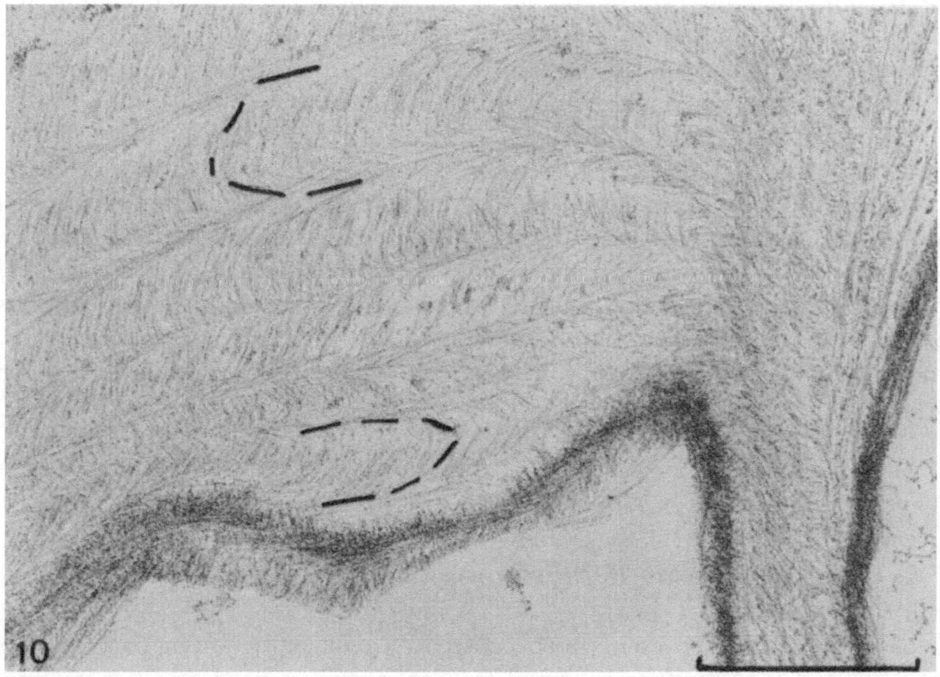

FIGURE 10. Regular bow-shaped arcs in an elongating wall. Epidermis of mung bean hypocotyl (zone of maximum elongation). Extraction with endopolygalacturonase and PATAg reaction. The arcs are sharply visualized throughout the thickness of the wall. They are due to straight and parallel subunits (underlined). Scale bar: 1 μm.

2.5. *Variability of Wall Texture within Organs*

As noted before, the cells do not reach the same stage of growth at the different levels within a growing organ. For example, mung bean hypocotyl, which has been extensively used as a material for growth experiments, presents a well-defined growth gradient (Prat, 1978, and bibliography therein). When the wall texture is analyzed step by step along the gradient, each level appears characterized by its own wall architecture. The more ordered zone—maximum number of layers, typical bow-shaped patterns—is the youngest zone of the hypocotyl, at the place where elongation starts (Roland and Vian, 1981 and unpublished results). Therefore, the maximum ordering corresponds to and is even necessary for the highest capacity for growth of a cell.

If controversy still exists concerning the occurrence and the relative importance of dispersed texture, it is probably in part because very early growth stages are transient and difficult to circumvent. In an elongating organ, some regions of wall may actually present a dispersed texture because of the rapid elongation. This always occurs in the outermost part of the wall, which can be considered as "senescent wall." In the innermost part, the cell still elaborates ordered and twisted patterns.

Another variability is expressed by a different complexity in wall architecture between internal and external tissues. The walls of internal tissues (mainly pith cells) remain consistently simple and nonlayered, like a typical three-ply construction. Conversely, the external tissues (mainly cortical parenchyma and epidermis) are more complex and correspond to multi-ply constructions. The ordering and the number of layers increase toward the outside of the organs, the epidermis being the most ordered wall (Prat and Roland, 1980). A direct relationship exists between the relative complexity of the morphogenesis of walls and their growth potentialities. Thus, the internal walls, in which morphogenesis is limited, seem to have a reduced capacity for mechanical resistance and support a nonpolarized and somehow passive extension, whereas at the same level, internal tissues increase in surface in a polarized manner by means of precise and selective events occurring in a three-dimensionally ordered wall.

3. *Positioning and Ordering of Wall Subunits*

3.1. *Wall Construction Results from a Secretory Process*

The cell wall represents a specialized type of extracellular covering. Its construction is relevant to a typical external secretion process, since it corresponds to a transport of products of synthesis from sites of formation in internal compartments to sites of release at the cell surface (Mollenhauer and Morré, 1966). The mechanisms of wall biosynthesis, studied by means of complementary techniques, have been often reviewed (for a complete discussion, see Morré and

Mollenhauer, 1976) and will not be dealt with here. The successive steps—polymerization, association, and positioning and orientation of wall subunits—are summarized only schematically in Fig. 11.

The diagram is in fact general, and many details are still to be provided, particularly concerning the exact sites of synthesis of the various wall subunits during the cytoplasmic steps of secretion.

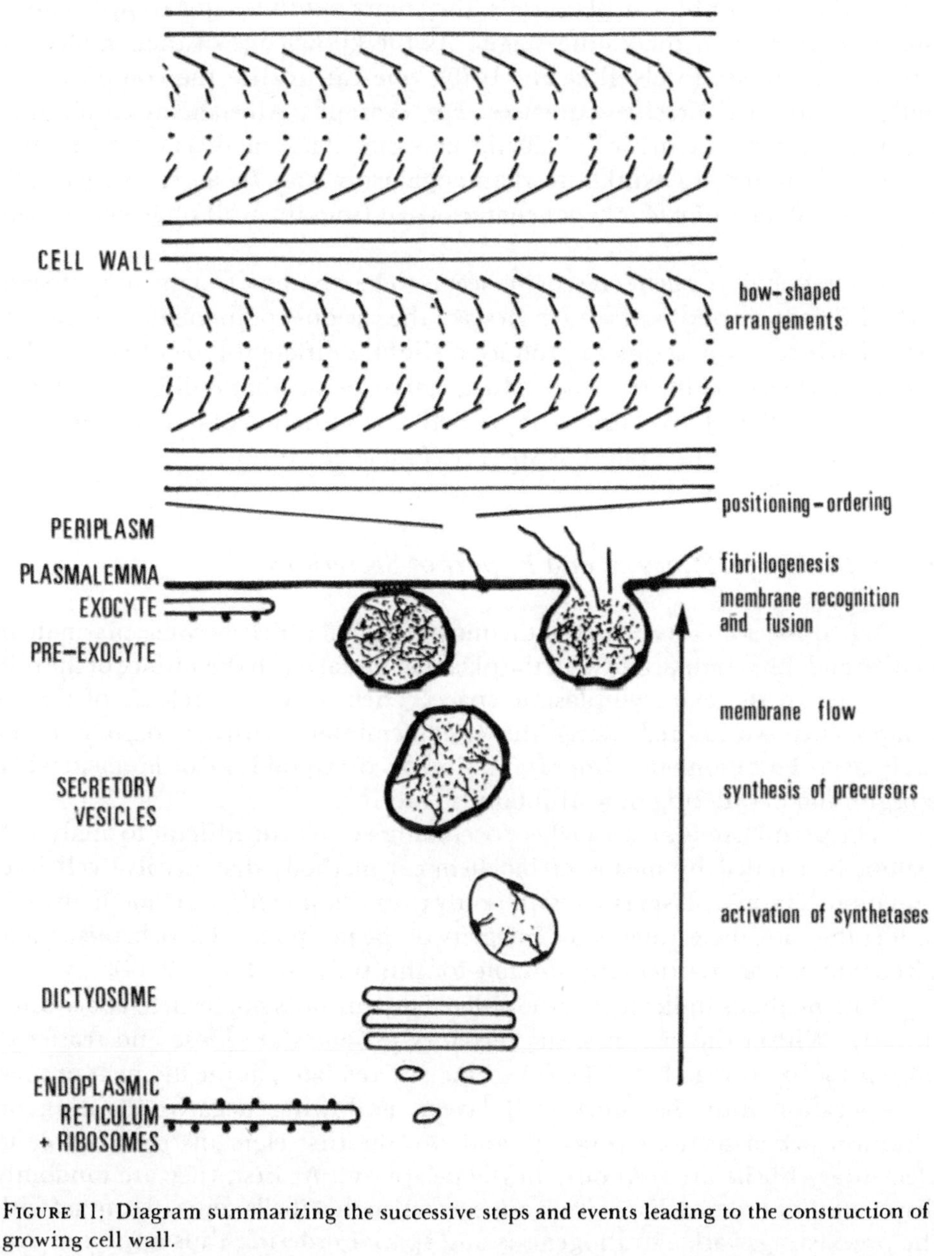

FIGURE 11. Diagram summarizing the successive steps and events leading to the construction of a growing cell wall.

Much attention has been focused on the sites of cellulose synthesis. It is accepted that in certain algae, cellulose synthesis is completed within Golgi cisternae and vesicles derived from the Golgi apparatus, i.e., before reaching the cell surface (see Brown and Willison, 1977). For higher plants, most authors accept the idea that cellulose deposition is a late event occurring at the surface of the plasmalemma, while pectic and hemicellulosic components are synthesized by the Golgi apparatus and derived secretory vesicles.

Much remains to be learned about the exact sites of synthesis of noncellulosic polysaccharides. In higher plant cells, they represent the major components of the wall, at least in the young stages. As the knowledge of their molecular structure increases (McNeil *et al.*, 1979), one can realize the complexity of building up such precise sequences. For example, Albersheim emphasized recently (see Albersheim *et al.*, 1980) the necessity of a great diversity of enzymes (about a hundred) to synthesize rhamnogalacturonan II, a very structurally complex polymer of pectic types characterized from the wall of growing plant cells.

Even though the general scheme seems to be common to a great number of cells, different chronologies of synthesis of the various components are likely to exist. Each cell type seems to produce a slightly different type of vesicle. For example, when cytochemistry is used as a marker of membrane differentiation, it indicates that the relative chronology of both accumulation of secretory products and membrane restructuration is specific (Prat *et al.*, 1977).

3.2. Opening of Exocytes and Export of Secretions

When the secretory vesicles, the membrane of which become plasmalemmalike and thus compatible with the plasmalemma, reach the surface of the cell, they open in the extracytoplasmic space. Then a massive release of matrix components synthesized along the endomembrane pathway occurs in the periplasm. This moment is important, since it corresponds to the moment when a beginning of ordering of wall subunits is seen.

The periplasm is an area where occurring events are difficult to analyze. It cannot be studied by means of biochemical methods that involve cell fractionation. The periplasm is accessible only to *in situ* investigation methods. The milder they are, the greater is the integrity of the periplasm. Cytochemistry and ultracryotomy are particularly suitable for this purpose (Figs. 12–14).

Both methods indicate that the ordered texture does not seem to be acquired directly. Within the exocytes, the secretory products are loose and scattered, appearing like thread balls. They become ordered later, at the moment of their incorporation into the inner wall layers. Following negative staining on ultrathin frozen sections (Figs. 12 and 13), the first elements comparable to elementary fibrils are seen only in the periplasm. At first, they are randomly disposed. The ordered disposition appears when the fibrils come in contact with the preexisting wall. Fibrillogenesis and spatial ordering thus appear as two subsequent steps, the physicochemical conditions encountered in the periplasm being determinant (see below).

3.3. *Role of the Plasmalemma and Cytoskeleton*

The literature concerning this point is abundant (extensive discussion and references can be found in Heath, 1974; Preston, 1974; Hepler, 1976; Brown and

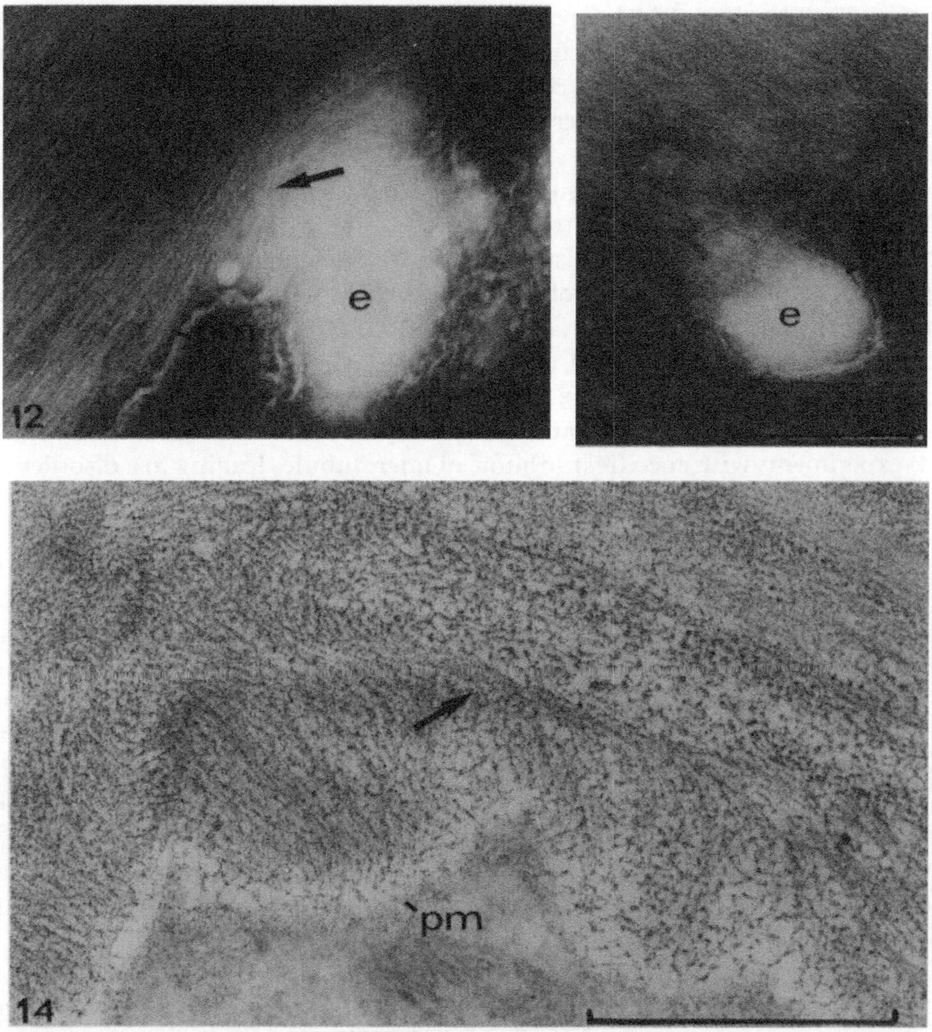

FIGURES 12–14. Opening of exocytes and release of wall subunits in the periplasm. Ultracryotomy and negative staining for Fig. 12 (pea root) and Fig. 13 (mung bean hypocotyl); cytochemistry (DMSO extraction, PATAg test) for Fig. 14 (mung bean hypocotyl). At the surface of the exocytes (e), the microfibrils are at first random. The arrows indicate the incorporation of polysaccharide chains into the inner layer of ordered wall. (pm) Plasmalemma. Scale bars: 0.25 μm (Fig. 12); 0.25 μm (Fig. 13); 1 μm (Fig. 14).

Montezinos, 1976; Giddings *et al.*, 1980; Mueller and Brown, 1980; Lloyd *et al.*, 1980). Moreover, the question is discussed by several authors in this volume (see Chapters 1, 4, 5, 6, 8, and 10) and concern algae, protoplasts, and higher plant cells.

Two main elements are generally taken into account in the cell-surface machinery engaged in the synthesis or the orientation, or both, of microfibrils: (1) the plasmalemma, the fluid structure of which is widely accepted and provides mobility for the presumptive integrated synthetases, and (2) the cytoplasmic structures lying underneath, i.e., the microtubules and the microfilaments. The more widely accepted scheme that could explain how oriented microfibril synthesis could occur in plant cells suggests that the plasmalemma-bound microfibril synthetase complexes are moving in the lipid bilayer via interactions with adjacent microtubules to which they are connected (Heath, 1974).

Different kinds of data support this hypothesis: (1) the occurrence of intramembrane particles or particle complexes of the plasmalemma ("terminal complexes",) (for a complete discussion, see Mueller and Brown, 1980); they are generally visualized by means of freeze-fracturing and appear to be involved in microfibril synthesis; (2) observations of various types of arrangement of those particles in rows, strings of particles, or more geometric and ordered arrays of particles (the so-called rosettes); (3) a correlation between the orientation of the microtubules in the cytoplasm and the orientation of the microfibrils in the wall; (4) experiments with specific inhibitors of microtubules leading to a disorder in wall texture.

In view of the three-dimensionally ordered texture of walls, especially elongating walls discussed in the previous sections, there is much concern with the particular question of how the wall subunits are oriented when deposited.

Alignments of particles have been generally described in algal or bacterial cells (see Preston, 1974; Brown and Montezinos, 1976). The recent results of Giddings *et al.* (1980) elegantly demonstrate an actual correlation between the structure of the complexes and the pattern of microfibril deposition in *Micrasterias.* Particle arrays have also been described in temporarily naked cells such as eggs of *Pelvetia* (Peng and Jaffe, 1976) or protoplasts of *Skimmia* (Robenek and Peveling, 1977). In higher plants grouped in tissues, the existence of such arrangements is still debated. The terminal complexes described in the freeze-fractured leaflets of the plasmalemma appear generally more randomly disposed (Brown and Willison, 1977; Mueller and Brown, 1980).

The exact involvement of microtubules in wall morphogenesis is also still a matter of discussion (Robinson, 1977). Evidence has been proposed for mutual orientation between cytoplasmic microtubules and microfibrils in different cases [e.g., secondary walls, tracheids, vessels, stomatal cells (for a review, see Hepler, 1976)]. In the case of actively elongating cells, no unequivocal evidence of microtubule organization that could explain the precise ordering of multi-ply constructions built up by the cells has yet been proposed. The exact wall–tubule interactions and in particular the way in which microtubules could impose a directional force on the microfibril-synthesizing machinery remains an open question.

3.4. Role of the Matrix: Gelation and Ordered Assembly

As previously pointed out, priority has often been given to the study of the skeletal fibrous wall components. They represent the polysaccharides that are crystalline (mainly cellulose) or that at least present some regularity in the way the molecules are packed. They may be studied by X-ray diffraction methods. The microfibrillar subunits are the only components left after the procedures necessary for shadow-casting. They have a structure visible in the electron microscope and represent the more easily visualized subunits.

However, the so-called matrix polysaccharides constitute an important proportion of cell walls, especially in young stages when an active secretion occurs. The importance of the matrix has long been underestimated and its role reduced to a sort of filling one. In particular, the study of the conformation, i.e., the shape in three dimensions that the molecular chains can adopt, should not be overlooked.

As with other macromolecules, primary, secondary, and tertiary structures may be described for polysaccharides to indicate the levels of organization. The primary structure corresponds to the covalent structure of monomer residues; the secondary structure represents the geometric arrangements that can be adopted in space (e.g., helix, ribbons, crumpled features); the tertiary structure is the way in which these arrangements pack together (Rees and Welsh, 1977). Little is known about the two latter structures. In this field, the investigations from a chemical viewpoint of Rees and his group are notable (Rees, 1977; Morris *et al.*, 1977, 1980) (for discussions see also Kirkwood, 1974; Preston, 1979; McNeil *et al.*, 1979).

One important property of matrix polysaccharides, mainly pectins, is their ability to form gels, i.e., an infinite three-dimensional network that can entrap the solvent and other molecules. The properties of polysaccharide gels have been studied mainly from polysaccharides of algal origin (Quatrano and Stevens, 1976; Rees, 1977). Gel-forming substances must have structural characteristics: the molecules must be long and contain periodic sequences capable of adopting ordered configurations, which are separated or "interrupted" by regions having a different type of regularity or no regularity at all. These so-called "junction zones" introduce a kink in the regular sequence and induce the chains to combine with a large number of partners. Thus, a three-dimensional gel network is established. The interruptions may be of different types (e.g., introduction of one or more different sugar units, chain branching). Figure 15 illustrates a generalized scheme for the mechanism of gelation and indicates some types of junction zones that have been recorded.

For pectins, the mechanism of gelation is less well understood. Pectins appear to gel by cooperative binding of divalent cations, mainly calcium. In the "egg-box" model proposed by Rees and his group (Rees, 1977), calcium fits

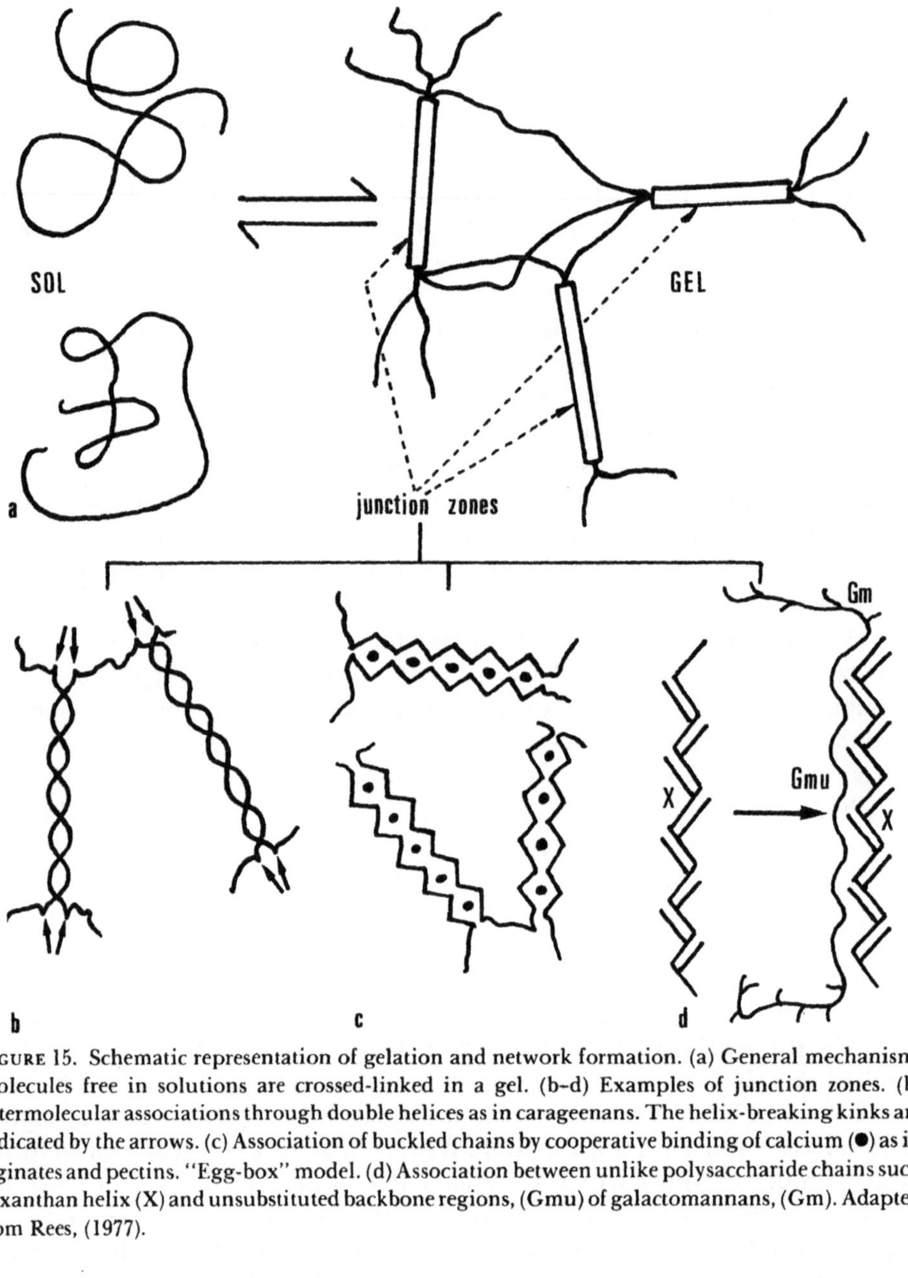

FIGURE 15. Schematic representation of gelation and network formation. (a) General mechanism. Molecules free in solutions are crossed-linked in a gel. (b–d) Examples of junction zones. (b) Intermolecular associations through double helices as in carageenans. The helix-breaking kinks are indicated by the arrows. (c) Association of buckled chains by cooperative binding of calcium (●) as in alginates and pectins. "Egg-box" model. (d) Association between unlike polysaccharide chains such as xanthan helix (X) and unsubstituted backbone regions, (Gmu) of galactomannans, (Gm). Adapted from Rees, (1977).

the cavities between polysaccharide chains. It is a cooperative association, since binding of the first cation between any pair of chains causes alignment that facilitates binding of the next, and so on along the sequence. Consequently, the array of ions between the chains looks like eggs held between the trays of an egg box. The result is both an increased rigidity and a cross-linking of polysaccharide chains.

Gelation phenomena are dependent on environmental conditions. Quatrano and Stevens (1976) have shown, from polysaccharides extracted from cell wall of *Fucus* zygotes (in this case, the major constituent is alginate), that the formation of clear, rigid gels with high mechanical strength depends mainly on the binding of Ca^{2+}-forming autocooperative junctions. A revised "domain model" has been proposed by Morris *et al.* (1980) as a common mechanism of polymer gelation. Specific chain–chain dimerization should be the primary mechanism of association in polyuronate systems such as pectins; further aggregation of such regions of ordered tertiary structure occurs only under favorable conditions of concentration and ionic strength.

In fact, the degree of interchain cross-linking via calcium ions is sensitive to the degree of methyl esterification of the galacturonan chains in the pectic polysaccharides. The egg-box mechanism seems to occur only for pectic polysaccharides of low methyl ester content. Thus, the strength of the association is affected by the distribution of those methyl galacturonan units. Pectins are known to be highly esterified. Results recently obtained by *in situ* extractions with purified endopolygalacturonase (Roland and Vian, 1981) have indicated an uneven distribution of methyl esterified pectins, nonsusceptible to enzyme attack, within a single elongating wall. A relationship likely exists between the ordering of the fibrillar units in bow-shaped arcs and the occurrence of a massive embedding of unesterified pectic polysaccharides. Mechanisms that control the gelling properties of the matrix, both at the time of synthesis and when it is in place, must exist. For example, the role of pectin esterases, the importance of which in growth processes is well established, should be thoroughly considered.

The idea that purely physicochemical phenomena may control wall morphogenesis is not new (see Roelofsen, 1959). In 1941, Frey-Wyssling (see Frey-Wyssling, 1953) had already postulated that within a gel, linear molecules with repetitive sequences could present ordered and even crystallized areas.

In vitro spontaneous assemblies are known to occur for the building up of protein structures such as viruses, microtubules, membranes, and others (Bouck and Brown, 1976; Sleytr, 1978). Typical assemblies have also been obtained *in vitro* from wall glycoprotein subunits of bacteria and of the alga *Chlamydomonas* (Catt *et al.*, 1978). In their study of the wall formation around the *Fucus* egg cell, Quatrano and Stevens (1976) have postulated that the wall assembly could be initiated by the self-assembly of alginates into a gel on the surface of the egg, triggered on fertilization and depending on the calcium contained in seawater.

Two series of data indicate that the morphogenesis of elongating walls in higher plants could in part result from self-assembly processes: (1) On the one hand, there is a strong correspondence between the twisted patterns observed in the elongating walls and the twisted patterns observed in other biological materials. It has been shown (see Bouligand, 1972; Neville and Luke, 1971; Neville *et al.*, 1976; Livolant *et al.*, 1978) that such systems are related to liquid crystals and that their morphogenesis implies self-assembly processes called mesomorphic states. (2) On the other hand, it has been shown that some wall fractions, mainly alkali-extracted fractions, have the ability to aggregate into fibrillar elements *in vitro* (Roland *et al.*, 1977; Reis, 1978). The aggregation

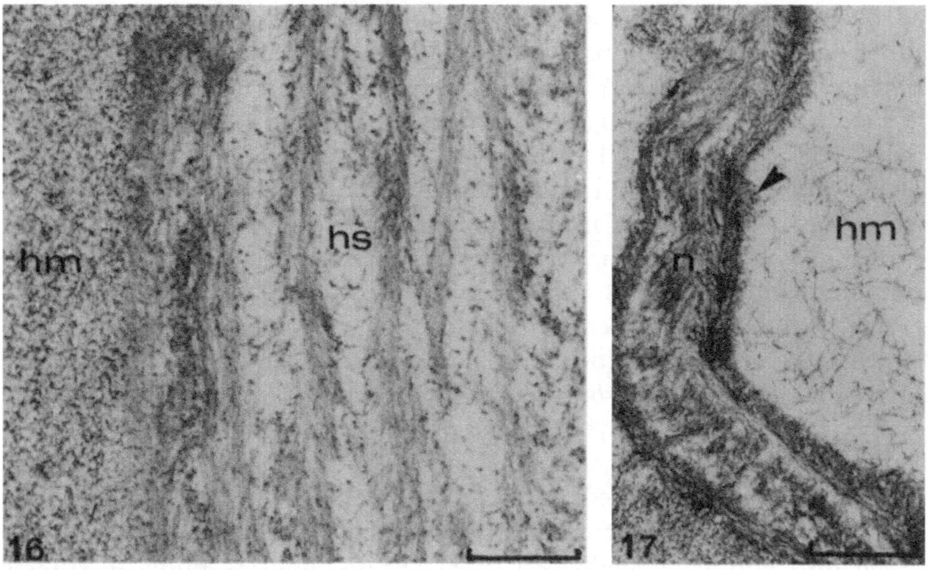

FIGURES 16 and 17. Gels assembled *in vitro* from alkali-extracted polysaccharides (walls of mung bean hypocotyl). Within a homogeneous gel (hm), subunits tend to aggregate in some areas (hs). (n) Multilayered nodule with wall-like texture. (▲) A zone where peripheral subunits are incorporated into the nodule. Scale bars: 1 μm (Fig. 16); 1 μm (Fig. 17). Figure 16 courtesy of D. Reis.

occurs either after a precipitation in ethanol or during the dialysis and gives rise to different types of gels. They can be homogeneous, but they often appear heterogeneous by the occurrence of birefringent nodules the structure of which evokes the *in vitro* texture of wall subunits (Figs. 16 and 17). Conditions of pH and ionic concentration appear to be important and to determine the capacity of aggregation of the subunits.

All these data led to the idea that *in vivo* arrangement results in part from a self-assembly process. When released into the periplasm, the polysaccharides could pass through mesomorphic states responsible for their characteristic orientation (for details, see Reis, 1978).

4. Concluding Remarks

From the data that have been discussed in this chapter, it appears that the assembly of the wall subunits cannot be explained simply in terms of a mechanical process, but rather is the result of a directed process. A defined morphogenetic program leading to the building up of a three-dimensionally ordered wall must exist.

We have emphasized the importance of the phenomena that occur at the surface of the cell, in the periplasmic area. It is obvious that our knowledge of these events is still fragmentary. In particular, new investigations will be required in order to gain more insight into the respective involvement of purely physicochemical phenomena (spontaneous assemblies of noncellulosic polysaccharides) and membrane phenomena (synthetic machinery) as directing forces whereby microfibril assembly and orientation are achieved. Experiments using plasmolyzed cells or naked protoplasts (Roland and Prat, 1973) indicate that the wall subunits built by the cells remain unorganized in the enlarged periplasm. They reveal the importance of the cell wall as a template for its own formation (Peng and Jaffe, 1976; Reis, 1978). The periplasmic area is a privileged zone wherein probably occur changes in pH and ionic strength that favor the formation and the breaking of bonds between the wall subunits. The three-dimensional assembly is also dependent on the concentration in this area and the chemical nature (repetitive or interrupted osidic sequences) of the released subunits. The latter one is ultimately subordinated to the activity of the synthetases along the endomembrane flow. It is expected that in the future we will greatly extend our knowledge of all these events.

References

Albersheim, P., 1975, The walls of growing plant cells, *Sci. Am.* **232**:81.

Albersheim, P., McNeil, M., Darvill, A. G., Valent, B., Hahn, M. G., Robertsen, B., Aman, P., Franzen, L. E., Desjardins, A., Ross, L. M., and Spellman, M., 1980, Recognition between plant cell and microbes is regulated by complex carbohydrates, 2nd Int. Congr. Cell Biol., Berlin, *Eur. J. Cell Biol.* **22**:230.

Bouck, G. B., and Brown, D. L., 1976, Self-assembly in development, *Annu. Rev. Plant Physiol.* **27**:71.

Bouligand, Y., 1972, Twisted fibrous arrangements in biological materials and cholesteric mesophases, *Tissue Cell* **4**:189.

Brown, R. M., and Montezinos, D., 1976, Cellulose microfibrils: Visualization of biosynthetic and orienting complexes in association with the plasma membrane, *Proc. Natl. Acad. Sci. U.S.A.* **73**:143.

Brown, R. M., and Willison, J. H. M., 1977, Golgi apparatus and plasma membrane involvement in secretion and cell surface deposition with special emphasis or cellulose, in: *International Cell Biology* (B. R. Brinkley and K. R. Porter, eds.), pp. 267–283, The Rockefeller University Press, New York.

Catt, J. W., Hill, G. J., and Roberts, K., 1978, Cell wall glycoproteins from *Chlamydomonas reinhardii* and their self-assembly, *Planta* **138**:91.

Erickson, R., and Silk, W. K., 1980, The kinematics of plant growth, *Sci. Am.* **242**:134.

Frey-Wyssling, A., 1953, *Submicroscopic Morphology of Protoplasm*, Elsevier, Amsterdam.

Frey-Wyssling, A., 1976, *The Plant Cell Wall*, Gebrüder Borntraeger, Berlin.

Giddings, T. H., Brower, D. L., and Staehelin, L. A., 1980, Visualization of particle complexes in the plasma membrane of *Micrasterias denticulata* associated with the formation of cellulose fibrils in primary and secondary cell walls, *J. Cell Biol.* **84**:327.

Heath, I. B. 1974, A unified hypothesis for the role of membrane bound enzyme complexes and microtubules in plant cell wall synthesis, *Planta* **146**:41.

Hepler, P. K., 1976, Plant microtubules, in: *Plant Biochemistry* (J. Bonner and J. E. Varner, eds.), pp. 141–187, Academic Press, New York.

Itoh, I., 1975, Application of freeze-etching technique for investigating cell wall organization of parenchyma cells in higher plants, *Wood Res.* **58**:20.

Kirkwood, S., 1974, Unusual polysaccharides, *Annu. Rev. Biochem.* **43**:401.

Livolant, F., Giraud, M. M., and Bouligand, Y., 1978, A goniometric effect observed in sections of twisted fibrous materials, *Biol. Cell.* **31**:159.

Lloyd, C. W., Slabas, A. R., Powell, A. J., and Lowe, S. B., 1980, Microtubules, protoplasts and plant cell shape, *Planta* **147**:500.

McNeil, M., Darvill, A. G., and Albersheim, P. A., 1979, The structural polymers of the primary cell walls of Dicots, in: *Progress in the Chemistry of Organic Natural Products*, pp. 191–249, Springer-Verlag, Berlin.

Mollenhauer, H. H., and Morré, D. J., 1966, Golgi apparatus and plant secretion, *Annu. Rev. Plant Physiol.* **17**:27.

Morré, D. J., and Mollenhauer, H. H., 1976, Interactions among cytoplasm, endomembranes, and the cell surface, in: *Transport in Plants III* (C. R. Stocking and V. Heber, eds.), pp. 288–344, Springer-Verlag, Berlin.

Morris, E. R., Rees, D. A. Thom, D., and Welsh, E. J., 1977, Conformation and intermolecular interactions of carbohydrate chains, *J. Supramol. Struct.* **6**:259.

Morris, E. R., Rees, D. A., and Robinson, G., 1980, Cation-specific aggregation of carrageenan helices: Domain model of polymer gel structure, *J. Mol. Biol.* **138**:349.

Mueller, S. C., and Brown, R. M., 1980, Evidence for an intramembrane component associated with a cellulose microfibril-synthesizing complex in higher plants, *J. Cell Biol.* **84**:315.

Mühlethaler, K., 1959, Plant cell walls, in: *The Cell* (J. Brachet and A. E. Mirsky, eds.), pp. 86–134, Academic Press, New York.

Neville, A. C., and Luke, B. M., 1971, A biological system producing a self-assembly cholesteric protein liquid crystal, *J. Cell Sci.* **8**:93.

Neville, A. C., and Gubb, D. C., and Crawford, R. M., 1976, A new model for cellulose architecture in some plant cell walls, *Protoplasma* **90**:307.

Northcote, D. H., 1972, Chemistry of the plant cell wall, *Annu. Rev. Plant Physiol.* **23**:113.

O'Brien, T. P., 1972, The cytology of cell wall formation in some eukaryotic cells, *Bot. Rev.* **38**:87.

Peng, H. B., and Jaffe, L. F., 1976, Cell wall formation in *Pelvetia* embryos: A freeze fracture study. *Planta* **133**:57.

Parameswaran, N., and Sinner, M., 1979. Topochemical studies on the wall of beech bark sclereids by enzymatic and acidic degradation. *Protoplasma* **101**:197.

Prat, R., 1978, Gradient of growth, spontaneous changes in growth rate and response to auxin of excised hypocotyl segments of *Phaseolus aureus, Plant Physiol.* **62**:75.

Prat, R., and Roland, J. C., 1980, Croissance différentielle des tissus et texture des parois, *Physiol. Veg.* **18**:241.

Prat, R., Vian, B., Reis, D., and Roland, J. C., 1977, Evolution of internal pressure, vacuolation and membrane flow during cell growth in mung bean hypcotyl, *Biol. Cell* **28**:269.

Preston, R. D., 1974, *The Physical Biology of Plant Cell Wall.*, Chapman and Hall, London.

Preston, R. D., 1979, Polysaccharide conformation and cell wall function, *Annu. Rev. Plant Physiol.* **30**:55.

Quatrano, R. S. and Stevens, P. T., 1976, Cell wall assembly in *Fucus* zygotes, *Plant Physiol.* **58**:224.

Rees, D. A., 1977, *Polysaccharide Shapes*, Chapman and Hall, London.

Rees, D. A., and Welsh, E. J., 1977, Secondary and tertiary structure of polysaccharides in solutions and gels, *Angew. Chem. Int. Ed. Engl.* **16**:214.

Reis, D., 1978, Precisions cytochimiques sur l'assemblage *in vitro* des hemicelluloses de l'hypocotyle de soja (*Phaseolus aureus* Roxb.), *Ann. Sci. Nat.* **19**:163.

Robenek, H., and Peveling, E., 1977, Ultrastructure of the cell wall regeneration of isolated protoplasts of *Skimmia japonica* Thunb., *Planta* **136**:135.

Robinson, D. G., 1977, Plant cell wall synthesis, *Adv. Bot. Res.* **5**:89.

Roelofsen, P. A., 1959, *The Plant Cell Wall*, Gebrüder Borntraeger, Berlin.

Roland, J. C., 1973, The relationship between the plasmalemma and plant cell wall, *Int. Rev. Cytol.* **36**:45.

Roland, J. C., 1978, General preparation and staining of thin sections, in: *Electron Microscopy and Cytochemistry of Plant Cells* (J. L. Hall, ed.), pp. 1–62, Elsevier, Amsterdam.

Roland, J. C., 1981, Comparison of arced patterns in growing and non-growing polylamellate cell walls of higher plants, in *Cell Walls '81, Proc. 2nd Cell Wall Meeting* (D. G. Robinson and H. Quader, eds.), pp. 162–170, Wissenschaftliche Verlagsgesellschaft, Stuttgart.

Roland, J. C., and Prat, R., 1973, in: *Protoplasts et Fusion de Cellules Somatiques Végétales, Colloq. Int. CNRS,* No. 212, pp. 243–271.

Roland, J. C., and Vian, B., 1979, The wall of growing plant cell: Its three dimensional organization, *Int. Rev. Cytol.* **61**:129.

Roland, J. C., and Vian, B., 1981, Use of purified endopolygalacturonase for a topochemical study of elongating cell walls at the ultrastructural level, *J. Cell Sci.* **68**:333.

Roland, J. C., Vian, B., and Reis, D., 1977, Further observations on cell wall morphogenesis and polysaccharide arrangements during plant growth, *Protoplasma* **91**:125.

Sleytr, U. B., 1978, Regular arrays of macromolecules on bacterial cell walls: Structure, chemistry, assembly, and function, *Int. Rev. Cytol.* **53**:1.

Vian, B., 1978, On the interpretation of twisted patterns in elongating plant cell wall: Informations obtained with ultracryotomy, *Protoplasma* **97**:379.

Vian, B., 1981, La technique des coupes au froid appliquée au matériel végétal, *Bull. Soc. Bot. Fr.* 128, *Actual Bot.* 89.

Vian, B., and Roland, J. C., 1980, Labelling of surface anionic sites with cationized ferritin or colloidal iron: Spacing of markers during cell growth and relation with wall architecture, *6th Int. Histochem. Cytochem. Congr. Brighton* (B. D. Lake, D. B. High, S. J. Holt, and P. J. Stoward, eds.), Royal Microsc. Soc., p. 396, Oxford.

Vian, B., Mueller, S., and Brown, R. M., 1978, A freeze-etching and replication study of wall deposition in elongating plant cells, *Cytobios* **22**:7.

Willison, J. H. M., 1976, An examination of the relationship between freeze-fractured plasmalemma and cell wall microfibrils, *Protoplasma* **88**:187.

3

Cell-Wall Formation in *Fucus* Zygotes: A Model System to Study the Assembly and Localization of Wall Polymers

RALPH S. QUATRANO

1. Introduction

Alterations in cell-wall structure and metabolism are closely coupled to the developmental processes of growth, differentiation, and morphogenesis. Although the structure and function of the cell wall have been studied extensively, relatively little attention has been devoted to the mechanisms by which a cell wall is initially assembled outside the plasma membrane. Understanding the steps and pattern in which various polymers are assembled into an extracellular structure will be helpful in elucidating a chemical and physical model of the cell wall. This developmental or inductive approach should complement that used by Albersheim, in which an existing cell wall is dissected with chemical and enzymatic techniques. From this, Albersheim's group can deduce the chemical architecture of an already existing structure (Keegstra *et al.*, 1973). Unfortunately, wall assembly in angiosperms is not amenable to biochemical experimentation, since most cases of initial wall construction occur within complex tissue masses and are difficult to isolate from other concurrent biochemical events. For example, in meristems, the construction of new cross-walls during telophase has been described cytologically, but to identify individual wall polymers and how they are linked to other polymers, as well as how the process of wall assembly is

RALPH S. QUATRANO • Department of Botany and Plant Pathology, Oregon State University, Corvallis, Oregon 97331.

directed to only the sites of new plasma-membrane formation (i.e., the "end walls"), is difficult to approach.

Regeneration of cell walls around enzymatically prepared protoplasts is an approach that attempts to create a natural system to study wall assembly around a wall-less cell. Although important information will be generated from studies using this approach (especially concerning optimal conditions for cell-fusion experiments), its usefulness as a model experimental system has some limitations. First, the lack of synchrony and absence of a homogeneous cell type in the regenerating system leads to the possibility that walls that are "different" (both temporally and qualitatively) are being regenerated in the population. Therefore, biochemical changes observed at any given time are in fact composites of an undefined number of individual wall regenerations occurring in the population. Second, since a heterogeneous mixture of enzymes is used to remove the wall, the extent of wall removal and membrane alterations cannot be precisely controlled or adequately measured. Cytologically and biochemically abnormal walls have been regenerated in such systems (Hanke and Northcote, 1974; Horine and Ruesink, 1972), and the question of whether such cells are still capable of normal growth or development has not always been adequately answered.

An ideal model system would be a natural one in which a large number of homogenous, wall-less cells are triggered to synchronously form cell walls under defined conditions. The zygote of the brown alga *Fucus* is such a system. Our long-term goal is to define the biochemical mechanisms and the regulatory steps involved in the processes by which specific wall polymers are transported, secreted, and finally assembled into an extracellular structure.

2. *Experimental System*

2.1. *Wall Formation in Fucus Zygotes*

Haploid eggs of the brown alga *Fucus* are spherical cells about 70 μm in diameter. They are released from the parent plant and fertilized in the open sea. If the same conditions are mimicked in the laboratory, gamete release and fertilization can occur reproducibly. Detailed procedures for the release of sperm and eggs, fertilization, and early development have been published elsewhere (Quatrano, 1980) and will not be discussed here. To highlight the developmental sequence, a cell wall is not present around the egg, but within minutes after fertilization an extracellular layer is evident. Using established laboratory culture procedures, every cell in a population of up to 10^8 zygotes possesses a cell wall by 30 min after fertilization. These zygotes develop synchronously with the increase in thickness of the cell walls for 10–12 hr in the absence of cell growth or cell division. At 14–16 hr, there appears a localized protuberance or rhizoid that is subsequently separated from the rest of the zygote by the first cell division at 20–24 hr after fertilization. Cell-wall assembly, rhizoid formation, and cell division occur synchronously in this homogeneous single-cell population. The cells are cultured aseptically, in a defined inorganic seawater medium. No light or external carbon source is required for early development.

In terms of studying cell-wall metabolism, this single-cell system offers two distinct and clearly separable stages that are experimentally approachable: (1) at fertilization, the assembly of the cell wall around a wall-less cell without concomitant growth or cell division, and (2) at rhizoid formation, localized wall extension. Both are under control of extracellular triggers; sperm activation of the egg is required for wall assembly, and the site of rhizoid formation can be determined by various gradients such as unilateral light (i.e., the rhizoid forms in the shaded hemisphere). Hence, the timing and localization of these processes are predictable and under experimental control (Quatrano, 1978).

Another distinct advantage of the *Fucus* system over angiosperms is that the number of polymers in the wall is relatively small each being distinct and well characterized. Greater than 95% by weight of the cell wall is composed of three major polysaccharides: cellulose, alginic acid, and the fucans or fucoidins (Quatrano and Stevens, 1976).

Alginic acid (alginate) is a binary, linear heteropolymer containing 1,4-linked β-D-mannuronic acid (M) and α-L-guluronic acid (G) residues in varying proportions. These monomers are arranged in a blockwise fashion within the same chain as $(MM)_n$, $(MG)_n$, and $(GG)_n$ stretches. It has been shown that the physical properties of alginate depend not only on the monomer composition but also on the relative proportion of these blocks. The ability of alginate to bind water and to form flexible gels is particularly dependent on the proportion of the various blocks in the polymer (Smidsrød, 1974).

The fucans are a complex array of fucose-containing, sulfated polysaccharides found only in brown algae. These polymers comprise a spectrum from high uronic acid/xylose and low sulfate content (F_1) to a relatively pure fucan sulfate (F_2). A more complete description of these polymers is given by Medcalf and Larsen (1977). Both alginate and the fucans are negatively charged, which offers the advantage of being amenable to separation, isolation, and characterization on the basis of charge.

It should also be pointed out that although the polymers in *Fucus* walls are composed of monomers different from those found in higher plants, the similarities are quite striking (Talmadge *et al.*, 1973; Quatrano and Stevens, 1976). For example, both sycamore and *Fucus* walls have about the same percentage of cellulose (23 vs. 20%) and the polyuronic acids (pectin in sycamore and alginate in *Fucus*) are similar in basic properties (e.g., Ca^{2+} binding, gel formation). The matrix polymers in sycamore (xyloglucans) and the fucan polymers in *Fucus* (xylofucan and galactofucan) are also found in similar percentages (21 vs. 20%). However, the sulfated property of the fucans is unique to the brown algae. Since the general properties and proportions of sycamore and *Fucus* walls are similar, extrapolation of results from this model algal system to higher plants can be made with more confidence.

2.2. *Composition and Structure of the Fucus Zygote Wall*

Levring (1952) was the first to describe the basic experimental system of cell-wall formation following fertilization in *Fucus*. Using light, polarizing, and electron microscopy, we essentially confirmed and extended his findings; i.e., in response to fertilization, an extracellular structure is formed *de novo*. In addition, we repeated the work of Nakazawa *et al.* (1969) showing by use of the fluorescent dye Calcofluor White (which binds to β-linked polysaccharides) that the rhizoid wall is also formed *de novo* and does not represent a local stretching of the preexisting zygote wall. We also developed a mild technique (homogenization in distilled water followed by sonication) to isolate walls intact and free of cytoplasmic contamination (Fig. 1). Only four major polymers were identified in cell walls from 24-hr zygotes: alginate (60%), cellulose (20%), and two electrophoretically distinct fucans, F_1 and F_2 (20%). The rate of carbohydrate deposition, the appearance of ordered molecules in the wall (birefringence), and the gross chemical composition of the wall during assembly are given in Fig. 2 and 3 (Quatrano and Stevens, 1976).

A wall can be detected within minutes by plasmolysis, electron microscopy, and Calcofluor White. Alginate is detected cytochemically by 15 min, and cellulose is found in homogenates of zygotes (but not in eggs) by 30 min. However, walls cannot be isolated until 60 min, and at that time only fragmented walls can be obtained. In such fractions, both alginate and cellulose are present. Birefringence was not observed until 60 min and is due to cellulose and not alginate. Since the birefringence is negative in the radial direction, it indicates that ordered cellulose microfibrils are arranged tangentially to the surface. F_1 is first detected in the wall at 60 min, corresponding to the time at which: (1) cellulose birefringence is first observed, (2) walls can be isolated in an intact condition, and (3) an increase in the internal pressure potential of the zygote can be measured. All this evidence points to a stable wall structure beginning at 60

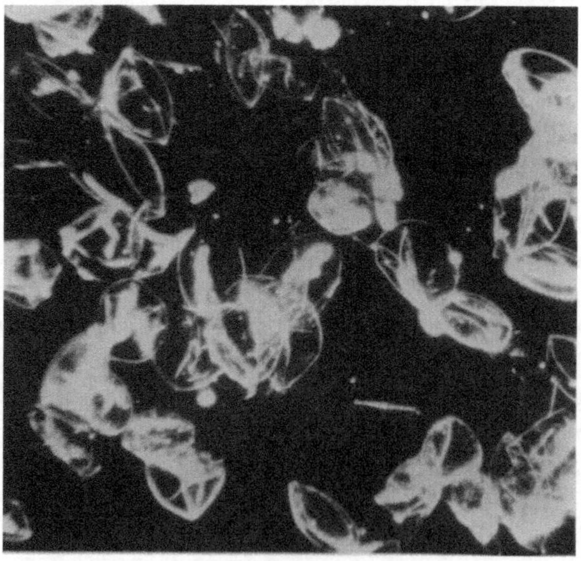

FIGURE 1. Dark-field image of a purified cell-wall preparation from 10-hr zygotes of *Fucus vesiculosus*. ×100. From Quatrano and Stevens (1976).

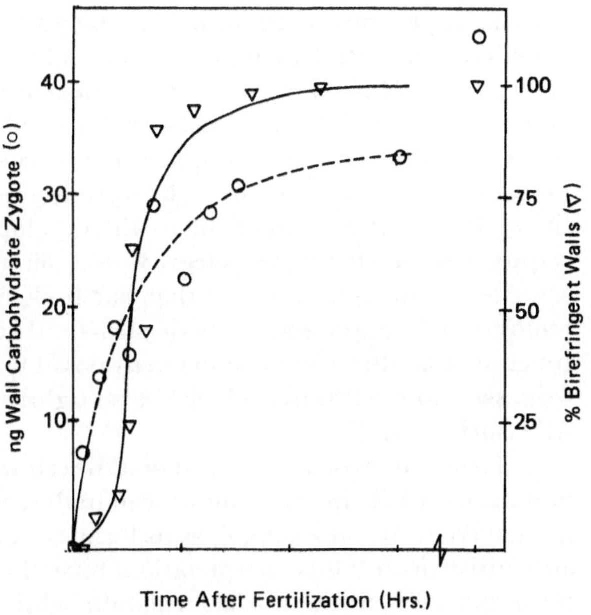

FIGURE 2. Time-course of the appearance of cell-wall birefringence and the accumulation of carbohydrate in isolated cell walls of *Fucus vesiculosus* zygotes. From Quatrano and Stevens (1976).

min after fertilization. In fact, 50–75% of a given population of isolated walls after 2–4 hr retain the characteristic shape of the zygote or embryo from which they were isolated (Quatrano and Stevens, 1976).

The delay in the appearance of birefringence (60 min) in relation to the presence of cellulose in isolated walls (15–30 min) could be due to the low density of cellulose microfibrils in the early wall. However, the possible interaction of F_1 and cellulose in microfibril orientation during *wall assembly* should not be overlooked. Although stabilization of the cell wall during isolation is coincident

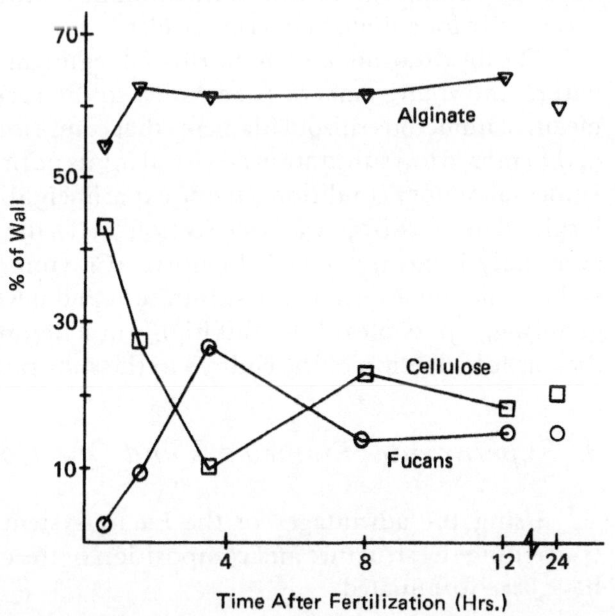

FIGURE 3. Changes in the major polysaccharide composition of isolated cell walls from *Fucus vesiculosus* at different times after fertilization. From Quatrano and Stevens (1976).

with the appearance of the fucan F_1, subsequent extraction of most of the fucans and all the alginate does not result in collapse of the characteristic shape of the embryo wall. Cellulose appears to be the main determinant of cell shape. But even in the most highly purified samples of the cellulose fraction from embryo walls, 5–15% of the neutral sugars can be accounted for by equal quantities of xylose and fucose, the remaining being only glucose. The facts that F_1 appears as the earliest *matrix* component in the developing wall (coincident with the acquisition of structural integrity and birefringence of the wall) tightly associated with cellulose and that harsh chemical treatment is required for removal of F_1 argue for the basic structural role of F_1 in wall assembly and function. Can this xylofucan in *Fucus* possibly be equated with the xyloglucan from sycamore cell walls, which is also tightly bound to cellulose (Valent and Albersheim, 1974)?

Several pieces of data suggest that the cell wall of the two-celled embryo is a mosaic, with F_2 being the major fucan in the rhizoid wall and the remainder of the embryo wall containing F_1 as its fucan component. Three pieces of evidence are consistent with this interpretation: First, the percentage of fucan in the wall remains at 20% during rhizoid formation, while the percentage of the monomer fucose in the wall *doubles* between 10 and 24 hr. Since F_2 has a higher fucose content than F_1, these results are consistent with F_2 substituting for F_1 in the rhizoid wall rather than F_2 *and* F_1 being added to the rhizoid cell wall (Quatrano and Stevens, 1976). Second, compared to F_1, F_2 is heavily sulfated and becomes preferentially labeled when 16-hr zygotes are incubated in $^{35}SO_4$ (Quatrano *et al.*, 1979). When two-celled embryos are specifically stained for sulfated polysaccharides or are analyzed autoradiographically after a $^{35}SO_4$ pulse, there is intense staining or an accumulation of localized silver grains associated with the rhizoid wall (Brawley and Quatrano, 1979). Finally, Vreugdenhil *et al.* (1976) have shown that the rhizoid wall contains a unique fibrillar layer that stains specifically for sulfated polysaccharides.

The localization of F_2 in the rhizoid region may have a specific function. In nature, the young embryo is anchored to the rocks in the intertidal region by means of numerous rhizoid filaments that comprise the holdfast. The attachment of the embryo to a substratum is critical for its normal development and survival. Under laboratory conditions, if one experimentally prevents F_2 from becoming localized in the rhizoid wall (see Section 3.1.1), the zygote develops a rhizoid and eventually forms a two-celled embryo (Crayton *et al.*, 1974). However, these embryos do not attach to any substratum and never develop a typically shaped germling. A possible role for this highly mucilaginous, sulfated polysaccharide, then, is for adhesion of the embryo to the substratum.

3. *Experimental Framework and Questions*

Using the advantages of the *Fucus* system and this basic information concerning the structure and composition of the cell wall, two major questions have been formulated:

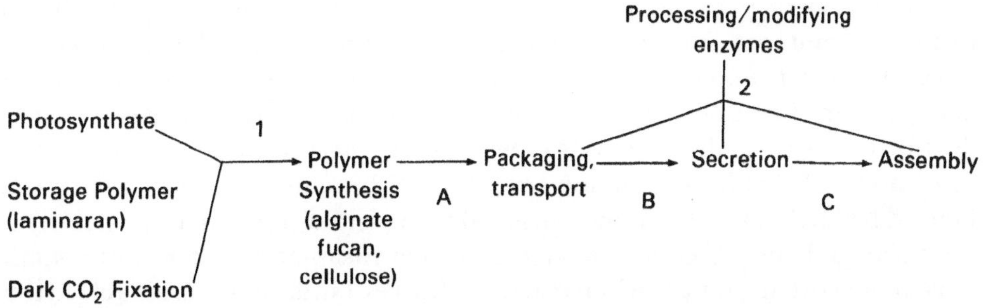

FIGURE 4. Probable biochemical and cellular pathways of wall polymers from synthesis to insertion into the cell wall. Biochemical experiments have focused on the synthetic (1) and processing (2) enzymes, while events indicated by A, B, and C are being approached cytologically.

1. What are the mechanisms by which specific polymers are assembled into the *Fucus* cell wall at fertilization and at what level of genetic expression are these controlled?
2. What is the mechanism by which a polysaccharide becomes incorporated into a localized region of the cell wall?

Figure 4 presents a general scheme of the steps to be elucidated from polysaccharide synthesis to assembly into the extracellular structure. In the context of this scheme and the two major questions posed above, the next section of this chapter will deal specifically with: (1) the role of enzymatic sulfation in the incorporation of the fucan F_2 into only the rhizoid wall of the two-celled embryo (Sites 2 and A, B, and C); (2) the control of cellulose synthesis (Site 1); and (3) the secretion and assembly of preformed alginate and other wall polymers in response to fertilization (Site C).

3.1. Localization and Deposition of Specific Polymers

3.1.1. Fucans

Although F_1 is incorporated into the wall by 4 hr after fertilization and may play an important role in the structural integrity of the wall (see Section 2.2), most of our effort has been focused on the local incorporation of F_2 into the rhizoid wall at 12–14 hr after fertilization. Acid extraction of zygotes or purified walls results in a fraction enriched in F_2, while a subsequent base extraction yields a fraction enriched in F_1 (Hogsett and Quatrano, 1975). Using acid extraction, we concluded that F_2 is present in an unsulfated form by 8 hr. It has not been clearly established yet what percentage of this F_2 is preformed in the egg or synthesized within the first 8 hr following fertilization. Beginning at about 10 hr, F_2 is enzymatically sulfated within Golgi-derived vesicles scattered at random in the cytoplasm. At the time of rhizoid formation, these vesicles accumulate and fuse with the plasma membrane at the rhizoid tip, where they secrete their contents of sulfated fucans into the wall (Quatrano and Crayton, 1973; Brawley and Quatrano, 1979; Quatrano *et al.* 1979).

The evidence that F_2 is present in an unsulfated form prior to rhizoid formation, and that sulfation occurs on a preexisting but unsulfated rather than F_2 synthesized *de novo*, comes from several sources. First, the amount of fucan per zygote in the acid-extracted fraction from whole embryos remains constant during the first 24 hr of development. Also, there is little metabolism or "turnover" of this fucan fraction from 8 to 36 hr, since 4-hr pulses with ^{14}C-labeled $NaHCO_3$, fucose, or mannose failed to yield significant radioactivity in the fucan polymer. Second, the acid-extracted fucan at 8 hr does not contain ester-linked sulfate, but at 16 hr it reaches a degree of sulfation comparable to that of fucan from mature plants. When zygotes are incubated with $^{35}SO_4$ for 60-min intervals during the first 24 hr of development, cells accumulate ^{35}S in the cytoplasm but do not incorporate it into ester-linked sulfate on the fucan chain of F_2 until 8–10 hr. At that time, there is a dramatic increase in sulfate incorporation into fucan, which is sustained for several hours. The data support the conclusion that there does not appear to be *de novo* synthesis of F_2 but rather a postsynthetic modification (i.e., sulfation) of a preformed fucan (Quatrano and Crayton, 1973).

This enzymatic sulfation of the fucan polymer appears to be required for localized insertion of F_2 into the rhizoid wall. Using cytochemical (Hogsett and Quatrano, 1978) and pulse-chase autoradiographic techniques (Brawley and Quatrano, 1979), we demonstrated that if F_2 is prevented from becoming sulfated (by growing zygotes in seawater lacking SO_4 but containing methionine to allow protein synthesis to continue), its presence in the rhizoid wall cannot be detected. Although the requirement for sulfation is thought to be responsible for transport to a specific cell-wall site [possibly driven by an intracellular electric field (refer to Jaffe and Nuccitelli, 1977)], its role in secretion, assembly, and rhizoid-wall function has not been thoroughly investigated. In addition to a negatively charged, sulfated F_2 being required, localization of F_2 is also dependent on a cytochalasin B (CB)-sensitive process, possibly a cytoskeletal structure such as microfilaments (Brawley and Quatrano, 1979). In the presence of CB, sulfation of F_2 occurs, but directional migration of the vesicles containing F_2 to the rhizoid site does not occur. Intracellular movement of vesicles containing wall material in other plant systems is also inhibited by CB (Mollenhauer and Morré, 1976; Pope *et al.*, 1979; Weiss and Wilbur, 1978). To summarize, preformed, unsulfated F_2 located at random in Golgi vesicles must be enzymatically sulfated and interact in some way with a CB-sensitive process or structure (possibly microfilaments) to become transported and secreted into the rhizoid, a site in the cell wall that is predetermined and fixed by some environmental gradient. A more complete discussion of the data and hypotheses concerning the mechanism for vesicle localization or transport in *Fucus* has been given in detail elsewhere (Quatrano, 1978; Brawley and Quatrano, 1979; Quatrano *et al.*, 1979). From the data presented here, it is clear that the processing and modifying enzymes of the F_2 polymer responsible for its sulfation are a critical factor in the assembly of F_2 into a specific wall location. The biochemical steps and the controls involved with fucan sulfation now become the central concern of our biochemical approach to understand the regulation of the local deposition of F_2.

TABLE I
Proposed Intermediates of Sulfate Transfer from Seawater to Fucoidin (F_2) and Their Presence (+) or Absence (−) in *F. distichus* Embryos at 8 and 16 Hours after Fertilization

Reaction[a]	Product[a]	8 Hours	16 Hours
$ATP + SO_4 \xrightarrow[\text{ATP sulfurylase}]{} APS + PPi$	APS	+	+
$APS + ATP \xrightarrow[\text{APS kinase}]{} PAPS + ADP$	PAPS	−	+
$PAPS + F_2 \xrightarrow[\text{Sulfotransferase}]{} F_2\text{-S} + PAP$	F_2-S	−	+

[a] (APS) Adenosine 5′-phosphosulfate; (PAPS) 3′-phosphoadenosine-5′-phosphosulfate; (PPi) pyrophosphate; (F_2) fucoidin (unsulfated); (F_2-S) fucoidin (sulfated).

Recently, we have determined that the sulfate intermediates adenosine 5′-phosphosulfate (APS) and 3′-phosphoadenosine-5′-phosphosulfate (PAPS) (Table I) are found in *F. distichus* embryos, and when pulsed with $^{35}SO_4$ at the time of F_2 sulfation both were labeled with ^{35}S. However, prior to sulfation (8 hr), the same amount of sulfate enters the zygote, but only APS is labeled and not PAPS (Table I). This would seem to indicate that the enzyme APS kinase and possibly the sulfotransferase are limiting and, coupled with previous results showing that new protein synthesis is required for sulfation (Quatrano and Crayton, 1973), suggests that these enzymes must be newly synthesized for F_2 to become sulfated and subsequently localized and inserted. Isolation, characterization, and purification of these enzymes are now in progress to obtain an *in vitro* sulfating system that should lead to a better understanding of the controls involved.

3.1.2. Alginic Acid (Alginate)

Alginate is found in vesicles and is deposited into the extracellular wall almost immediately after fertilization. Since alginate possesses a low M/G ratio and a block distribution consisting mainly of GG blocks (Quatrano and Stevens, 1976), the wall alginate is a transparent, flexible gel with high mechanical strength. The time of deposition suggests that it might already be present in the egg and on fertilization is released to the surface after fusion of stored vesicles with the plasma membrane. The high amount of G in the wall alginate and the very high probability of its forming a gel in the presence of Ca^{2+} suggest a mechanism for the initial wall assembly: the formation of an alginate gel at the egg surface when the intracellular soluble alginate is exposed to the Ca^{2+} in the seawater. To test this mechanism, we fertilized eggs in the absence of Ca^{2+}. No cell wall is evident when zygotes are stained with Calcofluor White, plasmolyzed, or viewed with polarized light. Fertilization occurs, however, since the immediate addition of Ca^{2+} to the washed eggs results in the entire population forming a wall, rhizoid, and eventually a two-celled embryo. As shown in Table II, the absence of Ca^{2+} in the artificial seawater medium also results in the release of wall polymers into the medium. The secretion of polymers is in part inhibited by KCN (10 mM) and is not evident in unfertilized eggs or eggs fertilized in the

TABLE II
Release of Wall Polymers from Fertilized or Unfertilized Eggs in the Presence or Absence of Ca^{2+} in an Artificial Seawater Medium

		Content (μg/μl)		
Eggs	Ca^{2+}	Hexose[a]	Fucose[a]	Uronic acid[a]
Unfertilized[b]	+ or −	0.85	0.08	0.21
Fertilized	+	0.65	0.06	0.21
Fertilized	−	5.00	0.20	0.42

[a] A 3.6-fold concentration (lyophilization) of a dialyzed seawater medium that had the fertilized or unfertilized eggs removed.
[b] No cell wall is detected in these treatments when assayed by plasmolysis.

presence of Ca^{2+}. Fertilized eggs grown in the absence of Ca^{2+} will now allow us to separate the processes of secretion from those of wall assembly. These preliminary results suggest that the initial wall is composed of Ca-alginate in the form of a matrix or gel, within which the subsequent fibrous wall components (e.g., cellulose) are deposited. Without the Ca^{2+}, alginate is not assembled into a structural gel and the polysaccharides normally inserted into the wall later are secreted into the medium. The polymers released have not yet been analyzed. Although the mechanism now appears to be a physical phenomenon (sol–gel transformation), it is still possible that fertilization results in the activation of an alginate epimerase, the activity of which has been shown to change M units into G units at the polymer level (Larsen and Haug, 1971). This polymer-processing or -modifying enzyme may be essential for the generation of more G units in the alginate stored in the egg in order for the alginate to have high gel-forming ability in the presence of Ca^{2+}. The M/G ratio of alginate in the egg cytoplasm compared to the M/G ratio in the zygote wall, as well as the activity of an epimerase following fertilization, should distinguish between these alternatives.

3.1.3. *Cellulose*

A hot, alkali-insoluble cell-wall component of *F. distichus* embryos was shown to be cellulose by chemical and enzymatic hydrolysis as well as by X-ray diffraction (Stevens and Quatrano, 1978). Cellulose was not found in eggs, but within 20–30 min following fertilization was present in cell walls of zygotes (Fig. 5). Incorporation of [^{3}H]glucose or $NaH^{14}CO_3$ confirms that cellulose in zygotes of *F. distichus* and *F. vesiculosus* is newly synthesized, with the highest rate of synthesis occurring within the first 60 min (Fig. 6). However, the increase in cellulose *does not* require protein or RNA synthesis. This is consistent with a posttranslational level of genetic control (Stevens and Quatrano, 1978).

The cellular site of cellulose synthesis in fucoid embryos has not been determined conclusively, but involvement of the plasma membrane is very likely. We do know that cellulose is assembled extracellularly (see Fig. 5) and is not formed within Golgi vesicles, as is the case with the cellulosic extracellular scales

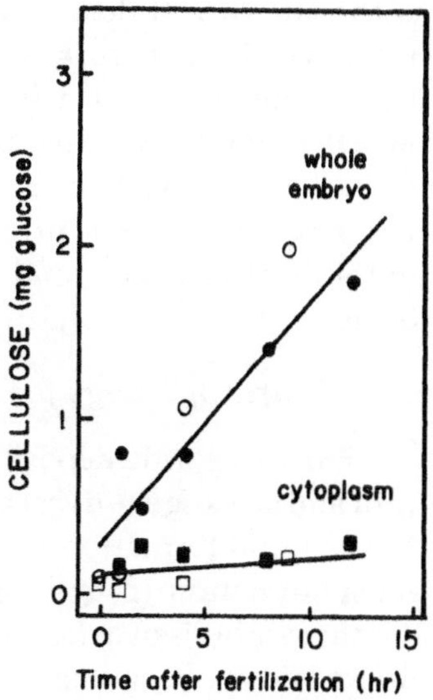

FIGURE 5. Amount of cellulose in whole embryos and cytoplasm from the same population of *Fucus distichus* at various times after fertilization. From Stevens and Quatrano (1978).

of the alga *Pleurochrysis* (Herth *et al.*, 1972) (see also Chapter 7). Vesicle–plasma membrane fusion may serve to bring molecules involved in cell-wall construction to the cell surface following fertilization (Brawley *et al.*, 1976). Strings of membrane particles on the plasma membrane, which form shortly after fertilization at the site and orientation where microfibrils later develop, also appear to be involved with cell-wall deposition (Peng and Jaffe, 1976), possibly

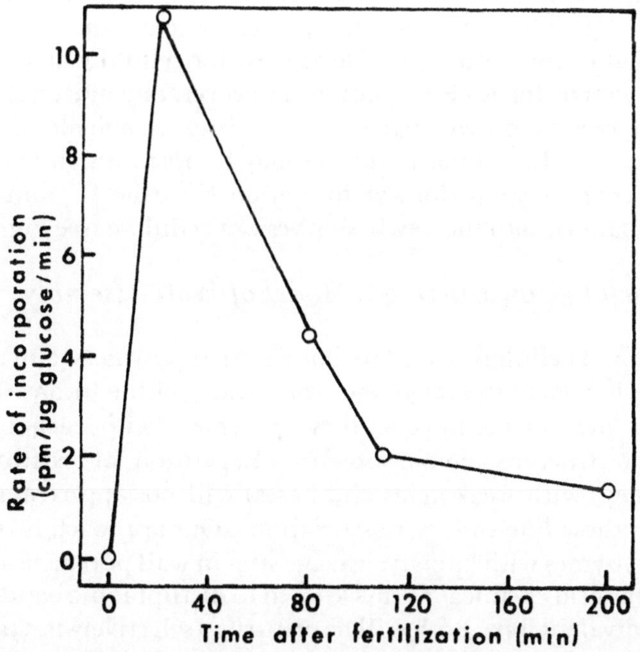

FIGURE 6. Rate of incorporation of ^{14}C from $NaH^{14}CO_3$ into the cellulose fraction of *Fucus vesiculosus* embryos at different times after fertilization. From Stevens and Quatrano (1978).

as the site of cellulose synthesis. However, the nature of these structures is unknown. Recent work with higher plants indicates that specialized particle rosettes are associated with the plasma membrane and cellulose microfibrils (Mueller and Brown, 1980). Thus, the mechanism of cellulose deposition into the developing wall of *Fucus* zygotes appears to involve the activation of preformed cellulose synthetase at the time of fertilization, the extracellular synthesis of cellulose, and its incorporation into the alginate matrix already formed.

3.1.4. Source of Energy for Wall Construction

Polysaccharide deposition into the cell wall takes place in the absence of light and an exogenous carbon source. An endogenous energy source is required for the rapid formation of the wall structure, since an uncoupler of oxidative phosphorylation (*p*-trifluoromethoxyphenylhydrazone) blocks wall assembly and the synthesis of at least one wall component, cellulose. The required carbon blocks and energy are most likely derived from the enzymatic hydrolysis of the storage β1,3-glucan (laminaran) that is found in *Fucus* eggs (see Fig. 1). Laminaran isolated from the cytoplasm of *Fucus* zygotes decreases during the first 7 hr of development. Zygotes from this stage of development also possess a glucanase that is found in a particulate fraction. When released *in vitro* by detergent, it is capable of hydrolyzing native laminaran to glucose. The enzyme appears to be an exo-β1,3-glucanase, since glucose and not laminarabiose or other oligosaccharides were found in incubation mixtures of the enzyme and native [^{14}C]laminaran (Quatrano and Stevens, 1976). Relatively large increases in respiration rate following fertilization of *Fucus* eggs have been reported (Whitaker, 1931). Thus, the zygotes possess a mechanism for laminaran breakdown that in turn could provide glucose for cellulose and other polymer syntheses, as well as for ATP formation. However, one must not overlook the evidence that certain brown algae fix CO_2 into carbohydrates in the dark (Nisizawa *et al.*, 1971). Similar pathways may be operating in *Fucus* zygotes to provide the carbon substrates for wall formation, since the ^{14}C from $NaH^{14}CO_3$ is rapidly incorporated into the newly synthesized cellulose (see Fig. 6).

3.2. Approach to Constructing a Model of Wall Assembly

Alginate and cellulose are well-characterized polymers, and several recent studies have elucidated the range and complexity of the fucans. However, the relationships and interactions of these polymers with each other in wall assembly and structure remain obscure. Repetition of existing extraction techniques (e.g., with weak acids and bases) will not appreciably extend our knowledge of these interpolymer associations. Our approach is twofold: (1) to use purified enzymes with known cleavage sites in wall polymers and (2) to treat cell walls with mild chemical agents known to disrupt ionic bonds, or H bonds or to remove divalent ions. Both will be utilized to selectively fractionate the cell wall so that the released oligomers can be characterized. Each of these released "subsets" of the 24-hr zygote wall will be used in the construction of a model of the mature wall in much the same way as done with higher plants (Keegstra *et al.*, 1973). Once several subsets have been fractionated and characterized from the 24-hr wall, we can trace their origins and ask when each was synthesized, secreted, and assembled into the wall.

As a start in this direction, we have recently isolated several strains of marine bacteria that are capable of growing in liquid culture containing a single purified wall polymer (e.g., alginate, fucoidin, *or* cellulose) as the sole carbon source (Quatrano and Caldwell, 1978). Alginases have been isolated both from the medium and from a sonicated extract. Several strains of bacteria have demonstrated the production of enzymes that selectively cleave G blocks in preference to M blocks, whereas others cannot distinguish between the different blocks (Quatrano *et al.*, 1981). Similar enzymes have been purified from a number of other sources (cf. Elyakova and Favorov, 1974). In addition, these partially purified enzymes can act on intact embryos and isolated cell walls, suggesting their potential usefulness in preparing protoplasts and selectively fractionating cell walls (Quatrano *et al.*, 1981).

We have developed in *Fucus* a method to isolate oligmers released from intact walls by the removal of noncovalent bonds, similar to the approach of Catt *et al.* (1978), with cell-wall glycoproteins of *Chlamydomonas*. *Fucus* walls from different stages of postfertilization development were isolated (20–50 mg) and suspended in an aqueous solution within a dialysis membrane. The membrane was immersed in a different solution containing, for example, the Ca^{2+} chelator EGTA (10 mM). With continuous stirring for several hours and several changes of the external solution, the contents within the bag were centrifuged (5000*g*, 10 min). The EGTA was removed from the supernatant by dialysis or ultrafiltration and the released oligmers concentrated by lyophilization. Using ^{14}C-labeled walls, EGTA can release about 30% of the radioactivity into a nondialyzable fraction that contains primarily uronic acids and significant amounts of fucose. The same method could be used on intact walls employing urea, guanidine, and/or high salt before and after enzymatic treatment.

References

Brawley, S. H., and Quatrano, R. S., 1979, Sulfation of fucoidin in *Fucus* embryos. IV. Autoradiographic investigations of fucoidin sulfation and secretion during differentiation and the effect of cytochalasin treatment, *Dev. Biol.* **73:**193.

Brawley, S. H., Wetherbee, R., and Quatrano, R. S., 1976, Fine-structural studies of the gametes and embryo of *Fucus vesiculosus* L. (Phaeophyta). II. The cytoplasm of the egg and young zygote, *J. Cell Sci.* **20:**255.

Catt, J. W., Hills, G. J., and Roberts, K., 1978, Cell wall glycoproteins from *Chlamydomonas reinhardii*, and their self-assembly, *Planta* **138:**91.

Crayton, M. A., Wilson, E., and Quatrano, R. S., 1974, Sulfation of fucoidin in *Fucus* embryos. II. Separation from initiation of polar growth, *Dev. Biol.* **39:**164.

Elyakova, L. A., and Favorov, V. V., 1974, Isolation and certain properties of alginate lyase VI from the mollusk *Littorina* so., *Biochim. Biophys. Acta* **358:**341.

Hanke, D. E., and Northcote, D. H., 1974, Cell wall formation by soybean callus protoplasts, *J. Cell Sci.* **14:**29.

Herth, W., Franke, W. W., Stadler, J., Bittiger, H., Keilich, G., and Brown, R. M., 1972, Further characterization of the alkali-stable material from the scales of *Pleurochrysis scherffelii:* A cellulosic glycoprotein, *Planta* **105**:79.

Hogsett, W. S., and Quatrano, R. S., 1975, Isolation of polysaccharides sulfated during early embryogenesis in *Fucus*, *Plant Physiol.* **55**:25.

Hogsett, W. S., and Quatrano, R. S., 1978, Sulfation of fucoidin in *Fucus* embryos. II. Required for localization in the rhizoid wall, *J. Cell Biol.* **78**:866.

Horine, R. K., and Ruesink, A. W., 1972, Cell wall regeneration around protoplasts isolated from *Convolvulus* tissue culture, *Plant Physiol.* **50**:438.

Jaffe, L. F., and Nuccitelli, R., 1977, Electrical controls of development, *Annu. Rev. Biophys. Bioeng.* **6**:445.

Keegstra, K., Talmadge, K. W., Bauer, W. D., and Albersheim, P., 1973, The structure of plant cell walls. III. A model of the walls of suspension cultured sycamore cells based on the interconnections of the macromolecular components, *Plant Physiol.* **51**:188.

Larsen, B., and Haug, A., 1971, Biosynthesis of alginate. III. Tritium incorporation with polymannuronic acid 5-epimerase from *Azotobacter vinelandii*, *Carbohydr. Res.* **20**:225.

Levring, T., 1952, Remarks on the submicroscopical structure of eggs and spermatozoids of *Fucus* and related genera, *Physiol. Plant.* **5**:528.

Medcalf, D. G., and Larsen, B., 1977, Fucose-containing polysaccharides in the brown algae, *Carbohydr. Res.* **59**:531.

Mollenhauer, H. H., and Morré, D. J., 1976, Cytochalasin B, but not colchicine, inhibits migration of secretory vesicles in root tips of maize, *Protoplasma* **87**:39.

Mueller, S. C., and Brown, R. M., 1980, Evidence for an intramembrane component associated with a cellulose microfibril-synthesizing complex in higher plants, *J. Cell Biol.* **84**:315.

Nakazawa, S., Takamura, K., and Abe, M., 1969, Rhizoid differentiation in *Fucus* eggs labeled with Calcofluor White and birefringence of cell wall, *Bot. Mag.* **82**:41.

Nisizawa, K., Akagawa, H., and Ikawa, T., 1971, Dark $^{14}CO_2$-fixation in brown algae, in: *Proceedings of the 7th International Seaweed Symposium*, pp. 532–536, John Wiley, New York.

Peng, H. B., and Jaffe, L. F., 1976, Cell-wall formation in *Pelvetia* embryos: A freeze-fracture study, *Planta* **133**:57.

Pope, D. G., Thorpe, J. R., Al-Azzawi, M. J., and Hall, J. L., 1979, The effect of cytochalasin B on the rate of growth and ultrastructure of wheat coleoptiles and maize roots, *Planta* **144**:373.

Quatrano, R. S., 1978, The development of cell polarity, *Annu. Rev. Plant Physiol.* **29**:487.

Quatrano, R. S., 1980, Gamete release, fertilization, and embryogeneis in the Fucales, in: *Handbook of Phycological Methods III: Developmental and Cytological Methods* (E. Gantt, ed.), pp. 59–68, Cambridge University Press, Cambridge.

Quatrano, R. S., and Caldwell, B. A., 1978, Isolation of a unique marine bacterium capable of growth on a wide variety of polysaccharides from macroalgae, *Appl. Environ. Microbiol.* **36**:979.

Quatrano, R. S., and Crayton, M. A., 1973, Sulfation of fucoidan in *Fucus* embryos. I. Possible role in localization, *Dev. Biol.* **30**:29.

Quatrano, R. S., and Stevens, P. T. 1976. Cell wall assembly in *Fucus* zygotes. I. Characterization of the polysaccharide components, *Plant Physiol.* **58**:224.

Quatrano, R. S., Brawley, S. H., and Hogsett, W. H., 1979, The control of the polar deposition of a sulfated polysaccharide in *Fucus* zygotes, in: *Determinants of Spatial Organization* (S. Subtelny and I. R. Konigsberg, eds.), pp. 77–96, Academic Press, New York.

Quatrano, R. S., Peterman, J. J., and Doubet, R. S., 1981, Deposition of polysaccharides in cell walls of early embryos of *Fucus*, in: *Pacific Seaweed Aquaculture* (I. A. Abbott, M. S. Foster and L. F. Eklund eds.) pp. 136–143, California Sea Grant Program, Inst. Marine Resources, Univ. California, La Jolla, CA.

Smidsrød, O., 1974, Molecular basis for some physical properties of alginates in the gel state, *Faraday Discuss. Chem. Soc.* **57**:263.

Stevens, P. T., and Quatrano, R. S., 1978, Cell wall assembly in *Fucus* zygotes. II. Cellulose synthesis and deposition is controlled at the post-translational level, *Dev. Biol.* **62**:518.

Talmadge, D. W., Keegstra, K., Bauer, W. D., and Albersheim, P., 1973, The structure of plant cell walls. I. The macromolecular components of the walls of suspension-cultured sycamore cells with a detailed analysis of the pectic polysaccharides, *Plant Physiol.* **51**:158.

Valent, B. S., and Albersheim, P., 1974, The structure of plant cell walls. V. On the binding of xyloglucan to cellulose fibers, *Plant Physiol.* **54:**105.

Vreugdenhil, D., Dijstra, M. L., and Libbenga, K., 1976, The ultrastructure of the cell wall of normal and apolar embryos of *Fucus vesiculosus*, *Protoplasma* **88:**305.

Weiss, R. E., and Wilbur, K. M., 1978, Effects of cytochalasin B on division and $CaCO_3$ extrusion in a calcareous alga, *Exp. Cell Res.* **112:**47.

Whitaker, D. M., 1931, On the rate of oxygen consumption by fertilized and unfertilized eggs. I. *Fucus vesiculosus*, *J. Gen. Physiol.* **15:**167.

4

Cell-Wall Regeneration by Protoplasts Isolated from Higher Plants

J. H. M. WILLISON AND A. S. KLEIN

1. *Introduction*

When first observed, the plant cell was effectively defined as a space between cell walls. In dead tissues, such spaces are empty, but the cell walls that delimit these spaces exist only because they were secreted by living protoplasts that once filled the spaces. The protoplast is the plant cell, as we now know it. Among higher plants particularly, protoplasts rarely exist naturally in the absence of their cell walls, and experimentalists have long sought to isolate them. Such isolated protoplasts were first prepared by cutting plasmolyzed tissues (von Klercker, 1892, Plowe, 1931). When the cut passed between the protoplast and the wall, the undamaged protoplast could be released. Only low yields of protoplasts could be obtained by this "mechanical" method, and it was not until the 1960s that higher yields were obtained as a result of the application to plants of the concept of protoplast isolation by enzymatic wall degradation (Cocking, 1960), a technique used previously for isolating bacterial and fungal protoplasts. Even higher yields were obtained once enyzmes of fungal origin were used (Gregory and Cocking, 1965; Takebe *et al.*, 1968), and protoplasts are now isolated by such means from many species for numerous experimental purposes. In principle, the technique is simple. The tissue is placed either within, or on the surface of, a hypertonic solution (commonly of mannitol) containing a suitable mixture of cell-wall

J. H. M. WILLISON • Biology Department, Dalhousie University, Halifax, Nova Scotia, Canada B3H 4J1. A. S. KLEIN • MSU-DOE Plant Research Laboratory, Michigan State University, East Lansing, Michigan 48824.

hydrolases. As the walls are degraded, the plasmolyzed protoplasts tend to become free and are ultimately liberated from the wall remnants using a coarse filter (such as a wire mesh). In practice, just as with any tissue culturing, the major problems are associated with maintaining sterility. For a detailed description of methods of protoplast isolation and culture, see Gamborg and Wetter (1975).

If suitably cultured, protoplasts will regenerate a new cell wall, whether isolated mechanically by cutting (Townsend, 1897; Prat and Roland, 1971) or enzymatically (Pojnar *et al.*, 1967; Nagata and Takebe, 1970). Once a wall is reestablished, the potential exists for the development of a callus and thence for the regeneration of plants (Nagata and Takebe, 1971), a potential that is now being exploited in revolutionary approaches to plant breeding (see Scowcroft, 1977). We are not concerned here, however, with cell-wall regeneration as a prerequisite for protoplast division [which is disputable (see Schilde-Rentschler, 1977; Herth and Meyer, 1978; Meyer and Herth, 1978)], but with cell-wall regeneration as an experimental approach toward a general understanding of cell-wall secretion. In this respect, isolated protoplasts have some intrinsic advantages over comparable systems. Because isolated protoplasts start without a wall, the process of the biosynthesis and secretion of the cell wall is separable, to some extent, from that of its assembly. Furthermore, the very nakedness of the cell permits observation of its surface, which would otherwise be impossible. However, as will become clear from this discussion, there are also some intrinsic experimental problems with isolated protoplasts in this context, notably that the process of protoplast isolation may so disturb the secretory process that some of the observations on cell-wall regeneration have only indirect relevance to normal wall secretion.

In this chapter we review cell-wall regeneration by protoplasts from the aspects of electron microscopy and biochemistry, concentrating on works published since 1976. Previous reviews have been those of Cocking (1970, 1972), Willison (1976), and Fowke (1978).

1.1. Sources of Protoplasts

Although protoplasts have been isolated from numerous higher plant sources using a variety of enzymes, certain tissues have been particularly favored and deserve introduction. Protoplasts were first isolated enzymatically in large numbers from the locule tissue of immature tomato fruit using a fungal pectinase of commercial origin (Gregory and Cocking, 1965). Protoplasts of this type (but isolated with different enzymes) were the first in which cell-wall regeneration was studied in any detail (Fig. 1) (Pojnar *et al.*, 1967; Willison and Cocking, 1972; Willison, 1973). These protoplasts do not undergo cell division, and no other protoplasts have yet been shown to regenerate comparable walls. A little later, Nagata and Takebe (1970) isolated protoplasts from tobacco mesophyll tissue using crude fungal cellulase and fungal pectinase prepared in Japan, and since then mesophyll protoplasts have been isolated from many species using similar enzymes. Although it has been possible to isolate

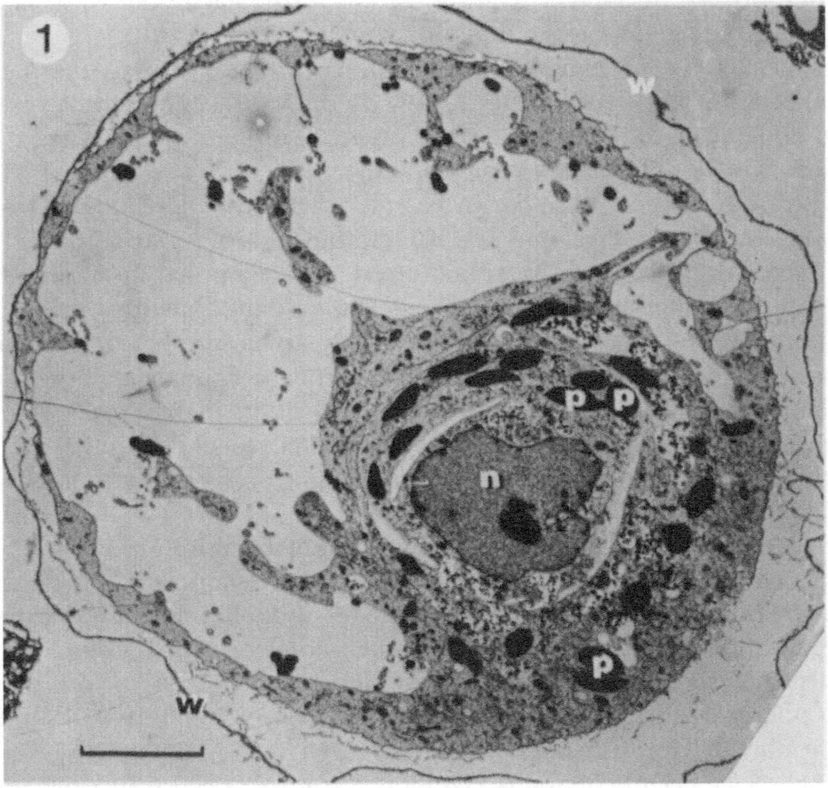

FIGURE 1. Isolated tomato fruit protoplast, cultured for about 4 days. The regenerated wall (multilamellar in this case) is clearly visible, particularly where it has tended to float away from the cell surface. (w) Wall; (n) nucleus; (p) plastid. Scale bar: 10 μm.

protoplasts from plant parts other than mesophyll, notably the meristematic apices (e.g., Binding and Nehls, 1978), mesophyll protoplasts have been most widely used in studies of cell-wall regeneration (e.g., Burgess and Fleming, 1974; Grout, 1975; Meyer and Abel, 1975; Willison and Cocking, 1975; Prat and Williamson, 1976; Kinnersley *et al.*, 1978). More recently, plants themselves have been discarded and cell-suspension cultures have been utilized (e.g., Hanke and Northcote, 1974; Williamson *et al.*, 1977; Takeuchi and Komamine, 1978; Davey and Mathias, 1979; Asamizu and Nishi, 1980; Wilkinson and Northcote, 1980), thereby simplifying isolation and culture procedures and allowing direct biochemical comparisons to be made between the preexisting wall and the regenerated wall.

2. *Studies of the Regenerated Wall: Perspective*

One framework in which to consider cell-wall regeneration is in comparison with the composition and structure of the native cell wall. Has the composition of cell-wall polymers been altered? Are the individual constituents present in the same proportions in the regenerating wall as in the native cell wall? Is there coordination in the synthesis of individual wall polymers during regeneration, or does the process of removing the cell wall perturb the synthesis of some wall components but not that of others? How do the various polysaccharides and proteins interact to form a new cell wall? There are no definitive answers to these questions, but we shall review progress to date.

There have been a number of recent reviews on the composition and structure of cell walls (Preston, 1974; 1979; Robinson, 1977; Darvill *et al.*, 1980). The composition of the native primary cell wall, compiled from analyses of tissue-culture material, is reviewed in Table I and includes operational definitions of the classes of noncellulosic polysaccharides based on chemical fractionation techniques. Pectins are thought to be components of the middle lamella, although they may also be constituents of the matrix, along with the hemicelluloses. Comprehensive compositional analyses of cell walls have been performed (e.g., Talmadge *et al.*, 1973); there is, however, no single unequivocal model for the three-dimensional organization of all the cell-wall polymers *in situ*. Furthermore, there is considerable variation in the composition of cell walls (e.g., between monocotyledonous plants and dicotyledonous plants and also within the same species during different stages of development). Nevertheless, characterization of the process of wall regeneration by isolated protoplasts may provide useful insights into the organization and function of the cell walls of higher plants because it furnishes another means by which cell-wall synthesis and assembly may be dissected.

There have been many more cytological studies of cell-wall regeneration than biochemical studies. Many cytological studies have been directed toward creating views of microfibrils as they develop on the previously naked protoplast surface, and we shall review these critically. In most studies, these microfibrils have been presumed to be cellulosic, but it is only recently that careful attempts to verify this have been made (Herth and Meyer, 1977; Klein *et al.*, 1981). While these studies have tended to confirm the presumption, they have done so with reservations (see later). Both biochemical and cytological studies have demonstrated that particularly during the early stages of regeneration, the regenerated wall is unusual. At the biochemical level, the wall is incomplete, due partly to loss of materials to the culture medium. Not unexpectedly, it may also contain materials typical of the response to wounding. At the cytological level, the microfibrils are more randomly and more loosely arranged than those of the walls of the parent tissue. In certain protoplasts, at least, there is also a temporary loss of order in the bundles of cortical cytoplasmic microtubules (Lloyd *et al.*, 1980).

3. Cellulose Microfibrils

3.1. Chemical Evidence for the Synthesis of Cellulose

The deposition of fibrillar material on the surfaces of isolated protoplasts is one of the characteristics of cell-wall regeneration and has been an important

TABLE I
Constituents of the Primary Cell Wall

Class	Percentage by weight	Method(s) of extraction from the isolated wall[a]	Examples and comments
Pectins	5–30	1. Hot H_2O 2. Hot H_2O + EDTA Additional extractions with ammonium or potassium oxalate	Rhamnogalacturonan (dicotyledonous plants) Arabinogalactan
Hemicelluloses	30–60	1. 4% KOH 2. 24% KOH (hot) or other combinations of alkali with borate	Xyloglucan (dicotyledonous plants) Arabinoxylan (monocotyledonous plants) Glucomannan Methylated and acetylated polysaccharides
α-Cellulose (not pure)	10–20	Insoluble in hot alkali	After alkali extraction, traces of sugars other than glucose (polysaccharides?) and protein are still present in this fraction.
Other carbohydrate		Usually isolated with one or more of the fractions listed above	Mixed-linked (β1,3 and β1,4) glucans (monocotyledons); *in Vinca*, homopolymers of galactose or arabinose. β1,3-Glucan (cotton) may be wound-induced.
Protein	3–10	Partially extracted by urea or LiCl; released by cellulase; much of this is bound to other cell-wall constituents.	Glycoproteins; enzymes
Lignin or its precursors	<1	—	Not normally found in primary cell walls.

[a] These extraction procedures do not produce "pure" cell-wall fractions; some polysaccharides are extracted with more than one fraction.

experimental system in cytological studies of cellulose synthesis (see later). As previously warned (Herth and Meyer, 1977), caution should be exercised before equating visualization of microfibrils with the presence of cellulose, since it has been observed that certain unbranched polysaccharides, particularly homopolymers, form fibrils *in vitro* [e.g., (1→4)-β-mannan and (1→4)-β-xylan of angiosperms], and some of the noncellulosic components of the cell wall could exist as fibrils *in situ* (see Preston, 1979). Just as Calcofluor fluorescence is an unreliable criterion for the presence of cellulose in regenerating walls (see Willison, 1976), so other criteria that have been used are also equivocal: cellulase digestibility is not a rigorous test for the presence of cellulose because most available cellulase preparations contain other polysaccharide-degrading enzymes (Anderson and Ray, 1978), and it is well established that alkali insolubility (once the basis of the method for isolating "pure" α-cellulose) is a property of other cell-wall polysaccharides. Nevertheless, in two reports on the chemical composition of the regenerating cell wall (Hanke and Northcote, 1974; Asamizu and Nishi, 1980), one or another of these biochemical procedures was used as the sole means to demonstrate the presence of cellulose in the regenerating cell wall.

To circumvent the problems outlined above, Klein *et al.* (1981) adopted the procedure of Updegraff (1969) to demonstrate the presence of newly synthesized cellulose in the regenerating walls of soybean protoplasts. Freshly isolated protoplasts were supplied with [^{14}C]glucose and, after being cultured for varying periods, incubations were terminated by the addition of 4 volumes of ethanol, which precipitates most polysaccharides and proteins. The quantity of cellulose synthesized was estimated by boiling the 80% ethanol precipitate, with added carrier cellulose, in acetic acid–nitric acid–water (8:1:2) for 1 hr and then determining the radioactivity remaining in the acetic–nitric acid resistant residue. Crystalline cellulose is the only polysaccharide cell-wall component thought to be resistant to acetic-nitric acid treatment (Updegraff, 1969); however, low-molecular-weight (1→4)-β-glucan (oligosaccharides) may be soluble under these conditions. The validity of the acetic–nitric acid procedure was verified by the following analyses: (1) Acetic–nitric acid resistant material was hydrolyzed to individual monosaccharides. Paper chromatography showed that glucose was the only monosaccharide present in the sample (see Saemann *et al.*, 1963). (2) Another sample was degraded enzymatically with a *Streptomyces* cellulase preparation (obtained from E. T. Reese); cellobiose and glucose (in an approximate ratio of 3:1) were the only small, radioactive saccharides released, indicating that the sample contained (1→4)-β-linked glucose. (3) Samples of acetic-nitric acid resistant material from protoplasts were permethylated and converted to alditol acetates. In most experiments, all the radioactivity cochromatographed with authentic 1,4,5-tri-*O*-acetyl-2,3,6-tri-*O*-methyl glucitol (i.e., the expected 1,4-linked glucose derivative). In some experiments, small amounts of radioactivity (less than 15%) cochromatographed with the derivative of 1,3-linked glucose. As a result of these careful tests, it can be concluded that the radioactive material in the acetic–nitric acid insoluble residue predominantly represents newly synthesized cellulose. In the same work (Klein *et al.*, 1981), this chemical evidence of synthesis of cellulose during cell-wall regeneration was correlated with visualization of fibrillar material on the surfaces of the protoplasts.

Potentially, chemical inhibitors of cellulose synthesis could be used to verify the relationship between cellulose synthesis and microfibril deposition, but results to date have been inconclusive for the following reasons: Coumarin (100 μM) and 2,6-dichlorobenzonitrile (DCB) (10 μM) specifically inhibit cellulose synthesis in cultured cotton ovules (Montezinos and Delmer, 1980). At higher concentrations, these compounds also inhibit the synthesis of other cell-wall polysaccharides. Treatment of soybean protoplasts with DCB (3.3 μM) inhibited 66% of the incorporation of [^{14}C]glucose into chemically characterized cellulose, as compared to controls (Klein and Delmer, unpublished results). In an ultrastructural study, Burgess and Linstead (1977) found that coumarin (0.5–1.5 mM) inhibited fibril deposition on tobacco mesophyll protoplasts, although after 4 days an aberrant wall was formed, and Meyer and Herth (1978) reported that coumarin (1 mM) and DCB (5 μM) totally inhibited cell-wall formation by similar protoplasts. Thus, while it is clear that DCB or coumarin inhibits both cellulose synthesis and microfibril deposition, these observations alone do not provide conclusive evidence that the microfibrils are composed of cellulose, since either the synthesis or assembly of other cell-wall components might also be affected.

3.2. Secondary Structure of Cellulose

Limited information is available on the secondary structure of cellulose synthesized during wall regeneration by isolated protoplasts. In one study, alkali-insoluble (20% NaOH) fibrous material was isolated from tobacco mesophyll protoplasts that had been cultured in saline medium for 6 days (Herth and Meyer, 1977), long after the first fibrillar material had been deposited at the cell surface. The sample contained trace quantities of sugars other than glucose and ninhydrin-positive material, and showed only a little similarity with the X-ray diffraction pattern of cotton cellulose. This was interpreted to indicate that the microfibrils were only weakly crystalline and might contain material other than cellulose. It should be noted that their X-ray diffraction pattern closely resembles that of alkali-insoluble 1,4-linked glucan microfibrils isolated from cultured rose cells, as described by Chanzy *et al.* (1979).

Asamizu *et al.* (1977) measured the distribution in the degree of polymerization (DP) of newly synthesized cellulose from protoplasts isolated from cultured carrot cells, by a method involving nitration of the cellulose to "nitrocellulose." After 24 hr of culture, more than 70% of the newly synthesized (protoplast) "nitrocellulose" had a DP of less than 200 glucose residues. This short-chain "nitrocellulose" was not characterized by any other chemical procedure, and it is possible that other noncellulosic polysaccharides were measured by their procedure in addition to the newly synthesized cellulose, thereby influencing the average DP. They reported that after 3 days in culture, the distribution of chain lengths shifted toward higher degrees of polymerization, and by 9 days more than

50% of the "nitrocellulose" had a DP greater than 1700 glucose residues, a size and distribution similar to that measured in the original carrot cells. It should be noted, however, that nitration causes chain degradation, leading to an underestimate of the actual lengths of cellulose chains from both protoplasts and intact cells (Billmeyer, 1971) (see also Chapter 15).

Although most of the fibrillar material visualized on the surfaces of protoplasts is probably cellulose, it is important to realize that in some instances the regenerated cell wall does not contain crystalline cellulose (e.g., Takeuchi and Komamine, 1978) (for more details, see Section 4.1).

3.3. Kinetics of Cellulose Synthesis: Biochemical Studies

Klein *et al.* (1981) studied the incorporation of [^{14}C]glucose into acetic-nitric acid resistant material [cellulose (see above)] by soybean protoplasts cultured for periods of 1–21 hr (Fig. 2). In parallel experiments, cytological observations were made at similar times. In samples fixed immediately after isolation, the surface of the plasma membrane was devoid of any fibrillar material. Cellulose synthesis was detected chemically within minutes after protoplasts were separated from the cell-wall-degrading medium, and continued over longer incubation periods. This lack of any lag phase (see Section 3.5), together with the absence of preexisting microfibrils, implies that the radioactive glucose was incorporated into new cellulose chains and therefore that cellulose-synthesizing machinery probably remains intact during the isolation of protoplasts from these rapidly growing cells. Comparable results have been reported by Asamizu and Nishi (1980), who found that carrot protoplasts incorporated [^{14}C]glucose into an alkali-insoluble "α-cellulose" fraction during the first 5 hr of culture.

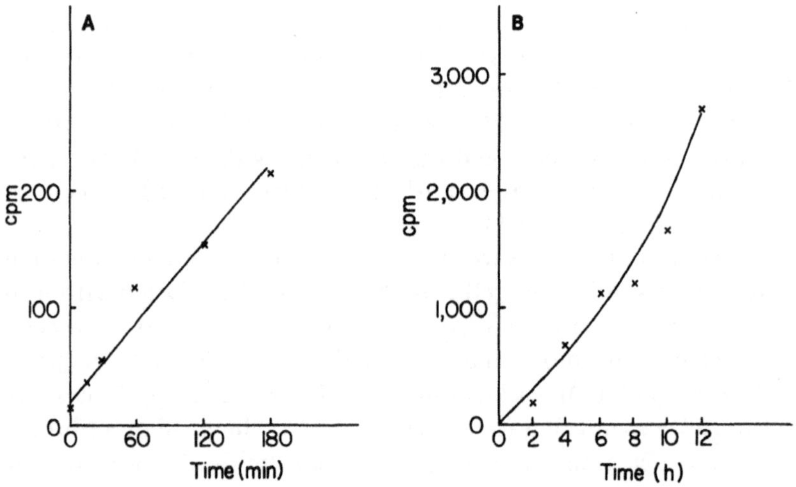

FIGURE 2. Incorporation of [^{14}C]glucose into acetic-nitric acid insoluble material (cellulose) by freshly isolated soybean protoplasts. Cellulose synthesis begins without an appreciable lag.

3.4. Microscopic Methods Used in Observing Protoplasts

Numerous microscopic techniques, both light optical and electron optical, have been employed in studying cell-wall regeneration by isolated plant protoplasts, and examples of these are listed in Table II, together with a brief summary of their relative advantages and disadvantages. While most studies have not been contentious, where dispute has arisen its source has usually been differences of opinion over image interpretation. Such problems of interpretation can be resolved only by comparing results obtained using different techniques and identifying common factors and by careful consideration of technical variables.

3.5. Kinetics of Microfibril Deposition: Cytological Studies

From the results of many studies (Prat and Williamson, 1976; Williamson *et al.*, 1977; Burgess *et al.*, 1978; Kinnersley *et al.*, 1978; Wilkinson and Northcote, 1980) (for earlier literature, see Willison, 1976), certain cautious generalizations can be made about the rates of cell-wall regeneration in relation to protoplast source, isolation technique, and culture medium. First, the rate of wall synthesis and the type of wall formed vary according to the source of the protoplasts: those isolated from rapidly growing tissue cultures and from meristematic cells tend to begin to regenerate walls immediately after isolation; mesophyll protoplasts generally display a short lag period (8–45 hr after transfer to the culture medium) before microfibrils can be detected, whereupon wall synthesis is vigorous, and for unexplained reasons some protoplasts (notably those from grasses and from aging callus cultures) are intrinsically very slow to commence wall regeneration, which is itself a relatively slow process. Second, although relatively little studied, the manner in which protoplasts are isolated has little (if any) influence on cell-wall regeneration, except that mesophyll protoplasts display a lag phase only after their initial isolation [i.e., if the regenerated wall is redigested, microfibrils reappear almost immediately, (Burgess *et al.*, 1978)]. Third, the vigor with which a new wall is deposited is influenced by the culture medium; most notably, it is inhibited in media in which the osmotic stabilizers are salts or are present in concentrations beyond those necessary to produce minimal plasmolysis (Meyer and Abel, 1975; Burgess *et al.*, 1978; see also Willison, 1976). When cell-wall regeneration is extremely slow, as in saline media, a complete cell plate is not formed during cell division. In such cases, the phragmoplast disintegrates, and cytokinesis occurs by budding (Herth and Meyer, 1978).

There are several possible explanations for the lag phase typical of protoplasts isolated from mesophyll and certain other differentiated tissues. These include repression of cellulose synthesis by the conditions required for protoplast isolation, damage to the cellulose-synthesizing machinery, or that these cells are not actively synthesizing cellulose at the time of protoplast isolation. The results of Burgess *et al.* (1978), described above, suggest that the last is probable.

TABLE II
Microscopic Techniques Used in Studying Cell-Wall Regeneration by Isolated Protoplasts

	Technique	Advantages	Disadvantages	Examples[a]
Light microscopy	1. Calcofluor fluorescence	Specimen living; method simple; some specificity of staining	Physiological effects problematical; staining specificity poor (see Section 3.2.1)	Nagata and Takebe (1970), Prat (1973)
	2. Immunofluorescence	Highly specific method for localizing defined antigens	Limited applicability	Lloyd *et al.* (1980)
Electron microscopy	1. Thin-sectioning, conventional staining	Much cytological detail; specimen not dried during processing	Interpretation difficult at high resolution; fixation, dehydration, and embedding artifacts; surface detail invisible	Burgess and Fleming (1974), Herth and Meyer (1977)
	2. Thin-sectioning, polysaccharide-specific staining	Some specificity of staining	Fixation, dehydration, and embedding artifacts; surfaces not imaged	Prat and Roland (1971), Prat (1973), Pearce *et al.* (1974)
	3. Single-stage, preshadowed replication	Surfaces of walls and membranes imaged	Prefixation usually necessary; drying artifacts	Pojnar *et al.* (1967), Williamson *et al.* (1977)
	4. Freeze-fracturing without etching	Plasma-membrane fracture faces exposed; chemical fixation may be avoided	Freezing artifacts; details of cell wall and cytoplasm obscure	Willison and Cocking (1975), Robenek and Peveling (1977), Davey and Mathias (1979), Schnabl *et al.* (1980), Wilkinson and Northcote (1980)
	5. Freeze-fracturing with etching	Plasma-membrane surfaces exposed in conjunction with fracture faces; resolution better than method 3	Chemical fixation usually necessary; freezing artifacts; less of surface visible than with methods 3 and 6	Willison (1973), Willison and Cocking (1975), Willison and Grout (1978), Davey and Mathias (1979)
	6. Scanning electron microscopy	Up to half of entire protoplast surface may be viewed	Drying artifacts; coating artifacts; resolution poorer than with methods 3–5	Burgess and Linstead (1976), Burgess *et al.* (1977, 1978)
	7. Negative staining	Simple; high resolution	Limited applicability (enables visualization of microfibrils stripped from protoplast surface only)	Herth and Meyer (1977), Burgess and Linstead (1979)

[a] While all the principal techniques that have been used are listed, this list of references is merely exemplary.

3.6. Cytological Observations of Microfibril Formation

In the context of this chapter, the most interesting cytological observations are those on the early stages of microfibril formation at protoplast surfaces. Unfortunately, although there have been several detailed reports in which almost every specimen-preparation technique has been used (see Table II), results have led neither to clear conclusions nor to a consensus. Before discussing all the available data, we will present typical results obtained using the deep-etching modification of the freeze-etching technique (see also Willison, 1973; Willison and Cocking, 1972, 1975). In this method, protoplasts are glutaraldehyde-fixed, washed several times with distilled water, and then frozen. After the frozen specimen is fractured, "etching" is obtained as a result of sublimation of some of the ice in which the protoplasts are embedded. In this way, protoplast surfaces are exposed, and replicas of these may be examined in the transmission electron microscope (for more details, see Willison and Rowe, 1980). Although the technique allows only a relatively small part of the surface of any protoplast to be viewed, much detail is resolved.

The surfaces of freshly isolated protoplasts are usually devoid of microfibrillar material. In rare instances, after mesophyll or tomato fruit protoplasts have been cultured for a few hours, microtubules appear beneath the plasma membrane surface (Fig. 3). Both mesophyll and tomato fruit protoplasts have a "lag phase" of several hours before microfibrils begin to appear, and impressions of microtubules (as in Fig. 3) have been observed only during this lag phase. In tobacco mesophyll protoplasts, when microfibrils first appear, they are long (Fig. 4), but ends may sometimes be observed (Figs. 4 and 5). Ends tend to be either tapered (Fig. 5) or associated with the edge of a neighboring microfibril (Fig. 4). Other forms are found, but these are always difficult to interpret (Fig. 6 and 10). After longer periods of culture, the quantity of microfibrils increases rapidly to levels at which the identification of ends becomes impossible (Fig. 7). Comparable results have been obtained with tomato fruit protoplasts, which are heterogeneous in their response to culture. Initially, many tomato fruit protoplasts produce a "multilamellar" wall (Fig. 1 and 8), the structure and insolubility in strong acids of which suggest a lipidic material of the cutin/suberin group, but chemical analysis has not been attempted. Microfibrils may appear between the protoplast and this multilamellar wall (Figs. 9 and 10), together with amorphous materials. This is a valuable finding, because the location of this microfibrillar material makes it impossible that it is a residue of the original wall, since any wall residue would lie outside the multilamellar wall (for evidence, see Willison, 1973). The micrograph in Fig. 10 is a good example of the problems associated with interpreting even the most promising micrographs (see the Fig. 10 caption for details). Although granules of various forms are evident in the micrograph, none is unequivocally associated with the end of a microfibril, and the identification of ends is somewhat problematical. Furthermore, since ice formation during freezing probably compresses a loose three-dimensional tangle of microfibrils against the surface of the membrane (for discussions, see Willison

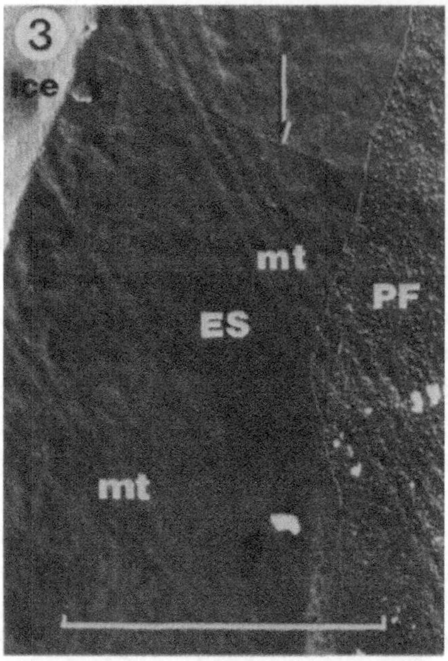
3
ice
mt
PF
ES
mt

4
ice
ES

5

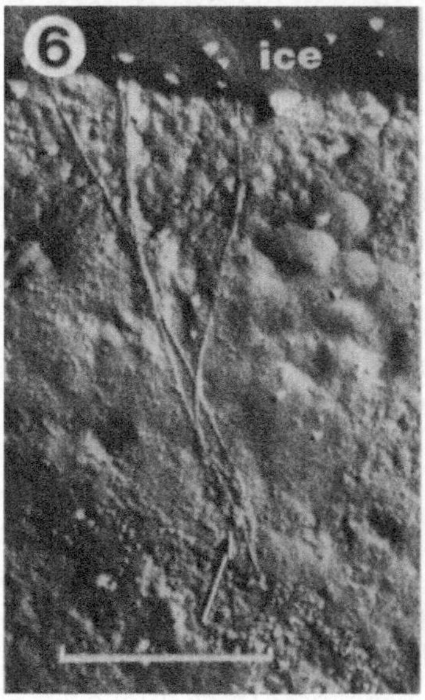
6
ice

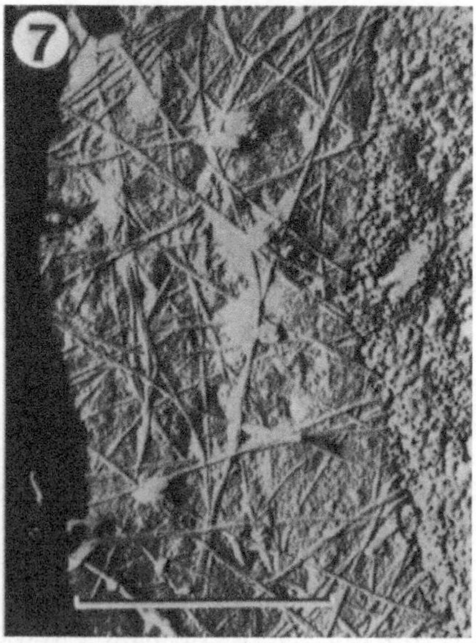
7

and Grout, 1978; Willison and Rowe, 1980), even the spatial relationships are suspect. Dehydration artifacts are not restricted to freeze-etching. In work with the scanning electron microscope (Burgess and Linstead, 1976, 1977), large "projections" had been shown to be associated with "fibers," and ontogenetic relationships between the two had been discussed critically. Later, more detailed study (Burgess *et al.* 1977; Willison and Grout, 1978; Burgess and Linstead, 1979) showed clearly that the relationship between the two structures was artifactual, however. The only conclusion that can be drawn with confidence from the freeze-etching results described above is that microfibrils are very long, although it does also seem probable that many of the ends taper to a thickness that cannot be resolved with this technique.

Generally, the results of all fine-structural studies of protoplast-wall regeneration are compatible with the results described above (for comparison, see Grout, 1975; Williamson *et al.*, 1977; Herth and Meyer, 1977; Willison and Grout, 1978; Burgess and Linstead, 1979; Klein *et al.*, 1981). That is, microfibrils are long and increase in quantity with time. Although prejudice might lead one to assume that the increase in quantity of microfibrils also involves an increase in length, no reliable estimates of the mean length of microfibrils in regenerated walls have ever been obtained, and therefore this has still to be demonstrated. Indeed, the negative staining results of Burgess and Linstead (1979) suggest that bundles of fibrillar material about 60 nm in length appear during the early stages of wall formation (together with normal microfibrils), but are not found later on. No direct evidence of a developmental relationship between the two was found, but a "quantum leap" hypothesis must be included among the possible hypotheses. Concern has been expressed about the apparent incompatibility of some cytological observations of microfibril formation at protoplast surfaces with the hypothesis of terminal-complex-mediated end synthesis of cellulose microfibrils (Willison and Grout, 1978; Burgess and Linstead, 1979), most notably the absence of any evidence of such complexes. It is becoming clear,

FIGURE 3. Part of the surface of an isolated tobacco leaf protoplast, cultured for a few hours only, showing the absence of microfibrils from the E surface (ES) of the plasma membrane: a "deep-etch" preparation. Impressions of microtubules (mt) are visible, however. The arrow indicates an ice segregation artifact, not a microfibril. (PF) P fracture face. Scale bar: 1 μm.

FIGURES 4 and 5. Microfibrils present on the surface of a tobacco mesophyll protoplast cultured for 30 hr. Note that even at this very low density, the microfibrils are long. Two ends are visible (arrows); one lies against the surface of another microfibril, the other is shown at higher magnification in Fig. 5. No distinctive structures are associated with these tips, but it must be noted that the material was glutaraldehyde-fixed and washed with several changes of distilled water. Scale bars: 1 μm (Fig. 4); 100 nm (Fig. 5).

FIGURE 6. Part of the surface of a tobacco mesophyll protoplast cultured for about 24 hr. Three microfibrils are visible, one of which ends within the field shown, possibly at a large granule (arrow). However, close by lie other similar granules but in the absence of microfibril tips. Scale bar: 500 nm.

FIGURE 7. Part of the surface of a tobacco mesophyll protoplast cultured for about 3 days. The microfibril density is such that all ends cannot be identified, and a mean length therefore cannot be estimated. Scale bar: 500 nm.

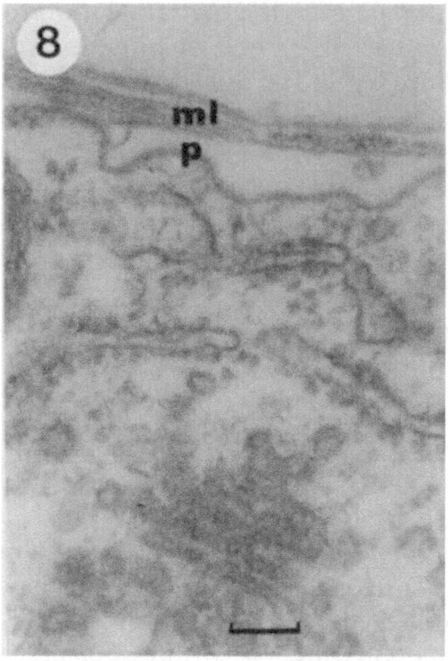

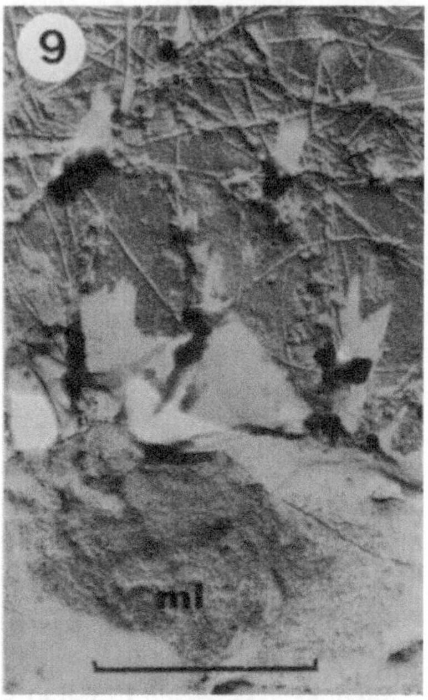

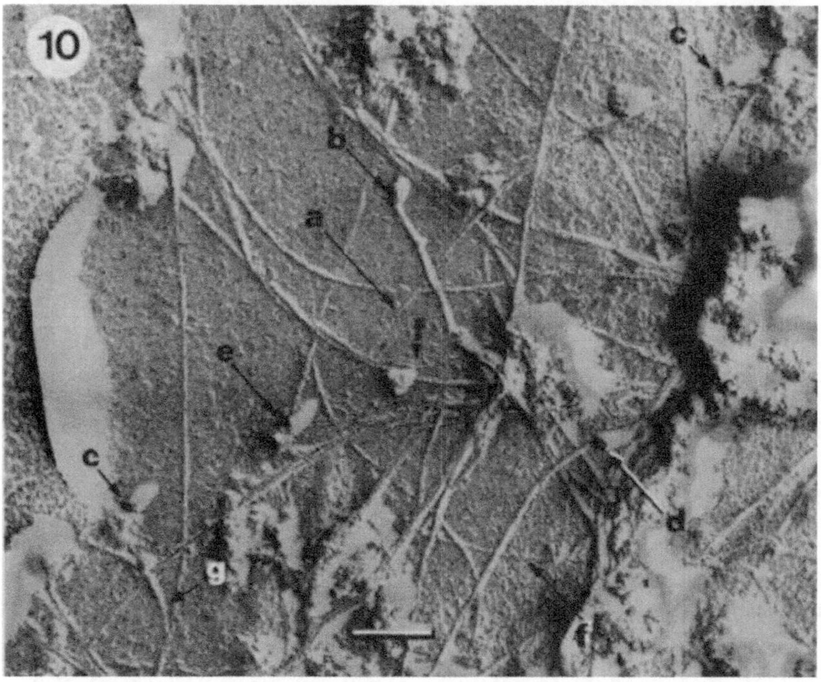

FIGURE 8. Part of an isolated tomato fruit protoplast, cultured for 2 days. (ml) Multilamellar wall; (p) plasmalemma. Scale bar: 100 nm.

FIGURE 9. "Deep-etched" tomato fruit protoplast, cultured for 4 days, showing the presence of microfibrils beneath the multilamellar wall (ml). Scale bar: 500 nm.

however, that although microfibrils look superficially very similar, there are important dissimilarities between classes of them. To date, the clearest ultrastructural evidence of complexes of particles associated with microfibril tips has come from cellulosic algal-cell walls and from secondary (or, more usually, unidentified) higher-plant-cell walls (R. M. Brown, Jr., and Montezinos, 1976; Mueller *et al.*, 1976; Willison and Brown, 1977; Willison and Grout, 1978; Mueller and Brown, 1980; Giddings *et al.*, 1980). In the case of primary walls, such particle complexes appear to be present (see, for example, Mueller and Brown, 1980), but some equivocation about their role and identity is justified. (The following doubts spring to mind: Can "primary walls" be identified with certainty in freeze-fracture preparations? Can one be sure that particle complexes associated with primary walls are active in the primary walls themselves? Are all "primary wall" microfibrils strictly comparable?) Isolated protoplasts regenerate a wall containing cellulose (see earlier), but the microfibrils probably contain cellulose cores that are "weakly crystalline and have dimensions in the range of 'subelementary fibrils'" (Herth and Meyer, 1977). These microfibrils are probably coated with noncellulosic glycans and are typical of primary walls. Therefore, the explanation for the absence of microfibril-terminal granules from isolated protoplasts might be that microfibrils of the primary-wall type are assembled by a quite different process from that of secondary-wall-type microfibrils. Compatibly with such a viewpoint, Grout (1975) and Burgess and Linstead (1979) have proposed that the microfibrils of protoplasts might arise by crystallization in the absence of terminal complexes (either at the plasma-membrane surface or beyond it). This proposal should not be dismissed simply because it is difficult to reconcile with findings from systems in which highly crystalline cellulose arises. Willison and Grout (1978), however, also note that the absence of microfibril-terminal granules can be explained by (1) loss of the complexes themselves due to glutaraldehyde fixation and (2) loss of impressions of complexes through insufficient compression of plasma membrane against wall.

The observation that microfibril tips commonly lie immediately against the surfaces of other microfibrils might be a dehydration artifact, but it is also possible that these are points of branching—that is, points where "fraying" (Burgess and Linstead, 1979) has led to the generation of two nuclei at which

FIGURE 10. "Deep-etched" tomato fruit protoplast, cultured for 4 days, showing a region of the surface of the plasma membrane lying beneath lamellae of the multilamellar wall (not visible in the micrograph). Only two microfibril tips are unequivocably visible; others may be present but are indistinct (e.g., being lost beneath nonmicrofibrillar material). Of these two microfibril tips, one (a) tapers to a point, and the other (b) appears to end in a granule. However, other granules of a size similar to that at (b) are not associated with microfibril tips (c). At first sight, a third microfibril tip (d) seems to end in a granule, but careful scrutiny reveals that it continues beyond the granule. Yet another doubtful case (e) probably represents a microfibril that rises sharply toward the viewer before ending (much clearer examples of this fracture artifact have been seen by the author). The two microfibrils (f) may be other examples of microfibrils tapering, but these tips lie too close to debris for this to be an unquestionable interpretation; similarly, microfibril (g) may terminate next to a second microfibril, but equally, it might pass just beyond. Scale bar: 100 nm.

crystallization, or end synthesis, can occur and from which microfibril growth continues.

It has been tacitly assumed throughout this discussion that the site of cellulose synthesis probably lies at the plasma membrane or beyond. No evidence from work on isolated protoplasts has supported the opposite contention, that microfibrils might be assembled intracellularly, although the existence of "wall-bodies" (Pearce *et al.*, 1974) deserves cautionary mention. Wall-bodies are large, membrane-bound, intracellular accumulations of cell-wall-like material, including microfibrils, that occur in certain (diverse) types of protoplasts. The contents may be released exocytotically from the protoplasts to reside in the extracellular space. While this may appear to be evidence of intracellular microfibril synthesis, it has not been interpreted as such (Willison, 1973; Pearce *et al.*, 1974). Instead, it is hypothesized that the membrane that limits the wall-body is plasma membrane that was taken up by the protoplast during plasmolysis [a phenomenon that is well established (Davey and Cocking, 1972)] and that this membrane retains the properties of plasma membrane, including the capacity to act as a site for microfibril assembly. It must also be noted that materials that would diffuse into the culture medium if released at the plasma membrane will accumulate within wall-bodies and may therefore appear in analyses of the chemical composition of fractions obtained from cultured protoplasts.

Evidence concerning the role of the plasma membrane in microfibril biosynthesis by isolated protoplasts is intriguing rather than illuminating. In a freeze–fracture study, Willison and Cocking (1975) showed that the particle partition coefficient (an index of the segregation of membrane-associated particles between the two complementary membrane fracture faces) differed between freshly isolated mesophyll protoplasts and those actively laying down microfibrils and that the change coincided with the onset of microfibril deposition. This suggests that a fundamental reorganization of the membrane occurred. Even less direct are the observations of Lloyd *et al.* (1980) using an immunofluorescent label directed against microtubules in cultured carrot (*Daucus corata*) cells and isolated protoplasts, which had been attached to glass surfaces and treated with detergent to leave a delipidated ghost into which the label could penetrate. The method clearly demonstrated the disposition of bundles of microtubules. In the elongated cells, these bundles tend to run as "hoops" lying at an angle normal to the longitudinal axis of the cell, just as do the microfibrils. In isolated protoplasts, however, the bundles become disordered, although the integrity of individual bundles is unaffected. Nevertheless, that the bundles of microtubules remain attached to the detergent-extracted skeleton of the plasma membrane suggests that there is a strong transmembrane link between microtubules and the outer surface of the plasma membrane. While their main conclusion was that the orientation of the microtubule "hoops" and the microfibrillar skeleton of the wall are interdependent, the results are also compatible with the hypothesis of Heath (1974) that cytoplasmic microtubules determine the orientation of wall microfibrils by determining the paths taken by the growing microfibril tips. A direct link between microtubule and microfibril tip-associated synthesizing complex is not the only possibility, however. The observations on microtubules described earlier (see Fig. 3) suggest that for primary-wall-type microfibrils deposited around isolated protoplasts, the effect of the microtubules might be to condition the plasma membrane somehow.

3.7. Arrays of Membrane-Associated Particles

Among the most-studied protoplast phenomena have been several distinctive types of aggregations of membrane-associated particles on freeze-fractured plasma membranes. These have been described on protoplasts isolated from stem callus of *Skimmia japonica* (Robenek and Peveling, 1975, 1977), carrot (*Daucus carota*) suspension-culture protoplasts and *Petunia* mesophyll protoplasts (Davey and Mathias, 1979), guard cell and mesophyll protoplasts of *Vicia faba* (Schnabl *et al.*, 1980), and protoplasts derived from a potato (*Solanum tuberosum*) callus culture (Wilkinson and Northcote, 1980). At least three types of particle array have been described: (1) loosely arranged, parallel strings of particles (Robenek and Peveling 1975, 1977; Davey and Mathias, 1979); (2) homogeneous arrays of closely hexagonally packed particles (Wilkinson and Northcote, 1980); (3) heterogeneous arrays of closely hexagonally packed particles (Robenek and Peveling, 1975, 1977; Davey and Mathias, 1979; Schnabl *et al.*, 1980; Wilkinson and Northcote, 1980).

Although membrane transport functions, associated either with microfibril synthesis (Robenek and Peveling, 1977) or with solute transport (Schnabl *et al.*, 1980), have been proposed for the hexagonal arrays, it has also been suggested that they are nonfunctional, being glycerol-induced (Davey and Mathias, 1979) or simply plasmolysis-induced (Wilkinson and Northcote, 1980). Robenek and Peveling (1975, 1977) suggested that the strings of particles might be involved with microfibril synthesis, but, again, Davey and Mathias (1979) found that such arrays arise only after prolonged treatment of the protoplasts with glycerol. If such arrays result simply from glycerol treatment (and for strings of particles this seems probable), then the term "artifact" seems appropriate, since they would play no physiological role in the functioning of normal protoplasts. However, since the hexagonally packed arrays may be plasmolysis-induced, it is possible that this type is physiologically influential. In this context, it is worth noting that the process of protoplast isolation has been reported to result in an increase in membrane permeability to certain ions (Taylor and Hall, 1976) and that plasmolysis alters the membrane potential (Racusen *et al.*, 1977). It seems improbable, however, that any type of particle array is essential to microfibril synthesis, because this continues in several types of protoplasts in the absence of ordered arrays of membrane particles (Willison and Cocking, 1972, 1975; Willison and Grout, 1978; Davey and Mathias, 1979).

Confidence in this conclusion can only be strengthened by bringing together the observations that, first, particle-array formation is induced by plasmolysis (Wilkinson and Northcote, 1980) and, second, cell-wall regeneration seems to be slowed by plasmolysis (Burgess *et al.*, 1978).

4. Cell-Wall Constituents Other Than Cellulose

4.1. Biochemistry

There have been a limited number of studies in which the noncellulosic components of regenerating cell walls have been characterized [soybean (Hanke and Northcote, 1974; Klein *et al.*, 1981), carrot (Asamizu and Nishi, 1980), *Vinca rosea* Takeuchi and Komamine, 1978; Tanaka and Uchida, 1979)]. It is difficult to compare the results directly, since the findings differed between experimental systems and even within the same system. Also, the composition and structure of the regenerating cell wall can only be compared with the cell wall of the parent tissue once the "regenerating cell wall" has been defined. Conventional methods for the isolation of cell walls from intact cells depend on the huge size difference between cytoplasmic organelles and pieces of the wall. In the cell wall, polymers are tightly interwoven, and thus the pieces of wall may approach the dimensions of intact cells, sedimenting at low centrifugal force or being retained by coarse filters. During the early stages of wall regeneration, wall fragments are small and loosely organized. Although growing cellulose chains may be anchored to the plasma membrane (see Mueller and Brown, 1980), during the early stages of wall regeneration, in the absence of a preexisting cellulosic network, much of the newly synthesized cell-wall material may float away from the surface of the protoplast (Prat and Roland, 1971; Hanke and Northcote, 1974; Burgess and Linstead, 1979). Furthermore, cultured plant cells normally secrete polysaccharides and proteoglycans into the extracellular milieu, a characteristic likely to be shared by protoplasts. These difficulties are compounded by the facts that cell-wall regeneration may not proceed synchronously and that the "regenerating cell wall" is likely to be compositionally dynamic. Thus, it is perhaps too early to build detailed models of the structure of the regenerating wall.

In a study of cell-wall formation by soybean protoplasts obtained from a cell culture, Hanke and Northcote (1974) incubated protoplasts in the presence of radioactive glucose. Following 20 hr of culture, radioactivity was incorporated into several materials including a pectic polysaccharide that was identified in the dialyzed filtrate (this filtrate represented both secreted material and the products of protoplast lysis). This pectin was characterized on the basis of its electrophoretic mobility as a weakly acidic pectinic acid, which was also shown to contain some galacturonic acid. After 40 hr of culture, two additional pectins were isolated from the dialyzed filtrate (at this point, not contaminated by cytoplasmic constituents from lysed protoplasts). All three pectic polysaccharides had electrophoretic characteristics similar to those of pectins isolated from the parent soybean callus, yet they were not found in the "washed protoplast fraction" captured by the filter, even after 40 hr of culture. The cell wall of the parent soybean callus contains classes of pectins in addition to the three identified as products of the synthetic activity of the protoplasts. Hanke and Northcote (1974) concluded that wall regeneration by soybean protoplasts was aberrant because some pectins were not synthesized and others were not assembled into the developing wall. It is important to appreciate, however, that it is not necessarily protoplast isolation *per se* that influences the biosynthetic and secretory activities of the cell. Plasmolysis of tobacco cells alters the pattern of biosynthesis of pectic polymers by comparison with unplasmolysed tissue (Boffey and Northcote, 1975). This is not a simple quantitative change, since incorporation of label from [^{14}C]glucose into galacturonic acid and rhamnose residues of the bulk pectic fraction was found to decrease, while incorporation into arabinose residues was virtually unaffected.

Asamizu and Nishi (1980) examined changes in the pattern of incorporation of [^{14}C]glucose and [^{3}H]myoinositol by carrot protoplasts into polysaccharide-containing polymers by supplying the label for 5 hr after varying periods of culture (during which the wall was forming). Radioactivity was found in all three classic cell-wall fractions (see Table I), but the proportion of label in pectin, hemicellulose, or "α-cellulose" fractions changed continuously throughout the first 100 hr of culturing. It is not clear whether the incipient cell wall, from protoplasts cultured for only short periods, was isolated by their conventional procedures. The overall neutral sugar composition of the regenerating walls from protoplasts, cultured for 24 hr or 7 days, differed from that of recently subcultured cells. Similarly, the extracellular, 80% ethanol-precipitable polymers from protoplasts, differed from those isolated from the medium of cultured cells. After 24 hr in culture, with continuous labeling from [^{14}C]glucose, the labeling pattern was generally similar to that obtained with carrot cells in suspension culture (Asamizu and Nishi, 1979), although it should be noted that the pectic fraction from the protoplasts contained glucose, which is not normally a constituent of pectic polysaccharides. These results are difficult to interpret meaningfully. Although carrot protoplasts appear to synthesize all the classic components of the cell wall, they probably do not produce all polymers simultaneously. The results do not allow us to discriminate among many more specific possibilities, such as that specific pectic or hemicellulosic polysaccharides are absent during certain stages of wall regeneration or that biosyntheses are differently coordinated during wall regeneration by comparison with normal wall secretion.

In a study comparable to that described above, Klein *et al.* (1981) isolated protoplasts from rapidly growing soybean suspension cultures and incubated the protoplasts with [^{14}C]glucose of high specific activity. A true cell-wall fraction could not be isolated during the first few hours of wall formation; thus, after 3 hr of culture, protoplasts were collected by centrifugation and two fractions of 80% ethanol-precipitable polymers prepared: those in whole protoplasts and those isolated from the cuture medium [extracellular material

(ECM)]. The ECM represents secreted polymers that have been released into the incubation medium [either because suitable well bonding sites were not yet present in the regenerating wall or because some of these polymers are identical to those normally found in the culture medium of suspension cultured cells (for details of the latter, see Moore, 1973)]. The distribution of radioactivity in the neutral sugars of the two fractions was distinct (Table III). Glucose had been rapidly converted to other neutral sugars, found as residues of noncellulosic polymers synthesized during the first 3 hr of wall regeneration. When similar samples were subjected to methylation analyses (Fig. 11), radioactivity could be detected in peaks that cochromatographed with differing pentoses and hexoses in diverse linkages. Terminal-arabinose, terminal-xylose, 4-linked hexose, and branched hexoses are found in pectic and hemicellulosic components of the native cell wall. The distribution of radioactivity in the ECM differed qualitatively from that of the protoplast fraction, indicating that different polymers may be present in the two fractions. This study, like previous studies, suggests that protoplasts synthesize a spectrum of polysaccharide-containing polymers that are, in many ways, similar to those of intact cells. It is clear, however, that more extensive analyses of regenerating walls are required before this conclusion can be stated without reservation. It must be noted, too, that there are important differences between regenerating walls and native walls. 3-Linked glucose was prominently labeled in both the protoplast and ECM fractions, suggesting the presence of callose (of which laminaribiose is the repeating unit). Callose is absent from the cell wall of cultured soybean cells, and its presence in the regenerating wall may represent a wound response induced by the act of protoplast isolation.

In a study of *Vinca rosea* protoplasts, Takeuchi and Komamine (1978) concluded that the regenerated wall, after 3 days in culture, was fully soluble in 24% KOH; thus, unlike the wall of the parent culture (Takeuchi and Komamine,

TABLE III
Distribution of Radioactivity in Neutral Sugars in Newly Synthesized Noncellulosic Polymers from Soybean Protoplasts[a]

Neutral sugar(s)	Protoplasts (mole%)	Extracellular material (mole%)
Rhamnose + fucose	7	8
Arabinose	22	21
Xylose	10	13
Mannose	24	14
Galactose	10	21
Glucose	27	23

[a] From Klein et al. (1981). Soybean protoplasts were cultured for 3 hr in wall-regeneration medium supplemented with [^{14}C(U)]glucose. At the end of the incubation, protoplasts were separated from the incubation medium. Polymers were collected by ethanol precipitation. The distribution of radioactivity in the alditol acetates derived from the neutral sugar residues in these polymers was determined and expressed as the mole% of label incorporated in the neutral sugars.

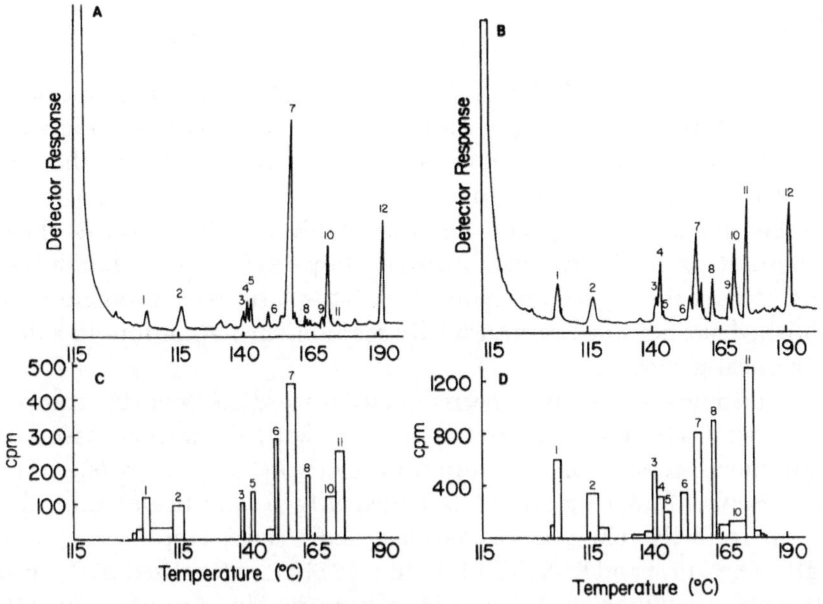

FIGURE 11. Gas-liquid chromatography elution profiles of: (A) permethylated derivatives of soybean cell walls; (B) polymers secreted into the culture medium of cell suspensions (2-day-old cultures). Elution profiles of permethylated derivatives of radioactively labeled polymers obtained from: (C) whole protoplasts; (D) extracellular molecules isolated from the culture medium of protoplasts cultured in the presence of [^{14}C]glucose for 3 hr. The peaks, indicated by numbers, represent variously methylated derivatives that are inferred to have arisen from sugar residues linked within the polymers as follows: (1) terminal-arabinose; (2) terminal-xylose; (3) 1,5 arabinose; (4) terminal-galactose; (5) 2- or 4-xylose; (6) 3-glucose; (7) 4-hexose; (8) 6-galactose; (9) 4,6-galactose; (10) 4,6-galactose + plasticizer; (11) 3,6-galactose; (12) inositol (internal standard). A spectrum of poly-saccharide-containing polymers was synthesized by the soybean protoplasts during the first 3 hr of wall regeneration. From Klein *et al.* (1981).

1977), it did not contain any crystalline cellulose. They found, instead, that the regenerated wall was composed of 3- and 4-linked glucans (which are also present in the cell wall of the *Vinca* culture from which the protoplasts were derived). Extracellular polysaccharides (including polyuronides and 3, 6-linked arabinogalactan) from protoplasts were found to be similar to those in the medium of the parent suspension culture. Hydroxyproline-containing glycoproteins from protoplasts isolated from a *Vinca* culture have also been examined (Tanaka and Uchida, 1979). Because protoplasts are easily disrupted, they are useful for the isolation of cell-wall precursors. High-molecular-weight (>100,000 daltons) labeled glycoproteins were isolated when protoplasts were fed [^{3}H]-proline and [^{14}C]arabinose, and conditions that inhibited protein synthesis or hydroxylation also inhibited the incorporation of [^{3}H]arabinose into this material. Tanaka and Uchida (1979) concluded that they had isolated precursors of "extensin" (a major structural protein of the cell wall, rich in hydroxyproline), but the possibility that these were other glycoproteins was not eliminated, since the material was not fully characterized (for details of plant glycoproteins, see Lamport, 1977; Brown and Kimmins, 1977; Clarke *et al.*, 1979).

4.2. Cytology

The studies described above, directed toward a fuller description of the molecular features of regenerating walls, have demonstrated that these walls vary in composition according to protoplast source. In many respects, the walls are similar to those of the parent tissue culture, but since they are sampled during their development, the composition changes with time (both because the loss of components to the medium is determined by the preexistence of suitable bonding sites and because the capacity to synthesize all the various components is neither reestablished at one time nor necessarily reestablished according to a developmentally ideal program).

No cytological studies have been directed toward the "matrix" components of the cell wall. Globular materials associated with the wall are visualized by most specimen-preparation techniques for electron microscopy (e.g., Fig. 10), but the presence of such materials has tended only to complicate studies directed toward microfibrils. The polysaccharide-specific stains employed by Prat and colleagues (e.g., Prat and Roland, 1971; Prat, 1973) have localized wall precursors in dictyosomes and their secretory vesicles, presumably en route to the cell wall. This tends to confirm the well-supported concept of the involvement of dictyosomes in the secretion of matrix substances. It has been hypothesized (Hanke and Northcote, 1974) that the phenomenon of protoplast budding [which sometimes is, and sometimes is not, involved with cell division (see Meyer and Abel, 1975; Willison, 1976)] is related to the loss of wall matrix materials to the culture medium, which results in microfibrils slipping past each other, creating microfibril-deficient regions of weakness in the wall where cell expansion is concentrated.

ACKNOWLEDGMENTS

A. S. K. is grateful to the following for making helpful comments on a preliminary draft of part of this work: T. Bacic, P. Carlson, D. P. Delmer, M. Jermyn, and D. T. A. Lamport. She also acknowledges the support of the Department of Energy (Contract DE-AC02-76ER01338) and the Biochemistry Department of Michigan State University. All the electron micrographs were obtained J. H. M. W. in the laboratory of Professor E. C. Cocking, Nottingham University.

References

Anderson, R., and Ray, P., 1978, Labeling of the plasma membrane of pea cells by a surface localized glucan synthetase, *Plant Physiol.* **61**:723.

Asamizu, T., and Nishi, A., 1979, Biosynthesis of cell wall polysaccharides in cultured carrot cells, *Planta* **146**:49.

Asamizu, T., and Nishi, A., 1980, Regenerated cell wall of carrot protoplasts isolated from suspension-cultured cells, *Physiol. Plant.* **48**:207.

Asamizu, T., Tanaka, K., Takebe, I., and Nishi, A., 1977, Change in molecular size of cellulose during regeneration of cell wall on carrot protoplasts, *Physiol. Plant.* **40:**215.

Billmeyer, F. W., Jr., 1971, *Textbook of Polymer Science*, 2nd ed., Wiley-Interscience, New York, p. 444.

Binding, H., and Nehls, R., 1978, Regeneration of isolated protoplasts of *Vicia faba* L., *Z. Pflanzenphysiol.* **88:**327.

Boffey, S., and Northcote, D. H., 1975, Pectin synthesis during wall regeneration of plasmolyzed tobacco leaf cells, *Biochem. J.* **150:**433.

Brown, R. G., and Kimmins, W. C., 1977, Glycoproteins, *Int. Rev. Biochem.* **13:**183.

Brown, R. M., Jr., and Montezinos, D., 1976, Cellulose microfibrils: Visualization of biosynthetic and orienting complexes in association with the plasma membrane, *Proc. Natl. Acad. Sci. U.S.A.* **73:**143.

Burgess, J., and Fleming, E. N., 1974, Ultrastructural observations of cell wall regeneration around isolated tobacco protoplasts, *J. Cell Sci.* **14:**439.

Burgess, J., and Linstead, P. J., 1976, Scanning electron microscopy of cell wall formation around isolated plant protoplasts, *Planta* **131:**173.

Burgess, J., and Linstead, P. J., 1977, Coumarin inhibition of microfibril formation at the surface of cultured protoplasts, *Planta* **133:**267.

Burgess, J., and Linstead, P. J., 1979, Structure and association of wall fibrils produced by regenerating tobacco protoplasts, *Planta* **146:**203.

Burgess, J., Linstead, P. J., and Harnden, J. M., 1977, The interpretation of scanning electron micrographs, *Micron* **8:**181.

Burgess, J., Linstead, P. J., and Bonsall, V. E., 1978, Observations on the time course of wall devélopment at the surface of isolated protoplasts, *Planta* **139:**85.

Chanzy, H., Imada, K., Mollard, A., Vuong, R., and Barnard, F., 1979, Crystallographic aspects of sub-elementary cellulose fibrils occurring in the wall of rose cells cultured *in vitro*, *Protoplasma* **100:**303.

Clarke, A. E., Anderson, R. L., and Stone, B. A., 1979, Form and function of arabinogalactans and arabinogalactan-proteins, *Phytochemistry* **18:**521.

Cocking, E. C., 1960, A method for the isolation of plant protoplasts and vacuoles, *Nature (London)* **187:**927.

Cocking, E. C., 1970, Virus uptake, cell wall regeneration, and virus multiplication in isolated plant protoplasts, *Int. Rev. Cytol.* **28:**89.

Cocking, E. C., 1972, Plant cell protoplasts—isolation and development, *Annu. Rev. Plant Physiol.* **23:**29.

Darvill, A., McNeil, M., Albersheim, P. A., and Delmer, D. P., 1980, The primary cell walls of flowering plants, in: *The Biochemistry of Plants* Vol. 1 (P. K. Stumpf and E. E. Conn, eds.), pp. 91–162, Academic Press, New York.

Davey, M. R., and Cocking, E. C., 1972, Uptake of bacteria by isolated higher plant protoplasts. *Nature (London)* **239:**455.

Davey, M. R., and Mathias, R. J., 1979, Close-packing of plasma membrane particles during wall regeneration by isolated higher plant protoplasts—fact or artefact, *Protoplasma* **100:**85.

Fowke, L. C., 1978, Ultrastructure of isolated and cultured protoplasts, in: *Frontiers of Plant Tissue Culture 1978* (T. A. Thorpe, ed.), pp. 223–233, International Association for Plant Tissue Culture, Calgary.

Gamborg, O. L., and Wetter, L. R., 1975, *Plant Tissue Culture Methods*, National Research Council, Ottawa.

Giddings, T. H., Jr., Brower, D. L., and Staehelin, L. A., 1980. Visualization of particle complexes in the plasma membrane of *Micrasterias denticulata* associated with the formation of cellulose fibrils in primary and secondary walls, *J. Cell Biol.* **84:**315.

Gregory, D. W., and Cocking, E. C., 1965, The large-scale isolation of protoplasts from immature tomato fruit, *J. Cell Biol.* **24:**143.

Grout, B. W. W., 1975, Cellulose microfibril deposition at the plasmalemma surface of regenerating tobacco mesophyll protoplasts: A deep etch study, *Planta* **123:**275.

Hanke, D. E., and Northcote, D. H., 1974, Cell wall formation by soybean callus protoplasts, *J. Cell Sci.* **14**:29.

Heath, I. B., 1974, A unified hypothesis for the role of membrane bound enzyme complexes and microtubules in plant cell wall synthesis, *J. Theor. Biol.* **48**:445.

Herth, W., and Meyer, Y., 1977, Ultrastructural and chemical analysis of the wall fibrils synthesized by tobacco mesophyll protoplasts, *Biol. Cellulaire* **30**:33.

Herth, W., and Meyer, Y., 1978, Cytology of budding and cleavage in tobacco mesophyll protoplasts cultivated in saline medium, *Planta* **142**:11.

Kinnersley, A. M., Racusen, R. H., and Galston, A. W., 1978, A comparison of regenerated cell walls in tobacco and cereal protoplasts, *Planta* **139**:155.

Klein, A. S., Montezinos, D., and Delmer, D. P., 1981, Cellulose and 1,3-glucan synthesis during the early stages of wall regeneration in soybean protoplasts, *Planta* **152**:105.

Lamport, D. T. A., 1977, Structure, biosynthesis and significance of cell wall glycoproteins, in: *Recent Advances in Phytochemistry*, Vol. II (F. A. Loewus and V. C. Runeckles, eds.), pp. 79–115, Plenum Press, New York.

Lloyd, C. W., Slabas, A. R., Powell, A. J., and Lowe, S. B., 1980, Microtubules, protoplasts and plant cell shape, an immunofluorescent study, *Planta* **147**:500.

Meyer, Y., and Abel, W. O., 1975, Budding and cleavage division of tobacco mesophyll protoplasts in relation to pseudo-wall and wall formation, *Planta* **125**:1.

Meyer, Y., and Herth, W., 1978, Chemical inhibition of cell wall formation and cytokinesis, but not of nuclear division, in protoplasts of *Nicotiana tabacum* L., cultivated *in vitro*, *Planta* **142**:253.

Montezinos, D., and Delmer, D. P., 1980, Characterization of inhibitors of cellulose synthesis in cotton fibers, *Planta* **148**:305.

Moore, T. S., Jr., 1973, An extracellular macromolecular complex from the surface of soybean cultures, *Plant Physiol.* **51**:529.

Mueller, S., and Brown, R. M., Jr., 1980, Evidence for an intramembrane component associated with a cellulose microfibril synthesizing complex in higher plants, *J. Cell Biol.* **84**:351.

Mueller, S., Brown, R. M., Jr., and Scott, T. K., 1976, Cellulose microfibrils: Nascent stages of synthesis in a higher plant cell, *Science* **194**:949.

Nagata, T., and Takebe, I., 1970, Cell wall regeneration and cell division in isolated tobacco mesophyll protoplasts, *Planta* **92**:301.

Nagata, T., and Takebe, I., 1971, Plating of isolated tobacco mesophyll protoplasts on agar medium, *Planta* **99**:12.

Pearce, R. S., Withers, L. A., and Willison, J. H. M., 1974, Bodies of wall-like material ("wall-bodies") produced intracellularly by cultured isolated protoplasts and plasmolysed cells of higher plants, *Protoplasma* **82**:223.

Plowe, J. Q., 1931, Membranes in the plant cell. 1. Morphological membranes at protoplasmic surfaces, *Protoplasma* **12**:196.

Pojnar, E., Willison, J. H. M., and Cocking, E. C., 1967, Cell wall regeneration by isolated tomato fruit protoplasts, *Protoplasma* **64**:460.

Prat, R., 1973, Contribution à l'étude des protoplastes végétaux. II. Ultrastructure du protoplaste isolé et régéneration de sa paroi, *J. Microsc. (Paris)* **18**:65.

Prat, R., and Roland, J.-C., 1971, Étude ultrastructurale des premiers stades de néoformation d'une envelope par les protoplastes végétaux séparés mécaniquement de leur paroi, *C. R. Acad. Sci. Ser. D* **273**:165.

Prat, R., and Williamson, F. A., 1976, Chronologie de la sécretion de parois par les protoplastes végétaux, *Soc. Bot. Fr. (Coll. Secret. Veget.)* **123**:33.

Preston, R. D., 1974, *The Physical Biology of Plant Cell Walls*, Chapman and Hall, London.

Preston, R. D., 1979, Polysaccharide conformation and cell wall function, *Annu. Rev. Plant Physiol.* **30**:55.

Racusen, R. H., Kinnersley, A. M., and Galston, A. W., 1977, Osmotically induced changes in electrical properties of plant protoplast membranes, *Science* **198**:405.

Robenek, H., and Peveling, E., 1975, Veränderungen des Plasmalemmas während der Zellwandregeneration an isolierten Protoplasten aus dem Sprosskallus von *Skimmia japonica* Thunb., *Planta* **127**:281.

Robenek, H., and Peveling, E., 1977, Ultrastructure of the cell wall regeneration of isolated protoplasts of *Skimmia japonica* Thunb., *Planta* **136**:135.

Robinson, D. G., 1977, Plant cell wall synthesis, *Adv. Bot Res.* **15**:89.

Saemann, J. F., Moore, W. E., and Millett, M. A., 1963, Sugar units present, in: *Methods in Carbohydrate Chemistry*, Vol. 3 (R. L. Whistler, ed.), pp. 54–69, Academic Press, New York.

Schilde-Rentschler, L., 1977, Role of the cell wall in the ability of tobacco protoplasts to form callus, *Planta* **135**:177.

Schnabl, H., Vienken, J., and Zimmermann, U., 1980, Regular arrays of intramembranous particles in the plasmalemma of guard cell and mesophyll cell protoplasts of *Vicia faba*, *Planta* **148**:231.

Scowcroft, W. R., 1977, Somatic cell genetics and plant improvement, *Adv. Agron.* **29**:39.

Takebe, I., Otsuki, Y., and Aoki, S., 1968, Isolation of tobacco mesophyll cells in intact and active state, *Plant Cell Physiol.* **9**:115.

Takeuchi, Y., and Komamine, A., 1977, Turnover of cell wall polysaccharides of a *Vinca rosea* suspension culture. 1. Synthesis and degradation of cell wall components, *Physiol. Plant.* **48**:271.

Takeuchi, Y., and Komamine, A., 1978, Composition of the cell wall formed by protoplasts isolated from *Vinca rosea* suspension culture, *Planta* **140**:227.

Talmadge, K., Keegstra, K., Bauer, W. D., and Albersheim, P., 1973, The structure of plant cell walls. I. Macromolecular components of the walls of suspension-cultured sycamore cells with a detailed analysis of pectic polysaccharides, *Plant Physiol.* **51**:158.

Tanaka, M., and Uchida, T., 1979, Heterogeneity of hydroxyproline-containing glycoproteins in protoplasts from a *Vinca rosea* suspension culture, *Plant Cell Physiol.* **20**:1295.

Taylor, A. R. D., and Hall, J. L., 1976, Some physiological properties of protoplasts isolated from maize and tobacco tissue, *J. Exp. Bot.* **27**:383.

Townsend, C. O., 1897, Der Einfluss der Zellkerns auf die Bildung der Zellhaut, *Jahrb. Wiss. Bot.* **30**:484.

Updegraff, D. M., 1969, Semi-micro determination of cellulose in biological materials, *Anal. Biochem.* **32**:420.

Von Klercker, J., 1892, A method for isolating living protoplasts, *Ofvers Vetensk. Akad. Forh. Stockholm* **49**:463.

Wilkinson, M. J., and Northcote, D. H., 1980, Plasma membrane ultrastructure during plant protoplast plasmolysis, isolation and wall regeneration: A freeze-fracture study, *J. Cell Sci.* **42**:401.

Williamson, F. A., Fowke, L. C., Weber, G., Constabel, F., and Gamborg, O., 1977, Microfibril deposition on cultured protoplasts of *Vicia hajastana*, *Protoplasma* **91**:213.

Willison, J. H. M., 1973, Fine structural changes occurring during the culture of isolated tomato fruit protoplasts, *Colloq. Int. CNRS* **212**:215.

Willison, J. H. M., 1976, Synthesis of cell walls by higher plant protoplasts, in: *Microbial and Plant Protoplasts* (J. F. Peberdy, A. H. Rose, H. J. Rogers, and E. C. Cocking, eds.), pp. 283–298, Academic Press, London.

Willison, J. H. M., and Brown, R. M., Jr., 1977, An examination of the developing cotton fiber: Wall and plasmalemma, *Protoplasma* **92**:21.

Willison, J. H. M., and Cocking, E. C., 1972, The production of microfibrils at the surface of isolated tomato-fruit protoplasts, *Protoplasma* **75**:397.

Willison, J. H. M., and Cocking, E. C., 1975, Microfibril synthesis at the surface of tobacco mesophyll protoplasts, a freeze-etch study. *Protoplasma* **84**:147.

Willison, J. H. M., and Grout, B. W. W., 1978, Further observations on cell-wall formation around isolated protoplasts of tobacco and tomato, *Planta* **140**:53.

Willison, J. H. M., and Rowe, A. J., 1980, *Replica, Shadowing and Freeze-Etching Techniques*, Vol. 8 of *Practical Methods in Electron Microscopy* (A. M. Glauert, ed.), North-Holland, Amsterdam.

5

Cellulose-Microfibril Assembly and Orientation in Higher Plant Cells with Particular Reference to Seedlings of *Zea mays*

SUSETTE C. MUELLER

1. *Introduction*

Cellulose-microfibril synthesis and deposition is one of the key events in plant-cell growth and differentiation. Microfibrils are deposited in highly organized layers to form the structural backbone of the cell wall (Preston, 1974). A complex mixture of other macromolecules interact with cellulose to give the cell wall strength, but yet allow it the property of extensibility (Cleland, 1971; Northcote, 1972). An important point of regulation of cell growth is the control of the orientation of microfibril deposition. In most growing cells, the overall axes of microfibril deposition are transverse with respect to the longitudinal cell axis (Roelofsen, 1965). This arrangement of microfibrils results in cellular extension along a longitudinal axis during growth (Green, 1974; Roelofsen, 1965).

A traditional approach to the study of microfibril deposition has been to analyze the orientations of microfibrils in various cell types by ultrathin sections of fixed material (Hepler and Palevitz, 1974). However, a major drawback to this technique is the inability to resolve adequately the most recently deposited microfibrils associated with the plasma membrane. In addition, this technique does not reveal the distributions of the cellulose-synthesizing enzymes that are thought to be plasma-membrane-bound (Brown and Willison, 1977; Delmer, 1977; Maclachlan, 1977). In recent years, freeze-fracture techniques have been

---

SUSETTE C. MUELLER • The Biological Laboratories, Harvard University, Cambridge, Massachusetts 02138.

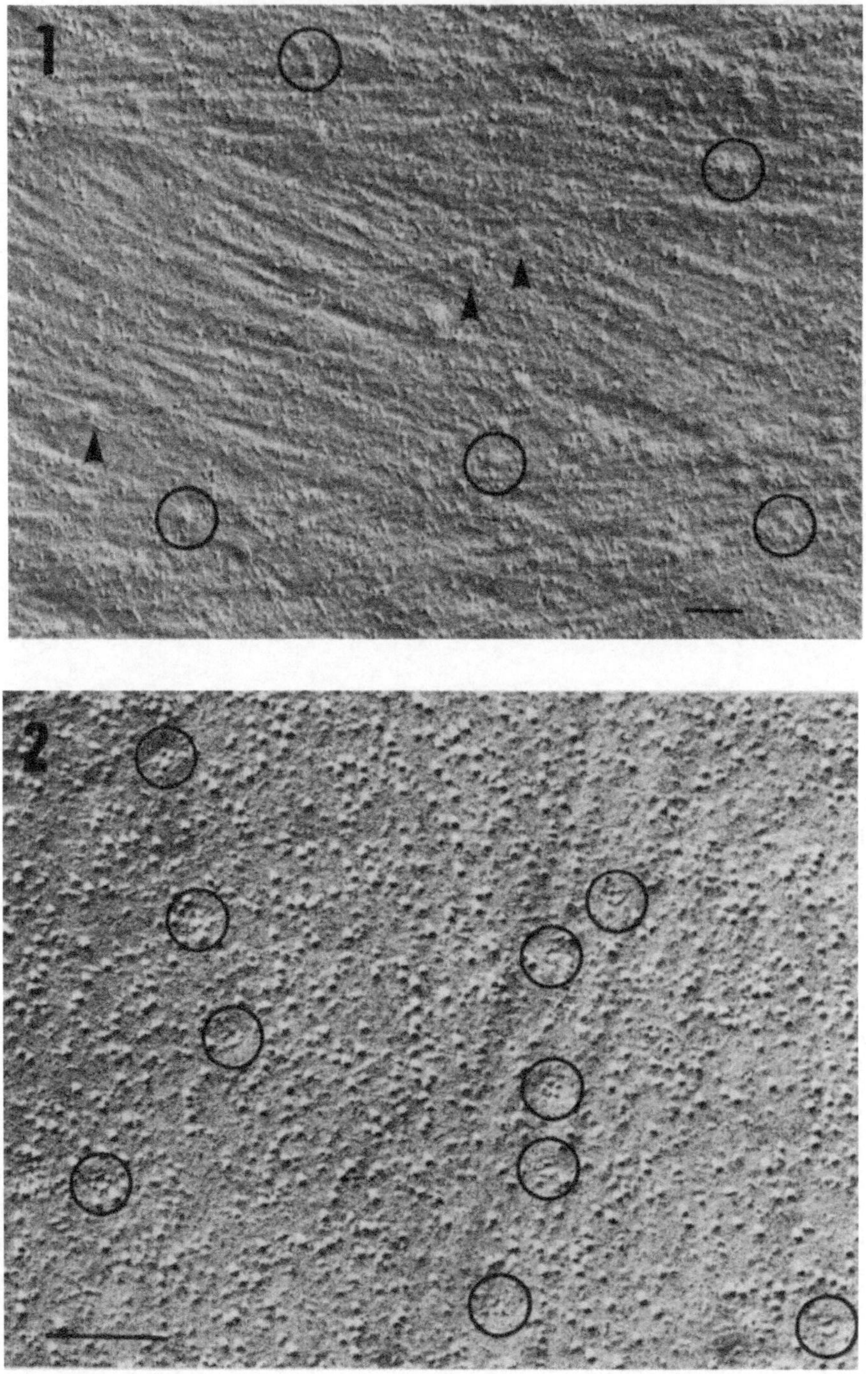

FIGURE 1. Mesocotyl plasma membrane prepared by freezing corn tissue between double-replica holders in nitrogen slush. An enlargement of an E fracture face on a longitudinal wall illustrates microfibril bands. Terminal complexes (circles) are concentrated in the membrane over the newly synthesized bands. Three terminal complexes are seen associated with the ends of microfibrils between the two bands (▲). ×76,800. Scale bar: 0.1 μm.

exploited in the search for a morphologically recognizable synthetase complex. The results have contributed significantly to our understanding of how microfibrils are synthesized and deposited (Brown and Willison, 1977; Brown, 1979).

As a result of recent freeze-fracture studies, presumptive cellulose-synthesizing structures have been observed in a bacterium (Brown *et al.*, 1976; Zaar, 1979), in algae (Brown and Montezinos, 1976; Giddings *et al.*, 1980; Willison and Brown, 1978), and in several higher plants (Mueller *et al.*, 1976; Mueller and Brown, 1980; Willison and Brown, 1977; Willison and Grout, 1978). Terminal synthesizing complexes have been observed on the E fracture face of the plasma membrane at the ends of microfibrils in corn seedling tissue (Mueller *et al.*, 1976). Terminal complexes are globular, about 25 nm in diameter, and may consist of subunits. Sometimes, during fracturing, terminal-complex-associated microfibrils are torn toward the cytoplasm from the wall through the outer leaflet of the membrane. When this happens, the central portion of the terminal complex (about 12 nm in diameter) remains attached to the microfibril (Mueller *et al.*, 1976; Mueller and Brown, 1980). A tear associated with a hole is observed in the outer leaflet of the membrane. A portion of the terminal complex appears to remain attached to the membrane and appears to surround the hole left by the central particle (Mueller *et al.*, 1976; Mueller and Brown, 1980).

Improvement of freezing techniques has subsequently allowed the visualization of the P fracture face (PF) component of the terminal synthesizing complex, termed a rosette (Mueller and Brown, 1980). This component consists of six intramembranous particles arranged in a rosette around a central nonshadowed depression (or hole) in the PF leaflet of the plasma membrane. Rosettes measure about 24 nm in diameter from the outer edges of particles on opposite sides of the unshadowed center. Rosette (PF) and terminal complex (EF) subunits are observed at the ends of individual microfibril impressions (Mueller and Brown, 1980, 1982a). Both rosettes and terminal complexes are associated with the most recently deposited bands of microfibrils (Figs. 1 and 2), and both are observed in greatly reduced numbers within pit fields where fewer microfibrils are deposited (Mueller and Brown, 1980, 1982a). The tendency of microfibrils to tear through the membrane might possibly reflect an affinity of the microfibril for the PF rosette (Mueller and Brown, 1982a). A model for the association of synthesizing complexes and microfibrils in the plasma membrane has been presented (Mueller and Brown, 1980; see also Giddings *et al.*, 1980). The synthesizing complex is embedded within the plasma membrane, and it was suggested that in fact it may be transmembranous (Mueller and Brown, 1980). As previously proposed for the terminal complex in *Oocystis* (Brown and Montezinos, 1976), terminal synthesizing complexes in corn cells might spin out microfibrils into the wall as they diffuse laterally through the membrane (Mueller and Brown, 1980).

FIGURE 2. Corn mesocotyl cell prepared as in Fig. 1. An enlargement of a portion of two microfibril bands illustrates that rosettes (circles) are also concentrated in the membrane over the microfibril bands on P fracture faces. ×159,000. Scale bar: 0.1 μm.

The EF component of the terminal complex has been observed in radish roots (Willison and Grout, 1978), cotton fibers (Willison and Brown, 1977), mung bean hypocotyls, and pine roots, and in corn roots, coleoptiles, and mesocotyls (Mueller and Brown, 1980). The rosette subunit of the terminal synthesizing complex has been observed in corn tissues, mung bean hypocotyls (Mueller and Brown, 1980), cotton fibers (Willison and Brown, unpublished micrographs), radish root hairs (Cooper and Brown, unpublished micrographs), and most recently in *Micrasterias*, an alga (Giddings *et al.*, 1980), and in *Adiantum* (Wada and Staehelin, 1981). Thus, it appears that the terminal complex-rosette structure is universally present in higher plants. The morphology of the terminal synthesizing complex is similar to that of higher plants in at least one alga.

Giddings *et al.* (1980) have obtained complementary double replicas of freeze-fractured *Micrasterias* that reveal particulate EF structures complementary to PF rosettes at microfibril termini (Giddings *et al.*, 1980). The observation that the width of each microfibril is positively correlated to the number of rosettes arranged in a linear sequence at the microfibril terminus provided evidence for the proposed role of rosettes in the synthesis of microfibrils.

Linear sequences of terminal synthesizing complexes of a different nature have been observed in several green algae (Brown and Montezinos, 1976; Montezinos and Brown, 1976; Willison and Brown, 1978), (see also Chapters 1 and 6). In *Oocystis*, the EF particles, or the terminal complexes, are thought to be responsible for microfibril synthesis; the PF particles, or granule bands, are proposed to be microfibril-orienting structures (Brown and Montezinos, 1976). In another green alga, *Glaucocystis*, which also has a linear terminal complex on the EF face of the plasma membrane, granule bands appear to be absent (Willison and Brown, 1978). There is no evidence to suggest that the rosettes in higher plants and *Micrasterias* are comparable to the granule bands in *Oocystis*, but rather it is thought that they might function in microfibril synthesis (Giddings *et al.*, 1980; Mueller and Brown, 1980).

2. *Factors That Affect the Visualization of Terminal Complexes*

The appearance of freeze-fractured membranes depends on the state of the membrane before freezing. For example, the use of fixatives or cryoprotectives or both can alter the lateral distribution of intramembranous particles (Willison and Rowe, 1980; see Rash and Hudson, 1979), and fixation can alter the distribution of intramembranous particles between E and P fracture faces (Satir and Satir, 1979). Montezinos and Brown (1976) found that terminal complex particles disappeared as a consequence of glutaraldehyde fixation. Willison (1976) showed that glycerol pretreatment of plant cells caused microfibril impressions on the plasma membrane to disappear because of the plasmolyzing effects of glycerol. Since terminal synthesizing complexes were not observed after glutaraldehyde and glycerol pretreatment, it was proposed that in higher plants the visualization of terminal complexes may also depend on the proximity of the plasma membrane to the cell wall (Willison, 1976). However, after glutaraldehyde fixation and glycerol cryoprotection, terminal complexes (Fig. 3) and rosettes (Fig. 4) can be observed on E and P fracture faces from pea stems, but are more difficult to recognize. Terminal complexes may lose their prominent profile after pretreatment as a consequence of the loss of cellular turgor pressure.

Slow freezing rates increase the likelihood that ice crystals will damage the membrane (Willison and Rowe, 1980). Membrane fracture faces prepared by freezing corn seedling tissue in liquid Freon 22 or by freezing in nitrogen slush between double-replica holders (Balzers Corp.) results in a general improvement in the preservation of membrane morphology with the latter technique (Mueller and Brown, 1980; Mueller, unpublished observations). For unfixed, noncryoprotected tissue, the frequency of membrane fracture faces judged to be more poorly preserved is greater in conventionally frozen material, although well-preserved membranes can also be observed on occasion. Poorly preserved membranes sometimes contain only depressions that lack distinctive rosette particles (see Mueller *et al.*, 1976, Fig. 6). Strings of intramembranous particles are frequently observed on conventionally frozen membranes, whereas they were rarely observed on membranes from tissues frozen between double-replica grids in nitrogen slush. Membranes containing particle strings and indistinct rosettes are interpreted to be damaged membranes (Mueller and Brown, 1980).

3. Oriented Microfibril Deposition

Microfibrils within each layer are frequently in parallel, linear arrays adjacent to the plasma membrane. The thickness of each layer is fairly constant, but the axis of microfibril orientation is different. In corn root parenchyma just beginning the rapid phase of elongation, microfibril orientation within each layer near the plasma membrane is nearly transverse with respect to the longitudinal cell axis (Mueller and Brown, 1982a). Older parenchyma cells that have completed the rapid elongation phase, by contrast, have layers of microfibrils that alternate between transverse and more longitudinal inclinations (Mueller and Brown, 1982a). Elongating mesocotyl and coleoptile parenchyma have transversely inclined axes of microfibril orientation, but frequently also contain longitudinal ribs of microfibrils at the corners of cells (Mueller and Brown, 1982a). These cells also have a higher proportion of layers with more longitudinally inclined microfibrils compared to rapidly elongating root parenchyma.

4. Unidirectional Microfibril Deposition

A striking feature of microfibril deposition is the pattern of terminal complex–microfibril tears on EF faces of the plasma membrane (Mueller and

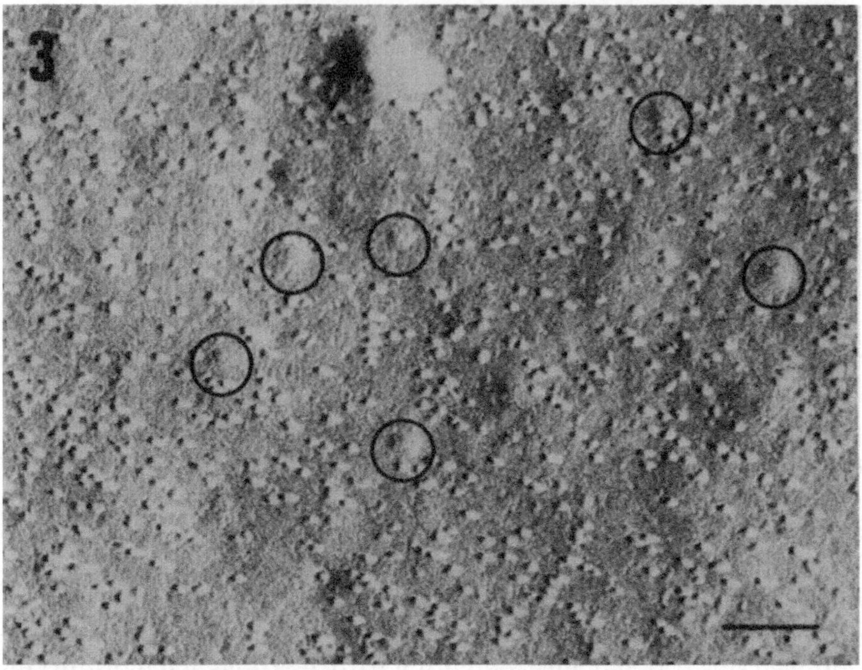

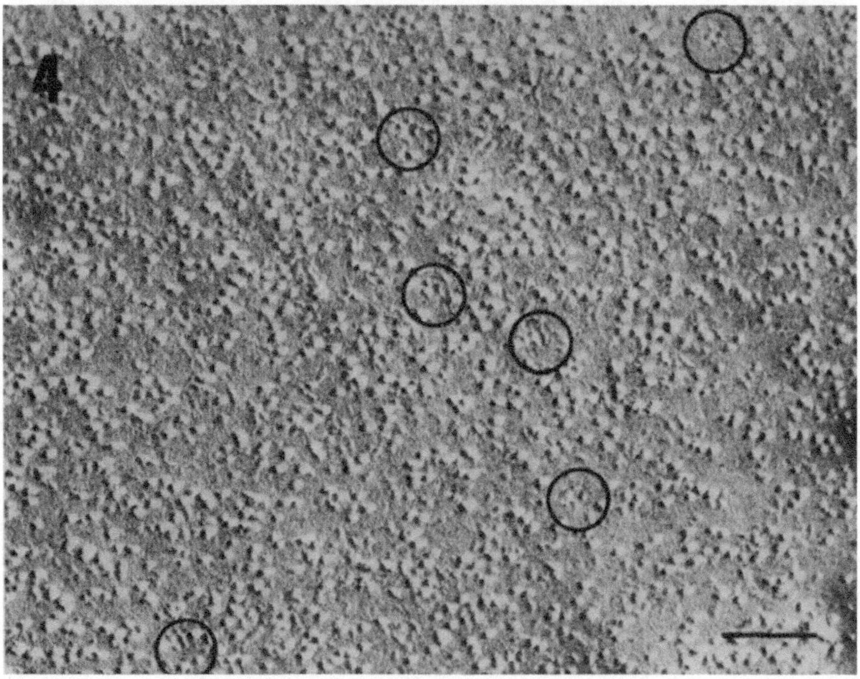

FIGURE 3. Longitudinal membrane of a parenchyma cell from a pea stem. The tissue was fixed 1 hr in 2% glutaraldehyde, 0.1 M cacodylate buffer, pH 7.4, and 0.1 M sucrose at room temperature, washed in buffer, and soaked in 30% glycerol in buffer for 1 hr. The pretreated tissue was then frozen in

Brown, 1982a). The tears were mentioned above in the discussion of the structure of the terminal synthesizing complexes. The orientation of the terminal complex hole in the membrane with respect to the position of the tear has been interpreted to be an indication of the direction of deposition (Mueller and Brown, 1982a). Often, tears cover large surfaces of the membrane, and they all point in the same direction (Mueller and Brown, 1982a). Thus, the tears suggest that at the time the cells were frozen, microfibril deposition was occurring in one direction across the plasma membrane.

When tears over the entire membrane of a longitudinal or transverse wall suggest that deposition is unidirectional, tears associated with small bands of microfibrils have approximately the same polarity. Sometimes a band of microfibril impressions terminates in association with synthesizing complexes and therefore appears to have been deposited in a unidirectional fashion. Occasionally on a transverse wall, bands of microfibrils associated with terminal complexes contain impressions in both polarities (Mueller and Brown, 1982a). In addition, one case of multidirectional terminal complex–microfibril has been observed in a stelar root cell of corn (Mueller and Brown, 1982a). In this cell, microfibril orientation appears disorganized.

Occasionally, curving configurations of terminal-complex-associated microfibrils are observed (Mueller and Brown, 1982a). On these membranes, microfibrils parallel to the main axis of deposition are associated with terminal complexes. In addition, a small population of terminal complexes is associated with the ends of microfibril impressions whose axis of orientation curves from the main axis to a new axis of orientation. A similar phenomenon is observed on a pine root plasma membrane, in which terminal complex–microfibril tears curve in the same direction with the same polarity (Mueller and Brown, 1982a).

It is hypothesized that terminal-complex-associated microfibril tears are an indication of the polarity of microfibril deposition. It might be argued that terminal-complex-associated tears result from a selective tearing process. This, however, seems unlikely for several reasons. Terminal complex–microfibril tears can be observed even after double-replica freeze–fracture when no knife is used (Mueller and Brown, 1982a). Furthermore, asymmetrical outlines of microfibrils around pit fields are observed on membranes that also contain unidirectionally oriented tears (Mueller and Brown, 1982a). If microfibrils are deposited toward the pit field from each side, then one would expect each end of the pit field to be a mirror image of the other. Thus, the observed asymmetrical pit field outlines suggest that microfibrils were deposited unidirectionally. Tears and impressions indicating the same polarity are usually associated with highly ordered microfibril configurations, whereas tears and impressions with many polarities

nitrogen slush. Despite the pretreatment with glutaraldehyde and glycerol, terminal complexes (circles) are observed on the E fracture face. ×122,800. Scale bar: 0.1 μm.

FIGURE 4. Longitudinal membrane adjacent to a pit field from a pea stem. The tissue was pretreated and frozen as in Fig. 3. Rosettes (circles) are observed on P fracture faces of fixed, glycerinated tissue. ×122,800. Scale bar: 0.1 μm.

have only been observed when the orientation of microfibril deposition appears to have been perturbed (Mueller and Brown, 1982a)

The association of terminal complexes and rosettes with flexuous, organized patterns of microfibrils suggests that microfibril impressions are actually records of the movement of synthesizing complexes through the plasma membrane. Patterns most suggestive of this are adjacent to pit fields (Fig. 5). The patterns of microfibrils around pit fields are similar to the patterns that are generated by particles on the surface of water flowing around a tree in the middle of a stream. The path taken by the terminal synthesizing complex diffusing laterally through the membrane would be similar to the path taken by the particle on the water, and the microfibril that is synthesized would become a permanent record.

Alternatively, microfibril patterns could result from the reorientation of

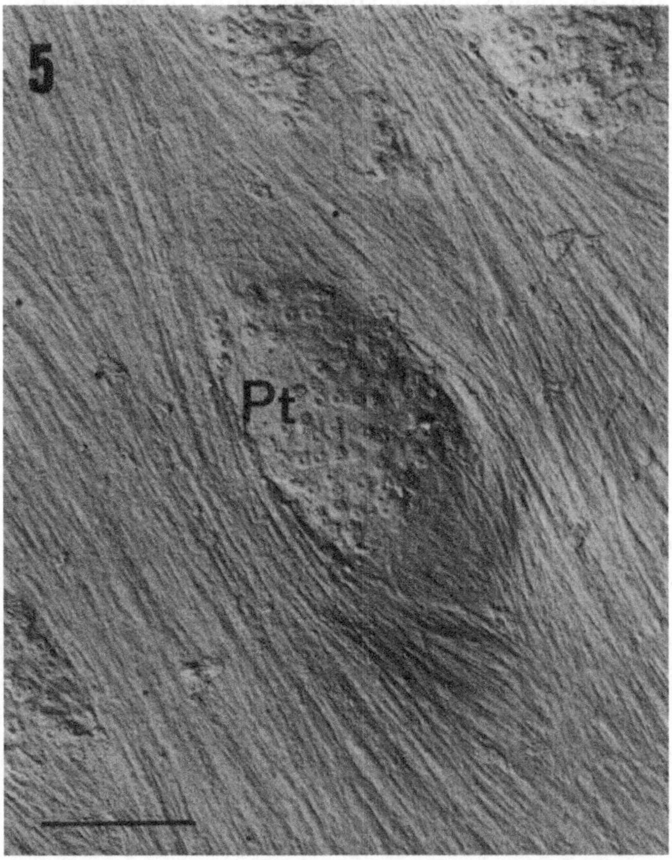

FIGURE 5. Longitudinal membrane of a parenchyma cell 3 mm from the corn root tip. Pieces of untreated tissue were frozen in Freon 22. The flexuous patterns of microfibril impressions around the pit field (Pt) suggest that the microfibrils record the movement in the membrane of synthesizing complexes. The pattern of microfibrils resembles a flow pattern. ×20,400. Scale bar: 0.1 μm.

microfibrils already deposited during the process of cell elongation. However, terminal-complex-associated microfibrils still display flexuous microfibril patterns around plasmodesmata on membranes of cells that are not undergoing expansion. These nonexpanding membranes are associated with older cells more than a centimeter from the root tip and membranes of transverse walls (Mueller and Brown, 1982a). Thus, as seems likely, if terminal complexes are responsible for the synthesis of microfibrils, then microfibril impressions on the plasma membrane could be a record of the recent movement of terminal complexes in the plane of the membrane.

5. *Experimental Procedures*

The freeze-etch relationship between cytoplasmic structures and the orientation of microfibrils have been studied to elucidate how terminal synthesizing complexes could be guided through the plasma membrane.

Corn (*Zea mays* L., Burpee's Snowcross) and pea (*Pisum sativum* L., Alaska variety) seeds were surface sterilized in 10% Clorox, rinsed in deionized water, and soaked for about 6 hr. Seeds were germinated in the dark 3 days for corn and 7–8 days for peas. Small pieces of tissue were excised from corn roots and mesocotyls and frozen either in Freon 22 at liquid nitrogen temperature or in nitrogen slush (Mueller and Brown, 1980). Segments of pea stems were excised from the region of elongation just below the subapical hook. The segments were frozen on 1-mm gold specimen holders (Balzers Corp., Hudson, New Hampshire) in nitrogen slush. Corn tissue frozen in Freon 22 and pea tissue was freeze-etched by conventional means as described in Mueller and Brown (1980). Corn tissue frozen between double-replica holders (Balzers Corp.) was fractured in a hinged double-replica device (Balzers Corp.).

Pea stem segments were pretreated before freezing as follows: fixation for 1 hr at room temperature in 2% glutaraldehyde, 0.1 M cacodylate buffer, pH 7.4, 0.1 M sucrose; rinsed in 0.1 M cacodylate buffer, pH 7.4; soaked in 30% glycerol in 0.1 M cacodylate buffer, pH 7.4, for 1 hr.

For colchicine treatment, corn seeds were germinated in 9-cm Petri plates. After 48 hr, 5 ml 1 mM colchicine (Sigma) in 10–20 mM citrate phosphate buffer, pH 7.0, or buffer alone was added with care not to disturb the seedlings. After 24 hr, pieces of tissue were excised from control and colchicine-treated roots 3–4 mm from the tip. The tissue was frozen in nitrogen slush between double-replica holders.

Tissue fractured in a double-replica device or with a razor blade (Mueller and Brown, 1980) was replicated immediately or etched for 30–60 sec before replicating. Fractured tissue was shadowed with platinum at an angle of 45° and then stabilized with a carbon film at an angle of 90°. Replicas were processed as described in Mueller and Brown (1980) and viewed in a Hitachi HU 11E-1 or a Philips 200 electron microscope.

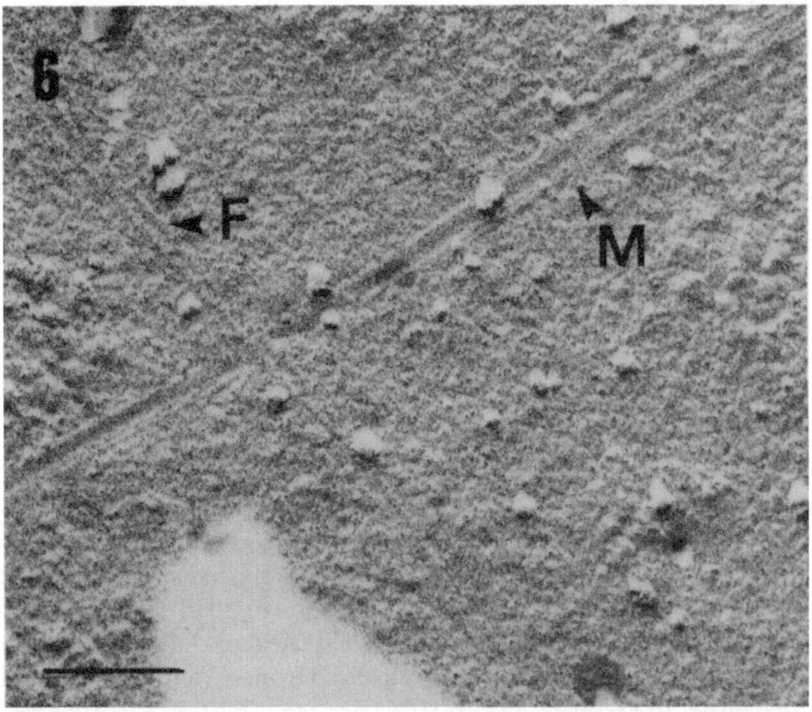
6
F
M

7
F
M

6. *Influence of Cytoplasmic Structures on Microfibril Deposition*

Fracture planes through the cytoplasm frequently reveal lengths of microtubules and bundles of microfilaments adjacent to the plasma membrane (Fig. 6) (Mueller and Brown, 1982b). In corn seedling tissue, these fracture planes contain microtubules parallel to the newest orientations of microfibrils, while microfilaments are generally parallel to the longitudinal cell axis (Fig. 7). The alignment of microtubules with new microfibrils supports the widely held theory that microtubules mediate microfibril orientation (Hepler and Palevitz, 1974). A direct association between a terminal-complex-associated band of microfibrils and a microtubule exactly opposite it has been shown (Mueller and Brown, 1982b). It is possible that the new orientation of microfibrils resulted from the interaction of terminal synthesizing complexes with the microtubule in the plane of the membrane, since the microtubule observed in the freeze-fractured cytoplasm appears close enough to the plasma membrane to be cross-linked. Furthermore, bands of microfibrils visualized commonly on E and P fracture faces associated with terminal complexes and rosettes may be the direct result of underlying parallel arrays of cortical microtubules (Hepler and Palevitz, 1974; Mueller and Brown, 1982b). Although the role of filaments, in particular actin microfilaments, is less clear (Hepler and Palevitz, 1974), it may be significant that the orientations between microtubules and microfilaments on freeze-etched transverse membranes (Fig. 6) are the same as on freeze-etched longitudinal membranes (Fig. 7) (Mueller and Brown, 1982b). Parallel orientations of microtubules and microfilaments are seldom observed.

One of the strongest lines of evidence for the existence of other microfibril-orienting mechanisms is the pattern of microfibrils produced in the presence of colchicine, a chemical that causes the depolymerization of microtubules and the disruption of the normal orientation of microfibril deposition (Hepler and Palevitz, 1974). After microtubule depolymerization, terminal complex-microfibril tears indicate that deposition is polarized in one direction over rather large surfaces of the plasma membrane, similar to the situation in control cells (Mueller and Brown, 1982b). Waves of unidirectionally-oriented microfibrils associated with terminal complexes at their tips are in thin, evenly spaced arrays.

The major effect of colchicine on microfibril orientation in higher plants appears to be mediated by its effect on cortical microtubules (Hepler and Palevitz, 1974). The initiation of new microfibril layers in untreated cells is

---

FIGURE 6. Freeze-etch preparation of corn tissue 6 mm from the root tip. The tissue was frozen as in Fig. 1 and fractured transversely across the root. The resulting cytoplasmic fracture plane is parallel and adjacent to the transverse plasma membrane. Microtubule cross-fractures (M) and microfilaments (F) are in the fracture plane. ×83,000. Scale bar: 0.1 μm.

FIGURE 7. Longitudinal membrane from a corn root cell 3 mm from the tip. The tissue was prepared as in Fig. 1. Terminal complexes (circle) are associated with microfibril impressions on the E fracture face of the plasma membrane. The impressions are parallel to microtubules (M) which are cross-fractured in the cytoplasm. Microfilaments (F) are usually parallel to the longitudinal cell axis. ×42,000. Scale bar: 0.2 μm.

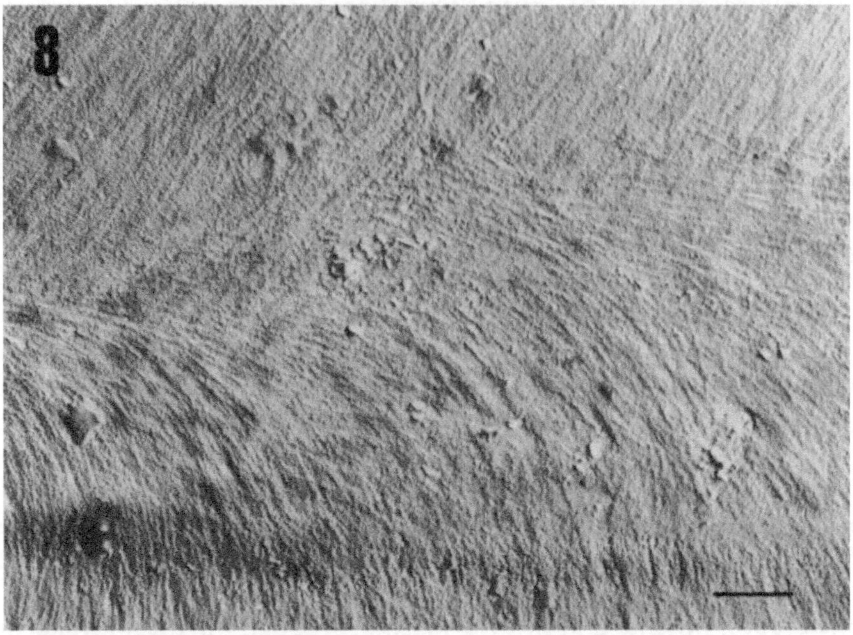

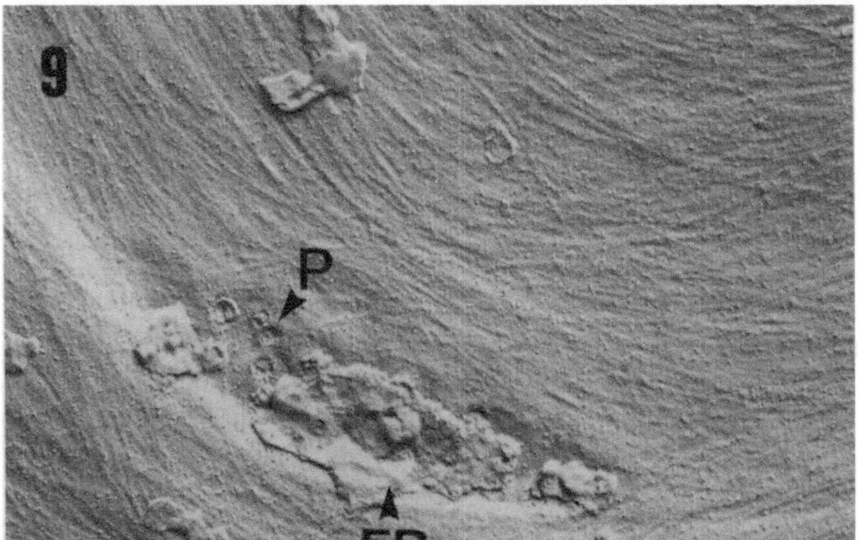

FIGURE 8. Intact corn roots treated with 1 mM colchicine for 24 hr and prepared for freeze-fracture as in Fig. 1. This P fracture face is from a parenchyma cell that became nearly spherical during colchicine treatment. Curving patterns of microfibril impressions are common and often appear at the junction between two axes of microfibril orientation. ×20,400. Scale bar: 0.5 μm.

characterized by the presence of parallel bands of microfibrils (Mueller and Brown, 1982b). Then, when microfibril deposition has been established in one orientation, older microfibril orientations are completely covered by terminal-complex-associated microfibrils with parallel orientations. In contrast, after colchicine treatment, microfibril bands are not observed in the newest layers; instead, evenly spaced terminal-complex-associated microfibrils are observed (Mueller and Brown, 1982b). When microfibril deposition is orderly, these microfibrils are in approximately parallel orientations.

Microfibril deposition is altered in several other ways during colchicine treatment. Often, instead of linear arrays of microfibrils, new layers of microfibrils deposited during colchicine treatment have swirling or curving configurations (Figs. 8 and 9). Microfibril impressions sometimes converge on a single point in the plane of the membrane (Mueller and Brown, 1982b). On a P fracture face, microfibril impressions are suggestive of patterns produced by two converging axes of microfibril orientations (Fig. 8). Ridges of microfibrils on the E fracture face also appear to be formed during colchicine treatment by the convergence of microfibril orientations (Mueller and Brown, 1982b). The ridges or depressions are associated with an increased concentration of terminal complexes or rosettes in the membrane. In all these examples, however, microfibril deposition is ordered in that the microfibril impressions are parallel, even if they are not linear.

The outlines of pit fields from colchicine-treated cells are very similar to pit field outlines from untreated cells (Fig. 9), but the long axes are usually curved following the major microfibril orientations. Since microtubules are not detected in the cytoplasm of colchicine-treated cells by thin section or freeze-fracture, the outlines of pit fields cannot be directly dependent on the presence of cortical microtubules.

Occasionally, freeze-fractured membrane surfaces of colchicine-treated cells also contain new terminal-complex-associated microfibrils with disoriented, multidirectional patterns (Mueller and Brown, 1982b). Multidirectional deposition on plasma membranes during colchicine treatment was observed to consist only of a very thin layer of terminal-complex-associated microfibrils. Previously deposited organized wall layers could be seen beneath the newly deposited disorganized layer. This suggests that the disruption of microfibril deposition was a transient phenomenon. Pit fields within regions of disoriented deposition are symmetrical in shape.

When intact seedlings are exposed to 1 μl/liter ethylene, microfibril deposition is altered in a different way. New layers of microfibrils have longitudinal orientations with respect to the longitudinal cell axis in mesocotyl and root parenchyma. Unidirectionally-oriented tears occur during ethylene

FIGURE 9. Corn roots prepared as in Fig. 8. A pit field on an E fracture face of a colchicine-treated cell is similar to a control cell pit field except that its long axis is parallel to the predominant microfibril orientation, which is curved. Normally, microfibrils are deposited in linear patterns. Exceptions in control cells are found near transverse membranes and adjacent to longitudinal ribs of microfibrils. (P) = Plasmodesma; (ER) = endoplasmic reticulum. ×30,100. Scale bar: 0.5 μm.

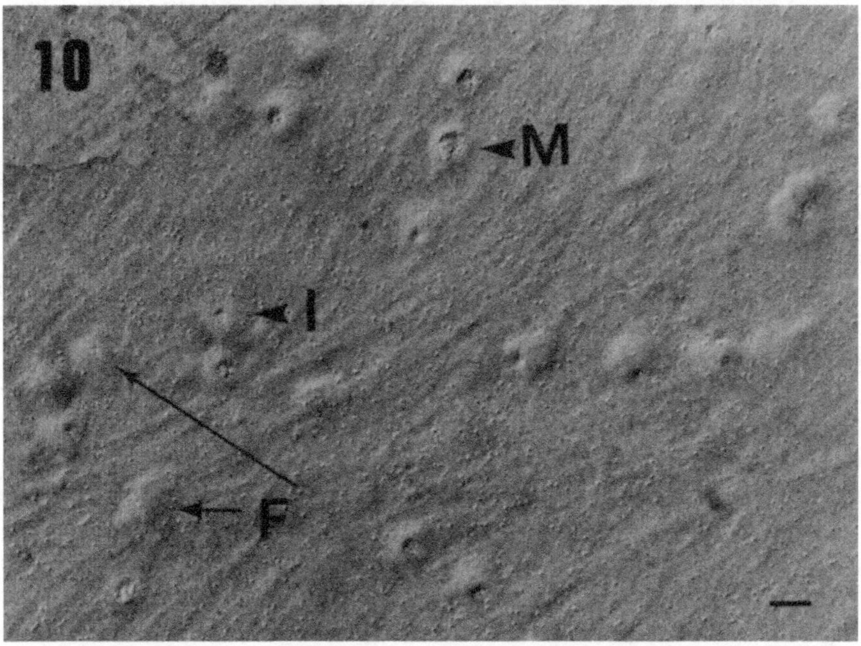

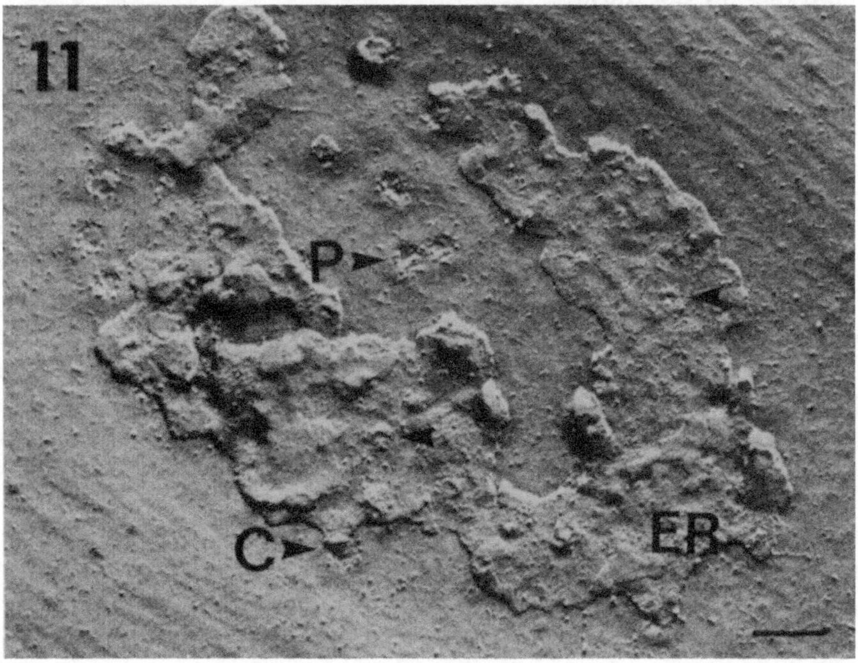

FIGURE 10. Protuberances on E fracture faces that may be plasmodesmata in various stages of formation. A mature plasmodesma (M) has a perforation surrounded by intramembranous particles in the center of the protuberance. A protuberance lacking the perforation may be an initial stage of formation (F) and a protuberance with a narrow central perforation but few particles surrounding it

treatment in association with a well-established longitudinal microfibril orientation (Mueller and Brown, 1982b). Some other root parenchyma, however, display new longitudinally inclined bands of microfibrils. These bands are small, containing only about 5–10 microfibrils each. Terminal complex-microfibril impressions from each group indicate that within bands, microfibril polarity is unidirectional. However, the bands are oriented with opposite polarities with respect to each other.

The results obtained from ethylene-treated roots and mesocotyls are consistent with those from control and colchicine-treated cells in that unidirectionally-oriented microfibrils are associated with ordered deposition. During ethylene treatment, microfibril deposition may be perturbed as a result of the restriction of microfibril deposition to a single axis of longitudinal orientation rather than alternating orientations of microfibrils. The patterns of microfibrils that are deposited in the presence of colchicine indicate that the control of microfibril deposition is complex, involving other cytoplasmic or plasma membrane structures in addition to microtubules. It seems clear that microtubules do have a direct effect on microfibril orientation, since bands of oriented microfibrils are present when microtubules are intact with or without ethylene treatment, but are no longer deposited in the presence of colchicine. Freeze–etch experiments need to be carried out to further characterize the time-course of colchicine- and ethylene-induced microfibril reorientations as well as the recovery of normal microfibril deposition on removal of these substances.

In addition to microtubules, endoplasmic reticulum membranes may be involved in the orientation of new microfibrils. Endoplasmic reticulum membranes are in close association with developing cell walls after cell division, and it is thought that they play an integral role in the development of plasmodesmata (Jones, 1976). On maturing walls, however, the formation of plasmodesmata might first be signaled by the appearance of an E fracture face protuberance with the shape and dimensions of a plasmodesma but lacking a central perforation (Fig. 10). Mature plasmodesmata viewed on EF faces have a central perforation surrounded by a ring of intramembranous particles (Figs. 10 and 11). The close associations between endoplasmic reticulum membranes and plasma membranes are also revealed by freeze–fracture (Fig. 11). The fractured endoplasmic reticulum membrane adjacent to the plasma membrane contains pits that appear to lead into the plasmodesmata (Fig. 11, C and arrowheads). Endoplasmic reticulum membranes could influence microfibril deposition by determining the sites of plasmodesmata formation and by interacting with the plasma membrane adjacent to the plasmodesma pore.

an intermediate stage (I). A transverse membrane of a corn root 2 mm from the apex prepared as in Fig. 1. ×49,200. Scale bar: 0.1 μm.

FIGURE 11. Pit field from a longitudinal membrane illustrating the EF appearance of mature plasmodesmata (P) and the continuity between the endoplasmic reticulum (ER) and the center of the plasmodesma (C). The arrowheads indicate pits in the ER membrane that may be associated with plasmodesmata in the adjacent plasma membrane and wall. ×90,200. Scale bar: 0.1 μm.

7. Conclusions

In summary, microfibril tears and impressions associated with terminal synthesizing complexes suggest that microfibril deposition occurs in one direction across the plasma membrane. Since microfibrils are anchored in the wall, terminal synthesizing complexes probably diffuse through the membrane during synthesis. The orientations of microfibrils observed on fracture faces of the plasma membrane can be partially accounted for by the presence of cytoplasmic microtubules that interact with the synthesizing complexes in the plane of the membrane. However, since polarized terminal complex–microfibril impressions or tears occur even in the absence of microtubules during colchicine treatment, the patterns of microfibrils must be a result of other orienting forces in the membrane. Microfilaments or other contractile proteins in the cytoplasm might influence the orientation and directionality of microfibrils during synthesis and deposition, since they are observed adjacent to the plasma membrane in freeze-fractured cells. We have proposed that since the patterns of microfibrils around pit fields resemble the patterns that water makes flowing around a tree in a stream, then microfibrils may be similarly oriented by a flowing plasma membrane (Mueller and Brown, 1982a; cf. Bretscher, 1974.) Membrane flow would account for the following observations: (1) unidirectional microfibril orientations even in the absence of microtubules; (2) flexuous patterns of microfibrils around pit fields; and (3) curving orientations of microfibrils adjacent to longitudinal ribs and transverse walls and in colchicine-treated cells.

The concomitant loss of orientated deposition and unidirectional orientations might be a consequence of the temporary disturbance of flow, for example, as occasionally happens during colchicine treatment. The motive force for a possible flowing plasma membrane is unknown. Alternatively, a few membrane proteins pulled or guided through the membrane by cytoplasmic elements might guide a large number of synthesizing complexes if these membrane proteins had a lateral affinity for synthesizing complexes. At present, it is not possible to distinguish among these possibilities.

References

Bretscher, M. S., 1974, Directed lipid flow in cell membranes, *Nature (London)* **260**:21–23.

Brown, R. M., Jr., 1979, Biogenesis of natural polymer systems, with special reference to cellulose assembly and deposition, in: *Structure and Biochemistry of Natural Biological Systems*, Proceedings of the Third Philip Morris Science Symposium (E. M. Walk, ed.), pp. 52–103, Philip Morris, New York.

Brown, R. M., Jr., and Montezinos, D., 1976, Cellulose microfibrils: Visualization of biosynthetic and orienting complexes in association with the plasma membrane, *Proc. Natl. Acad. Sci. U.S.A.* **73**:143–147.

Brown, R. M., Jr., and Willison, J. H. M., 1977, Golgi apparatus and plasma membrane involvement in secretion and cell surface deposition, with special emphasis on cellulose, in: *International Cell Biology, 1976–1977* (B. R. Brinkley and K. R. Porter, eds.), pp. 267–283, The Rockefeller University Press, New York.

Brown, R. M., Jr., Willison, J. H. M., and Richardson, C. L., 1976, Cellulose biosynthesis in

Acetobacter xylinum: Visualization of the site of synthesis and direct measurement of the *in vivo* process, *Proc. Natl. Acad. Sci. U.S.A.* **73**:4565–4569.

Cleland, R., 1971, Cell wall extension, *Annu. Rev. Plant Physiol.* **22**:197–222.

Delmer, D. P., 1977, The biosynthesis of cellulose and other plant cell wall polysaccharides, in: *Recent Advances in Phytochemistry* (F. A. Loewus and V. C. Runeckles, eds.), pp. 45–77, Plenum Press, New York.

Giddings, T. H., Jr., Brower, D. L., and Staehelin, L. A., 1980, Visualization of particle complexes in the plasma membrane of *Micrasterias denticula* associated with the formation of cellulose fibrils in primary and secondary cell walls, *J. Cell Biol.* **84**:327–339.

Green, P. B., 1974, Morphogenesis of the cell and organ axis—biophysical models, *Brookhaven Symp. Quant. Biol.* **25**:166–190.

Hepler, P. K., and Palevitz, B. A., 1974, Microtubules and microfilaments, *Annu. Rev. Plant Physiol.* **25**:309–362.

Jones, M. G. K., 1976, The origin and development of plasmodesmata, in: *Intercellular Communication in Plants* (B. E. S. Gunning and A. W. Robards, eds.), pp. 81–106, Springer-Verlag, New York.

Maclachlan, G. A., 1977, Cellulose metabolism and growth, in: *Plant Growth Regulation* (P. E. Pilet, ed.), pp. 13–20, Springer-Verlag, Berlin.

Montezinos, D., and Brown, R. M., Jr., 1976, Surface architecture of the plant cell: biogenesis of the cell wall with special emphasis on the role of the plasma membrane in cellulose biosynthesis, *J. Supromolecular Struct.* **5**:277 (229)–290 (242).

Mueller, S. C., and Brown, R. M., Jr., 1980, Evidence for an intramembrane component associated with a cellulose microfibril-synthesizing complex in higher plants. *J. Cell Biol.* **84**:315–326.

Mueller, S. C., and Brown, R. M., Jr., 1982a, The control of microfibril deposition in higher plants. I. Can directed membrane flow orient cellulose microfibrils? Indirect evidence from freeze-fractured plasma membranes, *Planta* (submitted).

Mueller, S. C., and Brown, R. M., Jr., 1982b, The control of microfibril deposition in higher plants. II. Freeze-fracture microfibril patterns following experimental alteration with colchicine and ethylene. *Planta* (submitted).

Mueller, S. C., Brown, R. M., Jr., and Scott, T. K., 1976, Cellulose microfibrils: Nascent stages of synthesis in a higher plant cell, *Science* **194**:949–951.

Northcote, D. H., 1972, Chemistry of the plant cell wall, *Annu. Rev. Plant Physiol.* **23**:113–132.

Preston, R. D., 1974, *The Physical Biology of Plant Cell Walls*, Chapman and Hall, London.

Rash, J. E., and Hudson, C. S., 1979, *Freeze–Fracture: Methods, Artifacts, and Interpretation*, Raven Press, New York.

Roelofsen, P. A., 1965, Ultrastructure of the wall of growing cells and its relation to the direction of growth, *Adv. Bot. Res.* **2**:69–149.

Satir, B., and Satir, P., 1979, Partitioning intramembrane particles during the freeze-fracture procedure, in: *Freeze-fracture: Methods, Artifacts, and Interpretations* (J. E. Rash and C. S. Hudson, eds.), pp. 43–50, Raven Press, New York.

Wada, M. and Staehelin, L. A., 1981, Freeze-fracture observations on the plasma membrane, the cell wall and the cuticle of growing protonema of *Adiantum capillus-venesis* L. *Planta* **151**:462–468.

Willison, J. H. M., 1976, An examination of the relationship between freeze-fractured plasmalemma and cell-wall microfibrils, *Protoplasma* **88**:187–200.

Willison, J. H. M., and Brown, R. M., Jr., 1977, An examination of the developing cotton fiber: Wall and plasmalemma, *Protoplasma* **92**:21–42.

Willison, J. H. M., and Brown, R. M., Jr., 1978, Cell wall structure and deposition in *Glaucocystis*, *J. Cell Biol.* **77**:103–119.

Willison, J. H. M., and Grout, B. W. W., 1978, Further observations on cell-wall formation around isolated protoplasts of tobacco and tomato, *Planta* **40**:57–58.

Willison, J. H. M., and Rowe, A. J., 1980, Replica, shadowing and freeze-etching techniques, in: *Practical Methods in Electron Microscopy*, Vol. 8 (A. M. Glauert, ed.), pp. 171–234, North-Holland, Amsterdam.

Zaar, K., 1979, Visualization of pores (export sites) correlated with cellulose production in the envelope of the gram-negative bacterium *Acetobacter*, *J. Cell Biol.* **80**:773–777.

6
Microfibril-Tip Growth and the Development of Pattern in Cell Walls

J. H. M. Willison

1. Introduction

No single model for the physical assembly of cellulose microfibrils is universally accepted at present, and it is likely that the subject will continue to be controversial for some time to come. Very important advances have been made since the publication of a major review of this subject by Shafizadeh and McGinnis (1971), but we seem no nearer to consensus (see, for example, Stöckmann, 1972; Atalla and Nagel, 1974; Preston, 1974; Palma *et al.*, 1976; Robinson, 1977a; Brown and Willison, 1977; Burgess, 1979; Colvin, 1980). Burgess (1979) has stated that there are two fundamental hypotheses for microfibril assembly; (1) that glucan chains are first polymerized in a soluble form, and these intermediate molecules are later assembled into microfibrils (perhaps extracellularly, separated from the plasma membrane; perhaps via an intermediate hydrated "nascent fibril"); (2) that polymerization of glucan chains occurs directly at the free ends of preexisting microfibrils (i.e., that polymerization and crystallization are simultaneous, or virtually simultaneous). This simple statement should not, of course, be allowed to hide the existence of numerous variations, extensions, and modifications of these hypotheses, many of which are as viable now as when they were proposed. Take, for example, the shear-stress solution crystallization hypothesis proposed (and rejected) by Stöckmann (1972). Stöckmann rejected the hypothesis on the basis that there is

J. H. M. Willison • Biology Department, Dalhousie University, Halifax, Nova Scotia, Canada B3H 4J1.

no stirring at the sites of microfibril assembly, citing the extracellular assembly of *Acetobacter* cellulose as a critical example. Yet we know now that plasma membranes, particularly their lipid components, stir very rapidly (see Singer, 1974) and that *Acetobacter* cellulose is assembled at the cell surface (Brown *et al.*, 1976; Brown and Willison, 1977). I am not proposing at present that a shear-stress solution crystallization hypothesis be revived (particularly for *Acetobacter*, where stirring is probably less vigorous than in eukaryotic plasma membranes), only that we have reason to be open-minded. I can see no compelling reason for believing that a single natural physical process of microfibril assembly exists; indeed, I would argue that *Acetobacter* may be an unsuitable model for the biogenesis of higher-plant cellulose and that reference to it in the development of unifying hypotheses has led to unnecessary controversy. Not only might *Acetobacter* be a special case, but also it seems feasible that the celluloses of primary walls and of secondary walls are assembled by differing processes (the products, at least, appear to have different properties).

Despite the foregoing reservations, most models for cellulose microfibril assembly acknowledge the probability that microfibrils lengthen exclusively, or predominantly, at their tips (probably from one end only, since two-ended growth would demand two sets of enzymes).Equally favored is that the site of assembly is either at the plasma membrane or just beyond it in the periplasmic space adjacent to the membrane. My concern here is not to argue a case for, or against, simultaneous polymerization and crystallization (although it does seem that this hypothesis best fits the structural data obtained to date), but to examine microfibril-tip growth in relation to the development of the architecture of the cell wall in two experimental systems, the unicellular alga *Glaucocystis* and the cotton fiber. I will assume that the microfibrils examined here grow only at their tips and that the site of assembly lies on the outer surface of the plasma membrane. Both these assumptions are reasonable given the present state of knowledge; both are compatible with the observations described here, but neither assumption can be considered proven. Two fundamental questions will be considered: (1) Can tip growth of microfibrils at the protoplast surface account for observed patterns of microfibrils in completed walls? (2) Are there sufficient microfibril ends for enzyme-mediated microfibril-tip growth to account for observed rates of cellulose deposition? Much of what is described here is drawn from three published works (Willison and Brown, 1977, 1978a,b).

2. *Glaucocystis nostochinearum*

Glaucocystis nostochinearum Itzigsohn is a unicellular alga with uncertain taxonomic affinities (for a recent discussion, see Kies, 1979). Its life cycle is similar to that of *Oocystis* (see Chapter 1) in that it usually reproduces by the formation of eight autospores. The architecture of its cell wall also bears superficial resemblance to that of *Oocystis*. Most of the microfibrils [which are cellulosic (see Robinson and Preston, 1971)] are ribbonlike (cross-sectional dimensions of 8 nm × 20 nm) and arranged in a "crossed polylamellate" fashion (Roelofsen, 1965), that is, as a series of lamellae distinguished by having microfibrils arranged about the cell alternatively as right-handed and left-handed helices. Except in the polar zones, microfibrils lie effectively parallel with each other within any one lamella, and at the equator those of adjacent lamellae make angles of close to 90° with each other. While crossed polylamellate walls have been well described for some time, the polar organization of the microfibrils has been elucidated only recently (see below). After cytokinesis, the autospores bounded by any single mother cell wall generate new cell walls virtually synchronously, but it is difficult to obtain reproductive synchrony between cells within cultures of the alga. Since the proportion of the life cycle during which the cell wall is deposited is relatively short, the difficulty of synchronizing cell cultures is serious for the experimenter who wishes to study cell-wall development. This problem was overcome by Richard Santos (who maintained the algal cultures) by spraying aqueous suspensions of aging cultures of the alga onto fresh plates of solidified culture medium, using a nebulizer. Although the reasons are unclear, autospore formation began almost synchronously within the small droplets produced by spraying. If cultures were sampled for electron microscopy within 2–3 days of spraying, the cells were found to be actively laying down the broad cellulosic microfibrils. Plentiful supplies of mother-cell walls, from which autospores had been released, were obtained by placing aged cultures in tubes of distilled water. As the cells settled, mother-cell walls formed a white horizon above the green sediment at the bottom. This white horizon was readily removed with a fine glass pipette.

2.1. *Architecture of the Microfibrillar Cell Wall*

The crossed polylamellate nature of equatorial regions of mother-cell walls was readily confirmed using differential interference contrast microscopy (Fig. 1) and electron microscopy of thin sections and surface replicas (Willison and Brown, 1978a). The polar organization was less tractable, however. *Glaucocystis* cells are ellipsoidal with flattened poles, and it is clearly impossible that there could be a unitary polar terminus for all microfibrils. In replicas, microfibrils in the polar region of the outer surface of the mother-cell wall were usually obscured by a coating of matrix materials, but when the inner surface was seen, it was clear that the polar organization was complex (Fig. 2). Possible models of the polar organization, drawn on hemispherical surfaces, were compared with several clear examples of poles from surface replicas (such as Fig. 2), and the best fit was selected (Fig. 3). In this model, microfibrils encircle three "rotation centers" at each pole, such that the microfibrils of any one wall lamella are interconnected with those of the adjacent lamella (for more details, see Willison and Brown, 1978a). The finding accounts for the greater number of lamellae, as compared with equatorial regions, found in thin sections of the poles. The model would imply, too, that wall layers are more strongly bonded to each other at the poles than in the equatorial zone, accounting for the finding that wall

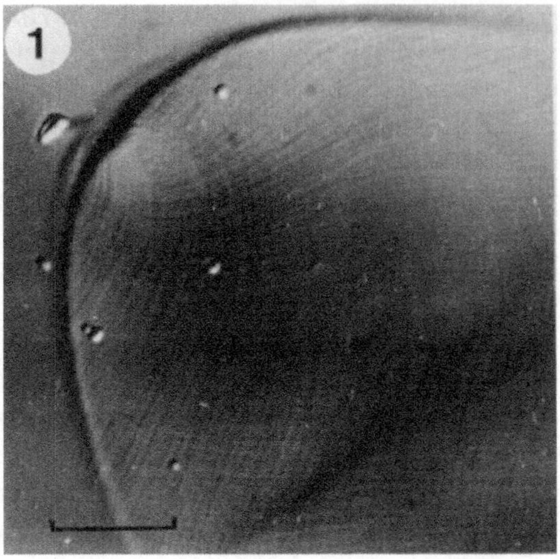

FIGURE 1. Isolated *Glaucocystis* mother-cell wall, showing one pole and the crossed helical organization of the wall. Differential interference contrast microscopy. Scale bar: 10 μm.

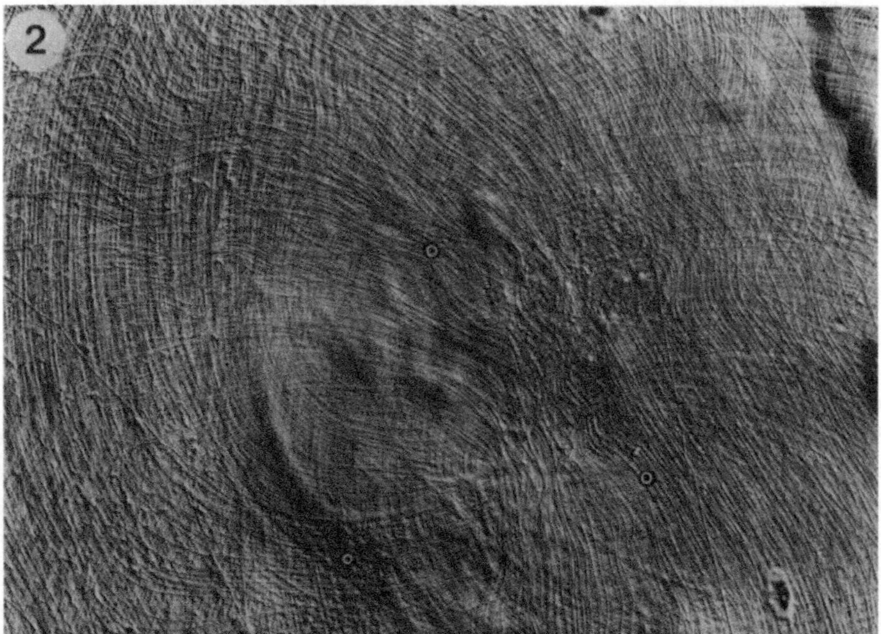

FIGURE 2. Inner aspect of a flattened pole of a mother-cell wall of *Glaucocystis*. The microfibrils encircle three "rotation centers" (the approximate centers of which are marked "o"). Platinum-carbon shadowed replica. Scale bar: 5 μm. Reprinted from Willison and Brown (1978a) with the permission of The Rockefeller University Press.

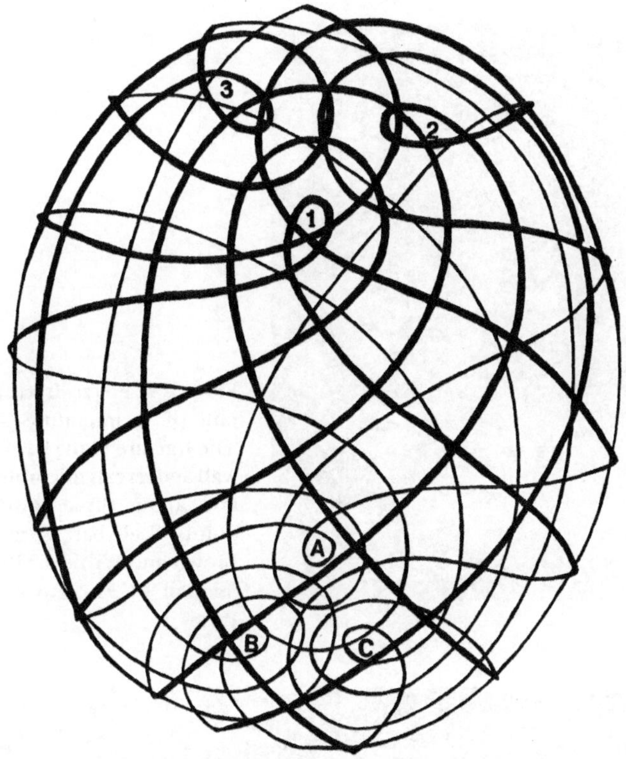

FIGURE 3. A model of the arrangement of the microfibrils in the *Glaucocystis* wall. This is an oblique view, and the ellipsoid is drawn as though transparent, with the heavier lines representing the front face. The three rotation centers at each of the poles are indicated (1, 2, and 3; A, B, and C). The three microfibrils that are indicated encircling each rotation center appear to make closed loops only because it is presumed in this model that the rotation centers of the various wall layers are superimposed (i.e, at some point in each loop, there is a step in the radial direction). Reprinted from Willison and Brown (1978a) with the permission of The Rockefeller University Press.

extension leads to rupture of microfibrils in the equatorial zone but not in the polar zones.

Surface replicas could not be used to confirm that the architecture of the cell wall at the time of deposition was equvalent to that in the mother wall, but all observations of freeze-fractured developing walls were completely compatible with the basic pattern shown in Fig. 2, including observations of the critically important poles (Figs. 4 and 5). We must presume, then, that cell extension leads to no greater reorganization of this cell wall than the circumferential spreading apart of microfibrils (hence an increase in girth), and an axial lengthening, by thinning, and eventual tearing of individual microfibrils (hence an increase in axial length).

2.2. *Plasma-Membrane Organization*

From a model-building point of view, it would be convenient if the boundary of the *Glaucocystis* protoplast were as simple as that of, for example, a typical mesophyll cell. However, in addition to the conventional protoplast, the

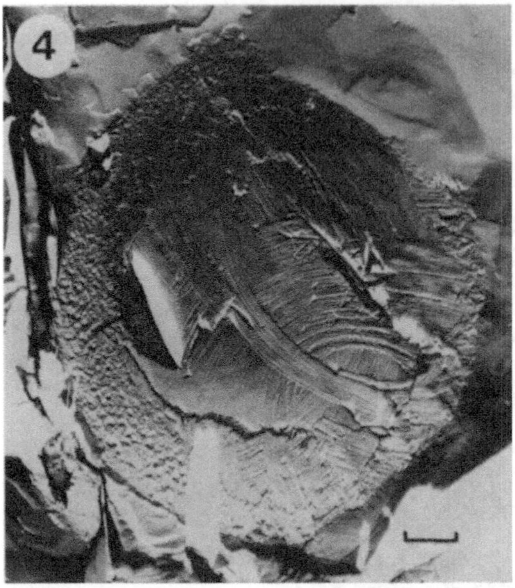

FIGURE 4. Freeze-fracture replica of the pole of an immature *Glaucocystis* cell. The fracture plane has passed within the wall and reveals microfibrils making three arcs about three equilaterally spaced points. Scale bar: 2 μm. Reprinted from Brown and Willison (1977) with the permission of The Rockefeller University Press.

FIGURE 5. Freeze-fracture replica of the pole of an immature *Glaucocystis* cell that was laying down the microfibrillar wall at the time of freezing. Impressions of microfibrils can be seen to encircle one "rotation center" (RC) at the pole. Ends of microfibrils within the "zone of synthesis" are visible (arrows). Scale bar: 1 μm.

plasma membrane covers a pair of flagella located equatorially in the periplasmic space. Also, closely subjacent to the plasma membrane are many flattened shield-shaped "lacunae" (Schnepf *et al.*, 1966) that cover the entire inner surface of the cell with the exception of a small area around the flagellar roots. The lacunae considerably complicate investigation of the plasma membrane by freeze-fracturing (the preferred technique for studying membrane organization) because their closely aligned margins lead to the creation of two additional fracture planes at the cell surface, reducing the probability that views of the plasma membrane itself will be obtained.

Despite the complications outlined above, several views of the plasma membrane of cells actively laying down the cellulosic wall were obtained. Microfibril ends could be seen as impressions in the fractured plasma membrane (Fig. 6). Although not comparable in detail with those of *Oocystis* (see Chapter 1), the plasma membrane was specialized in the region of the microfibril tips, suggesting the possibility that complexes of membrane proteins involved in microfibril biosynthesis are associated with microfibril tips (although a functional role for these tip-associated structures is a matter of hypothesis). In contrast with the situation in *Oocytsis*, however, microfibrils in *Glaucocystis* appear to be generated in one direction only and are indefinitely long (that is, both ends of any one microfibril were never identified). For reasons that will become clear later, we favor the hypothesis that microfibril tips (and therefore microfibril synthesis) are found within a restricted, though mobile, zone of the cell surface. Evidence that could be used to support this contention directly was

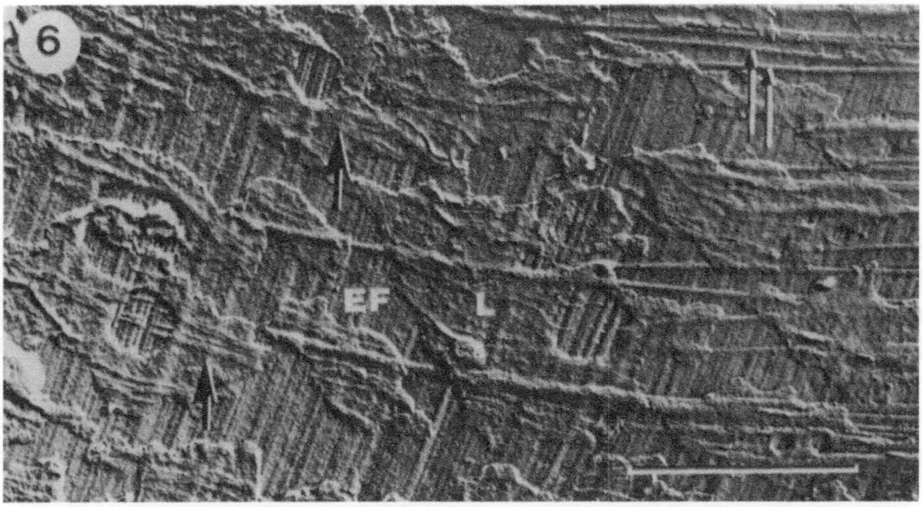

FIGURE 6. E fracture face of plasma membrane (EF), overlaid with parts of several lacunae (L), revealing (to right) impressions of microfibrils (small arrows), each of which ends within the compass of the micrograph. Fibrillar structures (large arrows) seem to lie between the lacunae and the plasma membrane, particularly ahead of the advancing microfibril tips. The orientation of the previously completed wall layer is also visible. Scale bar: 1 μm. Reprinted from Brown and Willison (1977) with the permission of The Rockefeller Univeristy Press.

found (i.e., regions of membrane where the tips appeared to be localized), but the fracture faces are so full of detail that ends could have been obscured, and the "evidence" is therefore tentative. Within zones where microfibril ends could be clearly identified, there were about six ends per square micrometer of cell surface (insufficient data were obtained to provide an estimate of error) (Willison and Brown, 1978b).

2.3. *Modeling the Pattern of Microfibril Deposition*

I have adopted the hypothesis that cellulosic microfibrils are generated by the passage of their growing tips over the cell surface. Since the *Glaucocystis* wall is highly ordered, there must be an orderly pattern of movement of microfibril tips that mirrors the architecture of the cell wall. Are there, then, enough microfibril tips? How fast must they move to generate the wall that is formed? To answer these questions, we need estimates of the major axial dimensions of the cell at the time of microfibril deposition and an estimate of the number of microfibrils in any one wall layer. By extrapolation from growth data (for details, see Willison and Brown, 1978b), the major axes of the unexpanded cell were found to be 27.5 μm (interpolar length) × 15 μm (equatorial diameter). By direct counting from surface replicas, it was estimated that there were approximately 2500 microfibrils per wall layer in the mother-cell wall. Given an equatorial circumference of 46 μm at the time of microfibril deposition, this translates into approximately 54 microfibrils per micrometer of equatorial circumference (a value that is reasonably consistent with the numbers of microfibrils found in thin sections of young cells, although accurate estimates cannot be made this way because the plane of sectioning is not known). If 2500 microfibril tips are distributed with a density of six per square micrometer (see above), then about 400 μm^2 of the cell surface is occupied by microfibril tips. An ellipsoid having the dimensions given above has a surface area of about 1300 μm^2, so there is clearly more than enough surface available for the observed density of microfibril tips, and the data support the supposition that microfibril ends are locally distributed in a zone of the cell surface (which will be called the "zone of synthesis" henceforth). From one large region of freeze-fractured plasma membrane, it appeared that the width of the zone of synthesis was about 10 μm; therefore, if there is one zone, its length is about 40 μm. This is similar to the 45° helical distance between poles, and if the zone of synthesis is a narrow band lying at right angles to the long axes of the microfibrils, then the zone stretches from pole to pole, making one spiral turn about the cell as it does so. There is no direct evidence to support such a contention, but it is the simplest single zone that is compatible both with the findings and with the assumptions made so far. Could such a zone of synthesis move over the surface of an ellipsoid to generate a wall having the observed architecture? The model shown in Fig. 7 suggests that it could, if the zone splits into three parts as it traverses the poles. Impressions of microfibrils circumventing a rotation center were revealed on the

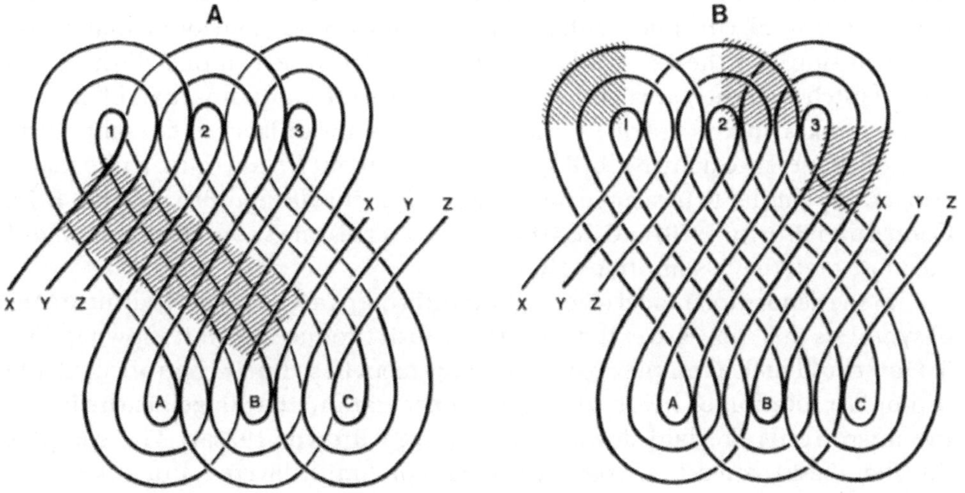

FIGURE 7. A model showing the proposed pattern of deposition of microfibrils in the *Glaucocystis* cell wall by a narrow "zone of synthesis" (hatched areas) that moves across the surface of the cell. One complete wall layer is shown in each diagram (A and B), the diagrams being drawn as though the ellipsoid had been cut vertically from pole to pole and the overlapping polar rotation centers separated. In (A), the zone of synthesis runs, roughly, from pole to pole; in (B), it has moved forward and split into three subzones, each encircling one of the rotation centers. After having encircled the upper set of rotation centers, the three subzones would recombine to move diagonally toward the lower rotation centers (A, B, and C). Reprinted from Willison and Brown (1978b) with permission of Springer-Verlag.

fracture face of plasma membrane pressed against a developing pole (Willison and Brown, 1978b), indirect evidence that microfibril tips pass around rotation centers in generating the polar architecture.

Before the deposition model (Fig. 7) is considered in more detail, the question of synthetic rate should be considered. Our model assumes that each of the microfibrils of any one wall layer is separately generated, but that all microfibrils are interconnected between wall layers (an assumption that is reasonable in view of the apparent lack of interweaving between wall layers). Thus, in estimating synthetic rate, we need consider only one microfibril, which passes without a break from one wall layer to the next, via the poles, contributing to each of the 12 wall layers as it goes. Such a microfibril is about 750 μm long when initially generated. Our first estimate of synthetic rate (Willison and Brown, 1978b) assumed that the period of cellulose synthesis was between 1 and 2 days, giving a microfibril growth rate of 0.25–0.5 μm/min. A wall deposition period of 1–2 days was assumed simply because the life cycle in our cultures was about 4 days and it was clear that cell wall was not deposited throughout the life cycle. In a later investigation (Colpitts and Brown, unpublished), time lapse microcinematography demonstrated that *Glaucocystis* protoplasts rotate unidirectionally within the wall with a rotation time of 21 ± 12 min (95% confidence level). Rotation starts shortly after cytokinesis is completed but ceases before much cell growth has occurred, a period that includes that during which we suspect, from thin-sections, that wall deposition occurs. Colpitts constructed a wooden ellipsoid, equipped with pins for rotation centers, and found that it took 16 revolutions for the 12 wall layers of the model to be completed (the mean helical pitch is slightly greater than the length of the longitudinal axis of the cell; hence, it takes an average of 1⅓ rotations for each wall layer to be completed). This translates into a microfibril growth rate of about 1.5 μm/min, a rate that is remarkably similar to measured rates in *Acetobacter xylinum* (Brown *et al.*, 1976; Cooper and Brown, 1979) and that therefore increases one's confidence in some of the more tentative assumptions above.

In the deposition model (Fig. 7), microfibril tips are contained within a zone of synthesis that moves over the cell surface so as produce the wall shown in Fig. 3. How could such a complex pattern of movement be so finely controlled? Could protoplast rotation be involved in pattern generation? The flagella arise before the microfibrils are laid down (Willison and Brown, 1978a). The complex flagellar bases consist, in part, of a pair of "multilayered structures" per flagellum (Robinson, 1977b) from which the cytoplasmic microtubules splay out (Fig. 8). These microtubules lie to the cytoplasmic side of the lacunae and are arranged roughly parallel with one (at least) of the orientations of the microfibrils (Robinson, 1977b; Willison and Brown, 1978a). We do not know whether the orientation of these cytoplasmic microtubules changes as wall layers are synthesized. Although there is a distance of perhaps 70–150 nm between the cytoplasmic microtubules and the outer surface of the plasma membrane, it is not impossible that some functional interrelationship exists between the two. Links between the plasma membrane and the lacunae are clearly visible in thin sections (Willison and Brown, 1978a), and freeze-fracturing reveals fibrous remnants in this same location (see Fig. 6). These fibrous remnants lie roughly parallel with the microfibrils at the leading edge of a developing wall lamella. Robinson (1977b) reports that cytoplasmic microtubules tend to lie parallel with the long axes of the lacunae (although it should be noted that lacunae have only a tendency to orderliness). Robinson (1977b) also found that applications of the drugs colchicine, colcemid, and podophyllotoxin (all of which usually interfere with microtubule assembly) to developing cells had no effect on the architecture of the wall generated, unlike that of *Oocystis*, but in *Glaucocystis* the drugs also had no effect on the cytoplasmic microtubules and the flagellar bases. The case for cytoplasmic control of microfibril orientation via cytoplasmic microtubules cannot therefore be ruled out, although it is clearly, at best, a long-distance control via some complex intermediaries.

The direct structural relationship between the cytoplasmic microtubules and the flagellar bases has some interesting implications. The timing of protoplast rotation and the data on microfibril growth rate imply that protoplast rotation occurs during microfibril deposition. Colpitts's model demonstrated that rotation could be instrumental in the generation of the architecture of the wall. Clearly, the helical architecture could arise simply as a product of two vectors—protoplast rotation and up-and-down movement of the zone of

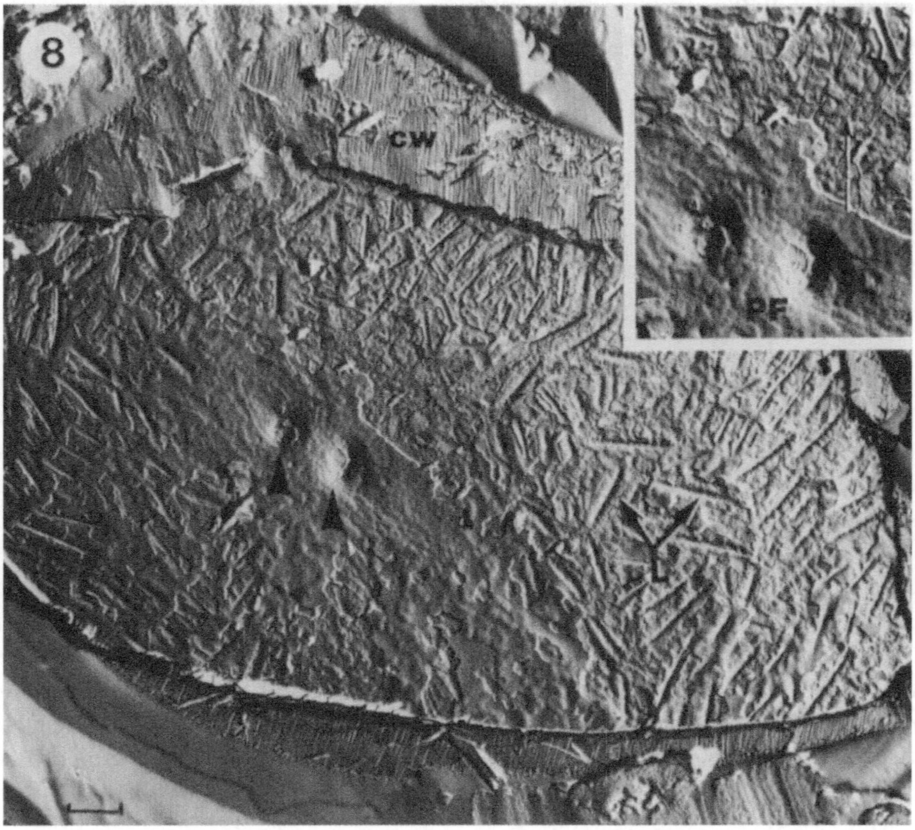

FIGURE 8. Equatorial zone of a *Glaucocystis* cell showing the cleaved bases of the flagella (arrowheads, center left). Striations, which are presumed to represent microtubules, radiate out from the bases of the flagella (see arrow, inset). (L) Lacunae; (cw) cell wall. Scale bar: 2 μm.

synthesis parallel to the longitudinal axis of the cell—but to do so, the rates and orientation must be perfectly attuned if the architecture of the wall is to remain the same from lamella to lamella and from individual to individual. A direct connection between the flagella (which might produce and control the protoplast rotation) and the cytoplasmic microtubules (which might control the orientation, though not the rate, of microfibril growth), as reported by Robinson (1977b), is therefore a critical finding. Such a model is not without its problems, however. It remains unclear how the zone of synthesis misses the flagella in its travels over the cell surface and yet generates a wall that lacks conspicuous gaps. It is not clear how the zone of synthesis can be so precisely divided and guided around the rotation centers. Even our reason for originally suspecting that the protoplast rotates raises a problem. Mature *Glaucocystis* cells contain an equatorial torus of rather amorphous, perhaps gelatinous, material (see Willison and Brown, 1978a). This material is presumed to be of dictyosome origin. The dictyosomes are concentrated near the flagella. A uniform torus is most readily formed by protoplast rotation (and for this reason, we looked for rotation), but since the torus lies inside the fully formed wall, it must be laid down after the microfibrillar wall has been completed. Is the observed protoplast rotation, then, concerned with the generation of the torus rather than the microfibrillar wall, or is the function of torus formation to interfere with wall synthesis and set the stage for the next developmental stage of the life cycle?

3. Cotton Fibers

Cotton fibers are single cells that develop from the epidermis of the ovule in the developing fruit of the cotton plant (*Gossypium hirsutum* L.). Because of their commercial importance, the structure of variously treated cotton fibers has been extensively studied in the past (see, for example, Rollins and Tripp, 1961), as has cotton fiber development (see, for example, Anderson and Kerr, 1938). The widely recognized "multinet growth hypothesis" for wall extension was based partly on observations of cotton fibers (see Roelofsen, 1958). More recently, there have been many studies of developmental changes in cotton fibers, including studies of cellular ultrastructure (Itoh, 1974; Westafer and Brown, 1976; Ramsey and Berlin, 1976; Willison and Brown, 1977; Ryser, 1979), cell-wall biochemistry and biophysics (Nowak-Ossorio *et al.*, 1976; Meinert and Delmer, 1977; Huwyler *et al.*, 1979; Montezinos and Delmer, 1980; Pillonel *et al.*, 1980), and growth characteristics (Beasley and Ting, 1973; Ryser, 1977; De Langhe *et al.*, 1978, Beasley, 1979).

The initially roughly isodiametric epidermal cells elongate to 20 mm, or more, in length over a period of about 30 days (see Schubert *et al.*, 1973). Elongation occurs throughout the length of the fiber, not just at the tip, although growth may be more rapid at the tip (Ryser, 1977). Most of this elongation occurs while the fiber is invested with a primary wall. Primary and secondary walls may be distinguished morphologically by their helical angles (Fig. 9), the primary wall having a very small helical angle (equivalent to "transverse" microfibrils) and the secondary wall having a helical angle of about 55° to the transverse (see Willison and Brown, 1977). Rollins and Tripp (1954) report the presence of a somewhat ephemeral "winding layer," separating the primary and secondary walls. Secondary wall deposition continues for about 30 days, forming a wall some 5–10 μm thick. The primary-wall composition varies according to developmental stage (Meinert and Delmer, 1977), but always has a relatively low cellulose content. By contrast, the secondary wall contains 95%, or more cellulose (Marx-Figini and Schulz, 1966; Beasley, 1979; Huwyler *et al.*, 1979), and this cellulose has a higher molecular weight than that of the primary wall (Marx-Figini and Schulz, 1966). Most studies have ignored possible differences between fiber types (notably, "lint" and "fuzz") found on ovules, and this may represent a source of confusion (Beasley, 1979).

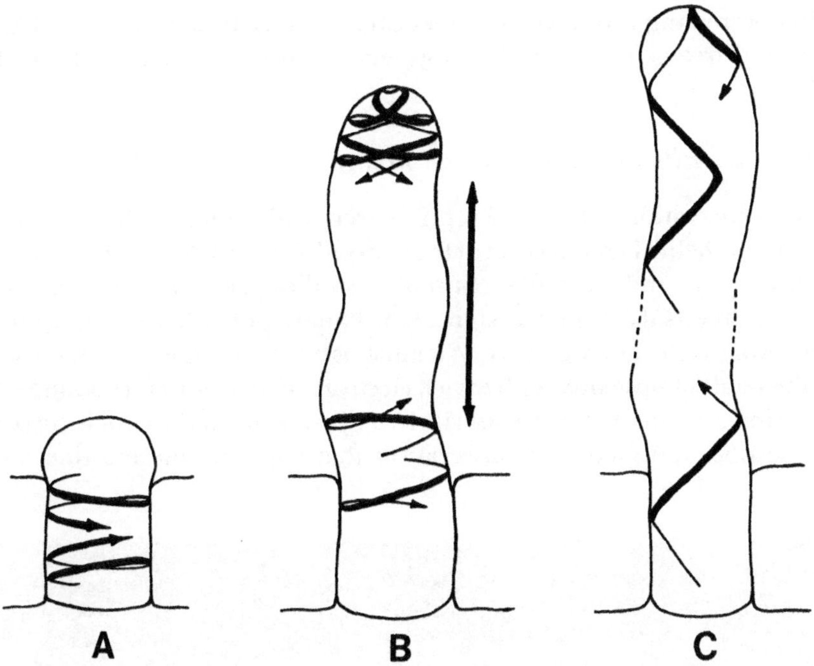

FIGURE 9. Models for the architecture of the inner (depositing) surface of the cell wall of the developing cotton fiber. (A) Before substantial elongation, microfibrils of the side walls are deposited as crossed slow helices. (B) During the period of elongation, primary wall microfibrils continue to be laid down as crossed helices. The pattern of deposition at the fiber tip has not been observed. In this model, left-handed and right-handed helices are continuous at the tip. (C) The secondary wall microfibrils are laid down as a fast helix. Periodic reversals of the helical sign occur (see top of fiber). The pattern of deposition at the tip has not been observed. In this model, the ascending helix is joined to the descending helix without a reversal of helical sign.

3.1. Methods

We applied two novel electron-microscopic approaches to studying the cotton fiber (Willison and Brown, 1977). First, to view the architecture of the inner surface (i.e., that next to the protoplast) of the wall while minimizing changes induced by the preparative technique, we froze unbroken masses of ovules, as removed from the carpels, in either liquid nitrogen or Liquid Freon 22 at its freezing point. The fibers were then fragmented in the frozen state, thawed and rinsed, and then air dried before preparation of surface replicas. The choice of quenchant made no difference in the appearance of the walls. Material selected carefully from the outer surfaces of the most rapidly frozen ovular masses was mounted and freeze-fractured without having been thawed. Effectively, therefore, the views of the freeze-fractured fibers shown here are of fibers that were completely undisturbed until they were very rapidly frozen (<0.5 sec from 20 to −100° C). After freezing, no changes could have taken place other than those introduced by the fracturing and replicating steps of the procedure, and the rapidity of freezing minimizes changes caused by ice formation. Full details of replica and freeze-fracture technology may be found in Willison and Rowe (1980).

3.2. *Wall Architecture and Plasma-Membrane Organization*

Our observations of surface replicas were fully compatible with previous reports of the helical architecture of the cotton-fiber wall. We did not observe the "winding layer" and, as in all other replica studies of which I am aware, we did not see the "reversals" of helical sign (right-handed or left-handed) that are quite evident from light micrographs. We must assume that these omissions result from the small sample sizes with which electron microscopists are bound to deal. We did find evidence that microfibrils lying just beneath the cuticle lie parallel with the fiber long axis, an observation that supports the multinet growth

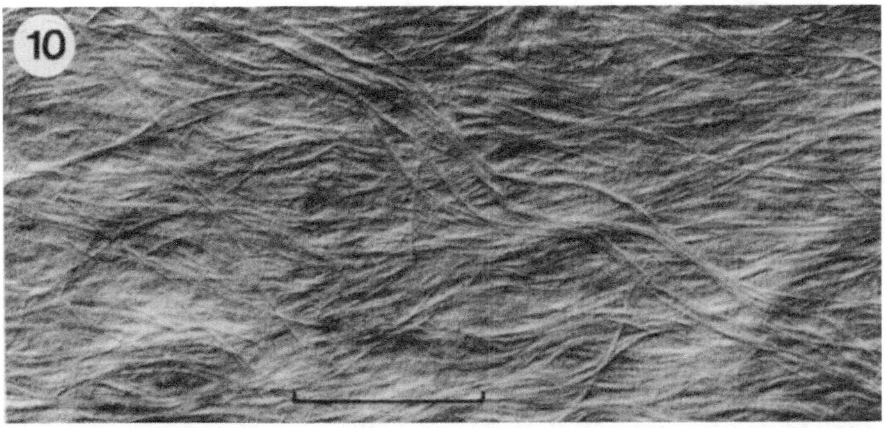

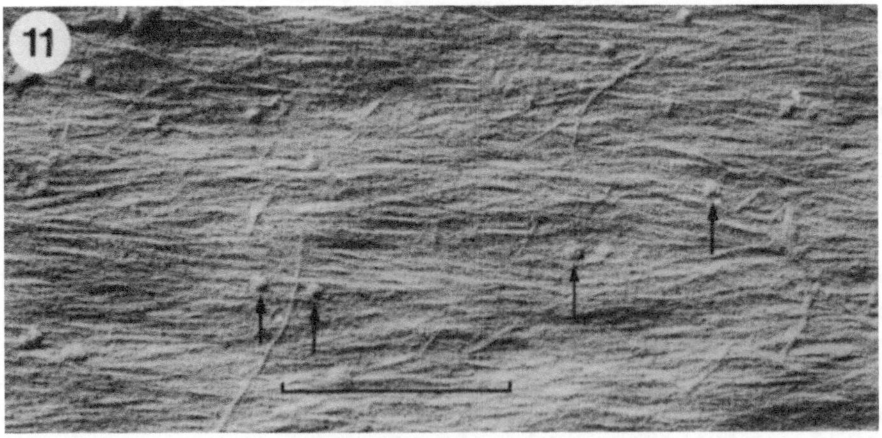

FIGURES 10 and 11. Replicas of the inner surfaces of the cell walls of cotton fibers, 15 days postanthesis, primary wall, (Fig. 10), and 22 days postanthesis, secondary wall (Fig. 11). Some of the 20–30 nm diameter globules are indicated by arrows (Fig. 11). Scale bars: 500 nm.

hypothesis for cotton fibers (for a critical review, see Roland and Vian, 1979). No microfibrils were straight, as are those of *Glaucocystis* (e.g., Fig. 6), but instead they were wavy, and the waviness was characteristically different between primary and secondary walls (Figs. 10, and 11). Inner surfaces of secondary walls were dotted with particles of a fairly uniform size, 20–30 nm (Fig. 12). These particles appeared to be loosely clustered, but enough uninterpretable debris was also present to lead to uncertainty about both their identity and their distribution. It was not possible to assign these particles to the ends of individual microfibrils.

Freeze-fracturing provided more novelty. Plasma-membrane fracture faces bore imprints of the walls pressed tightly against those membranes (Fig. 12). Most particularly, there were impressions of distinctive globules on E fracture faces associated with secondary walls (Fig. 13), of a size and appearance very similar to those described by Mueller *et al.* (1976) (see also Mueller and Brown, 1980; Willison *et al.*, 1980) associated with the ends of microfibrils. Again, in cotton, these particles were not visibly associated with microfibril ends. Counting only those that could be clearly identified, there were 5-10/μm^2 plasma membrane surface. It was interesting to note, too, that the fracture faces of plasma membranes associated with primary walls and secondary walls were different, in terms of their complements of membrane-associated particles (i.e., those particles that are typical of freeze-fractured unit membranes and that are

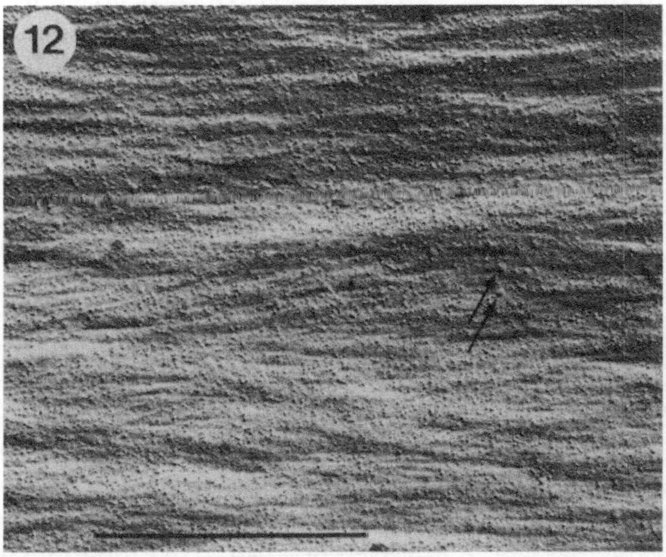

FIGURES 12 and 13. E fracture face of the plasma membrane of a 22-day-postanthesis cotton fiber, with secondary wall. At lower magnification (Fig. 12), the architecture of the inner surface of the cell wall is visible as impressions in the plasma membrane (cf. Fig. 11). A region of Fig. 12, shown at higher magnification in Fig. 13, reveals impressions of globules (arrows), of a size similar to those in Fig. 11, which are distributed with a frequency of 5–10/μm^2. Scale bars: 1 μm (Fig. 12); 100 nm (Fig. 13).

presumed to represent integral proteins), suggesting that the change from primary wall synthesis to secondary wall synthesis involves fundamental changes in the molecular composition of the plasma membrane.

3.3. *Are the Globules Involved in Cellulose Synthesis?*

Mueller and Brown (1980) have built a case for considering globules, similar to those described here (though clearly shown to be associated with microfibril termini and with the plasma membrane), as membrane-bound sites of microfibril assembly. Do the results from cotton fibers support this contention? Let us assume some typical cotton-fiber parameters and calculate how fast the cotton-fiber globules might move if they were agents of cellulose synthesis. Fiber dimensions: 20 mm long; 20 μm diameter; secondary wall thickness, 7 μm. Hence: wall volume, 8.82×10^5 nm^3; surface area of plasma membrane, 1.26×10^{12} nm^2. If we assume a microfibril diameter of 8 nm (an acceptable overestimate if the porosity of the wall is ignored), then the total length of microfibril in the wall of one fiber is 1.75×10^{14} nm. If we assume that the globules are distributed evenly, 5–10/μm^2 (as measured), then there are 6.3×10^6 to 1.26×10^7 per fiber. Hence, each globule would make 2.8×10^7 to 5.6×10^7 nm of microfibril. If the secondary wall is laid down over 30 days for 12 hr per day (light is required), then the rate of movement is 1.3–2.6 μm/min, a rate so close to the microfibril growth rate measured in *Acetobacter* (Brown *et al.*, 1976) that one is encouraged to consider these globules seriously. There are clearly several possible sources of error in the preceding calculation, and the absolute value obtained should be taken only as a rough approximation of the microfibril growth rate. Since no direct evidence of connection between globules and microfibrils was found, the calculation may be of more value in predicting the number of microfibril ends available for growth than as evidence that the globules might be involved with microfibril synthesis or assembly or both.

3.4. *Wall Architecture and Microfibril-Tip Growth*

Despite the detail shown in Fig. 9, there are several uncertainties in the finer details of the architecture of the wall of the cotton fiber. First, we know little about the detailed architecture of the tip or base. Only the outer surface of the tip has been seen in detail (and then after harsh pretreatment), where the microfibrils are "oriented at random" (Houwink and Roelofsen, 1954). Polarizing microscopy shows that this randomness is a surface feature and that inner layers, other than those in the center of the apex, are "transversely oriented. . .like those found on the inner side of the wall in the tubular part" (Houwink and Roelofsen, 1954). Polarizing microscopy shows only the predominant orientation. At the center of the apical dome, the wall is isotropic under the polarizing microscope, but in view of the far from random texture of the poles of *Glaucocystis*, we should beware of equating this isotropy with randomness, particularly on the inner surface of the wall. The side walls of the cotton fiber are better understood, but

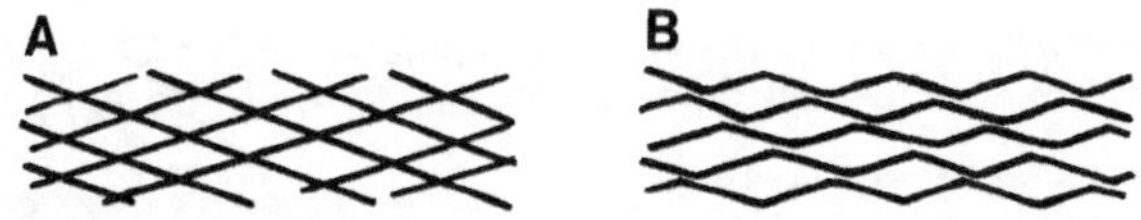

FIGURE 14. Representations of segments of tubes, cut along one side and opened out, consisting of (A) crossed slow helices and (B) wavy hoops in which the waves are phased so as to create "summative helices" having pitches similar to those in (A).

here too there are uncertainties. We have no idea of the length of microfibrils, and little idea of whether the "helices" (as I describe them) of the primary wall are a summative property or, as I suspect, a property of individual microfibrils (see Fig. 14 for clarification). More problematical is the organization in the radial direction. We do not know the extent to which the wall is layered. More particularly, we know little about the twisting and braiding of groups of microfibrils, of which there are sometimes hints. At a finer level, the extent to which compression due to drying or freezing might have flattened helical twisting of individual microfibrils to produce the characteristic flat waves (see Fig. 10) is unclear.

Excepting the reservations outlined above, the architecture of the side wall is compatible with the concept that microfibrils are generated at the protoplast surface by tip growth, the pattern of movement of the tips reflecting precisely that of the microfibrils in the inner surface of the wall. Those microfibril tips, following a helical path, which strayed into the apical dome of the fiber, would cross the dome and return as a reverse helix (see Fig. 9). Because of its curvature, as the tips approached closer to the center of the apical dome, their paths would approach closer to the radial direction, leaving the center of the dome relatively isotropic. It would be different for twists or braids to occur as a direct result of the pattern of tip movement unless the rate of movement was variable and precisely controlled with respect to the relative orientation of the constituent microfibrils of the twist or braid. Such fine control seems unlikely. The appearance of braiding, like that of weaving (Fig. 15), can be illusory, however, and I have never seen a convincing example of it. Certainly, there are bundles that "branch and rebranch with one another" (Anderson and Kerr, 1938), but these are quite compatible with a process in which microfibril tips aggregate, disaggregate, and reaggregate in new groups as they grow. Twisting of pairs, or bundles, of microfibrils (which are to be found) need not occur at the time of microfibril generation. Shearing at the wall-protoplast or wall layer-wall layer interface, as might occur during growth or specimen preparation, can clearly lead to twists that have no relevance to problems of microfibril synthesis.

3.5. Cytoplasmic Control of Microfibril Orientation

There is much evidence to support the idea that cellulose is synthesized in the plasma-membrane region (see Northcote, 1974). Pillonel *et al.* (1980) found that when labeled sucrose, but not uridine diphosphoglucose (UDPG) or guanosine diphosphoglucose (GDPG), was supplied from outside the cotton

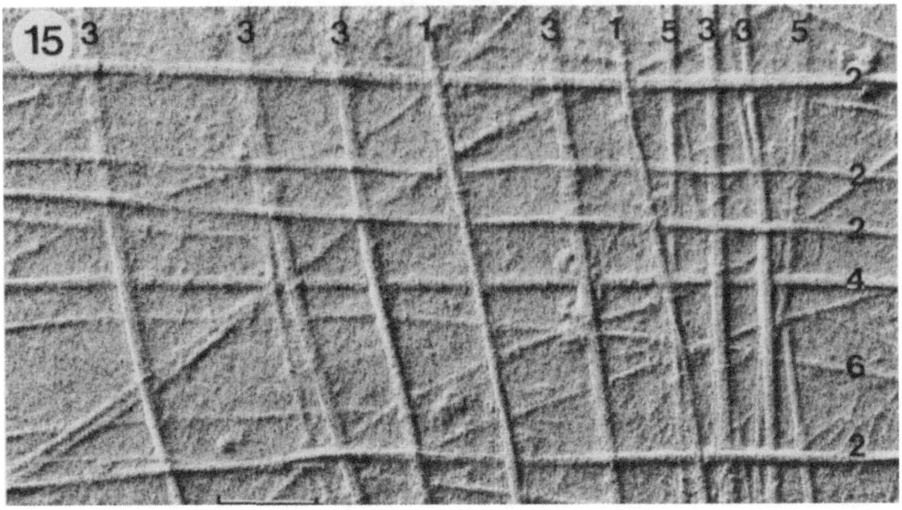

FIGURE 15. Replica of a much-expanded region of the mother cell wall of *Oocystis*. At first sight, the horizontal microfibrils labeled "2" are woven among the vertical microfibrils. Closer examination reveals that, rather than there being two layers of microfibrils (vertical and horizontal), there are six layers (not including the diagonals), as indicated by the numbering. Scale bar: 200 nm.

fiber, it would act as a precursor for secondary wall cellulose synthesis. This would seem to indicate that no nucleotide sugar glycosyl transferase system is available at the outer surface of the membrane and that the precursor must be fed from within the cell at a stage in the cellulose anabolic sequence that precedes this particular glycosyl transfer. This finding, together with the fine-structural findings described here and by Ryser (1979), is most readily interpreted in terms of a site of cellulose synthesis that is on the outer surface of the plasma membrane of the cotton fiber, yet is also integrated into the membrane.

For many years, microtubules have been considered the principal agents responsible for controlling microfibril orientation (see, for example, Heath, 1974; Palevitz, 1980; Galatis, 1980). Reasons for believing this include: parallelism between microtubules and microfibrils, colchicine-induced loss of cellular control of microfibril orientation, visibility of cross-bridges between microtubules and plasma membrane, and the influence of morphologically active hormones on patterns of cytoplasmic microtubules. In cotton fibers, the orientation of cytoplasmic microtubules lying just beneath the plasma membrane parallels that of microfibrils during both primary wall formation and secondary wall formation, despite the change in predominant microfibril orientation (Itoh, 1974; Westafer and Brown, 1976; Ryser, 1979), and links between microtubules and the plasma membrane have been recorded (Willison and Brown, 1977). The control of microfibril orientation can be understood in terms of the cytoplasmic microtubule acting as a track on which the membrane-bound microfibril-synthesizing complex travels (see Heath, 1974), although it is possible that microtubules act only to change microfibril orientations, not to maintain them (see Schnepf and Deichgräber, 1979).

In the model described above, we must presume that the membrane-bound microfibril-synthesizing complex is driven over the surface of the cell by the act of synthesis itself, the frictional resistance of the lipidic environment of the membrane against the terminal complex being negligible by comparison with the bonding of the microfibril to the wall. Nevertheless, there might be some resistance to movement of the terminal complex, and perhaps this gives rise to the waviness of the individual microfibrils in cotton fibers. Wavy microtubules or postsynthesis but precrystallization changes in microfibrils (for discussion of the latter, see Stöckmann, 1972) are other possible sources of waviness.

4. Conclusions

I conclude that: (1) microfibril-tip growth can account for the observed patterns of microfibrils in completed walls, but more critical observations, of the sort obtained for *Glaucocystis*, are required before microfibril deposition patterns can be used as evidence for tip growth in organisms other than *Glaucocystis* and *Oocystis*; and (2) if microfibril-tip growth is assumed, the microfibril growth rates obtained from data on the number of available ends and the rates of wall development are remarkably consistent, and are consistent with measured rates in *Acetobacter*.

References

Anderson, D. B., and Kerr, T., 1938, Growth and structure of cotton fibre, *Ind. Eng. Chem.* **30**:48.

Atalla, R. H., and Nagel, S. C., 1974, Cellulose: Its regeneration in the native lattice, *Science* **185**:522.

Beasley, C. A., 1979, Cellulose content in fibers of cottons which differ in their lint lengths and extent of fuzz, *Physiol. Plant* **45**:77.

Beasley, C. A., and Ting, I. P., 1973, The effects of plant growth substances on *in vitro* fiber development from fertilized cotton ovules, *Am. J. Bot.* **60**:130.

Brown, R. M., Jr., and Willison, J. H. M., 1977, Golgi apparatus and plasma membrane involvement in cell surface deposition, with special emphasis on cellulose biogenesis, in: *International Cell Biology 1976–1977* (B. R. Brinkley and K. R. Porter, eds.), pp. 267–283, The Rockefeller University Press, New York.

Brown, R. M., Jr., Willison, J. H. M., and Richardson, C. L., 1976, Cellulose biosynthesis in *Acetobacter xylinum*. 1. Visualization of the site of synthesis and direct measurement of the *in vivo* process, *Proc. Natl. Acad. Sci. U.S.A.* **73**:4565.

Burgess, J., 1979, Cellulose microfibrils—an end in sight?, *Nature (London)* **278**:212.

Colvin, J. R., 1980, The mechanism of formation of cellulose-like microfibrils in a cell-free system from *Acetobacter xylinum*, *Planta* **149**:97.

Cooper, K. M., and Brown, R. M., Jr., 1979, Cellulose assembly in *Acetobacter xylinum*: site of formation and variation of the rate of synthesis, *J. Cell Biol.* **83**:69a.

De Langhe, E., Kosmidou-Dimitropoulou, S., and Waterkeyn, L., 1978, Effect of hormones on nucleolar growth and vacuolaton in elongating cotton fibers, *Planta* **140**:269.

Galatis, B., 1980, Microtubules and guard-cell morphogenesis in *Zea mays* L., *J. Cell Sci.* **45**:211.

Heath, I. B., 1974, A unified hypothesis for the role of membrane bound enzyme complexes and microtubules in plant cell wall synthesis, *J. Theor. Biol.* **48**:445.

Houwink, A. C., and Roelofsen, P. A., 1954, Fibrillar architecture of growing cell walls, *Acta Bot. Neerl.* **3:**385.
Huwyler, H. R., Franz, G., and Meier, H., 1979, Changes in the composition of cotton fibre cell walls during development, *Planta* **146:**635.
Itoh, T., 1974, Fine structure and formation of cell wall of developing cotton fiber, *Wood Res.* **56:**49.
Kies, L., 1979, Zur systematischen Einordnung von *Cyanophora paradoxa*, *Gloeochaete wittrockiana* und *Glaucocystis nostochinearum*, *Ber. Dtsch. Bot. Ges.* **92:**445.
Marx-Figini, M., and Schulz, G. V., 1966, Uber die Kinetik und den Mechanismus der Biosynthese der Cellulose in den hoheren Pflanzen (nach Versuchen an den Samenhaaren der Baumwolle), *Biochim. Biophys. Acta* **112:**81.
Mienert, M. C., and Delmer, D. P., 1977, Changes in biochemical composition of the cell wall of the cotton fiber during development, *Plant Physiol.* **59:**1088.
Montezinos, D., and Delmer, D. P., 1980, Characterization of inhibitors of cellulose synthesis in cotton fibers, *Planta* **148:**305.
Mueller, S. C., and Brown, R. M., Jr., 1980, Evidence for an intramembrane component associated with a microfibril-synthesizing complex in higher plants, *J. Cell. Biol.* **84:**315.
Mueller, S. C., Brown, R. M., Jr., and Scott, T. K., 1976, Cellulose microfibrils: nascent stages of synthesis in a higher plant cell, *Science* **194:**949.
Northcote, D. H., 1974, Sites of synthesis of the polysaccharides of the cell wall, in: *Plant Carbohydrate Biochemistry* (J. B. Pridham, ed.), pp. 165–181, Academic Press, London.
Nowak-Ossorio, M., Gruber, E., and Schurz, J., 1976, Untersuchungen zur Cellulosebildung in Baumwollsamen, *Protoplasma* **88:**255.
Palevitz, B. A., 1980, Comparative effects of phalloidin and cytochalasin B on motility and morphogenesis in *Allium*, *Can. J. Bot.* **58:**773.
Palma, A., Buldt, G., and Jovanovic, S. M., 1976, Absolutes Molekulargewicht der nativen Cellulose des Alge *Valonia*, *Makromol. Chem.* **177:**1063.
Pillonel, C., Buchala, A. J., and Meier, H., 1980, Glucan synthesis by intact cotton fibers fed with different precursors at the stages of primary and secondary wall formation, *Planta* **149:**306.
Preston, R. D., 1974, *The Physical Biology of Plant Cell Walls*, Chapman and Hall, London.
Ramsey, J. D., and Berlin, J. D., 1976, Ultrastructure of early stages of cotton fiber differentiation, *Bot. Gaz.* **137:**11.
Robinson, D. G., 1977a, Plant cell wall synthesis, *Adv. Bot. Res.* 5:89.
Robinson, D. G., 1977b, Structure, synthesis and orientation of microfibrils. IV. Microtubules and microfibrils in *Glaucocystis*, *Cytobiologie* **15:**475.
Robinson, D. G., and Preston, R. D., 1971, Studies on the fine structure of *Glaucocystis nostochinearum* Itzigs. I. Wall structure, *J. Exp. Bot.* **22:**635.
Roelofsen, P. A., 1958, Cell-wall structure as related to surface growth, *Acta Bot. Neerl.* **7:**77.
Roelofsen, P. A., 1965, Ultrastructure of the wall in growing cells and its relation to the direction of growth, *Adv. Bot. Res.* **2:**69.
Roland, J.-C., and Vian, B., 1979, The wall of the growing plant cell: Its three-dimensional organization, *Int. Rev. Cytol.* **61:**129.
Rollins, M. L., and Tripp, V. W., 1954, Optical and electron microscopic studies of cotton fiber structure, *Textile Res. J.* **24:**345.
Rollins, M. L., and Tripp, V. W., 1961, The architecture of cellulose: A microscopic view, *For. Prod. J.* **11:**493.
Ryser, U., 1977, Cell wall growth in elongating cotton fibers: An autoradiographic study, *Cytobiologie* **15:**78.
Ryser, U., 1979, Cotton fibre differentiation: Occurrence and distribution of coated and smooth vesicles during primary and secondary wall formation, *Protoplasma* **98:**223.
Schnepf, E., and Deichgräber, G., 1979, Elongation of setae of *Pellia* (Bryophyta): Fine structural analysis, *Z. Pflanzenphysiol.* **94:**283.
Schnepf, E., Koch, W., and Deichgräber, G., 1966, Zur Cytologie und taxonomischen Einordnung von *Glaucocystis*, *Arch. Mikrobiol.* **55:**149.
Schubert, A. M., Benedict, C. R., Berlin, J. D., and Kohel, R. J., 1973, Cotton fiber development—kinetics of cell elongation and secondary wall thickening, *Crop Sci.* **13:**704.

Shafizadeh, F., and McGinnis, G. D., 1971, Morphology and biogenesis of cellulose and plant cell-walls, *Adv. Carbohydr. Chem. Biochem.* **26:**297.

Singer, S. J., 1974, The molecular organization of membranes, *Annu. Rev. Biochem.* **43:**805.

Stöckmann, V. E., 1972, Developing a hypothesis: Native cellulose elementary fibrils are formed with metastable structure, *Biopolymers* **11:**251.

Westafer, J. M., and Brown, R. M., Jr., 1976, Electron microscopy of the cotton fibre: New observations on cell wall formation, *Cytobios* **15:**111.

Willison, J. H. M., and Brown, R. M., Jr., 1977, An examination of the developing cotton fiber: Wall and plasmalemma, *Protoplasma* **92:**21.

Willison, J. H. M., and Brown, R. M., Jr., 1978a, Cell wall structure and deposition in *Glaucocystis*, *J. Cell Biol.* **77:**103.

Willison, J. H. M., and Brown, R. M., Jr., 1978b, A model for the pattern of deposition of microfibrils in the cell wall of *Glaucocystis*, *Planta* **141:**51.

Willison, J. H. M., and Rowe, A. J., 1980, *Replica, Shadowing and Freeze-Etching Techniques*, Vol. 8 of *Practical Methods in Electron Microscopy* (A. M. Glauert, ed.), North-Holland, Amsterdam.

Willison, J. H. M., Brown, R. M., Jr., and Mueller, S. C., 1980, A critical examination of the electron-microscopical evidence for swollen hydrated cellulose fibrils in green plants, *J. Microsc.* **118:**177.

7

The Role of the Golgi Apparatus in the Biosynthesis of Natural Polymer Systems with Particular Reference to Cellulose

DWIGHT K. ROMANOVICZ

1. Introduction

In its accepted role as the site of glycosylation and sulfation, the Golgi apparatus is involved in the biosynthesis of most natural polymers (for reviews, see Whaley, 1975; Northcote, 1979). Synthesis occurs in the context of a directed membrane flow throughout the endomembrane system (Morré *et al.*, 1971; Whaley *et al.*, 1971; Northcote, 1974)—from the nuclear envelope to the endoplasmic reticulum to the Golgi apparatus to the plasma membrane—and is accompanied by membrane differentiation that can be observed as changes in the thickness (Grove *et al.*, 1968) and composition (Keenan and Morré, 1970) of the membrane (see Morré, 1977). Consequently, the Golgi apparatus is a site for the synthesis and modification of membrane as well as a site for the synthesis of sugar-containing polymers. These two functions typically occur simultaneously, as illustrated in the formation of the cell plate from Golgi-derived vesicles: the contents of the vesicles form the middle lamella of the new transverse wall, while the vesicle membrane itself becomes the new plasma membrane of the daughter cells (see Whaley, 1975).

Within the scope of this volume, consideration may be restricted to those polymers that constitute the plant cell wall, typically classified into pectins,

DWIGHT K. ROMANOVICZ • Department of Biology, West Georgia College, Carrollton, Georgia 30118.

hemicelluloses, and cellulose. These widely used terms were originally based on solubility characteristics, but given the great variation in composition within each class, the implied specificity is somewhat misleading. It is more useful to identify wall components as simply cellulosic or noncellulosic, requiring demonstration of the β1,4 linkage and microfibrillar structure to establish the presence of cellulose.

There is currently wide acceptance of two separate pathways of cell-wall biosynthesis. The noncellulosic material is synthesized and secreted via the Golgi apparatus, while the synthesis of cellulose microfibrils occurs at the plasma membrane. The latter concept was proposed by Preston (1964), based on the observation that the complex microfibrillar structure of some plant-cell walls would require simultaneous crystallization and orientation of the microfibrils at the cell surface. Verification of the concept was provided by radioautographic studies which demonstrated that only noncellulosic polymers are synthesized in the Golgi apparatus, while incorporation of labeled glucose into cellulose occurs at the plasma membrane (Northcote and Pickett-Heaps, 1966; Wooding, 1968). The evidence from the cell-fractionation studies was not so conclusive, with β1,4-glucan synthetase activity demonstrated in Golgi fractions (Ray *et al.*, 1969), plasma-membrane fractions (Villemez *et al.*, 1968), or both (van der Woude *et al.*, 1974). These data were considered to be the result of synthetase activity that is in transit to the plasma membrane and is not active in the Golgi cisternae *in vivo* (Northcote, 1972, 1974; Shore and Maclachlan, 1975). Wright and Northcote (1976) demonstrated the presence of β1,4-glucan in Golgi-derived polysaccharide of root cap cells, but this material is in the form of very short microfibrils rendered soluble by a coating of noncellulosic material and is distinct from the long cellulose microfibrils of the secondary cell wall. Thus, the presence of β1,4-glucan synthetase in the Golgi apparatus is not in itself sufficient to refute the concept of cell-surface cellulose biosynthesis.

Morphological demonstration of cellulose microfibril synthesis provides even stronger support for the cell-surface interpretation. This visualization required the application of freeze-fracture to preserve the labile synthesizing complexes. With this technique, presumptive synthesizing complexes have been observed in association with microfibrils in the bacterium *Acetobacter* (Brown *et al.*, 1976), in several unicellular algae (Robinson and Preston, 1971, 1972; Brown and Montezinos, 1976; Giddings *et al.*, 1980), and in corn root (Mueller *et al.*, 1976; Mueller and Brown, 1980), although there is a conspicuous absence of such structures in some cellulose-synthesizing systems (Willison and Cocking, 1975) (see Willison, 1976). This work was the subject of a recent review (Brown and Willison, 1977) and is well covered in this volume.

There are, however, organisms that synthesize all components of the cell wall within the cisternae of the Golgi apparatus. These organisms possess a cell wall that is composed of discrete subunits called scales (see Moestrup, 1974a; Romanovicz, 1981). The distinctive ultrastructure and discrete nature of the scales permitted the observation that they are formed within the Golgi apparatus (Manton and Leedale, 1963; Parke and Rayns, 1964). In the Prymnesiophyceae

[Haptophyceae (see Hibberd, 1976)], the scales are constructed of microfibrils that resemble cellulose, so there was considerable interest in determining their composition. J. C. Green and Jennings (1967), analyzing isolated scales from *Chrysochromulina chiton*, were unable to demonstrate the presence of cellulose, finding only ribose and galactose as constituent monosaccharides. Brown and co-workers (Brown, 1969; Brown *et al.*, 1969, 1970), studying scale biogenesis in the related species *Pleurochrysis scherffelii*, established the polysaccharide nature of the scales and presented an accumulation of evidence indicating the presence of cellulose. This was followed by an investigation of alkali-purified *Pleurochrysis* scales (Herth *et al.*, 1972) that confirmed the presence of a cellulosic structural polysaccharide covalently linked to a peptide moiety. There was still considerable resistance to the notion of Golgi-derived cellulose (Colvin, 1972; Preston, 1974) until a cellulose I X-ray diffraction pattern was obtained from acid-purified *Pleurochrysis* scales (Romanovicz and Brown, 1976). This finding, coupled with a more thorough elucidation of the biosynthetic pathway (Brown *et al.*, 1973; Brown and Romanovicz, 1976), established beyond doubt the existence of Golgi-derived cellulose microfibrils. In the next section, *Pleurochrysis* will be discussed in more detail, followed by a consideration of its relationship to other cellulose-synthesizing systems.

2. *Pleurochrysis: A Model System for Golgi-Derived Cellulose*

2.1. *Scale Structure and Composition*

2.1.1. *Native Scales*

The scales of *Pleurochrysis* (Fig. 1) are composed of two microfibrillar subcomponents: a quadriradical network (radial microfibrils) to which is attached a band of microfibrils arranged in a spiral pattern (spiral microfibrils). Originally termed "concentric," this spiral arrangement was established by observation of partially degraded scales (Brown *et al.*, 1973) (Fig. 2). The diameter of the scales is 1.5–2.0 μm, varying proportionately with the length of the spiral microfibrils and the number of turns in the spiral. In nonmotile vegetative cells, the scales are coated with an amorphous polysaccharide that causes them to adhere and form a compact cell wall (Fig. 3).

2.1.2. *Noncellulosic Components*

When agar cultures of vegetative cells are flooded with seawater, the protoplasts escape into the liquid as naked, motile zoospores, leaving behind a cell-free scale preparation called scale fraction A (Brown *et al.*, 1970). Serial extraction of scale fraction A with increasing concentrations of trifluoroacetic acid (TFA) separated the individual subcomponents for chemical analysis (Romanovicz and Brown, 1976) (Table I). Several points about these data require discussion. First, the radial microfibrils are clearly noncellulosic, despite their

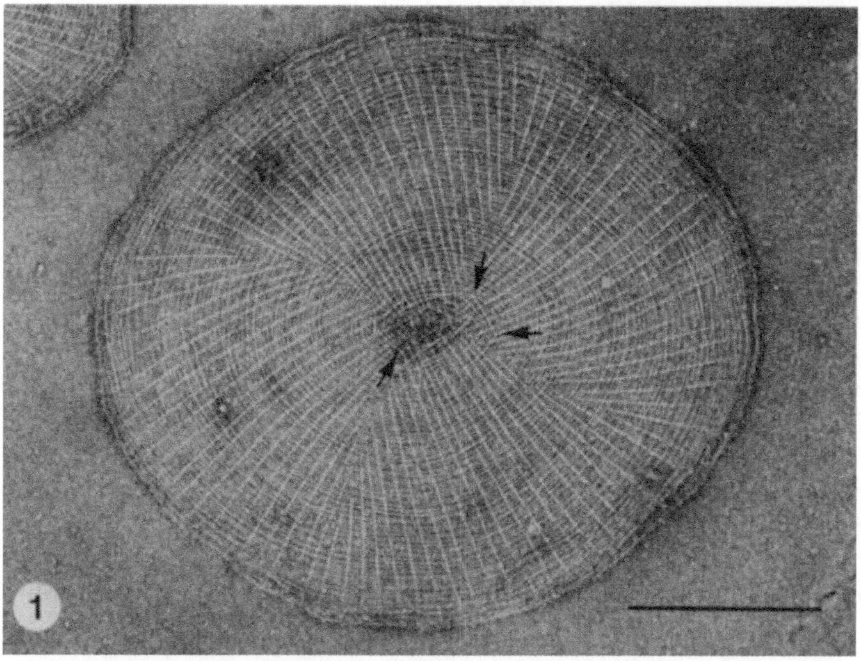

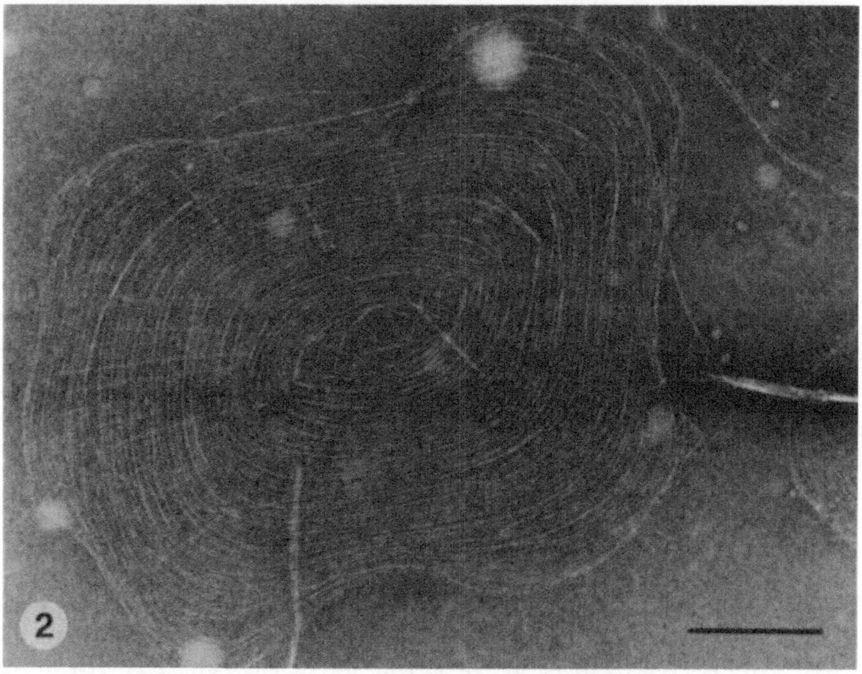

FIGURE 1. *Pleurochrysis* scale, negatively stained with uranyl acetate. The amorphous polysaccharide has been removed so that the quadriradial network and spiral microfibrils would be more evident. Note the "breakage loci" at the center of the scale (arrows). Scale bar: 0.5 μm.

FIGURE 2. *Pleurochrysis* scale, after extraction with 1 N HCl, negatively stained with phosphotungstic acid. The spiralling band pattern is apparent in the upper right quadrant. Scale bar: 0.25 μm.

microfibrillar structure. Second, there is some disagreement between these data and the results of other investigations of prymnesiophycean scales. Brown *et al.* (1970) found ribose in *Pleurochrysis* scales, and this sugar was also identified in the scales of other prymnesiophytes (Marker, 1965; J. C. Green and Jennings, 1967; Allen and Northcote, 1975). Ribose is an atypical component of plant-cell walls, and its presence is often interpreted as indicating cytoplasmic contamination of wall fractions. Romanovicz and Brown (1976) suggested that L-fucose had been erroneously identified as ribose. However, Allen and Northcote (1975) found both L-fucose and ribose in scales of *Chrysochromulina*, taking care to dispute the notion of cytoplasmic contamination. Two of the studies also indicated the presence of xylose (Marker, 1965; Allen and Northcote, 1975), but there was no indication of this monosaccharide in any of the other investigations. Ribose, xylose, and L-fucose are difficult to distinguish with commonly used chromatographic methods (Harris and Northcote, 1970), and it is clear that further analysis of prymnesiophycean scales is needed to convincingly establish the constituent monosaccharides of the radial and amorphous subcomponents. Third, the serial extraction revealed the presence of a coating substance on the spiral microfibrils, a coating that is responsible for maintaining the spiral arrangement of the microfibrils and conferring positive stainability in thin section. Finally, all three of the noncellulosic subcomponents contained small amounts of protein, a constituent of cell walls that has been receiving increased attention (Albersheim *et al.*, 1973; Keegstra *et al.*, 1973; Paradies *et al.*, 1977). Interestingly, the most abundant amino acid in the coating material is hydroxyproline. This fact, coupled with the role of the coating material in maintaining the spiral arrangement, suggests an analogy with Lamport's "extensin" (Lamport, 1970, 1973).

In an investigation of *Pleurochrysis*, Kavookjian and Brown (1978) reported a modified amorphous substance at the scale rim, recognizable only by its unique antigenic properties. Rim modifications are common among the Prymnesiophyceae and are considered to have phyletic significance (see Romanovicz, 1981). Allen and Northcote (1975) also observed an amorphous modification of the scale rim in *Chrysochromulina*, which they felt was responsible for scale adhesion. Little else is known about this rim modification; it is apparently a fruitful avenue for future study.

2.1.3. Scale Cellulose

If the spiral coating substance is hydrolyzed with 2 N TFA, the resultant pellet contains disorganized spiral microfibrils (Fig. 4), called acid-purified scale material. Further hydrolysis with 6 N HCl yields only glucose (Table I). When the acid-purified scale material is dried, it yields a cellulose I X-ray diffraction pattern (Romanovicz and Brown, 1976) (Table II), unequivocal proof of the presence of cellulose. The need for acid extraction and drying to obtain the cellulose I pattern indicates that scale cellulose has a low crystallinity with respect to other celluloses that yield such a pattern while wet and in the native state (Heyn, 1965; Caufield, 1971). This demonstration explained the inability of

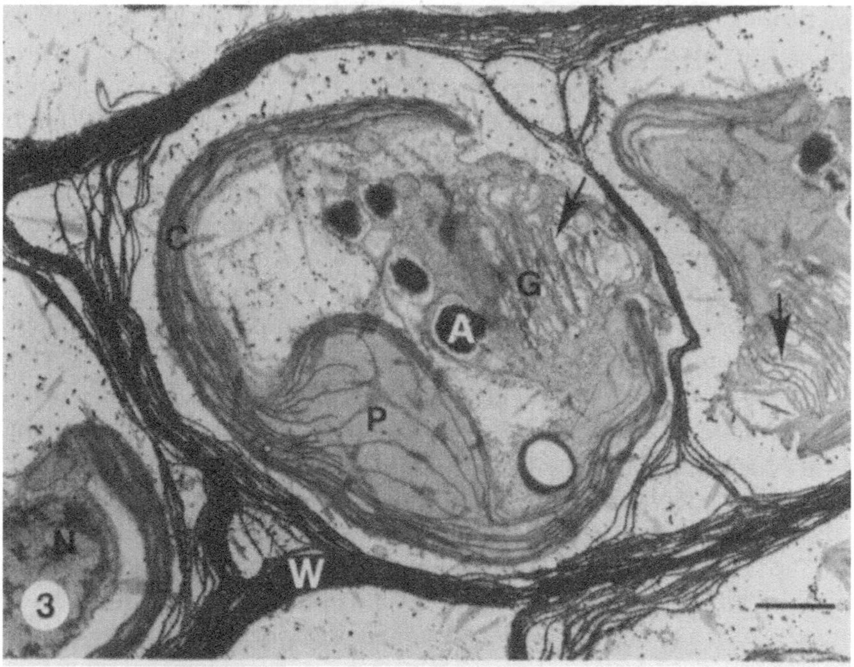

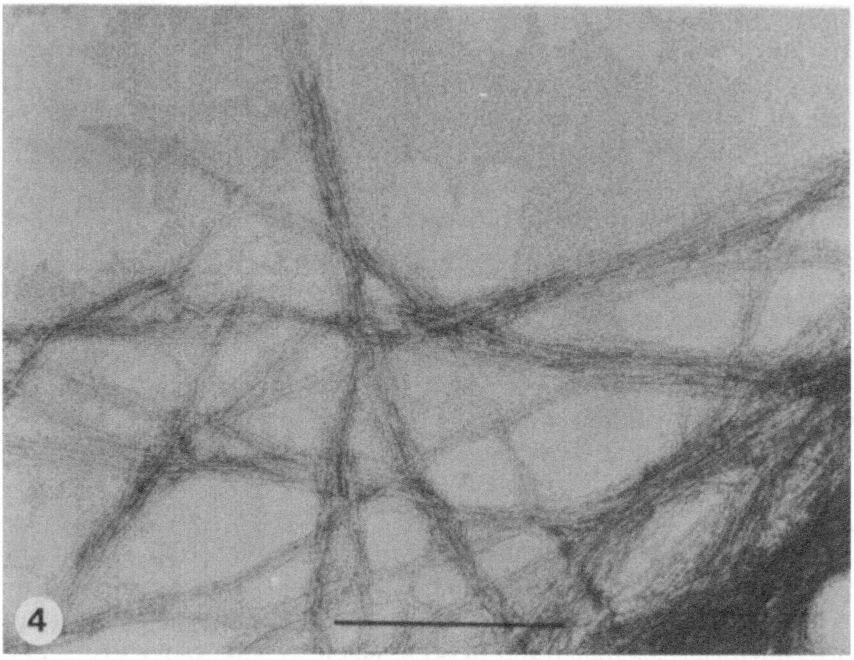

FIGURE 3. Glutaraldehyde-osmium fixed cells of *Pleurochrysis*, stained for polysaccharide localization with periodic acid-silver methenamine and counterstained with lead. Parietal chloroplast (C) with pyrenoid (P), autophagic vacuoles (A), nucleus (N), and Golgi apparatus (G) are visible in this section. Silver deposition is evident in the wall (W) and intracisternal scales (arrows). Scale bar: 1 μm.

TABLE I
Analysis of Scale Subcomponents

Scale subcomponent	Principal monosaccharides[a]
Amorphous polysaccharide	Galactose, glucose, L-fucose
Radial microfibrils	Galactose, arabinose, L-fucose
Spiral coating	Galactose, glucose, mannose
Spiral microfibrils	Glucose

[a] Thin-layer chromatography; butanol-pyridine-water (6:4:3).

J. C. Green and Jennings (1967) to obtain a cellulose diffraction pattern from *Chrysochromulina* scales: (1) they used native scales without acid extraction and (2) it is doubtful that scales from this organism contain much cellulose, since their micrographs reveal only a few spiral microfibrils confined to a narrow band at the rim.

Analysis of the infrared absorption spectra of native and acid-purified scale material (Fig. 5) supports several aspects of scale composition. The similarity between spectra of scale material and spectra of cotton fibers is immediately apparent, reaffirming the presence of cellulose. Low crystallinity is indicated by the single broad band in the 3000–3600 cm^{-1} region as compared with more highly crystalline forms of cellulose that yield as many as six distinct peaks in the region of absorption by hydrogen bonds (Marrinan and Mann, 1956; Liang and Marchessault, 1959; McCall *et al* ., 1971). Absorption in the 1600–1800 cm^{-1} region indicates the presence of protein (Mitchell and Scurfield, 1970; Paradies *et al*., 1977). That this absorption is still present in acid-purified scale material is no surprise, since scale cellulose has been shown to contain more than 30% bound protein (Herth *et al*., 1972; Romanovicz and Brown, 1976). Finally, the presence of sulfated polysaccharides in the noncellulosic subcomponents is indicated by

TABLE II
X-Ray Spacings (Å)[a]

Native scales, dried	Acid-purified scales[b] Wet	Acid-purified scales[b] Dried	Cotton
		6.01M	6.07M
		5.40M	5.49M
4.91W		4.32W	4.35M
4.07W	3.92M	3.99S	3.97S
3.25VW		3.21VW	3.21VW
2.59VW		2.60W	2.60W

[a] (S) Strong; (M) medium; (W) weak; (VW) very weak.
[b] Extracted with 2 N TFA, 3 hr, 121°C.

FIGURE 4. Acid-purified scale cellulose, negatively stained with uranyl acetate. Extraction of the coating material disrupted the spiral pattern. Scale bar: 0.25 μm.

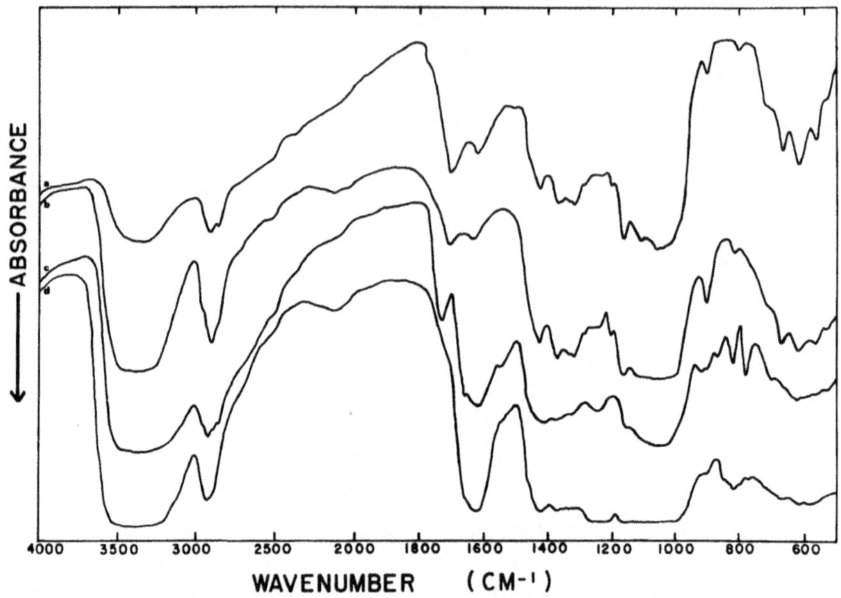

FIGURE 5. Infrared absorption spectra of acid-extracted cotton (a), acid-extracted *Pleurochrysis* scales (b), native cotton (c), and native *Pleurochrysis* scales (d). The overall similarity is evident. See the text for further interpretation.

the absorption at 1240 cm^{-1} and 820–850 cm^{-1} (Lloyd *et al.*, 1961) in native scale material and the loss of this absorption upon acid extraction.

The microfibrils of acid-purified scale cellulose are 3.7 × 4.9 nm in cross section (Romanovicz and Brown, 1976). These dimensions are identical to those of native spiral microfibrils as measured from intact scales, and are further supported by a crystallite size of 3.8 nm calculated from line-broadening in the X-ray diffraction pattern (Romanovicz and Brown, 1976). The agreement between native and acid-purified spiral microfibrils prevents any dispute of these measurements on the basis of disruptive isolation techniques (see Colvin, 1972; Preston, 1974), and since the measurements were made on individual microfibrils lying on the substrate, the dimensions cannot be said to be underestimated due to interference from overlapping images (see Preston, 1971).

It is clear from the preceding discussions that the spiral microfibrils of *Pleurochrysis* scales are cellulosic. Further, this constitutes the first demonstration of a 4- to 5-nm cellulose microfibril that is supported by both diffraction data and electron microscopy. In the next section, it will be demonstrated that the spiral microfibrils are synthesized and assembled within the cisternae of the Golgi apparatus, establishing beyond doubt the existence of Golgi-derived cellulose microfibrils.

2.2. *Scale Biogenesis*

2.2.1. *General Ultrastructure of the Endomembrane System*

The vegetative cell of *Pleurochrysis* contains a single, large Golgi apparatus composed of 30–40 individual cisternae. Polarization of the apparatus is evidenced by increasing distension of the cisternal lumen toward the secreting, or distal, face, which is always oriented toward the cell surface. The forming, or proximal, face of the apparatus is adjacent to an amplexus of the endoplasmic reticulum that arises as an extension of the nuclear envelope (Brown *et al.*, 1970). This entire endomembrane system is in a dynamic state of membrane flow, from the nuclear envelope to the plasma membrane. *Pleurochrysis* presents a particularly vivid example of this process, since the progression of each cisterna from the forming face to the secreting face is accompanied by the observable synthesis and assembly of the scale subcomponents. Viewed in static electron micrographs, the Golgi apparatus appears to be spatially differentiated into several regions, each of which conducts a specific portion of scale biogenesis. It should be remembered, however, that cisternal differentiation is actually temporal, in that each cisterna sequentially synthesizes a complete scale. Bearing this distinction in mind, scale biogenesis can be discussed in terms of spatially differentiated regions of the Golgi apparatus.

2.2.2. *Sequential Synthesis and Assembly*

The first visible product in the Golgi cisternae is a dense material found in some of the localized dilations characteristic of the proximal third of the Golgi apparatus (Manton, 1966; Brown *et al.*, 1970; Franke, 1970) (see Fig. 6). This dense material is a precursor for the radial microfibrils (Brown *et al.*, 1973; Brown and Romanovicz, 1976). Polymerization and microfibril formation are apparently separate events in radial microfibril synthesis, since the precursor is sufficiently polymerized to be retained through processing, yet amorphous in form.

The radial microfibrils first appear as two parallel rows of microfibrils, each row closely appressed to one of the opposing cisternal membranes (Fig. 6). The precursor material is apparently fed into the restricted lumenal space, where microfibril synthesis occurs next to the cisternal membrane. The close spatial relationship between the synthetic product and the cisternal membranes has been attributed to the membrane-bound nature of the synthetic enzymes (Northcote, 1974) or the attachment of the product to an integral membrane protein, providing a mechanism for the directed transport of the product within the endomembrane system (Northcote, 1979). Sulfation of the radial microfibrils occurs at this time, evidenced by the localization of arylsulfatase (Fig. 7) and acid phosphatase, two enzymes involved in the elaboration of sulfated products (Balasubramanian and Bachhawat, 1970; Hanker *et al.*, 1975). The timing of sulfation with respect to polymerization of the sulfated products is not well understood. It is clear in *Pleurochrysis* that at least some polymerization has already occurred prior to sulfation.

The transition from the double-row configuration to the quadriradial network has been described as an unfolding (Brown *et al.*, 1973; Brown and Romanovicz, 1976), based on the observation of bent or funnel-shaped radial microfibril configurations and the Z-stage Golgi cisterna in which the double row is turned 90° from its original orientation (Fig. 8). Bent intracisternal scales have also been noted in *Chrysochromulina* (Manton, 1972; Allen and Northcote,

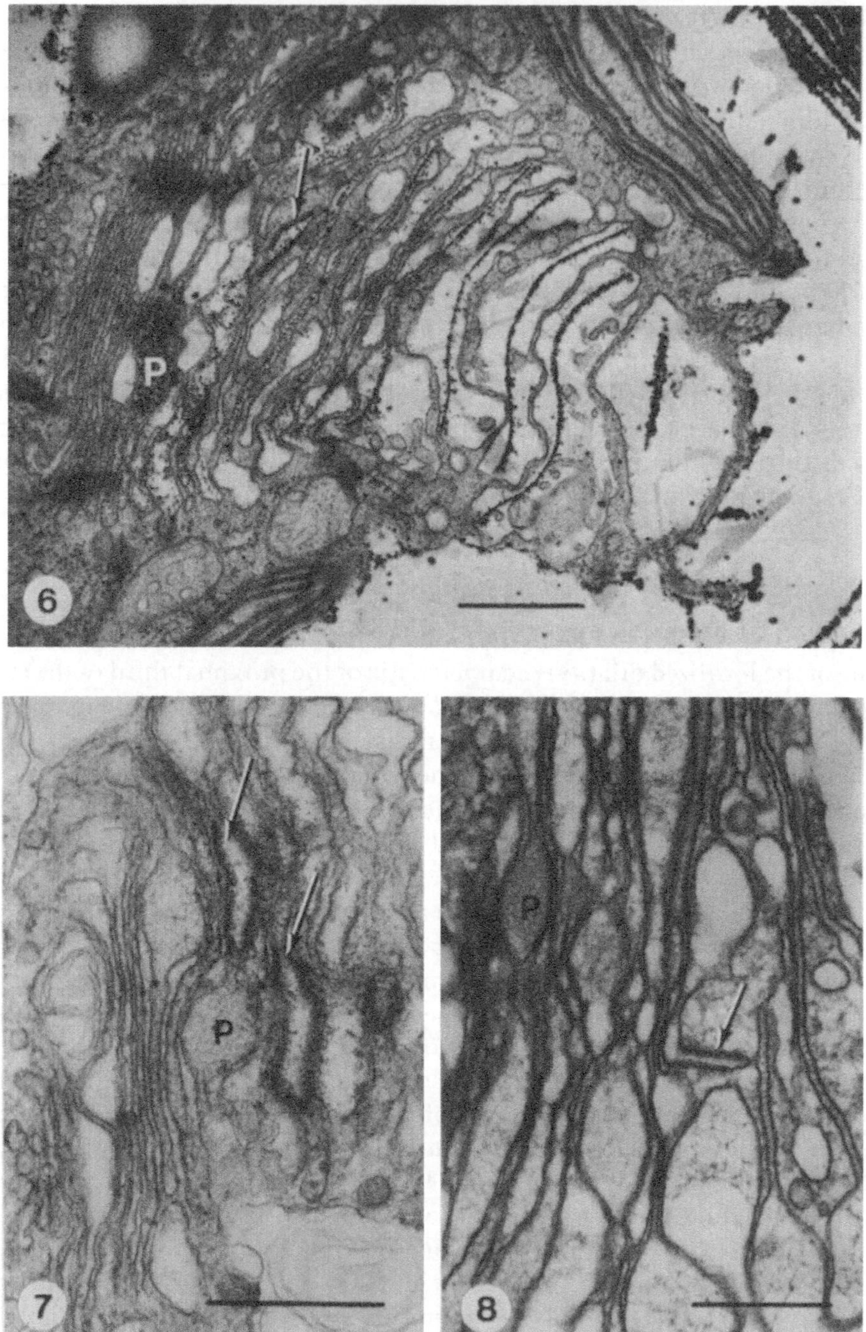

FIGURES 6–8. Thin sections through the Golgi apparatus of *Pleurochrysis* cells, with the secreting face toward the right. Periodic acid–silver methenamine reaction for polysaccharide localization, counterstained with lead citrate (Fig. 6). Note silver deposition in a radial precursor pool (P), double row of radial microfibrils (arrow), and intracisternal scales. All scales distal to the double row are unfolded. Localization of arylsulfatase activity (arrows) on a double row of radial microfibrils

1975). Further support for the unfolding concept comes from the novel work of Slaby and Slaby (1979), who envision the Golgi apparatus as a flattened, tightly spiralled tube within a cytoplasmic region that exhibits a spiral flow pattern. According to this hypothesis, the Z-stage cisterna is a point where the cytoplasmic flow shifts from an inward spiral to an outward one, separating the two rows of microfibrils since they are attached to opposing membranes (Slaby, pers. comm.). It must be mentioned that Pienaar (1969, 1971, 1976b) has consistently demonstrated a process of quadriradial formation in the related species *Hymenomonas* (*Cricosphaera*) *carterae* that does not involve a double row of microfibrils or unfolding. In this interpretation, the radial microfibrils are synthesized directly into the quadriradial network, with no separation of synthesis and orientation. However, a double row of microfibrils is evident in a micrograph from the study of another species of *Hymenomonas* (Pienaar, 1976a). Thus, it is likely that the particular morphological features used to support each of these two hypotheses are merely different views of the same process, so that a reevaluation of all the data is necessary to more accurately describe quadriradial formation.

Once the radial microfibrils are arranged in the characteristic quadriradial network, the cellulosic spiral microfibrils are added centripetally to the distal side, requiring several successive cisternal differentiations to complete the scale (Fig. 9). The cisternal membranes, which have been tightly appressed to the developing scale since radial microfibril synthesis, pull away from the scale with the addition of each turn of the spiral microfibrils. Brown and co-workers (Brown *et al.*, 1973; Brown and Romanovicz, 1976) have proposed two alternate hypotheses for the actual synthesis of the cellulose microfibrils in *Pleurochrysis*: (1) *in situ* synthesis at the point of attachment to the quadriradial network and (2) synthesis at a single locus, after which the microfibrils are fed into the lumen of the cisterna and deposited on the quadriradial network. The first hypothesis is consistent with the demonstration of cellulose microfibril synthesis by mobile enzyme complexes within the membrane, as discussed elsewhere in this volume. Although such complexes are not often demonstrated in thin section, presumably due to their labile nature, granular structures attached to the cisternal membrane have been observed in the region where spiral microfibrils first appear (Fig. 10). This hypothesis implies that polymerization, crystallization, and orientation of the glucan chains occur simultaneously. However, the central region of the scale always contains "breakage loci" (see Fig. 1) that are presumed to be the result of bending the cellulose microfibrils beyond the limit of their flexibility. The breakage loci have been taken to imply that crystallization and orientation are separate events in spiral microfibril synthesis, leading to the second hypothesis. This single-locus hypothesis is given additional support by the consistent presence of a central feeding tubule on the distal side of the

adjacent to a radial precursor pool (P) (Fig. 7). Z-stage Golgi cisterna with the double row of radial microfibrils (arrow) shifted 90° from its original orientation (Fig. 8). Note the precursor pool (P) proximal to the Z-stage cisterna. Scale bars: 0.5 μm.

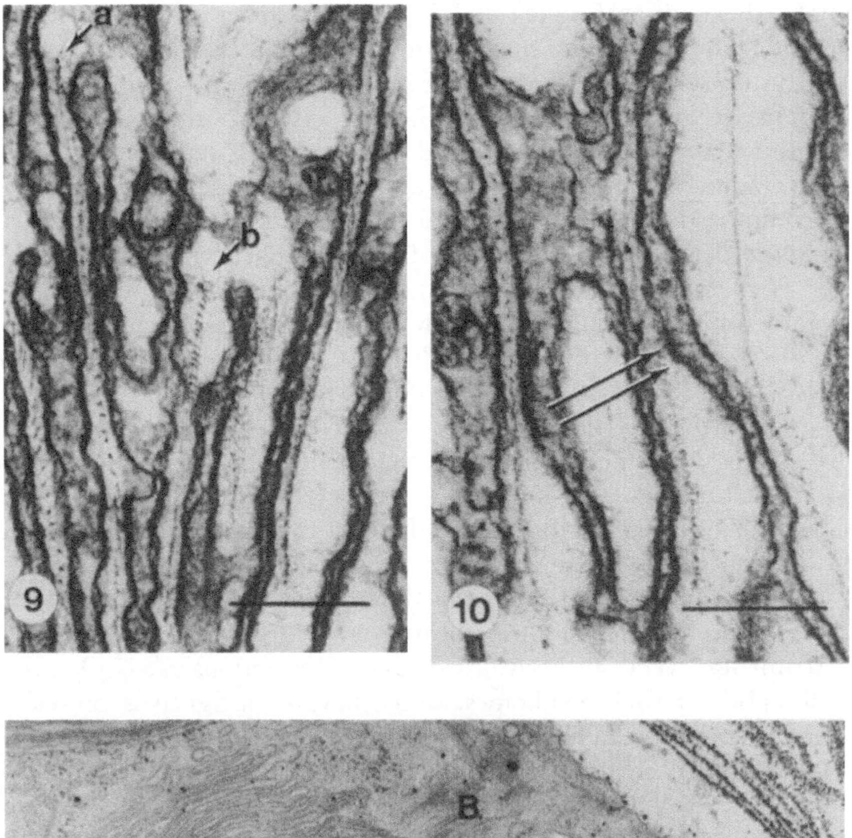

FIGURES 9–11. Thin sections through the Golgi apparatus of *Pleurochrysis* cells, with the secreting face toward the right. Portion of the distal Golgi region illustrating the centripetal addition of the spiral microfibrils (Fig. 9). An increase in the number of spiral microfibrils from a to b is evident. Note the proximity of the cisternal membranes at the point where the microfibrils are being added.

cisterna, leading to the suggestion that the microfibrils are made at this point and fed into the cisternal lumen. However, this hypothesis engenders even greater difficulty with respect to the mechanism of microfibril orientation. In view of the support that the mobile-enzyme-complex hypothesis has received from diverse cellulose-synthesizing systems, it is likely that a similar mechanism operates in *Pleurochrysis*. The question of microfibril orientation remains, and will be discussed in Section 3.2.

Synthesis and addition of the amorphous polysaccharide occurs immediately after the attachment of the spiral microfibrils at each point. Consequently, the rim of a particular scale is completed while the central portion consists only of the radial network. The process of amorphous addition can be observed as an increase in the density of polysaccharide staining that progresses centripetally over several successive cisternae (Fig. 11). A similar staining pattern is obtained from the localization of alkaline phosphatase activity (Brown and Romanovicz, 1976). Since the amorphous subcomponent is quite hydrated, its synthesis could account for the observed distension of the cisternal lumen at the distal region of the Golgi apparatus.

2.3. *Scale Exocytosis*

With scale synthesis completed, the distalmost cisterna fuses with the plasma membrane to release the scale into the extracellular space. It is likely that fusion is initiated at specific sites on the membranes involved (for a review, see Northcote, 1979). The formation of distinct particle complexes within the membrane prior to fusion has been demonstrated in some systems (Satir *et al.*, 1973; Beisson *et al.*, 1976, 1980) but not in others (da Silva and Nagueira, 1977). In *Pleurochrysis*, there is evidence for the involvement of microtubules in exocytosis, in that colchicine blocks scale exocytosis, but not scale synthesis. This is consistent with other demonstrations of the ability of colchicine to block the secretory process (see Hoffstein *et al.*, 1977). The Golgi apparatus of *Pleurochrysis* is flanked by an array of microtubules (Brown and Franke, 1971) and exhibits a distinctive orientation with respect to the basal bodies (see Fig. 11) that may be characteristic of the prymnesiophycean Golgi apparatus (see Manton 1964, 1967a,b, 1968).

According to Brown (1969), the maximal rate of exocytosis is one cisterna every 2 min, a rate that is consistent with data from other plants (see Chrispeels, 1976) and from rat goblet cells (Neutra and Leblond, 1966). If the cell is to avoid a rapid increase in surface area during scale biogenesis, much of the cisternal membrane must be recycled. Brown and Romanovicz (1976) have suggested a role

Scale bar: 0.25 μm. Region of the Golgi apparatus where spiral microfibrils are being added (Fig. 10). Three granules (arrows) are visible on the distal cisternal membrane at the point of microfibril addition. Scale bar: 0.25 μm. Periodic acid–silver methenamine reaction for polysaccharide localization, counterstained with lead (Fig. 11). The centripetal addition of the amorphous polysaccharide is indicated by the increase in staining density (a to d). Note the orientation of the Golgi cisternae with respect to the basal body (B). Scale bar: 0.5 μm.

for the extensive subsurface cisterna just below the plasma membrane, based on the localization of acid phosphatase activity in this structure and in the distalmost cisterna of the Golgi stack, similar to the Golgi–endoplasmic reticulum–lysosome (GERL) complex proposed by A. B. Novikoff and colleagues (P. M. Novikoff *et al.*, 1971; A. B. Novikoff and P. M. Novikoff, 1977). The hypothesis involves an endocytotic mechanism utilizing invaginations of the plasma membrane into the subsurface cisterna. Similar invaginations have been observed in other scale-forming species (see Klaveness, 1973), and endocytosis has been documented in *Chrysochromulina* (Manton, 1972), but Brown (personal communication) was unable to demonstrate the uptake of cationized ferritin by *Pleurochrysis*. Thus, the mechanism of membrane recycling remains unknown.

3. Relationship of Pleurochrysis to Other Systems

3.1. A General Model for the Secretory Process

Scale biogenesis exemplifies the features of the secretory process in a particularly vivid manner because the secretory product is clearly identifiable within the cell. The abrupt initiation and cessation of enzyme activity can be correlated with the synthesis of specific scale subcomponents the composition of which is known from chemical analysis of isolated scales. The role of cisternal membranes is illustrated by the direct contact between the scale and the membrane during synthesis.

A major question about the secretory process concerns the means by which genetic control is maintained. According to the endomembrane concept, glycosylating enzymes are synthesized and incorporated into the membranes at the endoplasmic reticulum where Golgi membrane originates. Thus the scale pattern is a reflection of the organization of these enzymes within the cisternal membrane, and sequential synthesis of scale subcomponents would require a mechanism for enzyme activation and deactivation. In *Pleurochrysis*, Brown and colleagues (Brown *et al.*, 1973; Brown and Romanovicz, 1976) observed a radial precursor pool still attached to the amplexus and suggested that topographic organization of Golgi cisternae may begin at this level. Such a mechanism might be especially important in those scale-forming organisms that simultaneously produce several different scale types within the same cisterna (Moestrup and Walne, 1979). In contrast, the association of polyribosomes with the Golgi apparatus (Brown *et al.*, 1973; Mollenhauer and Morré, 1974) and the increasing size of the cisternae toward the secreting face suggest that the addition of enzymes and membrane components to the cisternae does not occur exclusively at the forming face. The continued incorporation of enzymes into the differentiating cisternae could provide a mechanism for sequential synthesis of secretory products. Jaeken and Thines-Sempoux (1981) have suggested that the numerous small vesicles associated with Golgi cisternae transfer enzymes from one cisterna to the preceding one, making available the necessary enzymes at each level of the

Golgi apparatus. While insertion of membrane components could occur at cisternal regions not directly involved in synthesis, the incorporation of synthetic enzymes would have to be under precise control in order to maintain the specific scale pattern.

Scale biogenesis also involves highly specific changes in membrane configuration, as in the unfolding of the radial microfibrils. In other scale-forming organisms, the three dimensional shape of the scale is reflected in the cisternal membrane configuration before synthesis is initiated (Leadbeater, 1972) (see Romanovicz, 1981). These configurations are clearly under genetic control, but the means by which this control is achieved are completely unknown. Microtubules cannot be involved, since colchicine does not prevent scale synthesis; and there are no other observable structural elements in the Golgi region that could be invoked as transducers in this process.

Although a complete endomembrane system has been demonstrated in many algae and fungi, and in animal secretory tissue, there are some reservations when applying this concept to higher plants. From a serial-section analysis of Golgi apparatus in cultured sycamore cells and maize root-cap cells, Robinson (1980) concluded that the dictyosomes possess neither a polarity in terms of forming and secreting faces nor transitional elements from the endoplasmic reticulum. Secretory products are transferred from Golgi cisternae to the plasma membrane, but the regeneration of the Golgi from the endoplasmic reticulum (ER) cannot be supported. Kristen (1980) has proposed that the flow of membranes from ER to Golgi occurs only in conjunction with the synthesis and transport of proteinaceous secretory products, while secretion of polysaccharides does not require direct transitions between ER and Golgi. In addition, it has been suggested that, in some plant cells, a direct transfer may occur between the ER and the plasma membrane, without involving the Golgi apparatus (Bowles and Northcote, 1974). *Pleurochrysis* is thus an example of the general secretory pathway involving a complete endomembrane system producing a glycoprotein product, but as such may not pertain to those plant cells that secrete a product composed primarily of polysaccharide.

3.2. The Question of Microfibril Orientation

The complex and specific pattern of microfibrils in plant-cell walls makes it apparent that a genetically controlled orientation mechanism must exist. The most frequently invoked cellular structures are microtubules, often observed to underlie the plasma membrane in an orientation similar to that of the microfibrils in the wall (Newcomb, 1969; Heath, 1974). This hypothesis receives additional support from the demonstration that colchicine can disrupt the pattern of microfibril orientation in some organisms (P. B. Green, 1962; Pickett-Heaps, 1967). In contrast, there are other cellulose-synthesizing organisms in which microfibril orientation cannot be linked to any cytoplasmic structures (Kiermayer, 1968). The absence of microtubules in the Golgi region, and the inability of colchicine to disrupt spiral microfibril orientation, make it unlikely that microtubules are involved in cellulose microfibril orientation in *Pleurochrysis*.

Microfibril orientation of *Oocystis* has been linked to arrays of granules, called granule bands, that underlie the synthesizing complexes in the plasma membrane (Brown and Montezinos, 1976) (see also Chapter 1). The granule bands undergo a cyclic reorganization, resulting in a cell wall composed of layers with alternating microfibril orientations. It is interesting to note that colchicine prevents the reorganization of the granule bands, halting the alternation of wall layers, but does not interfere with the existing organization of the granule bands or their ability to orient microfibrils. Since the granule bands are observable only in freeze-fracture preparations, the demonstration of a similar mechanism in *Pleurochrysis* is unlikely. The probability of a fracture within the distal membrane of a cisterna in the region of spiral microfibril synthesis is quite low, a fact that points out of the unsuitability of *Pleurochrysis* to the current direction of research into cellulose biosynthesis.

It has been suggested repeatedly by Brown and colleagues that the restricted space of the cisternal lumen might function as a passive orientation mechanism in spiral microfibril synthesis. When combined with the concept of a mobile enzyme complex propelled by forces derived from the formation of the microfibrils, the result is a gradual centripetal spiral movement of the synthesizing complex(es) until the center of the scale is reached. This proposal also encompasses the existence of breakage loci, since it separates the processes of crystallization and orientation. A similar passive orientation mechanism has been suggested in the desmid *Micrasterias* (Giddings *et al.*, 1980), wherein the deposition of secondary wall cellulose microfibrils is essentially random, the overall shape of the semicell having been established by the primary wall, which then constrains the formation of the secondary wall.

Allen and Northcote (1975) described the structure of *Chrysochromulina* scales as "woven." If this were the case, a more complicated orientation mechanism would be required. However, their micrograph does not convincingly demonstrate a woven structure, and the scales of *Pleurochrysis* are clearly layered, not woven.

3.3. Golgi Involvement in Plasma-Membrane-Cellulose Synthesis

Considering the role of the endomembrane system in the synthesis and modification of membrane, one may infer the cellulose-synthesizing complexes on the plasma membrane have been transported via the Golgi apparatus. This has been demonstrated in *Micrasterias*, wherein rosettes of particles that arise in the Golgi cisternae become incorporated into the plasma membrane, where they synthesize cellulose microfibrils (Dobberstein and Kiermayer, 1972; Kiermayer and Dobberstein, 1973; Kiermayer and Sleytr, 1979; Giddings *et al.*, 1980). Consequently, the demonstration of Golgi-derived cellulose microfibrils in *Pleurochrysis* simply represents a difference in the timing of enzyme activation and is a confirmation of the membrane-flow hypothesis rather than a refutation of the cell-surface theory of cellulose biosynthesis. Synthesis within the Golgi apparatus results in a cell wall of discrete packets rather than the continuous structures made possible by cell-surface biosynthesis.

4. Concluding Remarks

In the years since *Pleurochrysis* was proposed as a model system for the biosynthesis of cellulose, three series of observations have been made: First, the investigation of several diverse organisms has left no doubt that cellulose is typically synthesized at the cell surface. Second, continued investigation of *Pleurochrysis* has established unequivocally that cellulose can be synthesized in the Golgi apparatus. Rather than being contradictory, these results illustrate the ubiquity of endomembrane involvement in the biosynthesis of polymers by eukaryotic cells. Synthesis within the Golgi apparatus permitted the development of elaborately structured scales, while synthesis at the cell surface led to continuous cell walls that are perhaps better suited to the needs of land plants. Last, scale formation has been demonstrated in an increasing number of diverse organisms, including the motile zoospores in the Charales and two Chlorophycean algae (see Romanovicz, 1981). The latter data led Moestrup (Moestrup, 1974b; Moestrup and Thomsen, 1974) to propose that scale-forming organisms are in fact ancestral to all land plants.

ACKNOWLEDGMENTS

The author is indebted to Bettina A. Bussard for typing the manuscript and helping in its preparation. The figures are reprinted from Brown and Romanovicz (1976) and Romanovicz and Brown (1976), courtesy of John Wiley and Sons, Inc.

References

Albersheim, P., Bauer, W. D., Keegstra, K., and Talmadge, K. W., 1973, The structure of the wall of suspension-cultured sycamore cells, in: *Biogenesis of Plant Cell Wall Polysaccharides* (F. Loewus, ed.), pp. 114–147, Academic Press, New York.

Allen, D. M., and Northcote, D. H., 1975, The scales of *Chrysochromulina chiton*, *Protoplasma* **83**:389.

Balasubramanian, A. S., and Bachhawat, B. K., Sulfate metabolism in brain, *Brain Res.* **20**:341.

Beisson, J., Lefort-Tran. M., Pouphile, M., Rossignol, M., and Satir, B., 1976, Genetic analysis of membrane differentiation in *Paramecium*: Freeze-fracture study of the trichocyst cycle in wild-type and mutant strains, *J. Cell Biol.* **69**:126.

Beisson, J., Cohen, J., Lefort-Tran, M., Pouphile, M., and Rossignol, M., 1980, Control of membrane fusion in exocytosis: Physiological studies on a *Paramecium* mutant blocked in the final step of the tricocyst extrusion process, *J. Cell Biol.* **85**:213.

Bowles, D. J., and Northcote, D. H., 1974, The amounts and rates of export of polysaccharides found within the membrane system of maize root cells, *Biochem. J.* **142**:139.

Brown, R. M., Jr., 1969, Observations on the relationship of the Golgi apparatus to wall formation in the marine Chrysophycean alga, *Pleurochrysis scherffelii* Pringsheim, *J. Cell Biol.* **41**:109.

Brown, R. M., Jr., and Franke, W. W., 1971, A microtubular crystal associated with the Golgi field of *Pleurochrysis scherffelii*, *Planta* **96**:354.

Brown, R. M., Jr., and Montezinos, D., 1976, Cellulose microfibrils: Visualization of biosynthetic and orienting complexes in association with the plasma membrane, *Proc. Natl. Acad. Sci. U.S.A.* **73**:143.

Brown, R. M., Jr., and Romanovicz, D. K., 1976, Biogenesis and structure of Golgi-derived cellulosic scales in *Pleurochrysis*. I. Role of the endomembrane system in scale assembly and exocytosis, *Appl. Polymer Symp.* **28**:537.

Brown, R. M., Jr., and Willison, J. H. M., 1977, Golgi apparatus and plasma membrane involvement in secretion and cell surface deposition, with special emphasis on cellulose biogenesis in: *International Cell Biology, 1976–1977* (B. R. Brinkley and K. R. Porter, eds.), pp. 267–283, The Rockefeller University Press, New York.

Brown, R. M., Jr., Franke, W. W., Kleinig, H., and Sitte, P., 1969, Cellulosic wall component produced by the Golgi apparatus of *Pleurochrysis scherffelii*, *Science* **166**:894.

Brown, R. M., Jr., Franke, W. W., Kleinig, H., Falk, H., and Sitte, P., 1970, Scale formation in Chrysophycean algae, I. Cellulosic and noncellulosic wall components made in the Golgi apparatus, *J. Cell Biol.* **45**:246.

Brown, R. M. Jr., Herth, W., Franke, W. W., and Romanovicz, D. K., 1973, The role of the Golgi apparatus in the biosynthesis and secretion of a cellulosic glycoprotein in *Pleurochrysis*: A model system for the synthesis of structural polysaccharides, in: *Biogenesis of Plant Cell Wall Polysaccharides* (F. Loewus, ed.), pp. 207–257, Academic Press, New York.

Brown, R. M., Jr., Willison, J. H. M., and Richardson, C. L., 1976, Cellulose biosynthesis in *Acetobacter xylinum*: Visualization of the site of synthesis and direct measurement of the *in vivo* process, *Proc. Natl. Acad. Sci. U.S.A.* **73**:4565.

Caufield, D. F., 1971, Crystallite sizes in wet and dry *Valonia ventricosa*, *Text. Res. J.* **41**:267.

Chrispeels, M. J., 1976, Biosynthesis, intracellular transport, and secretion of extracellular macromolecules, *Annu. Rev. Plant Physiol.* **24**:19.

Colvin, J. R., 1972, The structure and biosynthesis of cellulose, *CRC Crit. Rev. Macromol. Sci.* **1**:47.

Da Silva, P. P., and Nagueira, L., 1977, Membrane fusion during secretion: A hypothesis based on electron microscope observation of *Phytophthora palmisora* zoospores during encystment, *J. Cell Biol.* **73**:161.

Dobberstein, B., and Kiermayer, O., 1972, Das Auftreten eines besondern Typs von Golgivesikeln während der Sekundärwandbildung von *Micrasterias denticulata* Bréb, *Protoplasma* **75**:185.

Franke, W. W., 1970, Central dilations in maturing Golgi cisternae—a common structural feature among plant cells?, *Planta* **90**:370.

Giddings, T. H., Jr., Brower, D. L., and Staehelin, L. A., 1980, Visualization of particle complexes in the plasma membrane of *Micrasterias denticulata* associated with the formation of cellulose fibrils in the primary and secondary cell walls, *J. Cell Biol.* **84**:327.

Green, J. C., and Jennings, D. H., 1967, A physical and chemical investigation of the scales produced by Golgi apparatus within and found on the surface of cells of *Chrysochromulina chiton* Parke et Manton, *J. Exp. Bot.* **18**:359.

Green, P. B., 1962, Mechanism for plant cellular morphogenesis, *Science* **138**:1404.

Grove, S. N. Bracker, C. E., and Morré, D. J., 1968, Cytomembrane differentiation in the endoplasmic reticulum-Golgi apparatus-vesicle complex, *Science* **161**:171.

Hanker, J. S., Thornburg, L. P., Yates, P. E., and Romanovicz, D. K., 1975, The demonstration of arylsulfatases with 4-nitro-1,2-benzenediol mono(hydrogen sulfate) by the formation of osmium *blacks* at the site of copper capture, *Histochemistry* **41**:207.

Harris, P. J., and Northcote, D. H., 1970, Patterns of polysaccharide biosynthesis in differentiating cells of maize root tips, *Biochem. J.* **120**:479.

Heath, I. B., 1974, A unified hypothesis for the role of membrane-bound enzyme complexes and microfibrils in plant cell wall synthesis, *J. Theor. Biol.* **48**:4545.

Herth, W., Franke, W. W., Stadler, J., Bittiger, H., Keilich, G., and Brown, R. M., Jr., 1972, Further characterization of the alkali-stable material from the scales of *Pleurochrysis scherffelii:* A cellulosic glycoprotein, *Planta* **105**:79.

Heyn, A. N. J., 1965, Crystalline state of cellulose in fresh and dried mature cotton fiber from unopened bolls as studied by X-ray diffraction *J. Polym. Sci.* **A3**:1251.

Hibberd, D. J., 1976, The ultrastructure and taxonomy of the Chrysophyceae and Prymnesiophyceae (Haptophyceae): A survey with new observations on the structure of the Chrysophyceae, *Bot. J. Linn. Soc.* **72**:55.

Hoffstein, S., Goldstein, I. M., and Weissmann, G., 1977, Role of microtubule assembly in lysosomal enzyme secretion from human polymorphonuclear leukocytes, *J. Cell Biol.* **73**:242.

Jaeken, L., and Thines-Sempoux, D., 1981, A three-dimensional study of organelle interrelationships in regenerating rat liver. 6. Golgi apparatus, *Cell Biol. Intl. Reports* **5**:261.

Kavookjian, A. M. A., and Brown, R. M., Jr., 1978, Antigenic properties of the Golgi-derived scales of *Pleurochrysis scherffelii, Cytobios* **23**:45.

Keegstra, K., Talmadge, K., Bauer, W. D., and Albersheim, P., 1973, The structure of plant cell walls. III. A model of the walls of suspension-cultured sycamore cells based on the interconnections of macromolecular components, *Plant Physiol.* **51**:188.

Keenan, T. W., and Morré, D. J., 1970, Phospholipid class and fatty acid composition of Golgi aparatus isolated from rat liver and comparison with other cell fractions, *Biochemistry* **9**:19.

Kiermayer, O., 1968, The distribution of microtubules in differentiating cells of *Micrasterias denticulata* Bréb., *Planta* **83**:223.

Kiermayer, O., and Dobberstein, B., 1973, Membrankomplexe dictyosomaler Herkunft als "Matrizen" für die extraplasmatische Synthesis und Orientierung von Microfibrillen, *Protoplasma* **77**:437.

Kiermayer, O., and Sleytr, U. B., 1979, Hexagonally ordered "rosettes" of particles in the plasma membrane of *Micrasterias denticulata* Bréb. and their significance for microfibril formation and orientation, *Protoplasma* **101**:133.

Klaveness, D., 1973, The microanatomy of *Calyptrosphaera sphaeroidea,* with some supplementary observations of the motile stage of *Coccolithus pelagicus, Norw. J. Botn.* **20**:151.

Kristen, U., 1980, Endoplasmic reticulum–dictyosome interconnections in ligula cells of *Isoetes lacustris, Eur. J. Cell Biol.* **23**:16.

Lamport, D. T. A., 1970, Cell wall metabolism, *Annu. Rev. Plant Physiol.* **21**:235.

Lamport, D. T. A., 1973, The glycopeptide linkages of extensin: O-D-galactosyl serine and O-L-arabinosyl hydroxyproline, in: *Biogenesis of Plant Cell Wall Polysaccharides* (F. Loewus, ed.). pp. 149–164, Academic Press, New York.

Leadbeater, B. S. C., 1972, Fine structural observations on six new species of *Chrysochromulina* (Haptophyceae) from Norway with preliminary observations on scale production in *C. microcylindra* sp. nov., *Sarsia* **49**:65.

Liang, C. Y., and Marchessault, R. H., 1959, Infrared spectra of crystalline polysaccharides. I. Hydrogen bonds in native celluloses, *J. Polym. Sci.* **37**:385.

Lloyd, A. G., Dodgson, K. S., Price, R. G., and Rose, F. A., 1961, Infrared studies on sulfate esters. I. Polysaccharide sulfates, *Biochim. Biophys. Acta* **46**:108.

Manton, I., 1964, Observations with the electron microscope on the division cycle in the flagellate *Prymnesium parvum* Carter, *J. R. Microsc. Soc.* **83**:317.

Manton, I., 1966, Observations on scale production in *Prymnesium parvum, J. Cell Sci.* **1**:375.

Manton, I., 1967a, Further observations on the fine structure of *Chrysochromulina chiton* with special reference to the haptonema, "peculiar" Golgi structure and scale production, *J. Cell Sci.* **2**:265.

Manton, I., 1967b, Further observations on scale formation in *Chrysochromulina chiton, J. Cell Sci.* **2**:411.

Manton, I., 1968, Further observations on the microanatomy of the haptonema in *Chrysochromulina chiton* and *Prymnesium parvum, Protoplasma* **66**:35.

Manton, I., 1972, Observations on the biology and microanatomy of *Chrysochromulina megacylindra* Leadbeater, *Br. Phycol. J.* **7**:235.

Manton, I., and Leedale, G. F., 1963, Observations on the microanatomy of *Crystallolithus hyalinus* Gaarder and Morkali, *Arch. Mikrobiol.* **47**:115.

Marker, A. F. H., 1965, Extracellular carbohydrate liberation in the flagellates *Isochrysis galbana* and *Prymnesium parvum*, *J. Mar. Biol. Assoc. U. K.* **45**:755.

Marrinan, H. J., and Mann, J., 1956, Infrared spectra of the crystalline modifications of cellulose, *J. Polym. Sci.* **21**:301.

McCall, E. R., Morris, N. M., Tripp, V. W., and O'Conner, R. T., 1971, Low temperature infrared absorption spectra of cellulosics, *Appl. Spectrosc.* **25**:196.

Mitchell, A. J., and Scurfield, G., 1970, An assessment of infrared spectra as indicators of fungal cell wall composition, *Aust. J. Biol. Sci.* **23**:345.

Moestrup, Ø., 1974a, Scale formation: A special function of the Golgi apparatus in algae, in: *Proceedings of the Eighth International Congress on Electron Microscopy* (J. V. Sanders and D. J. Goodchild, eds.), pp. 564–565, Australian Academy of Science, Canberra.

Moestrup, Ø., 1974b, Ultrastructure of the scale-covered zoospores of the green alga *Chaetosphaeridium*, a possible ancestor of the higher plants and bryophytes, *Biol. J. Linn. Soc.* **6**:111.

Moestrup, Ø., and Thomsen, H. A., 1974, An ultrastructural study of the flagellate *Pyramimonas orientalis* with particular emphasis on Golgi apparatus activity and the flagellar apparatus, *Protoplasma* **81**:247.

Moestrup, Ø., and Walne, P. L., 1979, Studies on scale morphogenesis in the Golgi apparatus of *Pyramimonas tetrarhynchus* (Prasinophyceae), *J. Cell Sci.* **36**:437.

Mollenhauer, H. H., and Morré, D. J., 1974. Polyribosomes associated with the Golgi apparatus, *Protoplasma* **79**:333.

Morré, D. J., 1977, Membrane differentiation and the control of secretion: A comparison of plant and animal Golgi apparatus, in: *International Cell Biology, 1976–1977* (B. R. Brinkley and K. R. Porter, eds.), pp. 293–303, The Rockefeller University Press, New York.

Morré, D. J., Mollenhauer, H. H., and Bracker, C. E., 1971, Origin and continuity of Golgi apparatus, in: *Origin and Continuity of Cell Organelles* (J. Reinert and H. Ursprung, eds.), pp. 82–126, Springer-Verlag, New York.

Mueller, S. C., and Brown, R. M., Jr., 1980, Evidence for an intramembraneous component associated with a cellulose microfibril-synthesizing complex in higher plants, *J. Cell Biol.* **84**:315.

Mueller, S., Brown, R. M., Jr., and Scott, T. K., 1976, Cellulosic microfibrils: Nascent stages of synthesis in a higher plant cell, *Science* **194**:949.

Neutra, M., and Leblond, C. P., 1966, Synthesis of the carbohydrate of mucus in the Golgi complex as shown by electron microscope radioautography of goblet cells from rats injected with glucose H^3, *J. Cell Biol.* **30**:119.

Newcomb, E. H., 1969, Plant microtubules, *Annu. Rev. Plant Physiol.* **20**:253.

Northcote, D. H., 1972, Chemistry of the plant cell wall, *Annu. Rev. Plant Physiol.* **23**:113.

Northcote, D. H., 1974, Complex envelope system: Membrane systems of plant cells, *Philos. Trans. R. Soc. London B Ser.* **268**:119.

Northcote, D. H., 1979, The involvement of the Golgi apparatus in the biosynthesis and secretion of glycoproteins and polysaccharides, in: *Biomembranes*, Vol. 10 (L. A. Manson, ed.), pp. 51–76, Plenum Press, New York.

Northcote, D. H., and Pickett-Heaps, J. D., 1966, A function of the Golgi apparatus in polysaccharide synthesis and transport in the root cap cells of wheat, *Biochem. J.* **98**:159.

Novikoff, A. B., and Novikoff, P. M., 1977, Cytochemical contributions to differentiating GERL from the Golgi apparatus, *Histochem. J.* **9**:525.

Novikoff, P. M., Novikoff, A. B., Quintana, N., and Hauw, J. J., 1971, Golgi apparatus, GERL, and lysosomes of neurons in rat dorsal root ganglia, studied by thick and thin section cytochemistry, *J. Cell Biol.* **50**:859.

Paradies, H. H., Goke, L., and Werz, G., 1977, On the spatial structure of plant cell wall protein: Secondary structure of cell wall protein from *Acetabularia*, *Protoplasma* **93**:249.

Parke, M., and Rayns, D. G., 1964, Studies on marine flagellates. VII. *Nephroselmis gilva* sp. nov. and some allied forms, *J. Mar. Biol. Assoc. U. K.* **44**:209.

Pickett-Heaps, J. D., 1967, The effect of colchicine on the ultrastructure of dividing plant cells, xylem wall differentiation, and distribution of cytoplasmic microtubules, *Dev. Biol.* **15**:206.

Pienaar, R. N., 1969, The fine structure of *Hymenomonas (Cricosphaera) carterae.* II. Observations on scale and coccolith production, *J. Phycol.* **5**:321.

Pienaar, R. N., 1971, Coccolith production in *Hymenomonas carterae, Protoplasma* **73**:217.

Pienaar, R. N., 1976a, The microanatomy of *Hymenomonas lacuna* sp. nov. (Haptophyceae), *J. Mar. Biol. Assoc. U. K.* **56**:1.

Pienaar, R. N., 1976b, The rhythmic production of body covering components in the Haptophycean flagellate *Hymenomonas carterae*, in: *The Mechanisms of Mineralization in the Invertebrates and Plants* (N. Watabe and K. M. Wilbur, eds.), pp. 203–229, University of South Carolina Press, Columbia.

Preston, R. D., 1964, Structural and mechanical aspects of plant cell walls with particular reference to synthesis and growth, in: *Formation of Wood in Forest Trees* (M. H. Zimmerman, ed.), pp. 169–188, Academic Press, New York.

Preston, R. D., 1971, Negative staining and cellulose microfibril size, *J. Microsc.* **93**:7.

Preston, R. D., 1974, *The Physical Biology of Plant Cell Walls*, Chapman and Hall, London.

Ray, P. M., Shininger, T., and Ray, M., 1969, Isolation of β-glucan synthetase particles from plant cells and identification with Golgi membranes, *Proc. Natl. Acad. Sci. U.S.A.* **64**:605.

Robinson, D. G., 1980, Dictyosome–endoplasmic reticulum associations in higher plant cells? A serial-section analysis, *Eur. J. Cell Biol.* **23**:22.

Robinson, D. G., and Preston, R. D., 1971, Fine structure of swarmers of *Cladophora* and *Chaetomorpha*. I. The plasmalemma and Golgi apparatus in naked swarmers, *J. Cell Sci.* **9**:581.

Robinson, D. G., and Preston, R. D., 1972, Plasmalemma structure in relation to microfibril biosynthesis in *Oocystis, Planta* **104**:234.

Romanovicz, D. K., 1981, Scale formation in flagellates, in: *Cytomorphogenesis in Plants* (O. Kiermayer, ed.), pp. 27–62, Springer-Verlag, Vienna.

Romanovicz, D. K., and Brown, R. M., Jr., 1976, Biogenesis and structure of Golgi-derived cellulosic scales in *Pleurochrysis*. II. Scale composition and supramolecular structure, *Appl. Polymer Symp.* **28**:587.

Satir, B., Schooley, C., and Satir, P., 1973, Membrane fusion in a model system, *J. Cell Biol.* **56**:153.

Shore, G., and Maclachlan, G. A., 1975, The site of cellulose synthesis: Hormone treatment alters the intracellular location of alkali-insoluble β1,4-glucan (cellulose) synthesis, *J. Cell Biol.* **64**:557.

Slaby, F., and Slaby, M., 1979, A new theory of the structure and function of the Golgi complex in protein secreting cells: Each stack of cisternae is a rotating, spiralled membrane-bound tube, *J. Cell Biol.* **83**:431a.

Van der Woude, W. J., Lembi, C. A., Morré, D. J., Kindinger, J. I., and Ordin, L., 1974, β-Glucan synthetases of plasma membrane and Golgi apparatus from onion stem, *Plant Physiol.* **54**:333.

Villemez, C. L., Jr., McNab, J. M., and Albersheim, P., 1968, Formation of plant cell wall polysaccharides, *Nature (London)* **218**:878.

Whaley, W. G., 1975, *The Golgi Apparatus* (*Cell Biology Monographs*, Vol. 2), Springer-Verlag, New York.

Whaley, W. G., Dauwalder, M., and Kephart, J., 1971, Assembly, continuity, and exchanges in certain cytoplasmic membrane systems, in: *Origin and Continuity of Cell Organelles* (J. Reinert and H. Ursprung, eds.), pp. 1–45, Springer-Verlag, New York.

Willison, J. H. M., 1976, An examination of the relationship between freeze-fractured plasmalemma and cell-wall microfibrils. *Protoplasma* **88**:187.

Willison, J. H. M., and Cocking, E. C., 1975, Microfibril synthesis at the surfaces of isolated tobacco mesophyll protoplasts, a freeze-etch study, *Protoplasma* **84**:147.

Wooding, F. B. P., 1968, Radioautographic and chemical studies of incorporation into sycamore vascular tissue walls, *J. Cell Sci.* **3**:71.

Wright, K., and Northcote, D. H., 1976, Identification of β1,4-glucan chains as a part of a fraction of slime synthesized within the dictyosomes of maize root caps, *Protoplasma* **88**:225.

8

Interaction of Cell-Wall Formation and Cell Division in Higher Plant Cells

YVES MEYER AND WERNER HERTH

1. Introduction

In the typical process of cell division in higher plant cells, there is a rather strict coupling of the sequence of nuclear division and cell-wall formation processes.

From a developmental point of view, the two phenomena are linked in time as already described by Strasburger (1888) (Fig. 1). First the cell grows by longitudinal extension, with little cell-wall deposition, followed by nuclear division. A special structure, the phragmoplast, is formed between the nuclei; the cell plate is formed in the phragmoplast by coalescence of vesicles containing cell-wall material. This new cross wall is then reinforced by deposition of the primary wall. Later, the entire cell wall may be reinforced by secondary wall deposition.

From a morphogenetic point of view, the structure responsible for the maintenance of shape is the cell wall (Prat *et al.*, 1977). Since cytokinesis presumes at least a temporary passage through a nonspherical shape, which has to be stabilized by a wall, *a priori*, it seems difficult for a higher plant cell to divide without a cell wall.

From a functional point of view, the plasmalemma of animal cells is a particularly important structure in the control of mitosis. If such phenomena exist in higher plants, it is evident that the state of the cell wall, which is in close

YVES MEYER • Faculte des Sciences Exactes et Naturelles, Universite de Perpignan, 66025 Perpignan, France. WERNER HERTH • Zellenlehre, Universitat Heidelberg, D-6900 Heidelberg, Federal Republic of Germany.

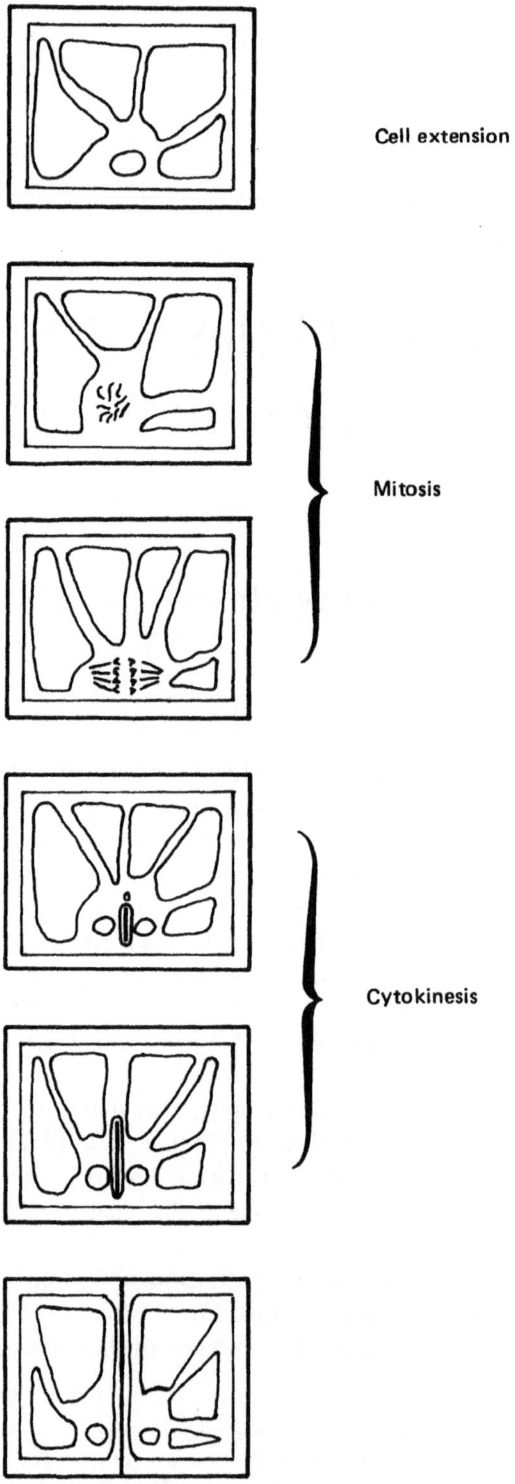

FIGURE 1. Normal division process of higher plant cells.

contact with the plasmalemma, might play a determining role in the control of the mitotic cycle.

The development of the technology of protoplast isolation and culture has recently allowed an experimental approach to these different points. In the original experiments with sugar media (e.g., Cocking, 1972), protoplasts regenerated a cell wall before nuclear division took place. It was therefore deduced that for isolated protoplasts, the cell wall is a prerequisite for nuclear division. Experiments with continuous digestion of the newly formed cell wall, leading to inhibition of nuclear division, were interpreted in the same manner (Schilde-Rentschler, 1977). However, our own experiments have meanwhile shown that the developmental sequence in isolated protoplasts is highly dependent on the composition of the culture medium (Meyer and Abel, 1975b). Indeed, with variations of the culture medium, and use of inhibitors, it was possible to use tobacco mesophyll protoplasts for detailed studies of the processes of cell-wall formation and cell division and their mutual dependence.

2. *Experimental Data with Tobacco Mesophyll Protoplasts*

2.1. *Dependence of Cell-Wall Formation on the Culture Medium*

It is relatively difficult to obtain protoplasts that regenerate a cell wall and divide. Very often, they degenerate in culture without cell-wall formation or entry into mitosis. This led to a search for less stringent culture media, and it was during such trials that we noticed that it is advantageous for protoplast survival to maintain the osmotic strength of the medium with a saline mixture (Meyer, 1974) instead of sugar or sugar-alcohols used previously (Table I). In addition,

TABLE I
Different Culture Media for Tobacco Mesophyll Protoplasts

Major nutrients (mg/liter)	RO.6	WO.6	Medium 2	Other nutrients common to both media (mg/liter)	
KNO_3	5,400	5,000	500	$FeSO_4 \cdot 7H_2O$	3.72
KCl	7,200	4,000	0		
NaCl	2,000	2,000	0	Na_2EDTA	2.78
$CaCl_2 \cdot 2H_2O$	10,800	10,800	1000	Minor nutrients according to	
KH_2PO_4	250	250	250	Murashige and Skoog (1962):	
$MgSO_4 \cdot 7H_2O$	1,500	1,500	25	Vitamins	
NH_4Cl	0	250	0	Mesoinositol	1 (g/liter)
				Thiamin-HCl	1
				Folic acid	1
Sucrose (g/liter)	1	30	10		
				Hormones	
				2, 4-D	1
Mannitol (g/liter)	0	0	130	6-BA	1
				Tween 80	0.02%

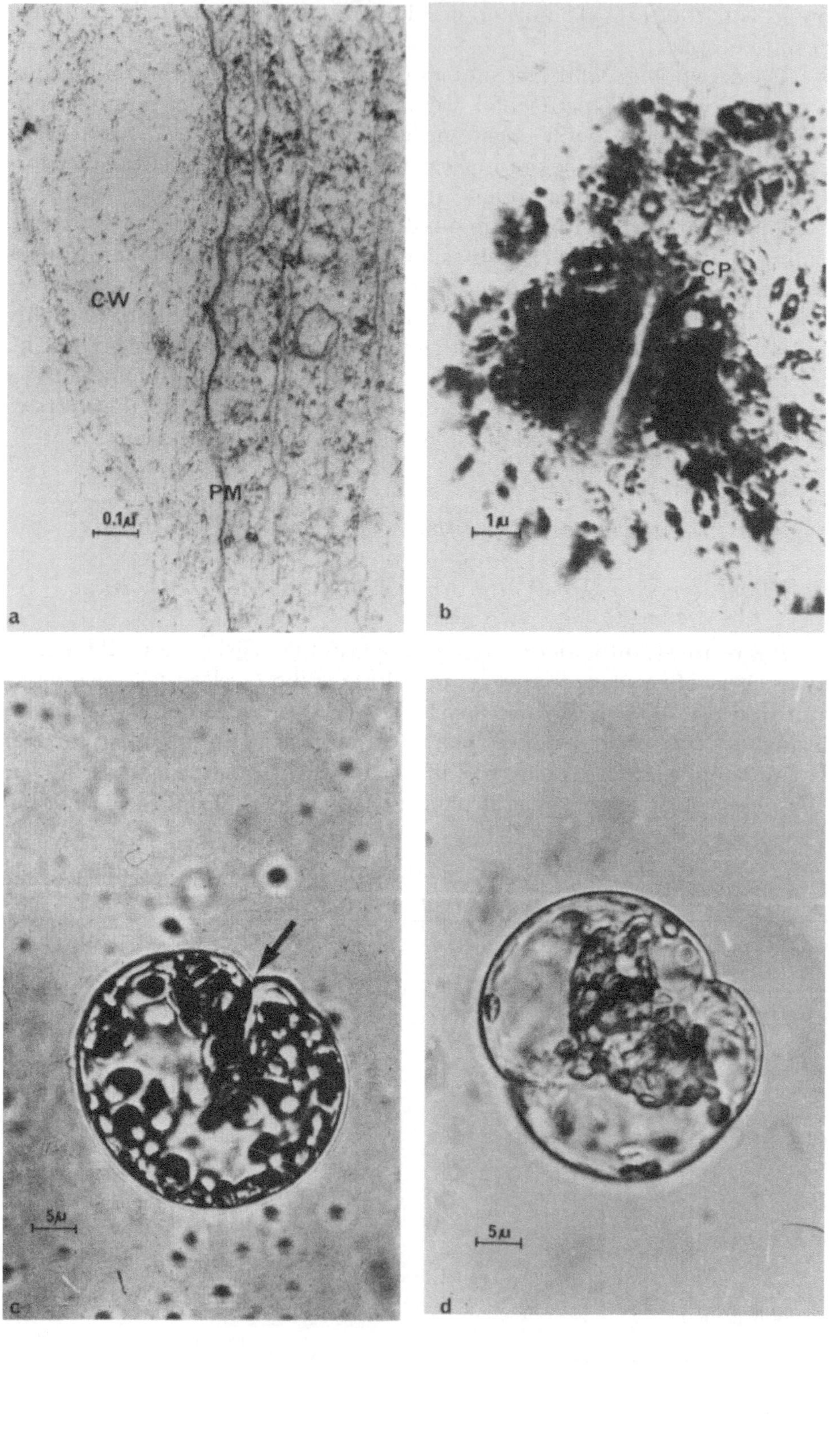
CW
PM
0.1μ
a
CP
1μ
b
5μ
c
5μ
d

this saline environment allows a simple control of wall synthesis (Meyer and Abel, 1975b); in the presence of ammonium in the saline medium (WO.6), the great majority of protoplasts reform a rigid cell wall, separable from the plasmalemma by plasmolysis (Fig. 2a). The composition of this cell wall resembles that of a typical primary wall (Herth and Meyer, 1977) (see also Chapter 4). In saline medium lacking ammonium (RO.6), the deposition of wall material is much more limited. Some material that can be stained with Calcofluor White (possibly cellulose) is present on the surface of protoplasts after several days of culture, However, plasmolysis does not show a rigid cell wall. Angular shapes that are observed indicate the presence of regions covered with wall material (Fig. 3a), and therefore rigid, alternating with naked zones. This irregular distribution of wall material is confirmed by electron microscopy.

2.2. *Effects of Saline Medium on Nuclear and Cell Division*

The presence of ammonium in the saline medium does not fundamentally affect the progress of mitosis. The entry into mitosis of protoplasts cultivated in ammonium takes place after 48 hr of culture, while those cultivated in its absence divide only after 72 hr. However, in both cases mitosis is normal. In contrast. cytokinesis is greatly affected by the medium. In medium WO.6 (+ NH_4^+), protoplasts have already regenerated a cell wall when mitosis begins. At the end of telophase, the phragmoplast develops between the nuclei (Fig. 2b), and the cell plate grows and cleaves the mother cell into two daughter cells (Fig. 2c, d). This type of cytokinesis corresponds precisely to normal cytokinesis in higher plants. In medium RO.6 (− NH_4^+), cytokinesis is very different and takes place by budding (Meyer and Abel, 1975a). The protoplast starts to form a bud often before nuclear division (Fig. 3b). The bud grows until a stage resembling figure 8 is attained (Fig. 3c). Then the nucleus divides close to the constriction, or one of the two nuclei from a previous nuclear division migrates toward the constriction. No remnants of the phragmoplast are found between the nuclei at this stage. One of the nuclei migrates into the bud; the other nucleus stays in the mother half. Then the constriction proceeds, and finally the daughter cells separate. All this takes place with almost no cell-wall deposition. The constriction region contains microtubules and some microfilaments as well as smooth and coated vesicles (Herth and Meyer, 1978). The arrangement of these elements resembles a cell-plate formation stage without deposition of cell-wall material (Fig. 3d). The processes of division by cleavage and by budding are shown schematically in Fig. 4.

FIGURE 2. Cleavage division of tobacco mesophyll protoplasts. (a) Electron micrograph showing microfibrils (cellulose) in the cell wall (CW) regenerated by the protoplast. (PM) Plasma membrane; (R) endoplasmic reticulum. (b) Cell plate (CP) formation at the and of mitosis. (c) Beginning of cleavage (arrow). (d) Two-celled stage.

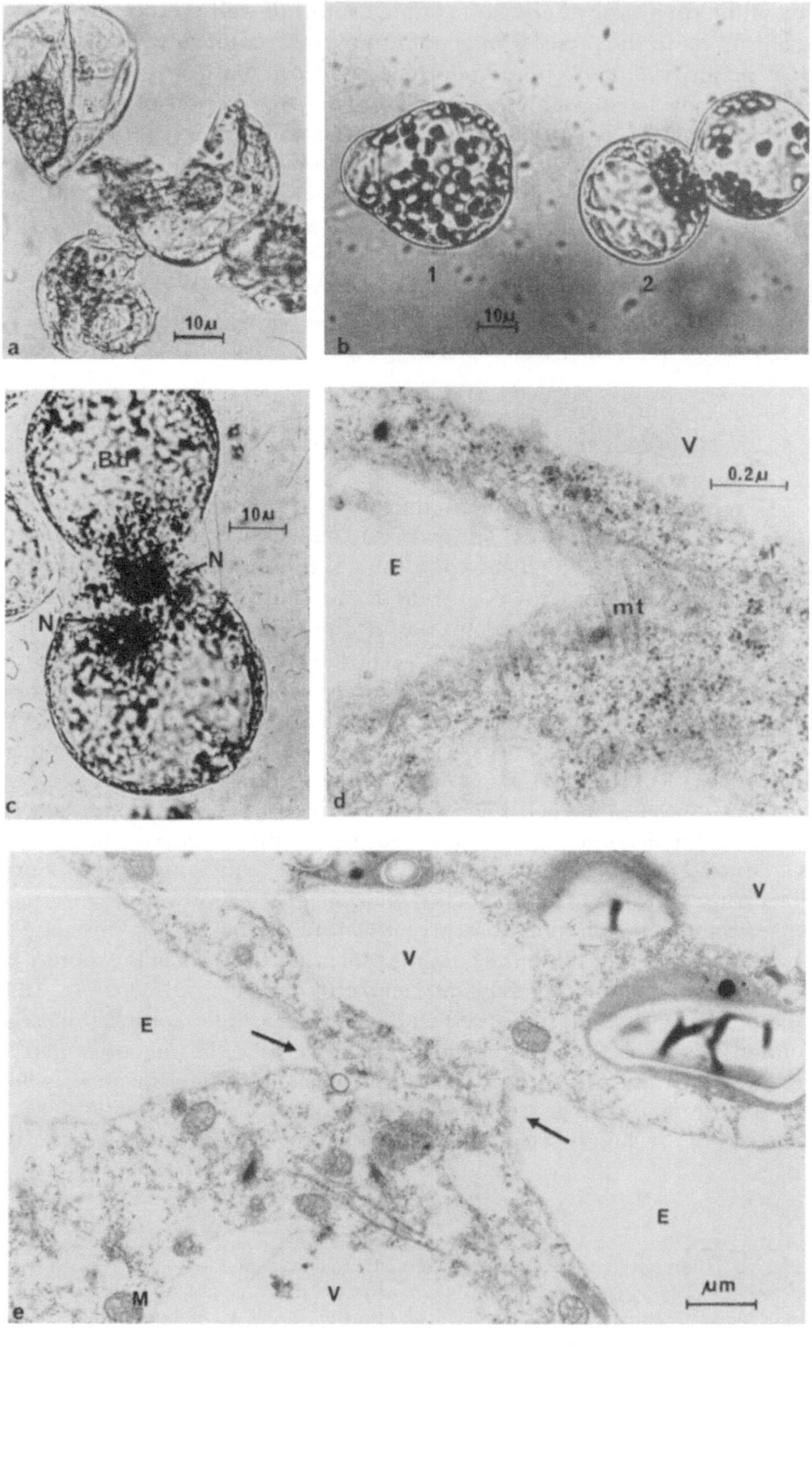
a
10μ
b
1
2
10μ
c
Bu
N
N
10μ
d
V
0.2μ
E
mt
e
V
V
V
E
E
V
M
μm

2.3. *Correlation between the Presence of a Cell Wall and Unlimited Divisions*

While cells regenerated from protoplasts are capable of unlimited division, it has never been possible to maintain a population of protoplasts in continuous culture, dividing by budding. Only two or three successive cycles are obtained with protoplasts isolated from tobacco leaf mesophyll, and only rarely is more than one cycle obtained with protoplasts isolated from tobacco cells cultivated *in vitro*. After this growth phase, cells are not irreversibly damaged, since transfer to a medium that induces cell-wall formation allows them to resume a normal rhythm of division and to form colonies. It seems, therefore, that during division by budding, certain substances present in freshly isolated protoplasts are no longer synthesized, leading to cessation of division after several cycles.

Under medium conditions slightly unfavorable to cell-wall formation, solidification of the culture medium with agar activates formation of the cell wall, probably by slowing the diffusion of wall components. One can then note that the percentage of protoplasts that form cell colonies (and that thus undergo an unlimited series of divisions) increases at the same time as the percentage of protoplasts forming a cell wall. All those colonies are from cells with walls.

A correlation therefore exists between the presence of a cell wall and the ability to undergo unlimited divisions. This coincidence does not necessarily imply a cause-and-effect relationship between the two phenomena.

2.4. *Hormonal Control*

Auxin is necessary from the beginning of culture to induce the mitotic program of tobacco mesophyll protoplasts. Cytokinin is necessary only for a later stage about 24 hr before mitosis, 4–5 hr before entry into the S phase. Auxin and cytokinin are also necessary for the formation of the cell wall. In their absence, the culture medium becomes toxic, probably due to the release of cell-wall products (Meyer and Cooke, 1979). Furthermore, the protoplasts undergo a morphological development characterized by cytoplasmic extrusions (Fig. 5). This does not reflect a completely perturbed metabolic activity, since the levels of synthesis of ribosomal RNA, polyadenylated RNA, and proteins are not noticeably affected by the presence of hormones (Cooke and Meyer, 1981), but the patterns of synthesis might be changed.

FIGURE 3. Budding division of tobacco mesophyll protoplasts. (a) Angular aspect of protoplasts plasmolyzed in 1 M NaCl after 3 days culture in medium RO.6. (b) Protoplasts (1) at the beginning of budding and (2) at a stage preceding the separation of daughter protoplasts. (c) Migration of a nucleus (N) into the bud (Bd) (orcein staining). (d) Presence of numerous microtubules (mt) close to the furrow (E). (V) Vacuole. (e) Late stage of budding process. Note the absence of wall material in the furrow (E). (V) Vacuole.

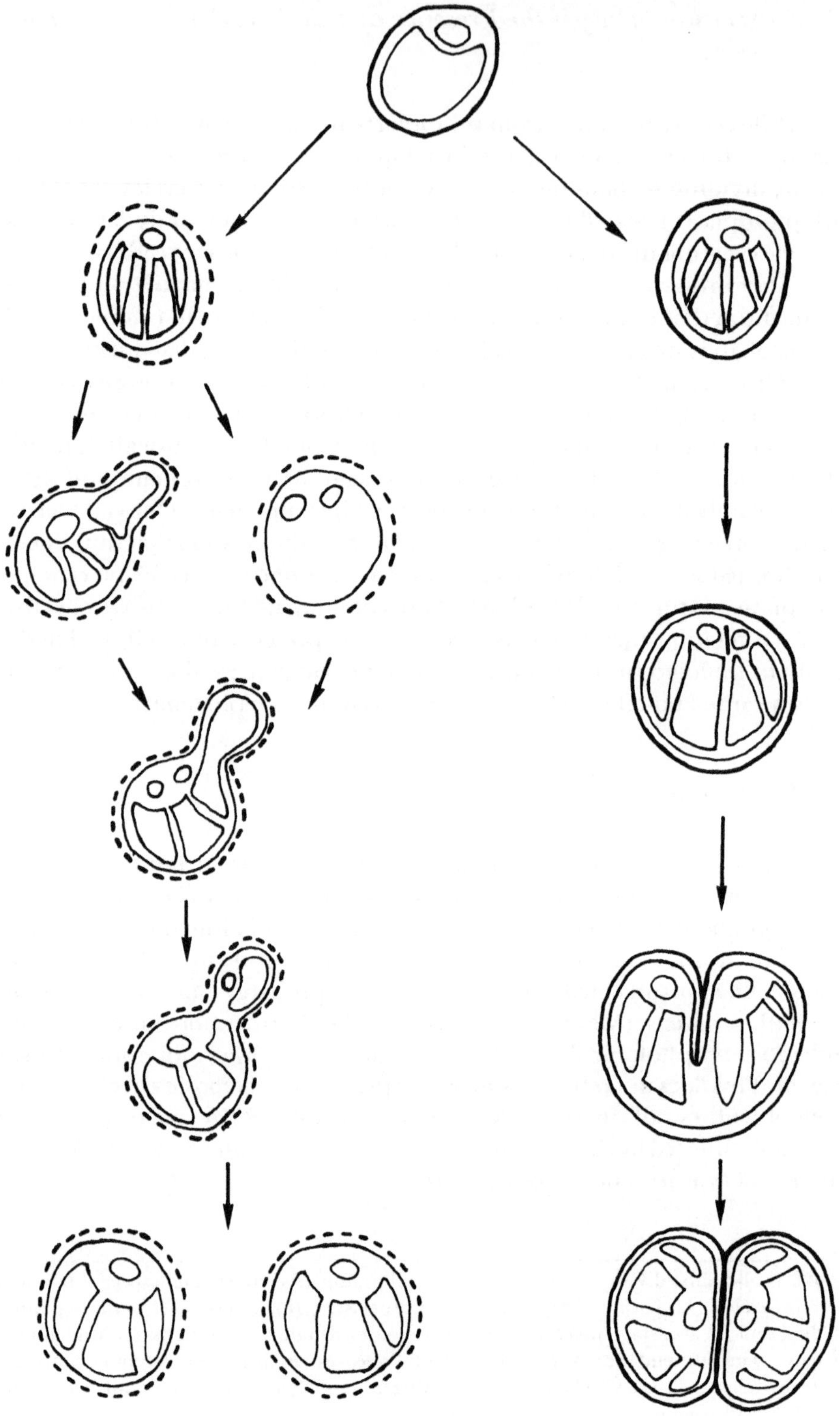

FIGURE 4. Schematic presentation of the different stages of budding or cleavage division of protoplasts in culture.

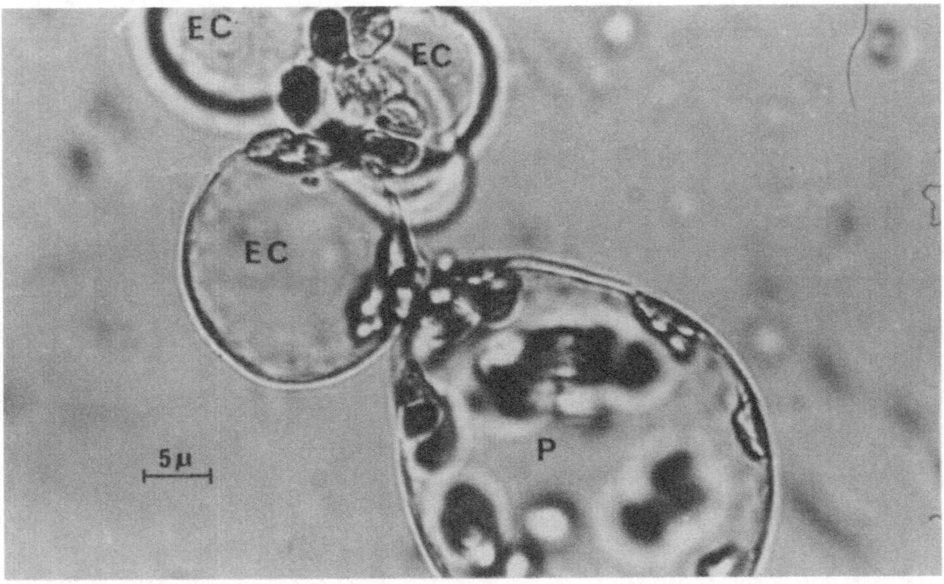

FIGURE 5. Protoplast cultivated for 5 days in medium WO.6 lacking hormones. Mitosis does not take place, but parts of the cytoplasm are extruded. (P) Protoplast; (EC) extruded cytoplasm.

2.5. *Inhibitor Studies*

We tested a variety of substances for their effects on cell-wall formation, nuclear division, and cytokinesis. The results on tobacco mesophyll protoplasts are summarized in Table II. Most substances tested turned out to have no specific effects at low concentrations; some were toxic or lethal at high concentrations. Only with cytochalasin B, colchicine, 2, 6-dichlorobenzonitrile (DCB), and coumarin was a total inhibition of cell-wall formation obtained. The effects of cytochalasin B and of colchicine were obviously accompanied by a general distortion of the cell metabolism leading to cell death with prolonged incubation. With the electron microscope, the cytochalasin- and colchicine-treated protoplasts both showed a very bad preservation of the cellular structures, indicating either damage to the cell or a different behavior toward fixation as compared to normal cells.

Coumarin had effects on cell-wall formation similar to those of DCB, but in addition partly inhibited mitosis. DCB inhibited cell-wall formation completely and did not inhibit mitosis. Treatment with this substance (chemical formula in Fig. 6) therefore led to the formation of multinucleate cell-wall free protoplasts (Fig. 7). These protoplasts look normal in ultrathin sections. After removal of the inhibitor, cell-wall formation starts within a few hours and is completed after 24 hr. Microtubules are not affected by DCB (Fig. 8). The first fibrillar cell-wall material to be formed is not obviously topographically correlated with these cortical microtubules (Fig. 8). These data clearly set apart the processes of nuclear division and cell-wall formation.

TABLE II
Effect of Various Inhibitors on Cell-Wall Formation and Mitosis

Substances tested	Concentrations tested (mg/liter)	Concentration inhibiting wall formation (mg/liter)	Concentration inhibiting mitosis (mg/liter)	Remarks
2,6-Dichlorobenzonitrile	0.1–10	1	10	—
Coumarin	10–1,000	200	(partial inhibition) 1000	—
Cytochalasin B	0.1–10	1	1	See the text for details.
Colchicine	100–10,000	2000	(total inhibition) 200	—
Caffeine	10–1,000	100	100	General inhibition of development.
Chlorogenic acid	10–500	50	50	At noninhibiting concentrations, black particles (possibly lignin) are produced.
Abscissic acid	0.1–20	None	None	—
Ethephon	10–500	200	200	General inhibition of development.
Cyclocel	10–500	200	200	General inhibition of development.
Morphactin	0.1–10	None	None	—
Gibberellic acid A_3	0.1–50	None	None	—
Fusicoccin	1–50	None	None	—
Chlorbufam	0.1–10	None	None	—
Fluorodeoxyuridine	0.025–25	0.25	0.25	General inhibition of development, reversible by 10^{-3} M thymidine.
Hadacidin	1–100	10	10	General inhibition of development, reversible by 10^{-3} M adenosine.
Benzimidazol	10–500	500	500	General inhibition of development; with 100 mg/liter, the protoplasts became cytokinin-independent.
Theophyllin	10–1,000	500	500	General inhibition of development; with 200 mg/liter, the protoplasts became cytokinin-independent.

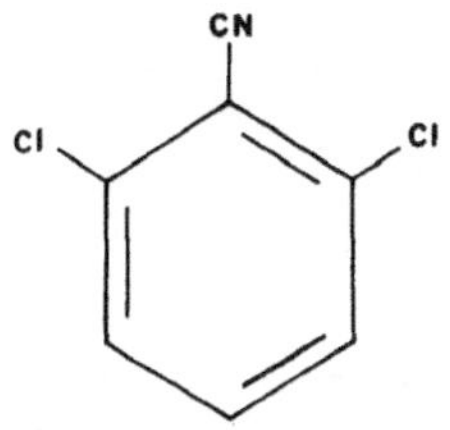

FIGURE 6. Chemical formula of 2,6-dichlorobenzonitrile (dichlobenil).

3. Discussion

The inhibition of cell-wall formation described above corresponds roughly with literature data on the effects of coumarin on cell-wall regeneration of tobacco and *Mougeotia* protoplasts (Harada *et al.*, 1972; Burgess and Linstead, 1977; Marchant, 1979; Marchant and Hines, 1979). It should be noted that these authors do not report nuclear division in the presence of coumarin, while we observe that 10–30% of tobacco protoplasts cultivated under our conditions undergo mitosis, and thus become polynucleate. Probably the inhibition of mitosis is due to a nonspecific effect of coumarin on the metabolic state of the

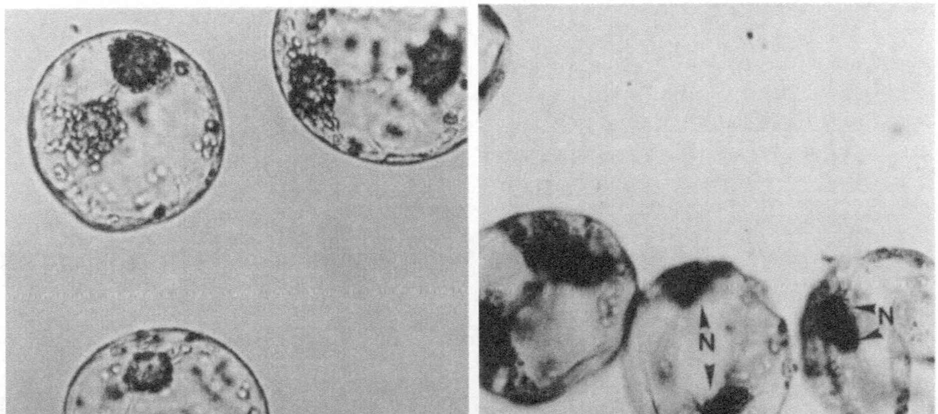

FIGURE 7. Protoplasts cultivated in the presence of dichlobenil. The regeneration of the cell wall as well as the formation of the cell plate are inhibited, whereas mitosis is not disturbed, leading to polynucleate protoplasts. (a) Living cells. (b) Cells fixed with glutaraldehyde and stained with orcein. (N) Nucleus.

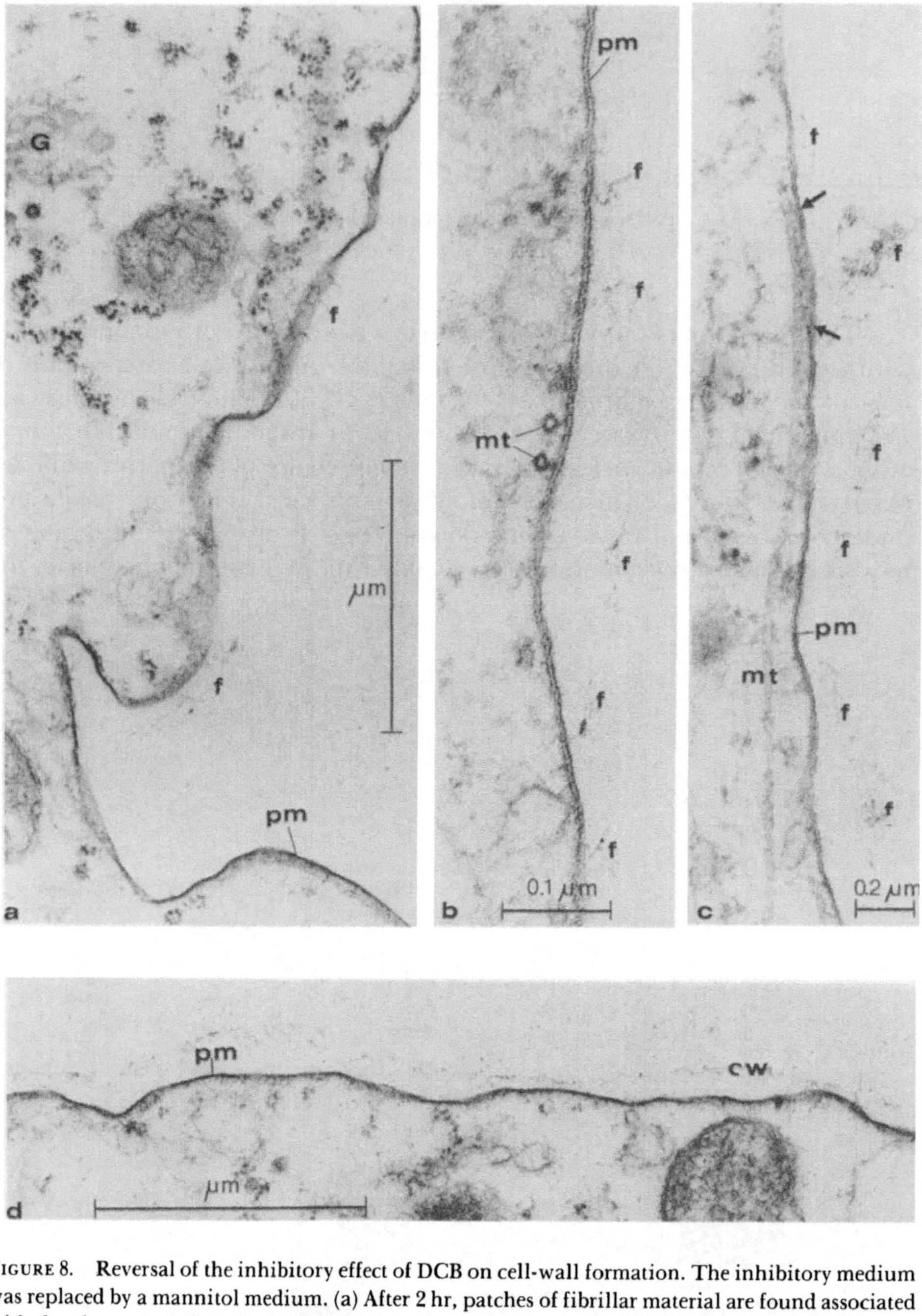

FIGURE 8. Reversal of the inhibitory effect of DCB on cell-wall formation. The inhibitory medium was replaced by a mannitol medium. (a) After 2 hr, patches of fibrillar material are found associated with the plasma membrane (pm). (G) Dictyosome. ×80,000. (b, c) After 6 hr of reversal, the fibrillar material is found in patches rather regularly over the cell surface. Section (b) is normal to a pair of cortical microtubules (mt); section (c) is parallel to a cortical microtubule. Between arrows in (c): zone of close contact between cortical microtubule and plasma membrane. (b)×200,000; (c)×100,000. (d) The newly synthesized cell wall after 16 hr of reversal forms a coherent cell wall (cw) parallel to the plasma membrane. Such protoplasts are osmotically stable. ×50,000.

protoplasts that is minimized in our saline medium. The use of dichlobenil has allowed the resolution of this ambiguity.

Until now, no other potent inhibitors of cell-wall formation in protoplasts have been discovered. In most of these studies concentrating on cell-wall formation, the question of a coupling with nuclear and cell division has not been investigated. Fowke and co-workers have performed detailed ultrastructural analysis (Fowke *et al.*, 1974, 1975) describing nuclear division, phragmoplast structure, and cytokinesis in protoplasts, as well as the presence or absence of the cell wall. These authors have interpreted the data as evidence for a normal process of cell division resembling typical higher plant cell division. However, Fowke *et al.* (1975), and Pearce and Cocking (1973) also suspected that the process may perhaps proceed in the absence of a complete cell wall.

Multinucleate protoplasts have sometimes been described (Reinert and Hellmann, 1971), but it has not been definitively established whether they were formed by mitosis or by fusion. It now seems clear that in some of these cases, nuclear division must also have occurred in the absence of cytokinesis.

The relationship between cell-wall formation and nuclear division has been studied in detail only by Schilde-Rentschler (1977). By culturing tobacco leaf protoplasts in the presence of purified cellulase, she shows that continuous digestion of the cell wall blocks the appearance of mitosis. The inhibition of cellulolytic activity by cellobiose reestablishes cell-wall formation and protoplast mitotic activity, which clearly demonstrates that the antimitotic action of cellulase is not due to a toxic contaminant. These results have been interpreted as showing that mitosis requires the presence of the cell wall. This interpretation is obviously incompatible with the obtention of polynucleate protoplasts in the presence of dichlobenil or coumarin. The results of Schilde-Rentschler could be due either to an intoxification of the medium following cell-wall degradation, comparable to that observed during hormonal deficiency, or to a reorientation of protoplast metabolism toward cell-wall synthesis, in order to compensate for the continuous destruction of the cell wall to the detriment of mitotic metabolism.

The clear-cut correlation of nuclear division, cytokinesis, and cell-wall formation in typical higher plant cells has natural exceptions:

Naked cells of certain endosperms undergo nuclear division without cell-wall formation; later, cell-wall formation may take place without nuclear division (Strasburger, 1888). In endosperm, lateral wall ingrowth may even occur in the absence of any phragmoplastlike structures (Schulz and Jensen, 1977). In pollen meiosis, a callosic wall is formed before division takes place, the pollen wall being formed only after the two divisions have been completed (Heslop-Harrison, 1964). In lactifer development of Euphorbiaceae, giant polynucleate cells are formed with continuous growth of the lactifer not directly related to nuclear divisions and phragmoplasts (Mahlberg and Sabharwal, 1968). In many cases, cells performing special functions become polyploid and grow to special forms, e.g., embryo suspensors, some hair cells, some secretory cells such as maize calyptra cells. These exceptions already demonstrate the normal sequence of events in cell division to be only one special program that may be altered in the case of special differentiations.

With inhibitors, namely, colchicine (Umetsu *et al.*, 1975a), cytochalasin B, DCB (Hogetsu *et al.*, 1974), and coumarin (Umetsu *et al.*, 1975b), cell-wall formation and cell division in higher plants are all susceptible to the same substances as protoplasts.

In addition, phragmoplast formation is sensitive to the action of caffeine (Röper, 1976; Lasselain, 1978), deoxyguanosine, and methyl-3-hydroxy-6-quinazoline dione 2, 4 (Deyson, 1969; Brulfert *et al.*, 1974). Caffeine does not inhibit cell-wall formation by protoplasts. In effect, they are stable only at high concentrations of calcium and magnesium, which are anatagonists of caffeine (Becerra and Lopez-Saez, 1978). To our knowledge, the effect of deoxyguanosine and of methyl-3-hydroxy-6-quinazoline dione 2, 4 on protoplasts has not been investigated.

4. Conclusions

The results presented in this review allow us to discuss the different points raised in the introduction.

First, the succession of events—mitosis, cytokinesis, cell-wall formation—which take place during normal cell division in higher plant cells is only one possible program. The events are not mechanically linked because, under certain conditions, it is possible to uncouple them and to obtain mitosis without cytokinesis or cytokinesis and cell-wall formation independent of mitosis. Protoplasts dividing by budding even provide an example wherein cytokinesis (budding) begins before mitosis. Perhaps there exists some kind of signal feedback mechanism to be able to perform the normal program, but the single steps of this program, such as nuclear division, phragmoplast formation, and cell-plate and cell-wall formation, seem, in principle, to be independent processes that may also be combined in another programmatic sequence, or may occur independent of each other.

The second point is that of the maintenance of cell shape. The prerequisite of the growth of higher plant cells is turgor. The cell contains a vacuole with a high solute concentration, hypertonic in relation to the medium, which confers a tendency to extension that is opposed by the cell wall. The shape of a plant cell containing a hypertonic vacuole is thus that of its cell wall. The obtention of cytokinesis by budding shows that higher plant cells, like animal and many lower plant cells, are capable of directly determining their shape in media in which the vacuole is not very hypertonic.

The third point is that of a possible control of mitosis by the cell wall via the plasmalemma. For multicellar animals, there exists a complex system of cell–cell interactions dependent on membrane receptors. For higher-plant cells, reactions involving the plasmalemma have been demonstrated using two fungal toxins, helminthosporoside and fusicoccin (Strobel and Hess, 1974; Dohrmann *et al.*, 1977). In both cases, receptors in the plasmalemma have been described, and Strobel suggests that the binding of helminthosporoside to receptors is sufficient to induce a conformational change in the plasmalemma that initiates the intracellular events of the host response, without the need for penetration of the toxin. Further, Albersheim and Anderson-Prouty (1975) develop the hypothesis that the majority of host-parasite relations can be triggered by a stimulation of the host plasmalemma by the cell wall or by polysaccharides released from the cell wall of the parasite. Such relationships between the plant-cell plasmalemma and its own cell wall have not been described, but are extremely likely. Do they play a role in the control of mitosis? Lloyd (1978) notes that the frequency of division in culture of mesophyll protoplasts (90–95% under good conditions) is superior to that of enzymatically isolated mesophyll cells (10–20%) and considers that this indicates that the mesophyll cell wall is responsible for their mitotic cessation. While this interpretation is attractive, it probably does not correspond to the natural situation. If tobacco mesophyll cells divide poorly in culture, this is because they are injured during isolation. In effect, protoplasts are isolated using a mixture of cellulase and pectinase, while cells are isolated using only pectinase at concentrations 10–50 times higher. It is known that pectinases, even when purified, are extremely toxic to plant cells. Further, other mesophyll cells isolated mechanically (asparagus, *Cattlaya*) divide well in culture as well as tissue disks [potato, Jerusalem artichoke (for a review, see Kahl, 1974)]. It therefore appears that the inhibition of mitosis in mature higher-plant tissues is not maintained by a defined structure of their cell wall. Obviously, this does not exclude the existence of a finer control mechanism between mitosis and the cell wall.

Note added in proof: The suggested existence of mechanisms of control of metabolism by cell wall polysaccharides recently has received decisive experimental confirmation. Bishop *et al.* (*Proc. Natl. Acad. Sci. USA* **78**:3536–3540, 1981) and Ryan *et al.* (*Plant Physiol.* **68**:616–618, 1981) have characterized the elusive "wound hormone" as being a group of polysaccharides liberated from the wall during wounding.

Acknowledgments

We should like to thank Dr. Richard Cooke for translating this chapter and Springer-Verlag for permission to reproduce the following material: Figures 2b, 3b, 3c, and 3e, previously published in Herth and Meyer (1978), and Table I and Fig. 7a, in Meyer and Herth (1978).

References

Albersheim, P., and Anderson-Prouty, A., 1975, Carbohydrates, proteins cell surface, and the biochemistry of pathogenesis, *Annu. Rev. Plant Physiol.* **26**:31–52.

Becerra, J., and Lopez-Saez, J. F., 1978, Effects of caffeine, calcium and magnesium on plant cytokinesis, *Exp. Cell Res.* **111**:301–308.

Brulfert, A., Clain, E., and Deysson, G., 1974, Deoxyguanosine, a potent cytokinesis inhibitor in plant cells, *Experientia* **30**:1010.

Burgess, J., and Lindstead, P. J., 1977, Coumarin inhibition of microfibril formation at the surface of cultured protoplasts, *Planta* **133**:267–273.

Cocking, E. C., 1972, Plant cell protoplasts—isolation and development, *Annu. Rev. Plant Physiol.* **23**:29–50.

Cooke, R., and Meyer, Y., 1981, Hormonal control of tobacco protoplast nucleic acid metabolism during *in vitro* culture, *Planta* **52**:1–7.

Deyson, G., 1969, Sur un nouvel inhibiteur de la cytodienese des cellules végétales, la methyl-3-hydroxy-6-quinazoline dione 2–4, *C. R. Seances Soc. Biol. Paris* **163**:37.

Dohrmann, U., Hertel, R., Pesci, P., Cocucci, S. M., Marre, E., Randazzo, G., and Ballio, A., 1977, Localization of "in vitro" binding of the fungal toxin fusicoccin to plasma-membrane rich fractions from corn coleoptile, *Plant Sci. Lett.* **9**:291–299.

Fowke, L. C., Bech-Hausen, L. W., Constabel, F., and Gamborg, O. L., 1974, A comparative study on the ultrastructure of cultured cells and protoplasts of soybean during cell division, *Protoplasma* **81**:189–203.

Fowke, L. C., Bech-Hausen, L. W., Gamborg, O. L., and Constabel, F., 1975, Electron microscope observations of mitosis and cytokinesis in multinucleate protoplasts of soybean, *J. Cell Sci.* **18**:491–507.

Harada, H., Ohyama, K., and Cheruel, J., 1972, Effects of coumarin and other factors on the modification of form and growth of isolated mesophyll cells, *Z. Pflanzenphysiol.* **66**:307–324.

Herth, W., and Meyer, Y., 1977, Ultrastructural and chemical analysis of the wall fibrils synthesized by tobacco mesophyll protoplasts, *Biol. Cellulaire* **30**:33–40.

Herth, W., and Meyer, Y., 1978, Cytology of budding and cleavage in tobacco mesophyll protoplasts cultivated in saline medium, *Planta* **142**:11–21.

Heslop-Harrison, J., 1964, Cell walls, cell membrane and protoplasmic connections during meiosis and pollen development, in: *Pollen Physiology and Fertilization* (H. F. Linskens, ed.), pp. 39–47, North-Holland, Amsterdam.

Hogetsu, T., Shibaoka, H., and Shimokoriyama, M., 1974, Involvement of cellulose synthesis in actions of gibberellin and kinetin on cell expansion: 2, 6-Dichlorobenzonitrile as a new cellulose-synthesis inhibitor, *Plant Cell Physiol.* **15**:389–393.

Kahl, G., 1974, Metabolism in plant storage tissue slices, *Bot. Rev.* **40**:263–314.

Lasselain, M. J., 1978, Contribution à l'etude du mécanisme de la cytodièrèse chez les végétaux supérierus: Évolution des constituants du phragmoplaste dan les conditions normales et sous l'influence d'inhibiteurs de divers types. Doctoral Thesis Faculty of Pharmacology, Paris.

Lloyd, C., 1978, The first division, *Nature (London)* **276**:562–563.

Mahlberg, P. G., and Sabharwal, P., 1968, Origin and early development of nonarticulated lactifers in embryos of *Euphorbia marginata*, *Am. J. Bot.* **55**:375–381.

Marchant, H. J., 1979, Microtubules, cell wall deposition and the determination of plant cell shape, *Nature (London)* **278**:167–168.

Marchant, H. J., and Hines, E. R., 1979, The role of microtubules and cell-wall deposition in elongation of regenerating protoplasts of Mougeotia, *Planta* **146**:41–48.

Meyer, Y., 1974, Isolation and culture of tobacco mesophyll protoplasts using a saline medium, *Protoplasma* **81**:363–372.

Meyer, Y., and Abel, W. O., 1975a, Importance of the wall for cell division and in the activity of the cytoplasm in cultured tobacco protoplasts, *Planta* **123**:33–40.

Meyer, Y., and Abel, W. O., 1975b, Budding and cleavage division of tobacco mesophyll protoplasts in relation to pseudo-wall and wall formation, *Planta* **125**:1–13.

Meyer, Y., and Cooke, R., 1979, Time course of hormonal control of the first mitosis in tobacco mesophyll protoplasts cultivated *in vitro*, *Planta* **147**:181–185.

Meyer, Y., and Herth, W., 1978, Chemical inhibition of cell wall formation and cytokinesis, but not of nuclear division, in protoplasts of *Nicotiana tabacum* cultivated *in vitro*, *Planta* **142**:253–262.

Murashige T., Skoog F., 1962, A revised medium for rapid growth and bioassays with tobacco tissue cultures. *Physiol. Plantarum* **15**:473–497.

Pearce, R. S., and Cocking, E. C., 1973, Behaviour in culture of isolated protoplasts from "Paul's Scarlet" rose suspension culture cells, *Protoplasma* **77**:165–180.

Prat, R., Vian, B., Reis, D., and Roland, J. C., 1977, Evolution of internal pressure, vacuolisation and membrane flow, during cell growth in mung bean hypocotyl, *Biol. Cellulaire* **28**:269–280.

Reinert, J., and Hellmann, S., 1971, Mechanism of the formation of polynuclear protoplasts from cells of higher plants, *Naturwissenschaften* **58**:419.

Röper, W., 1976, Nuclear fusion and irregular cytokinesis in binucleate and tetraploid cells of *Vicia faba* after caffeine treatment, *Experientia* **32**:1260.

Schilde-Rentschler, L., 1977, Role of the wall in the ability of tobacco protoplasts to form callus, *Planta* **135**:177–181.

Schulz, P., and Jensen, W. S., 1977, Cotton embryogenesis: The early development of the free nuclear endosperm, *Am. J. Bot.* **64**:384–394.

Strasburger, E., 1888, *Uber Kern and Zellteilung in Pflanzenreich nebst einem Anhang über Befruchtung*, 165, Fischer, Jena, Germany.

Strobel, G. A., and Hess, W. N., 1974, Evidence for the presence of the toxin binding protein on the plasma membrane of sugarcane cells, *Proc. Natl. Acad. Sci. U.S.A.* **71**:1413–1417.

Umetsu, N., Ojima, K., and Matsuka, K., 1975a, Enhancement of cell separation by colchicine in cell suspension cultures of soybean, *Planta* **125**:197–200.

Umetsu, W., Satoh, S., and Matsuda, K., 1975b, Effects of 2,6-dichlorobenzonitrile on suspension-cultured soybean cells, *Plant Cell Physiol.* **17**:1071–1075.

9

Callose-Deposit Formation in Radish Root Hairs

Kay M. Cooper

1. Introduction

Callose was identified as a substance of characteristic staining and solubility properties more than a century ago (Nägeli, 1861; see the review by Currier, 1957). Mangin (1890, 1892) emphasized that there is a wide distribution of this plant-cell constituent among various cells and tissues. The long list of callose deposition sites includes fungal and algal cell walls, pollen tube walls and as plugs in pollen tubes, sieve cells, guard-cell walls, cell plates, freely growing walls of endosperm, plasmodesmata, pit fields, and root hairs. Much callose formation is undoubtedly a part of normal cell differentiation or development. Perhaps the most conspicuous synthesis of callose is that which occurs rapidly on stimulation of plant cells by external factors. "Wound callose" is a phrase coined to describe callose synthesized in response to mechanical injury to plant tissue (Currier, 1957) and can describe equally well that callose formed on chemical damage, in response to fungal, bacterial, or viral infection, or in response to environmental stress.

Callose of grape sieve tubes was identified biochemically by Aspinall and Kessler (1957) as mostly β1,3-glucan. The sugar nucleotide uridine diphosphate (UDP) *D*-glucose has been shown to donate its sugar moiety *in vitro* to β1,3-glucan as well as to β1,4-glucan, glycolipids, and other plant polymers (see the reviews by Hassid, 1970; Shafizadeh and McGinnis, 1972; Nikaido and Hassid, 1972). It has been assumed by some (Shore and Maclachlan, 1975; Anderson and Ray, 1978) that exogenously supplied UDP-glucose donates its sugar moiety to β-glucans at the cell surface of intact plant cells. An attempt by this author to

Kay M. Cooper • Department of Botany, University of North Carolina at Chapel Hill, Chapel Hill, North Carolina 27514.

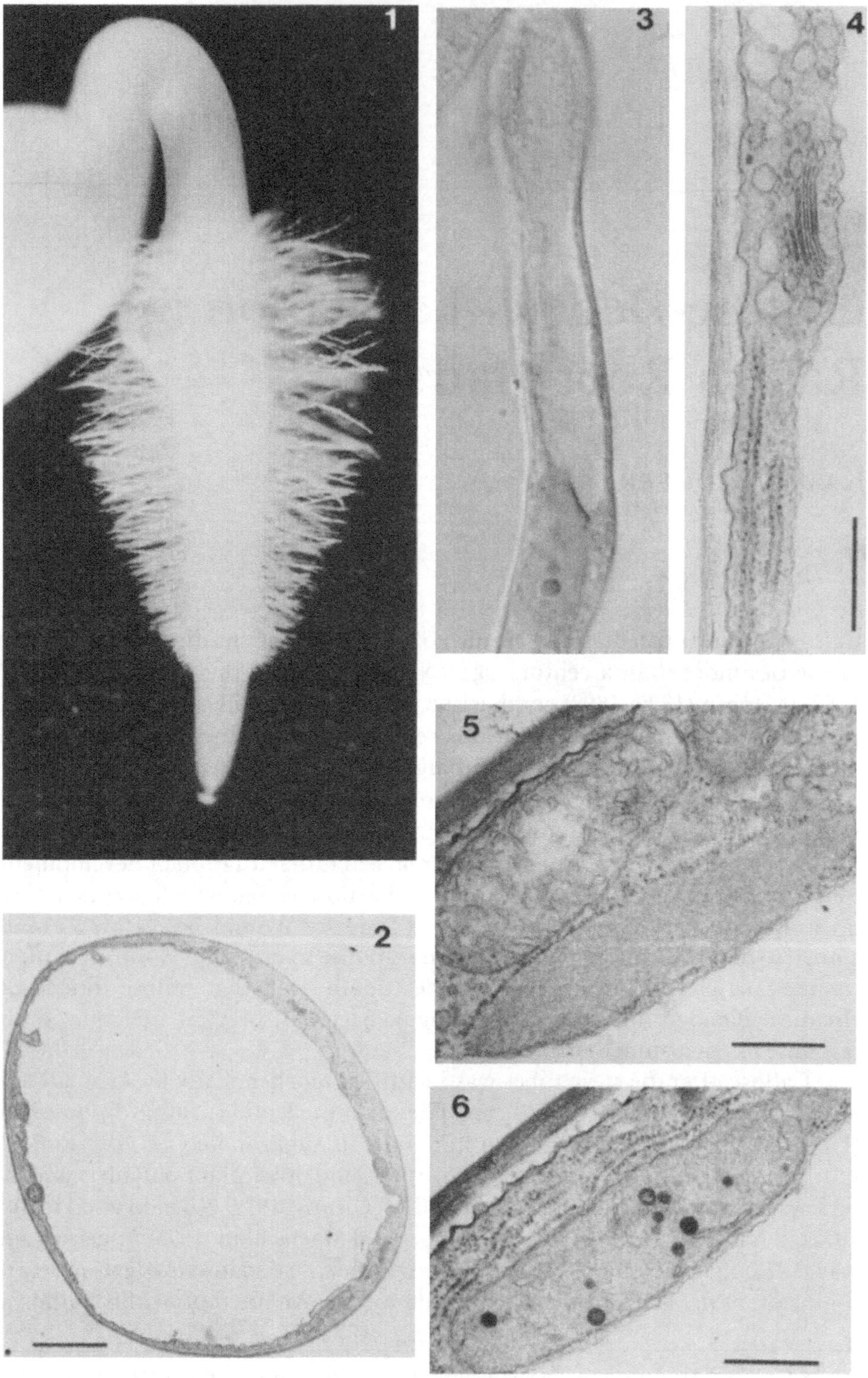

FIGURE 1. Intact 2-day-old radish seedling, with cone of elongating root hairs.

identify UDP-glucose-utilizing sites within intact *Rhaphanus sativus* roots by means of autoradiography was not successful, because little radioactive label was incorporated from the supplied UDP-[^{3}H]glucose by undamaged cells. It was noted, however, that many of the radish root hairs exhibited extremely heavy labeling by this radioactive sugar nucleotide. Furthermore, incubation of radish roots in UDP-glucose induced the formation of substance(s) having the staining reaction of callose in amounts far above what is normally observed in this cell type. These phenomena were studied further to identify the sites of incorporation of glucose moieties from exogenously supplied UDP-glucose in radish root hairs and to determine the morphology of the presumptive callose regions in these cells.

2. *UDP-Glucose-Incubated Root Hairs*

2.1. *Control Root-Hair Ultrastructure*

Root hairs over a certain length are highly vacuolate cells and therefore present difficult fixation problems (Leech *et al.*, 1963). Newcomb and Bonnett (1965) studied microtubule and microfibril orientation in the earliest stages of root-hair formation in *Raphanus sativus* and noted the difficulty of fixation of the older root hairs. Use of a primary fixative of low osmolarity for an extended period of time results in fairly good preservation of the ultrastructure of undamaged root hairs, as well as good preservation of other large vacuolate cells of the radish root (Cooper, 1978). Root hairs examined in this study were 1–2 mm long and were all more mature than those reported by Newcomb and Bonnett (1965) and Bonnett and Newcomb (1966). Older root hairs of this length are still elongating and form a "cone" around the root (Fig. 1). The unstressed root hairs contain large central vacuoles, with a thin layer of cytoplasm beneath the cell wall (Fig. 2). The nuclei of older hairs migrate out of the bases of the root hairs and are positioned at a rather uniform distance behind the growing root hair tips (Fig. 3), as noted for elongating root hairs by Haberlandt in 1887. Cell surfaces of mature unstressed large vacuolated root hairs are typically smooth and in close association with the cell walls (Fig. 4), although regions of cytoplasmic surface irregularities are also found, presumably in regions of ongoing cell-wall synthesis. Abundant dictyosomes with associated vesicles (Fig. 4), mitochondria (Fig. 5), and plastids with a sparse lamellar system (Fig. 6) may be observed in these cells, as well as axially aligned microtubules. A striking feature of these root

FIGURES 2–6. Morphology of control root hairs. Fig. 2. Cross section of root-hair region with thin layer of cytoplasm surrounding the large central vacuole. Scale bar: 2 μm. Fig. 3. Elongating end of root hair with nucleus in typical position for mature root hair. Fixed in 1% glutaraldehyde. Nomarski microscopy. Fig. 4. Root hair region with thin layer of cytoplasm containing a Golgi apparatus with associated vesicles, rough endoplasmic reticulum, and a relatively smooth cytoplasmic surface. Scale bar: 0.5 μm. Fig. 5. Mitochondrion and local dilation of the rough endoplasmic reticulum with typically dense matrix. Scale bar: 0.5 μm. Fig. 6. Plastid with sparse lamellar system. Scale bar: 1 μm

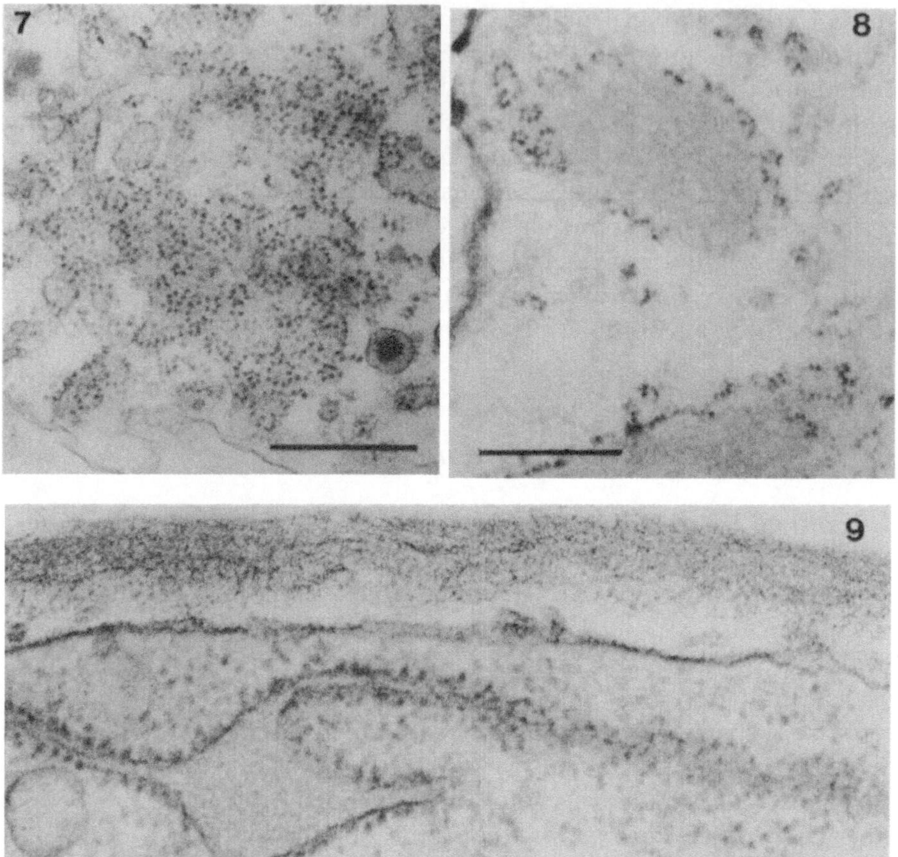

FIGURES 7–11. Morphology of control root hairs. Fig. 7. Polyribosomes on rough endoplasmic reticulum cisternae. Scale bar: 0.5 μm. Fig. 8. Local dilations of the rough endoplasmic reticulum with polyribosomes evident on their surfaces. Scale bar: 0.5 μm. Fig. 9. Local enlargement of the rough endoplasmic reticulum with connections to undilated reticulum. Scale bar: 0.2 μm. Fig. 10.

hairs is the extensive rough endoplasmic reticulum system. Polyribosomes are prominent both on the flat endoplasmic reticulum cisternae (Fig. 7) and on local endoplasmic reticulum dilations (Fig. 8), as described extensively by Bonnett and Newcomb (1965) for radish root hairs and hairless epidermal cells, as well as for cortical cells. These endoplasmic reticulum local dilations contain an electron-dense matrix in unstressed root hairs (Figs. 5, 8, and 9).

2.2. *UDP-Glucose Utilization by Damaged Root Hairs*

When root hairs are stressed by incubation in 10 mmole UDP-D-glucose or in 400 μm UDP-D-[1-^{3}H]glucose (2 mCi/ml, specific activity 4.85 Ci/mmole) in distilled water, the hairs appear to be in different stages of disorganization. Some hairs are essentially normal, with the same ultrastructure as the unstressed control root hairs. The majority of root hairs incubated in UDP-glucose are empty of cytoplasm or contain cytoplasm that is in a state of extreme disintegration. The endoplasmic reticulum forms into vesicles (Fig. 10), the mitochondria disintegrate, and the Golgi cisternae assume a whorled appearance (Fig. 11). These cells probably have suffered damage that is too severe to be repaired, with complete breaks of plasmalemma allowing contact of the cytoplasm with the external medium. This deteriorating cytoplasm is typical if dying cells (Trump and Arstila, 1971), with the endoplasmic reticulum comparable in appearance to the "mass of bubbles" seen by Mollenhauer *et al.* (1960) in excised root tips.

Most radish root hairs incubated in UDP-[^{3}H]glucose exhibit very heavy radioactive labeling of substances not removed by fixation procedures, as demonstrated by autoradiography (Fig. 12). Excessive labeling is found only in root hairs with cytoplasmic disorganization (Figs. 10, 11, and 13), while undamaged root hairs have little label (Fig. 13). Incubation of excised radish root in UDP-[^{3}H]glucose results in heavy labeling of the cut or otherwise damaged cells of all cell types (Fig. 14), with label comparable in quantity to the amount of labeling found in the damaged root hairs. Again, little label is located either in or at the surface of undamaged cells. This evidence supports the observation by Delmer *et al.* (1977) that cotton fibers require damage in order for enzymatic activity utilizing exogenously supplied UDP-glucose to occur and verifies the conclusions of Raymond *et al.* (1978) that only the exposed surfaces of cut or damaged cells utilize supplied UDP-glucose in large amounts.

Radioactive glucose label donated by UDP-[^{3}H]glucose in damaged root hairs to substances not removed by fixation is associated largely with vesicles that are often clearly of Golgi origin (Fig. 11 and 15). Label is associated to a lesser extent with vesiculated rough endoplasmic reticulum (Fig. 16). Heavy label is

◄ Autoradiogram of root hair incubated in UDP-[^{3}H]glucose for 10 min, with cytoplasmic disorganization and radioactively labeled vesicles. Scale bar: 2 μm. Fig. 11. Autoradiogram of root hair incubated in UDP-[^{3}H]glucose for 10 min, with whorled Golgi bodies (◄) having associated radioactive label. Scale bar: 1 μm.

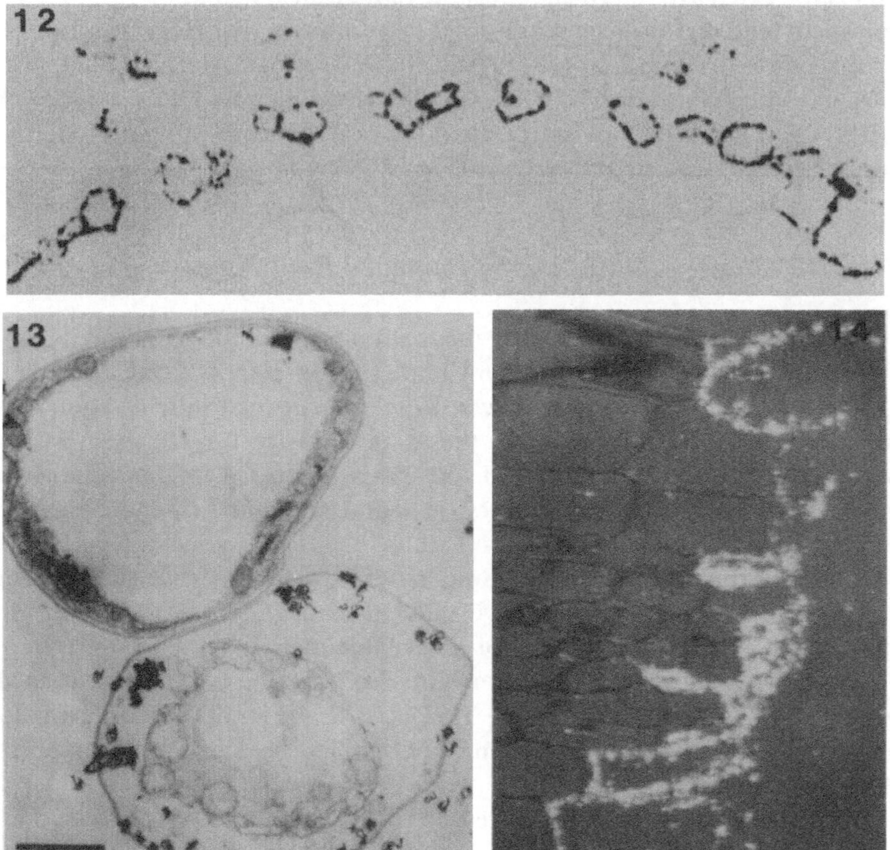

FIGURES 12–18. Root hairs incubated in UDP-[^{3}H]glucose. Fig. 12 Autoradiogram of 1-μm-thick section of radish root incubated in UDP-[^{3}H]glucose for 1 hr. with heavy radioactive label in some of the root hairs. Bright-field light microscopy. Fig. 13. Autoradiogram of root hairs incubated in UDP-[^{3}H]glucose for 10 min. Only the root hair with ultrastructural damage is radioactively labeled.

found in patches over the root-hair cell walls (Fig. 17), but whether this is due to activity by enzymes embedded in the wall or by plasma-membrane enzyme activity is not known. The synthesized radioactive polymers preserved during fixation are often membrane-bound (Fig. 18).

In vitro incubation of radish root hairs in 400 μm UDP-D-glucose plus 0.625 μm UDP-D-[U-^{14}C]glucose in 100 μl distilled water yields 70% ethanol-insoluble radioactive polymers that can be collected by centrifugation. Degradation by appropriate glucanases results in radioactive dimers that cochromatograph with laminaribiose and cellobiose, respectively, indicating the ability of these cells to synthesize glucans containing β1,3- and β1,4-linkages.

2.3. Aniline-Blue-Induced Fluorescence in Incubated Root Hairs

Staining of nonincubated control radish root hairs with aniline blue fluorochrome, followed by examination under UV light, reveals some tip callose and occasional callose deposits along the length of the root hairs. This occurrence of callose in root hairs has been noted previously (Ridgway, 1913; Roberts, 1916; Howe, 1921; Kumarasinghe and Nutman, 1977). After incubation of radish roots in UDP-glucose followed by aniline blue staining, UV light activates a homogeneous fluorescence appearing at the tips of the root hairs, with spotty fluorescence occuring along the length of the hairs (Fig. 19). Prolonged incubation in UDP-glucose results in homogeneous fluorescence along the whole length of some of the root hairs (Fig. 20). Stress in hypotonic solutions results in fluorescent material extruding from the hair tip (Fig. 21), along with other cell contents. Pelleting of this fluorescent root-hair cytoplasm reveals the fluorescent material to have a very vacuolar appearance (Fig. 22), comparable to the appearance of the disintegrating cytoplasm of UDP-glucose-incubated root hairs as seen in electron micrographs (see Fig. 10). Light microscopic autoradiography indicates a correlation of this pellet fluorescence and radioactive label donated by UDP-[^{3}H]glucose (compare Fig. 23 with Fig. 24). Exogenously supplied UDP-glucose evidently is donating its sugar moiety in damaged root hairs to glucans having a positive aniline blue staining reaction for callose. Isolated plasma-membrane fractions incubated with UDP-glucose fluoresce with aniline blue staining. Germinating pollen grains, incubated in an osmotically damaging solution of UDP-glucose, exhibit the same type of cytoplasmic staining response (Fig. 25). This phenomenon of synthesis of

Scale bar: 2 μm. Fig. 14. Autoradiogram of 2½-μm-thick section of the cut end of an excised radish root incubated in UDP-[^{3}H]glucose for 1 hr. Note that heavy radioactive label is associated only with cut or damaged cells. Polarization microscopy. Fig. 15. Autoradiogram of root hair incubated in UDP-[^{3}H]glucose for 10 min. Golgi vesicles are radioactively labeled. Scale bar: 0.5 μm. Fig. 16. Autoradiogram of root hair incubated in UDP-[^{3}H] glucose for 10 min, showing labeled vesiculated rough endoplasmic reticulum. Scale bar: 0.5 μm. Fig. 17. Autoradiogram of root hair incubated in UDP-[^{3}H]glucose for 10 min, with heavy concentration of radioactive label at the cell wall. Scale bar: 2 μm. Fig. 18. Autoradiogram of root hair incubated in UDP-[^{3}H]glucose for 1 hour. The radioactively labeled compound is apparently membrane-associated. Scale bar: 1 μm.

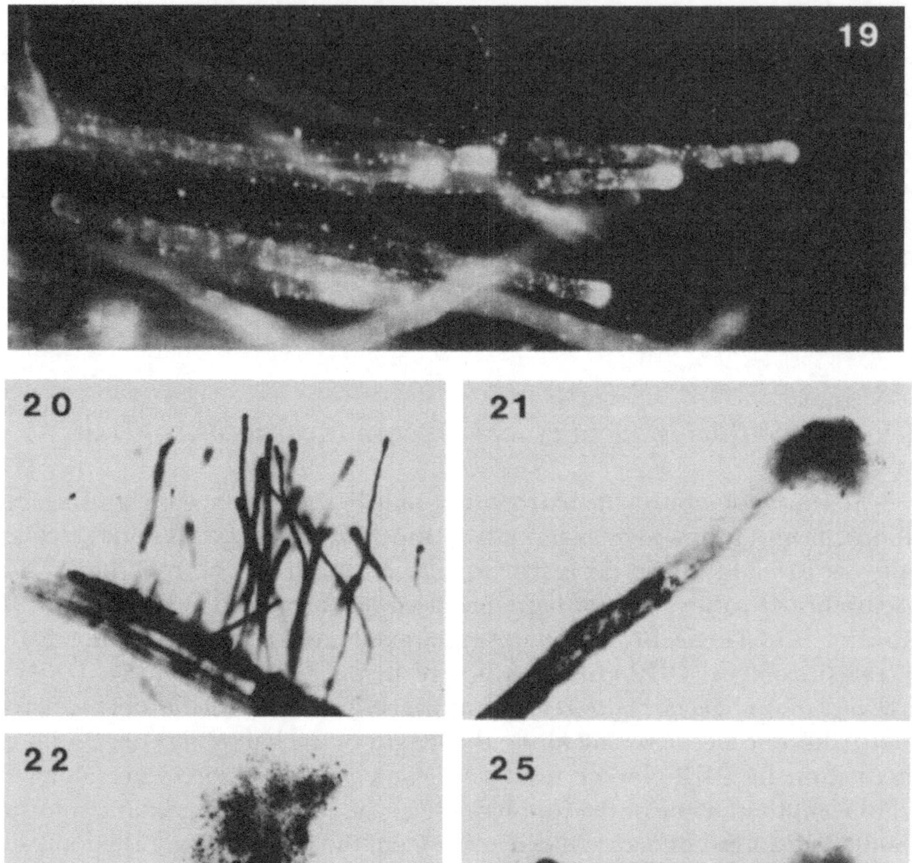

FIGURES 19–25. Material stained with aniline blue, then viewed with dark-field fluorescence microscopy (Fig. 19–23 and 25) or dark-field light microscopy (Fig. 24). Figure 19 is a positive print of a black-and-white negative. Figures 20–25 are negative contrast prints of Ektachrome color slides.

positive callose staining polymers may be a general *in vitro* response of damaged cells to UDP-glucose incubation.

2.4. Ultrastructure Changes in Intact Incubated Root Hairs

Some root hairs incubated in UDP-glucose are not severely damaged, but exhibit other ultrastructural changes, most notably the formation of deposits between the cytoplasm and the cell walls (Fig. 26). These deposits give a positive callose staining reaction (Fig. 26, inset). Little glucose from exogenously supplied UDP-[^{3}H]glucose is incorporated into these presumptive callose deposits. Apparently, the cells containing the callose deposits are not damaged to the extent that there is free access of the exogenously supplied UDP-glucose to the cytoplasmic or membrane-bound synthetases. It is possible that callose deposits in radish root hairs of this study were induced by the stress of incubation procedures, and the deposits may prevent or retard further cell damage by acting as substitute permeation barriers at lesions in the plasma membrane (Wheeler, 1974). These callose deposits are very heterogeneous with respect to electron density—unlike those described in *Rhizobium*-infected clover root hairs by Kumarasinghe and Nutman (1977)—with regions ranging from electron-opaque to electron-lucent, and contain various membranous inclusions in a matrix apparently continuous with the cell walls. Areas in these deposits also react in different degrees to cytochemical staining for periodate-reactive polysaccharides (Fig. 27) (Thiéry, 1967).

Short-term incubations of root hairs result in ultrastructural changes that give indications of how these callose deposits might be formed. After 5 min of incubation, the rough endoplasmic reticulum profiles are noticeably swollen (Fig. 28), as compared with the flat cisternae of rough endoplasmic reticulum found in unstressed control radish root hairs. The large local dilations of the rough endoplasmic reticulum found in unstressed root hairs lose their typically dense matrix, suggesting either a dilution or loss of contents (compare Fig. 28 and 29 with Fig. 8 and 9). This matrix possibly is transferred to the swollen rough endoplasmic reticulum with which the dilations are connected (Fig. 28). After incubation for 15 min, irregularly shaped smooth vesicles form in the root-hair

Fig. 19. Root hairs incubated in UDP-glucose for 10 min. The tips of the root hairs, as well as isolated regions along the length of the hairs, fluoresce bright yellow under UV light. Fig. 20. Root hairs incubated in UDP-glucose for 1 hr. The whole length of some of the root hairs fluoresces homogeneously. Fig. 21. Root hair incubated in UDP-glucose for 1 hr. Stress of the hypotonic staining solution causes the fluorescent cell contents to eject through the relatively weak tip of the root hair. Fig. 22. Cytoplasm of root hairs incubated in UDP-glucose for 1 hour. The fluorescent material has a marked vesicular appearance. Fig. 23. Autoradiogram of cytoplasm of root hair incubated in UDP-[^{3}H]glucose for 1 hr. Fluorescence microscopy reveals a brightly fluorescing area (▲). Fig. 24. Dark field view of Fig. 23, with silver grains evident over the fluorescent region (▲). Fig. 25. Germinating *Nicotiana* sp. pollen grain incubated in an osmotically damaging solution of UDP-glucose. The expelled cytoplasm fluoresces bright yellow in the same manner as the incubated root-hair cytoplasm (Fig. 21).

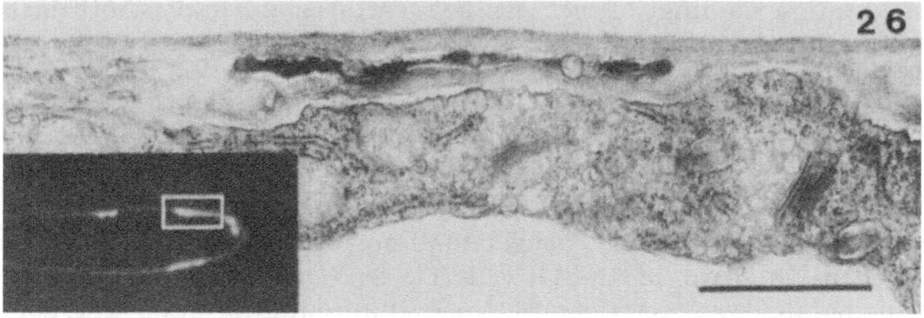

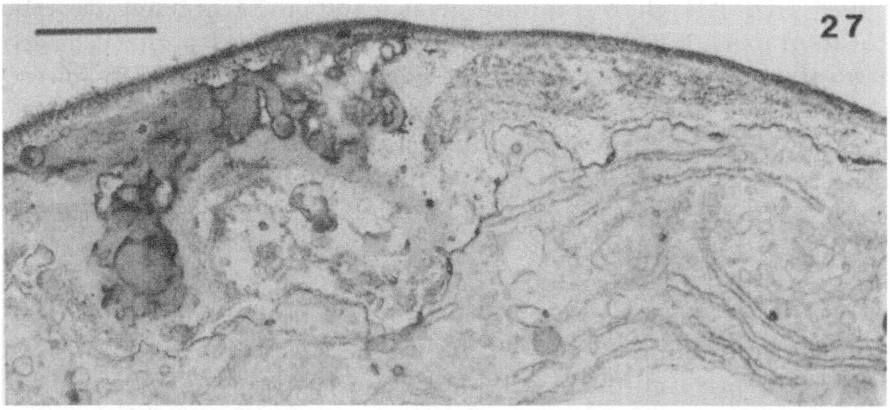

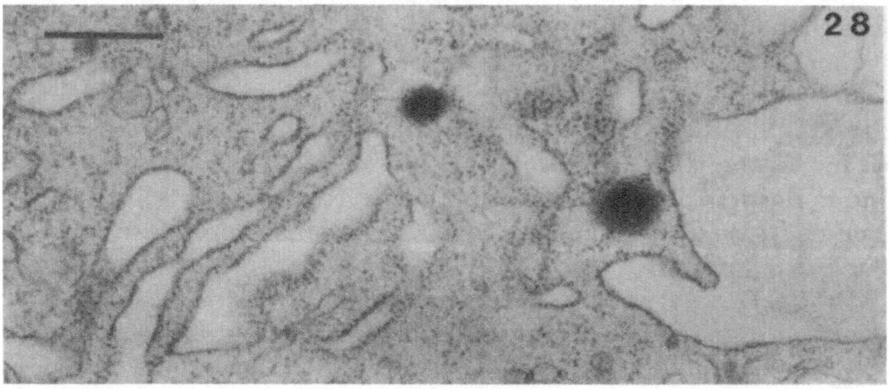

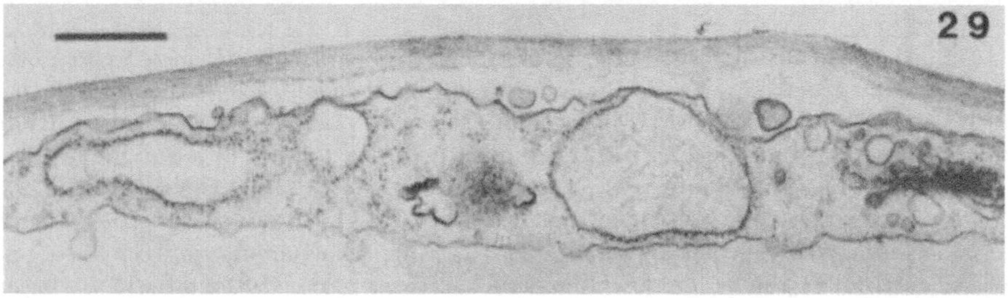

FIGURES 26–32. Morphology of intact root hairs incubated in UDP-glucose. Fig. 26. Root hair incubated in UDP-glucose for 1 hr, with callose deposits formed beneath the cell wall. Scale bar: 1 μm. *Inset*: Thick section (2½ μm) serial to that of Fig. 26, stained with aniline blue fluorochrome. The

cytoplasm. These often tubular vesicles are continuous with the rough endoplasmic reticulum (Fig. 30), as well as with membrane-bound tubular processes located at the cytoplasmic surfaces of the incubated cells in regions of forming callose deposits, as identified by aniline blue staining (Figs. 31 and 32). Similar tubular processes have been described by Newcomb and Bonnett (1965) in very young radish root hairs, with the unit membrane delimiting the tubules frequently identifiable as an extension of the plasmalemma; however, no function was postulated for them by these researchers. In older root hairs, these processes are seldom found in unstressed cells and may be part of the endoplasmic reticulum system formed to channel stored substances rapidly from the local rough endoplasmic reticulum dilations to the forming callose deposits at the cell surface.

3. Discussion

3.1. Callose-Deposit Composition

Arens (1949) first reported that water-soluble aniline blue fluoresced yellow in the presence of callose. Eschrich and Currier (1964) further established methods of identifying callose by its diachrome and fluorochrome reactions. Golden yellow fluorescence on UV irradiation following aniline blue staining has become an accepted mode for identifying the occurrence of this substance. Aspinall and Kessler (1957) chemically defined callose in *Vitis vinifera* sieve plates as β1,3-glucan containing a small amount of uronic acid residues. Since this one instance of chemical verification of callose composition, it has been widely assumed in the literature that callose, as identified by its aniline blue staining reaction, is synonomous with β1,3-glucan (Smith and McCully, 1978). Recent work questions both the specificity of aniline blue fluorochrome for β1, 3-glucans (Faulkner *et al.*, 1973; Reynolds and Dashek, 1976; Smith and McCully, 1978) and the criteria of alkali or acid solubility as a test for highly polymerized β1,3-glucans (Herth *et al.*, 1974; Heiniger and Delmer, 1977). Faulker *et al.* (1973) have demonstrated that aniline blue lacks specificity for linkage configuration (α or β) or for linkage group (-1,3-, -1,4-, -1,6-). Smith and McCully (1978), in a more detailed study on aniline blue specificity, found similar results. These authors concluded that the strong fluorescence found in such areas as sieve plates is probably due to localized structural packing of wall polymers, rather than to specific linkage. Evidently, the trend to assume aniline-blue-identified callose to be β1,3-glucan can lead to inaccuracies. In 1957, Currier (1957) suggested that

outlined fluorescent area corresponds to the region of callose deposition shown in Fig. 26. Dark-field fluorescence microscopy. Fig. 27. Callose deposit in a root hair incubated in UDP-glucose in 1 hr, stained cytochemically for periodate-reactive polysaccharides, showing a variety of staining intensities. Scale bar: 1 μm. Fig. 28 and 29. Root hairs incubated in UDP-glucose for 5 min. The usually flat cisternae of the rough endoplasmic reticulum dilate, and the matrix of the endoplasmic reticulum enlargements becomes less dense than in the control root hairs. Scale bars: 0.5 μm.

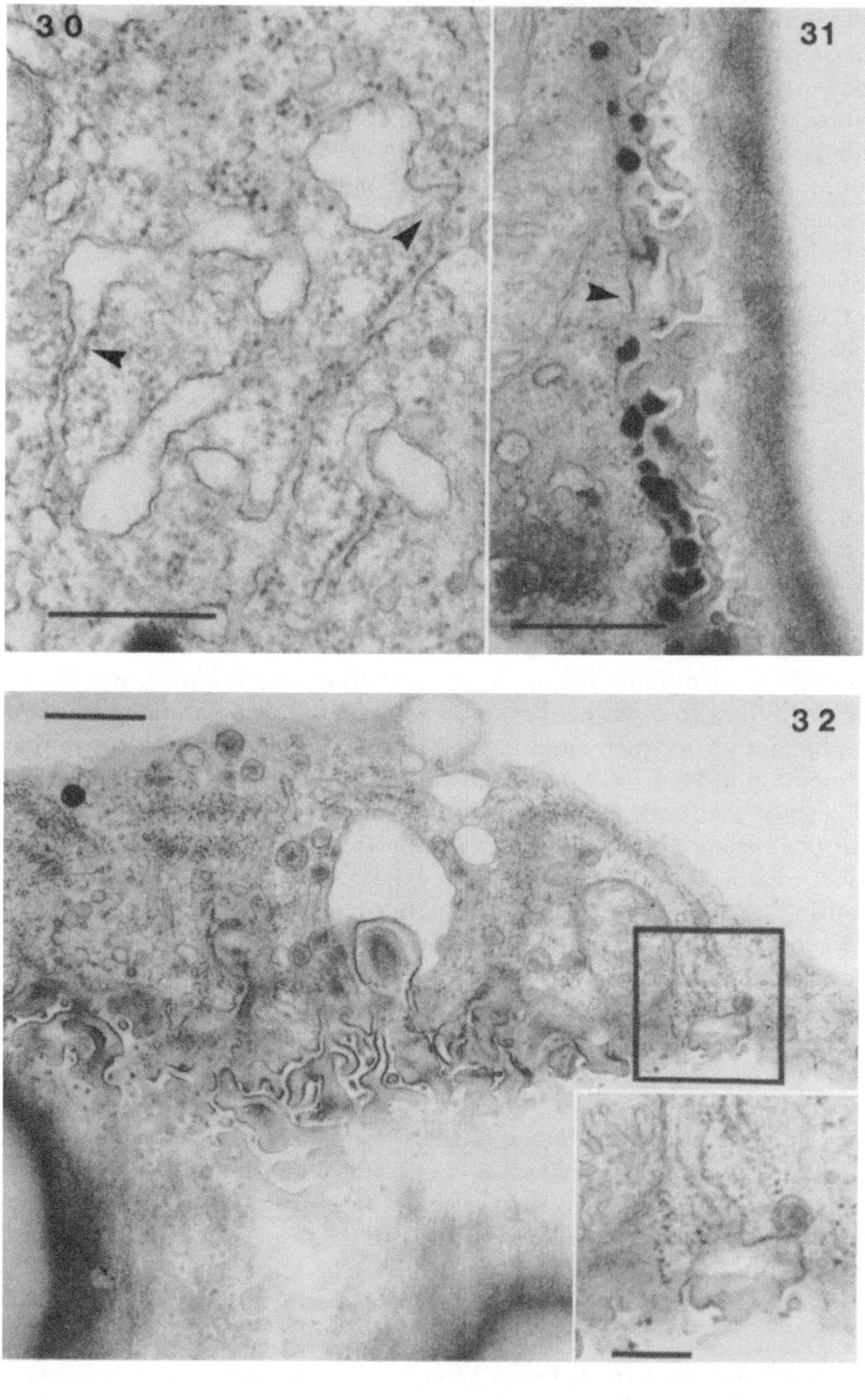

Fig. 30. Root hair in UDP-glucose for 15 min, with connection between rough endoplasmic reticulum and irregular smooth vesicles (◄). Scale bar: 0.5 μm. Fig. 31. Root hair incubated for 15 min in UDP-glucose, with membrane-bound tubular processes at the cell surface, and plasma

callose, as defined cytochemically, is in reality a heterogeneous substance with composition differences dependent on its source. This earlier idea may turn out to be the correct hypothesis.

Callose is usually marked by electron-transparency when viewed with the electron microscope, being neither osmiophilic nor stained by the usual electron-dense stains (Cresti and van Went, 1976) [see illustrations of phloem callose in Esau *et al.* (1962)]. Most callose deposits formed by reaction to infection or stress consist of material of varying electron density. Callose deposits, therefore, may be composed of the aniline-blue-positively staining polymers, plus a complex of other substances, and "callose-containing deposits" a more apt term for these regions. Nägeli (1861) and Salmon (1947) suggested sieve-tube callose to be proteinaceous in nature. Tsinger and Petrovskaya-Baranova (1967) reported proteins as well as lipids in callose pollen tube plugs. Lazarovits and Higgins (1975) found large osmiophilic globules embedded in callose layers formed by leaf cells of resistant tomato plants infected with *Cladosporium fulvum.* These workers suggested that the granules originate from the cytoplasm and may be similar to the osmiophilic globules observed by Kim *et al.* (1974), which are thought to be composed of phospholipid and protein. Lignin, as well as substantial amounts of cellulose as identified histochemically, is also found associated with callose deposition (Ride and Pearce, 1979). Cresti and van Went, (1976) assume that the electron-dense materials in callose plugs in pollen tubes are derived from disintegrated cytoplasm that becomes enclosed in the callose deposits during formation. Heath and Heath (1971) believe that plasmalemma is sequestered into the matrix of callose deposits in cowpea leaves. Hanchey and Wheeler (1971) demonstrated cell-wall fragments embedded in "cell-wall lesions" (probable callose deposits) and suggested that portions of the matrix are wall materials. Presumptive regions of callose deposition in clover root hairs stimulated by *Rhizobium* infection have a uniform texture and lack granules and membranes. However, the electron density is similar to that of the cell walls. While these deposits possibly contain β1,3-glucan as the aniline-blue-reactive agent, they presumably are composed of other substances as well.

Radish root-hair callose deposits contain areas unstained by electron dense stains (see Fig. 26) or by Thiéry's silver proteinate test for periodate-reactive polysaccharides (see Fig. 27). Because these cells have the capacity to synthesize β1,3-linked glucans, as demonstrated by *in vitro* incorporation studies, it is certainly possible that the nonstaining regions of these aniline-blue-reactive deposits contain β1,3-glucans. The membrane-bound tubules leading into the forming callose deposits contain moderately electron-dense substances (see Fig. 32) and are apparently partitioned off from the remaining cytoplasm by a newly formed plasma membrane (see Fig. 31). The tubules and their contents and possible

---

membrane apparently forming beneath the tubules (►). Scale bar: 0.5 μm. Fig. 32. Root hair incubated for 1 hr in UDP-glucose, with membrane-bound processes at the cell surface. Scale bar: 0.5 μm. *Inset*: Enlargment of the outlined area of Fig. 32, showing a rough endoplasmic reticulum connection to a smooth vesicle that presumably leads into a tubular process. Scale bar: 0.25 μm.

secretions therefore become a part of the callose deposits and may contribute to the membranous and electron-dense regions of these deposits. The enlarged dilations of rough endoplasmic reticulum were shown by Bonnett and Newcomb (1965) to stain positively for protein. It is interesting to speculate whether this protein matrix might be in the form of stored enzymes that contribute immediately on stress stimulation to the synthesis of callose-deposit substances.

3.2. *Evidence for Organelle Involvement in Callose Deposition in Other Cell Types*

Ultrastructural evidence is useful for postulating roles of apparently active organelles in regions of ongoing synthesis and deposition. The dilated rough endoplasmic reticulum in UDP-glucose-incubated root hairs may be a result of osmotic damage rather than an active cell response to stimuli. However, callose deposition in many plant cells besides root hairs is accompanied by dilation of the endoplasmic reticulum. A variety of cowpea immune to rust infection forms callose-containing deposits on the host cell wall around the point of penetration (Heath and Heath, 1971). Sheath formation around the invading haustoria results from the continued growth of callose deposits. These callose deposits of the haustorial sheath are of electron-translucent material containing variable amounts of darkly staining material, often with a membranous appearance, and are similar in appearance to the callose deposits described in radish root hairs in this study. The callosic sheath material apparently originates in irregular vesicles of dilated rough endoplasmic reticulum.

Lazarovits and Higgins (1976) studied the ultrastructural reactions of susceptible and resistant varieties of tomato to infection by *Cladosporium fulvum*. The highly resistant (immune) variety, Vinequeen, forms massive callose deposits in the mesophyll cells in infection lesions. Considerable ultrastructural changes occur in the cytoplasm of cells in the reacting tissues. In some cells, networks of swollen smooth endoplasmic reticulum give rise to large irregular vacuoles within the cytoplasm. Both smooth endoplasmic reticulum and Golgi vesicles are closely associated with the callose structures and may be involved in their deposition.

Cresti and von Went (1976) found that callose-plug formation in *Petunia* pollen tubes *in situ* is accompanied by swollen rough endoplasmic reticulum cisternae that contain material with an electron-transparency similar to that of the callose deposits. Intracisternal spaces of rough endoplasmic reticulum are continuous with callose grains, which, in turn, fuse with the forming callose plug. These workers suggest that callose is formed and accumulated in terminal swellings of the rough endoplasmic reticulum and that these callose-containing vesicles (callose grains) fuse with the plasmalemma, thereby adding their contents to the callose deposit. More recent work by Cresti *et al.* (1977) examined the ultrastructure of *in vitro* germinating *Lycopersicum peruvianum* pollen and noted the fusion of large Golgi vesicles with the plasma membrane in regions of callose deposition in the pollen grain and young pollen tubes. These researchers speculate that Golgi vesicles contain a callosic precursor that polymerizes on deposition between the plasma membrane and cell wall of the pollen tube or the intine of the germinating pollen. It was noted that the callose plugs in *Lycopersicum* are formed in the same manner as that described for *Petunia* pollen (Cresti and von Went, 1976), i.e., fusion of rough endoplasmic reticulum vesicles with the limiting membrane of the callose plug. This fact raises the interesting question of why the synthetic pathway for callose-plug formation (endoplasmic reticulum) should differ from the synthetic pathway for wall callose deposition (Golgi vesicles) in the same elongating pollen tube.

Studies on newly formed cell walls indicate that callose, as identified by aniline blue staining, is present in the developing cell plates and in recently formed walls (Waterkeyn, 1967; Fulcher *et al.*, 1976). Cell-plate formation at the time of cytokinesis is accompanied by definite distribution patterns of endoplasmic reticulum (Hepler and Newcomb, 1967). Of interest is the fact that the smooth-surfaced vesicles, which first designate the position of the cell plate, fuse to form aggregates of various sizes (Jones and Payne, 1977), some of which develop membranous arms similar in appearance to the endoplasmic-reticulum-derived tubular processes observed in radish root hairs. The majority of these vesicles that form the new cell plate are thought to originate from Golgi bodies, however (Dauwalder and Whaley, 1973), rather than from endoplasmic reticulum. Jones and Payne (1977) suggest that callose is deposited in the developing cell plate by these Golgi-derived vesicles.

Cell-fractionation studies have implicated the plasma membrane as a site for synthesis of β1,3-glucans. Van der Woude *et al.* (1974) showed that the capacity to synthesize both β1,3-glucans and β1,4-glucans resides in the plasma-membrane as well as the Golgi-apparatus fractions of onion stem. Anderson and Ray (1978) found β1,3-glucans, "presumably callose," bound to plasma membrane of slices of pea stem tissue that were supplied with UDP-glucose. These workers suggested that this cell-surface synthetase system could supply wound callose. Maltby *et al.* (1979) proposed that all β1,3 glucan in cotton fibers is located extracellularly, a location consistent with the aforedescribed hypothesis for cell-surface β1,3-glucan synthetases in pea stem tissue.

Autoradiography has indicated the sites of synthesis utilizing the glucose moiety of supplied UDP-glucose within nonintact radish root hairs to be Golgi vesicles, rough endoplasmic reticulum, and possibly plasma membrane; however, this technique does not correlate type of synthesized polymer with specific incorporation sites. Ultrastructural evidence implicates the direct involvement of the rough endoplasmic reticulum system of these cells in callose-deposit formation. It is possible, of course, that tubule membrane at forming callose deposits in root hairs has conformed to the membrane-flow theory postulated by Morré and Mollenhauer (1974) and differentiated into plasma membrane prior to synthesis of aniline-blue-reactive polymers at the cell surface.

4. Summary

Exogenously supplied UDP-glucose apparently does not cross the plasma membrane of undamaged cells of all cell types of radish seedling roots, nor is it utilized at the cell surfaces of intact cells. Damaged root hairs incorporate the glucose moiety of this sugar nucleotide into polymers not removed by fixation procedures. Sites of synthesis of these polymers include Golgi vesicles, rough endoplasmic reticulum, and possibly plasma membrane. The synthesized polymers give positive aniline blue staining reactions for callose. Undamaged root hairs stressed by incubation sometimes form callose deposits between the cell wall and the plasma membrane. The extensive rough endoplasmic reticulum system in these cells may contribute to these callose deposits.

References

Anderson, R. L., and Ray, P. M., 1978, Labeling of the plasma membrane of pea cells by a surface-localized glucan synthetase, *Plant Physiol.* **61**:723–730.

Arens, K., 1949, Provo de calose por meio da microscopia a luz fluorescente e appliçacoes do metodo, *Lilloa* **18**:71–75.

Aspinall, G. O., and Kessler, G., 1957, The structure of callose from the grape vine, *Chem. Ind. (London)* 1296.

Bonnett, H. T., Jr., and Newcomb, E. H., 1965, Polyribosomes and cisternal accumulations in root cells of radish, *J. Cell Biol.* **27**:423–432.

Bonnett, H. T., Jr., and Newcomb, E. H., 1966, Coated vesicles and other cytoplasmic components of growing root hairs of radish, *Protoplasma* **62**:59–75.

Cooper, K. M., 1978, Callose deposit formation and the utilization of UDP-glucose by the root hairs of *Rhaphanus sativus*, M.S. thesis, University of North Carolina, Chapel Hill.

Cresti, M., and van Went, J. L., 1976, Callose deposition and plug formation in *Petunia* pollen tubes *in situ*, *Planta* **133**:35–40.

Cresti, M., Pacini, E., Ciampolini, F., and Sarfatti, G., 1977, Germination and early tube development *in vitro* of *Lycopersicum peruvaianum* pollen: Ultrastructural features, *Planta (Berlin)* **136**(3):239–248.

Currier, H. B., 1957, Callose substance in plant cells, *Am. J. Bot.* **44**:478–488.

Dauwalder, M., and Whaley, W. G., 1973, Staining of cells of *Zea mays* root apices with the osmium-zinc iodide and osmium impregnation techniques, *J. Ultrastruct. Res.* **45**:279–296.

Delmer, D. P., Heiniger, U., and Kulow, C., 1977, UDP-glucose:glucan synthetase in developing cotton fibers. I. Kinetic and physiological properties, *Plant Physiol.* **59**:713–718.

Esau, K., Cheadle, V. I., and Risley, E. B., 1962, Development of sieve-plate pores, *Bot. Gaz.* **123**(4):233–243.

Eschrich, W., and Currier, H. B., 1964, Identification of callose by its diachrome and fluorochrome reactions, *Stain Technol.* **39**:303–307.

Faulkner, G., Kimmins, W. C., and Brown, R. G., 1973, The use of fluorochromes for the identification of $\beta(1 \rightarrow 3)$-glucans, *Can. J. Bot.* **51**:1503–1504.

Fulcher, R. G., McCully, M. C., Setterfield, G., and Sutherland, J., 1976, β-1,3-Glucans may be associated with cell plate formation during cytokinesis, *Can. J. Bot.* **54**:539–542.

Hanchey, P., and Wheeler, H., 1971, Pathological changes in ultrastructure: Tobacco roots infected with *Phytophthora parasitica* var. *nicotianae*, *Phytopathology* **62**:27–38.

Hassid, W. Z., 1970, Biosynthesis of sugars and polysaccharides, in: *The Carbohydrates*, 2nd ed., Vol. IIA (W. Pigman and D. Horton, eds.), pp. 301–373, Academic Press, New York.

Heath, M. C., and Heath, I. B., 1971, Ultrastructure of an immune and a susceptible reaction of cowpea leaves to rust infection, *Physiol. Plant Pathol.* **1**:277–287.

Heiniger, U., and Delmer, D. P., 1977, UDP-glucose: glucan synthetase in developing cotton fibers. II. Structure of the reaction product, *Plant Physiol.* **59**:719–723.

Hepler, P. K., and Newcomb, E. H., 1967, Fine structure of cell plate formation in the apical meristem of *Phaseolus* roots, *J. Ultrastruct. Res.* **19**:498–513.

Herth, W., Franke, W. W., Bittiger, H., Kuppel, A., and Keilich, G., 1974, Alkali-resistant fibrils of β-1,3- and β-1,4-glucans: Structural polysaccharides in the pollen tube wall of *Lilium longiflorum, Cytobiologie Z. Exp. Zellforsch.* **9**:344–367.

Howe, C. G., 1921, Pectic material in root hairs, *Bot. Gaz.* **72**:313–320.

Jones, M. G. K ., and Payne, H. L., 1977, Cytokinesis in *Impatiens balsamina* and the effect of caffeine, *Cytobios* **20**(78):79–92.

Kim, K. S., Fulton, J. P., and Scott, H. A., 1974, Osmiophilic globules and myelinic bodies in cells infected with two comoviruses, *J. Gen. Virol.* **25**:445–452.

Kumarasinghe, R. M. K., and Nutman, P. S., 1977, *Rhizobium*-stimulated callose formation in clover root hairs and its relation to infection, *J. Exp. Bot.* **28**(105):961–976.

Lazarovits, G., and Higgins, V. J., 1976, Ultrastructure of susceptible, resistant, and immune reactions of tomato to races of *Caldosporium fulvum, Can. J. Bot.* **54**(3/4):235–249.

Leech, J. H., Mollenhauer, H. H., and Whaley, W. G., 1963, Ultrastructural changes in the root apex, *Symp. Soc. Exp. Biol.* **17**:74–84.

Maltby, D., Carpita, N. C., Montezinos, D., Kulow, C., and Delmer, D. P., 1979, β-1,3-Glucan in developing cotton fibers, *Plant Physiol.* **63**:1158–1164.

Mangin, L., 1890, Sur la callose, nouvelle substance fondamentale existant dans la membrane, *C. R. Acad. Sci.* **110**:664–647.

Mangin, L., 1892, Observations sur la présence de la callose chez les Panérogames, *Bull. Soc. Bot.* **39**:260–267.

Mollenhauer, H. H., Whaley, W. G., and Leech, J. H., 1960, Cell ultrastructure responses to mechanical injury: A preliminary report, *J. Ultrastruct. Res.* **4**:473–481.

Morré, D. J., and Mollenhauer, H. H., 1974, The endomembrane concept: A functional integration of endoplasmic reticulum and Golgi apparatus, in: *Dynamic Aspects of Plant Ultrastructure* (A. W. Robards, ed.), pp. 84–137, McGraw-Hill, New York.

Nägeli, C. W., 1861, Ueber die Siebröhren von *Cucurbita, Sitzber. Bayerisch. Akad, Wiss.* **1**:212–238.

Newcomb, E. H., and Bonnett, H. T., Jr., 1965, Cytoplasmic microtubule and wall microfibril orientation in root hairs of radish, *J. Cell Biol.* **27**:575–589.

Nikaido, H., and Hassid, W. Z., 1972, Biosynthesis of saccharides from glycopyranosyl esters of nucleoside pyrophosphates ("sugar nucleotides"), *Adv. Carbohydr. Chem.* **27**:351–483.

Raymond, Y., Fincher, G. B., and Maclachlan, G. A., 1978, Tissue slice and particulate β-glucan synthetase activities from *Pisum* epicotyls, *Plant Physiol.* **61**:938–942.

Reynolds, J. D., and Dashek, W. V., 1976, Cytochemical analysis of callose localization in *Lilium longiflorum* pollen tubes, *Ann. Bot.* **40**:409–416.

Ride, J. P., and Pearce, R. B., 1979, Lignification and papilla formation at sites of attempted penetration of wheat leaves by non-pathogenic fungi, *Physiol Plant Pathol.* **15**(1):79–92.

Ridgway, C. S., 1913, The occurrence of callose in root hairs, *Plant World* **16**:116–122.

Roberts, E. A., 1916, The epidermal cells of roots, *Bot. Gaz.* **62**:488–506.

Salmon, J., 1947, Différenciation des tubes criblés chez les angiospermes: Recherches cytologique, *Rev. Cytol. Cytophysiol. Veg.* **9**:55–168.

Shafizadeh, F., and McGinnis, G. D., 1972, Morphology and biogenesis of cellulose and plant cell-walls, *Adv. Carbohydr. Chem.* **27**:297–349.

Shore, G., and Maclachlan, G. A., 1975, The site of cellulose synthesis: Hormone treatment alters the intracellular location of alkali-insoluble β-1,4-glucan (cellulose) synthetase activities, *J. Cell Biol.* **64**:557–571.

Smith, M. M., and McCully, M. E., 1978, A critical evaluation of the specificity of aniline blue induced fluorescence, *Protoplasma* **95**:229–254.

Thiéry, J.-P., 1967, Mise en évidence des polysaccharides sur coupes fines en microscopie électronique, *J. Microsc. (Paris)* **6**:987–1018.

Trump, B. F., and Arstila, A. U., 1971, Cell injury and cell death, in: *Principles of Pathobiology* (M. F. LaVia and R. B. Hill, Jr., eds.), pp. 9–96, Oxford University Press, London.

Tsinger, N. V., and Petrovskaya-Baranova, T. P., 1967, Function and physiological role of callose pollen tube plugs. *Soviet Plant Physiol.* **14**:404–410.

Van der Woude, C., Lembi, A., Morré, D. J., Kindinger, J. I., and Ordin, L., 1974, β-Glucan synthetases of plasma membrane and Golgi apparatus from onion stem, *Plant Physiol.* **54**:333–340.

Waterkeyn, L., 1967, Sur l'existance d'un "stade callosique" presente par la paroi cellulaire, au cours de la cytokinese, *C. R. Acad. Sci.* **265**:1792–1794.

Wheeler, H., 1974, Cell wall and plasmalemma modifications in diseased and injured plant tissue, *Can. J. Bot.* **52**(5):1005–1009.

10
Chitin-Fibril Formation in Algae

WERNER HERTH AND EBERHARD SCHNEPF

1. Introduction

Chitin is a fibrillar, crystalline polymer of *N*-acetylglucosamine. It belongs to the family of structural polysaccharides, as do cellulose, β1,3-glucan, β1,4-mannan, and β1,3-xylan. Chitin has not been found in higher plants, but is common in fungal cell walls ["fungine" of Braconnot (1811)] and in insect and arthropod cuticles ["chitin" from the Greek *χιτών*, the name for the tunic worn by ancient Greeks and Romans (Odier, 1843)]. Chitin is also found in lower animals, e.g., in bristles of polychaetes and hydrozoan thecae, and in the cyst walls of ciliates, as well as in the cell walls of bacteria [D-lactic-acid-substituted chitin in murein (for a review, see Muzzarelli, 1977)].

Earlier reports on the occurrence of chitin in algae have been based on staining reactions and often have not survived critical reexamination. However, more recent investigations have shown that at least in some centric diatoms and in the chrysoflagellate *Poterioochromonas*, there exist several cases with chitin clearly demonstrated by X-ray diffraction.

Fortunately, these chitin-forming algae are unicellular organisms that can be easily cultivated and are accessible to inhibitor studies and cytological investigations. This has allowed us to approach some of the open questions on the *in vivo* process of structural polysaccharide formation, namely, the question of the cytological site, of the assembly mechanism of the microfibril formation, and of the orientation mechanism. These investigations have been published in part elsewhere (Schnepf *et al.*, 1975, 1977; Herth *et al.*, 1977; Herth and Zugenmaier, 1977; Herth and Barthlott, 1979; Herth, 1978, 1979, 1980).

WERNER HERTH AND EBERHARD SCHNEPF • Zellenlehre, Universität Heidelberg, D-6900 Heidelberg, Federal Republic of Germany.

2. *β-Chitin-Fibril Formation of the Centric Diatoms Cyclotella and Thalassiosira*

It was already known to light microscopists at the end of the 19th century that certain centric diatoms form appendages of "diatom silk" to reduce the sinking velocity (Fig. 1) (Schütt, 1899). This parachutelike arrangement of fibrillar appendages radiates from the edge of the silica valves (Fig. 2). The appendages grow out of special openings, the pores or "strutted processes" (Anonymous, 1975) of the silica valve (Figs. 3 and 4), which are regularly distributed in the valve (for more details, see Herth and Barthlott, 1979). The appendages are ribbonlike with a conical proximal end and become gradually thinner toward the distal end (Figs. 5 and 6). They are rather stiff and show typical kinking sites after drying on the grid. They are resistant to alkaline extractions, but then have a tendency to disintegrate partially into smaller ribbonlike subunits (see Herth and Zugenmaier, 1977).

These appendages can be isolated (for the method, see Herth and Zugenmaier, 1977). They reveal an extremely sharp X-ray diffraction pattern typical for β-chitin (Dweltz *et al.*, 1968; Blackwell *et al.*, 1967) (see also Chapter 20). Chemical analyses also show that these fibrils are very pure *N*-acetylglucosamine polymers (Falck *et al.*, 1966). With electron diffraction, they yield point diagrams indicating their extremely high crystallinity (Zugenmaier *et al.*, 1981).

In sections of *Thalassiosira* or *Cyclotella* cells, the pores in the silica valve are easily recognized, but the chitin fibrils are almost invisible in conventional fixation.

When *Cyclotella meneghiniana* is fixed with tannic acid added to the glutaraldehyde fixative, the surface of the chitin fibrils is stained, and the fibril is seen to penetrate the pore channel longitudinally (Fig. 7). Its proximal end is found to originate from a conical groove of the plasma membrane. This region of the plasma membrane (Fig. 8) characteristically bears an extraplasmic and a cytoplasmic coat layer, both heavily stained with this tannic acid–glutaraldehyde–osmium tetroxide procedure. This coat is not observed in the other regions of the plasma membrane.

In *Cyclotella cryptica* with the standard glutaraldehyde–osmium tetroxide fixation, these coat layers are not so well preserved (Figs. 9 and 10). The chitin fibril is stained only at its proximal conical surface between the crystalline part and the plasma membrane (Figs. 9 and 10). The cytoplasmic coat layer without tannic acid fixation is more prominent than the extraplasmic coat layer. In oblique tangential section, the cytoplasmic coat appears as a polygonal network (Fig. 11) with meshes approximately 12 nm wide. In a cross section of the conical invagination, the coat consists of a cartwheel-like arrangement of projections (Fig. 12) 13–18 nm long.

In *Thalassiosira*, the invaginations of the plasma membrane are often less regular in outline, with undulated membrane and associated cytoplasmic vesicles (Fig. 13). The coat layer on the cytoplasmic face here consists of two layers (Fig. 13, arrows) and in part also appears polygonal in oblique tangential

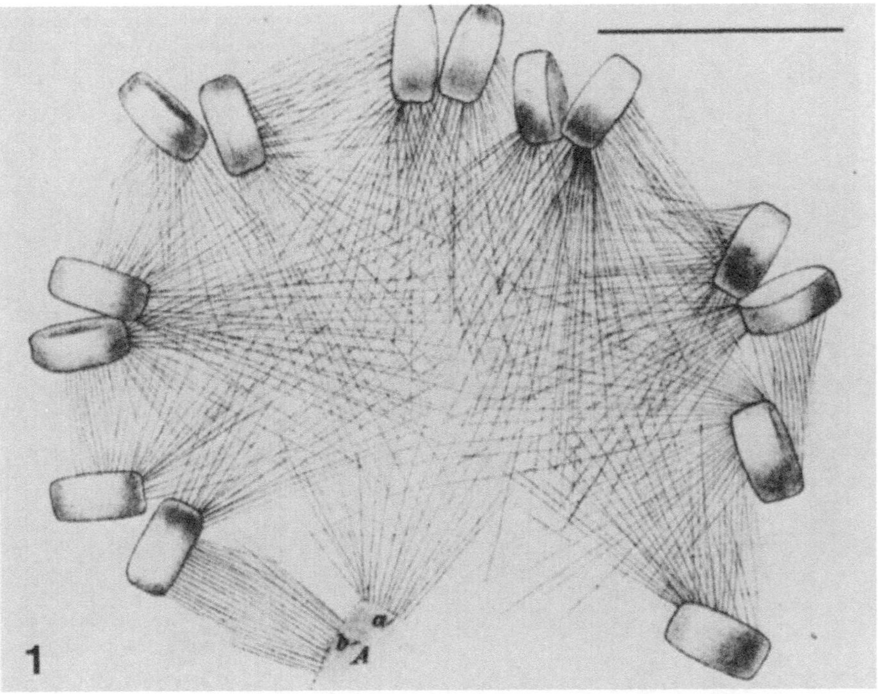

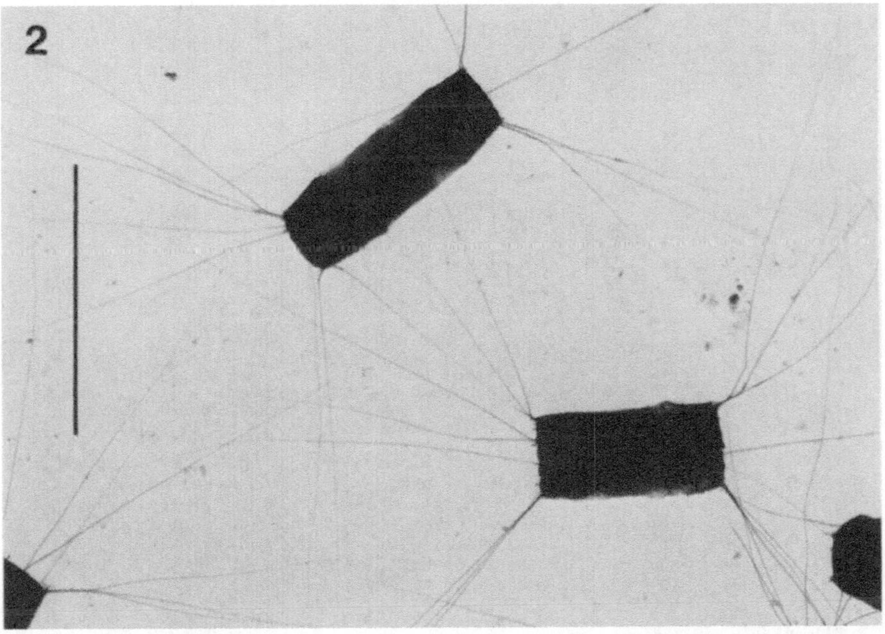

FIGURE 1. *Cyclotella socialis* Schütt. Reproduction of the original drawing of Schütt (1899). Scale bar: 50 μm.

FIGURE 2. *Cyclotella meneghiniana*. Electron micrograph of glutaraldehyde-fixed cells, with attached β-chitin appendages radiating from the edge of the valve. Scale bar: 10 μm.

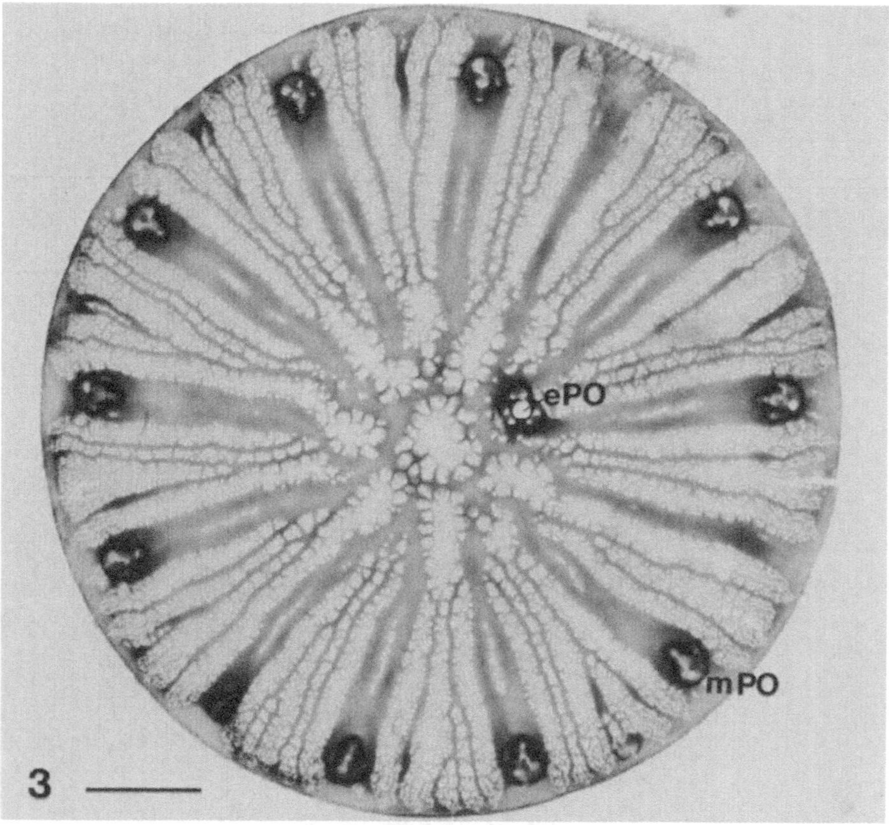

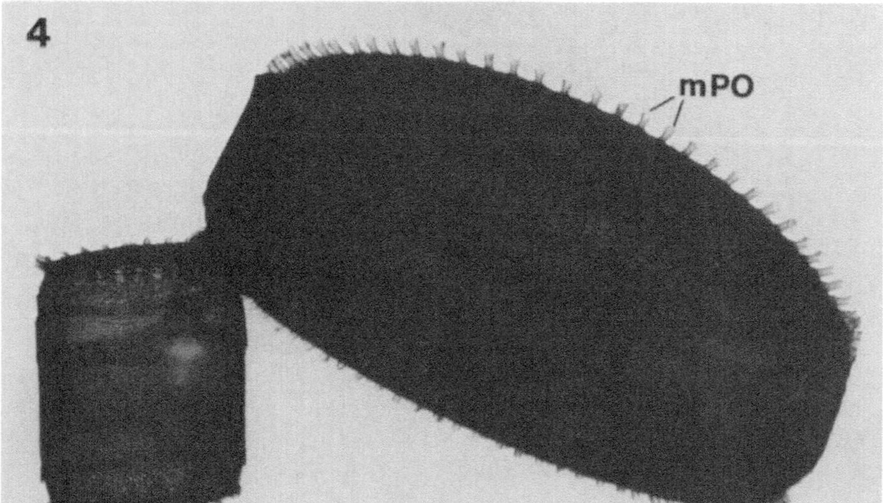

FIGURE 3. *Cyclotella cryptica*. H_2SO_4-cleaned valve (not yet fully silicified), with marginal pores (mPO) and eccentric pore (ePO). Electron micrograph. Scale bar: 1 μm.

FIGURE 4. *Thalassiosira fluviatilis*. H_2SO_4-cleaned whole cells. Note numerous chimneylike marginal pores (mPO). Scale bar: 10 μm.

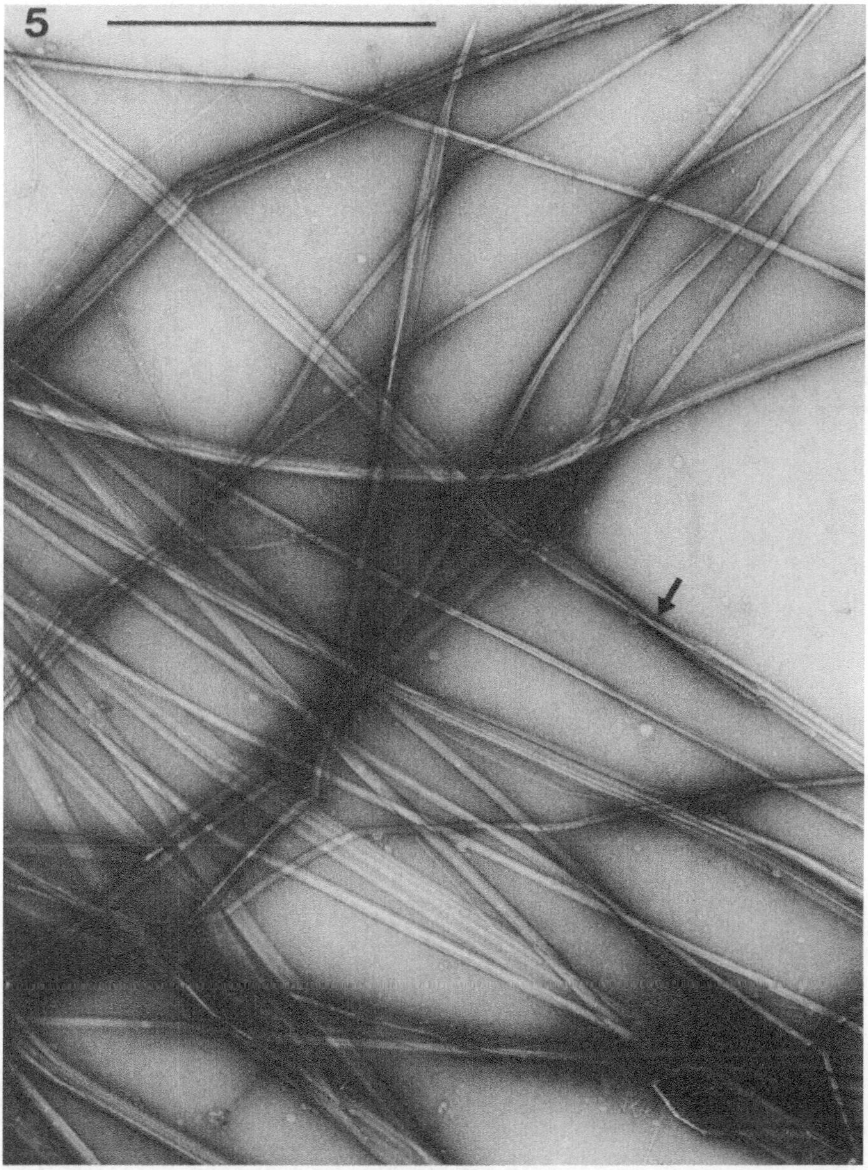

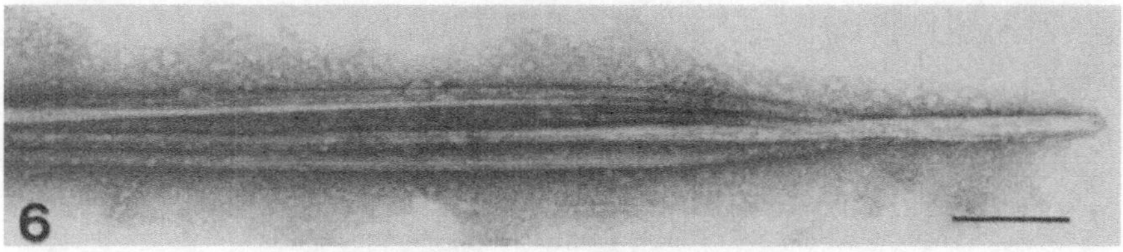

FIGURE 5. β-Chitin of *Thalassiosira fluviatilis* in negative stain, uranyl acetate. The aspect is ribbonlike (with small edge of a ribbon at the arrow), with kinking sites indicating high crystallinity; the ribbons are composed of several parallel subunits. Scale bar: 1 μm.

FIGURE 6. Higher magnification of the proximal conical end of a *Thalassiosira* chitin appendage. Scale bar: 0.1 μm.

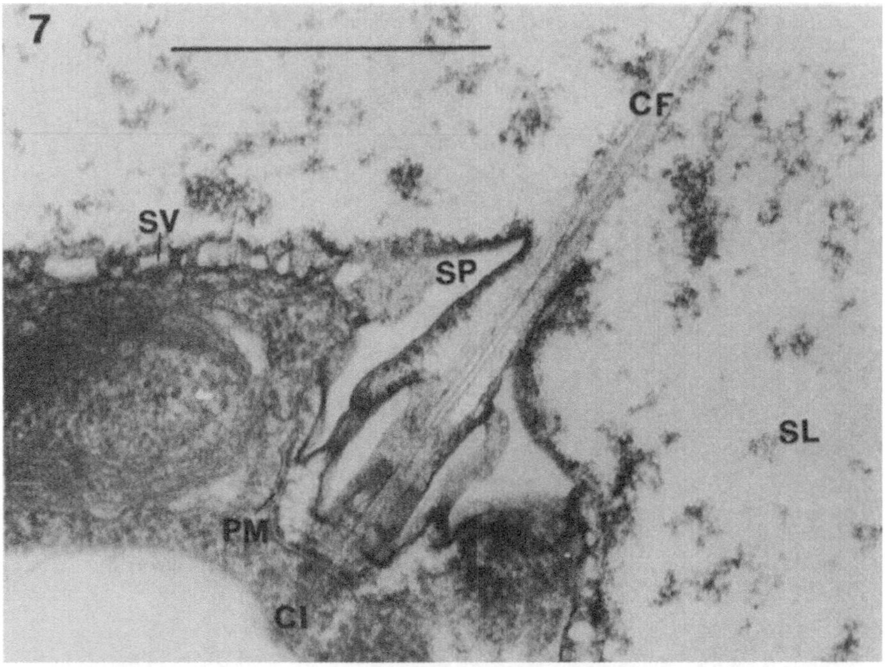

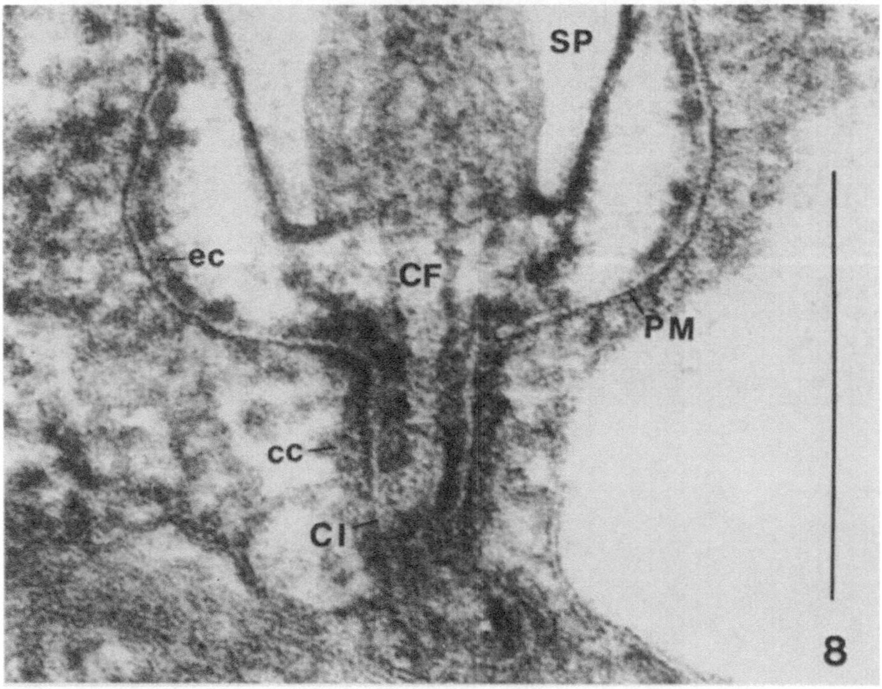

FIGURES 7 and 8. Sections of *Cyclotella meneghiniana*, fixed with glutaraldehyde-tannic acid and osmium tetroxide, then extracted with hydrofluoric acid to remove the silica. In Fig. 7, the chitin fibril (CF) is seen to penetrate the pore lumen, and to end in dense material. Scale bar: 0.5 μm. Figure 8 shows a high magnification of the point of origin of the chitin fibril (CF), a conical invagination (CI) of the plasma membrane (PM), with electron-dense extraplasmic coat (ec) and cytoplasmic coat (cc) layers. Scale bar: 0.2 μm.

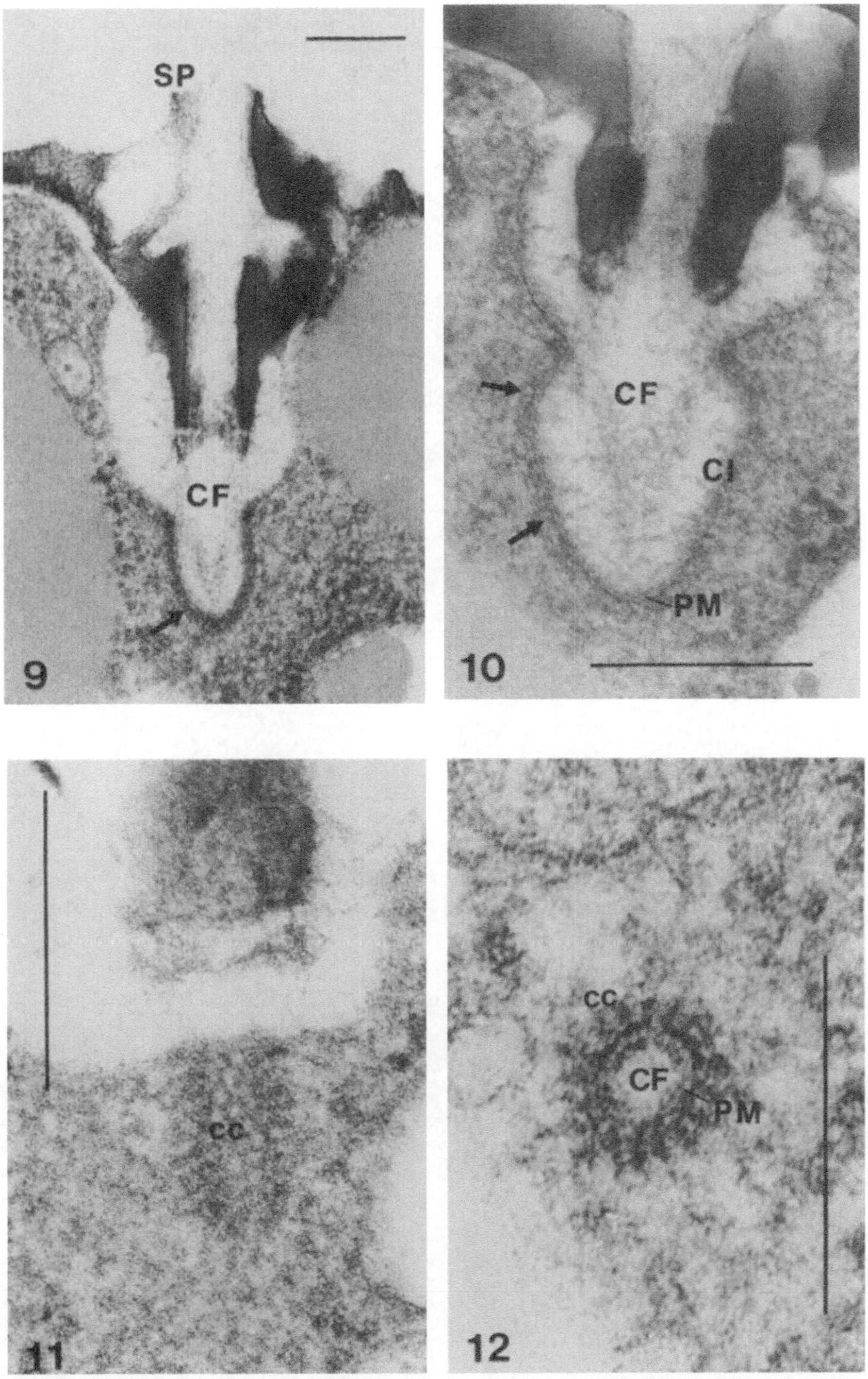

FIGURES 9–12. *Cyclotella cryptica*, glutaraldehyde-osmium tetroxide fixation. From Herth (1979) with the permission of Academic Press. Median longitudinal sections through the strutted process (SP), or pore apparatus (Figs. 9 and 10), show the chitin fibril (CF) to arise from a conical invagination (CI) of the plasma membrane (PM). Only the proximal end of the CF is positively stained. The PM invagination bears an electron-dense cytoplasmic coat (cc) layer (→). This layer is seen as a polygonal network in oblique tangential section (Fig. 11) and as a cartwheel-like arrangement in a cross section of the invagination (Fig. 12). Scale bar: 0.2 μm.

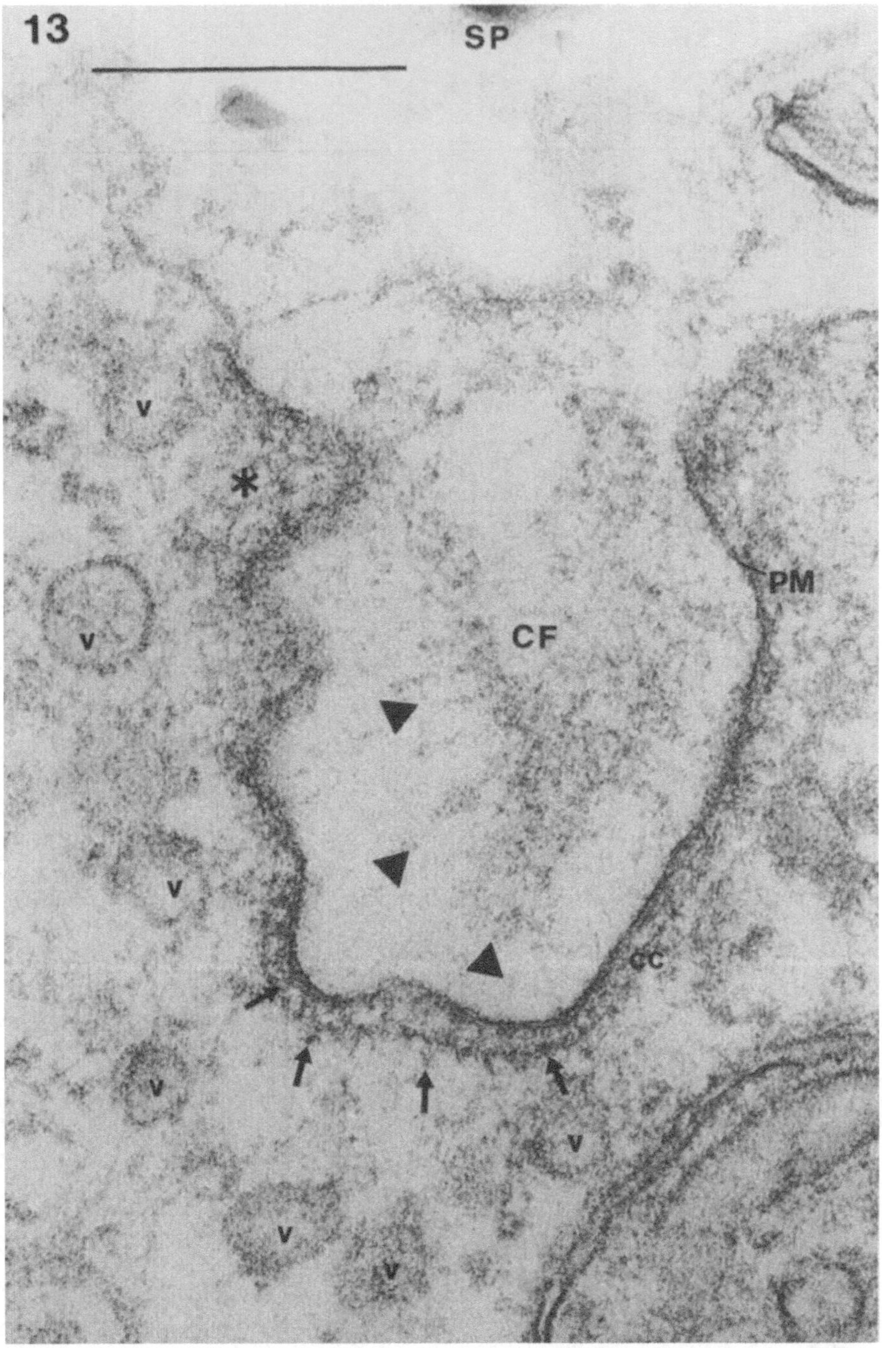

FIGURE 13. *Thalassiosira fluviatilis*. Fixation as in Figs. 9–12. High magnification of the plasma membrane (PM) invagination underlying a pore, median longitudinal section. The chitin fibril (CF) has positively stained projections connecting it with the plasma membrane (▶). Note the cytoplasmic coat layers (→) with a polygonal appearance in tangential section (*) and the associated cytoplasmic vesicles (V). (SP) Strutted process. Scale bar: 0.2 μm. From Herth (1979) with the permission of Academic Press.

section (*). In cross section, it looks less regular than the cartwheel-like arrangement in the *Cyclotella* species (Herth, 1979). There is a closer resemblance of the *Thalassiosira* coat layer to the cytoplasmic coat of typical coated vesicles in the same cell than in *Cyclotella*. In *Thalassiosira*, again, there is a stained region at the proximal end of the chitin fiber with fine projections from the crystalline nonstained part toward the plasma membrane (Fig. 13, arrowheads).

The chitin fibers of the centric diatoms investigated thus in all cases originate each from a special chitin-fibril-synthesizing apparatus, consisting of the pore in the silica valve and the underlying special region of the plasma membrane. This is summarized in Fig. 14. Only in *Thalassiosira* can cytoplasmic vesicles similar to chitosomes be observed, but the main process of chitin-fibril formation seems to be a transmembrane event with precursors on the cytoplasmic face, polymerization in the plasma membrane, and crystallization in the central lumen of the conical invagination of the plasma membrane. The coat layers observed may either, like clathrin, fix the synthetase enzymes in this region (e.g., Anderson *et al.*, 1977), or may be directly involved in the process of polymerization. The β-chitin appendages clearly grow at their proximal end. Considering the dimensions of the appendages, the time of formation, and the calculated surface area of the conical invaginations (for details, see Herth, 1979), in *Cyclotella cryptica* there may be 100,000 glycosidic linkages formed per minute and per square micrometer of specialized membrane region. The values for *C. meneghiniana* are 250,000, and for *Thalassiosira fluviatilis*, 700,000 glycosidic linkages formed per minute and per square micrometer of specialized membrane region.

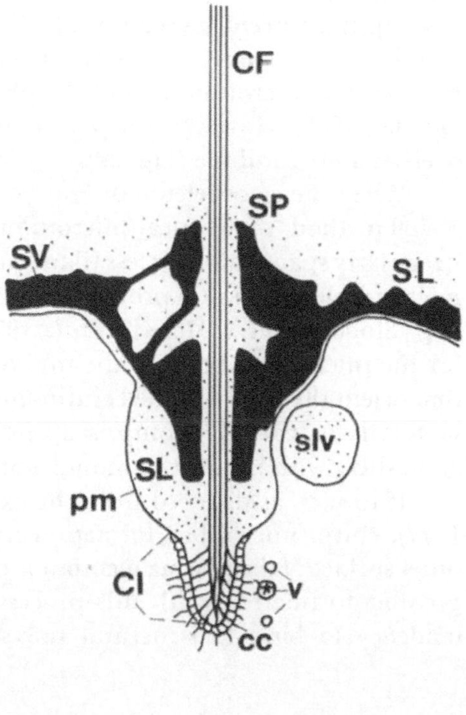

FIGURE 14. Schematic drawing summarizing the ultrastructural data on the chitin fibril (CF)-forming apparatus in centric diatoms. (cc) Cytoplasmic coat; (CI) conical invagination; (pm) plasma membrane; (SL) slime; (slv) slime vesicle; (SP) strutted process or pore apparatus; (SV) silica valve; (v) cytoplasmic vesicle. From Herth (1979) with the permission of Academic Press.

3. Lorica Microfibrils of *Poterioochromonas*

The chrysoflagellate alga *Poterioochromonas stipitata* (Fig. 15) (for cytological details and taxonomy, see Schnepf *et al.*, 1977) readily attaches to a substratum with its posterior end, forms a foot region by elongation of a posterior tail, secretes the hollow lorica stalk and later a cup, which then houses the cell (Figs. 16–18).

The lorica composition has been analyzed (Herth *et al.*, 1977). It has been shown by X-ray and sugar analysis that the alkali-resistant framework of this lorica consists of chitin microfibrils.

In the stalk region of the lorica, the microfibrils are oriented parallel to each other and form a helical arrangement around the cytoplasmic tail region where they are formed (Fig. 19). This arrangement corresponds to the distribution of the cortical microtubules in the cytoplasmic tail (Fig. 20). In the absence of these cortical microtubules, after colchicine application, this oriented arrangement of the stalk microfibrils may not be formed; only stalkless cups with less oriented individual microfibrils are secreted (Schnepf *et al.*, 1975). The microtubules therefore are thought not only to model the cell shape but also to exert an orienting influence on microfibril formation (Schnepf *et al.*, 1975).

It may be calculated from the dimensions of the cytoplasmic tail region involved in microfibril formation, from the dimension of the microfibrils formed, and from the rate of elongation of the stalk that in *Poterioochromonas* there is a rate of about 2,000,000 glycosidic linkages formed per minute and per square micrometer of plasma-membrane region involved (for details, see Schnepf *et al.*, 1975).

More detailed ultrastructural studies have now revealed that the cortical microtubules in the cytoplasmic tail of *Poterioochromonas* are embedded in an electron-translucent layer (Fig. 21). Between the microtubules and the plasma membrane, filamentous material is seen (Fig. 21). There is no direct contact between the microtubules and the plasma membrane (see also Robinson and Quader, 1980). However, the plasma membrane seems to be stabilized where it overlies a microtubule (Fig. 22).

When the cytoskeleton of *Poterioochromonas* is prepared with the Triton X-100 method, the cortical microtubules are seen to be associated with rather irregularly spaced more or less filamentous material (Fig. 23). This resembles the microtubule-associated protein (MAP) structures described for other objects (e.g., Sloboda *et al.*, 1976; Kim *et al.*, 1979). These structures could be responsible for the interaction between the microtubules and the membrane components that orient the newly formed chitin microfibrils. It is unclear whether they may be totally or partially comprised of contractile proteins. Attempts to localize actin there with indirect immunofluorescence have so far failed.

If in fact, as deduced from the experimental evidence (see Schnepf *et al.*, 1977), chitin microfibril formation in *Poterioochromonas* takes place on the outer surface of the plasma membrane of the cytoplasmic tail region, it should be possible to interfere with this process by addition of substances with a high tendency to bind to structural polysaccharides. There should be no access

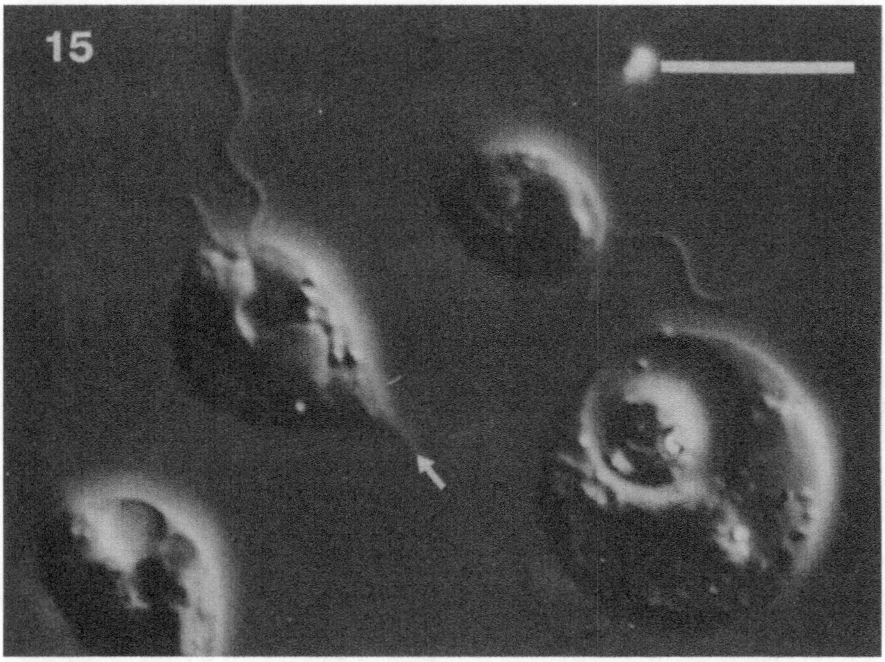

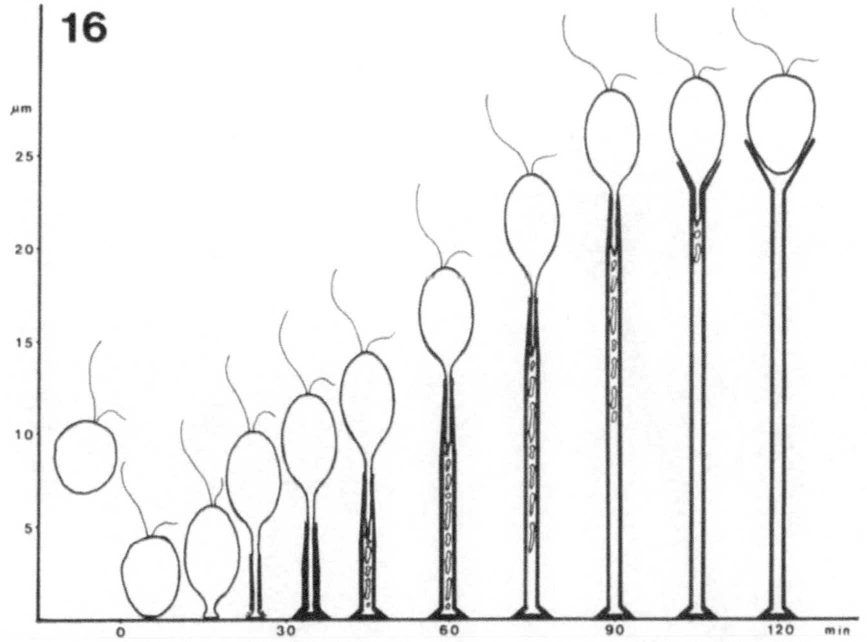

FIGURE 15. *Poterioochromonas stipitata*. Light micrograph, Nomarski differential interference contrast. Note the cytoplasmic tail (←), which serves to attach the cell to a substratum and is involved in lorica formation. Scale bar: 10 µm.

FIGURE 16. Schematic drawing summarizing the time-course of attachment, stalk, and lorica formation in *Poterioochromonas* (for details, see Schnepf *et al.*, 1975). Reproduced with the permission of Springer-Verlag.

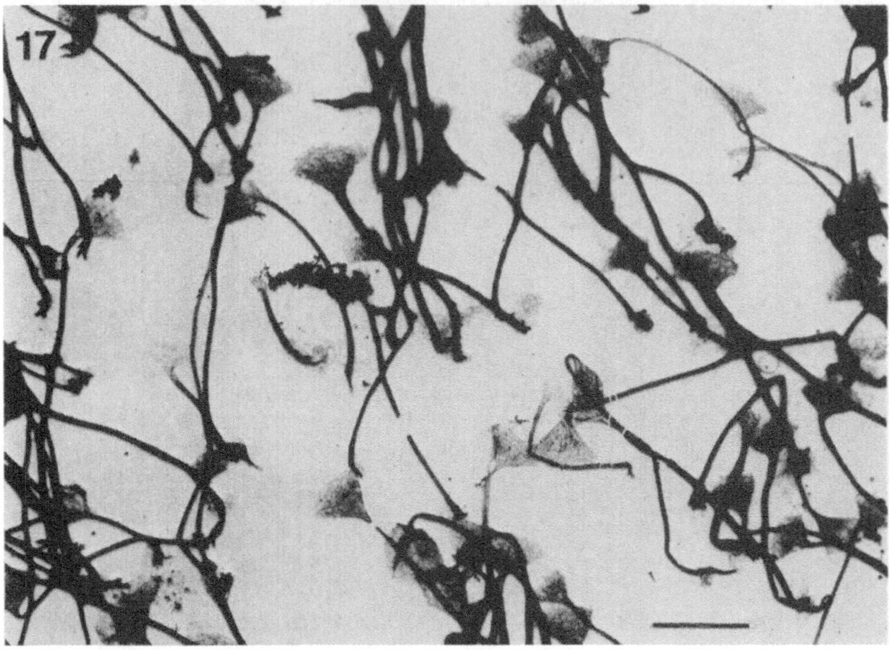

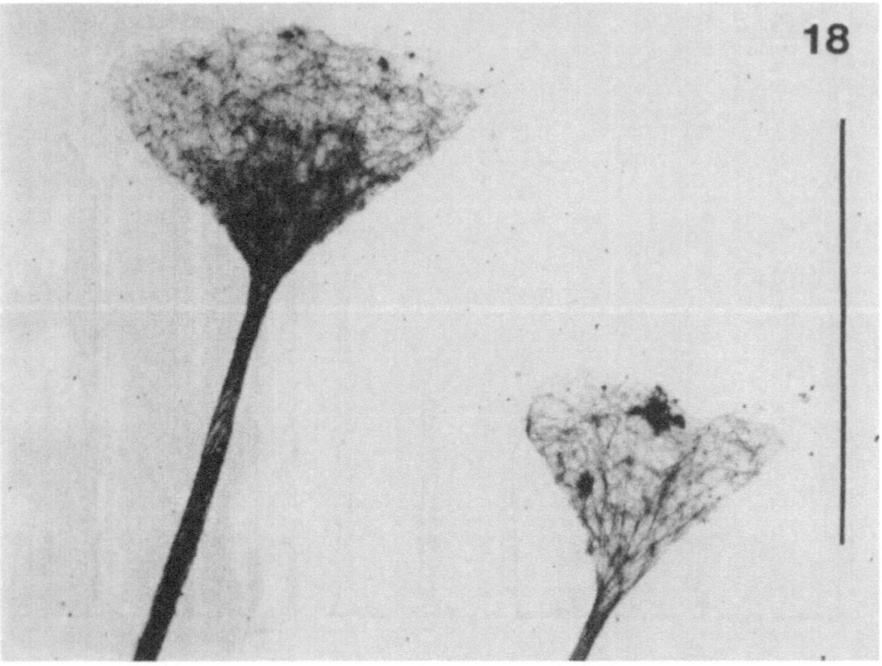

FIGURE 17. Loricae formed on the electron-microscope grid within 3 hr of exposure to the *Poterioochromonas* culture. Uranyl acetate stain. Scale bar: 10 μm.

FIGURE 18. Lorica stalk and cup composed of microfibrils. Uranyl acetate stain. Scale bar: 10 μm.

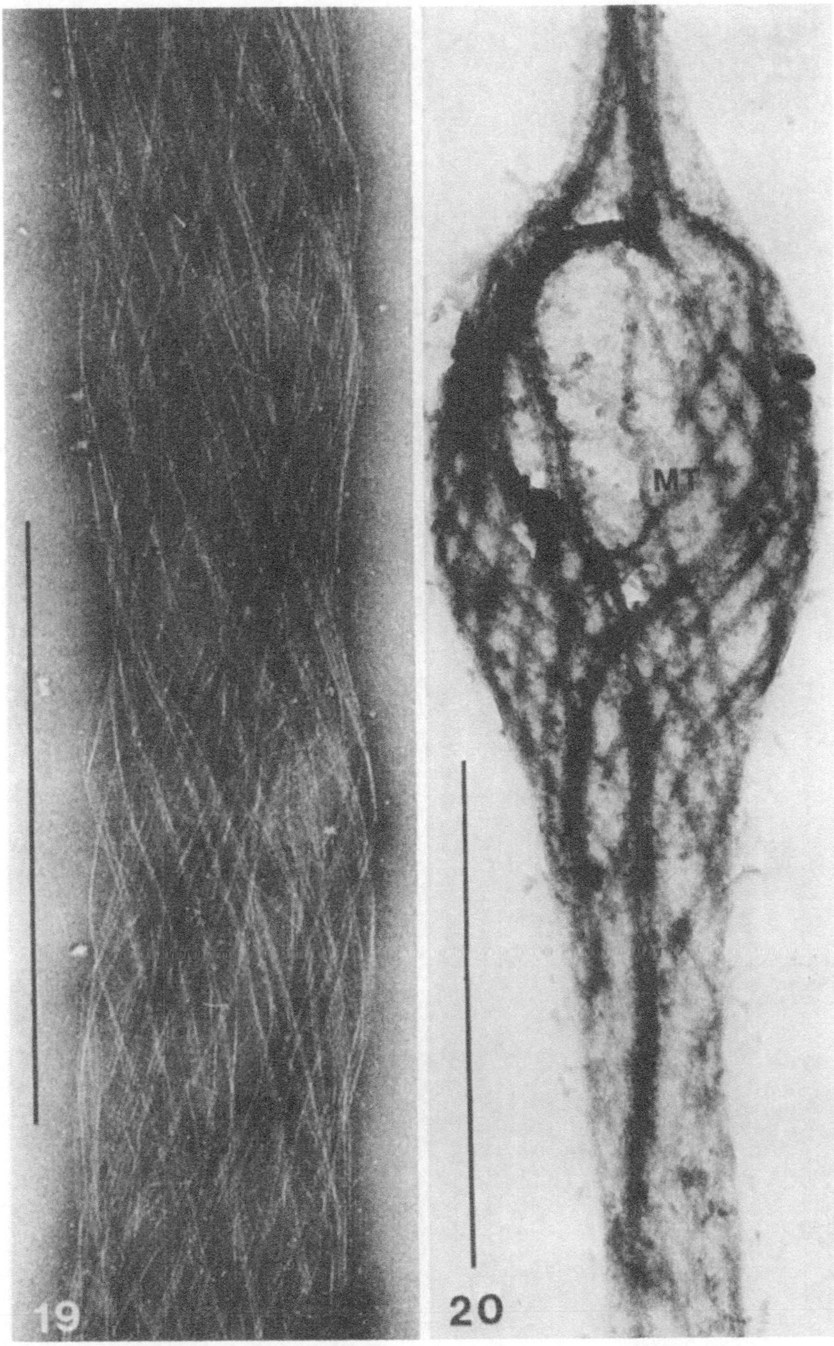

FIGURE 19. Helically ordered arrangement of chitin microfibrils in the lorica stalk region at high magnification. Negative stain with uranyl acetate. Scale bar: 1 μm.

FIGURE 20. Helical arrangement of microtubules (MT) in the remnant of the cytoplasmic tail of *Poterioochromonas* (detached from the cell by a water stream). Uranyl acetate stain. Scale bar: 1 μm.

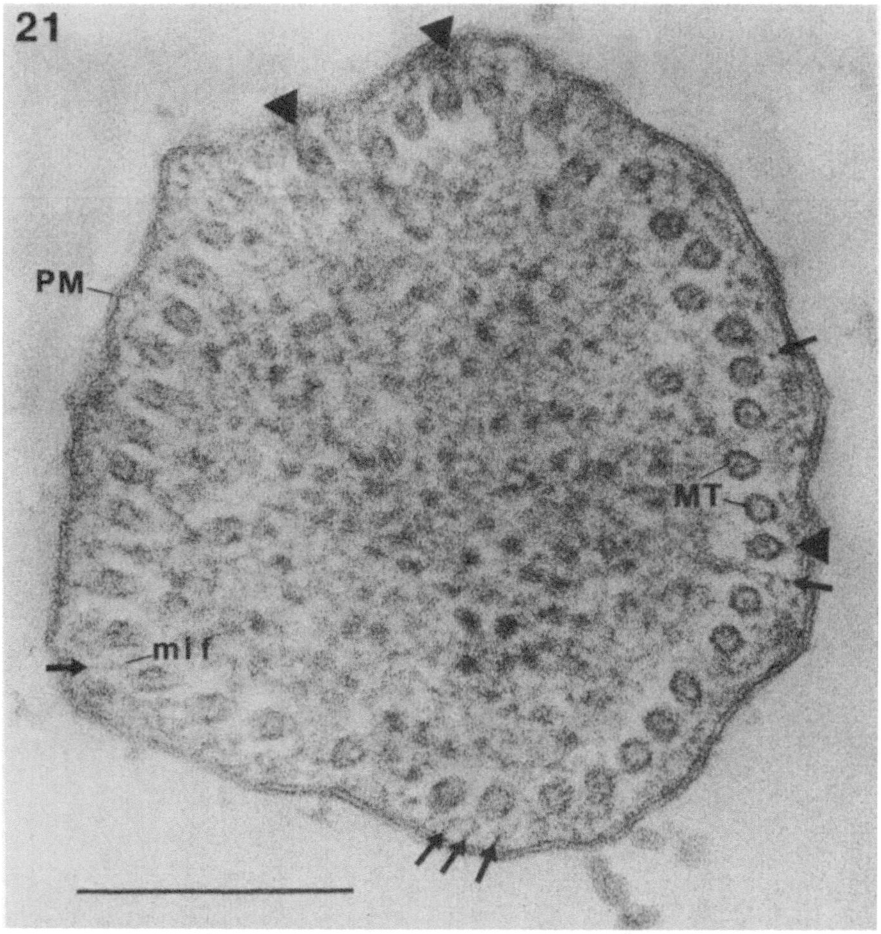

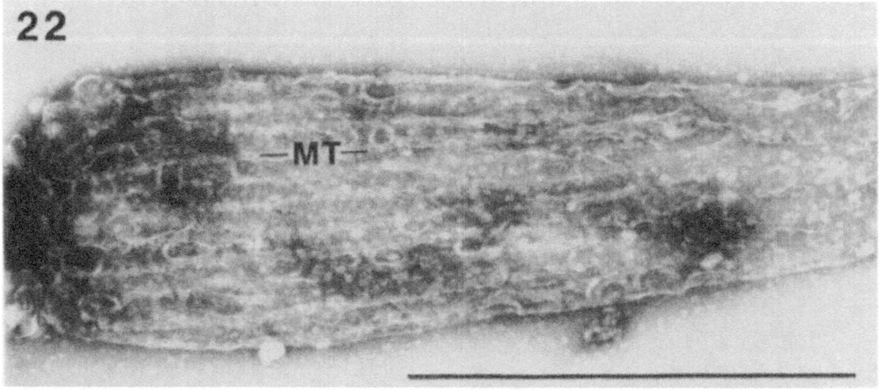

FIGURE 21. Cross section of the cytoplasmic tail of *Poterioochromonas*. The microtubules (MT) are embedded in a translucent zone and have no direct contact with the plasma membrane. Note 5-nm microfilaments (mif) between the microtubules and the plasma membrane (PM) (→) and cross-bridge-like structures (▶). Scale bar: 0.2 μm.

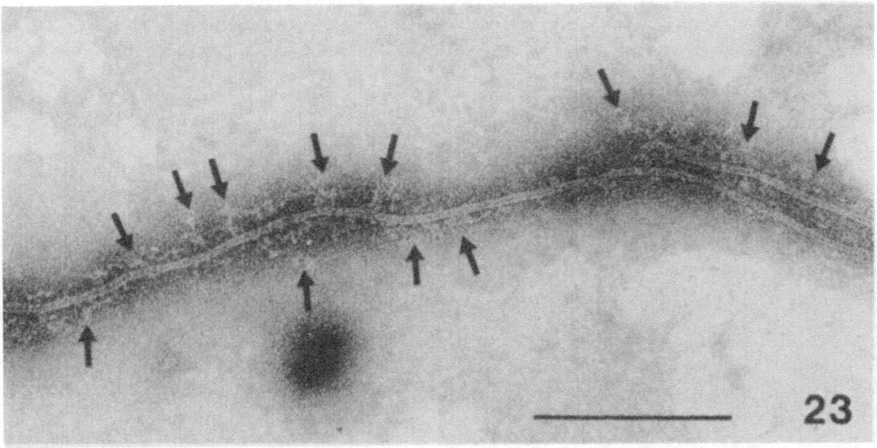

FIGURE 23. Microtubules from the cytoplasmic tail isolated with Triton X-100 and negatively stained with uranyl acetate. Note the irregularly associated filamentous to fuzzy projections from the microtubule surface (↓↓). Scale bar: 0.5 μm.

difficulties as in other objects surrounded by walls, since *Poterioochromonas* is a naked cell.

We tried to disturb or inhibit microfibril formation by addition of Calcofluor White ST or Congo Red, both substantial dyes for structural polysaccharides (see Herth, 1980). These two substances bind to polysaccharide chains in an oriented way (Herth and Schnepf, 1980), and with high affinity. Indeed, rather crude effects are observed when these two substances are applied to *Poterioochromonas* during lorica formation. Depending on the concentration applied (Figs. 24–27) (for details, see the figure captions), lorica formation is totally inhibited or malformed loricae are made. With Calcofluor White at high concentration, only patches of Calcofluor-positive material are secreted (Fig. 25). At medium concentration, thick stalks are formed (Fig. 26). At low concentration, complete but very short and very massive loricae are synthesized (Fig. 27). With Congo Red, similar effects are observed (for figures, see Herth, 1980). As seen in negatively stained preparations of the malformed loricae made in the presence of the inhibitors, chitin synthesis is undisturbed but the assembly of the microfibrils is affected. Depending on the concentration applied, microfibril formation may be totally suppressed, with formation only of short rods and aggregates of short rods (Figs. 28–30), or irregular individual subelementary fibrils (Figs. 31 and 32), which at low concentration of Calcofluor White may be intermeshed, but are only seldom parallel-associated as in the control microfibrils of the lorica stalk region (Fig. 33). The ultrastructural effects of Congo Red are also very similar.

FIGURE 22. Negative stain of a short cytoplasmic tail with parallel arrangement of the microtubules (MT). The plasma membrane is more vesiculated in the intermicrotubule zones, indicating a stabilization of the plasma membrane where it overlies a microtubule. Uranyl acetate stain. Scale bar: 1 μm.

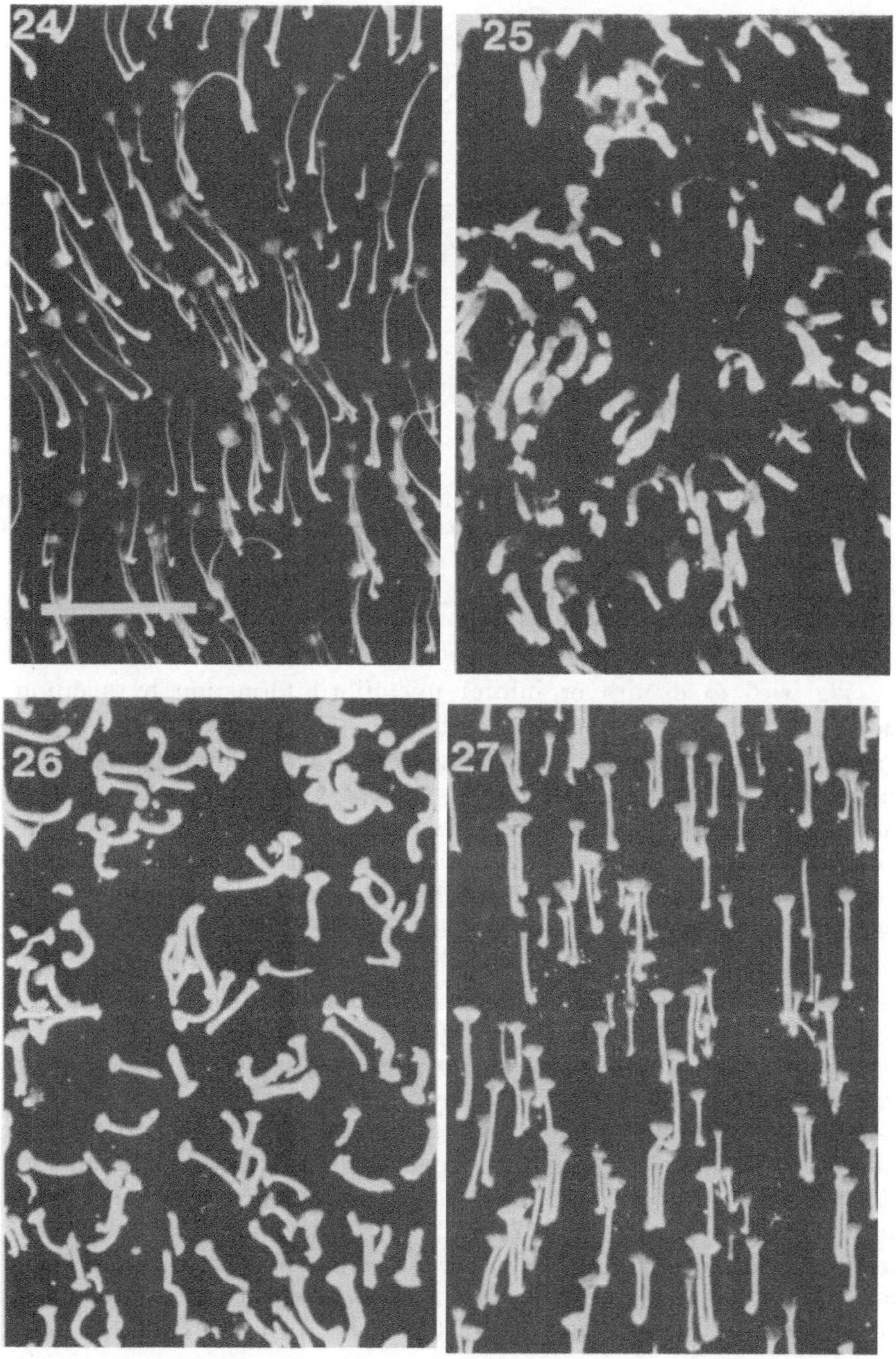

FIGURES 24–27. Calcofluor White (CW) experiments, light microscopy. From Herth (1980) with the permission of the Rockefeller University Press. Fig. 24. Control with numerous complete loricae after 2 hr of incubation without CW, stained for 5 min with 0.5% aqueous CW. Scale bar: 50 μm. Fig. 25. Incomplete loricae consisting mainly of stalks formed in the presence of 1 mg CW/ml. Fig. 26. Complete but very short and massive loricae formed in the presence of 0.5 mg CW/ml. Fig. 27. Loricae still very massive with 0.25 mg CW/ml.

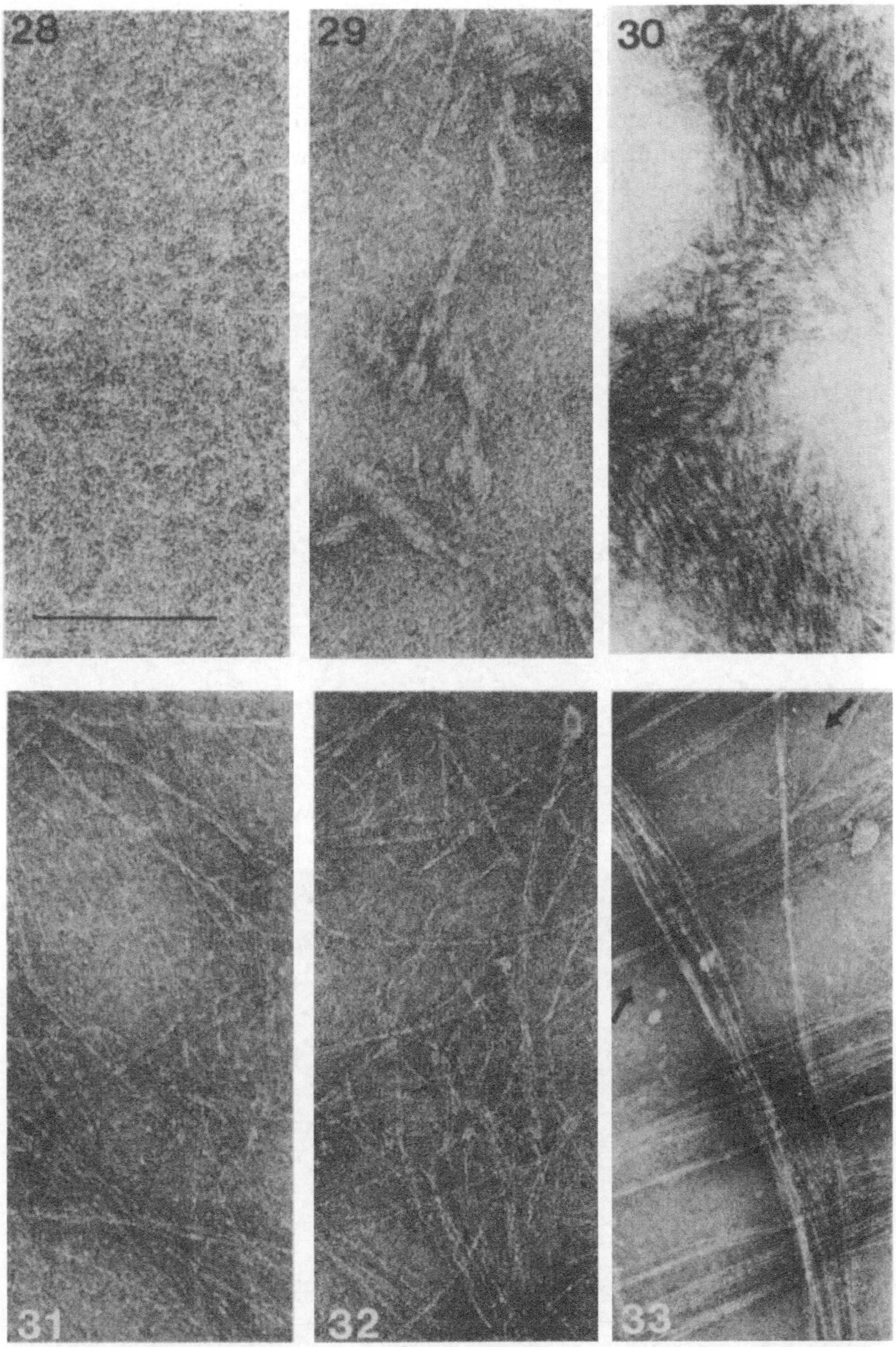

FIGURE 28–33. High magnification of negatively stained lorica material. From Herth (1980) with the permission of the Rockefeller University Press. Figs. 28–32. CW experiments. Fig. 33. Control. Fig. 28. Meshwork of short rods and short subelementary fibril pieces (1 mg CW/ml). Scale bar: 0.1 μm. Figs. 29 and 30. Aggregates of short rods (1 mg CW/ml). Fig. 31. Subelementary fibrils partially aggregated to microfibrils, and short rods (0.5 mg CW/ml). Fig. 32. Irregularly aggregated microfibrils (0.5 mg CW/ml) from the peripheral region of a patch. Fig. 33. Control microfibrils consisting of laterally associated subelementary fibrils and a few individual subelementary fibrils (→) in the range of 15 Å width.

The partial inhibition of microfibril formation has a second effect. The subelementary and intermeshed micro- and subelementary fibrils formed at lower Calcofluor White concentration (Fig. 34) are not oriented as in the controls (Fig. 35). This is not due to absence of microtubules in the cytoplasmic tail region (Herth, unpublished). Similar pictures are found with Congo Red application (Herth, 1980).

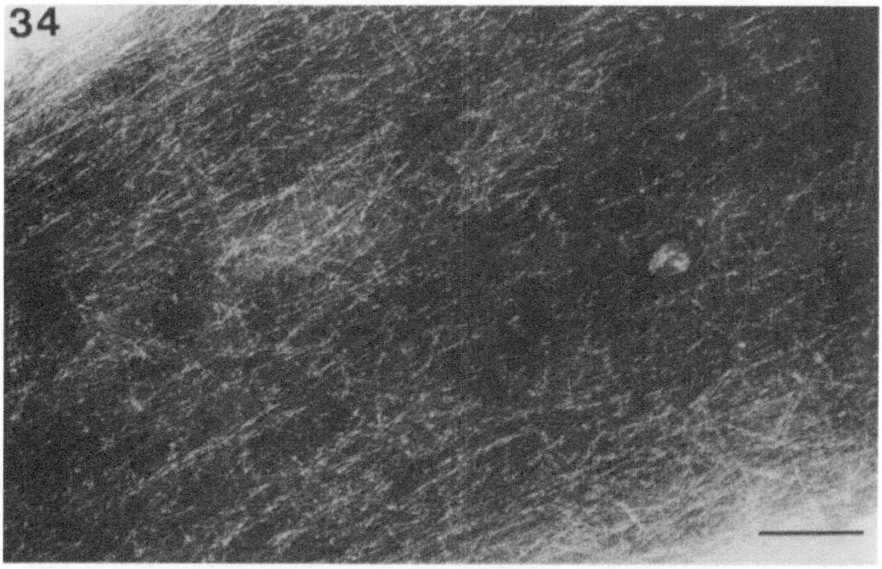

FIGURE 34 and 35. Loss of microfibril orientation with CW inhibition. From Herth (1980) with the permission of the Rockefeller University Press. Fig. 34. Lorica stalk region, formed at 0.25 mg CW/ml, with irregular fibril arrangement. Fig. 35. Control lorica stalk with parallel arrangement of microfibrillar ribbons consisting of laterally associated parallel subelementary fibrils. Scale bars: 0.1 μm.

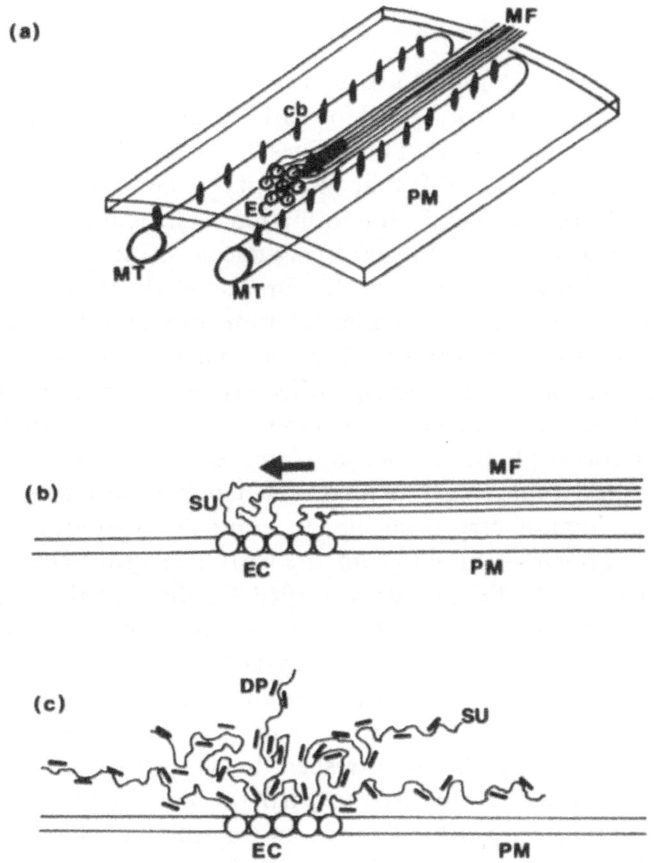

FIGURE 36. Model drawing. (a, b) The physical aggregation of polymerized subunits (SU) to microfibrils (MF) leads to a kinetic force (←) driving the enzyme complexes (EC) in a direction channeled by the cortical microtubules (MT). (cb) Cross bridge; (PM) plasma membrane. (c) With an intercalating dye such as Calcofluor White or Congo Red, the subunits are bound by the dye particles (DP) and therefore cannot regularly aggregate to microfibrils; the enzyme complexes cannot be moved regularly due to loss of the kinetic force of aggregation. From Herth (1980) with the permission of the Rockefeller University Press.

These results are interpreted as evidence for a two-step mechanism in microfibril formation. First, polymerization takes place, then microfibrils are formed by lateral aggregation of the polymers, via hydrogen bonding, and crystallize (Fig. 36a,b). The dyes obviously bind to the newly formed polymers and block the polymer surface against lateral aggregation (Fig. 36c).

The loss of orientation by inhibition of fibril aggregation shows that this reaction must play an essential role in microfibril orientation. We suppose that the lateral aggregation provides a kinetic force, which moves the synthetases laterally in the fluid plasma membrane (Fig. 36a,b). The microtubules, then, need not exert a kinetic force via contractile proteins, but would channel the movements of the synthetases in the membrane only by restricting the fluid membrane regions by interaction with the membrane via their associated proteins.

4. Conclusions

The investigation of the site of β-chitin-fibril formation in the centric diatoms demonstrates that microfibril formation is associated with a specialized region of the plasma membrane. This refines the general hypothesis of Preston (for a review, see Preston, 1974) that multienzyme complexes in the plasma membrane should be involved in the formation of structural polysaccharides. Only in *Thalassiosira* is this special invagination of the plasma membrane associated with "chitosome"-like vesicles (Bracker *et al.*, 1976; Ruiz-Herrera *et al.*, 1975), which in this case, as postulated for the fungi, might be synthetase en route to its destination. In contrast to most other objects investigated, the special membrane regions are cytologically differentiated in that they bear coat structures, perhaps a consequence of the special shape of the synthesizing site or its spatial restriction or both. In *Poterioochromonas*, the earlier results and the accessibility to the dyes also make an involvement of the plasma membrane of the cytoplasmic tail region very probable. It would be interesting to see these structures also in freeze-etching to compare with the recent freeze-etch data on cellulose formation (Mueller and Brown, 1980; Giddings *et al.*, 1980).

Both in the diatoms and in *Poterioochromonas*, there seems to be a short temporal lag between polymerization and crystallization of the microfibrils. The diatom appendages clearly grow at their proximal end. In *Poterioochromonas*, the zone of microfibril growth moves with respect to the newly formed microfibrils and possibly also with respect to the cell surface. This enables *Poterioochromonas* to direct the orientation of the newly formed microfibrils in either actively or indirectly influencing the movement of the synthetases and the aggregation of the polymers (see Chapter 1). This picture is only in partial agreement with the hypotheses of Heath (1974) and of Roland and Vian (1979) on the influence of microtubules on microfibril orientation, since it does not obligatorily need contractile MAPs.

The concept of a stepwise reaction in microfibril formation was already clearly formulated over a decade ago by Ben-Hayyim and Ohad (1968), but seems not to be accepted by the majority of workers, who seem to favor the concept of simultaneous polymerization and crystallization in a multienzyme complex (reviewed by Preston, 1974) (see also Robinson, 1977). Stepwise reactions would enable the cell to control the formation, aggregation, and orientation of the flexible polymer, and then to let it crystallize as supporting structure. It is tempting to generalize this hypothesis now shown for the cases of chitin-microfibril formation in the centric diatoms and *Poterioochromonas*, as well as cellulose formation in *Acetobacter xylinum* (Haigler and Brown, 1979) (see also Chapter 14), and also suggested from the cytological investigation of spike formation in chlorococcalean algae (Schnepf *et al.*, 1980, 1982).

ACKNOWLEDGMENT

This work was supported by the Deutsche Forschungsgemeinschaft.

References

Anderson, R. G. W., Brown, M. S., and Goldstein, J. L., 1977, Role of the coated endocytotic vesicle in the uptake of receptor-bound low density lipoprotein in human fibroblasts, *Cell* **10:**351–364.

Anonymous, 1975, Proposals for a standardization of diatom terminology and diagnoses, *Nova Hedwigia Z. Kryptogamenkd.*, Suppl. 53, pp. 323–354.

Ben-Hayyim, G., and Ohad, I., 1968, Synthesis of cellulose by *Acetobacter xylinum*. VIII. On the formation and orientation of bacterial cellulose fibrils in the presence of acidic polysaccharides, *J. Cell Biol.* **25:**191–206.

Blackwell, J., Parker, K. D., and Rudall, K. M., 1967, Chitin fibers of the diatoms *Thalassiosira fluviatilis* and *Cyclotella cryptica*, *J. Mol. Biol.* **28:**383–385.

Bracker, C. E., Ruiz-Herrera, J., and Bartnicki-Garcia, S., 1976, Structure and transformation of chitin synthetase particles (chitosomes) during microfibirl synthesis *in vivo*, *Proc. Natl. Acad. Sci. U.S.A.* **73:**4570–4574.

Braconnot, H., 1811, Sur la nature des champignons, *Ann. Chim. Phys.* **79:**265–304.

Dweltz, N. E., Colvin, J. R., and McInnes, A. G., 1968, Chitan (β-(1-4)-linked 2 acetamido-2-deoxy-D-glucan) fibers of the diatom *Thalassiosira fluviatilis*. 3. The structure of the chitan from X-ray diffraction and electron microscope observations, *Can. J. Chem.* **46:**1513–1521.

Falck, M., Smith, D. G., McLachlan, J., and McInnes, A. G., 1966, Studies on the chitin fibers of the diatom *Thalassiosira fluviatilis* Hustedt. 2. Proton magnetic resonance, infrared and X-ray study, *Can. J. Chem.* **44:**2269–2281.

Giddings, T. H., Brower, D. L., and Staehelin, L. A., 1980, Visualization of particle complexes in the plasma membrane of *Micrasterias denticulata* associated with the formation of cellulose fibrils in primary and secondary walls, *J. Cell Biol.* **84:**327–339.

Haigler, C., and Brown, R. M., 1979, The fluorescent brightener, Calcofluor White, alters cellulose synthesis in *Acetobacter xylinum*, *J. Cell Biol.* **83:**70a.

Heath, I. B., 1974, A unified hypothesis for the role of membrane bound enzyme complexes and microtubules in plant cell wall synthesis, *J. Theor. Biol.* **48:**445–449.

Herth, W., 1978, A special chitin-fibril-synthesizing apparatus in the centric diatom *Cyclotella*, *Naturwissenschaften* **65:**260.

Herth, W., 1979, The site of β-chitin fibril formation in centric diatoms. II. The chitin-forming cytoplasmic structures, *J. Ultrastruct. Res.* **68:**16–27.

Herth, W., 1980, Calcofluor White and Congo Red inhibit chitin microfibril assembly of *Poterioochromonas:* Evidence for a gap between polymerization and microfibril formation, *J. Cell Biol.* **87:**442–450.

Herth, W., and Barthlott, W., 1979, The site of β-chitin fibril formation in centric diatoms. I. Pores and fibril formation, *J. Ultrastruct. Res.* **68:**6–15.

Herth, W., and Zugenmaier, P., 1977, Ultrastructure of the chitin fibrils of the centric diatom *Cyclotella cryptica*, *J. Ultrastruct. Res.* **61:**230–239.

Herth, W., and Schnepf, E., 1980, The fluorochrome, Calcofluor White, binds oriented to structural polysaccharide fibrils. *Protoplasma* **105:**129–133.

Herth, W., Kuppel, A., and Schnepf, E., 1977, Chitinous fibrils in the lorica of the flagellate chrysophyte *Poterioochromonas stipitata* (syn. *Ochromonas malhamensis*), *J. Cell Biol.* **73:**311–321.

Kim, H., Binder, L. I., and Rosenbaum, J. L., 1979, The periodic association of MAP_2 with brain microtubules *in vitro*, *J. Cell Biol.* **80:**266–276.

Mueller, S. C., and Brown, R. M., 1980, Evidence for an intramembrane component associated with a cellulose microfibril-synthesizing complex in higher plants, *J. Cell Biol.* **84:**315–326.

Muzzarelli, R. A. A., 1977, *Chitin*, Pergamon Press, Oxford, New York, Toronto, Sidney, Paris, and Frankfurt.

Odier, A., 1843, Mémoire sur la composition chimique des parties cornées des insectes, *Mem. Soc. Hist. Nat. Paris* **1:**29–42.

Preston, R. D., 1974, *The Physical Biology of Plant Cell Walls*, Chapman and Hall, London.

Robinson, D. G., 1977, Plant cell wall synthesis, *Adv. Bot. Res.* **5:**89–151.

Robinson, D. G., and Quader, H., 1980, Topographical features of the membrane of *Poterioochromonas malhamensis* after colchicine and osmotic treatment, *Planta* **148**:84–88.

Roland, J. C., and Vian, B., 1979, The wall of the growing plant cell, *Int. Rev. Cytol.* **61**:129–166.

Ruiz-Herrera, J., Sing, V. O., VanDerWoude, W. J., and Bartnicki-Garcia, S., 1975, Microfibril assembly by granules of chitin synthetase, *Proc. Natl. Acad. Sci. U.S.A.* **72**:2706–2710.

Schnepf, E., Röderer, G., and Herth, W., 1975, The formation of the fibrils in the lorica of *Poteriochromonas stipitata:* Tip growth, kinetics, site, orientation, *Planta* **125**:45–62.

Schnepf, E., Deichgräber, G., Röderer, G., and Herth, W., 1977, The flagellar root apparatus, the microtubular system and associated organelles in the chrysophycean flagellate, *Poteriochromonas malhamensis* Peterfi (syn. *Poteriochromonas stipitata* Scherffel and *Ochromonas malhamensis* Pringsheim), *Protoplasma* **92**:87–107.

Schnepf, E., Deichgräber, G., Glaab, M., and Hegewald, E., 1980, Bristles and spikes in Chlorococcales: Ultrastructural studies in *Acanthosphaera, Micractinium, Pediastrum, Polyedriopsis, Scenedesmus*, and *Siderocystopsis. J. Ultrastruct. Res.* **72**:367–379.

Schnepf, E., Deichgräber, G., and Herth, W., 1982, Development of cell wall appendages in *Acanthosphaera zachariasi* (Chlorococcales): Kinetics, site of cellulose synthesis and of microfibril assembly, and barb formation, *Protoplasma*, in press.

Schütt, F., 1899, Centrifugales Dickenwachstum der Membran und extramembranöses Plasma, *Jahrb. Wiss Bot.* **33**:1–97.

Sloboda, R. D., Dentler, W. L., Bloodgood, R. A., Telzer, B. R. Granett, S., and Rosenbaum, J. L., 1976, Microtubule-associated proteins (MAPs) and the assembly of microtubules *in vitro*, in: *Cell Motility*, Vol. C (R. Goldmann, T. Pollard, and J. L. Rosenbaum, eds.), pp. 1117–1212, Cold Spring Harbor Laboratory, Cold Spring Harbor, New York.

Zugenmaier, P., Herth, W., and Lieser, G., 1981, Elektronenbeugungsuntersuchungen an Polymereinkristallen des β-Chitins. *Colloid & Polymer Sci.* **259**:472–473.

11

Synthesis of Chitin Microfibrils *in Vitro*

José Ruiz-Herrera

1. Introduction

Chitin is one of the most abundant organic compounds on earth. It is present in both animals and fungi, and characteristically it is absent in plants. In nature, the most abundant crystalline form of the polymer is α-chitin, which occurs in the form of microfibrils. Microfibrils have been observed by electron microscopy of fungal cell walls by use of different techniques: shadow-casting, heavy-metal replicas, negative staining, and oblique and tangential sections. In the cell walls of fungi, chitin microfibrils constitute the skeleton responsible for their rigidity (for a review, see Ruiz-Herrera, 1978).

Understanding the mechanism of chitin biosynthesis appears to be a prerequisite to comprehending how growth of the fungal cell wall and the exoskeleton of arthropods occurs. These studies underlie not only the biochemical mechanism of growth of the Gl_cNA_c-*N*-acetylglucosamine (poly-GlcNAc) chain but also how these are assembled to constitute a microfibril, the nature and location of the organelle(s) involved in microfibril biosynthesis, and how the process is timed and regulated.

During the last few years, *in vitro* systems for the biosynthesis of microfibrils have been developed and much progress in the purification, dissociation, and regulation of chitin synthetase has been made. In this chapter, I will try to summarize the work done in this field and offer a panoramic view of the present state of the art.

José Ruiz-Herrera • Instituto de Investigacion en Biologia Experimental, Facultad de Quimica, Universidad de Guanajuato, Apartado Postal N'187, Guanajuato, Gto. Mexico.

2. General Properties of Chitin Synthetase

Soon after the disclosure of the role of nucleotide sugars in the synthesis of disaccharides, their participation in the biosynthesis of several polysaccharides was established (for a review, see Leloir, 1972). One of the first polysaccharides synthesized *in vitro* was chitin (Glaser and Brown, 1957). The authors used particulate preparations from *Neurospora crassa* and UDPGlcNAc as glycosyl donor. Further studies have confirmed UDPGlcNAc as the precursor of chitin in both arthropods and fungi (e.g., Jaworski *et al.*, 1963; Porter and Jaworski, 1966; Camargo *et al.*, 1967; McMurrough *et al.*, 1971). There is no evidence that any other nucleotide of GlcNAc may be a precursor of chitin. Thus, only UTP was found involved in the formation of a chitin precursor in cell-free extracts from *Mucor rouxii* incubated with [^{14}C]-GlcNAc and several nucleotide triphosphates (McMurrough *et al.*, 1971). Likewise, cell-free extracts from *Cochliobolus miyabeanus* synthetized chitin only when incubated with [^{14}C]GlcNAc and UTP. GTP, ATP, and TTP were ineffective (Nanba and Kuroda, 1974). In nearly all systems studied, chitin synthetases are greatly stimulated by addition of a divalent metal. Mg^{2+} is most effective at concentrations of 2.5–25 mM (e.g., Glaser and Brown, 1957; McMurrough and Bartnicki-García, 1971; Keller and Cabib, 1971; López-Romero and Ruiz-Herrera, 1976). In *M. rouxii*, Mn^{2+} gave higher stimulations than Mg^{2+}, but the concentration range was narrower than that for Mg^{2+}, and Co^{2+} was almost as effective as Mg^{2+} (Ruiz-Herrera *et al.*, 1977). In *Cunninghamella elegans* (Moore and Peberdy, 1975) and *Mortierella vinacea* (Peberdy and Moore, 1975), no stimulatory effect of Mg^{2+} was observed. Possibly, these preparations contained a high level of Mg^{2+}. In *M. rouxii*, we have obtained variations in the activity of different samples incubated without Mg^{2+}, suggesting different levels of contamination with the metal. When the preparations were dialyzed against EDTA and the chelator was also eliminated by dialysis, there was no chitin synthetase activity in the absence of Mg^{2+}. But on addition of divalent metals, full activity was recovered (J. Ruiz-Herrera, unpublished data). Accordingly, it may be concluded that chitin synthetase has an absolute requirement for a divalent metal.

In all systems studied, chitin synthetase shows unconventional kinetics. Reciprocal plots of $1/v$ vs. $1/s$ are pronouncedly sigmoidal. This has been attributed to allosteric behavior of the enzyme (McMurrough and Bartnicki-García, 1971; Ruiz-Herrera *et al.*, 1977). De Rousset-Hall and Gooday (1975) reported a Hill coefficient of 4.0 for digitonin-dissociated enzyme from *Coprinus cinereus*, suggesting that the enzyme has four binding sites for UDPGlcNAc at the catalytic site. Ryder and Peberdy (1977) found a Hill coefficient of 1.0 for chitin synthetase from *Aspergillus nidulans*. For chitin synthetase from *M. rouxii*, either particulate, chitosomal, or dissociated with digitonin, a Hill coefficient of 1.0 has been found (J. Ruiz-Herrera, R. Hernández-Delgadillo, and E. López-Romero, unpublished data), suggesting that chitin synthetase had one binding site for substrate.

Free GlcNAc stimulates chitin synthetase to varying degrees. This stimulation, initially described by Glaser and Brown (1957) for *N. crassa*, has been confirmed for most systems analyzed (e.g., Porter and Jaworski, 1966; Camargo *et al.*, 1967; McMurrough and Bartnicki-García, 1971; Keller and Cabib, 1971; López-Romero and Ruiz-Herrera, 1976). It seems that GlcNAc acts as an allosteric activator (McMurrough and Bartnicki-García, 1971; Ruiz-Herrera *et al.*, 1977), since increasing concentrations of GlcNAc reduce the sigmoidal character of kinetic curves. The high concentrations of GlcNAc usually employed to obtain good stimulation of chitin synthetase are not likely to be found in the cytoplasm (thus, 4.5–12.5 mM GlcNAc is usually needed to attain half V_{max}). This stimulation is strictly dependent on substrate concentration: high stimulations are obtained at low levels of UDPGlcNAc and vice versa. Since UDPGlcNAc at much lower concentrations produces the same stimulatory effect as GlcNAc, it was concluded that GlcNAc probably has no role *in vivo*; it shows an effect *in vitro* because it mimics UDPGlcNAc, the real allosteric effector (McMurrough and Bartnicki-García, 1971; Ruiz-Herrera *et al.*, 1977). Gooday (1977), however, conjectured that a high concentration of GlcNAc may be found in the vicinity of chitin synthetase as a result of the action of chitinases and diacetyl chitobiase, and thus GlcNAc may play a regulatory function coordinating chitin breakdown and synthesis. There are conflicting reports about the effect of chitodextrins on chitin synthetase. Stimulation has been described for *N. crassa* (Glaser and Brown, 1957), *Venturia inaequalis* (Jaworski *et al.*, 1965) *Allomyces macrogynus* (Porter and Jaworski, 1966), *Blastocladiella emersonii* (Camargo *et al.*, 1967), *Mucor rouxii* (McMurrough and Bartnicki-García, 1971), *Cochliobolus miyabeanus* (Nanba and Kuroda, 1974), *Coprinus cinereus* (Gooday and De Rouset-Hall, 1975), and *Aspergillus flavus* (Moore and Peberdy, 1976). No effect was reported for *Saccharomyces carlsbergensis* (Keller and Cabib, 1971). Inhibition by chitodextrin was noted for chitin synthetase preparations from *M. vinacea* (Peberdy and Moore, 1975) and *C. elegans* (Moore and Peberdy, 1975). It is possible that chito dextrins serve as an alternative substrate for contaminating chitinases present in crude preparations and thus protect the newly synthesized product.

3. Effect of Nucleosides and Nucleotides

ATP at low concentrations stimulates chitin synthetase slightly in crude preparations (McMurrough and Bartnicki-García, 1971, Jan, 1974). UDP, a product of the chitin synthetase reaction (Porter and Jaworski, 1966; Keller and Cabib, 1971; Myers and Cantino, 1974), is inhibitory (Keller and Cabib, 1971; Hori *et al.*, 1974a; Peberdy and Moore, 1975; López-Romero and Ruiz-Herrera, 1976; Ruiz-Herrera *et al.*, 1977), but this effect is not specific; other nucleotides and nucleotide-sugars are also inhibitory (Keller and Cabib, 1971; Hori *et al.*, 1974a; Ruiz-Herrera *et al.*, 1977). A detailed kinetic analysis of inhibition brought about by nucleosides, nucleotides, and sugar-nucleotides was carried out by Hori *et al.*, (1974a). These authors found that all compounds tested were competitive inhibitors in *Piricularia oryzae.* UDP had the lowest K_i, 1.4 mM, followed closely by UMP and uridine, 1.73 and 1.92 mM, respectively. They concluded that chitin synthetase had a pyrimidine-specific site for UDPGlcNAc where the inhibitors competed. Competitive inhibition by UDP was also reported for *C. cinereus* (De Rousset-Hall and Gooday, 1975), *A. flavus* (López-Romero and Ruiz-Herrera, 1976), and *M. rouxii* (Ruiz-Herrera *et al.*, 1977). In the last two fungi, complex kinetics were observed, suggestive of allosteric interactions.

4. Polyoxins

Polyoxins are a family of antibiotics produced by *Streptomyces cacaoi* var. *asoensis* (Isono *et al.*, 1969) that bear a close structural resemblance to UDPGlcNAc. Polyoxins are strong and highly specific inhibitors of chitin synthetase. All chitin synthetase systems tested *in vitro* are competitively inhibited by polyoxins (Endo and Misato, 1969; Ohta *et al.*, 1970; Keller and Cabib, 1971; Bartnicki-Garcia and Lippman, 1972; López Romero and Ruiz-Herrera, 1976; Ruiz-Herrera *et al.*, 1977). The K_i values for polyoxin A are about 1000 times smaller than the K_m of chitin synthetase. Hori *et al.* (1974 a, b) concluded from kinetic data that polyoxins bind through their pyrimidine-nucleoside moiety to the uridine-binding site of chitin synthetase. Polyoxins can be effective *in vivo* (Endo *et al.*, 1970; Bartnicki-García and Lippman, 1972; Bowers *et al.*, 1974; Gooday and De Rousset-Hall, 1975; Benítez *et al.*, 1976), but generally at much higher concentrations. Apparently, polyoxin does not easily reach the site where chitin synthesis takes place *in vivo*. Another nucleoside-antibiotic that inhibits chitin synthetase is Nikkomycin, produced by *Streptomyces tendae* (Dahn *et al.*, 1976).

5. Activation and Inactivation of Chitin Synthetase

Cabib and Farkas (1971) reported that chitin synthetase from *Saccharomyces cerevisiae* and *S. carlsbergensis* was present mostly in an inactive (zymogenic) state that required activation by trypsin or by an endogenous "activating factor." In addition to trypsin, chymotrypsin, subtilisin, and papain-activated chitin synthetase (Ulane and Cabib, 1976). The activating factor was identified as a protease (Cabib and Ulane, 1973), purified (Ulane and Cabib, 1976) and shown to be identical with yeast proteinase B previously described (Lenney and Dalbec, 1967; Schott and Holzer, 1974). Acid protease (proteinase A) and carboxypeptidase C, the other two major proteases present in *Saccharomyces*, do not activate chitin synthetase (Hassilik, 1974). High concentrations of proteinase B inactivate chitin synthetase (Hassilik, 1974). Proteinase B is found in the cell vacuoles (Cabib and Ulane, 1973; Lenney *et al.*, 1974; for a review, see Holzer and Saheki, 1976). Activation of yeast chitin synthetase by proteinase B is blocked by the specific inhibitor of the protease (Cabib and Farkas, 1971; Ulane and Cabib, 1976) (see below). Activation of chitin synthetase by controlled proteolysis appears to be a general phenomenon in fungal enzymatic preparations, but the different systems vary in specificity. Thus, in *M. rouxii*, all proteases tested activated chitin synthetase, but a fungal acid protease was much more effective (Ruiz-Herrera and Bartnicki-Garcia, 1976; Ruiz-Herrera *et al*, 1977). Chitin synthetase from *A. flavus* was activated by fungal acid proteases only (López-Romero and Ruiz-Herrera, 1976), but enzyme from *A. nidulans* was activated by trypsin (Ryder and Peberdy, 1977). Chitin synthetases from *Agaricus bisporus* and *N. crassa* are activated by neutral but not by acid proteases (Bartnicki-García *et al.*, 1978; Arroyo-Begovich and Ruiz-Herrera, 1979). Chitin synthetases from *Phycomyces blakesleeanus* (Van Laere and Carlier, 1978), *Candida albicans* (Braun and Calderone, 1978), and *Schizophyllum commune* (Vermeulen *et al.*, 1979) are all activated by trypsin. Chitin synthetase from *C. cinereus* (Gooday and De Rousset-Hall, 1975) was not stimulated by proteases, possibly because the enzyme was already activated at the time of the assay. Crude chitin synthetase preparations from *M. rouxii* undergo self-activation and destruction owing to the presence of endogenous proteases (Ruiz-Herrera and Bartnicki-Garcia, 1976). Activation seems to be due to an acid protease since it is inhibited by pepstatin A, a specific inhibitor of acid proteases (Aoyagi and Umezawa, 1975), but not by neutral protease inhibitors: phenyl-methyl sulfanyl fluoride (PMSF), leupeptin, chymostatin. Inactivation of chitin synthetase was not blocked by any of the aforenamed protease inhibitors (J. Ruiz-Herrera and S. Bartnicki-García, unpublished observations). From the cytoplasm of the mycelium of *M. rouxii*, we have purified to homogeneity a protease that in very small concentrations effectively destroys chitin synthetase of the fungus (Fig. 1) (Diaz and Ruiz-Herrera, 1978).

Similarly to *M. rouxii*, chitin synthetase present in crude extracts of *N. crassa* undergoes a process of activation and destruction. Enzyme can be stabilized by incubation at alkaline pH. Protease inhibitors block activation but not destruction of chitin synthetase (Arroyo-Begovich and Ruiz-Herrera, 1979). The exact molecular change(s) effected by activating proteases on chitin synthetase and the physiological role of this process *in vivo* are not known. A previous hypothesis on the role of proteinase B in activating chitin synthetase at the onset of septum formation in yeast (Cabib and Farkas, 1971) is in conflict with the isolation of mutants lacking the protease, but able to bud normally (Wolf and Ehmann, 1978; Zubenko *et al.*, 1979).

6. *Autogenous Inhibitors of Chitin Synthetase*

Cells of *Saccharomyces* species contain a low-molecular-weight protein originally reported to inhibit chitin synthetase (Cabib and Keller, 1971). This claim was later amended when the inhibitory substance was found to block the action of the endogenous activating factor (Cabib and Farkas, 1971). This

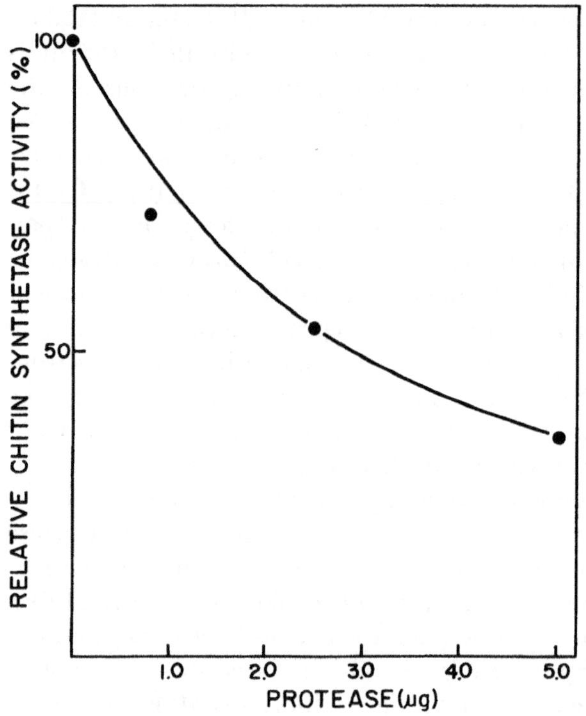

FIGURE 1. Inactivation of chitosomal chitin synthetase by purified acid protease from mycelial cells of *Mucor rouxii*.

protein turned out to be the specific inhibitor of proteinase B described by Lenney and Dalbec (1969). This inhibitor has been purified by several investigators (Ferguson *et al.*, 1973; Betz *et al.*, 1974; Ulane and Cabib, 1974; Lenney, 1975; for a review, see Holzer and Saheki, 1976). Proteinase B has a broad spectrum of action on yeast enzymes and is not a specific activator of chitin synthetase *in vitro* (Holzer and Saheki, 1976). A similar inhibitor has been described for a protease from *C. albicans* (Braun and Calderone, 1979).

The cytosol of *M. rouxii* contains a substance that inhibits chitin synthesis *in vitro* (McMurrough and Bartnicki, García, 1973). This substance was further characterized as a protein, and it was suggested that it inhibited chitin synthetase itself, and not its activation (López-Romaro *et al.*, 1978). However, it has been demonstrated more recently that the putative inhibitor is in fact a chitinase that preferentially degrades nascent chitin (E. López-Romero, S. Bartnicki-García, and J. Ruiz-Herrera, unpublished observations).

7. *Biosynthesis of Chitin Microfibrils*

Early studies on chitin biosynthesis were made with crude cell fractions, and the evidence for chitin synthesis rested exclusively on radioisotopic data. Identification of chitin relied on its susceptibility of chitinase, partial hydrolysis, and analyses of the acetylated dextrins and alkali insolubility of the product. Such studies could not establish the physical nature of synthesized product and left serious doubts as to whether chitin had been biosynthesized and whether UDPGlcNAc was the natural substrate for chitin synthetase or a poor substitute.

When a crude membrane-rich fraction of *M. rouxii* was incubated at low temperature with substrate and activators, a substantial part of chitin synthetase became "soluble"; i.e., it did not sediment when the suspension was centifuged at high speed (Ruiz-Herrera and Bartnicki-García, 1974). Preliminary findings mentioned at that time (Ruiz-Herrera and Bartnicki-García, 1974) and later expanded (Ruiz-Herrera *et al.*, 1975) warned that despite the transparency of the supernatant, the enzyme was sedimentable at high centrifugal forces, and it was not really soluble. With chitin synthetase present in these supernatants, polysaccharide microfibrils were synthesized *in vitro* for the first time. The product was identified as α-chitin by X-ray diffraction [this conclusion has been confirmed by electron diffraction (H. Chanzy and J. Blackwell, unpublished)]. Biosynthesis of chitin microfibrils was also obtained later with chitosomes from *S. cerevisiae, N. crassa, A. bisporus,* and *A. macrogynus* (Bartnicki-García *et al.*, 1978) (see below); γ-particles from *B. emersonii* (Mills and Cantino, 1978); chitin synthetase dissociated with digitonin from crude membrane-rich fraction of *S. cerevisiae* (Durán and Cabib, 1978) (see below); membrane-rich fractions of *S. commune* (Vermeulen *et al.*, 1979); and chitosomal subunits from *M. rouxii* (Ruiz-Herrera *et al.*, 1980) (see below).

8. *Chitosomes and Chitin Biosynthesis*

From the cytoplasm of the yeast form of *M. rouxii*, we have purified chitin synthetase in the form of microvesicles that we have named "chitosomes" (Bracker *et al.*, 1976; Ruiz-Herrera *et al.*, 1977). Chitosomes have a rather uniform size with an average diamter of 40–70 nm. No forms of chitin synthetase smaller than chitosomes have been detected in cell extracts from all fungi studied. Thus, chitosomes appear to be the elementary assemblies of chitin synthetase in fungal cells. Two forms of chitosomes are recognized, one impermeable to negative stains [proctoid (Fig. 2)], the other permeable to stains that reveal some internal structure and the presence of a delimiting shell 6–12 nm thick [cycloid (Fig. 3)]. Seemingly, the appearance is caused primarily by differences in the thickness of the negative stain and does not represent two developmental or functional states (C. E. Bracker, unpublished observations). In thin sections, chitosomes show only one type of morphology, that of a hollow sac with a tripartite shell 6.5–7.0 nm thick (Fig. 4) similar to a unit membrane, albeit thinner and more difficult to stain. High concentrations of trypsin induce a peculiar change in their morphology (Bartnicki-García *et al.*, 1978). The shell reveals two distinct layers, a thin fuzzy outer layer and a thick internal one. Also, an internal pouch is

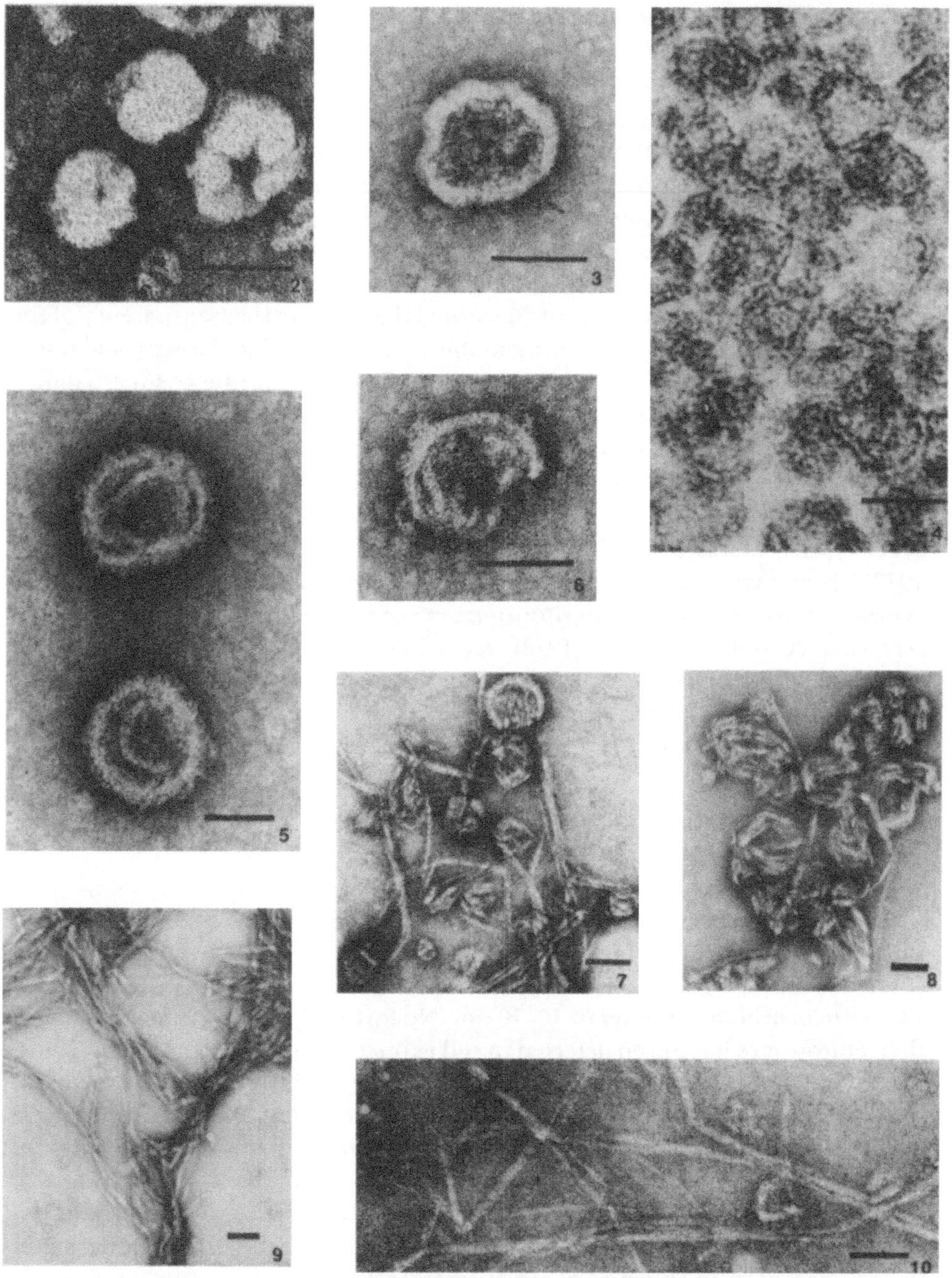

FIGURES 2–9. Chitosome structure and microfibril formation. Fig. 2. Proctoid forms of chitosomes. Fig. 3. Cycloid form of chitosomes. Fig. 4. Thin section view of isolated chitosomes. Fig. 5. Chitosomes after treatment with high concentration of trypsin. Fig. 6. Fibroid with rest of chitosomal shell still attached. Fig. 7. Fibroids associated with straight microfibrils. Observe the residual shell associated with fibroid. Fig. 8. Naked fibroids associated with short, straight microfibrils. Fig. 9. Long microfibrils synthesized by chitosomes. Scale bars: 50 nm. Fig. 10. Chitin synthesized by digitonin-dissociated chitin synthetase. Observe the arrangement of short needlelike microfibrils.

revealed (Fig. 5). On incubation with substrate (UDPGlcNAc) and activators, chitosomes undergo a seemingly irreversible series of transformations (Bracker *et al.*, 1976). Inside some chitosomes, thin and delicate microfibrils can be observed. Others show thicker coiled microfibrils that impart to the chitosomes an internal fibroid appearance. Often, these fibroids are continous with straight microfibrils. In most cases, the shell of the chitosome is either lost or broken but still adheres to the fibrillar content. In fewer cases, it remains intact. Images of these stages are shown in Fig. 6–8.

Fibroid coils are not known to occur in the cell walls of fungi. Presumably, their formation *in vitro* is a result of the crystallization of chitin within the confinement of the shell by the premature activation of chitosomes. The growing fibroid, however, probably causes the chitosome shell to split open, allowing the formation of long, straight microfibrils. Chitosomes are not exclusive to the yeast form of *M. rouxii*. We have applied the same procedure described for the isolation of chitosomes from *M. rouxii* to extracts from fungi belonging to different taxonomic groups (Bartnicki-García *et al.*, 1978). Analysis of the behavior of their chitin synthetase in sucrose gradients showed profiles of sedimentation almost identical to those of chitosomes from *M. rouxii*. Microscopic observation of the fractions containing chitin synthetase activity showed typical chitosomes. These behaved in the same fashion after either incubation with substrate or treatment with trypsin as those isolated from *M. rouxii* (Bartnicki-García *et al.*, 1978).

We have regarded chitosomes as the natural conveyors of chitin synthetase from its place of synthesis (conceivably the rough endoplasmic reticulum) to its site of action (probably the membrane–wall interphase). In this sense, chitosomes would be the natural packages of chitin synthetase, containing as many individual enzymes as are necessary for the formation of a microfibril (see below). One important question is whether chitosomes are real organelles or artifacts produced by vesiculation of larger membrane structures by the harsh ballistic method employed to break the cells. The following pieces of evidence support the true existence of chitosomes in the fungal cell: (1) The shell of chitosomes in thin sections appears to have a different width than that of the plasma membrane (Bracker *et al.*, 1976). (2) Chitosomes lack the ATPase characteristic of plasma membrane (M. Morales and J. Ruiz-Herrera, unpublished observations). (3) Chitosomes show important differences in their lipid composition as compared to whole cells; i.e., they lack phosphatidyl serine and have much higher amounts of glycolipids (Hernández, *et al.* in press). (4) Similar yields of chitosomes, morphologically indistinguishable, can be obtained by either osmotic lysis or ballistic disruption of both protoplasts from *M. rouxii* and slime mutant of *N. crassa* (Bartnicki-García, *et al.*, 1980). (5) Typical chitosomes capable of synthesizing chitin microfibrils and fibroids can be observed in cytoplasmic material manually extruded or removed by a microsyringe from the tip of sporangiophores of *P. blakesleeanus* (Figs. 11–13) (L. Herrera and J. Ruiz-Herrera, to be submitted). (6) Membrane-bound structures of the same size and the same membrane thinness as chitosomes can be observed in sections of whole

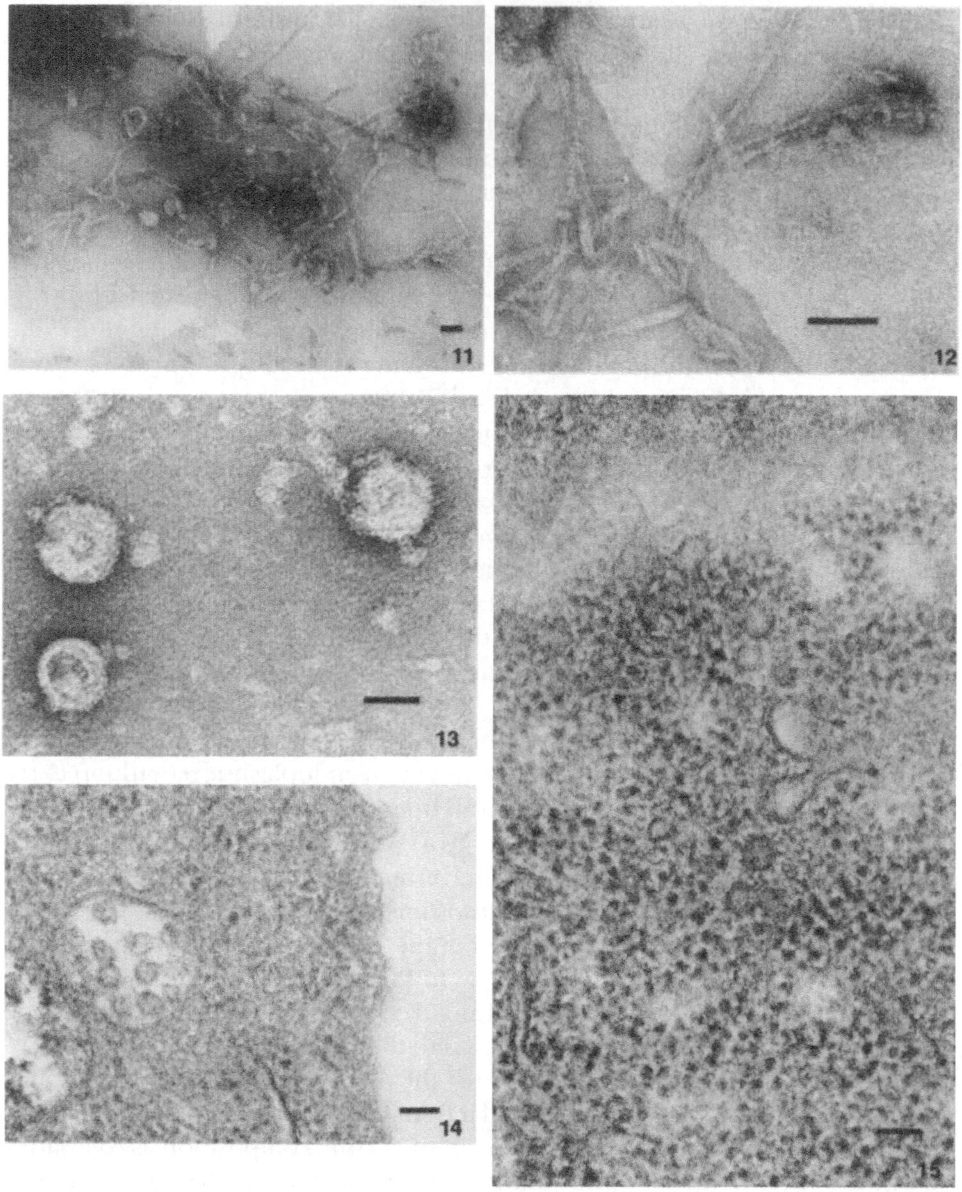

FIGURES 11–15. Fig. 11. Microfibrils synthesized by chitosomes present in extruded cytoplasmic material from the tip of *Phycomyces blakesleeanus*. Fig. 12. Fibroids continuous with straight microfibrils synthesized by chitosomes present in extruded material from the tip of *P. blakesleeanus*. Fig. 13. Chitosome treated with a high concentration of trypsin in cytoplasmic material removed from the tip of *P. blakesleeanus* with a microsyringe. Figs. 14 and 15. Thin section of intact cells of *M. rouxii* showing microvesicles similar to isolated chitosomes either inside multivesicular bodies (Fig. 14) or free (Fig. 15). Observe the proximity of microvesicles to the cell surface in both photographs. Scale bars: 50 nm.

cells; some are free and some inside the so-called multivesicular bodies (Figs. 14 and 15) (Bracker *et al.*, 1976).

*9. Dissociated Forms of Chitin Synthetase**

Glaser and Brown (1957) noted that chitin synthetase from *N. crassa* could be rendered "soluble" by treatment with butanol. These results have not been confirmed. Organic solvents and detergents are usually deleterious to chitin synthetase activity. When membrane-rich fractions of *M. rouxii* were treated with nonionic detergents and excess detergent was removed, chitin synthetase was partially recovered (Ruiz-Herrera and Bartnicki-García, 1976). Gooday and De Rousset-Hall (1975) successfully employed digitonin to dissociate chitin synthetase from *C. cinereus.* Digitonin had been previously utilized for the "solubilization" of several plant glycosyl transferases [e.g., β1,3-glucan synthetase from *Phaseolus aureus* (Feingold *et al.*, 1958) and β1,3 and β1,4 glucosyl transferases from *Avena sativa* (Tsai and Hassid, 1971)], but had been unsuccessfully tried by Glaser and Brown (1957) to dissociate chitin synthetase from *N. crassa.* Later on, the enzyme was dissociated with digitonin from crude membrane-rich fractions of *S. cerevisiae* (Durán and Cabib, 1978) and with cholate-digitonin from similar preparations of *C. albicans* (Braun and Calderone, 1979). In both cases, the enzyme aggregates showed a molecular weight in the vicinity of 500,000 as measured by gel filtration and were activated by trypsin (i.e., they were "zymogenic") and by phosphatidyl serine.

Treatment of chitosomes from yeast cells of *M. rouxii* with digitonin causes their disintegration and the release of chitin synthetase subunits that sediment as a sharp band on a sucrose-density gradient containing digitonin, with a sedimentation coefficient of 16 S. This value corresponds to that from a spherical protein with a molecular weight of about 500,000 (Ruiz-Herrera *et al.*, 1980). When digitonin is removed by centrifugation in a sucrose gradient, part of the enzyme is aggregated into laminar structures, resembling pieces of chitosomes (Bartnicki-García *et al.*, 1977). Presumably, the particles dissociated by digitonin represent the minimum subunits of chitin synthetase within the chitosomes and have the capacity to reassociate. Incubation of the chitosomal subunits with activators and substrate led to the synthesis of α-chitin identified by X-ray diffraction and undistinguishable by this criterion from the chitin produced by intact chitosomes. However, their appearance was different: chitin made by chitosomes appears in the form of long and narrow ribbons often associated with fibroids (see Figs 7 and 9), whereas chitin made by dissociated chitosomes appears as short needles (see Fig 10) resembling the "lozenge"-shaped particles of chitin synthesized by dissociated chitin synthetase from *S. cerevisiae* (Durán and Cabib, 1978). It is probable that the arrangement of subunits into the chitosome provides the spatial organization for the elaboration of typical (true) microfibrils.

*Considering the molecular size of the dissociated forms of chitin synthetase reported in the literature, we decided not to use the term soluble for such large subunits.

10. On the Mechanism of Chitin-Chain Formation

Several questions regarding the mechanism of chitin biosynthesis remain unanswered, namely: (1) Is there an acceptor for chain growth?, and if there is, what is its nature? (2) Does chitin biosynthesis involve a lipid carrier similar to those described for the biosynthesis of bacterial polysaccharides and eukaryotic glycoproteins?

In regard to the first question, Camargo *et al.* (1967) suggested that GlcNAc [which is a strong stimulator of chitin synthetase from diverse sources (see above)] could be used as a primer for chitin biosynthesis. This claim was not confirmed in other systems (Porter and Jaworski, 1966; McMurrough and Bartnicki-García, 1971; Ruiz-Herrera *et al.*, 1977), and it is now generally accepted that GlcNAc is only an allosteric activator of the enzyme. The stimulation of chitin synthetase by chitodextrins in *N. crassa* (Glaser and Brown, 1957), *V. inequalis* (Jaworski *et al.*, 1965), *A. macrogynus* (Camargo *et al.*, 1967), and *M. rouxii* (McMurrough and Bartnicki-García, 1971) was attributed to a primer effect, but no rigorous proof of this claim is available. Purified preparations of chitosomes synthesize chitin microfibrils when incubated with UDPGlcNAc in the absence of GlcNAc or chitodextrins. More relevant is a recent report by Mills and Cantino (1978). These authors incubated γ-particles from *B. emersonii* with UDP-[^{14}C]GlcNAc and demonstrated that GlcNAc was transfered to a lipid identified as diglycosyl diglyceride; as the reaction progressed, labeled lipid was incorporated into insoluble chitin. On the basis of these results, they concluded that the lipid served as an acceptor for the growth of the GlcNAc chain. On the other hand, Bartnicki-García *et al.* (1979) found that chitin microfibrils synthesized *in vitro* by chitin synthetase preparations from *M. rouxii* were covalently bound to a peptide fraction containing mostly basic amino acids and concluded that probably the peptide acted as acceptor for the initiation of the chitin chain. These data are not contradictory, since a sequential reaction involving a lipid and a protein acceptor is possible; alternatively, the mechanisms could be different for the two fungi. In any case, it is evident that much work remains to be done in this field.

In regard to the involvement of a lipid intermediate in chitin biosynthesis, there are no data, either positive or negative, reported in the literature. Chitin biosynthesis by crude fractions from *M. rouxii* was stimulated by lipid extracts obtained from that fungus (McMurrough and Bartnicki-García, 1971), and phosphatidyl serine was found to stimulate digitonin-dissociated chitin synthetase from yeasts (Durán and Cabib, 1978). These data probably reflect physical effects on the enzyme and not the action of lipid intermediate.

We have made an exhaustive study to detect a possible lipid carrier (J. Ruiz-Herrera, E. López, and T. Hernández, unpublished observations). Using different mixtures of solvents, chloroform–methanol (2 : 1), chloroform–methanol–H_2O (10 : 10 : 3), and butanol, under different conditions and at different times of incubation, we have been unable to extract a lipid containing [^{14}C]GlcNAc, when either chitosomes or digitonin-dissociated enzyme is incubated with labeled UDPGlcNAc. If we consider that a high-energy lipid intermediate is formed during chitin biosynthesis, the two following possibilities could be anticipated:

$$U^*DP\ GlcNAc + X\text{-}P \rightleftharpoons U^*DP + X\text{-}P\text{-}GlcNAc \quad (1)$$

or

$$U^*DP\ GlcNAc + X\text{-}P \rightleftharpoons U^*DP + X\text{-}P\text{-}O\text{-}P\ GlcNAc \quad (2)$$

Accordingly, incubation of chitin synthetase with unlabeled UDPGlcNAc and labeled UDP or UMP would lead to the formation of labeled UDPGlcNAc. UDP has been found to be the product of chitin synthetase (Porter and Jaworski, 1966; Keller and Cabib, 1971; Myers and Cantino, 1974), and therefore equation (1) would be more likely. Digitonin-dissociated chitin synthetase from *M. rouxii* was incubated with [^{3}H]UDP or [^{14}C]UMP either without or with 1 mM nonradioactive carrier. No label appeared in residual UDPGlcNAc. This experiment rules out the possibility of a high-energy intermediate in chitin biosynthesis. However, the possibility remained that the process could be divided into two phases: chain initiation, wherein a lipid carrier would be involved, and chain elongation, wherein transfer of GlcNAc moieties would be provided by the nucleotide-sugar in a fashion similar to glycoprotein biosynthesis (Parodi and Leloir, 1979). If this were the case, the sensitivity of our exchange assay would be below the resolution limits, since each chitin chain contains about 1800–2200 units (Calvo, 1981). We have thus prepared dolichol-P-O-P-O-GlcNAc and dolichol-P-O-P-O $(GlcNAc)_2$ labeled in the sugar moiety, employing membrane-rich fractions from *S. cerevisiae*. Incubation of either chitosomes of digitonin-dissociated chitin synthetase with cold UDPGlcNAc and the labeled lipid carriers led to biosynthesis of chitin containing no radioactivity. Therefore, it is likely that if a lipid carrier is involved in chain initiation, this must be different from the dolichols involved in glycoprotein biosynthesis in yeasts.

11. *Perspectives and Future Developments*

The last decade has witnessed marked interest in the study of chitin biosynthesis *in vitro*. Much progress on the requirements of chitin synthetase and on the effect of activators and inhibitors had been made. Also, chitin microfibrils were synthesized *in vitro* for the first time. These advances have opened a wide avenue to research on this subject. However, several important problems remain to be solved. The most important one is the purification of chitin synthetase. As described above, dissociated forms of the enzyme have been partially purified, but instability of the enzyme after a certain degree of purification is attained has hampered its purification. Therefore, it seems necessary to investigate the nature of the components that after removal during purification leave the enzyme so labile. Purification of chitin synthetase will permit its polypeptide analysis, preparation of antibodies, and detailed studies on the mechanism of proteolytic activation and on the binding of activators and inhibitors. Purification of chitin synthetase will also permit its reconstitution into model bilayers for the investigation of the process of fibril formation. Even with the present limitations, knowledge gained with *in vitro* systems can now be applied to the study of the mechanism of the process of fibrillogenesis as its occurs *in vivo*.

At this point, important questions await answers. Curiously enough, the first refers to the actual mechanism of microfibril assembly. It is accepted that GlcNAc chains in α-chitin are antiparallel (Blackwell *et al.*, 1978). Accordingly, synthesis and crystallization of chitin must be temporally separated; i.e., the microfibril cannot be assembled as it is synthesized. How this apparent problem is solved by the organism must await further studies with *in vivo* or model systems, but one is tempted to propose that there operates in the cell a mechanism that guarantees that several GlcNAc chains are synthesized together confined in a subcellular structure and then crystallize to assemble a microfibril.

Another question that must be solved in the near future regards the site of the cell where actual synthesis and assembly to chitin microfibrils take place. Several possible locations for these events have been suggested (see Chapter 10), but the problem remains unsettled.

Advances have also been made in the study of regulation of chitin biosynthesis *in vitro*, but the relevance of these processes to the living cell remains obscure. Study of these aspects will be helped by use of mutants lacking proteases and inhibitors. Some progress in this direction has been made (Wolf and Ehman, 1978; Zubenko *et al.*, 1979), but much remains to be done. By the same token, the genetic study of chitin synthetase is urgently needed for the understanding of the process of cell-wall growth. As yet, no mutants that lack chitin synthetase have been isolated. A mutant of *Aspergillus nidulans* that lacks chitin was unable to synthesize glucosamine (Katz and Rosenberger, 1970).

ACKNOWLEDGMENTS.

Original work from the author's laboratory was partially supported by a grant from PNCB, CONACYT, México. Thanks are given to Prof. S. Bartnicki-García for his critical review of the original manuscript, to Prof. C. E. Bracker for providing unpublished photographs, and to Mrs. Yolanda Aguirre for careful typing of the manuscript.

References

Aoyagi, T., and Umezawa, H., 1975, Structures and activities of protease inhibitors of microbial origin, in: *Proteases and Biological Control* (E. Reich, D. B. Rifkin, and E. Shaw, eds.), pp. 429-454, Cold Spring Harbor Laboratory, Cold Spring Harbor, New York.

Arroyo-Begovich, A., and Ruiz-Herrera, J., 1979, Proteolytic activation and inactivation of chitin synthase from *Neurospora crassa, J. Gen. Microbiol.* **113**:339.

Bartnicki-García, S., and Lippman, E., 1972, Inhibition of *Mucor rouxii* by polyoxin: Effects on chitin synthetase and morphological development, *J. Gen. Microbiol.* **71**:301.

Bartnicki-García, S., Bracker, C. E., and Ruiz-Herrera, J., 1977, Reassembly of solubilized chitin synthetase into functional structures resembling chitosomes, Abstracts of the Second International Mycological Congress, Tampa, Florida, p. 41.

Bartnicki-García, S., Bracker, C. E., and Ruiz-Herrera, J., 1978, Isolation of chitosomes from taxonomically diverse fungi and synthesis of chitin microfibrils *in vitro, Exp. Mycol.* **2**:173.

Bartnicki-García, S., Lippman, E., and Heick, J., 1979, Evidence for a polypeptide acceptor in fungal chitin biosynthesis, Abstracts of the Annual Meeting of the American Society for Microbiology, Los Angeles, p. 106.

Bartnicki-García, S., Bracker, C. A., and Ruiz-Herrera, J., 1980, Isolation of chitosomes from osmotically fragile cells of *Mucer rouxii* and *Neurospora crassa,* Abstracts of the Annual Meeting of the American Society of Microbiology, Los Angeles, p. 98.

Benítez, T., Villa, T. G., and García Acha, I., 1976, Effects of polyoxin D on germination, morphological development and biosynthesis of cell wall of *Trichoderma viride, Arch. Microbiol.* **108**:183.

Betz, H., Hinze, H., and Holzer, H., 1974, Isolation and properties of two inhibitors of proteinase B from yeast, *J. Biol. Chem.* **249**:4515.

Blackwell, J., Minke, R., and Gardner, K. H., 1978, Determination of the structures of α- and β-chitins by X-ray diffraction, in: *Proceedings of the First International Conference on Chitin/Chitosan* (R. A. A. Muzzarelli and E. R. Pariser, eds.), pp. 108–123, MIT Sea Grant Report MITSG 78-7, Boston.

Bowers, B., Levin, G., and Cabib, E., 1974, Effect of polyoxin D on chitin synthesis and septum formation in *Saccharomyces cerevisiae, J. Bacteriol.* **119**:564.

Bracker, C. E., Ruiz-Herrera, J., and Bartnicki-García, S., 1976, Structure and transformation and chitin synthetase particles (chitosomes) during microfibril synthesis *in vitro, Proc. Natl. Acad. Sci. U.S.A.* **73**:4570.

Braun, P. C., and Calderone, R. A., 1978, Chitin synthesis in *Candida albicans:* Comparison of yeast and hyphal forms, *J. Bacteriol.* **135**:1472.

Braun, P. C., and Calderone, R. A., 1979, Regulation and solubilization of *Candida albicans* chitin synthetase, *J. Bacteriol.* **140**:666.

Cabib, E., and Farkas, V., 1971, Control of morphogenesis: An enzymatic mechanism for the initiation of septum formation in yeast, *Proc. Natl. Acad. Sci. U.S.A.* **68**:2052.

Cabib, E., and Keller, F. A., 1971, Chitin and yeast budding: Allosteric inhibition of chitin synthetase by a heat-stable protein from yeast, *J. Biol. Chem.* **246**:167.

Cabib, E., and Ulane, R., 1973, Yeast chitin synthetase: Separation of the zymogen from its activating factor and recovery of the latter in the vacuole fraction, *J. Biol. Chem.* **248**:1451.

Calvo, C., 1981, Mechanisms de sinteus de quitosana en *Mucer rouxii. M. Sc. Thesis.,* Centro de Investigasion y Estudios Avanjados, Instituto Politicuico Nacinal, Mexico.

Camargo, P. E., Dietrich, C. P., Sonneborn, D., and Strominger, J. L., 1967, Biosynthesis of chitin in spores and growing cells of *Blastocladiella emersonii, J. Biol. Chem.* **242**:3121.

Dahn, U., Hagenmaier, H., Hohne, H., Konig, W. A., Wolf, G., and Zahner, H., 1976, Stoffwechselproduckte von Mikroorganismen. 154. Mitteilung. Nikkomycin, ein neuer Hemmstoff der Chitinsynthese bei Pilzen, *Arch. Microbiol.* **107**:143.

De Rousset-Hall, A., and Gooday, G. W., 1975, A kinetic study of solubilized chitin synthetase preparations from *Coprinus cinereus, J. Gen. Microbiol.* **89**:146.

Díaz, S. V., and Ruiz-Herrera, J., 1978, Purificación de una proteasa que inactiva la quitina sintetasa de *Mucor rouxii, Resúmenes, Reunión Bianual de Microbiologia,* Toluca, Mexico, p. B-11.

Durán, A., and Cabib, E., 1978, Solubilization and partial purification of yeast chitin synthetase: Confirmation of the zymogenic nature of the enzyme, *J. Biol. Chem.* **253**:4419.

Endo, A., and Misato, T., 1969, Polyoxin D, a competitive inhibitor of UDP-*N-acetyl-glucosamine: Chitin N*-acetyl glucosaminyl transferase in *Neurospora crassa, Biochem. Biophys Res. Commun* **37**:718.

Endo, A., Kakiki, K., and Misato, T., 1970, Mechanism of action of the antifungal agent polyoxin D, *J. Bacteriol.* **104**:189.

Feingold, D. S., Neufeld, E. F., and Hassid, W. Z., 1958, Synthesis of a β-1,3 linked glucan by extracts of *Phaseolus aureus* seedlings, *J. Biol. Chem.* **133**:783.

Ferguson, A. R., Katsunuma, T., Betz, H., and Holzen, H., 1973, Purification and properties of an inhibitor of the tryptophan-synthase inactivating enzymes in yeast, *Eur. J. Biochem.* **32**:444.

Glaser, L., and Brown, D. H., 1957, The synthesis of chitin in cell-free extracts of *Neurospora crassa, J. Biol. Chem.* **228**:729.

Gooday, G. W., 1977, Biosynthesis of the fungal wall: Mechanisms and implications, *J. Gen. Microbiol.* **99**:1.

Gooday, G. W., and De Rousset-Hall, A., 1975, Properties of chitin synthetase from *Coprinus cinereus, J. Gen. Microbiol.* **89**:137.

Hassilik, A., 1974, Inactivation of chitin synthetase in *Saccharomyces cerevisiae, Arch. Microbiol.* **101**:295.

Hernandez, J. Lopez-Romero, E., Cerbon, J., and Ruiz-Herrera, J. Lipid analysis of chitosomes, chitin synthesizing microvesicles, *Experimental Mycol.* in press.

Holzer, H., and Saheki, T., 1976, Mechanism and regulation of proteolytic processes in yeast, *Tokai J. Exp. Clin. Med.* **1**:115.

Hori, M., Kakiki, K., and Misato, T., 1974a, Further study on the relation of polyoxin structure of chitin synthetase inhibition, *Agric. Biol. Chem.* **38**:691.

Hori, M., Kakiki, K., and Misato, T., 1974b, Interaction between polyoxin and active center of chitin synthetase, *Agric. Biol. Chem.* **38**:699.

Isono, K., Asahi, K., and Suzuki, S., 1969, Studies on polyoxins, antifungal antibiotics. XII. The structure of polyoxins, *J. Am. Chem. Soc.* **91**:7490.

Jan, Y. N., 1974, Properties and cellular localization of chitin synthetase in *Phycomyces blakesleeanus, J. Biol. Chem.* **249**:1973.

Jaworski, E. G., Wang, L., and Marco, G., 1963, Synthesis of chitin in cell-free extracts of *Prodenia eridania, Nature (London)* **198**:790.

Jaworski, E. G., Wang, L. C., and Carpenter, W. D., 1965, Biosynthesis of chitin in cell-free extracts of *Venturia inaequalis, Phytopathology* **55**:1309.

Katz, D., and Rosenberger, R. F., 1970, A mutation in *Aspergillus nidulans* producing hyphal walls which lack chitin, *Biochim. Biophys. Acta* **208**:452.

Keller, F. A., and Cabib, E., 1971, Chitin and yeast budding: Properties of chitin synthetase from *Saccharomyces carlsbergensis, J. Biol. Chem.* **246**:160.

Leloir, L. F., 1972, Biosynthesis of polysaccharides seen from Buenos Aires, in: *Biochemistry of the Glycosidic Linkage* (R. Piras and H. G., Pontis, eds.), pp. 1–17, Academic Press, New York.

Lenney, J. F., 1975, Three yeast proteins that specifically inhibit yeast proteases A, B, and C, *J. Bacteriol.* **122**:1265.

Lenney, J. F., and Dalbec, J. M., 1967, Purification and properties of two proteinases from *Saccharomyces cerevisiae, Arch. Biochem. Biophys.* **120**:42.

Lenney, J. F., and Dalbec, J. M., 1969, Yeast proteinase B: Identification of the inactive form as an enzyme-inhibitor complex, *Arch. Biochem. Biophys.* **129**:407.

Lenney, J. F., Matile, P., Weimken, A., Schellenberg, M., and Meyer, J., 1974, Activities and cellular localization of yeast proteases and their inhibitors, *Biochem. Biophys. Res. Commun.* **60**:1378.

López-Romero, E., and Ruiz-Herrera, J., 1976, Synthesis of chitin by particulate preparations from *Aspergillus flavus, Antonie van Leeuwenhoek J. Microbiol. Serol.* **42**:261.

López-Romero, E., Ruiz-Herrera, J., and Bartnicki-García, S., 1978, Purification and properties of an inhibitory protein of chitin synthetase from *Mucor rouxii, Biochim. Biophys. Acta* **525**:338.

McMurrough, I., and Bartnicki-García, S., 1971, Properties of a particulate chitin synthetase from *Mucor rouxii, J. Biol. Chem.* **246**:4008.

McMurrough, I., and Bartnicki-García, S., 1973, Inhibition and activation of chitin synthesis by *Mucor rouxii* cell extracts, *Arch. Biochem. Biophys.* **158**:812.

McMurrough, I., Flores-Carreón, A., and Bartnicki-García, S., 1971, Pathway of chitin synthesis and cellular localization of chitin synthetase in *Mucor rouxii, J. Biol. Chem.* **246**:3999.

Mills, G. L., and Cantino, E., 1978, The lipid composition of the *Blastocladiella emersonii* γ-particle and the function of γ-particle lipid in chitin formation, *Exp. Mycol.* **2**:99.

Moore, P. M., and Peberdy, J. F., 1975, Biosynthesis of chitin by particulate fractions from *Cunninghamella elegans, Microbios* **12**:29.

Moore, P. M., and Peberdy, J. F., 1976, A particulate chitin synthetase from *Aspergillus flavus* Link: The properties, location, and levels of activity in mycelium and regenerating protoplast preparations, *Can. J. Microbiol.* **22**:915.

Myers, R. B., and Cantino, C. E., 1974, *The Gamma Particle,* S. Karger, Basel.

Nanba, H., and Kuroda, H., 1974, Studies on fungicides. X. Biosynthesis of β-glucan and chitin-like substance of cell wall from *Cochliobolus miyabeanus, Chem. Pharm. Bull.* **22**:610.

Ohta, N., Kakiki, K., and Misato, T., 1970, Studies on the mode of action of polyoxin D. Part II. Effect of polyoxin D on the synthesis of fungal cell wall chitin, *Agric. Biol. Chem.* **34**:1224.

Parodi, A. J., and Leloir, L. F., 1979, The role of lipid intermediates in the glycosylation of proteins in the eucaryotic cell, *Biochim. Biophys. Acta* **559**:1.

Peberdy, J. F., and Moore, P. M., 1975, Chitin synthetase in *Mortierella vinacea:* Properties, cellular location and synthesis in growing cultures, *J. Gen. Microbiol.* **90**:228.

Porter, C. A., and Jaworski, E. G., 1966, The synthesis of chitin by particulate preparations of *Allomyces macrogynus, Biochemistry* **5**:1149.

Ruiz-Herrera, J., 1978, The distribution and quantitative importance of chitin in fungi, in: *Proceedings of the First International Conference of Chitin/Chitosan* (R. A. A. Muzzarelli and E. R. Pariser, eds.), pp. 11-21, MIT Sea Grant Report MITSG 78-7, Boston.

Ruiz-Herrera, J., and Bartnicki-García, S., 1974, Synthesis of cell wall microfibrils *in vitro* by a "soluble" chitin synthetase from *Mucor rouxii, Science* **186**:357.

Ruiz-Herrera, J., and Bartnicki-García, S., 1976, Proteolytic activation and inactivation of chitin synthetase from *Mucor rouxii, J. Gen. Microbiol.* **97**:241.

Ruiz-Herrera, J., Sing, V. O., Van Der Woude, W. J., and Bartnicki-García, S., 1975, Microfibril assemby by granules of chitin synthetase, *Proc. Natl. Acad. Sci. U.S.A.* **72**:2706.

Ruiz-Herrera, J., López-Romero, E., and Bartnicki-García, S., 1977, Properties of chitin synthetase in isolated chitosomes from yeast cells of *Mucor rouxii, J. Biol. Chem.* **252**:3338.

Ruiz-Herrera, J., Bartnicki-García, S., and Brackner, C. E., 1980, Dissociation of chitosomes by digitonin into IGS subuncts with chitin syntetase activity. *Biochem. Biophys. Acta.* **629**:201.

Ryder, N. S., and Peberdy, J. F. 1977, Chitin synthase in *Aspergillus nidulans:* Properties and proteolytic activation, *J. Gen. Microbiol.* **99**:69.

Schott, E. H., and Holzer, H., 1974, Purification and some properties of tryptophan synthase inactivase II from yeast, *Eur. J. Biochem.* **42**:61.

Tsai, C. M., and Hassid, W. Z., 1971, Solubilization and separation of uridine diphospho-D-glucose: $\beta - (1 \rightarrow 4)$ glucan and uridine diphospho-D-glucose: $\beta - (1 \rightarrow 3)$ glucan glucosyl transferases from coleoptiles of *Avena sativa, Plant Physiol.* **47**:740.

Ulane, R. E., and Cabib, E., 1974, The activating system of chitin synthetase from *Saccharomyces cerevisiae:* Purification and properties of an inhibitor of the activating factor, *J. Biol. Chem.* **249**:3418.

Ulane, R. E., and Cabib, E., 1976, The activating system of chitin synthetase from *Saccharomyces cerevisiae:* Purification and properties of the activating factor, *J. Biol. Chem.* **251**:3367.

Van Laere, A. J., and Carlier, A. R., 1978, Synthesis and proteolytic activation of chitin synthetase in *Phycomyces blakesleeanus* Burgeff, *Arch. Microbiol.* **116**:181.

Vermeulen, C. A., Raeven, M. B. J. M., and Wessels, J. G. H., 1979, Localization of chitin synthase activity in subcellular fractions of *Schizophyllum commune* protoplasts, *J. Gen. Microbiol.* **114**:87.

Wolf, D. H., and Ehman, C., 1978, Isolation of yeast mutants lacking proteinase B activity, *FEBS Lett.* **92**:121.

Zubenko, G. S., Mitchell, A. P., and Jones, E. W., 1979, Septum formation, cell division, and sporulation in mutants of yeast defficient in proteinase B, *Proc. Natl. Acad. Sci. U.S.A.* **76**:2395.

12 Cellulose Synthesis in Detached Cotton Fibers

NICHOLAS C. CARPITA

1. *Introduction*

The mechanism of biosynthesis of cellulose and other cell-wall polymers has been researched and reviewed for decades, and there have been many conflicting reports on the mode of cellulose synthesis in cell-free systems from higher plants. However, more recent reports (Delmer, 1977; Franz, 1978; Ray, 1980) indicate that this plethora of controversy on the mechanism of cellulose synthesis reflects the notion that the *in vitro* synthesis has not been demonstrated unequivocally. The purpose of this chapter is to evaluate critically the reports of *in vitro* cellulose synthesis, to outline some of the major problems to be overcome for such an *in vitro* synthesis, and to discuss some physiological considerations for cellulose synthesis *in vivo*.

1.1. *Incorporation of Glucose from GDP-Glucose into Alkali-Insoluble Glucans*

Alkali-insoluble polysaccharides containing predominantly β1,4-linkages were first synthesized with enzyme preparations from mung bean seedlings (Elbein *et al.*, 1964) and elongating cotton fibers (Barber *et al.*, 1964). Many nucleotide glucose substrates were tested, but only GDP-glucose was substrate in this system. Other workers found that UDP-glucose could serve as substrate for alkali-insoluble polymers in other plant systems (Brummond and Gibbons, 1965; Ordin and Hall, 1967; Villemez *et al.*, 1967), yet critical examination of the

NICHOLAS C. CARPITA • Department of Botany and Plant Pathology, Purdue University, West Lafayette, Indiana 47907.

reaction products revealed a large proportion of β1,3-linkages (Feingold *et al.*, 1958; Chambers and Elbein, 1970; Heiniger and Delmer, 1977). In fact, only when GDP-glucose was used as substrate did the reaction products contain predominantly β1,4-glucan (Flowers *et al.*, 1969; Chambers and Elbein, 1970; Shore and Maclachlan, 1975). As a consequence, GDP-glucose was accepted as the precursor for cellulose for many years. The issue is still not resolved, however. Even in an original report of *in vitro* cellulose synthesis, GDP-mannose could be used as substrate for an alkali-insoluble polysaccharide (Barber *et al.*, 1964). GDP-mannose could stimulate the incorporation of glucose from GDP-glucose into an alkali-insoluble glucomannan (Elbein and Hassid, 1966; Villemez and Heller, 1970), and the sites of synthesis of this polymer were in vesicles derived from the Golgi complex (Shore and Maclachlan, 1975; Dürr *et al.*, 1979). Although Dürr *et al.* (1979) proposed that the synthesis of β1,4-glucan in the Golgi complex could represent newly synthesized cellulose synthetase en route to the plasma membrane, it is more probable that noncellulosic polysaccharide was synthesized in these vesicles.

1.2. Incorporation of Glucose from UDP-Glucose into Alkali-Insoluble Glucans

More recently, support has shifted in favor of UDP-glucose as the substrate for cellulose. UDP-glucose has long been recognized as a precursor for bacterial cellulose synthesized by *Acetobacter xylinum* (Glaser, 1958). UDP-glucose, and not GDP-glucose, is the major nucleotide-sugar in cotton fibers engaged actively in cellulose synthesis (Franz, 1969; Carpita and Delmer, 1980). Furthermore, the levels of UDP-glucose increase nearly 5-fold just before the onset of cellulose synthesis in secondary wall formation, and the UDP-sugar pool turns over in fibers cultured *in vitro* at a rate sufficient to account for the formation of both noncellulosic and cellulosic polysaccharide (Carpita and Delmer, 1981).

Reevaluation of the reaction products from cell-free systems indicate that glucose from UDP-glucose is incorporated into β1,4-linkages as well as β1,3-linkages. In fact, several reports present evidence that the type of linkage depends on the concentration of UDP-glucose in the reaction mixture; high concentrations of UDP-glucose yield predominantly β1,3-linkages, while low concentrations increase the proportion of β1,4-linkages (Péaud-Lenoël and Axelos, 1970; Smith and Stone, 1973; Musolan and Kindinger, 1978). Yet, as observed with GDP-glucose, the rate of β1,4-glucan synthesis from UDP-glucose can be increased by additions of other nucleotide-sugars. Recently, Ray (1980) has reported that the rates of polysaccharide formation from UDP-glucose:glucan synthetase associated with the Golgi apparatus are increased by adding UDP-xylose, and the product synthesized is a xyloglucan. The glucans synthesized from grass tissues often are comprised of mixed-linked glucans (polymers with β1,4- and β1,3-linkages) (Ordin and Hall, 1967; Villemez *et al.*, 1967; Tsai and Hassid, 1971). Thus, even more rigorous analyses of these reaction products claimed to be cellulose must be performed before confirmation of the *in vitro* synthesis of cellulose.

2. *Criteria for Cell-Free Synthesis of Cellulose*

Although there are a number of reports that claim that the *in vitro* synthesis of cellulose has been solved, the reaction products are probably not cellulose, but noncellulosic glucans. Confusion stems from the fact that both GDP-glucose and UDP-glucose can serve as the substrate for β1,4-glucan, and such linkages are found in so many noncellulosic polysaccharides such as the glucomannans, xyloglucans, and mixed-linked glucans. The lack of bona fide cellulose synthesis in enzyme preparations from higher plants may reflect the lability of the synthetase system. Only when two criteria are fulfilled will cellulose synthesis *in vitro* be demonstrated unquestionably:

1. The presence of β1,4-glucan must be verified. The presence of 1,4-linked hexose is best verified by separation of partially methylated alditol acetates by gas-liquid chromatography (Hakamori, 1964; Sandford and Conrad, 1966; Albersheim *et al.*, 1967). The linkages of the reaction product should be susceptible to enzymatic hydrolysis by cellulase isolated from several sources such as *Trichoderma verde*, *Penicillium verruculosum*, or *Streptomyces* strains (Reese, *et al.*, 1959, 1961; Bauer *et al.*, 1973) and not by an α-amylase (usually isolated from porcine pancreas to avoid β-hydrolases) or endo-β1,3-glucanase from *Rhizopus* (Reese, 1977). The enzymatic hydrolysis should yield glucose, cellobiose, cellotriose, and cellodextrins only, and not laminaribiose, laminaritriose, or mixed-linked trisaccharides. Such products can be separated easily by thin-layer or paper chromatography (Lederer and Lederer, 1957; Bartnicki-García and Lippman, 1967; Zweig and Whitaker, 1971). In most reports to date, the reaction products had been characterized sufficiently to fulfill this criterion.

2. The reaction products must be crystalline polymers—proven either by X-ray crystallography (Mann, 1963) or by resistance to hydrolysis in hot acetic-nitric acid (Updegraff, 1969). Since cellulose synthesis can be separated from the crystallization (Haigler *et al.*, 1980), preservation of both the glycosyl transfer and crystallization steps must be considered for *in vitro* studies to verify that the reaction product is indeed cellulose.

3. *Why Is Cellulose Synthesis Lost on Homogenization of Plant Tissues?*

No such proof yet exists for any reaction products reported as cellulose; consequently, the synthesis of cellulose has not been verified in particulate systems derived from higher plants. The synthetic activity may actually be lost totally on homogenization. An indication that cellulose synthetase may be more labile than expected is the observation that over 90% of the cellulose synthesis is immediately lost on detachment of the cotton fiber cells from the ovule; further homogenization of the fibers results in total loss of cellulose synthesis (Carpita and Delmer, 1980).

3.1. *Site of Cellulose Synthesis*

The ease of obtaining active synthesis of noncellulosic polysaccharides in particulate systems compared to the difficulties encountered in obtaining cellulose synthesis may reflect the difference in the subcellular location of the synthesis. Noncellulosic polysaccharide synthesis is generally localized with an internal membrane system (Dürr *et al.*, 1979; Ray *et al.*, 1969), whereas evidence based on electron and freeze–etch replicate microscopy indicates that cellulose synthesis occurs at the plasma membrane–cell wall interface (Brown and Montezinos, 1976; Willison and Brown, 1977; Giddings *et al.*, 1980). There are two exceptions to this statement: First, cellulose synthesis in some algae such as *Pleurochrysis* occurs in a Golgi complex, and the large scale of cell-wall material then fuses with the plasma membrane (Brown *et al.*, 1969). Second, a β1,3-glucan synthetase is localized on the plasma membrane in pea stem cells (Anderson and Ray, 1978) and probably in cotton fibers (Maltby *et al.*, 1979).

3.2. *Special Consideration Relevant to an Extracellular Synthesis of Cellulose*

If cellulose synthesis is indeed an extracellular event, then such a synthesis site creates some biological problems. First, the synthetase complex probably moves within the plasma membrane while the product is deposited onto the cell wall.* Second, a pathway of high-energy substrate(s) that provide(s) glucose must then traverse the plasma membrane. Third, the pressure potential gradient across the plasma membrane generated by the plant cell may impose critical orientation problems for the activity of the cellulose synthetase. Loss of cellulose synthesis on disruption of a cell may result from disorientation of the synthetase caused both by the sudden relaxation of the cell wall and by the extensive membrane disruption that occurs on release of turgor. Furthermore, some cellular activities are controlled by turgor (Cleland, 1977; Pierce and Raschke, 1980; Zimmerman, 1978), and since the synthetase is proposed to exist in a region that is capable of perceiving turgor, one must also consider the possibility that cellulose synthesis itself may be turgor-dependent.

4. *Protection of Cellulose Synthesis in Detached Cotton Fibers*

Carpita and Delmer (1980) have shown that polyethylene glycol (PEG) 4000 was able to protect cellulose synthesis in cotton fibers cut from ovules. This

*Observations based on freeze–etch analysis have revealed the appearance of complexes at the termini of microfibrils; these complexes are the proposed cellulose synthetases. There is no evidence *a priori* as to whether the complex moves ahead of the newly crystallized microfibrils or the complexes are fixed and the microfibrils move away from the complexes. It is assumed by most researchers, though, that the physical restraints imposed on the movement of such a large polymer would favor the movement of the complexes through the plasma membrane ahead of the growing microfibril.

report gives the complete details on experimental procedures and illustrates the attempts to optimize the incubation conditions. The work reported in this review will iterate some of the important features regarding the nature of the protection of cellulose synthesis by PEG in such fibers.

Cotton fibers from unfertilized ovules cultured *in vitro* were used. This system is optimal for the study of secondary wall cellulose synthesis for several reasons:

1. These fibers, like plant-grown fibers, expend most of their carbon budget in synthesis of polysaccharides. Mutsaers (1976) indicated that over 80% of the translocated sugar accumulated in wall polymers, and Carpita and Delmer (1981) observed that a high proportion of the carbon passing through the metabolite pool of glucose accumulates in cellulose and β1,3-glucan alone.
2. The fiber cells are large (nearly 2 cm long), and ovules cultured with associated fibers can be manipulated easily. Furthermore, the fiber cells can be removed from the ovule and analyzed separately.

A high-molecular-weight osmoticum unable to penetrate the cell wall (Carpita *et al.*, 1979) was used (1) to reduce turgor before detachment and (2) to mimic turgor on detachment; the inability to penetrate the cell-wall capillaries theoretically results in tension keeping the plasma membrane firmly appressed to the cell wall (Fig. 1).

4.1. Protection of Cellulose Synthesis by Polyethylene Glycol

The unfertilized ovules were excised from the ovary the day of anthesis and transferred aseptically to a medium devised by Beasley and Ting (1974) (BT medium). The ovules and associated fibers were incubated 15–16 days in darkness at 32° C. At this stage, the fibers had reached their ultimate length and rates of cellulose synthesis were maximal (Meinert and Delmer, 1977). The ovules were transferred to the medium of choice, usually containing 1 mM UDP-[^{14}C]-glucose or 20 mM [^{14}C]glucose, 0.10 M pentaerythritol, a buffer or BT medium, and PEG 4000 or mannitol as an osmoticum. The submerged fibers were cut from the ovules with surgical scissors, leaving only the detached fibers in the incubation medium. Polysaccharide synthesis in such fibers was compared to control incubations in which ovules and fibers were left intact during the incubation but were excised and the ovules and aerial fibers removed at the end of the incubation.

Polysaccharide material was purified as described in a previous report (Carpita and Delmer, 1980), and the radioactivity accumulating into the cellulose remaining was assayed by liquid scintillation spectroscopy.

When fibers were detached from ovules, the rate of [^{14}C]cellulose synthesis from [^{14}C]glucose decreased to 10% of that observed in intact, attached fibers (Table I). When PEG 4000 was added to the reaction mixture, however, over 50% of the capacity for cellulose synthesis was maintained compared to intact controls without PEG (Table I); mannitol, at concentrations equiosmolar to

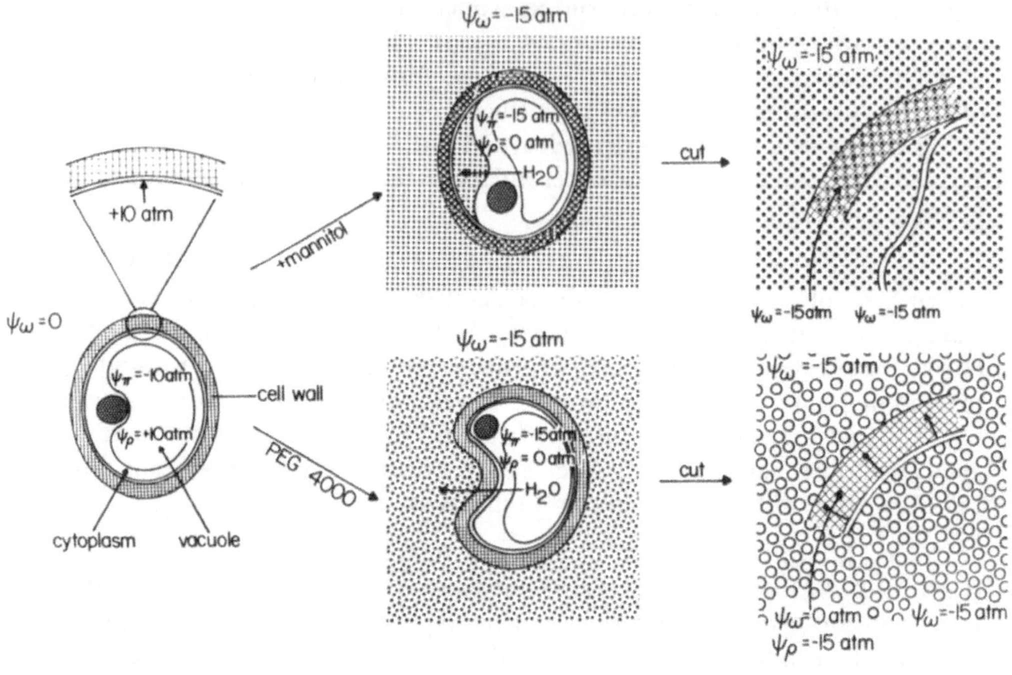

FIGURE 1. Representation of the theoretical influence of low- and high-molecular-weight osmotica on membrane–cell wall interaction of a plant cell. The osmotic potential developed by dissolved solutes within the protoplast of the cell causes water uptake. The resistance to water influx by the cell wall results in a turgor pressure pushing the plasma membrane tightly against the cell wall and a tension resulting in the elastic stretching of the cell wall fibrils. The addition of a low-molecular weight osmoticum, such as mannitol, at a soluble concentration higher than that of the cell results in movement of water out of the cell. The loss of protoplast volume diminishes turgor, and since mannitol is capable of penetrating the cell-wall capillaries, "plasmolysis" occurs, that is, the pulling away of the plasma membrane from the cell wall. Chopping the tissues would yield pieces of cell wall with the plasma-membrane fragments free to separate from the cell wall. When a high-molecular-weight osmoticum such as PEG is added. the inability of the osmoticum to penetrate the capillaries results in "cytorrhysis," or the collapse of the cell wall around the shrinking protoplast. Chopping the cell will yield pieces of cell wall with the plasma-membrane fragment firmly appressed to the wall. Since the osmotic potential of water within the cell-wall region is nearly zero, a tension will keep the plasma membrane tightly against the wall, thus mimicking turgor.

those of PEG 4000, was about half as effective as PEG. Both osmotica stimulated the incorporation of glucose into noncellulosic polysaccharides in detached fibers. A methylation analysis of the reaction products revealed that much of the radioactivity incorporated into polysaccharide was 1,3-linked and 1,4-linked hexose (Fig. 2); after digestion in acetic–nitric acid, however, only 1,4-linked glucose (cellulose) remained. The pH optimum was 6.0 (Carpita and Delmer, 1980); the optimal PEG concentration was 0.06–0.07 molal (Fig. 3). Concentrations of PEG higher than 0.07 molal resulted in severe inhibition of cellulose synthesis in both detached and intact fibers. Buffers were then tested to replace the BT medium. Potassium 2(*N*-morpholino)ethanesulfonate (MES) buffer was the most suitable of the buffers, yet not as suitable as the BT medium (Table II).

TABLE I

Incorporation of [^{14}C]Glucose into Polysaccharides by Intact or Detached Cotton Fibers[a]

Type of fibers incubated	Medium supplement	Polysaccharide synthesized (cpm/mg cellulose in fibers per hr): Cellulose	Noncellulosic polysaccharide
Intact	—	6090	3420
Intact	0.06 molal PEG 4000	8270	4200
Detached	—	490	1350
Detached	0.06 molal PEG 4000	3340	9700
Detached	0.20 M Mannitol	1520	9250

[a] Reproduced from Carpita and Delmer (1980) by permission of *Plant Physiology*. Incubations were for 1 hr at 32° C in BT medium, pH 6.0, supplemented with 10 mM potassium phosphate. "Intact" refers to fibers still associated with their ovules during the incubation; "detached," to fibers cut from their ovules prior to incubation.

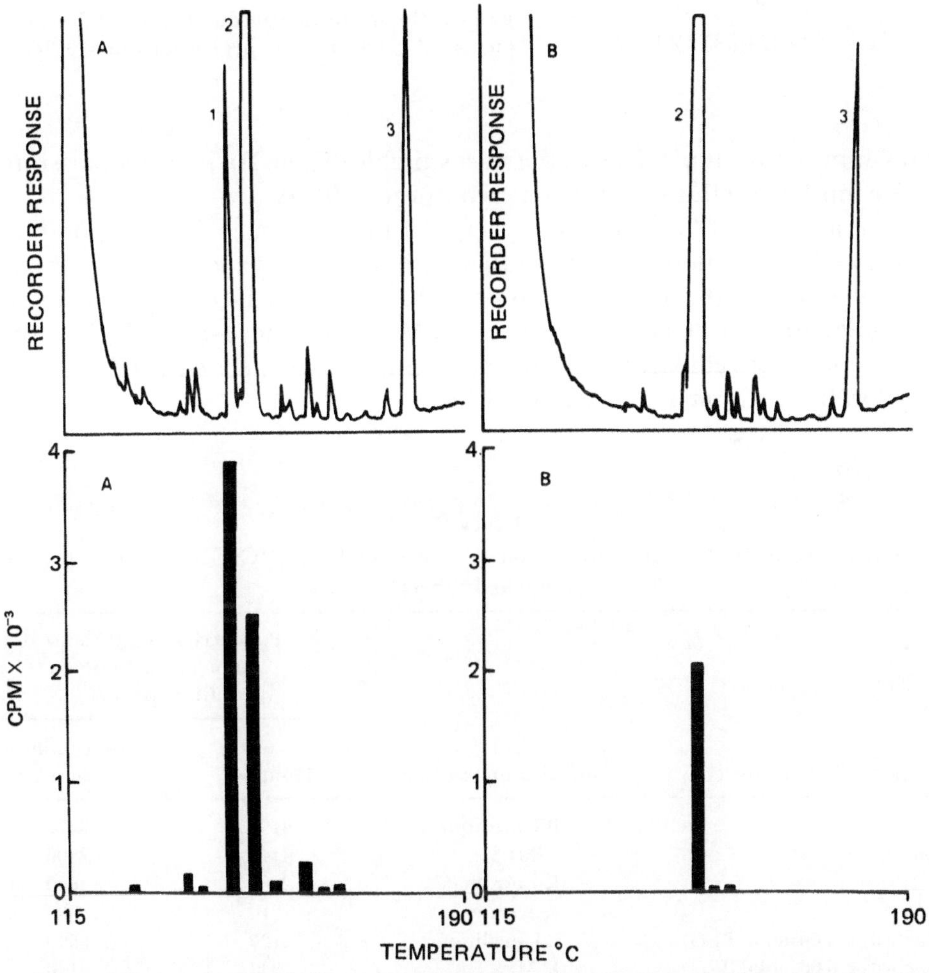

FIGURE 2. Separation of partially methylated alditol acetates and analysis of the radioactivity in the derivated sugars of the polysaccharides of cotton fibers detached in a medium containing 0.06 molal PEG. (A) Before acetic-nitric digestion; (B) after acetic-nitric digestion. Peaks: (1) 1,3-linked glucose; (2) 1,4-linked glucose; (3) inositol (internal standard). Reproduced from Carpita and Delmer (1980) by permission of *Plant Physiology*.

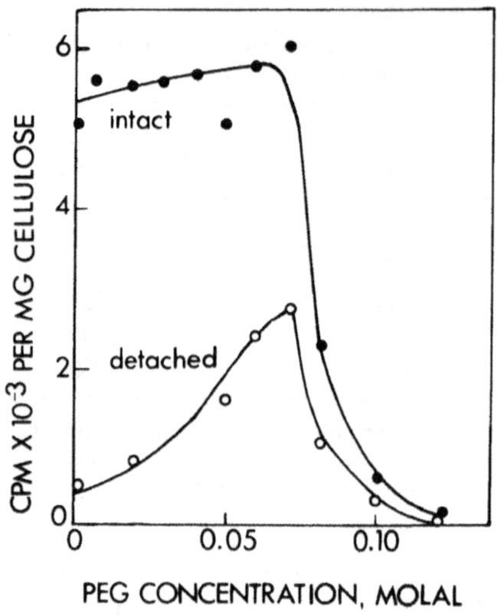

FIGURE 3. Comparison of [^{14}C]glucose incorporation into cellulose and noncellulosic polysaccharides in detached fibers as a function of the concentration of PEG in the incubation medium. The ovules were incubated in BT medium, pH 6.0, for 1 hr at 32° C.

Potassium acetate at pH 6.0 was the least suitable (Table II); in fact, potassium acetate inhibited cellulose synthesis even in intact fibers.

The addition of Ca^{2+} and Mg^{2+} synergistically stimulated the incorporation of glucose into cellulose, indicating that both ions may be necessary to protect cellulose synthesis (Carpita and Delmer, 1980). Under such conditions (10 mM Tris-MES, pH 6.0, 0.06 molal PEG, 3 mM $CaCl_2$, and 3 mM $MgCl_2$), the incorporation of radioactivity into cellulose approximately equaled that obtained in BT medium and 0.06 molal PEG.

TABLE II
Effect of Different Buffers on Polysaccharide Synthesis from [^{14}C]Glucose by Intact or Detached Fibers[a]

Type of fibers incubated	Incubation medium	Polysaccharide synthesized (cpm/mg cellulose in fibers per hr)	
		Cellulose	Noncellulosic polysaccharide
Intact	BT medium	4190	3470
Intact	K-MES	4140	2700
Intact	K-phosphate	2260	1850
Intact	K-acetate	210	580
Detached (+ 0.06 molal PEG)	BT medium	2120	8690
Detached (+ 0.06 molal PEG)	K-MES	1030	4140
Detached (+ 0.06 molal PEG)	K-phosphate	300	2970
Detached (+ 0.06 molal PEG)	K-acetate	10	300

[a] All buffers were at pH 6.0 and 50 mM with respect to the potassium ion. Incubations were for 1 hr at 32° C.

5. *Role of Polyethylene Glycol in the Protection of Cellulose Synthesis*

Since the detached fibers were unable to accumulate vital dyes such as neutral red (Fig. 4A,D), the cut ends were assumed to be open and accessible to potential precursors to cellulose such as glucose-1-P (G-1-P) or UDP-glucose,

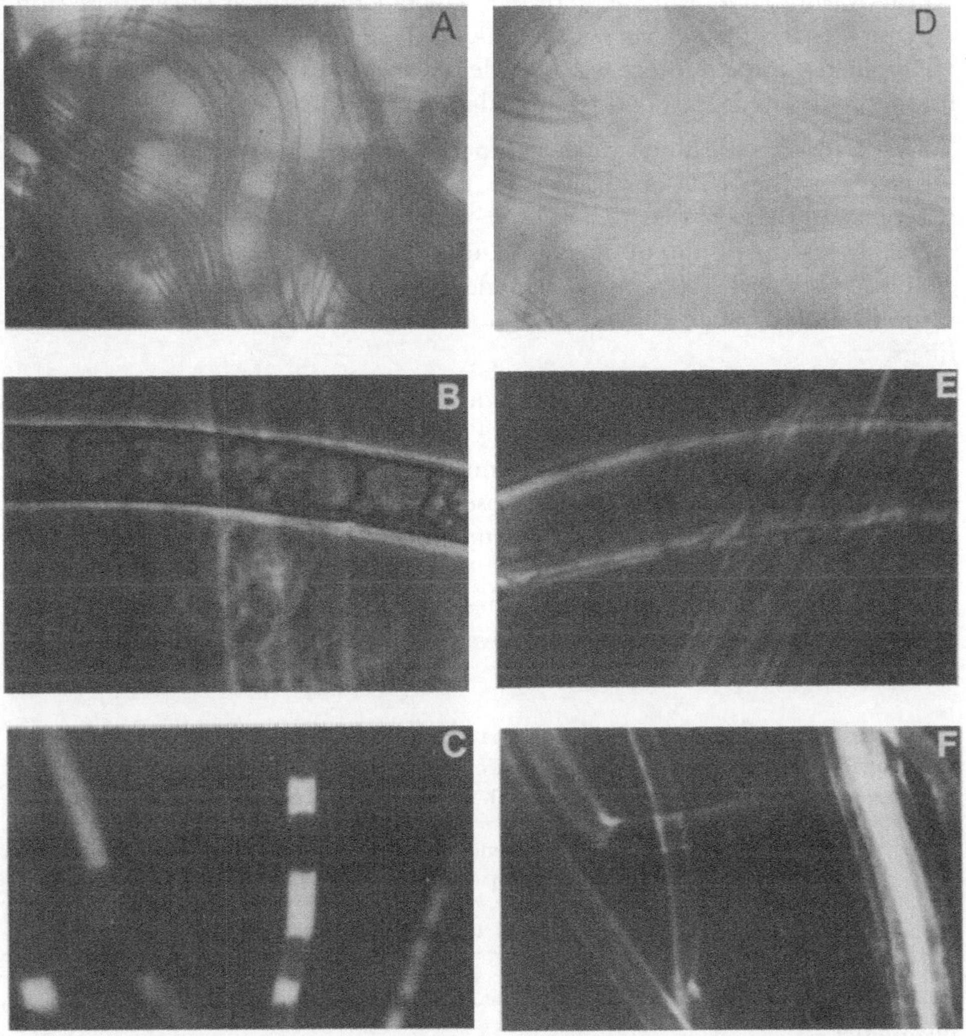

FIGURE 4. Microscopic examination of intact fibers or fibers detached in the presence of absence of PEG. The fibers were detached in 10 mM Tris-MES, pH 6.0 + 0.1 M pentaerythritol, 120 mM glucose, and 3mM each of $CaCl_2$ and $MgCl_2$ ± 0.06 molal PEG. (A) Intact fibers incubated for 5 min in 0.25% neutral red and observed by bright-field microscopy; (B) fibers detached in the absence of PEG and observed using phase-contrast microscopy; (C) fibers detached in the absence of PEG and incubated 5 min in 0.01% fluorescein diacetate, and observed; (D, E, F) similar in A, B, C, respectively, except that the fibers were detached in the presence of 0.06 molal PEG. (B, C, E, F) Reproduced from Carpita and Delmer (1980) by permission of *Plant Physiology*.

but little radioactivity from [^{14}C]-G-1-P or UDP-[^{14}C]glucose was incorporated into cellulose. The UDP-[^{14}C]glucose was used as substrate for the synthesis of the noncellulosic glucan (Table III). With PEG in the incubation medium, the capacity to synthesize noncellulosic glucan was greater when glucose rather than UDP-[^{14}C]glucose was used as a substrate (Table III). When the cut fibers were chopped progressively into smaller pieces, the capacity to synthesize both cellulose and noncellulosic polysaccharides from glucose decreased, while the synthesis from UDP-glucose increased. The rate of synthesis of cellulose from glucose was diminished, yet substantial, while no significant incorporation of UDP-glucose into cellulose was ever demonstrated (Table IV). This puzzling finding indicates at least two possibilities:

1. UDP-glucose is not a substrate for cellulose synthesis. This conclusion is unlikely, on the basis of analyses of UDP-glucose turnover in such fibers (Carpita and Delmer, 1981).
2. Resealing of the cut fibers is required for reinitiation of the synthesis of cellulose. Resealed fibers would exclude the UDP-glucose and would confine utilization of UDP-glucose to β1,3-glucan synthetase accessible in unsealed fibers.

Increasing the number of cuts would increase the amount of accessible glucan synthetase, while decreasing the numbers of resealed fibers. The role of PEG in protecting cellulose synthesis must be evaluated rigorously to assess this possibility. First of all, PEG was proposed to: (1) reduce turgor prior to cutting and (2) cause an artificial turgor on detachment (see Fig. 1). Mannitol, a small molecule able to penetrate the wall capillaries, could fulfill proposal (1) but not (2); mannitol did protect cellulose synthesis significantly, but was not as effective as PEG. Such experiments, however, have not verified that protection of turgor is

TABLE III
Comparison of the Incorporation of Radioactivity from [^{14}C]Glucose or UDP-[^{14}C]-Glucose into Polysaccharides of Intact or Detached Fibers as a Function of Medium pH[a]

Treatment	Cellulose	Noncellulosic polysaccharides	Cellulose	Noncellulosic polysaccharide
	(cpm/mg cellulose in fibers per hr)			
	pH 6.0		pH 7.5	
[^{14}C]Glucose				
Intact	5180	3520	1330	1940
Detached	490	1410	70	2370
Detached in PEG	2880	4350	530	3150
UDP-[^{14}C]-glucose				
Intact	60	190	10	410
Detached	20	1340	80	4530
Detached in PEG	50	440	20	1440

[a] Reproduced from Carpita and Delmer (1980) by permission of *Plant Physiology*. Incubations were for 1 hr at 32°C in BT medium ± 0.06 molal PEG.

TABLE IV
Incorporation of Radioactivity from [^{14}C]Glucose and UDP-[^{14}C]Glucose into Polysaccharides of Fibers Detached in Polyethylene Glycol as a Function of Degree of Cutting of Fibers[a]

Treatment	Polysaccharide synthesized (cpm/mg cellulose in fibers per hr)	
	Cellulose	Noncellulosic polysaccharides
[^{14}C]Glucose		
Detached (1 cut)	2470	8000
Detached (4 cuts)	1630	7990
Detached (7 cuts)	1480	6530
Detached (finely chopped)	1070	5990
UDP-[^{14}C]-glucose		
Detached (1 cut)	10	830
Detached (4 cuts)	20	1330
Detached (7 cuts)	20	1830
Detached (finely chopped)	20	3590

[a] Reproduced from Carpita and Delmer (1980) by permission of *Plant Physiology*. Fibers of about 20 mm were detached and incubated without further excision (1 cut) or cut with scissors 3 or 6 additional times. Finely chopped fibers were excised further into pieces 1–2 mm in length. All incubations were in BT medium, pH 6.0, plus 0.06 molal PEG for 1 hr at 32° C.

required for cellulose biosynthesis. In retrospect, alternative roles for PEG are apparent. The PEG may stabilize the cellulose synthetase directly. There is a precedent for such a stabilization; Halper and Srere (1977) have observed that PEG stabilized a specific interaction between mitochondrial malate dehydrogenase and citrate synthase. It should be pointed out, though, that this is an unlikely possibility in the detached fiber system, since PEG would be unable to penetrate to the proposed site of synthesis. Alternatively, the PEG could promote the resealing of the plasma membrane. What evidence is there for promotion of resealing?

1. Judged by phase-contrast microscopy, PEG prevents the disruption of the plasma membrane as well as the internal membrane system; excluding PEG from the medium results in extensive membrane disruption on detachment (Fig. 4B,E).

2. Although the detached fibers were unable to accumulate neutral red (Fig. 4A,D), such fibers were able to accumulate hydrolyzed fluorescein when fluorescein diacetate was added to the incubation medium (Fig. 4C,F). Fluorescein diacetate is utilized to check viability of protoplasts (Larkin, 1976); it works on the principle that in living cells, fluorescein diacetate diffuses into the cell, whereupon it is hydrolyzed by plant esterases to fluorescein and acetate. The charged fluorescein is unable to escape if the plasma membrane is intact. The discrepancy between the neutral red uptake and fluorescein entrapment is still unclear. Since neutral red is accumulated by vacuoles, the tonoplast may not be intact in the detached fibers, resulting in the inability to accumulate this vital dye.

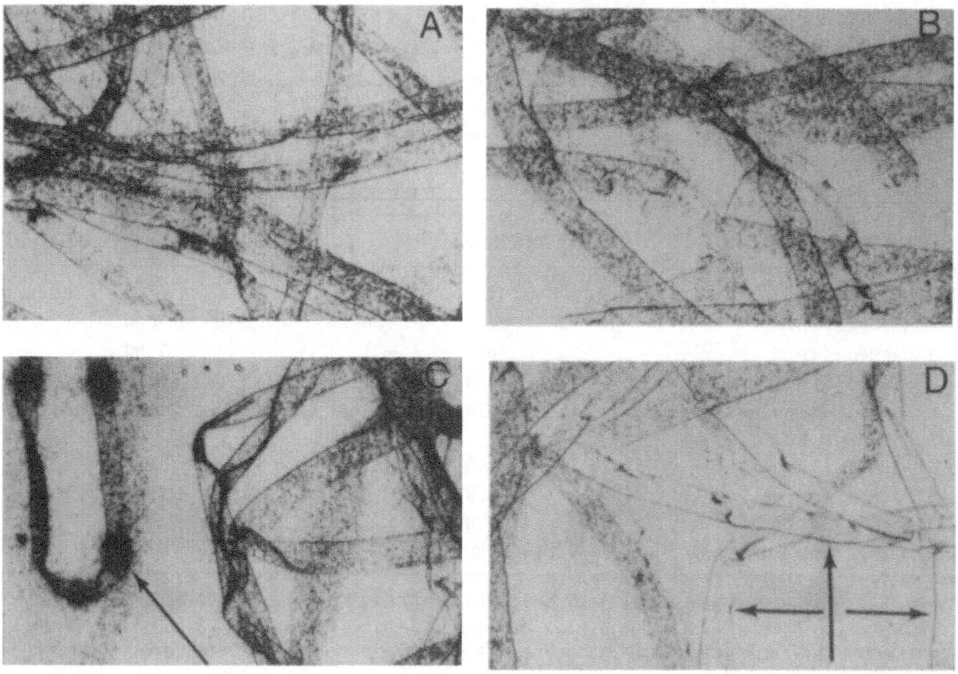

FIGURE 5. Autoradiograms of fibers intact or detached in 0.06 molal PEG after [^{14}C]glucose incorporation into polysaccharides. All incubations were in 10 mM Tris-MES, pH 6.0, + 0.1 M pentaerythritol, 3 mM $CaCl_6$, and 20 mM [^{14}C]glucose.(A) Polysaccharides of intact fibers before acetic-nitric digestion; (B) intact fibers after acetic-nitric digestion; (C) detached fibers (+PEG) before acetic-nitric digestion; (D) detached fibers (+PEG) after acetic-nitric digestion. Reproduced from Carpita and Delmer (1980) by permission of *Plant Physiology*.

3. The inability of UDP-glucose to serve as a substrate for cellulose. As described earlier, since UDP-glucose is unable to permeate the plasma membrane, it is substrate only for β1,3-glucan synthesis in unsealed fibers.

4. The influence of PEG in the promotion of protoplast fusion in other plant systems has long been recognized (Kao and Michayluk, 1974).

5. Both intact and detached cotton fibers in PEG accumulate radioactive glucose into polysaccharides in virtually *all* fibers, indicating that the energy-generating system is protected. On the other hand, cellulose synthesis was evident in only about *half* the detached fibers in PEG (Fig. 5). This experiment clearly indicates that besides the protection of the energy-generating system, a second requirement is necessary specifically for cellulose synthesis. The requirement may indeed be membrane resealing.

6. *Does Cellulose Synthesis Require a Membrane Potential?*

Could the intact membrane be required for generating a membrane potential across the plasma membrane? First of all, it is important to reiterate that cellulose synthesis in higher plants is probably an extracellular synthesis

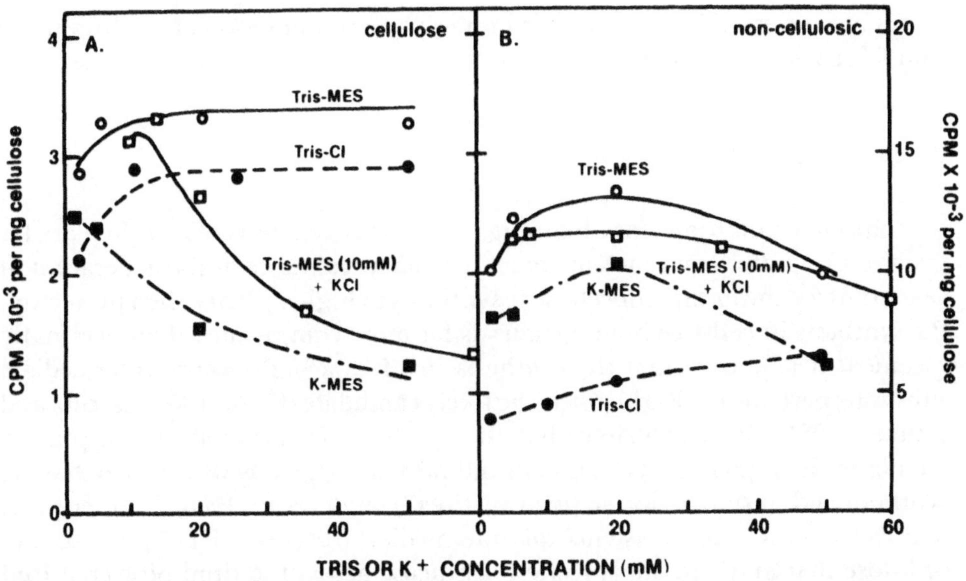

FIGURE 6. Influence of K^+ on cellulose synthesis in detached fibers in 0.06 molal PEG represented as the incorporation of [^{14}C]glucose into cellulose as a function of the ionic strength of the incubation media. All incubations were at pH 6.0 with buffer + 0.1 M pentaerythritol, 0.06 molal PEG, 3 mM $CaCl_2$, 3 mM $MgCl_2$, and 20 mM [^{14}C]glucose, for 1 hr, at 32°C. Reproduced from Carpita and Delmer (1980) by permission of *Plant Physiology*.

(Willison and Brown, 1977), and thus the cellulose synthetase would reside at a site where a potential could be perceived.* If such a potential were important for the synthesis, certainly an intact plasma membrane would be required. Also pertinent are reports of research with other biological systems revealing that detergents permeate only energized membranes (Komor *et al.*, 1979), and changes in membrane potentials cause changes in membrane rigidity of lipid vesicles (Lelkes, 1979). Such observations indicate a physiological significance for the physical state of the membrane. In cotton fibers, such potentials may be necessary for orienting the cellulose synthetase at the plasma membrane–cell wall interface or possibly for moving substrates across the lipid bilayer.

What is the evidence that such potentials could be important? First, K^+ inhibits cellulose synthesis in the detached fibers at concentrations that reduce potentials in other plant systems (Fig. 6) (Cocucci *et al.*, 1976). Second, treatments such as cold shock, detergents, or organic solvents, which are used to render the plasma membrane permeable to UDP-glucose, always increase incorporation of glucose into β1,3-glucan but inhibit cellulose synthesis (Delmer, personal communication). Third, direct evidence now demonstrates that β-glucan synthetase activity in partially purified membrane vesicles is

*The two most abundant biopolymers on earth are cellulose and chitin; both are long polymers of a β1,4-glycan. The *in vitro* chitin biosynthesis in cell-free systems of *Mucor rouxii* has been demonstrated (Bracker *et al.*, 1976), but this system does not require an intact membrane for the synthesis (Ruiz-Herrera *et al.*, 1980).

stimulated severalfold by potentials induced by a combination of a valinomycin and K^+ (Bacic *et al.*, 1980).

7. *Where Do We Go from Here?*

Optimal conditions for obtaining the *in vitro* synthesis of cellulose are far from resolved. If factors such as intact plasma membranes or the generation of potentials are important for cellulose synthesis in higher plants, then protecting the synthesis in cell-free homogenates is far more complicated than originally assumed. It is known that the synthesis involves a high-energy intermediate substrate; perhaps UDP-glucose is the likely candidate (Franz, 1969; Carpita and Delmer, 1981). It is probable that the synthesis is localized at the plasma membrane in higher plants (Willison and Brown, 1977). It is also known that the synthesis and crystallization steps are distinct (Haigler *et al.*, 1980). Furthermore, one can envision the existence of intermediates between UDP-glucose and cellulose that may provide a route for glucose across the lipid bilayer. Lipid intermediates may be involved in cellulose synthesis in *Acetobacter* (Colvin, 1959; Khan and Colvin, 1961; García *et al.*, 1974; Lennarz and Scher, 1972). Potential involvement of lipid and protein intermediates in polysaccharide synthesis has been proposed but not substantiated for higher plants (Forsee and Elbein, 1973; Pont Lezica *et al.*, 1976, 1978; Hopp *et al.*, 1978). Repeated attempts to detect such substances in cotton fibers have not been successful (D. P. Delmer, personal communication). The small surface area of the plasma membrane compared to that of an internal membrane system coupled with a rapid turnover of the lipid intermediate could make such detection extremely difficult. The movement of lipids across the bilayer ("flip-flop") is at present being explored by a few investigators (Cullis and deKruijff, 1978; Gerritson *et al.*, 1980), but I have found no reports on the influence of membrane potentials on such lipid movement.

Given these factors, one must emphasize the frustration of a researcher in cellulose biosynthesis. The requirement of a membrane potential for cellulose synthesis may indeed explain the lack of a bona fide synthesis in cell-free systems. Thus, a paradox may exist—integrity of the plasma membrane must be maintained, yet disruption of the membrane is necessary to obtain the synthetase. The situation is not hopeless. Techniques are available to induce potentials in isolated membrane vesicles, and reconstitution experiments with systems derived from plant, animal, and bacterial systems have been successful (Jones and Plewis, 1974; Schonfeld *et al.*, 1979; Haaker and Racker, 1979; Papazian *et al.*, 1979; Racker, 1979; Nelson *et al.*, 1980). Such experiments are now in progress to attempt to restore the capacity for cellulose synthesis under such conditions.

References

Albersheim, P., Nevins, D. J., English, P. D., and Karr, A., 1967, A method for the analysis of sugars in plant cell-wall polysaccharides by gas-liquid chromatography, *Carbohydr. Res.* **5**:340–345.

Anderson, R. L., and Ray, P. M., 1978, Labeling of the plasma membrane of pea cells by a surface-localized glucan synthetase, *Plant Physiol.* **61:**723–730.

Bacic, A., Klein, A. S., and Delmer, D. P., 1980, Stimulation of glucan synthesis *in vitro* by induction of membrane potentials, *Plant Physiol* (*Suppl.*) **65:**105.

Barber, G. A., Elbein, A. D., and Hassid, W. Z., 1964, The synthesis of cellulose by enzyme systems from higher plants, *J. Biol. Chem.* **239:**4056–4061.

Bartnicki-García, S., and Lippman, E., 1967, Enzymatic digestion and glucan structure of hyphal walls in *Phytophthora cinnamoni, Biochim. Biophys. Acta* **136:**533–543.

Bauer, W. D., Talmadge, K. W., Keegstra, K., and Albersheim, P., 1973, The structure of plant cell walls. II. The hemi-cellulose of the walls of suspension-cultured sycamore cells, *Plant Physiol.* **51:**174–187.

Beasley, C. A., and Ting, I. P., 1974, Effects of plant growth substances on *in vitro* fiber development from unfertilized cotton ovules, *Am. J. Bot.* **61:**188–194.

Bracker, C. E., Ruiz-Herrera, J., and Bartnicki-García, S., 1976, Structure and transformation of chitin synthetase particles (chitosomes) during microfibril synthesis *in vitro, Proc, Natl. Acad. Sci. U.S.A.* **73:**4570–4574.

Brown, R. M., and Montezinos, D., 1976, Cellulose microfibrils: Visualization of biosynthetic and orienting complexes in association with the plasma membrane, *Proc. Natl. Acad. Sci. U.S.A.* **73:**143–147.

Brown, R. M., Jr., France, W. W., Kleinig, H., Falk, H., and Sitte, P., 1969, Cellulosic wall component produced by the Golgi apparatus of *Pleurochrysis scherffelii, Science* **116:**894–896.

Brummond, D. O., and Gibbons, A. P., 1965, Enzymatic cellulose synthesis from UDP-D-glucose-^{14}C by *Lupinus alba, Biochem. Z.* **342:**308–318.

Carpita, N. C., and Delmer, D. P., 1980, Protection of cellulose synthesis in detached cotton fibers by polyethylene glycol, *Plant Physiol.* **66:**911–916.

Carpita, N. C., and Delmer, D. P., 1981, Concentration and metabolic turnover of UDP-glucose in developing cotton fibers. *J. Biol. Chem.* **256:**308–315.

Carpita, N. C., Sabularse, D., Montezinos, D. M., and Delmer, D. P., 1979, Determination of the pore size of the cell walls of living plant cells, *Science* **205:**1144–1147.

Chambers, J., and Elbein, A. D., 1970, Biosynthesis of glucans in mung bean seedlings: Formation of β-(1-4)-glucan from GDP-glucose and β-(1-3)-glucans from UDP-glucose, *Arch. Biochem. Biophys.* **138:**620–631.

Cleland, R., 1977, The control of cell enlargement, in: *Integration of Activity in the Higher Plant*, Proceedings of the 31st Symposium of the *Society for Experimental Biology*, (D. Jennings, ed.), pp. 101–115, Cambridge University Press.

Cocucci, M., Marre, E., Ballerin-Denti, A., and Scacchi, A., 1976, Characteristics of fusicoccin-induced changes in trans-membrane potentials and ion uptake in maize root segments, *Plant Sci. Lett.* **6:**143–156.

Colvin, J. R., 1959, Synthesis of cellulose in ethanol extracts of *Acetobacter xylinum, Nature* (*London*) **183:**1135–1136.

Cullis, P. R., and deKruijff, D. E. 1978, Polymorphic phase behavior in lipid mixtures as detected by ^{31}P NMR: Evidence that cholesterol may destablize bilayer structure in membrane systems containing phosphatidylethanolamine, *Biochem. Biophys. Acta* **507:**207–218.

Delmer, D. P., 1977, The biosynthesis of cellulose and other plant cell wall polysaccharides, in: *Recent Advances in Phytochemistry,* Vol. 11 (F. Loewus and V. C. Runeckles, eds.), pp. 45–77, Plenum Press, New York.

Dürr, M., Bailey, D. S., and Maclachlan, G., 1979, Subcellular distribution of membrane-bound glycosyltransferases from pea stems, *Eur. J. Biochem.* **97:**445–453.

Elbein, A. D., and Hassid, W. Z., 1966, The enzymatic synthesis of a glucomannan, *Biochem. Biophys. Res. Commun.* **23:**311–318.

Elbein, A. D., Barber, G. A., and Hassid, W. Z., 1964, The synthesis of cellulose by an enzyme from a higher plant, *J. Am. Chem. Soc.* **86:**309.

Feingold, D. S., Neufeld, E. F., and Hassid, W. Z., 1958, Synthesis of a β-1,3-linked glucan by extracts of *Phaseolus aureus* seedlings, *J. Biol. Chem.* **233:**783–788.

Flowers, H. M., Batra, K. K., Kemp. J., and Hassid, W. Z., 1969, Biosynthesis of cellulose *in vitro* from guanosine diphosphate D-glucose with enzymatic preparations from *Phaseolus aureus* and *Lupinus albus*, *J. Biol. Chem.* **244**:4969–4974.

Forsee, W. T., and Elbein, A. D., 1973, Biosynthesis of mannosyl- and glycosyl-phosphoryl-polyprenols in cotton fibers, *J. Biol. Chem.* **248**:2858–2867.

Franz, G., 1969, Soluble nucleotides in developing cotton hair, *Phytochemistry* **8**:737–741.

Franz, G., 1978, Biosynthesis of cellulose and hemicelluloses, *Papier* **32**(12):521–525.

Garcia, R. C., Recondo, E., and Dankert, M., 1974, Polysaccharides biosynthesis in *Acetobacter xylinum*: Enzymatic synthesis of lipid diphosphate and monophosphate sugars, *Eur. J. Biochem.* **43**:93–105.

Gerritson, W. J., deKruijff, B., Verkleij, A. J., deGier, J., and vanDeenen, L. L. M., 1980, Ca^{2+}-induced isotropic motion and phosphatidylcholine flip-flop in phosphatidylcholine-cardiolipin bilayers, *Biochem. Biophys, Acta.* **598**:554–560.

Giddings, T. H., Jr., Brower, D. L., and Staehlin, L. A., 1980, Visualization of particle complexes in the plasma membrane of *Microsterias denticulata* associated with the formation of cellulose fibrils in primary and secondary cell walls, *J. Cell Biol.* **84**:327–339.

Glaser, L., 1958, The synthesis of cellulose in cell-free extracts of *Acetobacter xylinum*, *J. Biol. Chem.* **232**:627.

Haaker, H., and Racker, E., 1979, Purification and reconstitution of the Ca^{2+}-ATPase from plasma membranes of pig erythrocytes, *J. Biol. Chem.* **254**:6598–6602.

Haigler, C. H., Brown, R. M., Jr., and Benziman, M., 1980, Calcofluor White ST alters the *in vivo* assembly of cellulose microfibrils, *Science* **210**:903–905.

Hakamori, S., 1964, A rapid permethylation of glycolipid and polysaccharide catalyzed by methysulfinyl carbanion in dimethyl sulfoxide, *J. Biochem.* **55**:205–208.

Halper, L. A., and Srere, P. A., 1977, Interaction between citrate synthase and mitochondrial malate dehydrogenase in the presence of polyethylene glycol, *Arch. Biochem. Biophys.* **184**:529–534.

Heiniger, U., and Delmer, D. P., 1977, UDP-glucose:glucan synthetase in developing cotton fibers. II. Structure of the reaction product, *Plant Physiol.* **59**:719–723.

Hopp, H. E., Romero, P. A., and Pont Lezica, R., 1978, On the inhibition of cellulose biosynthesis by coumarin, *FEBS Lett.* **86**:259–262.

Jones, O. T. C., and Plewis, K. M., 1974, Reconstitution of light dependent electron transport in membranes from a bacteriochlorophyll-less mutant of *Rhodopseudomonas spheroides*, *Biochim. Biophys. Acta* **357**:204–214.

Kao, K. N., and Michayluk, M. R., 1974, A method for high-frequency intergeneric fusion of plant protoplasts, *Planta* **115**:355–367.

Khan, A. W., and Colvin, J. R., 1961, Synthesis of bacterial cellulose from labeled precursor, *Science* **133**:2014–2015.

Komor, E., Weber, H., and Tanner, W., 1979, Greatly increased susceptibility of nonmetabolizing cells towards detergents, *Proc. Natl. Acad. Sci. U.S.A.* **76**:1814–1818.

Larkin, P. J., 1976, Purification and viability determinations of plant protoplasts, *Planta* **128**:213–216.

Lederer, E., and Lederer, M., 1957, *Chromatography: A Review of Principles and Applications*, Elsevier, New York, pp. 237–268.

Lelkes, P. I., 1979, Potential dependent rigidity changes in lipid membrane vesicles. *Biochem. Biophys. Res. Commun.* **90**:656–662.

Lennarz, W. J., and Scher, M. G., 1972, Metabolism and function of polyisoprenol sugar intermediates in membrane-associated reaction, *Biochim. Biophys. Acta* **265**:417–441.

Maltby, D., Carpita, N. C., Montezinos, D., Kulow, C., and Delmer, D. P., 1979, β-1,3-Glucan in developing cotton fibers: Structure, localization, and relationship of synthesis to that of secondary wall cellulose, *Plant Physiol.* **63**:1158–1164.

Mann, J., 1963, Crystallinity of cellulose, in: *Methods in Carbohydrate Chemistry*, Vol. III (R. L. Whistler, ed.), pp. 114–119, Academic Press, New York.

Meinert, M. C., and Delmer, D. P., 1977, Changes in biochemical composition of the cell wall of the cotton fiber during development, *Plant Physiol.* **59**:1088–1097.

Musolan, C., and Kindinger, J., 1978, The synthesis of β-1,3- and β-1,4-glucosyl glycosidic linkages at different concentrations of UDP-α-D-glucose by *Phaseolus vulgaris* leaves, *Rev. Roum. Biochim.* **15**:53–61.

Mutsaers, H. J. W., 1976, Growth and assimilate conversion of cotton bolls (*Gossypium hirsutum* L.). I. Growth of fruits and substrate demand, *Ann. Bot.* **40**:301–315.

Nelson, N., Anholt, R., Lindstrom, J., and Montal, M., 1980, Reconstitution of purified acetylcholine receptors with functional ion channels in plasma lipid bilayers, *Proc. Natl. Acad. Sci. U.S.A.* **77**:3057–3061.

Ordin, L., and Hall, M. A., 1967, Studies of cellulose synthesis by a cell-free oat coleoptile enzyme system: Inactivation by airborne oxidants, *Plant Physiol.* **42**:205–215.

Papazian, D., Rahamimoff, H., and Goldin, S. M., 1979, Reconstitution and purification by "transport specificity fractionation" of an ATP-dependent calcium transport component from synaptosome-derived vesicles, *Proc. Natl. Acad. Sci. U.S.A.* **76**:3708–3712.

Péaud-Lenoël, C., and Axelos, M., 1970, Structural features of the β-glucans enzymatically synthesized from uridine diphosphate glucose by wheat seedlings, *FEBS Lett.* **8**:224–228.

Pierce, M. J., and Raschke, R., 1980, Correlation between loss of turgor and accumulation of abscisic acid in detached leaves, *Planta* **148**:174–182.

Pont Lezica, R., Romero, P. A., and Dankert, M. A., 1976, Membrane-bound UDP-glucose, lipid glucosyltransferases from peas, *Plant Physiol.* **58**:675–680.

Pont Lezica, R., Romero, P. A., and Hopp, H. E., 1978, Glucosylation of membrane-bound proteins by lipid-linked glucose, *Planta* **140**:177–183.

Racker, E., 1979, Reconstitution of membrane processes, *Methods Enzymol.* **55**:699–711.

Ray, P. M., 1980, Cooperative action of β-glucan synthetase and UDP-xylose:xylosyltransferase of Golgi membranes in the synthesis of xyloglucan-like polysaccharide, *Biochim, Biophys. Acta* **629**:431–444.

Ray, P. M., Shiniger, T. L., and Ray, M. M., 1969, Isolation of β-glucan synthetase particles from plant cells and identification with Golgi membranes, *Proc. Natl. Acad. Sci. U.S.A.* **64**:605–612.

Reese, E. T. 1977, Degradation of polymeric carbohydrates by microbial enzymes, in: *Recent Advances in Phytochemistry*, Vol. 11 (F. Loewus and V. C. Runeckles, eds.), pp. 311–367, Plenum Press, New York.

Reese, E. T., Smakula, E., and Perlin, A. S., 1959, Enzymatic production of cellotriose from cellulose, *Arch. Biochem. Biophys.* **85**:171–175.

Reese, E. T., Parrish, F. W., and Manders, M., 1961, β-D-1,2-Glucanases in fungi, *Can. J. Microbiol.* **7**:309–317.

Ruiz-Herrera, J., Bartnicki-García, S., and Bracker, C. E., 1980, Dissociation of chitosomes by digitonin into 16 S subunits with chitin synthetase activity, *Biochim. Biophys. Acta* **629**:201–216.

Sandford, P. A., and Conrad, H. E., 1966, The structure of *Aerobacter aerogenes* A3(S1)polysaccharide. I. A reexamination using improved procedures for methylation analysis, *Biochemistry* **5**:1508–1517.

Schonfeld, M., Montal, M., and Feher, G., 1979, Functional reconstitution of photosynthetic reaction centers in planar lipid bilayers, *Proc. Natl. Acad. Sci. U.S.A.* **79**:6351–6355.

Shore, G., and Maclachlan, G. A., 1975, Hormone treatment alters the intracellular location of alkali-insoluble β-1,4-glucan (cellulose) synthetase activities, *J. Cell Biol.* **64**:557–571.

Smith, M. M., and Stone, B. A., 1973, β-Glucan synthesis by cell-free extracts from *Lolium multiflorum* endosperm, *Biochim. Biophys. Acta* **313**:72–94.

Tsai, C. M., and Hassid, W. Z., 1971, Solubilization and separation of uridine diphospho-D-glucose:β-(1-3)-glucan glucosyltransferases from coleoptiles of *Avena sativa, Plant Physiol.* **47**:740–744.

Updegraff, D. M., 1969, Semimicro determination of cellulose in biological materials, *Anal. Biochem.* **32**:420–424.

Villemez, C. L., and Heller, J. S., 1970, Is guanosine diphosphate-D-glucose a precursor of cellulose?, *Nature (London)* **227**:80–81.

Villemez, C. L., Jr., Franz, G., and Hassid, W. Z., 1967, Biosynthesis of alkali insoluble

polysaccharide from UDP-D-glucose with particulate enzyme preparations from *Phaseolus aureus*, *Plant Physiol.* **42**:1219–1223.
Willison, J. H. M., and Brown, R. M., Jr., 1977, An examination of the developing cotton fiber wall and plasmalemma, *Protoplasma* **92**:21–41.
Zimmerman, U., 1978, Physics of turgor and osmoregulation, *Annu. Rev. Plant Physiol.* **29**:121–148.
Zweig, G., and Whitaker, J. R., 1971, *Paper Chromatography and Electrophoresis*, Academic Press, New York, pp. 152–171.

13

The Control of Molecular Weight and Molecular-Weight Distribution in the Biogenesis of Cellulose

MARIANNE MARX-FIGINI

1. Introduction to the Concepts of Molecular Weight and Molecular-Weight Distribution and Their Relationships to the Polymerization Mechanism

General references to which the reader is referred for further discussion of the subjects covered in this section are Flory (1953, 1969), Tanford (1967), Billmeyer (1971), Vollmert (1973), Elias (1972), Hoffman *et al.* (1977), and Cantow (1967).

1.1. Molecular Weight and Degree of Polymerization

Cellulose belongs to the class of macromolecular compounds. Therefore, in order to determine the molecular properties of cellulose or the kinetic reaction mechanisms of, for example, the biopolymerization of cellulose, theoretical treatments and experimental methods developed for investigation on polymers must be employed.

The molecules of a macromolecular compound, whether a synthetic polymer or a natural one, consist of repeating small chemical units that are linked by covalent bonds. The number of such repeating monomeric units combined in a polymer molecule is called the degree of polymerization (DP).

MARIANNE MARX-FIGINI • Division Macromoleculas, Instituto de Investigaciones Fisicoquimicas Teoricas y Aplicadas, Universidad Nacional de La Plata. Conicet y CIC, La Plata, Argentina.

The molecular weight M of a macromolecular compound results, then, as the product of the DP and the molecular weight M_0 of the repeating unit:

$$M = DP \cdot M_0$$

The DP, and with it M, is one of the decisive properties of a macromolecular substance because of its strong influence on the mechanical properties of the polymer and because of its intimate relationship to the kinetics of the respective polymerization or polycondensation process. The molecular-weight determination is therefore one of the most informative procedures that must be carried out to characterize a polymer or to investigate the reaction mechanisms of a polymerization or degradation process.

1.1.1. Definition of the Different Averages

Whereas a low-molecular-weight compound consists of molecules having the same molecular weight throughout, a macromolecular substance represents a polymolecular mixture. This means that the macromolecules comprising a discrete polymer differ in their DP and M. As a consequence, molecular-weight determinations on a polymer yield only average values.

There can be distinguished different kinds of average values, the most important of which are as follows:

a. The *number average* M_n and DP_n, respectively, defined as:

$$M_n = \frac{\sum_{i=1}^{i=\infty} n_i \cdot M_i}{\sum_{i=1}^{i=\infty} n_i} \tag{1}$$

$$DP_n = \frac{\sum_{i=1}^{i=\infty} n_i \cdot DP_i}{\sum_{i=1}^{i=\infty} n_i} \tag{2}$$

b. The *weight average* M_w *and* DP_w, respectively, defined as:

$$M_w = \frac{\sum_{i=1}^{i=\infty} w_i \cdot M_i}{\sum_{i=1}^{i=\infty} w_i} \tag{3}$$

$$DP_w = \frac{\sum_{i=1}^{i=\infty} w_i \cdot DP_i}{\sum_{i=1}^{i=\infty} w_i} \tag{4}$$

c. The *viscosity average* M_η and DP_η, respectively, defined as

$$M_\eta = \frac{\left(\sum_{i=1}^{i=\infty} w_i \cdot M_i^a\right)^{1/a}}{\sum_{i=1}^{i=\infty} w_i} \tag{5}$$

$$DP_\eta = \frac{\left(\sum_{i=1}^{i=\infty} w_i \cdot DP_i^a\right)^{1/a}}{\sum_{i=1}^{i=\infty} w_i} \tag{6}$$

in which n_i is the number of molecules having the molecular weight M_i and the degree of polymerization DP_i, respectively; w_i is the total mass of molecules having the molecular weight M_i and the degree of polymerization DP_i, respectively; and a is the exponent of the relationships between $[\eta]$ and M or DP, respectively (see also Section 1.1.3). The terms n_i and w_i are related by the intrinsic viscosity

$$w_i = n_i \cdot m_i$$

where m_i is the mass of a single molecule having the molecular weight M_i and the degree of polymerization DP_i, respectively.

It always holds that

$$M_n < M_v < M_w \quad \text{and} \quad DP_n < DP_v < DP_w$$

The quotient M_w/M_n is a measure of the polymolecularity, called polydispersity, of a polymer. It is obvious from the foregoing that in monodisperse polymers (in most cases these are biopolymers), this quotient is near 1. In synthetic polymers, M_w/M_n lies mostly between 2 and 5, with some exceptions of smaller polydispersities or of polydispersities up to 30.

1.1.2. Principal Techniques of Molecular-Weight Determination

Two kinds of methods can be distinguished: absolute and relative (indirect). The absolute methods result directly in the determination of molecular weight based in all cases on well-founded theoretical laws for the relationship between the molecular structure and the measured phenomenon. Relative methods, on the contrary, demand a preceding calibration, since the measured phenomenon depends only parametrically on the molecular weight.

The principal techniques of absolute methods are: membrane osmometry, light scattering, X-ray small-angle scattering, sedimentation equilibrium, sedimentation velocity (in combination with the determination of the diffusion coefficient), and end-group determination. The most important indirect methods are viscometry, gel-permeation chromatography (GPC), and vapor-pressure osmometry. Of the methods mentioned, viscometry is the most frequently used technique because of its simplicity, rapidity, and economy.

The averages of molecular weight differ according to the method used. All techniques that measure colligative properties as well as end-group determinations yield the number-average M_n, whereas the scattering and sedimentation methods result in the weight-average M_w or in a value close to it. The average obtained by viscometry also lies near M_w, the difference being smaller the more monodisperse the polymer. In the case that exponent a of equations (8a) and (8b) (see Section 1.1.3) $= 1$, $M = M_w$. The resulting molecular weight, however, is defined only when the calibration has been carried out by use of an absolute method that yields the weight-average. GPC supplies both number- and weight-averages.

1.1.3. Molecular-Weight Determination on Cellulose

Because unsubstituted cellulose dissolves only in solutions of alkaline complexes of heavy metals such as cuprammonium hydroxide (CuO), cupriethylenediamine hydroxide (CuEn), cadmium-ethylenediamine hydroxide (Cadoxen), or iron–sodium-tartrate complex FeTNa), and because the respective solutions are more or less unstable in the presence of the oxygen of air, absolute molecular-weight determinations cannot be carried out on cellulose. Molecular-weight determinations in these solvents are therefore performed only by viscometrically, determining the intrinsic viscosity $[\eta]$ and converting it into M or DP by means of a calibration-function elaborated on cellulose nitrates and adapted to unsubstituted cellulose (Marx-Figini and Schulz, 1962; Marx-Figini, 1978). Derivatives of cellulose such as the nitric ester or the carbanilate dissolve in organic solvents, the respective solutions being stable. Therefore, molecular-weight determinations are often conducted on the cellulose derivatives instead of the unsubstituted cellulose. Basic requirements for these determinations are that the cellulose must have been converted polymeranalogously and homogeneously to its derivative (Marx-Figini, 1961; Marx-Figini, 1962).

The intrinsic viscosity $[\eta]$ of cellulose in (CuEn) and of cellulose nitrates in acetone or ethylacetate can be elaborated from measurements at a single concentration by means of the general equation of Schulz and Blaschke (1941):

$$[\eta] = \frac{\eta_{sp}/c}{1 + k_\eta \cdot \eta_{sp}} \tag{7}$$

using the corresponding values of k_η (Marx, 1955; Marx-Figini and Schulz, 1959).

The relationship between $[\eta]$ and M and DP, respectively, obeys the general equations of Staudinger, Mark and Howink:

$$[\eta] = K_M \cdot M^a \tag{8a}$$

$$[\eta] = K'_{DP} \cdot DP^a \tag{8b}$$

where K and a are constants that depend on the polymer and the solvent. The respective values valid for cellulose and cellulose nitrate in CuEn and acetone are compiled in Table I.

To use these constants the following experimental conditions must be met (Marx-Figini and Schulz, 1962):

1. Shear gradients of approximately $G = 300\ \text{sec}^{-1}$ for CuEn and $G = 1200\ \text{sec}^{-1}$ for acetone.
2. The range of applied concentrations must result in values of specific viscosity between $\eta_{sp} = 0.3$ and $\eta_{sp} = 0.6$.
3. The cellulose nitrate must exhibit a degree of substitution of $S = 2.90 \pm 0.02$, i.e, % N = 13.85 ± 0.05.

1.2. Relationship between Conversion and Molecular Weight in Polymerization and Biopolymerization Reactions

In polymerization or polycondensation processes, there exists generally a closed relationship between the molecular weight and the conversion (quantity of the synthesized polymer). To study the mechanism of a polymerization

TABLE I

Constants K and a of Equation 8b for Cellulose Nitrate and Unsubstituted Cellulose[a]

Material	Solvent	Range of *DP*	*K*	*a*
Cellulose nitrate[b]	Acetone	<950	0.82	1.0
		>950	4.46	0.76
Unsubstituted cellulose	CuEn	<950	0.42	1.0
		>950	2.29	0.76

[a] From Marx-Figini and Schulz (1962) and Marx-Figini (1978).
[b] Degree of substitution $S = 2.90 \pm 0.02$.

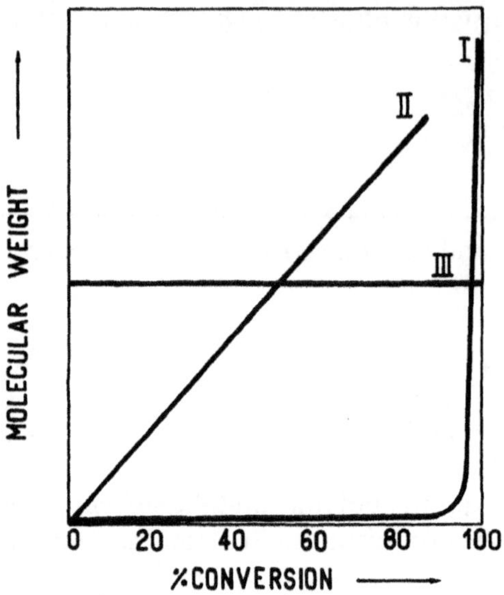

FIGURE 1. Relationship between molecular weight and percentage of conversion for a statistical (I), a nonstatistical (II), and a structure-correlated (III) step polymerization.

reaction it is therefore necessary to investigate, in addition to the conversion kinetics, the molecular weight and its distribution (see Section 1.3.1) as a function of the time of synthesis and of the conversions, respectively.

In principle, polymerization and polycondensation reactions (called also chain or addition polymerization and step polymerization, respectively) can be roughly classified as chance-correlated and structure-correlated processes. Among the chance-correlated reactions are distinguished statistical and non-statistical ones. All chain polymerizations involving radicals as well as most step polymerizations are statistical processes. Some anionic polymerizations of vinyl monomers, in contrast, are examples of nonstatistical ones. Structure-controlled reactions predominate in the field of biopolymerization.

Because nearly all biopolymerization processes are polycondensations (i.e., processes that occur by the combination of bifunctional monomers with the elimination of small molecules such as water), there is here briefly described the molecular-weight behavior of only step polymerization reactions. These relationships are represented in summary in Fig. 1.

1.2.1. *Statistical Step Polymerization*

Here, the aggregation step of a monomeric unit on a growing molecule is randomly distributed. Only low-molecular-weight polymers (oligomers) are produced at high values of conversion. A pronounced increase in the DP does not occur until conversions higher than 99%. Such high conversions, however, are reached only when the low-molecular-weight compound released by the condensation steps is continously removed.

1.2.2. *Nonstatistical Step Polymerization*

This type of reaction takes place according to a Poisson mechanism as it is known in chain polymerization. The Poisson mechanism is characterized by the facts that considering a discrete polymerization process, all molecules start simultaneously, all monomers have the same probability of being added onto a growing chain, and the polymerization does not occur in termination steps.

The behavior of the molecular weight is therefore characterized by an increase of the DP proportional to the conversion. A change in the reaction conditions that provokes a change in reaction rate and conversion consequently produces variations in the molecular weight.

A nonstatistical step polymerization seems to take place, for example, in the enzymatic *in vitro* synthesis of amylose from Cori-Ester using "primars" as starting material (Husemann *et al.*, 1958; Husemann, 1960).

1.2.3. Structure-Correlate Mechanism

This type of polymerization mechanism occurs by the involvement of a structure (e.g., a template) that governs the number of monomeric units to be combined in each polymer molecule. As a consequence, molecules with the same molecular properties are always produced. The best-known examples of this type of polymerization are protein, DNA, and RNA synthesis, wherein the controlling "structure" governs, however, not only the molecular weight but also the chemical structure.

The relationship between degree of polymerization and conversion is accordingly characterized by a complete constancy of the DP during the whole course of synthesis. Changes in the reaction conditions will produce changes in the polymerization rate and conversion, but not in the DP.

1.3. Molecular-Weight Distribution

As described in Section 1.1.1, most macromolecular compounds are polydisperse, which means that the molecular weight is not homogeneous but shows a certain distribution.

The molecular-weight distribution of a polymer can be expressed by a frequency function (mole-fraction distribution) or by a mass function (weight-fraction distribution) according to the number or to the mass of molecules. The frequency distribution indicates for each degree of polymerization or molecular weight, respectively, the number of molecules having this degree of polymerization as related to the total number of molecules. The mass distribution signifies consequently the weight of molecules having this degree of polymerization as related to 1 g of polymer. Of the two functions, the mass function is the more frequently used.

The distribution curves as such can be represented as an integral or as a differential distribution curve. The integral curve of a weight distribution indicates the weight percentage of all molecules having a degree of polymerization up to DP_i, DP_i being a discrete degree of polymerization, whereas the differential curve gives the weight fraction belonging to each degree of polymerization, normalized to 1 (Fig. 2).

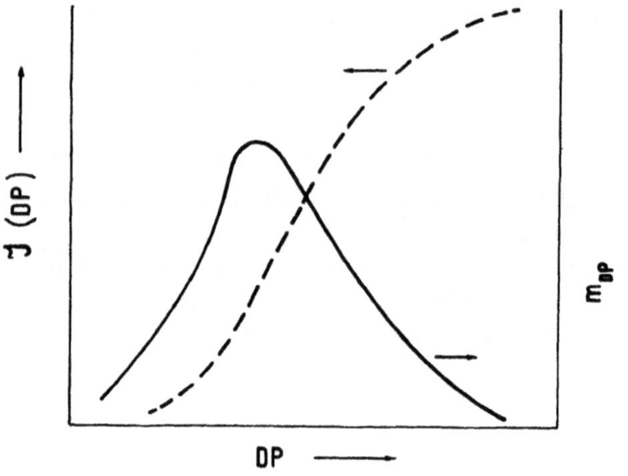

FIGURE 2. Example of an integral (- - -) and a differential (—) molecular-weight distribution curve (expressed in terms of *DP*).

1.3.1. Relationship between Polymerization Kinetics and Molecular-Weight Distribution

The molecular-weight distribution of a polymer is strictly related to the reaction kinetics of the polymerization. Therefore, in most studies of the mechanism of a polymerization process, the determination of the molecular-weight distribution is an indispensable step.

Statistical chain and step polymerizations result in the "Schulz–Flory distribution," also called "normal distribution" or "most probable distribution" (Schulz, 1935; Flory, 1936, 1953). Here the weight fraction m_{DP} of the polymer molecules having the degree of polymerization *DP* is given by

$$m_{DP} = \frac{DP}{\overline{DP}_n^{\,2}} \cdot e^{-DP/\overline{DP}_n} \tag{9}$$

The distribution becomes broader the higher the conversion, but maintains throughout the relationship $DP_w/DP_n = 2$. A typical Schulz–Flory distribution for several extending stages 9 of step polymerization is given in Fig. 3.

In contrast to the statistical polymerization processes, the nonstatistical ones as described in Section 1.2.2 result in extremely narrow molecular-weight distributions. In an ideal case, there arises Poisson distribution (Flory, 1940), which is distinguished by its extreme homogeneity, the polydispersity quotient being DP_w/DP_n:

$$DP_w/DP_n = \frac{1}{DP_n} + 1$$

According to the existing proportionality between *DP* and conversion, the distribution curve shifts with the conversion to a higher degree of polymerization, but maintains its homogeneity (Fig. 4).

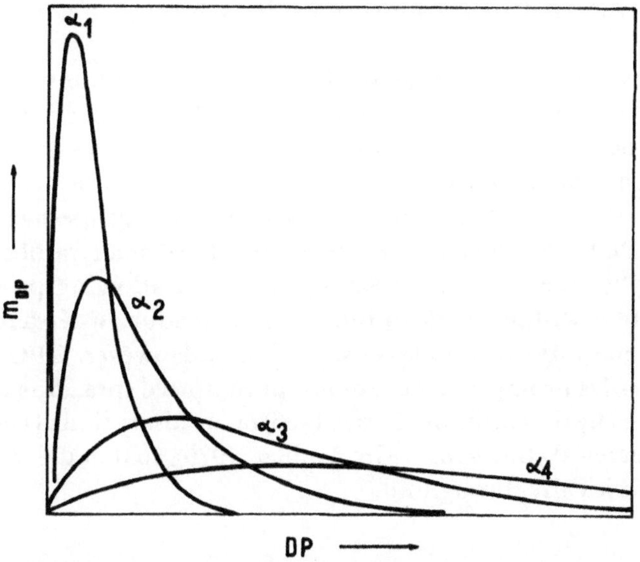

FIGURE 3. "Most probable distribution" of molecular weight (expressed in terms of *DP*) at increasing stages (α_1-α_4) of a statistical step polymerization.

Completely monodisperse molecular-weight distributions appear in structure-correlated polymerization mechanisms. Due to the constancy of the degree of polymerization with increasing conversion, no shift takes place with the conversion or the time of synthesis.

1.3.2. Principal Techniques of Determination of Molecular-Weight Distribution

Most of the methods applied in the determination of molecular-weight distribution are based on the dependence on the molecular weight of either the

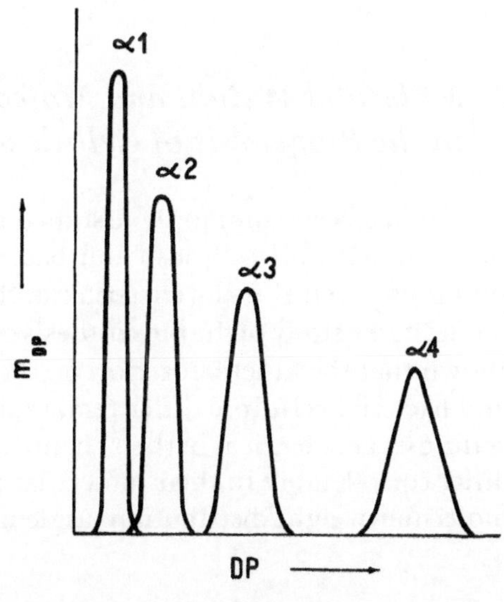

FIGURE 4. "Poisson distribution" of molecular weight, expressed in terms of *DP* at increasing stages (α_1-α_4) of a nonstatistical polymerization.

solubility or the molecular dimensions in solution of a macromolecular substance.

In the first case, use is made of the fact that the solubility of a polymer decreases with increasing molecular weight. The respective fractionation processes can be carried out by means of a precipitation or a solution technique.

A fractionation according to the dimensions of the dissolved macromolecules in the respective solvent is achieved by gel-permeation chromatography (GPC), a technique that can be considered as a type of size exclusion chromatography. GPC has the great advantage of allowing production of a distribution curve within several minutes to several hours, whereas fractionation by precipitation or dissolution takes several days. However, GPC still presents some problems, for example, a more or less pronounced spreading of the elution curves produced by the columns (Tung, 1966) or an adsorption of the polymer on the packing material (Dawkins, 1975; Ambler, 1976), so that the method cannot be applied uncritically throughout.

1.3.3. Determination of Molecular-Weight Distribution on Cellulose

As mentioned in Section 1.1.3, determinations of the molecular-weight distribution on cellulose samples must be carried out on cellulose nitrates or carbanilates. The most reliable fractionation method for cellulose nitrate is still the precipitation technique, using acetone as solvent and mixtures of water and acetone as precipitant (Marx-Figini, 1959; Marx-Figini and Schulz, 1963).

The application of GPC on cellulose nitrate or carbanilate is still limited to low-molecular-weight celluloses because of the insufficient separation capacity of the columns for higher-molecular-weight celluloses. Furthermore, it must be taken into account that an eventual adsorption of cellulose nitrate on the packing material is what makes the application of this method on cellulose nitrate of higher molecular weight still doubtful (Meyerhoff, 1970).

2. Molecular Weight and Molecular-Weight Distribution in the Biogenesis of Cellulose

In has been commonly assumed that the reaction mechanisms of biosynthesis of cell-wall cellulose and bacterial cellulose are identical and that the synthesis of cellulose by bacteria can therefore be considered as a representative model in the study of the biosynthesis of cellulose. Recently, however, it has been shown that the kinetic reaction mechanisms of biopolymerization of cell-wall and bacterial cellulose differ remarkably. As a consequence, although the two celluloses are identical in their chemical and supermolecular fine structure, they differ considerably in their molecular properties such as molecular weight and molecular-weight distribution, as demonstrated in the following sections.

2.1. *Cell-Wall Cellulose*

2.1.1. *Research Materials*

The best material for kinetic studies on the biosynthesis of cell wall cellulose is cotton cellulose. The seed hairs of cotton bolls represent a well-defined system because it is possible to establish exactly the postanthesis time of synthesis before the cotton boll is harvested. Furthermore, cotton seed hairs consist of nearly 95% cellulose, and the small portion of accompanying noncellulosic compounds can be eliminated without application of harsh methods.

Among the lower plants such as algae, a suitable material for this kind of investigation is the alga *Valonia* because it contains a relatively large proportion of unmodified cellulose, and it is possible to distinguish cells of different ages.

Before the determination of its molecular properties, the cellulose must be isolated from the noncellulosic compounds by applying techniques that avoid any change in the molecular properties. Such methods consist in successive extractions of the raw material with organic solvents (acetone and benzene) and with an aqueous solution of 2% NaOH. The precautions that must be followed to preserve the native molecular properties consist of a continuous bubbling of oxygen-free nitrogen and an addition of a reducing agent, e.g., sodium dithionite (Schulz and Marx, 1954; Marx-Figini and Schulz, 1963).

The isolation of cellulose from the cell walls of lower plants requires an additional extraction with boiling water between the solvent and the alkaline extraction (Marx-Figini, 1969b).

The compounds that dissolve in hot aqueous NaOH are principally noncellulosic constituents of the primary wall, most of which are probably other polysaccharides (Fig. 5). In lower plants such as the alga *Valonia*, the production of alkali-soluble material also continues during secondary wall synthesis and

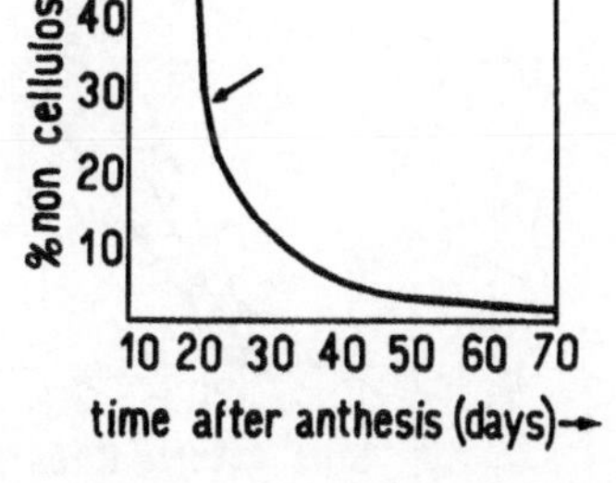

FIGURE 5. Cell-wall cellulose of higher plants. Percentage of noncellulosic compounds in the seed hairs of *Gossypium herbaceum* as a function of time after anthesis. (←) Onset of secondary wall synthesis.

TABLE II
Percentage of Noncellulosic Material Referred to Dry Weight per Cell in the Alga *Valonia*

Dry weight per cell (mg)	Noncellulosic material (%)
0.27	50.4
1.80	53.7
1.94	52.5
2.50	50.0
5.70	50.0

does not vary significantly with the age of the cell (Table II) (Marx-Figini, 1969a and b).

2.1.2. *Conversion Kinetics*

In the conversion kinetics of the biosynthesis of cell-wall cellulose, there can be distinguished two successive periods that differ remarkably in their reaction rates and therefore in the yield of cellulose per unit of time (Fig. 6) (Marx-Figini, 1963, Marx-Figini and Schulz, 1966). During the first period, only a small portion of cellulose is synthesized, whereas during the second period, nearly 95% of the whole mass of cellulose is produced. In the second period, conversion takes place at a constant rate, indicating the occurrence of a zero order reaction with respect to the conversion kinetics.

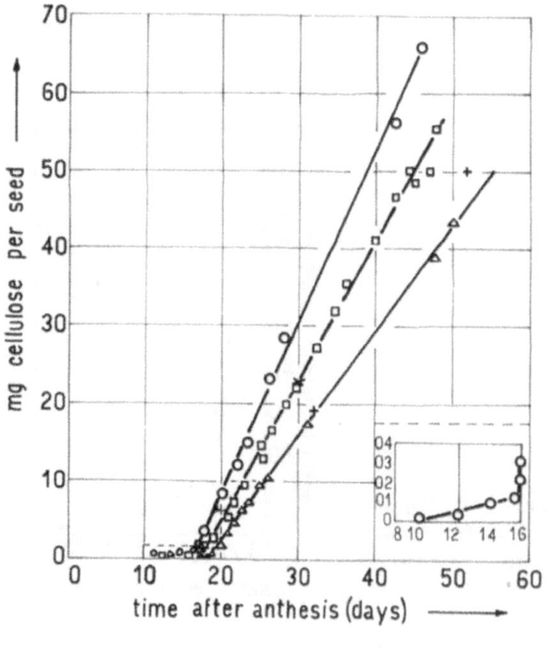

FIGURE 6. Cell-wall cellulose of higher plants. Conversion kinetics at different growth conditions: *Gossypium herbaceum* grown at 60,000 lux (○), 15,000 lux (□), and 4000 lux (△); *Gossypium hirsutum* grown at 15,000 lux (×) and 4000 lux (+).

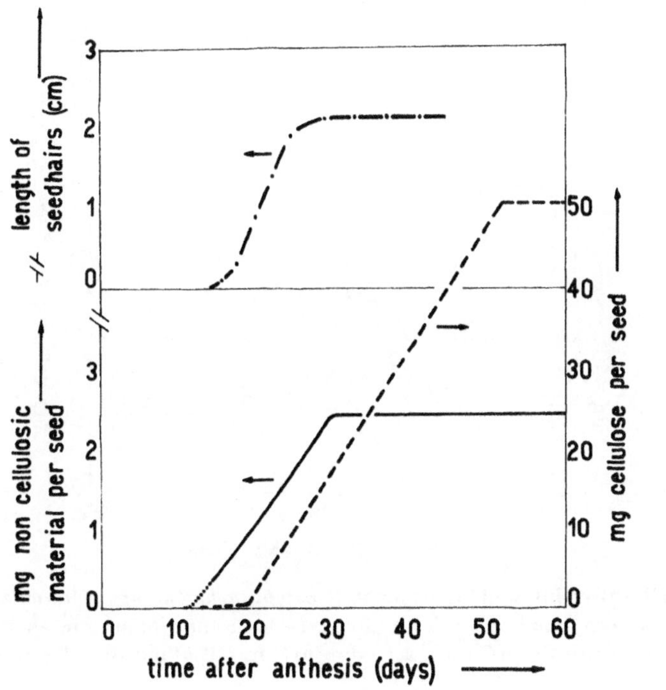

FIGURE 7. Cell-wall cellulose of higher plants (*Gossypium herbaceum*). Conversions of cellulose (- - -) and noncellulosic material (—), and cell size (— · — ·) as a function of time after anthesis.

It is reasonable to coordinate the two kinetic periods to the synthesis of primary and secondary wall cellulose, which is verified by the following facts (Marx-Figini and Schulz, 1966): (1) during the first, slow, period, hemicellulosic compounds are simultaneously and principally synthesized (see also Fig. 5); (2) the simultaneous synthesis of a large amount of noncellulosic compounds and little cellulose occurs only during cell elongation (Fig. 7); and (3) the seed hairs in formation show balloon swelling, typical of cellulose, only after the beginning of the rapid period, i.e., after the onset of the main production of cellulose.

The conversion-time kinetics of cellulose synthesis depend on the kinetics of the individual reaction conditions, which are given by the growth conditions of the respective plants (see Fig. 6) (Marx-Figini, 1966b). Detailed experimental data concerning the conversion kinetics of lower plant cellulose are still not available. As shown in Table II, more or less equal quantities of cellulose and noncellulosic compounds are continuously produced. The respective portions corresponding to the primary and secondary walls, however, are unknown.

2.1.3. *Molecular Weight*

As shown in Section 1.2, conclusions about a polymerization mechanism can be drawn only when the behavior of the molecular weight and its distribution, and their relationship to the conversion kinetics, are known. The corresponding studies on cell-wall cellulose in higher and lower plants have revealed the following facts (Marx-Figini, 1963, 1966a, 1969a,b; Marx-Figini and

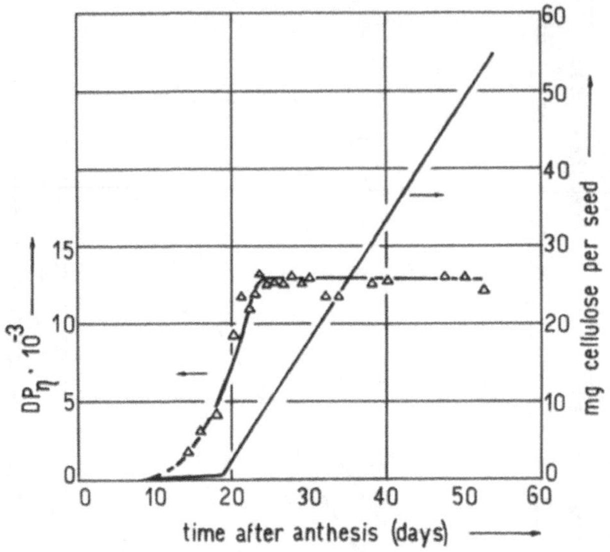

FIGURE 8. Cell-wall cellulose of higher plants (*Gossypium herbaceum*). Molecular weight (in terms of *DP*) (Δ) and corresponding yield of cellulose (—) as a function of time after anthesis. *DP* was determined viscometrically in CuEn[•] according to Marx (1955) and Marx-Figini and Schulz (1962).

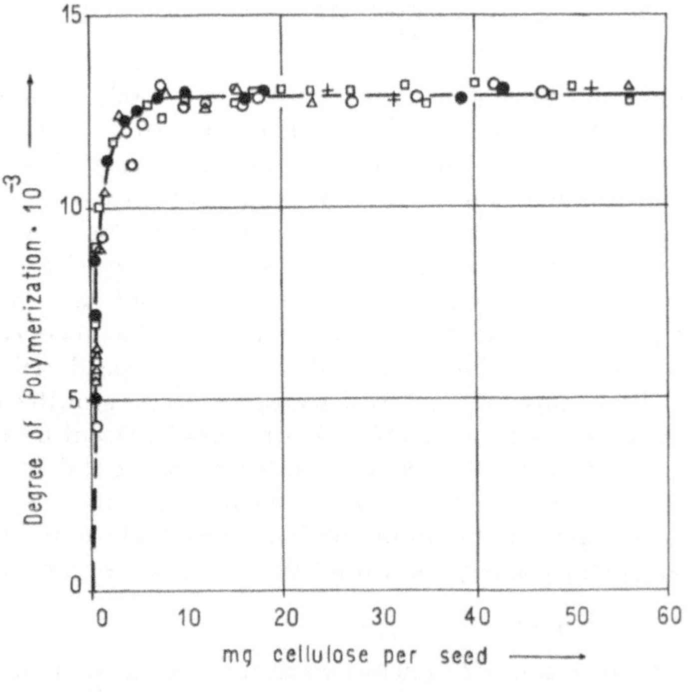

FIGURE 9. Cell-wall cellulose of higher plants (*Gossypium herbaceum* and *Gossypium hirsutum*). Molecular weight (in terms of *DP*) as a function of conversion at different growth conditions. ○ grown at 10–25°C, 1000 lux (see also Fig. 8); *Gossypium herbaceum*. □ Δ ● grown at 20–35°C and at different light intensities □ at 1500 lux, Δ at 6000 lux, and ● at 4000 lux, (see also Fig. 6); *Gossypium herbaceum*. + *Gossypium hirsutum*.

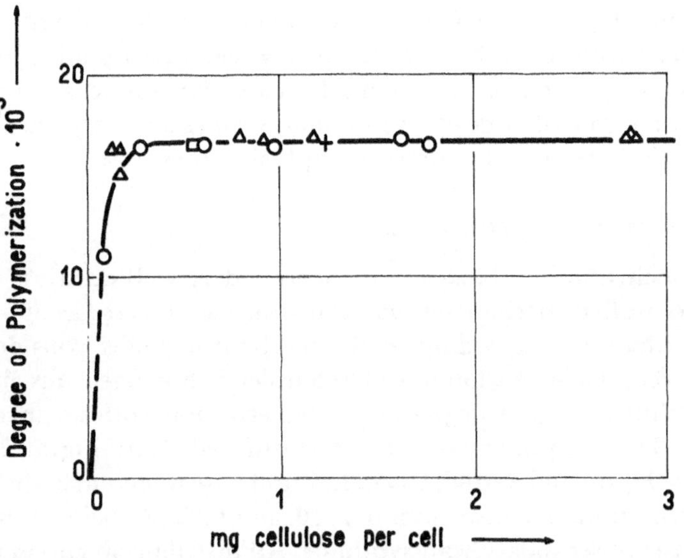

FIGURE 10. Cell-wall cellulose of lower plants (*Valonia*). Molecular weight (in terms of *DP*) as a function of conversion. DP was determined as indicated for Fig. 8. *Valonia* was grown at different places and therefore under different conditions. Δ Bay of Naples, + Adriatic Sea, □ North Sea, O cultivated in a seawater medium according containing extract of garden mold, 5%, 0.006% $Na_2HPO_4 \cdot H_2O$ and 0.01% $NaNO_3$ in each 100 ml. sea water, at 18°C, 2,000 lux, and PH 8.0.

Schulz, 1966): An increase of the chain length with the time of synthesis and the conversion occurs only during the period of primary wall cellulose synthesis (Fig. 8). However, during the entire period of synthesis of secondary wall cellulose, the degree of polymerization remains constant and does not exhibit variation, either with time (Fig. 8) or the conversion or with the growth conditions and the botanical species (Fig. 9).

In lower plants such as the alga, *Valonia,* even though actual conversion kinetics have not been determined, the molecular-weight behavior can be correlated very well with the quantity of cellulose produced by each cell. It is evident from Fig. 10 that the secondary wall cellulose of lower plants exhibits the same features as that of higher plants (see Fig. 9).

These results have shown that synthesis of primary and secondary wall cellulose differs not only in reaction rates but also in polymerization mechanisms. This is of special importance in relation to the independence of chain length from growth conditions, botanical species, and cell shape. All these facts have forced a conclusion that synthesis of secondary wall cellulose may occur by a "structure correlated" mechanism. This means that in cellulose biosynthesis, a certain compound of cell structure that acts as a "template" may be involved. This template may control the number of successive additions of activated glucose units to a growing cellulose chain. It is a reasonable assumption, therefore, to consider the molecular properties of native cellulose as "genetically controlled."

According to the molecular weights determined for native cell-wall celluloses of other plants (Timell, 1957; Goring and Timell, 1962; Marx-Figini and Schulz, 1963), cotton cellulose can be regarded as representative for all. The somewhat smaller values found for native wood celluloses, flax, ramie, and others can be considered to be caused by degradations that occur as a consequence of the more drastic isolation methods that must be applied in these cases.

2.1.4. Molecular-Weight Distribution

The conclusion that biosynthesis of secondary wall cellulose occurs by a structure-controlled mechanism was confirmed by the respective molecular-weight distributions. According to the mechanism under consideration, the molecular-weight distribution must be completely homogeneous and must not suffer any shift to higher degrees of polymerization with an increase of the conversion. These requirements are in fact fulfilled (Marx-Figini, 1964, 1966a, 1969a; Marx-Figini and Penzel, 1965). In Fig. 11 are represented the molecular-weight distributions at conversions of 2, 20, and 100%, 2% being a stage slightly after the onset of secondary wall synthesis. All distribution curves possess one maximum at both extremes. The main maximum at the high-molecular-weight end corresponds to the secondary wall cellulose. It is evident that its mass increases with the conversion. Its position, however, remains fixed at DP = 14,000, and its monodispersity does not broaden. The width of the maximum can

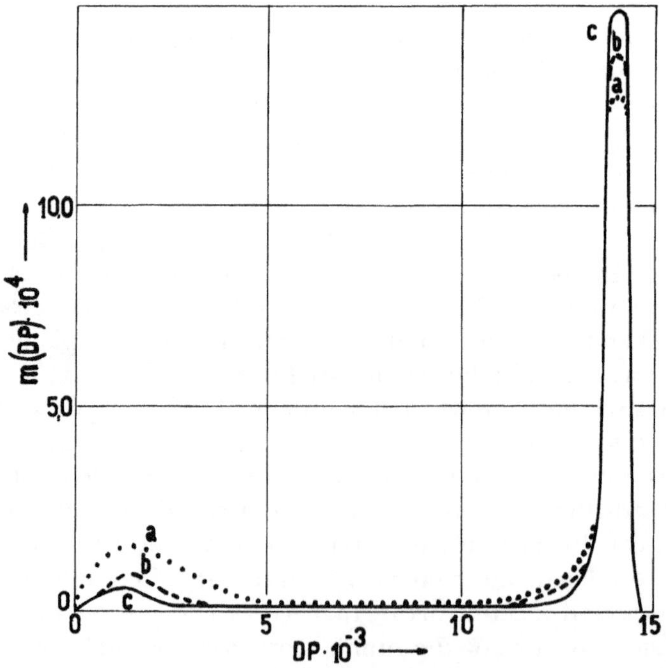

FIGURE 11. Cell-wall cellulose of higher plants (*Gossypium herbaceum*). Molecular-weight distribution (in terms of chain-length distribution) at increasing conversions: 2% (a), 20% (b), and 100% (c). The method of fractionation used was precipitation on the respective nitrates, according to Marx-Figini and Penzel (1965).

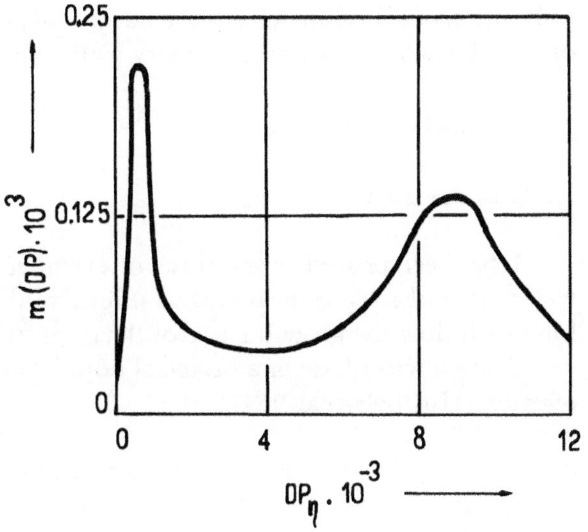

FIGURE 12. Cell-wall cellulose of higher plants (*Gossypium herbaceum*). Molecular-weight distribution (in terms of chain-length distribution) of primary wall cellulose. Conversion: 0.2%. Experimental technique as indicated in the Fig. 11 caption.

be considered as given by the limits of error of the fractionation method. In reality, the *DP* may be completely monodisperse. The small maximum at the low-molecular-weight end corresponds to the primary wall cellulose, the distribution of which is very much broader than that of the secondary wall cellulose (Fig. 12) as a consequence of its different polymerization mechanism (Marx-Figini, 1969a).

The same features as the cell-wall cellulose of higher plants are exhibited by the cell-wall cellulose of the alga *Valonia* (Fig. 13) (Marx-Figini, 1969b). The

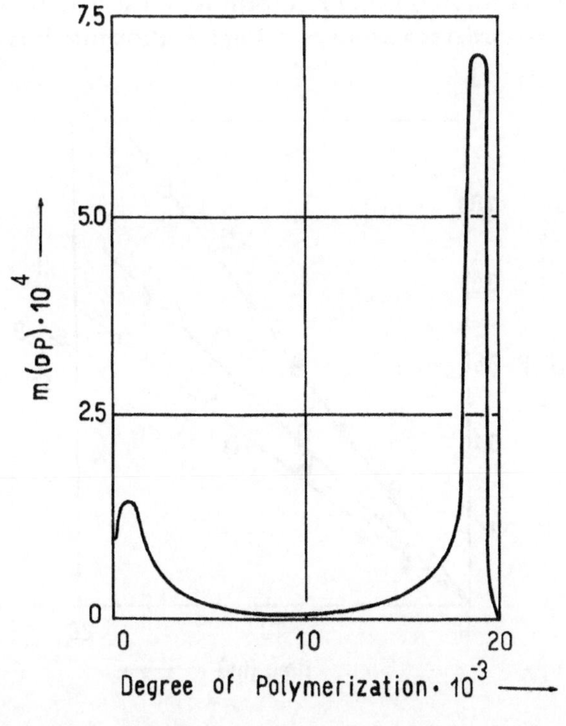

FIGURE 13. Cell-wall cellulose from lower plants (*Valonia*). Molecular-weight distribution (in terms of chain-length distribution) at a conversion of about 35%. Experimental technique as indicated in the Fig. 11 caption.

only difference is a somewhat higher degree of polymerization (*DP* 18,000 instead of *DP* 14,000 of the cotton secondary wall cellulose).

2.2. *Bacterial Cellulose*

2.2.1. *Conversion Kinetics*

It has been proved by celluloses of *Acetobacter acetigenum* (Brown, 1962) as well as *Acetobacter xylinum* (Marx-Figini and Pion, 1974) that the conversion kinetics follow the growth curve of the respective bacterial populations.

The growth phase of a bacterial population is known to obey a first-order reaction (Hinshelwood, 1947):

$$N_t = N_0 \cdot e^{k_N t} \tag{10}$$

Accordingly, the yield of cellulose with time can be described by the equation

$$C_t = C_0 \cdot e^{k_c t} \tag{11}$$

Growth of the bacterial population and quantity of cellulose produced coincide not only in the reaction order but also in the rate constants, i.e., $K_c = K_N$ (Fig. 14) (Marx-Figini and Pion, 1974).

The rate constant K_N may also be expressed as the mean generation time (or interdivision time) τ according to the general relationship between k and τ for first-order reactions, $\tau = \log 2/k$. Plotting the values of $\log (c_t/c_0$ as a function of

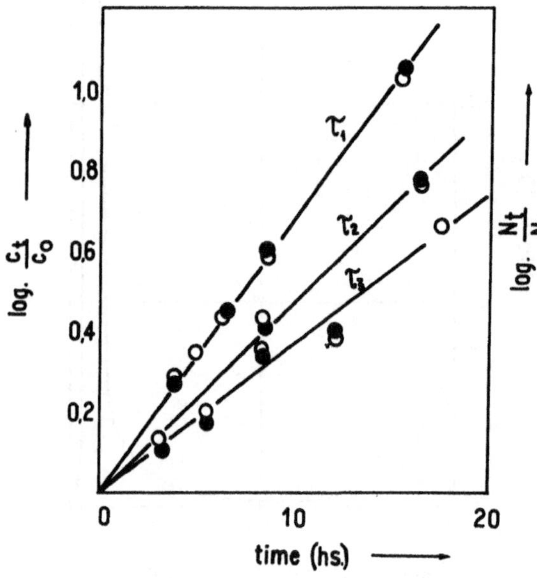

FIGURE 14. Bacterial cellulose (*Acetobacter xylinum*). Number of viable bacteria per milliliter (●) and cellulose produced (○) in terms of a first-order reaction as a function of time. There were three different cultures, having different generation times τ: $\tau_1 = 4.8$ hr; $\tau_2 = 6.2$ hr; $\tau_3 = 8.0$ hr.

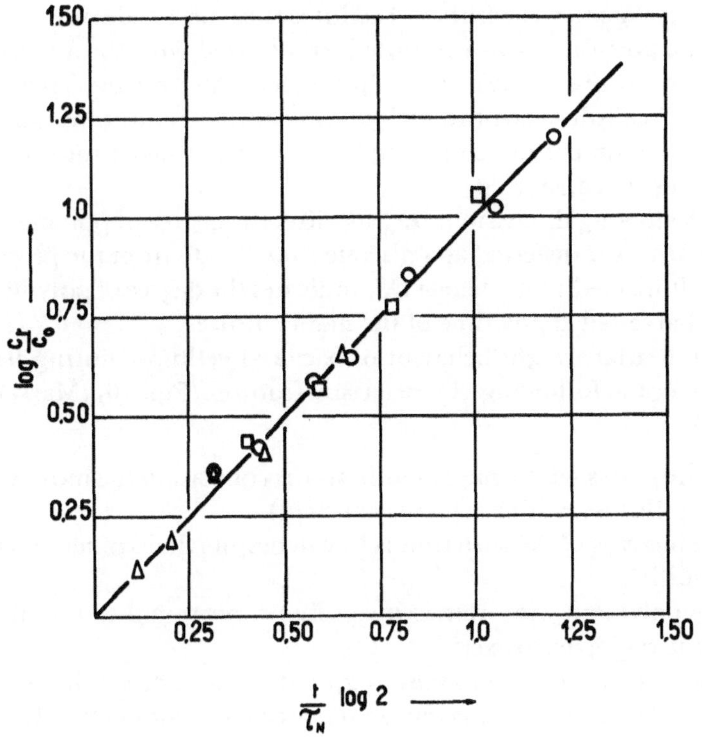

FIGURE 15. Bacterial cellulose (*Acetobacter xylinum*). Cellulose produced in terms of a first-order reaction as a function of the reduced time t/τ (number of generations). Experimental data are the same as in Fig. 14.

the respective reduced times t/τ, a straight line is formed that passes through the origin (Fig. 15).

In contrast to the exponential growth phase of the bacterial population, the lag phase does not produce cellulose. When cell division is prevented by the use of washed bacteria in a culture medium lacking nitrogen compounds, there is metabolic but no synthetic activity observed.

The strong agreement between the rate constant of bacterial population growth and product formation also exists in the synthesis of other compounds, probably noncellulosic polysaccharides, that are deposited together with the cellulose and amount to approximately 50% of the crude material.

From these, before determination of its molecular properties, the cellulose has to be isolated by successive extractions with water and a 2% aqueous solution of NaOH, observing the same precautions as mentioned in Section 2.1.1 to avoid degradation.

2.2.2. *Molecular Weight*

According to general microbiological techniques, cultures of bacteria in a liquid medium are induced in two stages: a preculture and a culture in mass, hereafter called the main culture. The main culture is started by inoculation with bacteria from the preculture. Because, in the case of *Acetobacter xylinum* or

Acetobacter acetigenum, synthesis of cellulose also occurs during the preculture period, and a portion of this cellulose is transferred with the attached bacteria into the main culture medium, the cellulose harvested during a main culture is composed of cellulose synthesized at preculture conditions and cellulose synthesized at main culture conditions. This must be taken into account when kinetic studies are considered.

In the following discussion, DP_0 signifies the degree of polymerization of a cellulose that was transferred at a discrete time ($t = 0$) from the preculture into the main culture medium, whereas DP_t indicates the degree of polymerization of a cellulose harvested at any time of the main culture.

The molecular-weight behavior of bacterial cellulose during its synthesis exhibits, then, the following characteristic features (Fig. 16) (Marx-Figini and Pion, 1974):

1. *DP* increases with time of synthesis (in contrast to the molecular weight of cell-wall cellulose, which remains constant).
2. The increase of *DP* with time is linear, despite the exponential increase of the cellulose yield.
3. The polymerization rate is inversely proportional to the interdivision time τ_N of the respective bacteria.
4. When the bacterial population growth ceases, the synthesis of cellulose stops and the degree of polymerization remains constant (Table III).

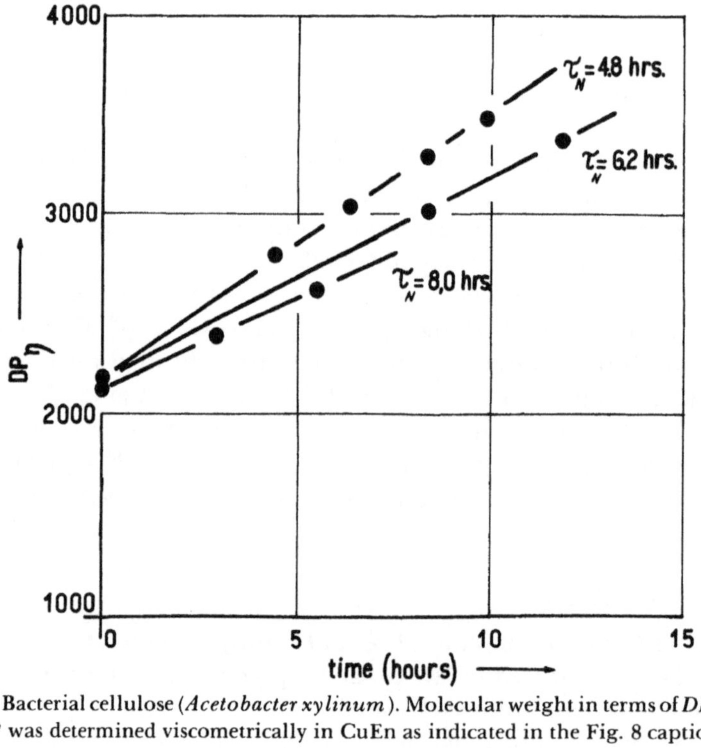

FIGURE 16. Bacterial cellulose (*Acetobacter xylinum*). Molecular weight in terms of *DP* as a function of time. *DP* was determined viscometrically in CuEn as indicated in the Fig. 8 caption.

TABLE III
Number of Viable Bacteria per Milliliter (N), Yield of Cellulose, and Corresponding Degree of Polymerization (DP) for a Representative Main Culture of *Acetobacter xylinum*[a]

Time (hr)	$N \cdot 10^{-6}$	Cellulose (mg)	DP
0	0.56	5.0	2190
8.5	1.30	14.5	3020
12.2	2.21	19.9	3400
16.4	3.50	32.0	3240
21.1	5.80	60.3	3410
25.1	2.0	84.8	3470
30.2	0.2	84.1	3470

[a]The culture time was extended up to the death phase of the bacteria.

At more prolonged times of synthesis, a slight decrease of the DP_η occurs, followed by a renewed increase (see also Table III). This decrease of DP_η had been found independently of two groups of authors (Husemann and Werner, 1963; Marx-Figini and Pion, 1974). The first group has interpreted the decrease of DP_η as the action of a degrading enzyme that is released into the culture medium and that would have its optimum activity at a PH between 6.5 and 4.0. The action of a cellulase in cultures of *Acetobacter xylinum*, however, could not be confirmed by other authors. A more precise interpretation of the interim decrease can be given considering the polymerization mechanism described below (see Fig. 17) and taking into account a certain synchronization of the bacteria.

The cited findings have led to the following conclusions:

1. There exists a strict control of the polysaccharide synthesis by the bacteria, as indicated by the agreement of the reaction order as well as of the rate constants between growth of bacterial population and increase in the yield of cellulose.
2. This control must consist of an initiation of a defined and identical number of cellulose molecules by each bacterium in a given culture at a defined time within its life cycle, which is probably the time of cell division or an event related to it.
3. The polymerization process occurs without a termination step and is indicated by the increase in DP_η that also continues when $t/\tau >> 1$. If the polymerization mechanism also comprises a termination step, a constant level of DP_η should appear just after one generation.
4. Cellulose synthesis must take place in permanent contact with the cell, as indicated by the inverse proportionality between polymerization rate and the interdivision time τ.

The conclusions pointed out in (2) and (3) assume that the polymerization mechanism of bacterial cellulose is a nonstatistical one of the type of a Poisson mechanism.

Recently, the relationship between molecular weight, reaction rate, time of

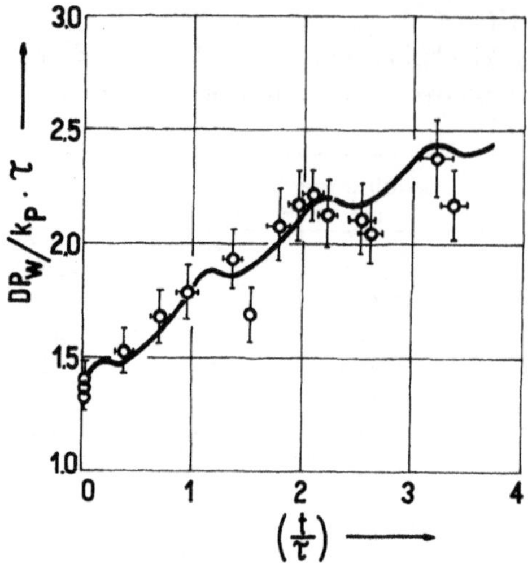

FIGURE 17. Bacterial cellulose. Reduced molecular weight (in terms of DP) as a function of the generation time t/τ, theoretically derived, whereby k_p signifies a first-order chain-propagation rate constant and $k_p \cdot \tau = 1570$. (Φ) Experimentally observed values.

synthesis, and generation time of the bacteria in bacterial polymerization processes has been theoretically derived assuming a Poissonlike polymerization mechanism and taking into account an exponential increase of chain initiation and a linear chain growth. The theoretical approach (Figini, 1974) confirmed the assumption of a Poissonlike mechanism (Fig. 17) and revealed furthermore that the experimentally observed interim decrease of DP_η is a consequence of a certain synchronization of the bacteria and the nonstatistical polymerization mechanism.

2.2.3. *Molecular-Weight Distribution*

According to the characteristic features of the polymerization mechanism of bacterial cellulose, it was expected that the chain-length distributions at polymerization times $t/\tau \gg 1$ would show an increasing number of maxima and therefore an increasing broadening with time. The experimentally determined distribution curves confirmed the prediction and in this way proved once again the aforestated mechanism (Fig. 18) (Marx-Figini and Pion, 1976).

In both distribution curves, there exists a pronounced maximum at nearly the same position of the low-molecular-weight end that corresponds to the

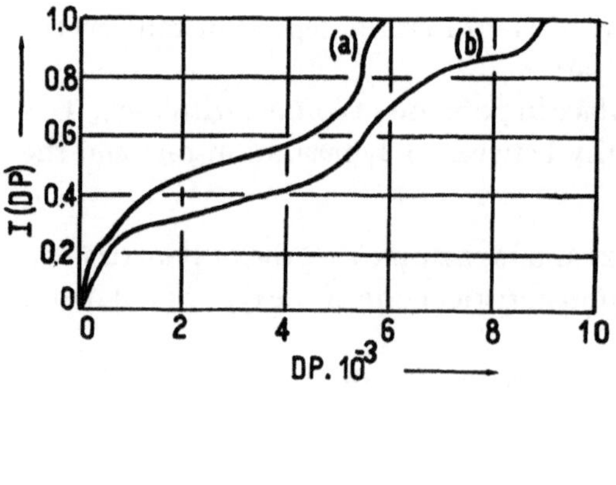

FIGURE 18. Bacterial cellulose (*Acetobacter xylinum*). Integral chain-length distribution curves at different numbers of generation: (a) $t/\tau = 2.0$, $DP_\eta = 2640$; (b) $t/\tau = 5.2$, $DP_\eta = 3800$. The fractionation technique was as indicated in the Fig. 11 caption.

recently initiated molecules. The high-molecular-weight end, in constrast, shifts with increasing number of generations to higher degrees of polymerization, increasing simultaneously the number of maxima.

Chain-length distributions of bacterial cellulose with more than one maximum have been found (Husemann and Werner, 1963). In these investigations, however, the influence of the growth of the bacterial population was not studied. It was therefore impossible to interpret them correctly.

3. Discussion

3.1. Comprehensive Comparison of the Differences in the Controlling Features and Their Consequences on the Polymerization Mechanisms and the Molecular Properties of Cell-Wall and Bacterial Cellulose

According to the present state of knowledge, it is well established that biopolymerization of both cell-wall and bacterial cellulose occurs under continuous control of the living cell and therefore in continuous contact with it. It differs, however, in the way in which the control is exerted.

In the case of cell-wall synthesis, the cell governs the molecular weight directly through the involvement of a "templatelike" compound or cell structure that "counts" the number of condensation steps occurring in the polymerization of each cellulose molecule. The molecular and supermolecular structure of this template must be genetically controlled. In this way, nature assures that cell-wall cellulose always possesses the same molecular properties independently from the particular growth conditions or variations of botanical species.

In the biogenesis of bacterial cellulose, on the other hand, the cell can actuate directly the number of chain initiations produced by each bacterium. The cell can control the molecular weight only indirectly by regulating the time of polymerization (given by the lifetime of the bacterium), and the rate of chain propagation.

As a consequence of the differences in the controlling features, both celluloses differ in the kinetic mechanisms of their polymerization and therefore in their molecular properties. Cell-wall cellulose is always distinguished by monodisperse molecular weight that may be altered only by posterior degradation reactions. Bacterial cellulose, on the other hand, shows monodispersity only during one synchronized generation of the bacterium, although its polymerization mechanism corresponds to a "Poisson"-like one. By the same measure, an increase in the number of generations also increases the polydispersity of the cellulose produced because of the small *DP* of the recently initiated molecules and the continuous chain growth of the previously initiated ones.

These characteristic differences between cell-wall and bacterial cellulose are illustrated in summary in Figs. 19 and 20.

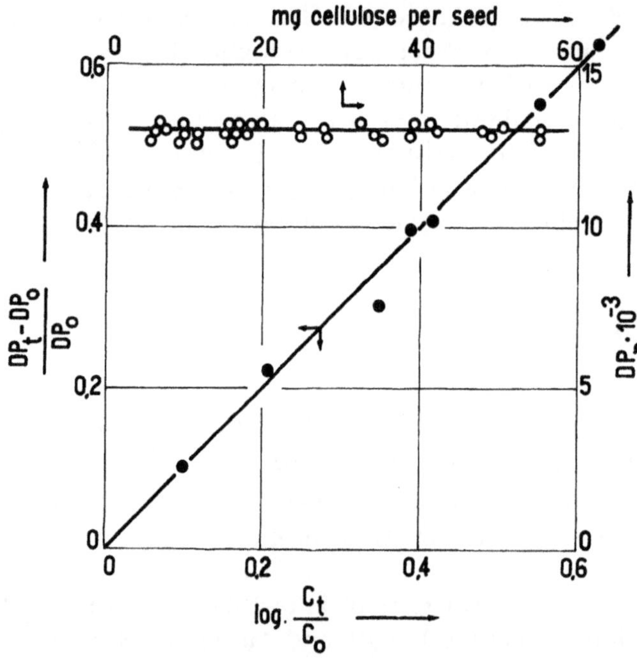

FIGURE 19. Comparison of the molecular-weight behavior at increasing conversions in biogenesis of cell-wall cellulose (○) and bacterial cellulose (●).

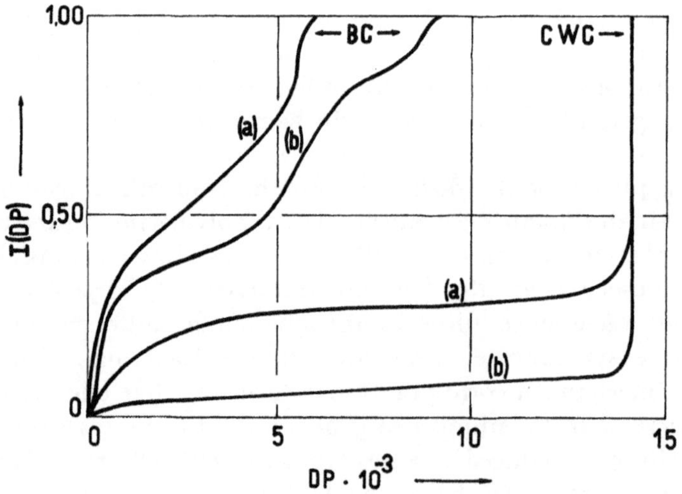

FIGURE 20. Comparison of integral molecular-weight distributions at increasing conversions in biogenesis of cell-wall cellulose (CWC) and bacterial cellulose (BC). Early stages of synthesis (a). Later stages of synthesis (b).

3.2. Consequences with Respect to the Formation of the Supermolecular Structure

Despite the different kinetics of their polymerization processes and the resulting differences in the molecular properties of cell-wall and bacterial cellulose, the two celluloses coincide in their supermolecular features. They are built up of the same morphological units (Frey-Wyssling and Mühlethaler, 1946, 1963) and show the same crystal lattice (Meyer and Misch, 1937; Meyer and Mark, 1930; Khonvine *et al.*, 1932), the so-called cellulose I. Until recently, this crystal lattice had been found exclusively in cellulose samples that had not undergone dissolution or special chemical treatments. For a long time, cellulose I was considered to be a particularly native property that originates as a consequence of simultaneous polymerization and crystallization, for example, epitatic polymerization on a recently preformed crystal lattice. According to the kinetic results, it is difficult, however, to understand that two polymerization processes that are directed by the existence of the same preformed crystal lattice would obey different kinetic mechanisms. Moreover, electron-diffraction studies (Macchi *et al.*, 1968; Macchi and Palma, 1969) have shown that the crystal lattice I can also be formed on dissolved and reprecipitated celluloses. This means that the crystal lattice I is not a property typical for native cellulose only. Rather, it suggests that the supermolecular structure of cellulose *is* a consequence of its molecular properties and that polymerization and crystallization are mechanisms that occur *independently* of each other. This does not exclude that crystallization may take place just when chain propagation is still going on.

3.3. Conclusions That Can Be Drawn from the Kinetics and Molecular-Weight Behavior with Respect to the Site of Synthesis

Although there does not yet exist any insight into or idea of the nature of the controlling structure, one can derive from the native degree of polymerization of cell-wall cellulose that a cellulose molecule requires as its site of synthesis a structure or organelle with a dimension of about 14,000 × 5.14 Å (unless molecules of smaller but constant chain length were first synthesized and then combined to produce a *DP* of about 14,000; this mechanism, however, would involve two controls: that of the chain length and that of the combination of a constant number of these chains). Of the organelles, only the microtubules and the plasma membrane fulfill this condition. Electron-microscopic findings seem to show that microtubules are accumulated during secondary wall synthesis and found frequently in proximity to growing cellulose fibers (Ledbetter and Porter, 1963; Nagai and Rebhun, 1966; Newcomb and Hepler, 1964; Wooding and Northcote, 1964; Cronshaw and Bouck, 1965; Robards and Kidwai, 1972). This in connection with their length of several micrometers led to the suggestion that the microtubules are directly connected with cellulose synthesis. To prove this hypothesis Marx-Figini (1971) cultured *Valonia* in the presence of colchicine, which is known to provoke disappearance of the microtubules. Later determination of the number of new cells formed as well as of the yield and degree of polymerization of the cellulose at different stages of cell-wall synthesis revealed that colchicine had affected only the number of new cells. The yield and *DP* of the cellulose, in contrast, were the same as in the corresponding blank tests. It can be concluded, therefore, that the microtubules do not participate directly in biosynthesis of cellulose molecules. Rather, the plasma membrane may be considered as a site of synthesis and a controlling structure, as indicated by electron-microscopic findings and sedimentation experiments (Wardrop and Foster, 1964; Mühlethaler, 1967; Bauer, 1967; Villemez *et al.*, 1968; Robinson and Preston, 1972; Brown, and Montezinos, 1976). According to some electron-microscopic results (Wardrop and Foster, 1964), the plasma membrane may be successively replaced by a new one, leaving the older one in the newly synthesized cell wall. This observation agrees with the detection of proteins within the cell wall (Lamport, 1965; Thomson and Preston, 1967) and with analytical results on the chemical composition of cell-wall cellulose at different stages of cotton seed hair development. These indicate a nearly constant relationship between protein and glucose constant during a considerable period of secondary wall formation (Huwyler, 1978).

The often unsuccessful attempts at cell-free *in vitro* synthesis of actual cellulose (poly-β1,4-D-glucose) using cell homogenates as synthetase source (Barber and Hassid, 1965; Shore and Maclachlan, 1975; Shore *et al.*, Delmer *et al.*, 1977) can be taken as a further proof for the requirement *in vivo* of something more than only the synthetases and the activated glucose, namely, that of the "structure" controlling the number of addition steps in the synthesis of the cellulose molecules (Marx-Figini, 1966c).

Concerning bacterial cellulose, kinetic investigations have revealed an involvement of cell division (or an event related to it) in the initiation of the cellulose chains. One may argue, therefore, that chain initiation is connected with the appearance of a new part of the cell membrane or the cell envelope. The observed inverse proportionality of chain propagation and interdivision time also suggests that chain propagation occurs in continuous contact with the cell. This conclusion has been confirmed by biochemical and electron-microscopic studies (Cooper and Manley, 1975a–e; Brown and Montezinos; Brown, *et al.* 1977). The most probable site of synthesis is the cell surface as was visualized by Brown *et al.* (1976). There, synthesis takes place, including chain initiation and the addition steps, whereas the recently synthesized part of each chain may be loosened from the cell while polymerization is still going on, and undergo arrangement of its supermolecular structure.

References

Ambler, M. R., 1976, Gel permeation chromatography of low molecular weight polymers, *J. Polym. Sci. Polym. Lett. Ed.* **14**:683.

Barber, G. A., and Hassid, W. Z., 1965, Synthesis of cellulose by enzyme from the developing cotton ball, *Nature (London)* **207**:295.

Bauer, H., 1967, Ultrastruktur und Zellwandbildung von *Acanthamoeba* sp., *Vierteljahresschr. Naturforsch. Ges. Zuerich* **112:**173.

Billmeyer, F. W., Jr., 1971, *Textbook of Polymer Science*, Wiley-Interscience, New York.

Brown, R. M., 1962, The mechanism of cellulose biosynthesis by *Acetobacter acetigenum*, *J. Polym. Sci.* **59:**155.

Brown, R. M., Jr., and Montezinos, D., 1976, Cellulose microfibrils: Visualization of the biosynthetic and orienting complexes in the plasma membrane, *Proc. Natl. Acad. Sci. U.S.A.* **73:**143.

Brown, R. M., Jr., Willison, J. H. M., and Richardson, C. L., 1976, Cellulose biosynthesis in *Acetobacter xylinum*, visualization of the site of synthesis and direct measurement of the *in vivo* process, *Proc. Natl. Acad. Sci. U.S.A.* **73:**4565.

Cantow, M., 1967, *Polymer Fractionation*, Academic Press, New York and London.

Cooper, D., and Manley, R. St. J., 1975a, Cellulose synthesis by *Acetobacter xylinum*. I. Low molecular weight compounds present in the region of synthesis, *Biochim. Biophys. Acta* **381:**78.

Cooper, D., and Manley, R. St. J., 1975b, Cellulose synthesis by *Acetobacter xylinum*. II. Relation between cellulose synthesis and cell envelope components, *Biochim. Biophys. Acta* **381:**97.

Cooper, D., and Manley, R. St. J., 1975c, Cellulose synthesis by *Acetobacter xylinum*. III. Matrix, primer, and lipid requirements and heat stability of the cellulose forming enzymes, *Biochim. Biophys. Acta* **381:**109.

Cronshaw, J., and Bouck, B., 1965, The fine structure of differentiating xylem elements, *J. Cell Biol.* **24:**415.

Dawkins, J. V., 1975, Gel permeation chromatography with cross linked polystyrene gels and poor and theta solvents for polystyrene, 1-3 *Makromol. Chem.* **176:**1777.

Delmer, D. P., Heiniger, U., and Kulow, C., 1977, UDP-glucose: Glucan synthetase in developing cotton fibers, *Plant Physiol.* **59:**713.

Elias, H. G., 1972, *Makromoleküle*, Hüthig und Wepf Verlag, Heidelberg.

Figini, R. V., 1974, Kinetics and molecular weight of macromolecules by polymerization in organized systems, Abstracts IUPAC—Symposium on Macromolecules, Río de Janeiro.

Flory, P. J., 1936, Molecular size distribution in linear condensation polymers, *J. Am. Chem. Soc.* **58:**1877.

Flory, P. J. 1940, Molecular size distribution in ethylene oxide polymers, *J. Am. Chem. Soc.* **62:**1561.

Flory, P. J., 1953, *Principles of Polymer Chemistry*, 1st ed., Cornell University Press, Ithaca and London.

Flory, P. J., 1969, *Principles of Polymer Chemistry*, 2nd ed., Cornell University Press, Ithaca and London.

Frey-Wyssling, A., and Mühlethaler, K., 1946, Submicroscopic structure of cellulose gels, *J. Polym. Sci.* **1:**172.

Frey-Wyssling, A., and Mühlethaler, K., 1963, Die Elementarfibrillen der Cellulose, *Makromol. Chem.* **62:**25.

Goring, D. A., and Timell, T. E., 1962, Molecular weight of native celluloses, *Tappi* **45:**454.

Hinshelwood, C. N., 1947, *The Chemical Kinetics of Bacterial Cells*, 1st ed., Oxford University Press, London.

Hoffmann, M., Krömer, H., and Kuhn, R., 1977, *Polymeranalytik I*, Georg Thieme Verlag, Stuttgart.

Husemann, E., 1960, Ueber den enzymatischen Auf- und Abbau von Amylose, *Makromol. Chem.* **35:**239.

Husemann, E., and Werner, R., 1963, Cellulosesynthese durch *Acetobacter xylinum*, *Makromol. Chem.* **59:**43.

Husemann, E., Fritz, B., Lippert, R., and Pfannemuller, B., 1958, Ueber natürliche und synthetische Amylose, *Makromol. Chem.* **26:**199.

Huwyler, H., 1978, Die Zellwandkomponenten und Glukoseaktivitäten in Baumwollhaaren verschiedener Entwicklungsstadien (Dissertation), Diss-Print, Fribourg.

Khonvine, Y., Champetier, G., and Sutra, R., 1932, X-ray study of the cellulose of *Acetobacter xylinum*, *C. R. Acad. Sci.* **194:**208.

Lamport, D., 1965, The protein component of primary cell wall, *Nature (London)* **213:**684.

Ledbetter, M.C., and Porter, K. R., 1963, Morphology of microtubules of plant cells, *J. Cell Biol.* **19:**239.

Macchi, E., and Palma, A., 1969, Morphological studies on precipitated cellulose, *Makromol. Chem.* **123**:208.

Macchi, E., Marx-Figini, M., and Fischer, E., 1968, Elektronenbeugungs untersuchungen an nativer und umgefällter Cellulose, *Makromol. Chem.* **120**:235.

Marx, M., 1955, Viskosimetrische Molekulargewichtsbestimmung von Cellulose in Kupfer-Äthylendiamin, *Makromol. Chem.* **16**:157.

Marx-Figini, M., 1959, Methodisches zur Bestimmungder Molekulargewichtsverteilung von Cellulose, *Makromol. Chem.* **32**:233.

Marx-Figini, M., 1961, Ueber die Abhängigkeit des Polymerisationsgrads und des Substitutionsgrads von den Nitrierbedingungen für Cellulosenitrate verschiedener Herkunft, *Makromol. Chem.* **50**:196.

Marx-Figini, M., 1962, Ueber den Einfluss des Substitutionsgrads auf die Viskositätszähl von Cellulosenitraten, *Makromol. Chem.* **52**:133.

Marx-Figini, M., 1963, Kinetische Untersuchungen zur Biosynthese der Cellulose in der Baumwolle. I, *Makromol, Chem.* **68**:227.

Marx-Figini, M., 1964, Kinetische Untersuchungen zur Biosynthese der Cellulose in der Baumwolle. II, *Makromol. Chem.* **80**:235.

Marx-Figini, M., 1966a, Zur Kinetik der Biosynthese der Cellulose in höheren Pflanzen: Molekulargewichtsverteilung bei 2% Umsatz, *Makromol. Chem.* **97**:202.

Marx-Figini, M., 1966b, Kinetics of the biosynthesis of cellulose of cotton balls by different light intensities, *Nature (London)* **210**:755.

Marx-Figini, M., 1966c, Comparison of biosynthesis of cellulose *in vitro* and *in vivo* in cotton balls, *Nature (London)* **210**:754.

Marx-Figini, M., 1969a, On the biosynthesis of cellulose in higher and lower plants, *J. Polym. Sci. C* **28**:57.

Marx-Figini, M., 1969b, Untersuchungen zur Biosynthese der Cellulose in der Alge *Valonia*, *Biochim. Biophys. Acta* **177**:27.

Marx-Figini, M., 1971, Investigations on biosynthesis of cellulose: DP and yield of cellulose of the alga *Valonia* in the presence of colchicine, *Biochim. Biophys. Acta* **237**:75.

Marx-Figini, M., 1978, Significance of the intrinsic viscosity ratio of unsubstituted and nitrated cellulose in different solvents, *Angew. Makromol. Chem.* **72**:161.

Marx-Figini, M., and Penzel, E., 1965, Absolutes Molekulargewicht und Molekulargewichtsverteilung von nativer, unbelichteter Baumwollcellulose, *Makromol. Chem.* **87**:507.

Marx-Figini, M., and Pion, B. G., 1974, Kinetic investigations on biosynthesis of cellulose by *Acetobacter xylinum*. 1, *Biochim. Biophys. Acta* **338**:382.

Marx-Figini, M., and Pion, B. G., 1976, Kinetic investigations on biosynthesis of bacterial cellulose. 2, *Makromol. Chem.* **177**:1013.

Marx-Figini, M., and Schulz, G. V., 1959, Methodisches zur Bestimmung der Viskositätszahl (Staudinger-Index) von Cellulosen und Cellulosenitraten. *Makromol. Chem.* **31**:140.

Marx-Figini, M., and Schulz, G. V., 1962, Die Viskosimetrische Molekulargewichtsbestimmung von Cellulosen und Cellulosenitraten unter Standarbedingungen, *Makromol. Chem.* **54**:102.

Marx-Figini, M., and Schulz, G. V., 1963, Neure Untersuchungen über Grösse und Grössenverteilung der β-glukosidischen Ketten nativer Cellulosen, *Makromol. Chem.* **62**:49.

Marx-Figini, M., and Schulz, G. V., 1966, Über die Kinetik und den Mechanismus der Biosynthese der Cellulose in den höheren Pflanzen, *Biochim. Biophys. Acta* **177**:27.

Meyer, K. H., and Mark, H., 1930, *Der Aufbau der hochpolymeren organischen Naturstoffe*, Akademische Verlagsgeselzschaft, Leipzig.

Meyer, K. H., and Misch, L., 1937, Positions des atomes dans le nouveau modele spatial de la cellulose, *Helv. Chim. Acta* **20**:232.

Meyerhoff, G., 1970, Anwendung der Gelchromatographie auf Cellulosenitrate, *Makromol. Chem.* **134**:129.

Mühlethaler, K., 1967, Ultrastructure and formation of plant cell walls, *Annu. Rev. Plant. Physiol.* **18**:1.

Nagai, R., and Rebhun, L. J., 1966, Cytoplasmic microfilaments in streaming *Nitella* cells, *J. Ultrastruct. Res.* **14**:571.

Newcomb, E. H., and Hepler, P. K., 1964, Microtubules and fibrils in the cytoplasm of *Coleus* cells undergoing secondary wall deposition, *J. Cell Biol.* **20:**529.

Robards, A. W., and Kidwai, P., 1972, Microtubules and microfibrils in xylem fibres during secondary cell wall formation, *Cytobiology* **6:**1.

Robinson, D. G., and Preston, R. D., 1972, Plasmalemma structure in relation to microfibre biosynthesis in *Oocystis*, *Planta* **104:**234.

Schulz, G. V., 1935, Über die Beziehung zwischen Reaktionsgeschwindigkeit und Zusammensetzung des Reaktionsproduktes bei Makropolymerisationsvorgämgen, *Z. Phys. Chem. Abt. B* **30:**379.

Schulz, G. V., and Blaschke, F., 1941, Molekulargewichtsbestimmungen an Makromolekularen Materialien. IX. Eine Gleichung zur Berechnung der Viscositätszähl bei kleinen Konzentrationen, *J. Prakt. Chem.* **158:**130.

Schulz, G. V., and Marx, M., 1954, Über Molekulargewichte und Molekulargewichtsverteilung nativer Cellulose, *Makromol. Chem.* **14:**52.

Shore, G., and Maclachlan, G. A., 1975, The site of cellulose synthesis: Hormone treatment alters the intracellular location of alkali-insoluble β-1,4-glucan (cellulose) synthetase activities, *J. Cell Biol.* **64:**557.

Tanford, C., 1967, *Physical Chemistry of Macromolecules*, John Wiley, New York.

Thomson, E. W., and Preston, R. D., 1967, Proteins in the cell walls of some green algae, *Nature (London)* **213:**684.

Timell, T. E., 1957, Molecular weight of native celluloses, *Sven. Papperstidn.* **60:**836.

Tung, L. H., 1966, Method of calculating molecular weight-distribution functions from gel-permeation chromatograms, *J. Appl. Polym. Sci.* **10:**375.Villemez, C. L., McNab, J. M., and Albersheim, P., 1968, Formation of plant cell wall polysaccharides, *Nature (London)* **218:**878.

Vollmert, B., 1973, *Polymer Chemistry*, Springer-Verlag, Berlin, Heidelberg, and New York.

Wardrop, A. B., and Foster, R. C., 1964, A cytological study of the oat coleoptile, *Aust. J. Bot.* **12:**135.

Wooding, F. B., and Northcote, D. H., 1964, The development of the secondary wall of the xylem in *Acer pseudoplatanus*, *J. Cell Biol.* **23:**327.

14
Biogenesis of Cellulose I Microfibrils Occurs by Cell-Directed Self-Assembly in *Acetobacter xylinum*

CANDACE H. HAIGLER AND MOSHE BENZIMAN

1. Introduction

The biogenesis of native cellulose is a complex process requiring the biosynthesis of high-molecular-weight β1,4-glucans; the formation of highly crystalline, metastable cellulose I microfibrils; and, in many cases, the organization and orientation of these microfibrils into cell walls (Colvin, 1964). Proposed mechanisms of cellulose biogenesis have historically been intimately related to the contemporary understanding of the physical structure of cellulose (Ranby, 1952; Preston, 1964). It is evident that any proposed mechanism of cellulose biogenesis must be consistent with the ultimate physical properties of the macromolecule. Since almost all native cellulose is fibrillar cellulose I, it is probable that there is a common biosynthetic mechanism among all organisms that produce crystalline cellulose I microfibrils (Sarko and Muggli, 1974).

Native cellulose I is now generally thought to exist in a crystalline lattice of extended, parallel glucan chains. It has been proposed that cellulose II, which is infrequently found in nature, is composed of antiparallel glucan chains and is

CANDACE H. HAIGLER • Department of Botany, University of North Carolina at Chapel Hill, Chapel Hill, North Carolina 27514. MOSHE BENZIMAN • Department of Biological Chemistry, Institute of Life Sciences, The Hebrew University of Jerusalem, 91904 Jerusalem, Israel.

more thermodynamically stable than cellulose I (Sarko and Muggli, 1974; Gardner and Blackwell, 1974a). An understanding of the molecular mechanisms by which organisms synthesize and assemble crystalline, metastable cellulose I fibrils is not yet complete. No hypothesis has heretofore been advanced that adequately unifies knowledge of the crystal structure of cellulose, the substructure of the microfibrils, and the biosynthetic mechanism. There are at least three basic unanswered questions concerning cellulose microfibril biogenesis:

1. What is the time–space relationship of polymerization of β1,4-glucans and crystallization into microfibrils? Are polymerization and crystallization simultaneous or consecutive processes (Stockman, 1972)? What is the intermediate between enzyme and microfibril—an activated glucose molecule, an oligomeric residue, or a free high polymer (Colvin, 1972)?

2. Is there an ultimate unit of cellulose microfibril structure? Is it the 3.5-nm elementary microfibril (Mühlethaler, 1967)? Is it a 1.5-nm subelementary fibril such as the fibrils seen by Hanna and Cote (1974) in wood cambium and by Franke and Ermen (1969) in quince slime? Are divisions of cellulose microfibrils below their macroscopic level merely artifacts of specimen preparation (Nieduszynski and Preston, 1970)?

3. What cytological processes regulate the formation of microfibrils from polymerized chains (Preston, 1964)? Are polymerizing enzymes specifically organized within membranes to facilitate crystallization (Haigler *et al.*, 1980)?

Recent experiments have supported the ordered-granule hypothesis first proposed by Preston (1964), which asserts that organized groups of enzymes determine the orientation of crystalline microfibrils. Organized granules that could function in this way have been visualized by freeze–etching in the plasma membranes of algae and higher plants (Brown and Montezinos, 1976; Willison and Brown, 1978; Mueller and Brown, 1980; Giddings *et al.*, 1980). Macromolecular mechanisms of cellulose biosynthesis have been proposed for several of these organisms on the basis of such morphological evidence. Until recently, however, there has been little experimental evidence concerning the molecular dynamics of polymerization and crystallization of cellulose. Further, attempts to synthesize cellulose I *in vitro* have been inconclusive (Ben-Hayim and Ohad, 1965; Cooper and Manley, 1975; Delmer, 1977; Colvin, 1980). Therefore, the molecular mechanisms that mediate polymerization of β1,4-glucans and their crystallization into microfibrils have remained elusive.

Recently, Calcofluor White ST, a fluorescent brightener that associates with linear β1,4-polysaccharides (Maeda and Ishida, 1967; Rattee and Breuer, 1974), and carboxymethylcellulose (CMC), a substituted derivative of cellulose that can cocrystallize with native cellulose (Ben-Hayim and Ohad, 1965), have been used to probe the relationship between polymerization and crystallization during cellulose microfibril biogenesis in *Acetobacter xylinum* (Haigler *et al.*, 1980; Benziman *et al.*, 1980). These studies have shown that in this bacterium, which synthesizes a ribbon of cellulose, polymerization and crystallization are coupled, consecutive processes, and crystallization limits the rate of polymerization. Our results suggest that organized extrusion pores in the lipopolysaccharide (LPS) layer of *A. xylinum* mediate the crystallization of parallel-chain glucan aggregates into microfibrils and the hierarchical assembly of microfibrils into a cellulose ribbon at the cell surface. Therefore, cellulose-fibril biogenesis in *A. xylinum* appears to occur by a cell-directed, self-assembly process as will be described below. Such a mechanism may also apply to cellulose synthesis in other organisms.

2. *Acetobacter xylinum as a Model System for Cellulose Biogenesis*

In *A. xylinum*, crystalline 3.5-nm microfibrils are assembled in association with rows of particles (presumed to be groups of multienzyme complexes) that are situated below extrusion pores in the LPS layer of the bacterium (Fig. 1) (Brown *et al.*, 1976; Zaar, 1979). These microfibrils associate by hydrogen bonding (Gardner and Blackwell, 1974b) into bundles that in turn form a twisting ribbon (Fig. 2). The microfibril bundles can be detected within the typical, striated ribbons (Fig. 3). As seen by dark-field microscopy, the bacterial cell turns on its axis as it is propelled forward by the elongating ribbon. This suggests an inherent kinetic force in the crystallization process. Ribbons from many bacteria intertwine into a tough pellicle of cellulose on the surface of the culture medium. The composite ribbon is analogous to microfibrils in walls of other cellulose-synthesizing organisms. *Acetobacter xylinum* is easily maintained in culture, and conditions of synthesis are readily manipulated; therefore, this bacterium is an excellent model for studying the molecular mechanisms of cellulose polymerization and crystallization.

3. *Effect of Calcofluor White ST on Cellulose Biogenesis*

3.1. *Morphology of the Cellulose Synthesized in the Presence of Calcofluor*

When Calcofluor (Fig. 4) is added to active cultures of *A. xylinum*, it disrupts ribbon assembly by associating with the subunits of the ribbon and preventing their normal aggregation. The planar Calcofluor molecule binds along the glucan chains (Fig. 5) through hydrogen bonding (Maeda and Ishida, 1967) and dipolar interactions (Rattee and Breuer, 1974). In the presence of Calcofluor, the bacteria synthesize broad bands that are extruded perpendicularly to the longitudinal cell axis instead of normal twisting ribbons (Fig. 6). The bands synthesized in the presence of Calcofluor fluoresce brilliantly when viewed with dark-field fluorescence microscopy (Fig. 6, inset). Once Calcofluor is bound to cellulose, it cannot be completely removed by washing or boiling in alkali, indicating that the association is very strong.*

*Similar alterations in ribbon assembly have been induced by many direct cellulose dyes such as Congo Red, Trypan Blue, Benzo Orange R, Amidine Red, Amidine Black, Diamine Violet, Chrysamine, and Sirus Red, and with other fluorescent brighteners, but the discussion here will be limited to the effects of Calcofluor.

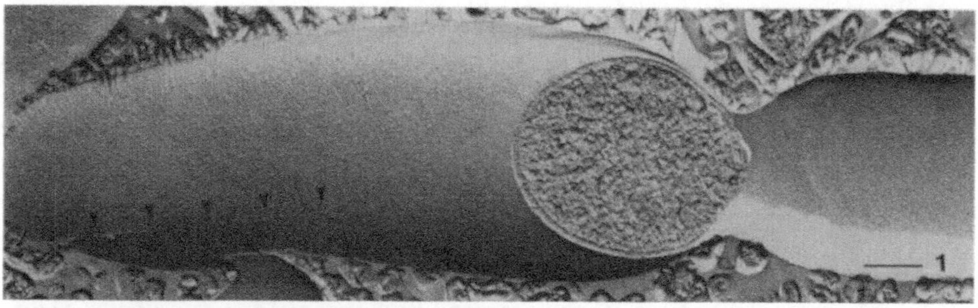

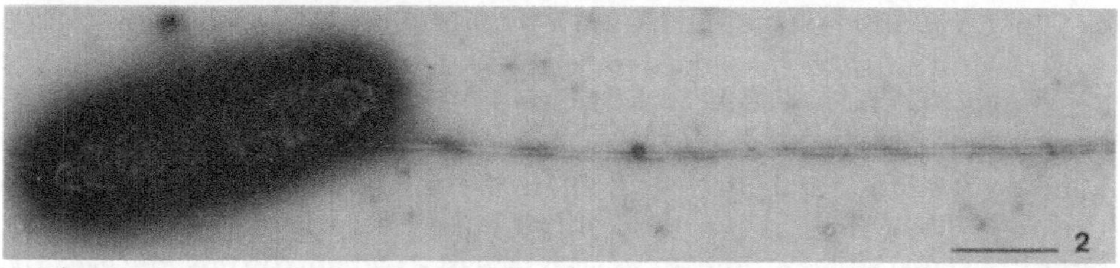

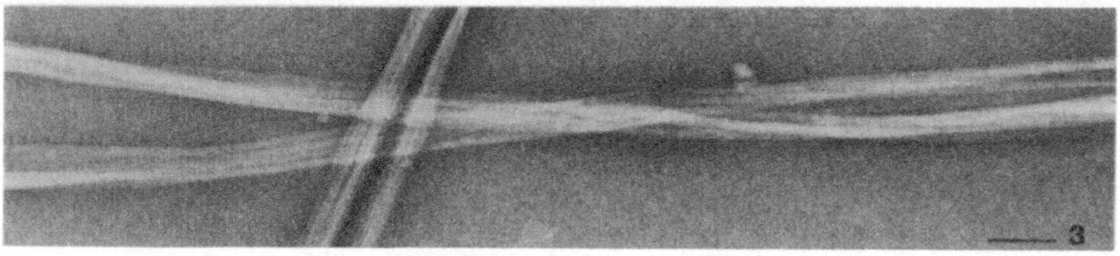

FIGURE 1. Freeze-etching of *A. xylinum* cells that are actively synthesizing cellulose. The micrograph reveals a row of particles in the PF face of the LPS layer (arrows) that are presumed to be groups of multienzyme complexes involved in cellulose polymerization. The outer layer of the LPS containing the associated extrusion pores has been fractured away. Scale bar: 0.2 μm.

FIGURE 2. Negatively stained preparations showing the typical twisting cellulose ribbon in association with the cell. Scale bar: 1 μm.

FIGURE 3. Higher-magnification micrograph of negatively stained cellulose ribbons. This micrograph reveals that the ribbon is striated. Here, four separate ribbons cross each other. The large striations correspond to the bundles of microfibrils. Suggestions of individual microfibrils can also be detected. Scale bar: 0.1 μm.

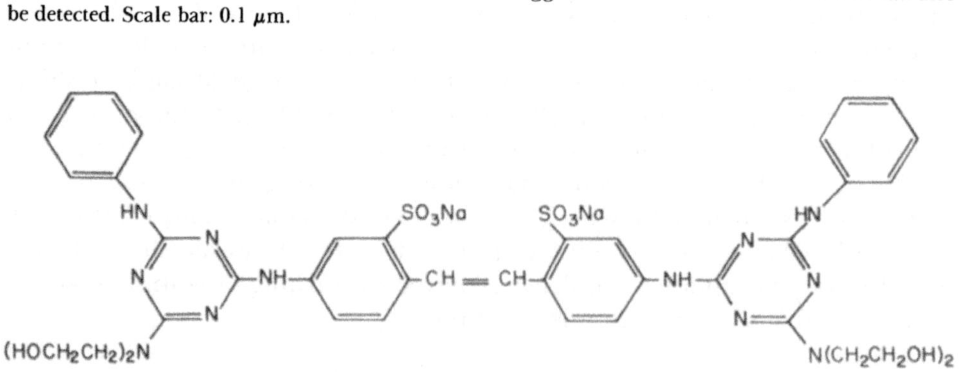

FIGURE 4. Structure of Calcofluor White ST (4,4′-bis[4-anilino-6-bis-(2-hydroxyethyl)amino-*s*-triazin-2-ylamino]-2-2′-stilbene-disulfonic acid). Calcofluor is a highly planar derivative of stilbene that associates with straight-chain polysaccharides with free hydroxyl groups.

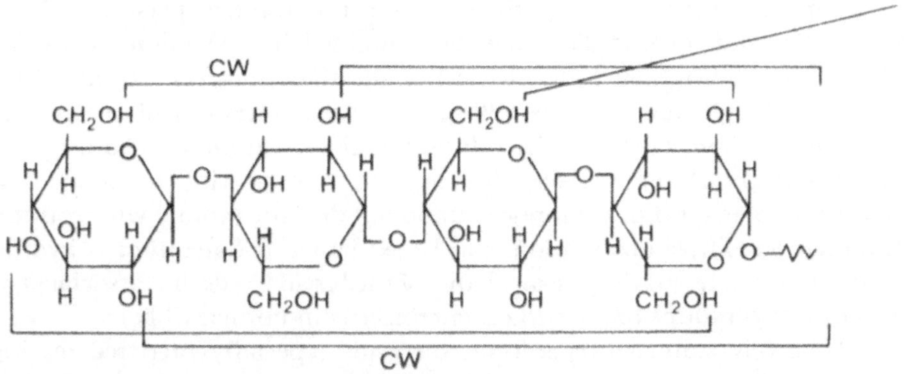

FIGURE 5. Binding of Calcofluor to cellulose. Calcofluor (CW) binds along the glucan chains through hydrogen bonding and dipolar interactions, thereby preventing normal microfibril formations and ribbon assembly.

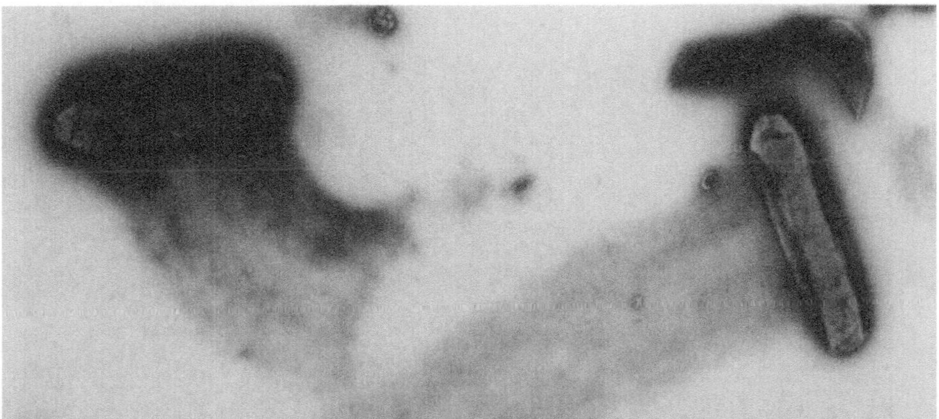

FIGURE 6. Negative staining of cells producing cellulose in the presence of 0.25 mM Calcofluor for 30 min. The micrograph shows that the morphology of the product is dramatically altered. The bent fibrils originate all along the longitudinal cell surface. Scale bar: 1 μm. *Inset:* Production of band cellulose observed *in vivo* with dark-field fluorescence microscopy. Scale bar: 5 μm.

Continued illumination with UV light causes rapid dispersion of the blue bands. This UV degradation may be explained by: (1) photosensitization, whereupon the brightener absorbs energy and then transfers it to its cellulose substrate (Baugh and Phillips, 1954), or (2) isomerization of planar *trans*-Calcofluor to nonplanar *cis*-Calcofluor, which would induce chain breakage and dispersion (Lanter, 1966). In contrast, normal ribbons stained with Calcofluor after synthesis fluoresce but do not degrade rapidly with continued illumination. These observations can be explained if Calcofluor is hydrogen bonded to virtually all the glucan chains of the broad bands, but associated only with hydroxyl groups of the surface microfibrils in normal ribbons.

Negatively stained preparations on grids especially prepared for high-resolution electron microscopy (White and Brown, 1981) reveal that the bands are composed of bent fibrils that originate along the longitudinal cell surface (Fig. 7). The smallest measurable fibrils average about 1.5 nm in diameter and are smaller than any fibrils commonly observed in normal *A. xylinum* cellulose. The 1.5-nm fibrils can sometimes be seen to curve and undulate, suggesting low crystallinity. Larger fibrils measuring 3.0, 4.5, and 6.0 nm in diameter appear to arise by fasciation of the smaller fibrils. These larger fibrils show "cracking sites," suggesting rigidity and high crystallinity. It is possible that each 1.5-nm

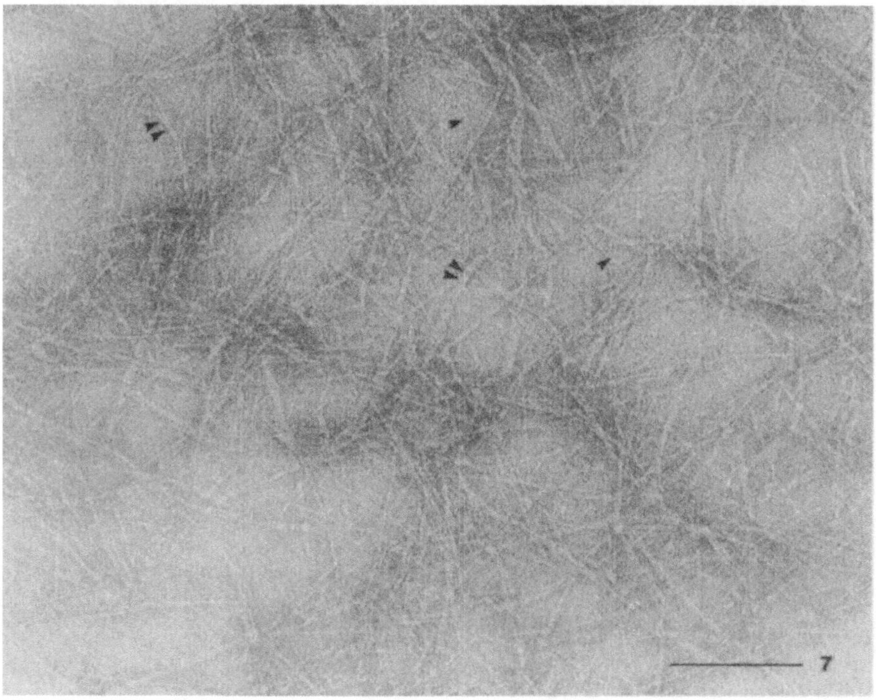

FIGURE 7. High-resolution micrograph of cellulose produced in the presence of Calcofluor showing a range of fibril widths. The smallest resolvable fibrils measure about 1.5 nm (arrows). Larger fibrils appear to arise by fasciation of the smaller fibrils (double arrows). Scale bar: 0.1 μm.

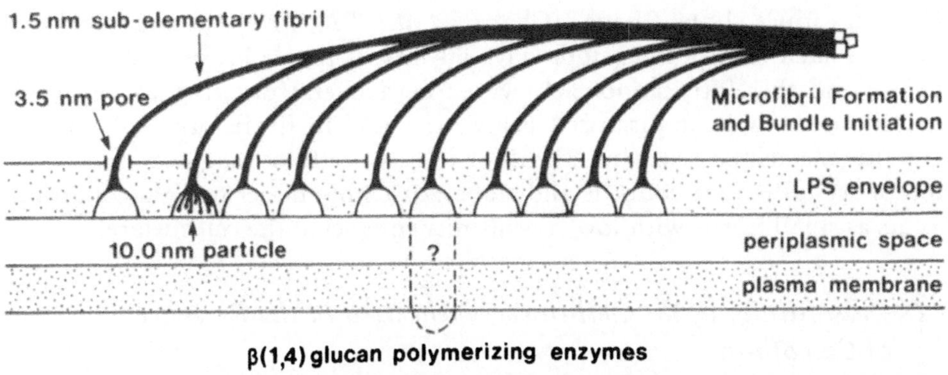

FIGURE 8. Proposed model of normal cellulose assembly. Glucan-chain aggregates from organized multienzyme complexes and extrusion pores crystallize into microfibrils, which then assemble into bundles and the normal ribbon at the cell surface (see the text for further discussion).

fibril observed in negative staining corresponds to an aggregate of glucan chains synthesized by the multienzyme complexes below one pore. Calcofluor White ST may hydrogen bond with such glucan aggregates as they are extruded, preventing formation of the 3.5-nm microfibrils commonly observed in *A. xylinum* cellulose. Diagrams of normal ribbon assembly and Calcofluor disruption of microfibril and ribbon assembly are shown in Fig. 8 and 9.

The small fibrils in the Calcofluor-induced altered cellulose are very similar

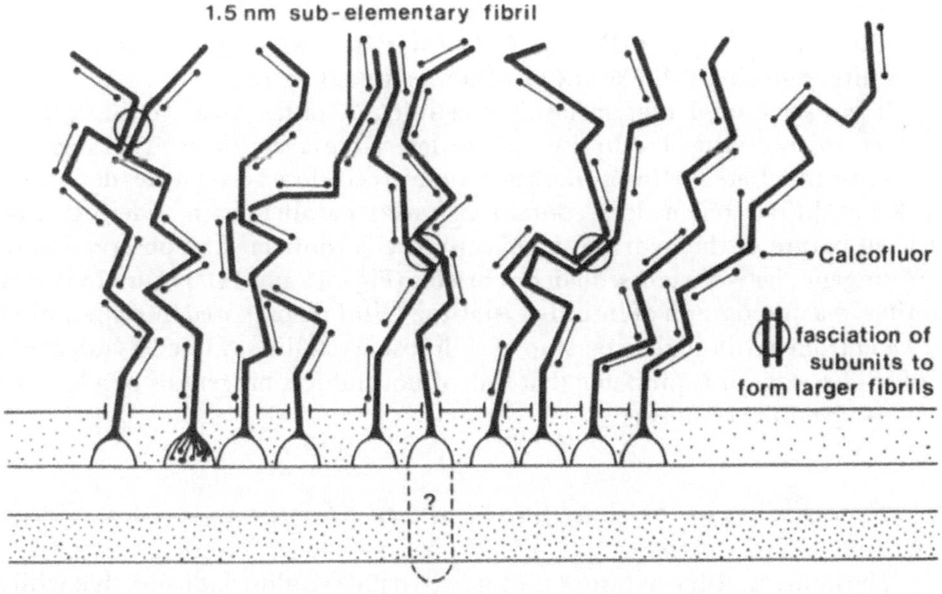

FIGURE 9. Proposed model of disrupted microfibril and ribbon assembly in the presence of Calcofluor. Calcofluor associates with glucan-chain aggregates as they are extruded from the pores, preventing the cellulose-to-cellulose hydrogen bonding required for normal microfibril crystallization.

to 1.5-nm fibrils of apparent low crystallinity observed by Franke and Ermen (1969) in quince slime. Chanzy (Chanzy *et al.*, 1978, 1979) has suggested that fibrils of such small dimensions have 60–80% of their glucan chains on the surface and therefore cannot achieve enough lateral ordering to be highly crystalline. As will be discussed below, the 1.5-nm fibrils are probably not crystalline. Their presence in Calcofluor-induced cellulose suggests that formation of crystalline microfibrils and ribbon assembly under normal conditions occurs as small fibrils with low crystallinity interact at the cell surface.

3.2. *Crystallinity of the Cellulose Synthesized in the Presence of Calcofluor*

There is ample evidence that this band product is cellulose that differs from native bacterial cellulose only in its crystallinity. The band product is rapidly degraded by purified cellulases; [^{14}C]glucose is incorporated into an alkali-insoluble product and into a band adjacent to the cell as visualized by autoradiography (Fig. 10); the degree of polymerization is at least 1680, which is in the same range as cellulose from control pellicles not treated with Calcofluor; and X-ray diffraction confirms that the Calcofluor-induced product crystallizes into cellulose I after drying.

In contrast to native cellulose, however, the band cellulose has no detectable crystallinity in the wet state. Native *A. xylinum* cellulose has crystallite sizes of 6.5 and 7.5 nm in the wet and dry states, respectively. Crystallites obtained on drying of Calcofluor-induced cellulose vary in size depending on the initial concentration of Calcofluor used to induce the alteration, normal 6.9-nm crystallites resulting if 0.25 mM Calcofluor is used (Fig. 11) and 2.8-nm crystallites resulting if 1–5 mM Calcofluor is used (Fig. 12).

It is exceptional that metastable cellulose I rather than cellulose II is regenerated by drying of cellulose with no detectable crystallinity. This suggests that while the altered cellulose does not contain cellulose I crystallites detectable by X-ray diffraction, it does contain extensive parallel-chain ordering. The ordered nature of the hydrated band cellulose is confirmed by observation of birefringence between crossed Nichol prisms (Figs. 13 and 14). Figure 13 is very similar to a micrograph of liquid-crystalline cellulose prepared by evaporation of water from a colloidal dispersion of cellulose crystallites (Marchessault *et al.*, 1959). Therefore, it is probable that Calcofluor induces biogenesis of a kind of liquid-crystalline cellulose.

3.3. *Reversibility of the Calcofluor-Induced Alteration*

The noncrystalline nature of the undried band cellulose indicates that while polymerization of β1,4-glucans continues in the presence of Calcofluor, normal *in vivo* microfibril crystallization is prohibited. Since Calcofluor acts by binding to glucan bundles only after they exit the pores in the LPS layer (Benziman *et al.*,

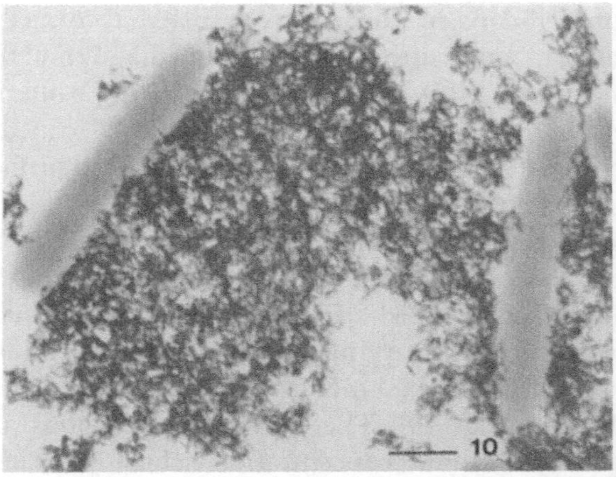
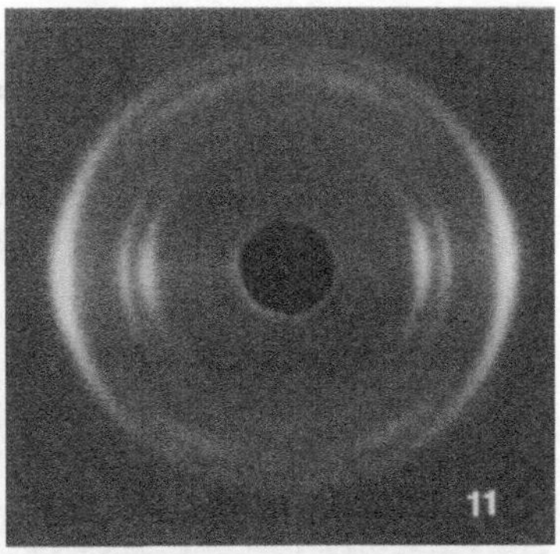
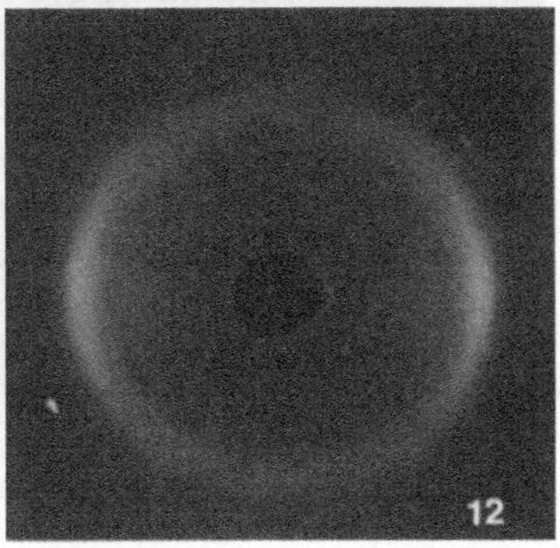
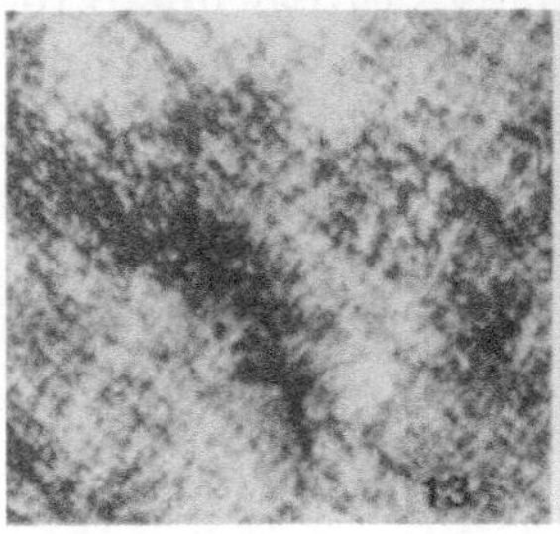
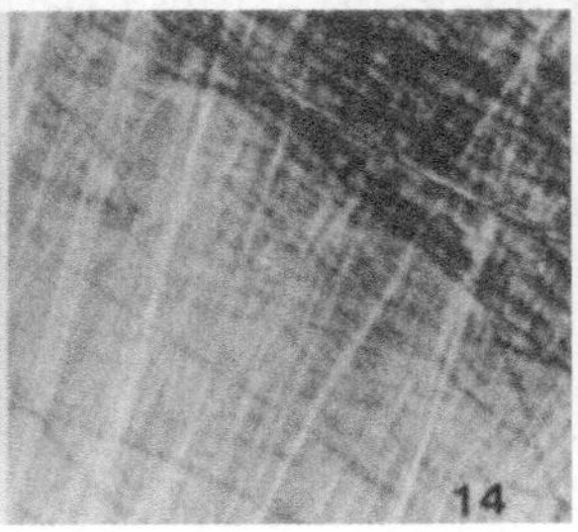

FIGURE 10. Electron-microscope autoradiograph showing that [^{14}C]glucose is incorporated into a band next to the cell. Scale bar: 0.5 μm.

FIGURE 11. X-ray diffractogram representative of normal *A. xylinum* cellulose without Calcofluor, wet or dry, indicating a crystallite size of 6.5–7.5 nm. The same kind of pattern results from altered cellulose produced in the presence of 0.25 mM Calcofluor and then dried. The fiber axis is vertical.

FIGURE 12. X-ray diffractogram representative of altered *A. xylinum* cellulose after drying when 1–5 mM Calcofluor is used to induce the alteration. A 2.8-nm crystallite size can be calculated from this pattern. The fiber axis is vertical.

FIGURE 13. Birefringence of hydrated band cellulose observed between crossed Nichol prisms with polarization microscopy. The micrograph indicates that while the band cellulose is not crystalline, it is highly ordered.

FIGURE 14. More fibrillar ordering than the kind of birefringence shown in Fig. 13 observed in the same preparations. These variations may be due to stretching under the coverslip or to partial dehydration.

1980), microfibril crystallization must occur extracellularly under normal conditions. However, ordered ribbons are typically visualized in direct association with the cell envelope, so crystallization of microfibrils and ribbon assembly, while they are extracellular processes, must still occur in association with the cell.

The control of crystallization and ribbon assembly at the cell surface is supported by the observed reversibility of the Calcofluor-induced alteration.

When cells synthesizing normal ribbons in glucose are transferred to media containing glucose plus Calcofluor, band synthesis begins immediately and the ribbon is pushed away from the cell (Fig. 15). When cells synthesizing bands in Calcofluor are subsequently washed in buffered glucose and transferred to media containing glucose alone, ribbon synthesis begins immediately at the cell surface and the band is pushed away from the cell (Fig. 16).

When gradual reversion from band synthesis to ribbon synthesis occurs in large cultures by depletion of Calcofluor as it binds to newly synthesized cellulose, products with morphologies intermediate between ribbons and bands can be observed (Fig. 17). Experiments with varying Calcofluor concentrations have shown that a threshold concentration of 0.1 mM is required to consistently induce synthesis of band cellulose (Fig. 18). The intermediate ribbons are presumably synthesized near the threshold concentration where sufficient interchain hydrogen bonding occurs to achieve partially normal microfibril aggregation. Thus, the morphology of the cellulose is critically dependent on the amount of free Calcofluor in the medium that can interfere with normal hydrogen bonding and thereby alter the association of ordered glucan aggregates at the cell surface.

3.4. Increase of Rate of Polymerization in the Presence of Calcofluor

The proposal that crystallization occurs at the cell surface is substantiated by kinetic data that show that: (1) polymerization and crystallization are coupled processes that can be separated by Calcofluor and (2) crystallization is rate-limiting in the biogenesis of cellulose microfibrils (Benziman *et al.*, 1980). The rate of cellulose synthesis as measured by [^{14}C]glucose incorporation increases 200–400% in the presence of Calcofluor, while the rate of glucose oxidation increases no more than 15% (Fig. 19). This slight increase in the rate of glucose oxidation is probably due to lack of aggregation in the presence of Calcofluor concentrations greater than 0.1 mM. Aggregation occurs as the cells become intertwined in normal cellulose ribbons; thus, when normal ribbon assembly is prevented by the presence of Calcofluor, cells have maximum accessibility to oxygen during the entire course of the incubation.

The rate of polymerization is dramatically increased when Calcofluor prevents normal crystallization; therefore, polymerization and crystallization must be tightly coupled. Since polymerization of β1,4-glucans occurs in association with the cell (see Chapter 17), the observed rate enhancement can be readily explained if the process of crystallization is also cell-associated, occurring consecutively with polymerization. The observed rate enhancement is inconsistent with the hypothesis that microfibrils crystallize from prepolymerized oligomers or high polymers far from the cell.

The rate of cellulose synthesis returns to that of the control only at aggregation time, whereupon normal ribbon production resumes (Fig. 20). The amount of cellulose synthesized and the amount of Calcofluor disappearing from the medium up to aggregation time are approximately constant (Fig. 21).

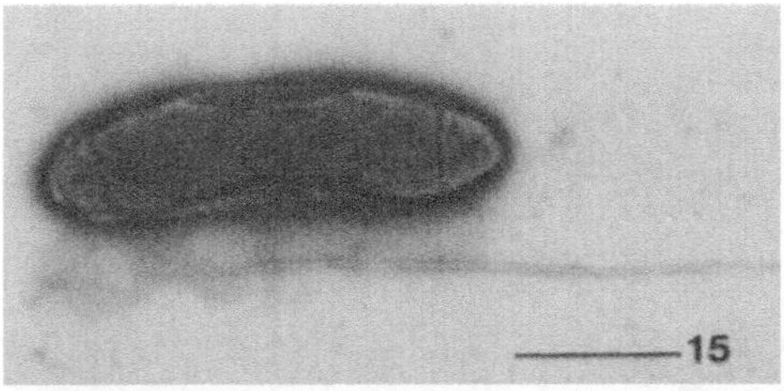

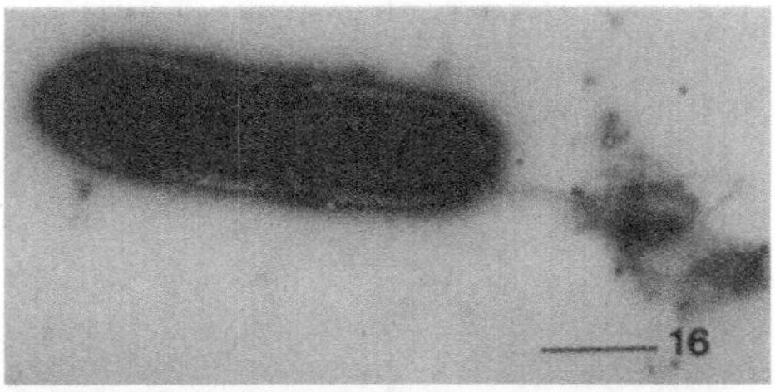

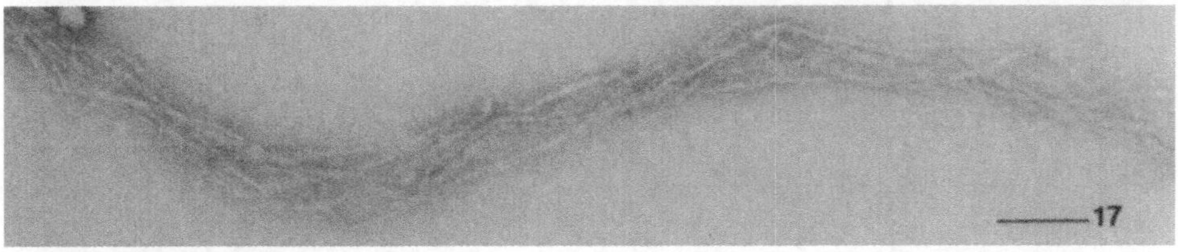

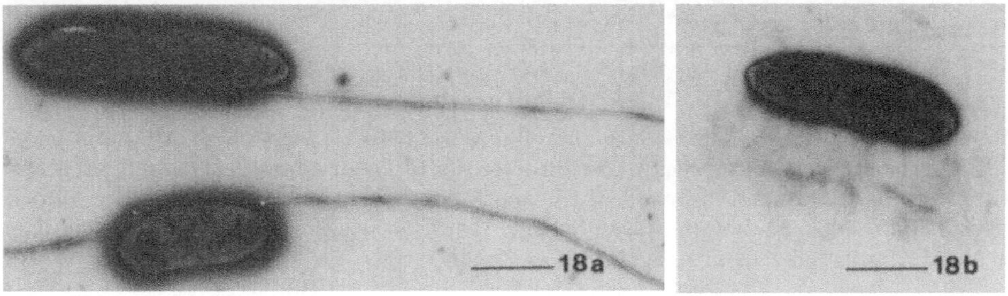

FIGURE 15. Induction of abrupt reversion from normal ribbon production to band production by a cell transferred from media without Calcofluor to media containing Calcofluor. Cells were incubated for 1 min in each solution. Scale bar: 1 μm.

FIGURE 16. Resumption of normal ribbon production when Calcofluor is washed away from a cell synthesizing band cellulose and incubation is continued in Calcofluor-free glucose. Cells were incubated for 1 min in each solution. Scale bar: 1 μm.

FIGURE 17. Cellulose products with morphology intermediate between bands and normal ribbons. Such products are presumably synthesized near the threshold concentration of Calcofluor required to induce band synthesis (see Fig. 18), where Calcofluor can only incompletely prohibit the hydrogen bonding required for normal crystallization. Scale bar: 0.2 μm.

FIGURE 18. Cells synthesizing normal ribbons in the presence of 0.025 mM Calcofluor (a) and band material in the presence of 0.1 mM Calcofluor (b). Concentrations of Calcofluor around 0.05 mM induce a mixture of product morphologies, possibly because of the tendency of Calcofluor to form micelles at pH 6.0–7.0. Therefore, we consider 0.1 mM to be the threshold concentration required to consistently induce band cellulose. Scale bars: 1 μm.

The amount of free Calcofluor remaining at aggregation time is approximately 0.1 mM, which is very close to the observed threshold concentration of Calcofluor required to induce production of band material. These data confirm that Calcofluor acts as a reactant in the process of cellulose polymerization, binding stoichiometrically to cellulose as it is synthesized and extruded.

As Calcofluor concentration increases, the rate of polymerization increases up to a saturation concentration of about 1.0 mM, after which increasing Calcofluor concentration cannot further increase the rate of polymerization (Fig.

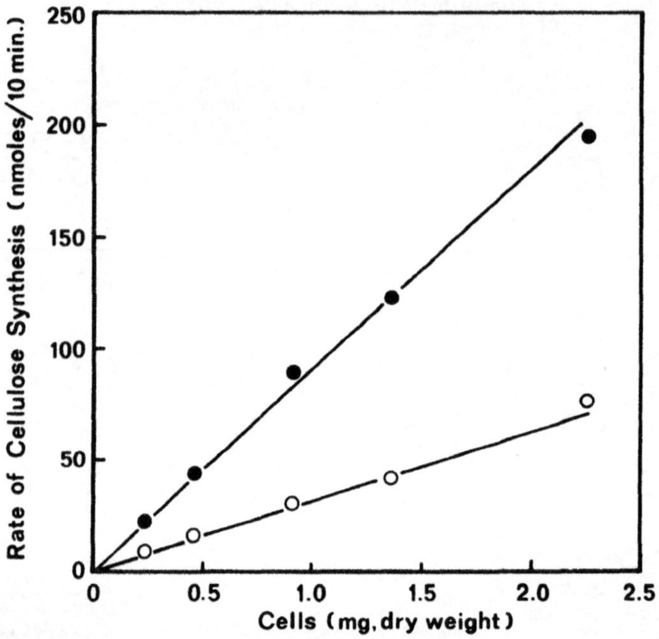

FIGURE 19. Effect of Calcofluor on the rate of cellulose synthesis as a function of cell concentration. Indicated amounts of cells (dry weight) were incubated for 10 min in the absence (○) and presence (●) of 0.57 μmole Calcofluor. The standard reaction mixture contained 40 mM D-[^{14}C(U)]glucose (125,000 cpm/μm) in 50 mM phosphate buffer (pH 7.0); incubation was carried out at 30°C in stoppered scintillation vials with shaking (80 oscillations/min).

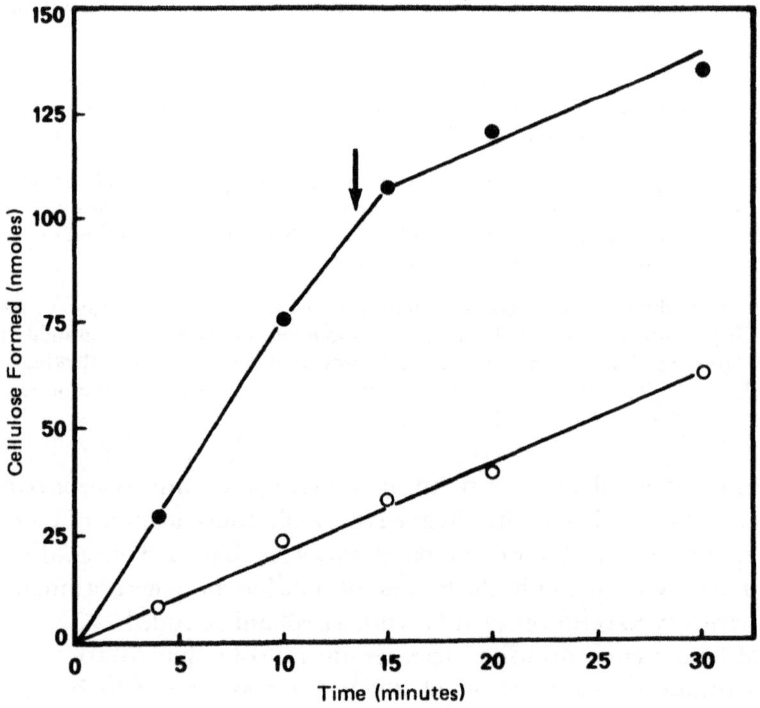

FIGURE 20. Time-course of cellulose synthesis in the absence (○) and presence (●) of 0.43 μmole Calcofluor. Cells (0.25 mg dry weight) were incubated in standard reaction mixtures. (↓) Time of aggregation.

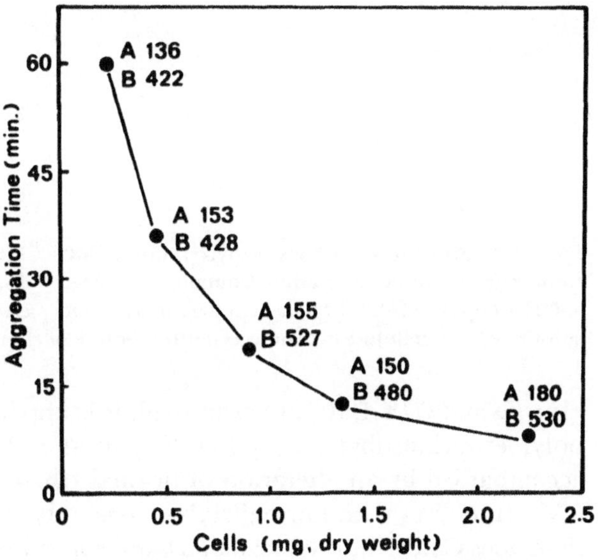

FIGURE 21. Aggregation time as a function of cell concentration of 0.57 μmole Calcofluor. Reactions were stopped when macroscopic cellulose aggregates appeared in the reaction mixture. Amount (μmole) of cellulose synthesized at aggregation time (A) was determined from incorporation data. Amount (μmole) of Calcofluor utilized at aggregation time (B) was calculated by determining the concentration of Calcofluor remaining in solution by fluorometry. Note that (A) and (B) have a ratio of 1/3 that is independent of cell concentration.

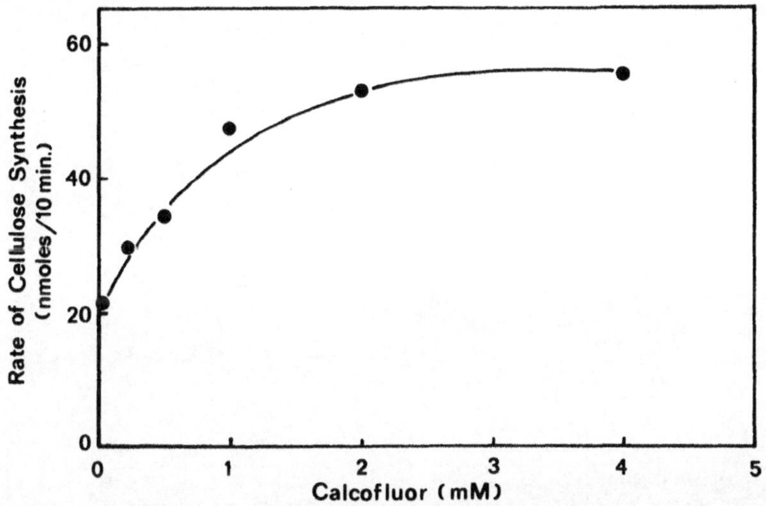

FIGURE 22. Effect of Calcofluor concentration on the rate of cellulose synthesis. Cells (0.4 mg dry weight) were incubated for 10 min in the standard reaction mixture in the presence of the indicated concentrations of Calcofluor.

22). Therefore, the degree of rate enhancement depends on the extent to which Calcofluor can compete for the hydrogen-bonding sites involved in normal crystallization.

4. *Effect of Carboxymethylcellulose on Cellulose Biogenesis*

Ben-Hayim and Ohad (1965) observed an increase in rate of glucose polymerization and an increase in pellicle orientation when CMC (Fig. 23) was added to cellulose-synthesizing cultures of *A. xylinum*. When cells incubated in

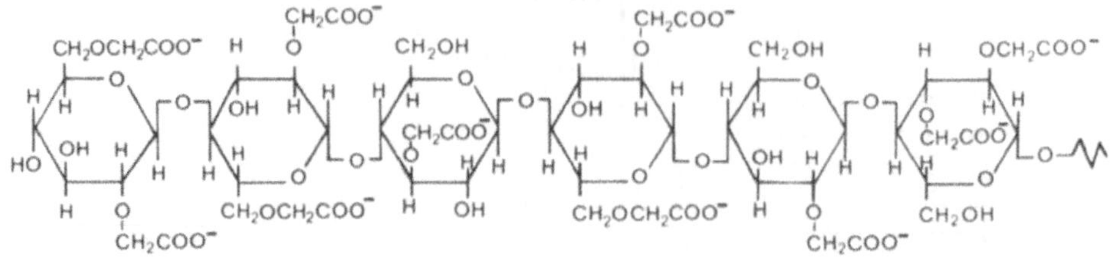

FIGURE 23. Structure of carboxymethylcellulose (CMC). CMC is a water-soluble cellulose ether that forms a gel or colloid in solution. Commonly, its degree of substitution of $-OCH_2COONa$ for the $-OH$ groups is 0.1–1.2. CMC is manufactured from mercerized, noncrystalline cellulose, and, once substituted, the cellulose cannot recrystallize because of electrostatic repulsion.

glucose and 0.1% CMC are examined, it is apparent that the increase in rate of polymerization observed by Ben-Hayim and Ohad and verified in our lab is accompanied by an alteration of normal ribbon assembly (Fig. 24). However, CMC interferes with normal ribbon assembly at a higher level of organization than does Calcofluor. Instead of a band composed of bent 1.5- to 6.0-nm fibrils, long 10- to 20-nm-wide bundles originate along the cell surface at approximately

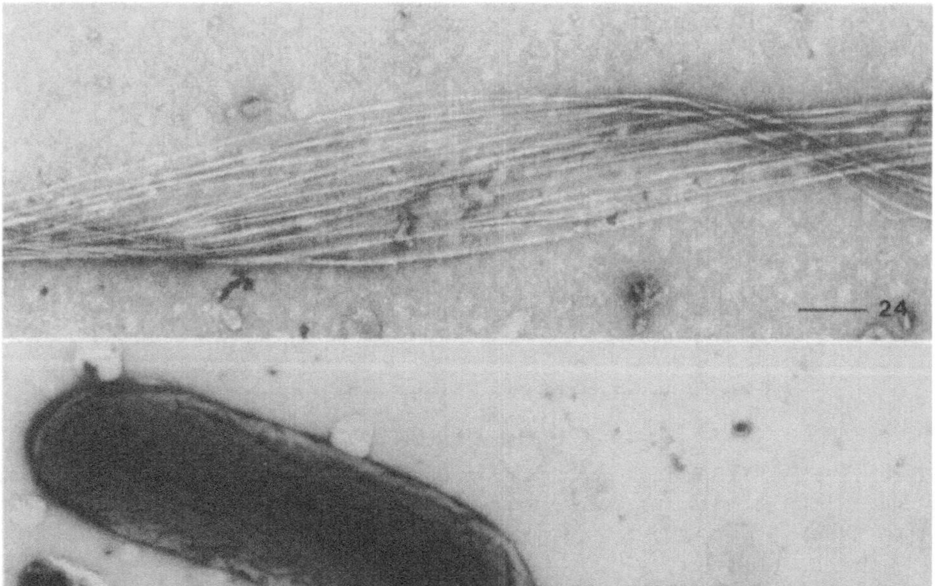

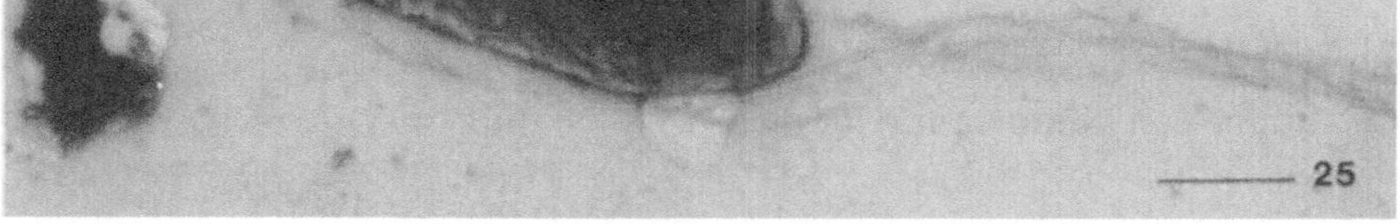

FIGURE 24. Bundle subunits of the ribbon prevented from aggregating in the presence of CMC. Compare this CMC-altered ribbon with the normal ribbon at about the same magnification in Fig. 3. Scale bar: 0.1 μm.

FIGURE 25. Individual bundles seen leaving the cell surface at regular intervals along the longitudinal axis after cellulose synthesis in the presence of CMC. They then form a loosely intertwining aggregate of bundles. Scale bar: 0.5 μm.

regularly spaced intervals in the presence of CMC (Fig. 25). Since CMC is a high-molecular-weight polymer, it is reasonable that it interferes with ribbon assembly at a higher level than does Calcofluor. CMC must associate with the subunits of the ribbon and, as a consequence of its negatively charged carboxyl groups, prevent the fasciation of the bundles.

The finding of two probes that can interfere with ribbon assembly at two different levels has important consequences. It suggests that the synthesizing apparatus may be organized so that ribbons are assembled hierarchically. Such a hierarchical mechanism may involve cocrystallization of ordered aggregates of glucan chains into microfibrils (Calcofluor can intervene here), aggregation of microfibrils into bundles, and fasciation of bundles to form the twisting ribbon (CMC can interfere here). Such a hierarchical assembly mechanism is also consistent with our findings that CMC alone can increase rate of polymerization only about 30%. Whereas Calcofluor can prevent both microfibril crystallization and ribbon assembly, CMC can hinder only the final assembly of bundles into ribbons. Ribbon assembly from crystallized bundles may then be slightly rate-limiting, but not nearly so much as the initial crystallization of microfibrils and formation of bundles from the ordered glucan aggregates.

5. *Cellulose Microfibrils Are Synthesized by Cell-Directed Self-Assembly in Acetobacter xylinum*

5.1. *A Proposed Mechanism of Hierarchical Assembly*

Experimental use of Calcofluor and CMC has suggested that polymerization and crystallization are coupled processes, that crystallization is rate-limiting, and that the extracellular ribbon is assembled hierarchically from ordered glucan aggregates at the cell surface. This concept of cell-directed self-assembly of cellulose in *Acetobacter xylinum* is described in the following sections.

5.1.1. *Polymerization of β1,4 Glucans*

The number of multienzyme complexes below each pore determines the number of glucan chains produced at each extrusion site. If the basic unit of synthesis is the 1.5-nm fibril frequently seen in Calcofluor-induced cellulose, one would expect it to include 12–15 glucan chains. It is possible that the basic glucan aggregate is even smaller, since suggestions of smaller fibrils have been seen in negatively stained preparations.

5.1.2. *Parallel Glucan Chains Form Nondissociable Aggregates*

It is probable, since wet Calcofluor-induced cellulose has no detectable crystallinity, that the pores in the cell envelope of *A. xylinum* are not the primary agents of microfibril crystallization as has been previously proposed (Brown *et al.*, 1976; Zaar, 1979). Nevertheless, the pores probably facilitate later microfibril crystallization by aggregating the glucan chains synthesized by the multienzyme complexes as they are extruded (Zaar, 1979). Therefore, each group of multienzyme complexes and their associated extrusion pore forms a nondissociable aggregate of parallel glucan chains. Formation of such ordered aggregates minimizes entropy of the glucan chains and facilitates the formation of cellulose I, since the chains are held in parallel and hindered from folding. The pores, therefore, would promote the extended chain conformation and high packing efficiency that Stockmann (1972) maintains are essential for biogenesis of cellulose I.

5.1.3. *Adjacent Glucan Aggregates Crystallize into Microfibrils*

We have already discussed evidence indicating that each pore extrudes a nascent fibril that is too small to crystallize into true cellulose I. As first suggested by Zaar (1979), it is probable that ordered glucan aggregates from more than one extrusion site fasciate before final crystallization to produce the typical 3.5-nm microfibril in *A. xylinum* cellulose. There is morphological evidence to suggest that the synthesis–extrusion sites in the LPS layer are sometimes grouped together (Fig. 26). Such spatial organization may facilitate microfibril formation from glucan aggregates extruded from adjacent pores.

It is possible that groups of two to four particles are preferred, reflecting the enhanced crystalline stability of a fibril containing at least 35 glucan chains (Chanzy *et al.*, 1979). Several groups of three particles can be seen in Fig. 26, and smaller and larger groups can be seen throughout the micrograph. The apparent redundancy of the 3.5-nm microfibril in natural systems (Blackwell and Kolpak, 1976) may be related to its crystalline stability. The 2.8-nm crystallite measured after drying of cellulose synthesized in the presence of 1–5 mM Calcofluor would correspond well to the dimensions of the 3.5-nm microfibril. The driving force to form this 2.8-nm crystallite is evidently great enough to displace Calcofluor on drying, while the more extensive chain aggregation required to form larger crystallites cannot occur in cellulose synthesized in 1–5 mM Calcofluor.

5.1.4. *Microfibrils from Discontinuous Segments of Extrusion Pores Aggregate into Bundles*

Negative staining of the bacterial cell surface shows that the ribbon is formed by accumulation of several tapering bundles that start at defined points along the cell surface (Fig. 27). Correspondingly, freeze–fracture replicas indicate that the row of particles presumed to be synthetases is divided not only into small groups but also into discontinuous or overlapping segments (Fig. 28). Since a crystallite larger than 3.5 nm exists in *A. xylinum* cellulose and in many other celluloses, it is apparent, even if the 3.5-nm microfibril is a fundamental unit of cellulose I crystallinity, that microfibrils must associate into a higher crystalline order. It is possible that the discontinuous segments of particles observed in the LPS layer facilitate the heirarchical assembly of microfibrils into larger bundles and crystallites.

A minimum requirement of a 2.8-nm crystallite or a 3.5-nm microfibril for

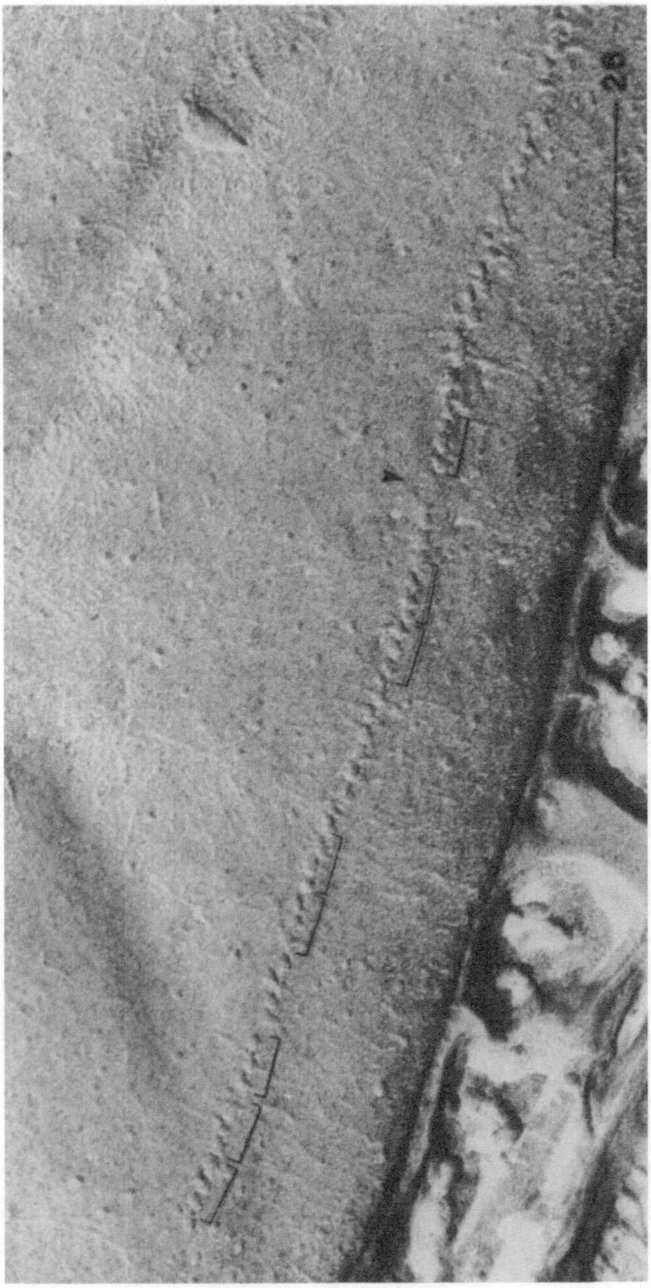

FIGURE 26. Freeze-etch replica of the fractured bacterial LPS layer showing that the particles (presumed synthetases) are sometimes segregated into small groups (brackets). Such spatial proximity of the synthetases may facilitate microfibril crystallization from adjacent glucan aggregates. Also note the large gap in the row of particles (arrow), which may facilitate the formation of separate bundles. Scale bar: 0.1 μm.

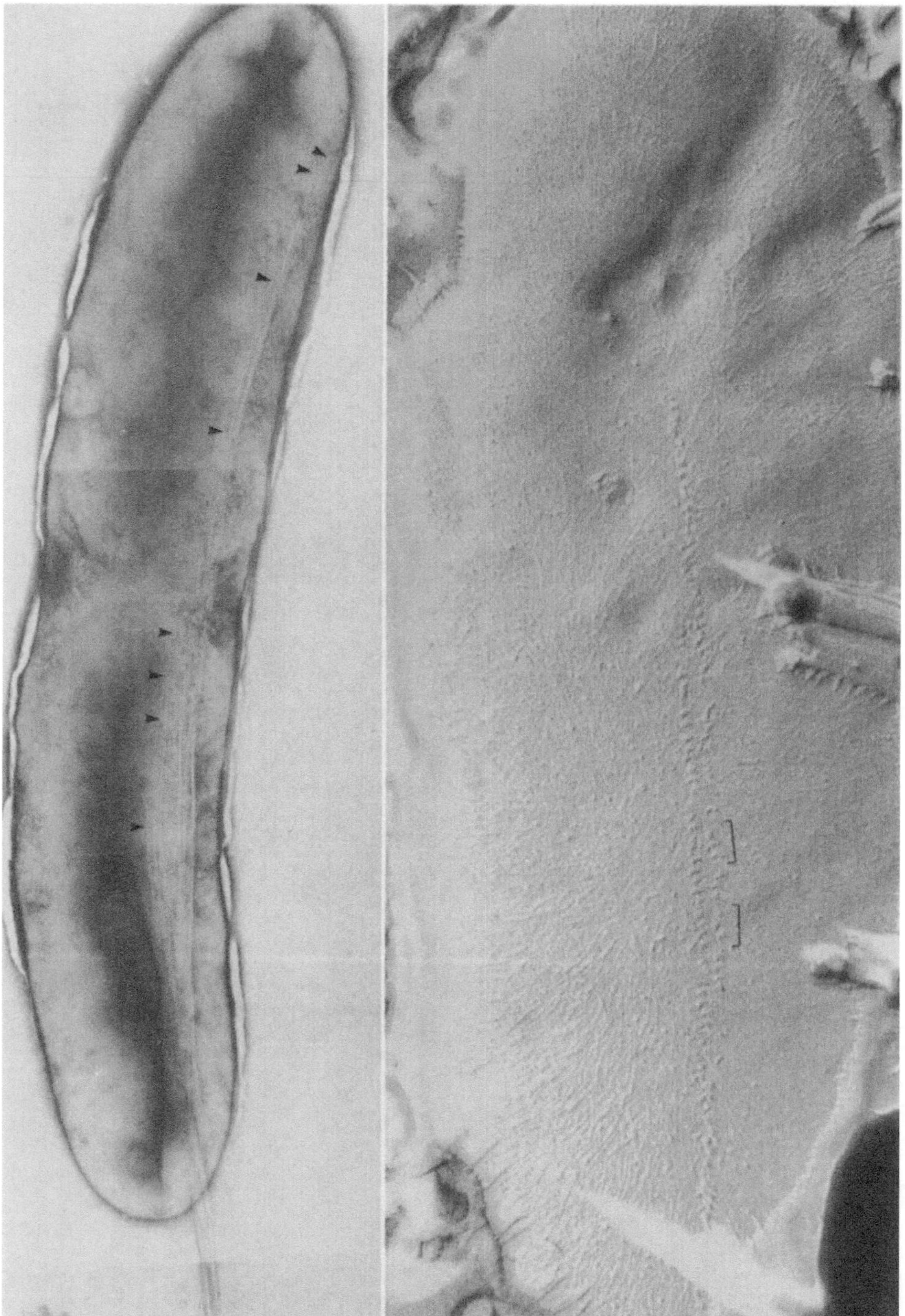

27
28

true cellulose I crystallinity is not necessarily inconsistent with the existence of larger cellulose crystallites. Frey-Wyssling and Mühlethaler (1963) proposed that elementary microfibrils fasciated to form larger microfibrils. Nieduszynski and Preston (1970), though they argue against the existence of 3.5-nm elementary fibrils, state that larger crystallites could possibly be formed by sufficiently close packing of 3.5-nm microfibrils. If the 3.5-nm elementary microfibril does represent a fundamental unit of cellulose I crystallinity that can exist as a subunit of a larger crystallite, it is reasonable that 3.5-nm fibrils would be revealed by mechanical disruption of larger fibrils as has been frequently observed (Frey-Wyssling *et al.*, 1966; Manley, 1971; Blackwell and Kolpak, 1976). The existence of 3.5-nm elementary microfibrils as subunits of cellulose I within larger crystallites may begin to explain the long-standing discrepancy between microfibril size as measured by electron microscopy and crystallite size as measured by X-ray diffraction.

5.1.5. Bundles of Microfibrils Fasciate to Form the Twisting Ribbon

Degradation of *A. xylinum* ribbons by cellulases suggests that hydrogen bonding between bundles in the ribbon is not as extensive as that within and between the 3.5-nm microfibrils making up the bundles, since the initial stage of degradation is the splaying of ribbons into bundles (White and Brown, 1981) (see also Chapter 23). A 6.9-nm crystallite could be easily contained within an individual bundle; therefore, it is probable that the bundles fasciate by limited contact along their surfaces to form the ribbon. Consequently, they are readily separated by the presence of CMC during synthesis or by cellulase degradation after synthesis. The characteristic twisting of the ribbon is also evident in small cellulose microfibrils and in the bundles and may result from relief of thermodynamic strain during cellulose I lattice formation (Stockmann, 1972).

In summary, we hypothesize that the proximity and organization of the synthesizing apparatus and extrusion pores (implying cell direction) are fundamental to the crystallization of uniformly sized microfibrils and the hierarchical assembly of the bundles and ribbon from ordered glucan aggregates at the cell surface (implying self-assembly). Stockmann (1972) proposed that covalent bonds and hydrogen bonds must be set simultaneously if extended-chain cellulose I microfibrils are formed in nature. He was assuming, however, that activated oligomers are added to growing ends of microfibrils far from the cell. If the biogenetic mechanism includes the mediation of the cell surface, as our results with Calcofluor and CMC indicate, cellulose I can be formed when

FIGURE 27. Close observation of the cell surface of the bacterium shows that the ribbon grows thicker along the cell length as discrete bundles with separate points of origin (arrows) accumulate. This cell is about to divide and is producing two ribbons in the same direction from two separate sets of synthesizing sites. Scale bar: 0.5 μm.

FIGURE 28. Freeze–etch replica of the fractured bacterial LPS showing a region of overlap of two separate rows of synthesizing sites. Offset rows of sites such as these could mediate the formation of separate bundles of cellulose like those visible in Fig. 27 and in CMC-induced cellulose. Groups of pores are also visible in this micrograph (brackets). Scale bar: 0.1 μm.

polymerization and crystallization are consecutive, not simultaneous, processes. The two processes remain tightly coupled and coordinated, however, through the synthesis and extrusion of ordered glucan aggregates that cocrystallize while attached to their forming enzyme complexes.

5.2. *Native Band Material May Be Synthesized When Cell-Directed Crystallization Is Disrupted*

Cellulose with aberrant morphology is sometimes observed in cultures of *A. xylinum* without addition of interfering agents to the incubation medium. These aberrant forms, which we have called native band material, are most often seen in old cultures or in cultures stored for extended periods in the cold (Fig. 29). While the induction of such altered products is not completely understood, it is probable that they are synthesized when the membrane organization required for normal microfibril and ribbon assembly is disrupted (Zaar, 1977). Further implicating the importance of precise coordination of the synthesizing appa-

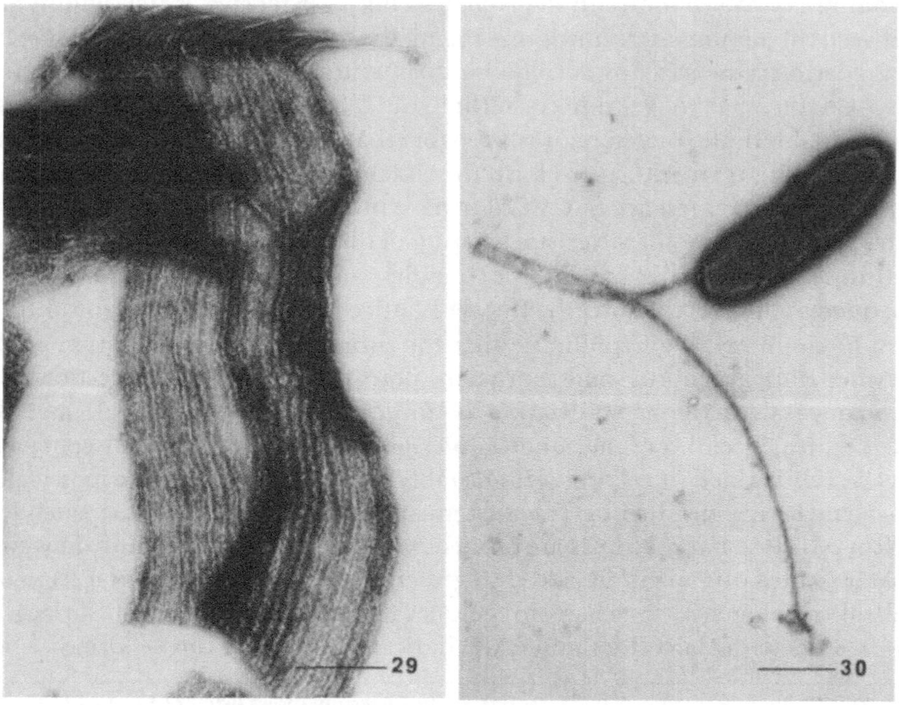

FIGURE 29. Native band material. This material is sometimes synthesized by *A. xylinum* without addition of interfering agents. These altered products appear nonfibrillar, suggesting that the cellular organization required for cell-directed self-assembly of normal microfibrils and ribbons has become disrupted. Scale bar: 0.5 μm.

FIGURE 30. Reversal in direction of ribbon synthesis accompanied by a brief disorganization of microfibril and ribbon assembly at the time of reversal. Scale bar: 1 μm.

ratus, we have observed that reversal in direction of synthesis sometimes occurs and is accompanied by brief disorganization of microfibril and ribbon assembly (Fig. 30).

5.3. How Is Cell-Directed Crystallization Rate-Limiting?

If microfibril crystallization and ribbon assembly require that parallel glucan-chain aggregates from adjacent sites interact as we have described, the time required for the proper interaction to occur could be rate-limiting. When Calcofluor binds to nascent fibrils immediately after extrusion, the time lag required for adjacent aggregates to come into proper proximity is removed, allowing the polymerizing enzymes to synthesize β1,4-glucans at their maximum rate.

Furthermore, if a perfect ribbon is assembled from the product of 100 or more synthesizing complexes, the individual complexes must be synthesizing cellulose at the same rate. Therefore, each enzyme would be restrained to the rate of the slowest ones. When Calcofluor is present, the synthesizing complexes are autonomous, and each can proceed at its maximum efficiency.

It may also be true that the crystallized ribbon is harder to move away from the cell than the more diffuse altered product. The kinetic force normally associated with formation of the twisting ribbon suggests, however, that crystallization of cellulose releases energy and does not require some "pushing away" on the part of the cell, which would be rate-limiting. Therefore, it seems more likely that the rate of polymerization is limited by the time required for the cell-directed self-assembly of crystalline microfibrils.

6. Does Cell-Directed Crystallization Occur in Other Organisms?

The concept of cell-directed crystallization presented here may have important implications for mechanisms of cellulose biogenesis in other organisms. Freeze–fracture studies of corn root (and pine and mung bean seedlings) indicate that organized membrane-bound particles, called rosettes, and globular complexes at the ends of microfibrils are involved in wall synthesis of higher plants (Mueller and Brown, 1980). No groups of rosettes are observed in corn, and the fibrils arising from each terminal globule and single rosette are small and nonstriated. The terminal globule may represent a group of enzymes organized to polymerize the glucan chains making up each separate microfibril. It does not appear, however, that several globules and rosettes are grouped together to form composite fibrils in corn.

Interestingly, similar rosettes have been observed in the plasmalemma of the alga *Micrasterias denticulata* (Giddings *et al.*, 1980). During primary wall synthesis, isolated rosettes are seen in association with single, nonstriated 5.0-nm microfibrils. During secondary wall synthesis, however, 6–175 of these rosettes have been shown to be aggregated in association with large, striated microfibrils in which the number of striations and overall size of the fibril can be correlated with the pattern of rosette organization. Similar correlations of terminal globule organization and microfibril size have been shown in the alga *Oocystis* (Brown and Montezinos, 1976). This is compelling evidence that the same kind of hierarchical assembly of cellulose occurs in organisms other than *Acetobacter*.

7. *Theoretical Implications of Cell-Directed Self-Assembly*

It may well prove true that synthesis of extended parallel-chain cellulose I microfibrils requires organized multienzyme complexes that polymerize at least 30–35 glucan chains in close association so that the parallel-chain conformation required for cellulose I is established and maintained directly at the site of polymerization. Further, membrane integrity and spatial organization of integral proteins may be required to facilitate chain aggregation, fasciation, and crystallization. The size of the fibrils and the size of the crystallites may depend on the number and arrangement of multienzyme complexes and associated membrane structures that are in a proximity to facilitate interactions of glucan aggregates before final crystallization and fibril formation. Very small fibrils of low crystallinity such as are observed in quince slime (Franke and Ermen, 1969), primary walls of rose and cotton (Chanzy *et al.*, 1978, 1979), and wood parenchyma (Hanna and Cote, 1974) may well be formed by small groups of dispersed enzymes. The presence of hemicelluloses in the slime and primary wall may, analogously to CMC, regulate the extent of fibril aggregation and crystallization that is possible. Such a mechanism fits the theoretical argument made by Sarko (1974) that the biosynthetic mechanisms of cellulose synthesis in different organisms should be identical, with different crystallite sizes arising by small differences in synthetic mechanism—that is, we would propose, by juxtaposition of different numbers of multienzyme complexes.

The mechanism of cell-directed self-assembly is attractive because it allows the same basic biogenetic machinery to operate in different ways depending on time in the cell cycle. Synthesis of the primary wall and the secondary wall in *Micrasterias* seems to be a good example of differential organization of the same basic enzymatic machinery. It is interesting to speculate that a relatively unorganized primary wall can be synthesized at a maximum rate, fulfilling the cell's immediate requirement for a containing wall. If small microfibrils are laid down in a matrix of hemicelluloses, the cell may be able to rapidly produce a wall with high expansibility. When rigidity is the most important consideration, as it is in secondary wall synthesis, enzyme complexes may become organized, and a highly crystalline wall can be produced, although possibly at a slower rate. Further, it is easy to see that environmental demands may have required the development of a more extensive hierarchical assembly mechanism in algae (to produce larger fibrils and crystallites) than in higher plant cells that are surrounded by other cells.

The mechanism of cell-directed crystallization wherein polymerization and crystallization are coupled processes at the cell surface also circumvents many problems inherent in other models (Colvin, 1972) of microfibril biogenesis. Production of metastable cellulose I *in vivo* is guaranteed through production of parallel-chain glucan aggregates by the organized multienzyme complexes and extrusion pores. No diffusion of extracellular intermediates need be hypothesized, since an activated glucose or oligomer can be transferred to the growing fibril in association with the cell. Microfibril initiation occurs at the site of the enzyme, and therefore, mechanisms for acellular nucleation need not be advanced. Anastomoses, or twisting without fusion of separate microfibrils, can be easily accounted for by the separation of groups of synthesizing sites similarly to the discontinuous array of extrusion pores that gives rise to the bundles in *A. xylinum* ribbons. The apparent inconsistency between crystallite size of elementary microfibrils may possibly be explained.

It is apparent that use of these interfering probes is a significant step forward in the study of cellulose biogenesis. Data we have obtained with Calcofluor and with CMC have allowed us to propose a mechanism of cellulose I microfibril biogenesis that begins to unify the crystal structure of cellulose, the substructure of the microfibril, and the observed organization of membrane-bound complexes in cellulose-synthesizing organisms. It is doubtful that native crystalline cellulose I microfibrils will be successfully formed *in vitro* unless the membrane integrity and the enzyme organization that facilitate consecutive, coupled polymerization and crystallization can be preserved *in vitro*.

Acknowledgments

We thank Anatole Sarko and C. Chuah (Department of Chemistry, College of Environmental Science and Forestry, State University of New York, Syracuse) for performing the X-ray diffraction; Sam Hudson (School of Textile Chemistry, North Carolina State University, Raleigh) for determining the degree of polymerization; R. Malcolm Brown, Jr., for valuable discussions; Alan White for preparing high-resolution grids; Kay Cooper for assistance with autoradiography and for performing the freeze-etching; Susan Sizemore for artwork; and Richard Santos for technical assistance. Figures 2, 6 (inset), 13, and 17 are reproduced with the permission of *Science*.

References

Baugh, P. J., and Phillips, G. O., 1954, Degradation of cellulose and its derivatives. D. Photochemical degradation, in: *Cellulose and Cellulose Derivatives*, Vol. 5, No. 5 (T. E. Timell, ed.), p. 1066, Wiley-Interscience, New York.

Ben-Hayim, G., and Ohad, I., 1965, Synthesis of cellulose by *Acetobacter xylinum*. VIII. On the formation and orientation of bacterial cellulose fibrils in the presence of acidic polysaccharides, *J. Cell Biol.* **25**:191.

Benziman, M., Haigler, C. H., Brown, R. M., Jr., White, A. R., and Cooper, K. M., 1980, Cellulose biogenesis: Polymerization and crystallization are coupled processes in *Acetobacter xylinum*, *Proc. Natl. Acad. Sci. U.S.A.* **77**:6678.

Blackwell, J., and Kolpak, F. J., 1976, Cellulose microfibrils as disordered arrays of elementary fibrils, in: *Applied Polymer Symposia*, 28 (T. E. Timell, ed.), pp. 751–761, John Wiley, New York.

Brown, R. M., Jr., and Montezinos, D., 1976, Cellulose microfibrils: Visualization of biosynthetic and orienting complexes in association with the plasma membrane, *Proc. Natl. Acad. Sci. U.S.A.* **73**:143.

Brown, R. M., Jr., Willison, J. H. M., and Richardson, C. L., 1976, Cellulose biosynthesis in *Acetobacter xylinum:* Visualization of the site of synthesis and direct measurement of the *in vivo* process, *Proc. Natl. Acad. Sci. U.S.A.* **73**:4565.

Chanzy, H., Imada, K., and Vuong, R., 1978, Electron diffraction from the primary wall of cotton fibers, *Protoplasma* **94**:299.

Chanzy, H., Imada, K., Mollard, A., Vuong, R., and Barnard, F., 1979, Crystallographic aspects of sub-elementary cellulose fibrils occurring in the walls of rose cells cultured *in vitro*, *Protoplasma* **100**:317.

Colvin, J. R., 1964, The biosynthesis of cellulose, in: *The Formation of Wood in Forest Trees* (M. H. Zimmerman, ed.), pp. 189–201, Academic Press, New York.

Colvin, J. R., 1972, The structure and biosynthesis of cellulose, *CRC Crit. Rev. Macromol. Sci.* **1**:47.

Colvin, J. R., 1980, The mechanism of formation of cellulose-like microfibrils in a cell-free system from *Acetobacter xylinum*, *Planta* **149**:97.

Cooper, D., and Manley, R. St. John, 1975, Cellulose synthesis by *Acetobacter xylinum*. III. Matrix, primer and lipid requirements and heat stability of the cellulose forming enzymes, *Biochim. Biophys. Acta* **381**:109.

Delmer, D. P., 1977, The biosynthesis of cellulose and other plant cell wall polysaccharides, in: *Recent Advances in Phytochemistry 11* (F. A. Leowus and U. C. Runeckles, eds.), pp. 45–77, Plenum Press, New York.

Franke, W. W., and Ermen, B., 1969, Negative staining of plant slime cellulose: An examination of the elementary fibril concept, *Z. Naturforsch.* **24**:918.

Frey-Wyssling, A., and Mühlethaler, K., 1963, Die Elementarfibrillen der Cellulose, *Makromol. Chem.* **62**:25.

Frey-Wyssling, A., Mühlethaler, K., and Muggli, R., 1966, Elementarfibrillen als Grundbausteine der nativen Cellulose, *Holz Roh- Werkst.* **24**:443.

Gardner, K. H., and Blackwell, J., 1974a, The structure of native cellulose, *Biopolymers* **13**:1975.

Gardner, K. H., and Blackwell, J., 1974b, The hydrogen bonding in native cellulose, *Biochim. Biophys. Acta* **343**:232.

Giddings, T. H., Jr., Brower, P. L., and Staehelin, L. A., 1980, Visualization of particle complexes in the plasma membrane of *Micrasterias denticulata* associated with the formation of cellulose fibrils in primary and secondary cell walls, *J. Cell Biol.* **84**:327.

Haigler, C., Brown, R. M., Jr., and Benziman, M., 1980, Calcofluor White alters the *in vivo* assembly of cellulose microfibrils, *Science* **210**:903.

Hanna, R. B., and Cote, W. A., Jr., 1974, The sub-elementary fibril of plant cell wall cellulose, *Cytobiologie* **10**:102.

Lanter, J., 1966, Properties and evaluation of fluorescent brightening agents, *J. Soc. Dyers Colour.* **82**:125.

Maeda, H., and Ishida, N., 1967, Specificity of binding of hexopyranosyl polysaccharides with fluorescent brightener, *J. Biochem.* **62**:276.

Manley, R. St. John, 1971, Molecular morphology of cellulose, *J. Polym. Sci. Polym. Phys. Ed.* **9**:1025.

Marchessault, R. H., Morehead, F. F., and Walter, N. M., 1959, Liquid crystal systems from fibrillar polysaccharides, *Nature (London)* **184**:632.

Mueller, S. C., and Brown, R. M., Jr., 1980, Evidence for an intramembrane component associated with a cellulose microfibril-synthetic complex in higher plants, *J. Cell Biol.* **84**:315.

Mühlethaler, K., 1967, Ultrastructure and formation of plant cell walls, *Annu. Rev. Plant Physiol.* **18:**1.

Nieduszynski, I., and Preston, R. D., 1970, Crystallite size in natural cellulose, *Nature (London)* **225:**274.

Preston, R. D., 1964, Structural plant polysaccharides, *Endeavor* **23:**153.

Ranby, B. G., 1952, The mercerisation of cellulose I: A thermodynamic discussion, *Acta Chem. Scand.* **6:**101.

Rattee, I. D. and Breuer, M. M., 1974, *The Physical Chemistry of Dye Adsorption*, pp. 180–182, Academic Press, New York.

Sarko, H., and Muggli, R., 1974, Packing analysis of carbohydrates and polysaccharides. III. *Valonia* cellulose and cellulose II, *Macromolecules* **7:**486.

Stockmann, V. E., 1972, Developing a hypothesis: Native cellulose elementary fibrils are formed with metastable structure, *Biopolymers* **11:**251.

White, A. R., and Brown, R. M., Jr., 1981, Enzymatic hydrolysis of cellulose: Visual characterization of the process, *Proc. Natl. Acad. Sci. U.S.A.* **78:**1047.

Willison, J. H. M., and Brown, R. M., Jr., 1978, Cell wall structure and deposition in *Glaucocystis*, *J. Cell Biol.* **78:**103.

Zaar, K., 1977, The biogenesis of cellulose by *Acetobacter xylinum*, *Cytobiologie* **16:**1.

Zaar, K., 1979, Visualization of pores (export sites) correlated with cellulose production in the envelope of the gram-negative bacterium *Acetobacter xylinum*, *J. Cell Biol.* **80:**773.

15

A Study of the Polymerization Kinetics of Bacterial Cellulose through Gel-Permeation Chromatography

Gerard J. F. Ring

1. Introduction

The importance of cellulose as a structural material in the plant kingdom is easily recognized by the fact that it is the most abundant organic material known (Ward, 1954). Naturally occurring cellulose is rapidly polymerized from nucleoside diphosphoglucose into a mechanically strong yet flexible matrix. But, depending on the source, it is also associated to varying degrees with other noncellulosic components that necessitate purification by chemical extraction. Consequently, the elucidation of the structure and formation of cellulose has been hampered by the consideration that all known isolation techniques alter its natural form to some extent and may even degrade the polymers (Mark, 1954). Thus, any purified natural cellulose is reliant on its isolation procedure as an integral part of its description.

A study of the kinetics of cellulose biosynthesis is particularly vulnerable to isolation techniques that will remove low-molecular-weight polymers that have not yet been consolidated into microfibrillar structures. The solubilization of this material will have the detrimental effect of reducing the total mass and increasing the apparent number-average degree of polymerization. But, at the present state of the art, it is impossible to avoid solubilizing low-molecular-weight material (Colvin, 1972). Therefore, a study of cellulose polymerization

GERARD J. RING • The Institute of Paper Chemistry, Appleton, Wisconsin 54912.

kinetics will not be representative of the total production of cellulose, but rather only of the production of isolatable cellulose.

Kinetic studies involving the remarkably pure extracellular cellulose produced by the bacterium *Acetobacter xylinum* have demonstrated that the yield of isolatable bacterial cellulose as a function of time follows the same rate of increase as the culture's total-mass accumulation (Marx-Figini and Pion, 1974; A. M. Brown, 1962) (see Also Chapter 13). This observation has been interpreted as being indicative of an essentially complete cellulose isolation. Furthermore, the relatively mild isolation procedures that were utilized have been demonstrated to be nondegradative (Marx-Figini and Schultz, 1963). Therefore, bacterial cellulose represents an almost ideal source wherein the molecular-weight distribution and the molecular-weight averages of cellulose have undergone a minimum of modification.

The significance of being able to accurately measure the molecular-weight averages with time as more and longer molecules are produced is that the mechanism of polymerization can be elucidated if the proper types of averages are measured. The previous investigations, however, have been able to obtain only the weight-average molecular weight and not the number-average molecular weight, which is necessary to estimate the actual increase in the number of polymers. Therefore, these studies could not prove with certainty a particular mechanism for cellulose production.

Accurate number-average molecular weights are difficult to obtain because the high molecular weights attained by bacterial cellulose are generally above the sensitivity limits of membrane osmometers. Furthermore, the relatively short synthetic intervals involved preclude obtaining sufficient material for more than one absolute measurement; hence, a weight average is usually preferred. In this study, however, these drawbacks were eliminated by the utilization of gel-permeation chromatography (GPC). The technique of GPC yields both number-average and weight-average molecular weights simultaneously without recourse to duplicate samples. But, because it is a relative technique rather than an absolute technique, the measurement of accurate molecular-weight averages relies on the determination of an accurate calibration. An accurate calibration depends on the proper assignment of molecular-weight averages to their proper positional values on the calibration curve. When an accurate calibration is obtained, the molecular-weight averages calculated from every GPC chromatogram must be corrected for the effects of instrumental spreading or peak dispersion, which tend to raise the value of the weight-average and lower the value of the number-average molecular weight.

When there is difficulty in obtaining calibration standards for the type of polymer that is being studied, however, it is customary to utilize an alternative series of calibration standards and translate the calibration so obtained via the Mark–Houwink constants for both polymers into the calibration desired. This is known as a universal calibration. Peak dispersion is also a factor when universal calibration is used, but the dispersion correction must then compensate for two different polymers. Since it was necessary to incorporate a universal calibration for the measurement of bacterial-cellulose molecular-weight averages, special precautions were taken in this study to avoid dispersional errors in these determinations. This required the development of a dispersionally correct universal calibration. Because the complete details of this technique are beyond the scope of this presentation, the reader is referred to Ring (1980) and Ring *et al.* (1982).

The primary objective of this study was to determine the existence or nonexistence of unconsolidated, soluble, polymeric cellulose precursors as proposed by Colvin and Leppard (1977). Because a polymeric precursor would have a distinctly more discrete molecular-weight distribution, it was realized that the effects of a polymeric precursor could be detected by observing the change in the molecular-weight distribution with synthesis time. This would also mean that the molecular-weight averages would vary in a different manner as opposed to a monomeric polymerization. Thus, GPC was utilized as the experimental method to obtain the molecular-weight distributions and their molecular-weight averages. A Poisson polymerization function was selected to determine the theoretical variations in the molecular-weight averages due to precursor molecular weight because it ideally modeled the hypothesized property of soluble precursors randomly attaching to any previously consolidated cellulose chain, and it has also been previously proposed for bacterial-cellulose biosynthesis (Marx-Figini and Pion, 1974).

2. *Bacterial-Cellulose Production*

2.1. *Description of Acetobacter xylinum*

The genus *Acetobacter* refers to a group of gram-negative bacteria that have the ability to oxidize ethanol to acetic acid. In particular, the bacterium *Acetobacter xylinum*, noted for its ability to produce vinegar, is usually found in wine vats as ". . .a sort of moist skin, swollen, gelatinous and slippery. . ." (Pasteur, 1868). It has been referred to in the past as "mother-of-vinegar" and was the basis of the early vinegar industry. The gelatinous skin is actually a polysaccharide matrix within which the bacterial cells are enmeshed and is more commonly known as a pellicle.

Because *A. xylinum* is an obligate aerobe requiring a constant supply of oxygen, it can be speculated that the function of the polysaccharide pellicle is to provide a buoyant environment at the air–liquid interface. This idea finds some support in the observation of "pellicle balloons" filled with carbon dioxide in constantly agitated cultures.* However, the pellicles of *A. xylinum* are com-

*Pellicle balloons, found after 48 hr of shaking in a 250-ml Erlenmeyer inoculation flask, were approximately 2–2.5 inches in diameter. When placed on a paper towel, they would quickly deflate to compact circular pellicles. The internal volume was sufficient to contain the first knuckle of a thumb. Until deflation, they were rather turgid and at times were difficult to remove through the neck of the inoculation flask.

monly found in a membranelike condition in nonagitated cultures. The polysaccharide of the pellicle is predominantly cellulose derived biochemically from various possible sugar analogues (A. J. Brown, 1886), and its biosynthesis has been the subject of numerous investigations (Hassid, 1971).

2.2. *Microfibril Formation*

Bacterial cellulose has been found to have the same native crystal structure as cotton cellulose (Barsha and Hibbert, 1934). In contrast to cellulose from green plants, however, it is not deposited into multilayered walls surrounding the cell. Rather, it is formed into numerous ribbons, each consisting of several microfibrils, interlaced into a matted pellicle structure (Sowden and Colvin, 1978). The individual ribbons are of various lengths and are subject to continued elongation from the addition of glucose monomers (Hestrin and Schramm, 1954). It is known that live, biosynthetically active bacterial cells are responsible for this elongation process (Marx-Figini and Pion, 1974); however, the exact role that the cells play in this process is still a controversial subject (Colvin, 1972). The primary polymerization of glucose into polyglucosan chains (β1,4-polyanhydroglucose) is believed to be performed by a synthetase enzyme complex, situated within the bacterial cell envelope (Glaser, 1958), which extrudes these chains extracellularly through pores in the surrounding outer membranes (R. M. Brown, Jr., *et al.*, 1976; Colvin and Leppard, 1977; Zaar, 1979). But the manner in which these chains are consolidated into microfibrils and then microfibrillar ribbons has not as yet been agreed on.

One proposal (R. M. Brown, Jr., *et al.*, 1976; Zaar, 1979), based on electron-microscopic evidence, postulates that the polymerization enzymes are arranged in a linear array that allows the connected polymer chains to immediately crystallize into about 46 continuous microfibrils, and that these microfibrils are then consolidated into a single, unified ribbon extruding from the polar region of the cell. During cellular division, this linear array is bisected such that the mother and daughter cells will each receive half the synthetase sites, but with no curtailment in the activity of these sites (R. M. Brown, Jr., 1979). Initially, the newly divided cells will produce cellulose ribbons of a reduced width until the biosynthesis of new enzyme complexes restores their original number per cell. Thus, as each new complex is activated, the incorporation of new microfibrils will increase the width of the already established, microfibrillar ribbon. In this manner, the process of cellular replication does not interfere with the ongoing process of cellulose biosynthesis.

In contrast to this proposal, it has been suggested that the synthetase enzymes are not attached to the growing cellulose microfibrils but, instead, produce oligomeric precursors that are excreted into the external milieu (Colvin and Leppard, 1977). These precursors then form nascent, highly hydrated fibrils via a presumably simple but random polymer-association process, followed by an end-to-end covalent bonding of the precursors to form substantially longer molecules after they have associated into nascent microfibrils. Finally, this second proposal also assumes that the initiation of new microfibrils is a random process, which occurs from the chance association of several precursors to form an "acceptor," identical to the growing tip of a previously formed microfibril.

2.3. *Bacterial-Cellulose Kinetics*

When bacterial cellulose is harvested by extracting the noncellulosic components with NaOH under nitrogen, the yield of the isolated cellulose has been found to increase with time at the same rate as the total dry mass (A. B. Brown, 1962; Marx-Figini and Pion, 1974). This rate is described by an exponential function that is typical of the logarithmic growth phase of bacterial cultures (Oginsky and Umbreit, 1959). This function, known as the law of bacterial growth, explicitly describes the total mass M of a culture after time t and has the form:

$$M = M_0(k_{ic} + k_{as})e^{\alpha t} \tag{1}$$

where M_0 is the initial mass at $t = 0$, α is the first-order rate constant, and k_{ic} and k_{as} are the weight fractions for the isolated cellulose and the alkali-soluble material, respectively. Theoretically, during the logarithmic growth phase, the bacterial cells are experiencing maximum enzymatic activity for the prevailing reaction conditions due to a nonlimiting supply of nutrients and oxygen. These investigations (Marx-Figini and Pion, 1974; A. M. Brown, 1962) have also demonstrated that the increase in the number of cells conforms to the law of bacterial growth with approximately the same rate constant α as the rate of mass accumulation [within 3% for values of $\alpha = 0.08$–0.15 (Marx-Figini and Pion, 1974)]:

$$N = N_0e^{\alpha t} \tag{2}$$

where N_0 is the initial number of cells at time $t = 0$, and N is the total number after time t.

It has been shown experimentally that the ratio k_{ic}/k_{as} may range from a value of 0.65 to a value of 1.33 for various cultures (Marx-Figini and Pion, 1974); however, it is also known that alkali extraction will remove a polyglucosan of relatively low molecular weight (Hestrin and Schramm, 1954; Ben-Hayyim and Ohad, 1965; Kjosbakken and Colvin, 1975). Therefore, it is possible to assume that the parameter k_{ic} does not represent the entire cellulose polymerization production and that k_{as} represents some soluble cellulose as well as noncellulose material. But, because the isolated cellulose does follow the law of bacterial growth, the mass of isolated cellulose is proportional to the total cellulose polymerization product:

$$w = w_0\,(k_i + k_s)e^{\alpha t} \tag{3}$$

where w_0 and w are the initial and final masses, respectively, of the total cellulose polymerization product; k_i is the weight-fraction of the isolatable cellulose; and k_s is the weight-fraction of the soluble, or nonisolatable, cellulose. Thus, the observation that the yield of isolated cellulose follows the law of bacterial growth is very significant, because it demonstrates that alkali-extracted bacterial cellulose can be considered as a distinct enzymatic product regardless of whether or not it represents the total cellulose polymerization product.

With first-order kinetics, an average period during which the number of cells will double, τ_{cell}, can be calculated from the first-order rate constant α:

$$\tau_{cell} = \ln 2/\alpha \tag{4}$$

This average time is interpreted as the time necessary for a typical cell to reproduce and is commonly referred to as the generation time. During each generation, the appearance of new cells is accompanied by the production of new cellulose molecules. However, it is also possible that at the time of cellular replication, the polymerization of older molecules may be terminated. If this were true, then the measurable average length or degree of polymerization of the cellulose molecules experiencing such a termination would rapidly reach a constant value. But it has been observed that *A. xylinum* does not produce a constant degree of polymerization (DP); rather, the DP slowly increases over a period of several generations (Marx-Figini and Pion, 1974; Husemann and Werner, 1963). Thus, the evidence for *A. xylinum* suggests that no such termination reaction is present and that the cellulose molecules, once initiated, continue to grow for the life of the cells. For this reason, it has been proposed that a possible mechanism for bacterial cellulose polymerization is a stochastic "living-polymer" process described by a Poisson function (Marx-Figini and Pion, 1974).

3. Theoretical and Experimental Objectives

A Poisson probability function can be defined as a mathematical expression that predicts the likelihood that a certain number of random events will occur during any particular time period, given that the occurrence of each separate event is independent of the occurrence of the other events and that it is impossible for two events to occur simultaneously at the same position (Brownlee, 1965). As a consequence, the Poisson function is discrete, disallowing fractional events.

The utilization of a Poisson function as a predictor of the chainlengths of isolated bacterial cellulose requires that an "event" be defined in terms of the degree of polymerization. Specifically, a Poisson function implies that during synthesis, each molecule of cellulose that is incorporated into the mass of isolated cellulose may experience a random number of events, whereby its polymeric length would be stochastically increased by subunits of equal size or molecular weight (Flory, 1940). If an event is defined in terms of the incremental change in

DP, then the magnitude of this incremental change becomes important, particularly in light of the controversy that exists about the mode of microfibril formation. For the case of simultaneous polymerization and crystallization, an event would correspond to the polymerization of cellulose by a subunit of glucose or perhaps cellobiose (Hassid, 1969). However, for the case of oligomeric precursors, an event would correspond to the bonding of an oligomer onto either another oligomer or a cellulose chain composed of repeated oligomers. This then postulates that if a Poisson function is applicable to the formation of isolated bacterial cellulose, each polymer would have a length that could be measured in multiples of the subunit's size or degree of polymerization.

Because of the important implications regarding the mechanism of microfibril formation, it is the purpose of this investigation to document the simultaneous rates of mass accumulation and DP attainment by measuring the cellulose yield and the number-average and weight-average DPs as functions of time, and to compare these experimental values against the theoretically calculated values for a Poisson-regulated growth pattern. This requires deriving explicit expressions for the experimental parameters in terms of the Poisson function. It also requires defining the independent variables for each expression, including the molecular weight of the event subunit, in such a manner that they can be measured unambiguously.

4. Poisson Polymerization Function

The process of polymer introduction is of key importance to the derivation of a polymerization distribution for a synthetic system that is continually producing new molecules with every generation. Previously, it has been assumed that for *A. xylinum* the number of polymers produced per cell is a constant and that the first-order rate equation for the increase in the number of cells with time could be used in conjunction with the polymer/cell proportionality to predict the total number of polymeric molecules isolated at any time (A. M. Brown, 1962). This expression for $n_i(t)$, the total number of polymers isolated after time t, is written as

$$n_i(t) = N_0 K e^{\alpha t} \tag{5}$$

where N_0 is the original number of cells, K is equal to the number of polymers per cell, and α is the same first-order rate constant defined earlier. However, equation (5) assumes only that the number of polymers isolated is always proportional to the number of cells present. The definition of the total number of polymers present in any sample is the ratio of the mass to the number-average molecular weight (Elias, 1977). If $\overline{M}_n$ is the number-average for the isolated bacterial cellulose, then the number of polymers that have been isolated is

$$n_i(t) = w_0 k_i \mathrm{A}\, e^{\alpha t} / \overline{M}_n \tag{6}$$

where A is Avogadro's number. Because equations (5) and (6) are equal only when their coefficients ($N_0 K$) and ($w_0 k_i \mathrm{A}/M_n$) are also equal, the implication is that $\overline{M}_n$ must be assumed constant with time. However, for a synthetic material that is increasing in average molecular weight, this assumption cannot be made.

An alternative assumption is that $\overline{M}_n$ varies with time, similar to equation (5), by the empirical function

$$\overline{M}_n(t) = K'e^{\alpha' t} \tag{7}$$

where K' and α' are analogous, respectively, to K and α. Equation (6) may then be rewritten as

$$n_i(t) = n_0 e^{(\alpha - \alpha')t} = n_0 e^{\beta t} \tag{8}$$

where $n_0 = w_0 k_i \mathrm{A}/K'$ and the symbol $\beta = \alpha - \alpha'$ is the rate constant for the increase in the number of polymers with synthesis time, which, depending on the value of α', will not generally be equal to α. Therefore, no simplifying assumptions can be made about the rate constants for the increase in the number of polymers and the mass of cellulose with time. Both α and β will have to be measured experimentally.

To measure α, a series of cultures initiated at the same time but harvested at various intervals is required. From these samples, the number of cells, total dry mass, or yield of isolated cellulose has to be obtained. A plot of the logarithm of these parameters against the elapsed time between initiation and harvesting results is a straight line. For the yield of isolated cellulose, $w_k(t)$, this line would be described by

$$\ln w_k(t) = \ln (w_0 k_i) + \alpha t \tag{9}$$

Thus, the slope of such a line is a measure of the rate constant.

To measure β, it is necessary to obtain both the yield and the number-average molecular weight $\overline{M}_n$. Thus, plotting the logarithm of the number of polymers against time will also result in a straight line with β as its slope if equation (8) is valid:

$$\ln n_i(t) = \ln[w_k(t)/\overline{M}_n] = \ln n_0 + \beta t \tag{10}$$

In much the same manner as equation (4), a doubling or generation time for the number of polymers, τ_{poly}, can be defined:

$$\tau_{\mathrm{poly}} = \ln 2/\beta = \delta/\beta \tag{11}$$

where $\delta = \ln 2$. The parameter τ_{poly} can be used to define a new dimensionless time-parameter γ in units of generation or doubling time:

$$\gamma = t/\tau_{\text{poly}} = \beta t/\delta \tag{12}$$

that permits a simplified notation for the kinetic expressions. Equation (8) can be restated simply as

$$n_i(t) = n_0 e^{\beta t} = n_0 e^{\delta\gamma} = n_0 2^{\gamma} \tag{13}$$

The number of polymers that are introduced at any given time γ' between 0 and γ will be

$$d[n_i(t)]/d\gamma' = n_0\,\delta\,e^{\delta\gamma'} \tag{14}$$

Those polymers introduced at γ' will polymerize for a duration of $\gamma - \gamma'$.

If polymerization proceeds through the sequential addition of cellulose precursors having a degree of polymerization equal to z, then the molecular weight of a polymer introduced as a precursor with k number of further additions is

$$M_k = x\,z\,(k + 1) \tag{15}$$

where $x = 162$, the molecular weight of anhydroglucose.

The proportion of polymers introduced at γ' with k number of polymerization events after $\gamma - \gamma'$, according to a Poisson probability function, is given as

$$p[k,\lambda,(\gamma - \gamma')] = e^{[-\lambda(\gamma-\gamma')]}[\lambda(\gamma - \gamma')]^k/k! \tag{16}$$

where λ is the constant, average rate that an event will occur per generation.

The combined products of equations (14) and (16) for every value of γ' yield the number distribution for a Poisson probability function and a first-order polymer introduction rate:

$$n(k,\lambda,\gamma) = n_0 p(k,\lambda,\gamma) + n_0\delta \int_0^{\gamma} e^{\delta\lambda'} p[k,\lambda,(\gamma - \gamma')]d\gamma' \tag{17}$$

which on integration yields

$$n(k,\lambda,\gamma) = n_0 e^{-\lambda\gamma}\frac{(\lambda\gamma)^k}{k!} + \frac{n_0\lambda^k\delta e^{\delta\gamma}}{(\delta + \lambda)^{k+1}}\left\{1 - e^{-(\delta+\lambda)\gamma}\sum_{i=0}^{k}\frac{[(\delta + \lambda)\gamma]^i}{i!}\right\} \tag{18}$$

The number distribution is utilized to define the polymeric mass, and the number-average and the weight-average degrees of polymerization, respectively:

$$w(\lambda,\gamma) = \sum_{k=0}^{\infty} M_k\,n(k,\lambda,\gamma) = x\,z\,u_1(\lambda,\gamma) \tag{19}$$

$$\overline{DP}_n(\lambda,\gamma) = (1/x) \sum_{k=0}^{\infty} M_k\, n(k,\lambda,\gamma) / \sum_{k=0}^{\infty} n(k,\lambda,\gamma) = z\, u_1(\lambda,\gamma)/u_0(\lambda,\gamma) \tag{20}$$

and

$$\overline{DP}_w(\lambda,\gamma) = (1/x) \sum_{k=0}^{\infty} M_k^2\, n(k,\lambda,\gamma) / \sum_{k=0}^{\infty} M_k\, n(k,\lambda,\gamma) = z\, u_2(\lambda,\gamma)/u_1(\lambda,\gamma) \tag{21}$$

where the function, with $j = 0, 1, 2$

$$u_j(\lambda,\gamma) = \sum_{k=0}^{\infty} (k+1)^j n(k,\lambda,\gamma) \tag{22}$$

is referred to as the jth moment of the number distribution.

As can be seen, z has the role of a simple scaling factor and can be computed directly once the moments $u_0(\lambda,\gamma)$ $u_1(\lambda,\gamma)$, and $u_2(\lambda,\gamma)$ have been determined. However, to be able to calculate all three moments, it is necessary to first determine the average occurrence rate, λ. This is accomplished by solving simultaneously the polydispersity relationship, which is independent of z, for several values of γ. Thus

$$\overline{DP}_w(\lambda,\gamma)/\overline{DP}_n(\lambda,\gamma) = u_2(\lambda,\gamma)\, u_0(\lambda,\gamma)/u_1^2(\lambda,\gamma) \tag{23}$$

In summary, the experimental procedure suggested by the preceding theoretical derivation is: (1) Measure yield and the molecular weight averages (w_k, $\overline{DP}_n$, and $\overline{DP}_w$) as functions of time t. (2) If $\overline{DP}_n$ is constant, calculate α according to equation (9). If $\overline{DP}_n$ is time-dependent, calculate β according to equation (10). (3) Calculate τ_{poly} and convert all time values from t to γ. (4) Determine the constant λ from a simultaneous solution of equation (23) for all values of γ. (5) Finally, calculate z from equations (19)-(21).

5. *Experimental Procedures*

5.1. *Acetobacter xylinum Cultures*

5.1.1. *Culture Conditions*

Acetobacter xylinum, American Type Culture Collection srain number 12733, was grown in a sterile liquid medium consisting of (in g/liter): glucose, 20; Bactopeptone, 5; yeast extract, 5; disodium phosphate, 2.7; and citric acid, 1.15. The pH of the medium was 6.0 throughout the duration of the experiments. Rapid growth rates were maintained by pipetting 2 ml of a 24-hr-old culture into a 250-ml Erlenmeyer flask containing 50 ml of medium at 30° C. All inoculation cultures were kept under a constant 125 cycle/min agitation to insure sufficient homogeneity in the aliquots removed for inoculation.

5.1.2. *Growth and Harvest of Bacterial Cellulose*

Six 4000-ml Erlenmeyer flasks were each filled with 450 ml of the liquid medium described above for a surface-to-volume ratio of 0.7 cm^{-1}. Every flask was simultaneously inoculated with 2 ml of inoculum from a 24-hr inoculation culture. Cultures were grown at 30° C with no agitation. Each flask was harvested after at least 72 hr of growth by draining the cellulose pellicles on a coarse sintered-glass filter, storing them overnight in a deep-freeze, then freeze-drying them.

5.1.3. *Purification of Bacterial Cellulose*

The freeze-dried pellicle was added to 100 ml 2% nitrogen-saturated NaOH and autoclaved for 60 min at 121° C. The alkali containing the cellular debris was then drained and replaced with 1% acetic acid and then placed on a shaker for 4 hr. After this period, fresh acid was added and the shaking continued overnight. The acid was finally replaced with distilled water that was also replaced every 24 hr for 3 days. Finally, after freeze-drying, the purified cellulose was measured for its yield on an analytical balance.

5.2. *Determination of the Molecular-Weight Averages*

5.2.1. *Preparation of Cellulose Tricarbanilates*

Due to the difficulty involved in directly carbanilating dried bacterial cellulose, the methylol derivative was first prepared in dimethylsulfoxide (DMSO) (Johnson *et al.*, 1976). An oil bath, maintained at 120°C by a combination hot plate-stirrer, was used to prepare the methylol cellulose derivatives. The oil bath consisted of a Pyrex crystallizing dish, 120 × 60 mm, filled with silicone fluid. A 2-inch Teflon-covered magnetic stirring bar was utilized to minimize the formation of temperature gradients, and the temperature was controlled to within ±0.5°C with the aid of an immersion probe connected to the hot plate.

The cellulose-DMSO solutions were prepared by initially mixing the bacterial cellulose with 40 ml DMSO in 100-ml beakers that were suspended in the silicone bath directly above the stirring bar. The size of these beakers was selected to allow the level of the DMSO to be sufficiently below the level of the silicone fluid to prevent the occurrence of polyoxymethylene films that would form if the surface of the solutions were allowed to cool excessively. A 1-inch Teflon-covered stirring bar was added to the mixture and operated by induction from the larger stirring bar in the silicone fluid. The temperature of the mixture was monitored by a suspended thermometer. Once the mixture had attained the same temperature as the bath, the thermometer was removed and 2.4 g paraformaldehyde (Tridom Chemical Co.) was added directly to the mixture. This addition produced the immediate result of formaldehyde gas evolution accompanied by the dissolution of the cellulose to produce a clear solution. After complete dissolution, the methylol cellulose solution was heated until no further gas evolution could be observed. The solution was then transferred to a dropping funnel.

The bacterial cellulose was regenerated by slowly introducing the methylol cellulose solution from the dropping funnel into 200 ml of a 5% sodium sulfite solution (10 g Na_2SO_3 + 200 ml H_2O) (Schroeder *et al.*, 1982). The aqueous regeneration solution was kept at 80° C and constant stirring to facilitate rapid formation of the soluble formaldehyde–sulfite addition complex (Walker, 1975). After 20 min, the cellulose mixture was centrifuged, decanted, and washed with five 100-ml aliquots of distilled H_2O. Freeze–drying subsequently produced a highly divided foamlike material. A cotton-linter cellulose (Hercules type N-30, $\overline{DP}_n$ = 816) that could be derivatized without this procedure demonstrated a $\overline{DP}_n$ = 797 after treatment, indicating very little degradation.

After freeze-drying, 4.6–62.4 mg bacterial cellulose was placed into a 100-ml narrow-necked, screw-top derivatization bottle and vacuum-desiccated overnight. Into this bottle, 100 ml pyridine and 7.2 ml phenylisocyanate was then pipetted, after which the bottle was sealed with a Teflon-lined cap. The bottles were then placed in a reactor vessel that was kept slowly revolving at 80° C for 48 hr in an oil bath (Schroeder and Haigh, 1979). After completion of the reaction, 3.3 ml methanol was added to each bottle to destroy any excess phenylisocyanate.

The derivative was precipitated by mixing the pyridine solution with 130 ml dioxane and slowly dripping this new mixture into a rapidly stirred solution of 800 ml methanol and 5 ml glacial acetic acid in a 2000-ml beaker. After all the pyridine–dioxane solution had been added, the tricarbanilate was allowed to settle overnight. The methanol solution was then siphoned off, and the precipitate was collected in a centrifuge bottle. The precipitate was washed in succession with mixtures of 200 ml methanol and 1.25 ml glacial acetic acid, 200 ml distilled H_2O and 1.25 ml glacial acetic acid, and finally washed with 200 ml distilled H_2O. The precipitate was then freeze-dried.

5.2.2. GPC Analysis

To prepare the samples for GPC, approximately 9 mg bacterial cellulose tricarbanilate was dissolved in 5 ml stabilized tetrahydrofuran (THF). To this solution, 0.03 ml of a 2.5% THF solution of methyl-*N*-phenylcarbamate was added as a low-molecular-weight,total-permeation internal standard. Styragel columns (Waters Associates) having permeability ranges of 10^2, 10^3, 10^4, 10^5, and 10^6 nm were used to obtain the chromatograms for the bacterial-cellulose tricarbanilates. The elution solvent was purified THF (Perrin *et al.*, 1966), the elution rate 2 ml/min, and the spectrophotometric detector was operated at 235 nm. A dispersion-compensated universal calibration, utilizing a linear polystyrene calibration curve and cellulose tricarbanilate standards, was used to calculate the number-average and weight-average degrees of polymerization (Ring, 1980; Ring *et al.*, 1982).

The chromatograms were digitalized at 2-ml intervals and analyzed by numerical integration techniques for the molecular-weight averages. McCrackin's computer program GPC was utilized for this analysis (McCrackin, 1977). Base lines were justifiably assumed to be linear across the first and last points entered in the program and were automatically subtracted from the data set.

6. *Results and Discussion*

The moments of the number distribution for a Poisson polymerization function, as defined by equation (21), have definite physical interpretations. The first moment $u_1(\lambda,\gamma)$ is proportional to the predicted yield $w(\lambda,\gamma)$ as demonstrated by equation (18), and the zeroth moment $u_0(\lambda,\gamma)$ is proportional to the predicted number of polymers $w(\lambda,\gamma)/\overline{DP}_n(\lambda,\gamma)$, from equations (19) and (20). The second moment $u_2(\lambda,\gamma)$, which is equal to the product of $w(\lambda,\gamma)$ and $\overline{DP}_w(\lambda,\gamma)$ from equations (19) and (21), corresponds to the manner in which the polymers are spread about the average value and can thus be regarded as a measure of the width of the distribution.

The actual experimental values of $w_{ic}(\gamma)$, $\overline{DP}_n(\gamma)$, and $\overline{DP}_w(\gamma)$, measured at different synthesis times γ, can also be used to define experimental moments $v_j(\gamma)$ analogous to the Poisson distribution moments $u_j(\lambda,\gamma)$, where

$$v_0(\gamma) = w_{ic}(\gamma)/\overline{DP}_n(\gamma) \tag{24}$$

$$v_1(\gamma) = w_{ic}(\gamma) \tag{25}$$

and

$$v_2(\gamma) = w_{ic}(\gamma)\,\overline{DP}_w(\gamma) \tag{26}$$

These experimental moments would then be proportional to $u_j(\lambda,\gamma)$ if the Poisson-predicted values of $w(\lambda,\gamma)$, $\overline{DP}_n(\lambda,\gamma)$, and $\overline{DP}_w(\lambda,\gamma)$ were equal to the experimental values of $w_k(\gamma)$. $\overline{DP}_n(\gamma)$, and $\overline{DP}_w(\gamma)$. Thus, by substituting equations (24)-(26) into equations (19)-(21)

$$v_0(\gamma) = x\ u_0(\lambda,\gamma) \tag{27}$$

$$v_1(\gamma) = x\ z\ u_1(\lambda,\gamma) \tag{28}$$

and

$$v_2(\gamma) = x\ z^2\ u_2(\lambda,\gamma) \tag{29}$$

What must now be demonstrated is whether or not unique values of λ and z can be obtained to satisfy equations (27)-(29).

The results of four experiments determining the yield $w_k(t)$ and the degrees of polymerization [$\overline{DP}_n(t)$ and $\overline{DP}_w(t)$] from six simultaneously inoculated cultures are presented in Table I. In agreement with previous results (Marx-

TABLE I
Experimental Data

Expt.-Sample	t (hr)	w_i (mg)	$\overline{DP}_n$	$\overline{DP}_w$	$\overline{DP}_w/\overline{DP}_n$
7-1	72.57	6.1	1310	2510	1.92
7-2	78.35	7.7	1270	2730	2.14
7-3	84.55	9.4	1190	3040	2.56
7-4	90.82	14.5	1690	5130	3.04
7-5	96.60	19.0	1470	4490	3.05
7-6	102.55	22.3	1290	3320	2.57
8-1	72.07	4.6	905	2240	2.48
8-2	77.73	8.5	976	2770	2.84
8-3	85.08	10.6	1170	2850	2.44
8-4	90.37	11.3	1250	3740	3.01
8-5	95.37	12.4	1200	3590	3.07
8-6	101.65	12.5	1080	2480	2.30
9-1	71.20	8.6	1360	3850	2.84
9-2	74.03	10.9	1490	4410	2.95
9-3	75.92	13.1	1650	5510	3.34
9-4	77.95	12.8	1650	5647	3.42
9-5	80.73	19.8	1720	5770	3.36
9-6	82.63	34.1	2210	7290	3.30
10-1	73.35	8.4	1100	3440	3.13
10-2	77.73	15.5	1670	5360	3.21
10-3	93.03	36.4	1210	4520	3.75
10-4	95.10	30.7	1670	6000	3.60
10-5	97.23	45.1	1700	6010	3.53
10-6	101.13	62.4	1350	5470	4.04

Figini and Pion, 1974; A. M. Brown, 1962), the yields for Experiments 7, 9, and 10 demonstrated first-order kinetics. By utilizing equation (9) as a linear regression model, one can calculate the first-order rate constant α (Table II) for each experiment, and the three sets of data can be plotted colinearly as functions of $\ln(w_{ic}/w_0)$ vs. t/τ_{cell} as in Fig. 1. (Experiment 8 did not demonstrate linearity

TABLE II
Determination of First-Order Rate Constants α and β

Expt.	Slope (sec^{-1})	Intercept	Correlation coefficient (r)
7	$\alpha = 0.046$	$w_0 = 217.0\ \mu g$	0.993
	$\beta = 0.042$	$n_0 = 1.33\ \mu mol$	0.988
9	$\alpha = 0.110$	$w_0 = 2.77\ \mu g$	0.974
	$\beta = 0.076$	$n_0 = .167\ \mu mol$	0.981
10	$\alpha = 0.067$	$w_0 = 73.3\ \mu g$	0.990
	$\beta = 0.063$	$n_0 = .467\ \mu mol$	0.981

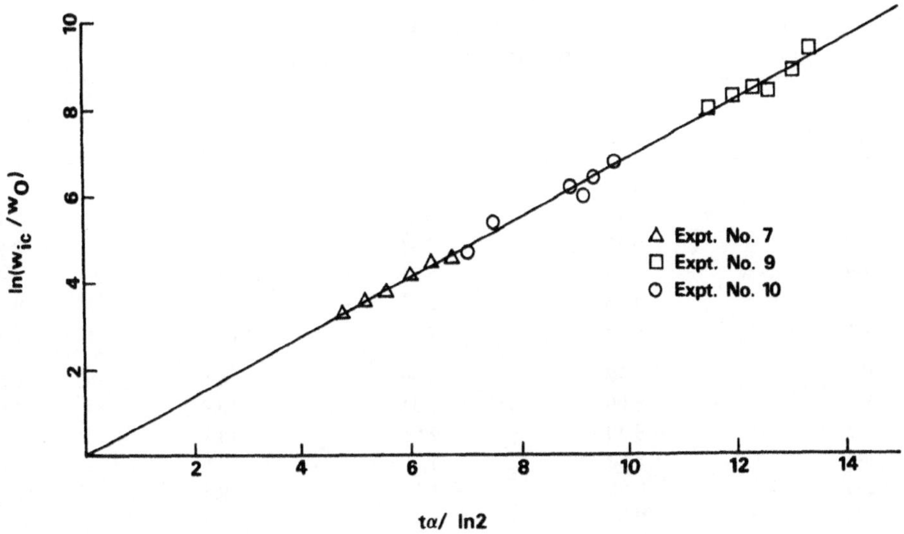

FIGURE 1. Yields of isolated bacterial cellulose from cultures harvested after 72 hr of synthesis.

because it became contaminated with a competing organism during inoculation and apparently entered its death phase prematurely. The first six experiments were preliminary in nature and were utilized to establish inoculation and cellulose-derivatization techniques.)

The weight-average DPs increased with time, as had been previously reported, until about 95 hr, where a decrease in these values can be seen to occur. However, such decreases are not attributed to degradation since the continued introduction of newer, low-molecular-weight material will gradually offset the effects of the higher-molecular-weight material and reduce the weight-average DP (Marx-Figini and Pion, 1974).

For Experiments 7, 9, and 10, the number of polymers $[n_i(t) = w_{ic}(t)/\overline{DP}_n(t)]$ also demonstrated first-order kinetics. The first-order rate constants, β, for these experiments were calculated using equation (10) and are presented together with the respective rate constants, α, in Table II. What is notable is that the values of α for each experiment are greater than the values for β, justifying the precautions taken in equation (8). The significance of α being greater than β is that the effective number of polymers per cell is not constant, but rather decreases with time, as can be seen by

$$n/N = (n_0/N_0)e^{(\beta - \alpha)t} \tag{30}$$

where n is the number of polymers and N is the number of cells.

In Table III, the experimental moments $v_j(\gamma)$ calculated from the experimental data are presented for Experiments 7, 9, and 10 along with the values of γ calculated by multiplying t with $\beta/\ln 2$. Table IV documents that for each culture, all three moments demonstrate at least approximate first-order kinetics. This was expected for $v_0(\gamma)$ and $v_1(\gamma)$, but the first-order behavior for $v_2(\gamma)$ further emphasizes the inherent appropriateness of monitoring the extent of polymerization through the second experimental moment rather than $\overline{DP}_w$,

TABLE III
Moments from Experimental Data

Expt.-Sample	γ	$v_0(\gamma)$	$v_1(\gamma) \times 10^{-3}$	$v_2(\gamma) \times 10^{-7}$
7-1	4.42	4.66	6.1	1.35
7-2	4.77	6.04	7.7	2.10
7-3	5.15	7.91	9.4	2.86
7-4	5.53	8.59	14.5	7.44
7-5	5.88	12.9	19.0	8.53
7-6	6.24	17.3	22.3	7.40
9-1	7.78	6.34	8.6	3.31
9-2	8.09	7.31	10.9	4.80
9-3	8.30	7.94	13.1	7.22
9-4	8.52	7.76	12.8	7.23
9-5	8.83	11.5	19.8	11.43
9-6	9.03	15.4	34.1	24.85
10-1	6.62	7.66	8.4	2.89
10-2	7.02	9.29	15.5	8.31
10-3	8.40	30.2	36.4	16.44
10-4	8.59	18.4	30.7	18.42
10-5	8.78	26.5	45.1	27.12
10-6	9.13	46.1	62.4	34.12

since the conflicting effects of increasing polymer mass are avoided. Expressed in time units of γ, the slopes for $v_0(\gamma)$ vs. γ are approximately equal to the constant $\delta = \ln 2$. The slopes for the $v_1(\gamma)$ and $v_2(\gamma)$ relationships, however, are not equal from one experiment to the next, but for each experiment, the magnitude of the slopes for $v_0(\gamma)$, $v_1(\gamma)$, and $v_2(\gamma)$ increases in that respective order. These results indicate that based on the rate of polymer introduction [$v_0(\gamma)$ vs. γ], the rate of mass accumulation [$v_1(\gamma)$ vs. γ] and the relative polymer elongation rate [$v_2(\gamma)$ vs. γ] can vary from condition to condition and are apparently not identically

TABLE IV
First-Order Kinetic Relationships for $v_0(\gamma)$, $v_1(\gamma)$, and $v_2(\gamma)$

Expt.	Moment	Slope	Intercept	Correlation coefficient (r)
7	$v_0(\gamma)$	0.695	−1.54	0.988
	$v_1(\gamma)$	0.752	5.37	0.933
	$v_2(\gamma)$	1.280	10.80	0.973
9	$v_0(\gamma)$	0.690	−3.59	0.980
	$v_1(\gamma)$	1.020	1.06	0.972
	$v_2(\gamma)$	1.470	5.79	0.974
10	$v_0(\gamma)$	0.693	−2.58	0.981
	$v_1(\gamma)$	0.737	4.30	0.990
	$v_2(\gamma)$	0.870	11.70	0.965

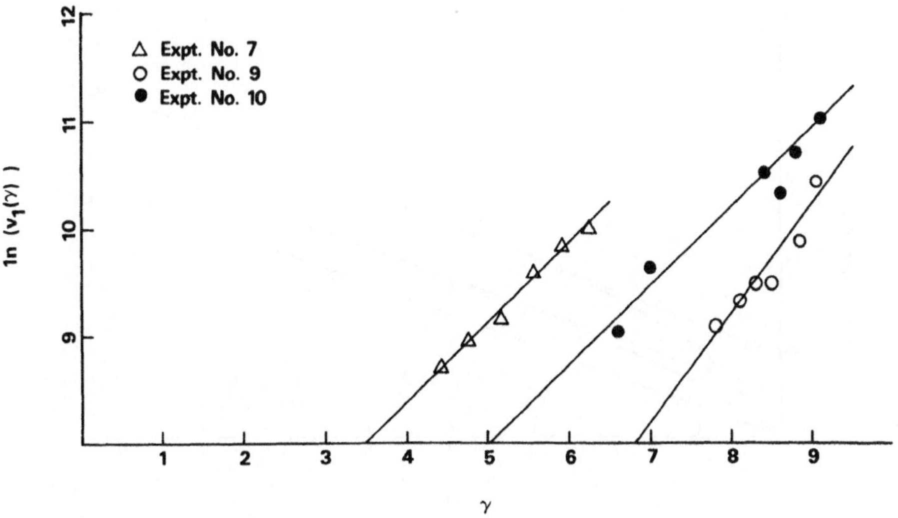

FIGURE 2. Experimental first moment as a function of polymer-number generation time.

synchronized with the cellulose replication rate. Figure 2 depicts the moments $v_1(\gamma)$ as a function of γ and Fig. 3 the $v_2(\gamma)$-vs.-γ relationship for Experiments 7, 9, and 10.

It is now obvious from the semilogarithmic forms of equations (27)–(29) that the slopes for $u_j(\lambda,\gamma)$ would be equal to the slopes for $v_j(\gamma)$ if the experimental results were, in fact, a consequence of a random Poisson polymerization process. Furthermore, the intercepts for $u_1(\lambda,\gamma)$ and $u_1(\lambda,\gamma)$ would differ from the intercepts of $v_1(\gamma)$ and $v_2(\gamma)$ by the values of $\ln z$ and $\ln z^2$, respectively. Thus, the desired value for λ has to satisfy a severe slope-intercept criterion. Crucial to this whole question is whether or not the functions of $u_j(\lambda,\gamma)$ are linear on a semilogarithmic basis. Figure 4 demonstrates for $\lambda = 1$ that $u_0(\lambda,\gamma)$ has a

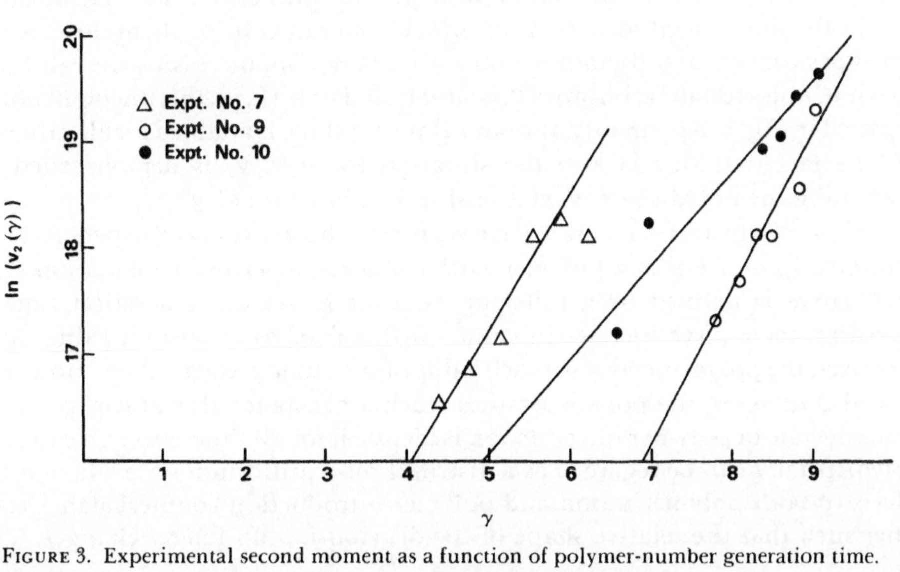

FIGURE 3. Experimental second moment as a function of polymer-number generation time.

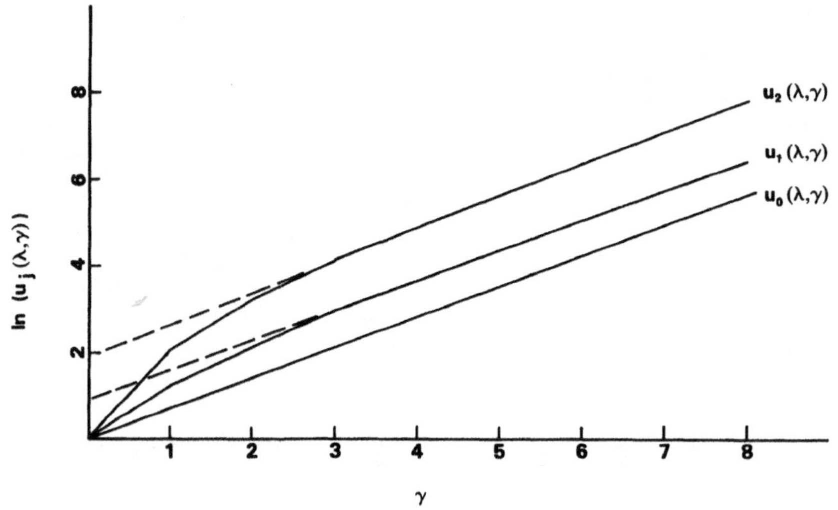

FIGURE 4. Theoretical Poisson moments $u_0(\lambda,\gamma)$, $u_1(\lambda,\gamma)$, and $u_2(\lambda,\gamma)$, where $\lambda = 1$.

constant slope equal to $\delta = \ln 2$, as expected, but that $u_1(\lambda,\gamma)$ and $u_2(\lambda,\gamma)$ have only approximate linearity at values of γ greater than 3 generations. In this linear range, however, the slopes of the functions $u_1(\lambda,\gamma)$ and $u_2(\lambda,\gamma)$ are practically equal to the slope for $u_0(\lambda,\gamma)$. But because the experimental slopes for $v_j(\gamma)$ had significantly different values, a value of λ equal to unity will not be applicable to these values. Therefore, the possibility of whether or not a different value for λ could produce different values for the slopes of $u_1(\lambda,\gamma)$ and $u_2(\lambda,\gamma)$ has to be investigated.

The functions $u_1(\lambda,\gamma)$ and $u_2(\lambda,\gamma)$ vary directly in value with λ. Thus, the value of z will vary inversely with λ, since larger $u_1(\lambda,\gamma)$ and $u_2(\lambda,\gamma)$ values require a smaller correction to equal $v_1(\gamma)$ and $v_2(\gamma)$ through equations (28) and (29), respectively. Choosing values for λ less than unity will not significantly change the slope for either $u_1(\lambda,\gamma)$ or $u_2(\lambda,\gamma)$, but rather bring them closer to the limiting situation of superimposition with $u_0(\lambda,\gamma)$. But increasing the value of λ does not, unfortunately, improve this situation. For $u_1(\lambda,\gamma)$, this phenomenon is depicted in Fig. 5. Apparently, the same limiting slope is attained for all values of λ. The fact that this is also the situation for $u_2(\lambda,\gamma)$ is demonstrated by examining the polydispersity relationship as a function of γ.

Figure 6 illustrates a family of curves for the theoretical polydispersity-vs.-γ relationship of a Poisson polymerization process, described by equation (23). Each curve is defined by a different value of λ. As the generation time γ increases, each curve increses in value until is reaches a constant value. As λ increases, the polydispersities at each value of γ assume greater values. However, as λ also increses, the polydispersities reach a constant value at earlier times. Since the rate of polymer introduction is identical for all these curves, a constant polydispersity can be regarded as a distribution-equilibrium state wherein the effects of both polymerization and polymer introduction counterbalance each other such that the relative shape of the distribution no longer changes. The

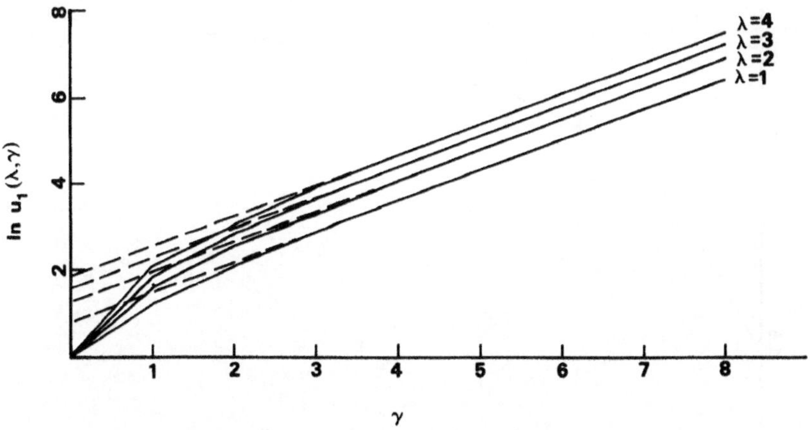

FIGURE 5. Theoretical Poisson moment $u_1(\lambda,\gamma)$ as a function of generation time and λ.

experimental relationship between the equilibrium polydispersity π_e and its respective value of λ was found to be a hyperbola

$$(\pi_e - 2.000)\,(\lambda + \ln 1.994) = -\ln 1.999 \approx -\delta \tag{31}$$

as demonstrated by Fig. 7. The maximum value of π_e, which corresponds to an infinitely large value for λ and an infinitely small value for the event size z, is 2. Therefore, monomeric stochastic polymerization of isolatable cellulose cannot exceed this polydispersity, while oligomeric polymerizations would have reduced values for π_e, dependent on their respectively greater values of z, and lesser values for λ. A value of $\lambda = 0$ would be indicative of no polymerization, only polymer introduction, and would result in $\pi_e \approx 1$.

As is evidenced in Table I, the polydispersities for Experiments 7, 9, and 10

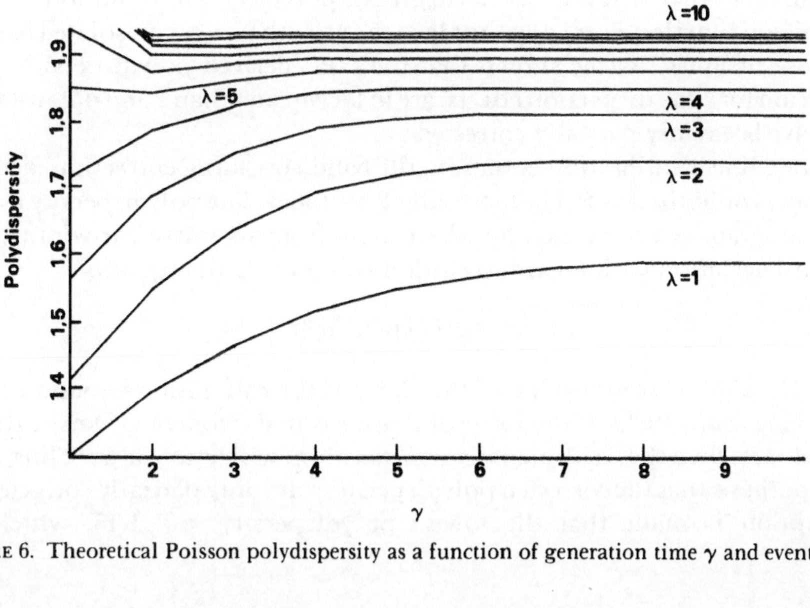

FIGURE 6. Theoretical Poisson polydispersity as a function of generation time γ and event rate λ.

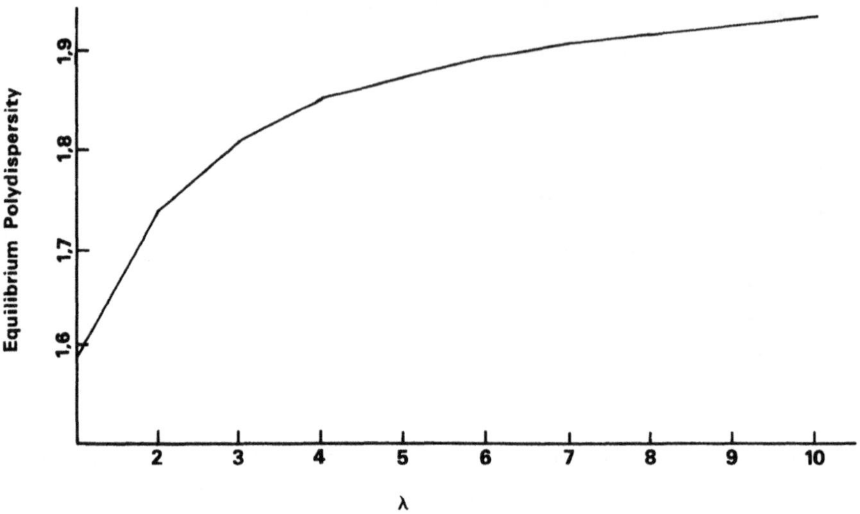

FIGURE 7. Theoretical Poisson equilibrium polydispersity vs. event rate λ.

are, with the exception of only one value, greater than 2. Because no value of λ exists that will produce this result for a Poisson polymerization mechanism, it must be assumed that an entirely different mechanism is operating for the production of isolatable bacterial cellulose. The rate of polymer introduction, defined by $u_0(\lambda,\gamma)$-vs.-γ, is constant regardless of the value of λ, since polymerization is an independent phenomenon. Hence, the polymerization-sensitive moments, $u_1(\lambda,\gamma)$ and $u_2(\lambda,\gamma)$, do not properly predict the observed polymerization distributions. The actual empirical moment for $v_2(\gamma)$ appears to have a value exceeding what is predicted by $u_2(\lambda,\gamma)$, thus accounting for the larger polydispersities. Therefore, it is questionable that isolated cellulose is the result of a polymerization process whereby each cellulose molecule experiences a random event that will increase its degree of polymerization. But to prove that the Poisson function is not applicable to the growth process of isolated bacterial cellulose, it must first be shown that the GPC-derived polydispersities, after correction for peak dispersion effects, are in fact greater than 2 and not just values that have been only partially corrected.

For a linear calibration equation, the band-spreading correction factor is a constant, symbolized as P (Hamielec and Ray, 1969). The polydispersity π of any chromatogram is determined approximately from its statistical width σ_v (the standard deviation with regard to elution volume) via the equation

$$\pi = P^2 \exp(D_2{}^2\sigma_v{}^2) \tag{32}$$

where D_2 is the absolute value of the slope of the calibration equation (Ring, 1980; Ring *et al.*, 1982). If the value of P is measured erroneously larger than its true value P_t, then the resulting polydispersities, π, will be too large. Thus, to test the hypothesis that the corrected polydispersities are only partially corrected, the assumption is made that the lowest polydispersity, $\pi = 1.15$, which was obtained from the narrowest cellulose calibration standard utilized, was actually unity or perfectly monodisperse. The correction factor that was necessary to produce this result was

$$P_t^2 = 0.87\ P^2 \tag{33}$$

However, the highest polydispersity that will be reduced by this factor to $\pi = 2$ or below is 2.3. Since the majority of polydispersities for the bacterial celluloses listed in Table I are above this value, it must be concluded that true polydispersities above 2 actually exist for the experimental samples. Therefore, a Poisson polymerization mechanism is not applicable for isolatable bacterial cellulose.

The significance of specifying isolated bacterial cellulose is that only experimentally isolated material can be analyzed for its molecular-weight distribution. However, it may be hypothesized that although isolated cellulose does not demonstrate Poisson polymerization kinetics, the unaffected, total, product may, because the isolation technique utilized may expose the total polymerization product to several possible degradation reactions that are functions of alkali (NaOH) concentration, oxygen concentration, temperature, and time. In alkali, cellulose will swell and become uniformly more accessible to degradation that is a direct function of all four of the aforementioned variables. Thus, to minimize severe degradation, it is necessary to avoid oxygen and excessive temperatures while limiting the exposure of cellulose to the lowest practical concentration of alkali for the shortest period. In the presence of oxygen, chain cleavage will increase the total number of molecules while reducing their average length. As a result, the polydispersity of cellulose undergoing oxidation rapidly decreases (Dyer and Phifer, 1971). In the absence of oxygen, rapid degradation will occur at temperatures above 140°C (Rydholm, 1965; Best, 1968), but between this high temperature and 50°C, degradation is limited to the peeling reaction (Whistler and BeMiller, 1958). Since peeling is expected to remove only as much as 68 glucose units per chain before the reaction is stopped (Lai and Sarkanen, 1967), the polydispersity will be reduced only at a rate equal to the approximate decrease in $\overline{DP}_w$, since DP_n tends to remain constant due to the very small change in the number of molecules (Dyer and Phifer, 1971). Finally, the removal of low-molecular-weight material through the various sample-handling procedures results in an increased $\overline{DP}_n$, which also results in a decrease of the polydispersity. Therefore, the only expected outcome of an alkaline isolation procedure is a reduction in the polydispersity, making it very unlikely that the relatively large polydispersities observed were actually increased from lower values due to the sample preparation procedures. Thus, it may be concluded that for both the total polymerization product and the isolatable cellulose, a Poisson polymerization function does not apply.

Further evidence for this conclusion can be obtained from an examination of the GPC chromatograms of the bacterial cellulose cultures. Figure 8 depicts the six consecutive chromatograms for Experiment 7. The observable growth pattern is that the greatest accumulation of mass occurs predominantly at the

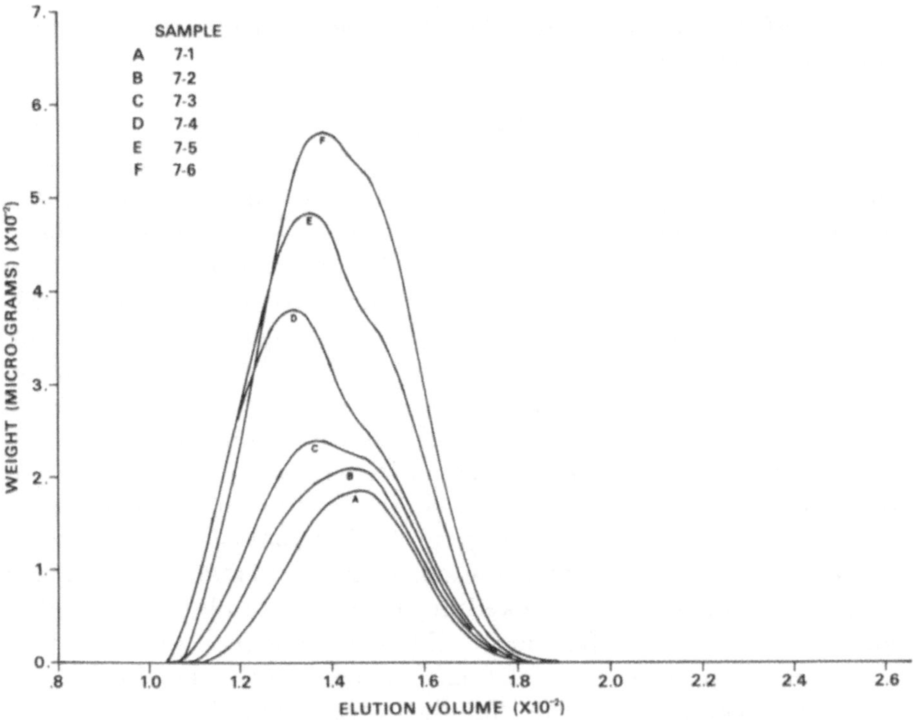

FIGURE 8. Yield-normalized GPC chromatograms for Expt. 7 (weight-biased).

lower elution-volume values. Because elution volume in GPC is inversely related to the molecular weight, this noticeable mass accumulation corresponds to molecules with high molecular weights. When these same chromatograms are converted from a mass basis for their ordinate to a molar basis, as in Fig. 9, it becomes apparent that the greatest accumulation is due to a relatively few high-molecular-weight molecules. Therefore, the basic assumption for a Poisson mechanism, that all polymers have equal probabilities for further polymerization, was clearly demonstrated by both Figs. 8 and 9 to be not valid for the actual production of isolatable bacterial cellulose.

In regard to soluble polymeric precursors proposed by Colvin and Leppard (1977), which have not yet been isolated (Colvin *et al.*, 1979), the experimental conclusions discussed above are not sufficient to disprove their existence. It only suggests that they will bond preferentially to established chains in a fashion not described by a Poisson distribution. However, the second important aspect of their precursor model is that each soluble precursor is capable of initiating a new cellulose chain. Thus, because an enzyme or cell is continually producing precursors throughout the synthesis duration, both the number of polymers per enzyme and the number per cell increase when a precursor initiates a new chain. Under these hypothetical conditions, equation (30) would then demonstrate a positive exponent, meaning that β was greater than α. But since the opposite was found experimentally, that α was greater than β, it must be further concluded that a polymeric precursor is not capable of randomly introducing a new cellulose chain. Therefore, if a polymeric precursor exists, its incorporation into cellulose chains or microfibrils cannot be referred to as a simple polymeric-association process, but rather as a process wherein much greater control of the destinies of the precursors is exercised (see Chapter 14).

Although the experimental results are in contradiction to the proposed mechanism of Colvin and Leppard (1977), no such contradiction exists for the proposed mechanism of R. M. Brown, Jr., *et al.* (1976) and Zaar (1979). The high relative rates of polymer elongation are consistent with the concept of an enzyme continually connected to its cellulose chain, because the entire production of each enzyme would then serve to elongate only one single cellulose chain, rather than being distributed over several chains. Furthermore, the fact that the rate constant α was determined to be greater than β is compatible with their observation that a microfibrillar ribbon does not reach its maximum width until after cellular replication, producing a time delay in the introduction of new isolatable cellulose molecules that would result in an effective decrease in their numbers per cell with time. From Table II, it is apparent that this time delay may be relatively short, as demonstrated by Experiments 7 and 10, in which α is approximately equal to β, or relatively long, as demonstrated by Experiment 9, in which α is much greater than β.

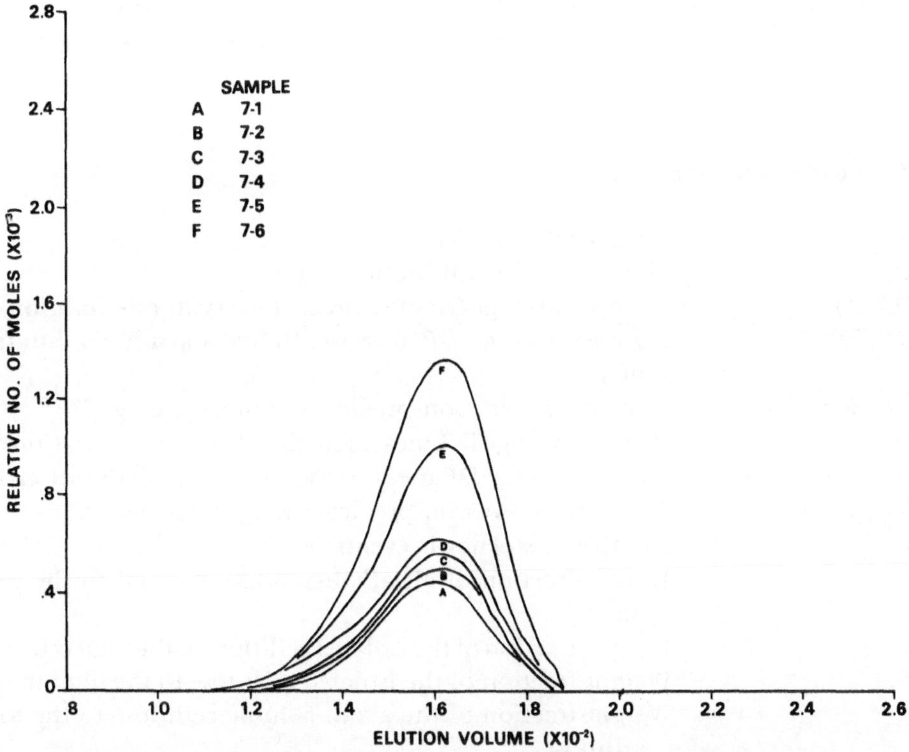

FIGURE 9. GPC chromatograms for Expt. 7 with ordinate converted to mole number.

7. Conclusions

This study has demonstrated that the production of bacterial cellulose exhibits approximate first-order kinetics for the increase in the number of polymers, in the mass of the polymers, and in the relative length of the polymers measured by the second moment of the number distribution, for several generations. It has also demonstrated that the number of polymers per cell effectively decreases for the entire culture during synthesis.

Contrary to what has previously been proposed, however, these results are not compatible with a Poisson polymerization mechanism (Marx-Figini and Pion, 1974). The experimental polydispersities, measured from the GPC chromatograms of the bacterial-cellulose samples, exhibit values significantly above 2, while the predicted Poisson polydispersities, on the other hand, will never exceed this value. The experimental polydispersities are much greater than the predicted theoretical values because the second distribution moment is much larger than what would be theoretically predicted.

Furthermore, because the number of polymers per cell decreases rather than increases, the results are in contradiction to the mechanism proposed by Colvin and Leppard (1977). In this mechanism, soluble polymeric precursors randomly associate together either to form new cellulose molecules or to elongate previously established molecules. The results, however, are not in contradiction to the mechanism proposed by R. M. Brown, Jr., *et al.* (1976) and Zaar (1979) that each cellulose chain remains attached to its synthesizing enzyme. Apparently, the bacterial cell rigorously controls both the introduction of new polymers and their subsequent polymerization.

8. Nomenclature

A	Avogadro's number
D_2	Slope of a GPC calibration curve
$\overline{DP}_n(t)$	Number-average *DP* measured after a synthesis duration of *t*
$\overline{DP}_n(\gamma)$	Number-average *DP* measured after a synthesis duration of γ
$\overline{DP}_n(\lambda,\gamma)$	Theoretical, Poisson-predicted number-average *DP*
$\overline{\mathrm{DP}}_{\mathrm{w}}(t)$	Weight-average *DP* measured after a synthesis duration of *t*
$\overline{DP}_w(\gamma)$	Weight-average *DP* measured after a synthesis duration of γ
$\overline{DP}_w(\lambda,\gamma)$	Theoretical, Poisson-predicted weight-average *DP*
k	Number of stochastic events
k_{as}	Weight-fraction of all alkali-soluble material to the total mass
k_{i}	Weight-fraction of the isolated cellulose to the total cellulose
k_{ic}	Weight-fraction of the isolated cellulose to the total mass
k_{s}	Weight-fraction of the alkali-soluble cellulose to the total cellulose

K	Ratio of the number of polymers to the number of cells
K'	Empirical coefficient for the exponential increase in the experimentally determined number-average degree of polymerization
M	Total mass of a bacterial culture after a duration of synthesis
M_k	Molecular weight of a cellulose polymer with k number of stochastic polymerization events
M_0	Initial mass of a bacterial culture on inoculation at time $t = 0$
$\overline{M}_n$	Number-average molecular weight
n	Total number of cellulose polymers
$n_i(t)$	Number of cellulose polymers isolated after an alkali extraction and a synthesis duration of t
$n_i(\gamma)$	Number of cellulose polymers isolated after an alkali extraction and a synthesis duration of γ
n_0	Initial number of polymers inoculated into a new medium
$n(k,\lambda,\gamma)$	Total number of cellulose polymers predicted by a Poisson distribution, a number distribution
N	Total number of bacterial cells
N_0	Initial number of bacterial cells inoculated into a new medium
$p[k,\lambda,(\gamma - \gamma')]$	Poisson polymerization function
P	A GPC dispersion correction factor determined experimentally
P_t	True GPC correction factor for dispersion
$u_j(\lambda,\gamma)$	Moments of a Poisson polymerization distribution
$u_j(\gamma)$	Moments of an experimentally determined polymerization distribution
w	Total mass of a bacterial culture after a duration of synthesis
$w_{ic}(t)$	Total mass of cellulose isolated after an alkaline extraction and a synthesis duration of t
$w_{ic}(\gamma)$	Total mass of cellulose isolated after an alkaline extraction and a synthesis duration of γ
$w(\lambda,\gamma)$	Total mass predicted by a Poisson polymerization function
w_0	Total initial mass of a bacterial culture after inoculation
x	Molecular weight of anhydroglucose, equal to 162 daltons
z	Degree of polymerization of the cellulose precursor
α	First-order rate constant for the increase in the total number of bacterial cells, the total mass of the bacterial culture, or the total mass of isolatable cellulose
α'	Empirical first-order rate constant for the increase in the number-average molecular weight
β	First-order rate constant for the increase in the total number of cellulose polymers after extraction by alkali
δ	ln 2

γ	A dimensionless time parameter determined from the ratio of real time, t, to the average time necessary for the total number of isolatable cellulose polymers to double
γ'	Time at which a particular group of polymers was introduced into the bulk of isolatable cellulose
λ	Poisson event rate
π	Polydispersity, $\overline{DP}_w/\overline{DP}_n$
π_e	Equilibrium polydispersity, the polydispersity achieved as a limit in a Poisson polymerization given a constant value of λ
σ_v	Empirical width or standard deviation of a GPC chromatogram
τ_{cell}	Average time necessary for the number of bacterial cells to double during the logarithmic growth phase of a bacterial culture
τ_{poly}	Average time necessary for the number of cellulose polymers to double during the logarithmic growth phase of a bacterial culture

References

Barsha, J., and Hibbert, H., 1934, Studies on reactions relating to carbohydrates and polysaccharides. XLVI. Structure of the cellulose synthesized by the action of *A. xylinum* on fructose and glucerol, *Can. J. Res.* **10**:170.

Ben-Hayyim, G., and Ohad, I., 1965, Synthesis of cellulose by *Acetobacter xylinum*. VIII. On the formation and orientation of bacterial cellulose fibrils in the presence of acidic polysaccharides, *J. Cell. Biol.* **25**:191.

Best, E. V., 1968, Kinetics of hot alkaline cleavage of the glycosidic bonds of methyl β-D-glucoside and methyl β-cellobioside, Ph.D. dissertation, The Institute of Paper Chemistry, Appleton, Wisconsin.

Brown, A. J., 1886, The chemical action of pure cultivations of *Bacterium aceti*, *Trans. Chem. Soc. (London)* **49**:172.

Brown, A. M., 1962, The mechanism of cellulose biosynthesis by *Acetobacter acetigenum*, *J. Polym. Sci.* **59**:155.

Brown, R. M., Jr., 1979, Biogenesis of natural polymer systems with special reference to cellulose assembly and deposition, Proceedings of the Third Philip Morris Science Symposium, Philip Morris, Inc., Richmond, Virginia.

Brown, R. M., Jr., Willison, J. H. M., and Richardson, C. L., 1976, Cellulose biosynthesis in *Acetobacter xylinum:* Visualization of the site of synthesis and direct measurement of the *in vivo* process, *Proc. Natl. Acad. Sci. U.S.A.* **73**:4465.

Brownlee, K. A., 1965, *Statistical Theory and Methodology in Science and Engineering*, John Wiley, New York.

Colvin, J. R., 1972, The structure and biosynthesis of cellulose, *CRC Crit. Rev. Macromol. Sci.* **1**:47.

Colvin, J. R., and Leppard, G. G., 1977, The biosynthesis of cellulose by *Acetobacter xylinum* and *Acetobacter acetigenus*, *Can. J. Microbiol.* **23**:701.

Colvin, J. R., Sowden, L. C., Daoust, V., and Perry, M., 1979, Additional properties of a soluble polymer of glucose from cultures of *Acetobacter xylinum*, *Can. J. Biochem.* **57**:1284.

Dyer, J., and Phifer, L. H., 1971, Application of GPC to studies of the viscose process. II. The effects of steeping and alkali-crumb aging, in: *Gel Permeation Chromatography* (K. H. Altgelt and L. Segal, eds.), pp. 481-491, Marcel Dekker, New York.

Elias, H., 1977, *Macromolecules*, Vol. 1, *Structure and Properties*, Plenum Press, New York.

Flory, P. J., 1940, Molecular size distribution in ethylene oxide polymers, *J. Am. Chem. Soc.* **62**:1561.

Glaser, L., 1958, The synthesis of cellulose in cell-free extracts of *Acetobacter xylinum*, *J. Biol. Chem.* **232**:627.

Hamielec, A. E., and Ray, W. H., 1969, An analytical solution to Tung's axial dispersion equation, *J. Appl. Polym. Sci.* **13**:1319.

Hassid, W. Z., 1969, Biosynthesis of oligosaccharides and polysaccharides in plants, *Science* **165**:137.

Hassid, W. Z., 1971, Biosynthesis of cellulose. A. Biosynthesis of cellulose and related plant cell-wall polysaccharides, in: *Cellulose and Cellulose Derivatives*, Part IV, Vol. 5 (N. M. Bikales and L. Segal, eds.), pp. 679-693, Wiley-Interscience, New York.

Hestrin, S., and Schramm, M., 1954, Synthesis of cellulose by *Acetobacter xylinum*. 2. Preparation of freeze-dried cells capable of polymerizing glucose to cellulose, *Biochem. J.* **58**:345.

Husemann, E., and Werner, R., 1963, Cellulosesynthese durch *Acetobacter xylinum*. I. Über Molekulargewicht und Molekulargewichtsverteilung von Bakteriencellulose in Abhangigkeit von der Synthesedauer, *Makromol. Chem.* **59**:43.

Johnson, D. C., Nicholson, M. D., and Haigh, F. C., 1976, Dimethyl sulfoxide/paraformaldehyde: A nondegrading solvent for cellulose, *Appl. Polym. Symp.* **38**:931.

Kjosbakken, J., and Colvin, J. R., 1975, New evidence for an intermediate polymer of glucose in cellulose biosynthesis by *Acetobacter xylinum*, *Can. J. Microbiol.* **21**:111.

Lai, Y.-Z., and Sarkanen, K. V., 1967, Kinetics of alkaline hydrolysis of glycosidic bonds in cotton cellulose, *Cell. Chem. Technol.* **1**:517.

Mark, H., 1954, Structures and properties of cellulose fibers. A. Interaction and arrangement of cellulose chains, in: *Cellulose and Cellulose Derivatives*, Part I, Vol. 5 (E. Ott, H. M. Spurlin, and M. W. Grafflin, eds.), pp. 217-230, Wiley-Interscience, New York.

Marx-Figini, M., and Pion, B. G., 1974, Kinetic investigations on biosynthesis of cellulose by *Acetobacter xylinum*, *Biochim. Biophys. Acta* **338**:382.

Marx-Figini, M., and Schulz, G. V., 1963, Neure Untersuchungen über Grosse und Grosseinverteilung der β-glukosidischen Ketten nativer Cellulosen, *Makromol: Chem.* **62**:49.

McCrackin, F. L., 1977, Calibration of gel permeation chromatography columns using polydisperse polymer standards, *J. Appl. Polym. Sci.* **21**:191.

Oginsky, E. L., and Umbreit, W. W., 1959, *Bacterial Physiology*, Freeman, San Francisco.

Pasteur, L., 1868, *Etudes sur le Vinaigre*, Paris.

Perrin, D. D., Armarego, W. L. F., and Perrin, D. R., 1966, *Purification of Laboratory Chemicals*, Pergamon Press, Oxford.

Ring, G. J. F., 1980, The molecular weight distributions of bacterial cellulose as a function of synthesis time, Ph.D. dissertation, The Institute of Paper Chemistry, Appleton, Wisconsin.

Ring, G. J. F., Stratton, R. A., and Schroeder, L. R., 1982, Dispersion compensated calibration equations: Theory and utilization for direct and universal calibration (in press).

Rydholm, S., 1965, *Pulping Processes*, John Wiley, New York.

Schroeder, L. R., and Haigh, F. C., 1979, Cellulose and wood pulp polysaccharides: Gel permeation chromatographic analysis, *Tappi* **62** No. (10):103.

Schroeder, L., Wabers, B., and Ring, G. J. F., 1982 (in prep.).

Sowden, L. C., and Colvin, J. R., 1978, Morphology, microstructure, and development of colonies of *A. xylinum*, *Can. J. Microbiol.* **24**:772.

Walker, J. F., 1975, *Formaldehyde*, Robert E. Krieger Publishing Co., Huntington, New York.

Ward, K., Jr., 1954, Occurrence of cellulose, in: *Cellulose and Cellulose Derivatives*, Part I, Vol. 5, (E. Ott, H. M. Spurling, and M. W. Grafflin, eds.), pp. 9-27, Wiley-Interscience, New York.

Whistler, R. L., and BeMiller, J. N., 1958, Alkaline degradation of polysaccharides, in: *Advances in Carbohydrate Chemistry*, Vol. 13 (M. L. Wolfrom and R. S. Tipson, eds.), pp. 289-329, Academic Press, New York.

Zaar, K., 1979, Visualization of pores (export sites) correlated with cellulose production in the envelope of the gram-negative bacterium *Acetobacter xylinum*, *J. Cell. Biol.* **80**:773.

16

Does β-Glucan Synthesis Need a Primer?

GORDON A. MACLACHLAN

1. Introduction

The title of this chapter is a polemical one intended to draw attention to the fact that primer requirements for the synthesis of β-glucans (e.g., cellulose, callose) have not received the attention that the subject merits. Studies on primer requirements for synthesis of other polymers, e.g., α-glucans, DNA, glycoproteins, have been extraordinarily informative in elucidating mechanisms of synthesis and regulation and in understanding their biological functions. Such studies with enzymatic systems that generate β-glucans could be expected to be similarly useful, with implications for questions in cell biology about the localization of these events and for questions in physiology about their relation to and control of growth and development.

The textbook definition of a primer as an essential acceptor molecule, which is elongated by glycosyl transfer during polysaccharide synthesis, was first proposed as central to the mechanism of phosphorylase action by the Coris for glycogen and by Hanes for starch biosynthesis. A similar mechanism appears to operate during synthesis of these products from sugar nucleotides by glucan synthases. In these instances, it has been demonstrated that synthesis occurs by transfer of single monomer units to the nonreducing ends of accessible chains of preformed α-glucan. As would be expected, the introduction of new nonreducing chain ends into the primer, either by partial hydrolysis with amylases or by addition of side chains with the appropriate branching enzyme, often results in an enhancement of the priming capacity of the system.

The definition of the term primer has been widened in recent years to

GORDON A. MACLACHLAN • Biology Department, McGill University, Montreal, Quebec, H3A 1B1 Canada.

include acceptors for the synthesis of other polymers, notably DNA (Kornberg, 1974) and glycoproteins (Kornfield and Kornfield, 1980). In such systems, the donors are not necessarily monomer units but may include partially assembled oligomers, which are transferred *en bloc* to acceptor nascent chains. Completion of the polymer may then depend on the presence of a ligase to join nascent chains together. Primers for polysaccharide synthesis may also include glycoproteins as initial acceptors for glycosyl transfer (Whelan, 1976). Elongation of the glycoprotein by repeated transfer to the nonreducing end of the glycone could result in the protein components remaining at the core of the completed polysaccharide. Such a mechanism has been proposed for both starch and glycogen biosynthesis (Tandecarz and Cardini, 1978).

The possibility of "unprimed" synthesis of α1,4-glucans has been repeatedly raised over the past ten years, but this remains in dispute (Boyer and Preiss, 1979). It is often difficult to be certain that an apparently purified enzyme preparation does not contain small but crucial amounts of tightly adsorbed primer polysaccharide (Hawker *et al.*, 1974).

A radically different glycoprotein-dependent mechanism has been proposed (Robyt, 1979) as a result of studies on the biosynthesis of certain bacterial α-dextrans and fructans by dextran and fructan sucrase. These enzymes themselves appear to contain covalently linked nascent glycones. They have at least two active sites, both capable of accepting hexose from the activated donor (sucrose) and then "inserting" one sugar unit between the protein catalyst and the reducing end of the other sugar. Very long polysaccharides may then be assembled by repeated transfer of the growing chain to the monoglycosylated site. This offers an explanation for apparently "unprimed" synthesis, i.e., instances wherein the addition of putative nascent acceptor oligomers to a synthetase system cannot be shown to enhance the rate of synthesis.

It is usually taken for granted that callose and cellulose biosynthesis requires a primer in the same sense as starch or glycogen biosynthesis. Yet the challenging suggestion by Marx-Figini (1976), that cellulose synthesis in secondary walls may be directed by a template, and proposals (Maclachlan, 1976) for an "insertion mechanism" of synthesis, have not received biochemical testing. This chapter surveys the evidence for primers, most particularly the results that have been interpreted as *indirect* evidence for the existence of required acceptors, and *direct* attempts to demonstrate a role for acceptors during *in vitro* synthesis. It is concluded that there is not sufficient decisive evidence as yet to justify acceptance of any one biochemical model for the mechanism of β-glucan synthesis.

2. *Results*

2.1. *Background Studies: Indirect Evidence for β-Glucan Primer Requirements*

A massive literature has accumulated in the past 25 years on the capacities of isolated plant-membrane preparations to synthesize water-insoluble products from glucose nucleotides. Such products would include insoluble β-glucans, but when these observations are compared to data for the synthesis and deposition of callose or cellulose by living cells and tissues, it is evident that the cell-free systems are relatively inefficient (Maclachlan *et al.*, 1979). Maximum rates of synthesis of insoluble products from UDP- or GDP-glucose *in vitro* are generally at least an order of magnitude less than the rates of synthesis of cellulose or callose *in vivo*. This generalization holds whether one calculate the relative rates on a membrane-protein, cell, or fresh-weight basis. Moreover, most of the membrane preparations will continue to generate insoluble products for only a few minutes, whereas most plant cells continue to deposit cellulose during their entire life span, and the capacity to generate callose appears to be retained by cells, even during senescence.

There are many possible reasons for the relatively low activity of β-glucan synthetase systems in isolated membrane preparations. One reason may be that synthetases are particularly sensitive to destruction by hydrolases, e.g., proteases (Raymond *et al.*, 1978). Soluble fractions of tissue homogenates rapidly bring about the inactivation of membrane-bound synthetases (Thomas *et al.*, 1969; Chao and Maclachlan, 1978) due to the presence of endogenous enzymatic inactivator(s). Thus, during tissue homogenization and membrane isolation, unless care is taken to perform the manipulations rapidly under conditions where hydrolases are inactive, membrane synthetases can be expected to be labile.

Another reason for inefficiency of isolated membrane-bound β-glucan synthetases may be that the key membranes that are most enriched in synthetase complexes are not completely recovered by current isolation procedures. It is probable that homogenization results in part of the plasma membrane remaining bound to cell-wall fragments and part vesiculating in such a way that receptor sites for glucose nucleotides are not fully accessible to added substrate. Indeed, fractions of homogenates containing cell-wall fragments often retain more β-glucan synthetase activity than the particulate membrane fraction (Maclachlan *et al.*, 1979). Washing to remove the membranes that adhere to cell walls, or more drastic homogenization techniques, e.g., sonication, merely lead to further net losses of activity.

Apart from lability of synthetase complexes and inadequate recovery techniques, efficient glycosyl transfer to β-glucans may depend on the integrity of association between a membrane complex and preformed polysaccharides. In normal turgid plant cells, the plasma membrane is pressed against the wall, with particularly close physical proximity at plasmodesmata. If this structure contains transmembrane complexes that catalyze the vectorial transfer of glucose from nucleotides in the cytoplasm to preexisting primers in the wall, physical disruption of the friable association between its elements could result in synthetase inactivation.

To be sure, plasmolyzed cells are still able to generate a new wall around protoplasts at a distance from the old walls, given time, and isolated protoplasts that are visibly free of wall fragments at the ultrastructural level are nevertheless able to form a new wall, which includes cellulose microfibrils (Herth and Meyer, 1977). Nevertheless, there is always a lag period before new wall material begins to be laid down by protoplasts, and the lag is often substantial, i.e., hours or even days. The lag period could represent recovery time during which synthetase complexes eventually become reassociated with a limiting amount of primer that is retained by the plasma membrane during isolation. Such membrane-associated primers may not be visible in the electron microscope, since they are not necessarily crystalline or microfibrillar.

Indirect evidence for a requirement for primers in β-glucan synthesis, and for a dependence on physical integrity for maximum efficiency, may be summarized as follows:

1. β-glucan synthetase activity in homogenates of whole cells is greatly reduced by the mere act of separating the slurry by centrifugation into wall, particulate, and supernatant fractions (Maclachlan *et al.*, 1979). Some of the original activity can be recovered by recombining the subfractions, suggesting that limited reconstruction of physical integrity can take place, i.e., spontaneous reassembly.

2. Slices of cells, e.g., cotton hairs (Franz and Meier, 1969; Delmer *et al.*, 1974), or tissue, e.g., pea stems (Shore and Maclachlan, 1975; Shore *et al.*, 1975), in which wall–plasma membrane associations remain at the inner surfaces of cut cells, will utilize UDP-glucose much more rapidly to form insoluble β-glucans than the corresponding preparations of membrane homogenates (see Table II) (see also Chapter 12). The synthesis occurs exclusively in cut cells, and not in adjacent living cells (Fig. 1) (see also Chapter 9), suggesting that UDP-glucose is utilized only on the inner surfaces of exposed plasma membrane. Radioautographs at the electron-microscopic level confirm that synthesis is confined to regions where the plasma membrane remains in contact with the wall (Mueller and Maclachlan, 1980). Although the tissue-slice system contains living cells, the fact that only the cut cells show activity means that it should be regarded as an *in vitro* system.

3. The molecular weight of insoluble β-glucans formed from [^{14}C]glucose *in vivo* is relatively high [degree of polymerization (DP) up to 10,000], particularly in older tissue (Spencer and Maclachlan, 1972). It is also high in such products formed from UDP-[^{14}C]glucose by tissue slices, but not in products formed by particulate membrane preparations (Gordon and Maclachlan, 1980). This suggests that systems containing intact wall–plasma membrane structures can lengthen existing wall glucans by transglucosylation to form radioactive polymers of substantial chain length, whereas isolated membranes without endogenous primers can generate only products of relatively low molecular weight.

4. Addition of purified endocellulase for a brief period to tissue slices before adding UDP-glucose can markedly stimulate the subsequent incorporation of glucose into alkali-insoluble products (Wong *et al.*, 1977). No such stimulation is observed by cellulase pretreatment on the incorporation into soluble products or on the synthesis by isolated membranes. This suggests that the availability of exposed chain ends in preexisting cellulose (which is increased by endocellulase activity) may be a limiting factor in the rate of glucosyl transfer by adjacent membrane.

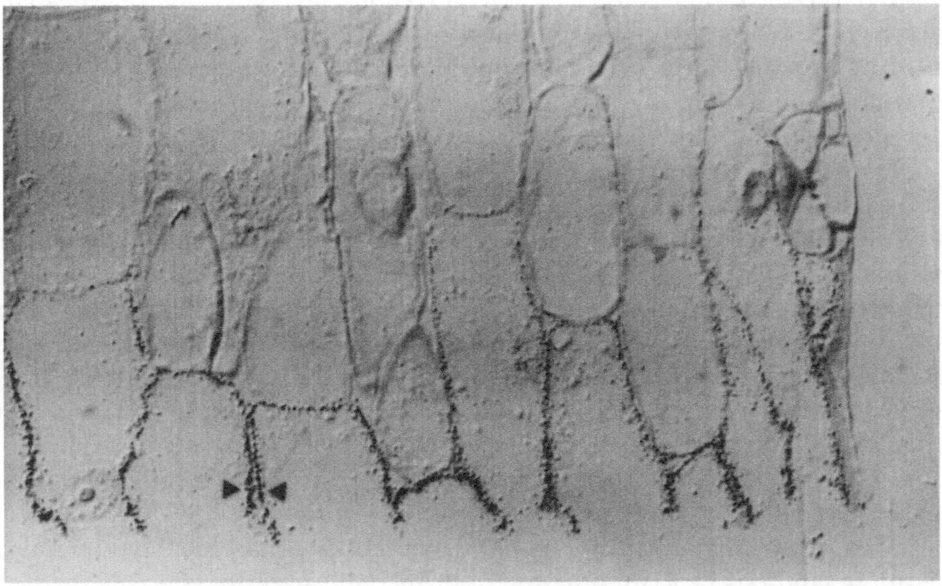

FIGURE 1. Radioautograph of pea tissues after floating on UDP-[^{3}H]glucose. Freshly cut thin slices (0.5 mm thick) of growing regions of etiolated stems were floated on 5 μm substrate for 30 min. Slices were extraced with chloroform–methanol–water (10:10:3) and hot water, embedded, and sectioned longitudinally, to show intact cells inside the slice and cut cells at the surface that was exposed to substrate. The radioautograph shows that only cut cells incorporated label into insoluble products. (►◄) Point at which both surfaces of exposed cell walls were clearly capable of synthesis.

5. Preincubation of membrane preparations with UDP-glucose, under conditions where some products are synthesized, may result in an enhancement of the subsequent capacity of the preparations to generate insoluble products from GDP-glucose (Hopp *et al.*, 1978). This was interpreted as indicating that a primer formed from UDP-glucose is required for glucosyl transfer from GDP-glucose. Earlier studies, which all showed only an apparent inhibition of incorporation from one glucose nucleotide by the other (e.g., Shore and Maclachlan, 1973), may have been conducted with membrane preparations that were already saturated with primer.

2.2. *Direct Biochemical Studies on Primer Requirements*

2.2.1. *Bacteria*

In the pioneer study by Glaser (1958) on cell-free cellulose synthesis by membranes from *Acetobacter*, it was reported that cellodextrins were required for synthesis from UDP-glucose, and it was suggested that they were needed as primers. Subsequent studies on the system (e.g., Ben-Hayyim and Ohad, 1965) were unable to reproduce any direct indication for such a dependency. It has usually been considered that membrane systems that are fully active in forming β-glucan without added acceptors are probably already saturated with endogenous acceptor chains at the sites of synthesis. Evidence for the existence of natural products that could act as endogenous acceptors or "intermediate polymers" has been published (Kjosbakken and Colvin, 1975; King and Colvin, 1976). The paradoxical stimulation of bacterial-wall synthesis by hydrolytic enzyme activities (autolysins) has also been interpreted as evidence for a dependence on available endogenous acceptors for polymer chain-lengthening (Thompson, 1971; Shockman, 1965).

Cooper and Manley (1975) reinvestigated the direct effects of a cellodextrin series on cell-free cellulose synthesis from UDP-[^{14}C]glucose by *Acetobacter* membranes. Using particulate preparations, the addition of dextrins larger than cellotetraose resulted in as much as a 15-fold enhancement in the yield of alkali-insoluble labeled products. The increase occurred, in part, at the expense of alkali-soluble products, which was interpreted as indicating that cellodextrins evoked an increase in molecular weight of newly formed β-glucan. Evidence that alkali-soluble products of such a system may be an intermediate stage in cellulose synthesis has also been provided by experiments with membranes from *Acanthamoeba* (Potter and Weisman, 1976).

Cooper and Manley (1975) concluded that cellodextrins probably stimulated cellulose synthesis by a combination of primer and activator effects. However, the optimum concentration for cellotetraose (15 mM) was considered to be far in excess of that expected for efficient primer action and enzyme saturation. Labeled cellodextrins were not employed to establish unambiguously whether or not any part of them was incorporated into the products. Thus, at present, it is impossible to conclude with certainty that primer action has in fact been demonstrated directly in bacterial cell-free systems.

It should also be noted that soluble carboxymethylcellulose (CMC) has been reported to stimulate the secretion of cellulose microfibrils from *Acetobacter* cells (Ben-Hayyim and Ohad, 1965). But in this case, studies on the incorporation of [^{14}C]-CMC and effects on aggregation time led to the conclusion that CMC was coprecipitating with newly formed fibrils and not forming covalent associations with new cellulosic chains. There is no indication, therefore, that CMC can act as an external acceptor molecule for surface glycosyltransferase in this organism.

2.2.2. *Fungi*

A highly active β1,3-glucan synthetase system in yeast membranes has recently been studied (Shematek *et al.*, 1980), and it was observed that remarkably uniform products (DP 60–80) can be generated *in vitro* from UDP-glucose. There was no evidence of a requirement for lipid intermediates, but the enzymatic activity was firmly bound to the plasma membrane and had the ability to catalyze transglycosylation at the inner (protoplasmic) surface only. Chemical reduction of reducing terminals, hydrolysis, and recovery of radioactive sorbitol from the reaction medium indicated that the polymers were all labeled at the chain end where initiation usually takes place. This suggests that the labeled products were assembled *de novo* without primer requirements. The endogenous insoluble glucan (from cell-wall residues) that was found in the particulate enzyme preparation did not serve as a primer, though it is possible that some other acceptor, e.g., a protein, was an intermediate in the process. Protein acceptors had been suggested earlier by Balint *et al.* (1976), who showed that a substantial part of the β-glucan formed by yeast membranes from GDP-glucose or UDP-glucose could be solubilized by conditions conducive to β-elimination. These β-glucans appeared to contain variable proportions of 1,3- and 1,6-linkages, depending on reaction conditions. The presence of glucan-protein fractions in yeast cell walls has long been suspected.

The present conclusions are therefore tentative, but they do not support a requirement for a classic polysaccharide primer in β-1,3-glucan synthesis.

2.2.3. *Higher Plants*

Repeated attempts have been made in this laboratory to increase the rate of wall synthesis or reduce the lag period observed in the regeneration of a new wall by isolated protoplasts, by adding potential saccharide primers to the medium. Tests have been conducted with protoplasts suspended in a slurry of cellulose or laminarin powders, in gels of carboxymethyl derivatives of β1,4- or β1,3-glucans, and in solutions of cellodextrins or laminaridextrins. The results have been entirely negative, forcing the conclusion that protoplasts will use none of these supplied saccharides as acceptors for transglycosylation.

Nevertheless, many authors have reported that various sugars, and in particular cellobiose, have stimulatory effects on the synthesis of β-glucans from glucose nucleotides in membrane systems from higher plants. In most instances, e.g., with lupine membranes (Thomas and Stanley, 1968), the stimulation can be duplicated with a variety of polyols, including glycerol and sucrose, and the effective concentrations are much too high to justify any conclusion that these additives are acting as acceptor molecules for transglycosylation. Nevertheless, the suspicion lingers that cellobiose may have a specific role in these biosynthetic systems, and it has often been included in reaction mixtures as insurance in case such a dependency turns out to be demonstrable.

In the case of pea stem membranes, labeled products with various solubilities are generated from both radioactive UDP- and GDP-glucose, and unlabeled cellobiose stimulates synthesis from the latter substrate only (Table I). Similar results have subsequently been obtained in this laboratory using pea membranes plus authentic GDP-[^{14}C]glucose in the presence and absence of cellohexaose. This dextrin increased synthesis of neutral lipids and water-soluble products as well as insoluble β-glucans. Insoluble cellulose has no significant effect on synthesis in these systems, but the soluble derivative CMC results in a change in the solubility of labeled products from buffer-insoluble to soluble fractions (see Table I). Such a result could have come about if CMC had successfully competed for endogenous primers and acted as an acceptor molecule for a limited amount of glycosyl transfer. However, the further addition of insoluble cellulose alters the solubility back to a product distribution close to that of the control system, i.e., to a preponderance of insoluble labeled product. Accordingly, the most likely explanation of these results is not primer action by

TABLE I
Effects of Potential Primers on Yields and Solubility of Products Formed by Pea Membranes from Glucose Nucleotides[a]

	Incorporation into products (pmol · hr^{-1} · mg protein^{-1})			
Conditions	Buffer-soluble	Alkali-soluble	Alkali-insoluble	Total
UDP-[^{14}C]glucose alone	15	18	74	107
+ Cellobiose	16	11	73	100
+ Cellulose	12	10	85	107
+ CMC	61	13	25	99
+ CMC + cellulose	38	14	60	112
GDP-[^{14}C]glucose alone	54	21	106	181
+ Cellobiose	69	82	164	315
+ Cellulose	48	49	97	195
+ CMC	163	3	23	189
+ CMC + cellulose	113	25	86	224

[a] Adapted from Spencer *et al.* (1971). Substrates were used at 4 mm and additives at 0.8% (wt. vol.).

CMC but, rather, effects of cellulose and CMC as adsorbants of biosynthesized products. Products that adsorb to CMC are rendered soluble, but the addition of excess cellulose results in precipitation. Thus, these data cannot be taken as evidence for primers.

The problem of distinguishing between newly synthesized products that are insoluble because of their high molecular weight and those that are insoluble because of their adsorption to other materials is illustrated by observations on properties of plant xyloglucans. These hemicelluloses have β1,4-linked glucan as the main chain of comparatively low molecular weight. They are found in many dicot cell walls firmly adsorbed by multiple hydrogen bonding to insoluble cellulose. The collective strength of these bonds leads to such a tight association that only reagents such as alkali, urea, or β-glucanase are sufficient to partially solubilize them (Bauer *et al.*, 1973). This is one basis for a theory on the potential role of xyloglucan: cellulose complexes as mediators or primary cell-wall rigidity–extension during growth (Valent and Albersheim, 1974).

In cell-free membrane preparations from pea stem, none of the water-soluble cellodextrins (cellobiose to cellohexaose) has any significant effect on the yield of water-insoluble products formed from UDP-[^{14}C]glucose (Table II). However, when [^{14}C]cellodextrins are added to cell-free or tissue-slice systems, they are incorporated into water-insoluble fractions in amounts that increase with increasing chain length (Table III). The bulk of the incorporation in the pea system is into the alkali-soluble fraction, but part is alkali-insoluble, i.e., the residue after several extractions with hot 1 M NaOH. A plot of incorporation vs. chain length yields a logarithmic relationship. This is different from results of similar tests on the primer capacity of a maltodextrin series for sustaining α-glucan synthesis, where dextrins with more than four glucose units are all equally effective as primers. Thus, cellodextrins may not have acted as primers in the tests described above.

TABLE II
Cellodextrins Do Not Alter Rate of Synthesis of β-Glucan from UDP-Glucose

Additive (5 mM) plus UDP-[^{14}C] glucose (5 mM)	Incorporation from UDP-[^{14}C]glucose[a] (μmol Glc · 7 min^{-1} · 5 mm seg.$^{-1}$) (± S.E.)	
	Particulate fraction	Tissue slices
None (control)	1.60 ± 0.46	20.97 ± 3.34
Cellobiose	1.64 ± 0.37	24.99 ± 3.39
Cellotriose	1.20 ± 0.20	21.34 ± 3.09
Cellotetraose	1.50 ± 0.17	17.80 ± 1.22
Cellopentaose	1.81 ± 0.30	20.04 ± 2.80
Cellohexaose	1.89 ± 0.21	17.65 ± 2.79

[a] Products are residues after extraction with lipid solvents and boiling water.

It has proved necessary to conclude that the results in Table III are in fact a consequence of adsorption of added [^{14}C]cellodextrins to endogenous insoluble materials, rather than incorporation by covalent linkages into β-glucan, for two reasons: (1) the incorporation into insoluble fractions was not dependent on concurrent synthesis from UDP-glucose (or GDP-glucose); and (2) apparent incorporation was greater in boiled denatured enzyme preparation (Table IV), where sites for adsorption would increase.

3. *Discussion and Conclusions*

An apparent conflict exists between the various kinds of *indirect* evidence for a primer requirement for β-glucan synthesis and the failure of numerous *direct* attempts to demonstrate such a requirement in experiments *in vitro*.

TABLE III
Incorporation of Radioactive Cellodextrins into Insoluble Fractions during Synthesis from UDP-Glucose

Added cellodextrin	Incorporation from [^{14}C]cellodextrin[a] (pmol dextrin · 30 min^{-1} · 50 mm seg.$^{-1}$)			
	Particulate fraction		Tissue slices	
	Alkali-soluble	Alkali-insoluble	Alkali-soluble	Alkali-insoluble
Cellobiose	83	4	383	143
Cellotriose	171	7	914	217
Cellotetraose	227	12	2656	231
Cellopentaose	491	32	5009	394
Cellohexaose	1009	85	9365	1094

[a] Reaction mixtures contained unlabeled UDP-glucose (5 mM) plus radioactive [^{14}C]cellodextrin (5 mM). After 30-min incubation, they were extracted with lipid solvents and boiling water before being extracted with hot 1 M NaOH (alkali-soluble fraction).

TABLE IV
[14C]Cellohexaose Adsorbs Strongly to Slices of Pea Tissue

	Recovery of [^{14}C]cellodextrin[a] (pmol · 30 min^{-1} · 5 mm seg.$^{-1}$)	
Conditions	Alkali-soluble	Alkali-insoluble
Tissue slices + [^{14}C]cellohexaose	3480	376
+ UDP-glucose (5 mM)	3780	447
Boiled slices + [^{14}C]cellohexaose	6204	1311

[a] Reaction mixtures were extracted with lipid solvents and boiling water before being extracted with hot 1 M NaOH.

The problem may lie in difficulties in bringing the proposed acceptor molecule into a proper spatial relationship with the membrane-bound synthetases. *In vivo,* the synthetase–primer connection may be close and seldom broken, except perhaps under plasmolyzing conditions or hormonal regimes where the orientation of membrane complexes and the direction of growth are altered. *In vitro,* it is perhaps expecting too much to think that a water-soluble potential primer, e.g., cellohexaose, or an insoluble polymer, could present acceptor chain ends to the active site of a transglycosylase that may be buried in the hydrophobic environment of a membrane. Yet that active site certainly appears to be accessible to water-soluble glucose nucleotide substrates and even to macromolecules like proteases that inactivate it. This is a paradox that may perhaps not be resolved until means are found to remove from the membrane and solubilize the transglucosylase responsible for chain elongation of β-glucans. If such a preparation could be purified and made to retain its activity, primer requirements and specificities would be relatively simple to determine.

The problem may also lie in the fact that only poly- and oligosaccharides have been tested to date for their ability to act as acceptor molecules in β-glucan synthesis. If the concept of primer is extended to include glycolipids or glycoproteins as acceptors for chain elongation, it may be that one of these species is the key requirement for initiating β-glucan synthesis.

Whelan (1976) suggested that a protein primer may be a universal requirement for synthesis of polysaccharides, including cellulose. Franz (1976) was the first to provide evidence that the synthesis of products resembling β-glucans by plant membranes could be enhanced by the addition of extracted proteins and that the latter were glucosylated. More recently, Pont-Lezica *et al.* (1978) summarized evidence for a 1,4-linked oligoglucan covalently attached to polypeptide in pea membranes, which was synthesized from UDP-glucose. These observations need further study and corroboration, but they are suggestive that β-glucan synthetase complexes may recognize only a particular glycoprotein as acceptor molecule. It is conceivable that β-glucan synthesis proceeds by a mechanism that requires glycoprotein as the enzymatic catalyst itself, where new saccharides are inserted at the reducing end of a convalently attached chain (Robyt, 1979). In this event, synthetase systems would appear to be "unprimed" and unresponsive to addition of oligosaccharides.

Colvin (1972) was the first to suggest that glycolipids may be required for the biosynthesis of β-glucan, and many authors have since described the synthesis of glucosylated lipids such as polyisoprenylphosphates by plant membranes (Elbein, 1979). These could turn out to be complex intermediates for the synthesis of glycoproteins (e.g., the "G"-oligosaccharide containing mannose and *N*-acetylglucosamine, "capped" with glucose) (Bailey *et al.*, 1979). But it is also possible that oligosaccharides containing glucose only are assembled on lipid-linked acceptors (Pont-Lezica *et al.*, 1978; Torossian and Maclachlan, 1980). Whether these elongate to a certain size and then are transferred *en bloc* to a true acceptor, as in synthesis of *N*-linked glycoproteins, or whether they can be elongated indefinitely to form a polysaccharide with the lipid primer as core, remains to be investigated. Again, it would be helpful to have a glucose-nucleotide-polyisoprenylphosphate transglucosylase that is solubilized and purified in active form, in order to properly study such questions.

It is concluded that the question of the nature of a primer for synthesis of β-glucans like cellulose and callose, and even the question of whether one is required, in unanswerable in biochemical terms with the present state of our knowledge of synthetase properties. The fact that young growing plant cells generate cellulose with an average molecular weight that increases during cell development (Spencer and Maclachlan, 1972), as does the DP of cellulose that is newly deposited around protoplasts (Asamizu *et al.*, 1977), strongly suggests that at least in young tissues, cellulose synthesis proceeds by chain elongation of preexisting cellulose. Whether the mechanism involves terminal additions of single of multiple glucose units, or an insertion via lipid or other intermediates, is not yet clear. It would seem to be a waste of time to continue studies of crude membrane-bound synthetase activities if problems like these are to be answered satisfactorily. Techniques are available for solubilizing and purifying hydrophobic enzymes in active form, and it is high time that these were applied to dissect and characterize the essential components of β glucan synthetase systems.

Acknowledgments

Thanks are due to Drs. Fred Spencer, Yves Raymond, and Susette Mueller, and Ms. Su Chen, for permission to quote results from their studies in this laboratory. The work was supported by grants from the Natural Sciences Research Council of Canada and the Ministry of Education of Quebec.

References

Asamizu, T., Tanaka, K., Takebe, I., and Nishi, A., 1977, Change in molecular size of cellulose during regeneration of cell wall on carrot protoplasts, *Physiol. Plant.* **40**:215.

Bailey, D. S., Dürr, M., Burke, J., and Maclachlan, G. A., 1979, The assembly of lipid-linked oligosaccharides in plant and animal membranes, *J. Supramol. Struct.* **11**:123.

Balint, S., Farkas, V., and Bauer, S., 1976, Biosynthesis of β-glucans catalysed by a particulate enzyme preparation from yeast. *FEBS Lett.* **64**:44.

Bauer, W. D., Talmadge, K. W., Keegstra, K., and Albersheim, P., 1973, The hemicellulose of the walls of suspension cultured sycamore cells, *Plant Physiol.* **51**:174.

Ben-Hayyim, G., and Ohad, J., 1965, Synthesis of cellulose by *Acetobacter xylinum, J. Cell Biol.* **25**:191.

Boyer, C. D., and Preiss, J., 1979, Properties of citrate-stimulated starch synthesis catalysed by starch synthase I of developing maize kernels, *Plant Physiol.* **64**:1039.

Chao, H.-Y., and Maclachlan, G. A., 1978, Soluble factors in *Pisum* extracts which moderate β-glucan synthetase activity, *Plant Physiol.* **61**:943.

Colvin, J. R., 1972, The structure and biosynthesis of cellulose, *Crit. Rev. Macromol. Sci.* **1**:47.

Cooper, D., and Manley, R. St. J., 1975, Cellulose synthesis by *Acetobacter xylinum.* III. Matrix, primer and lipid requirements and heat stability of the cellulose-forming enzymes, *Biochim. Biophys. Acta* **381**:109.

Delmer, D. P., Beasley, C. A., and Ordin, L., 1974, Utilization of NDP-glucoses in developing cotton fibers, *Plant Physiol.* **53**:149.

Elbein, A. D., 1979, The role of lipid-linked saccharides in the biosynthesis of complex carbohydrates, *Annu. Rev. Plant Physiol.* **30**:239.

Franz, G., 1976, The dependence of membrane-bound glucan synthetases on glycoprotein which can act as acceptor molecules, *Appl. Polym. Symp.* **28**:611.

Franz, G., and Meier, H., 1969, Biosynthesis of cellulose in growing cotton hairs, *Phytochemistry* **8**:579.

Glaser, L., 1958, The synthesis of cellulose in cell-free extracts of *Acetobacter xylinum, J. Biol. Chem.* **232**:627.

Gordon, R., and Maclachlan, G. A., 1980, Molecular weight distribution of products of pea β-glucan synthetase activity, *Plant Physiol. Suppl.* **65**:106.

Hawker, J. S., Ozbun, J. L., Ozaki, H., Greenberg, E., and Preiss, J., 1974, Interaction of spinach ADP-glucose α-1,4,-glucan glycosyl transferase and α-1,4-glucan,α-1,4-glucan glycosyl transferase in synthesis of branched α-glucan, *Arch, Biochem. Biophys.* **160**:530.

Herth, W., and Meyer, Y., 1977, Ultrastructural and chemical analysis of the wall fibrils synthesized by tobacco mesophyll protoplasts, *Biol. Cellulaire* **30**:33.

Hopp, H. E., Romero, P. A., Daleo, G. R., and Pont-Lezica, R., 1978, Synthesis of cellulose precursors: The involvement of lipid-linked sugars, *Eur. J. Biochem.* **84**:561.

King, G. G. S., and Colvin, J. R., 1976, Intermediate polymers of cellulose biosynthesis, *Appl. Polym. Symp.* **28**:623.

Kjosbackken, J., and Colvin, J. R., 1975, New evidence for an intermediate polymer of glucose in cellulose biosynthesis by *Acetobacter xylinum, Can. J. Microbiol.* **21**:111.

Kornberg, A., 1974, *DNA Synthesis,* Freeman, San Francisco.

Kornfeld, R., and Kornfield, S., 1980, Structure of glycoproteins and their oligosaccharide units, in: *The Biochemistry of Glycoproteins and Proteoglycans* (W. J. Lennarz, ed.), pp. 1–34, Plenum Press, New York.

Maclachlan, G. A., 1976, A potential role for endo-cellulase in cellulose biosynthesis, *Appl. Polym. Symp.* **28**:645.

Maclachlan, G. A., Dürr, M., and Raymond, Y., 1979, Towards the isolation of effective membrane-bound β-glucan synthetase complexes, *Methodol. Surveys, Biochem.* **9**:147.

Marx-Figini, M., 1976, Comparative study of cell wall and bacterial cellulose biosynthesis and its structural aspects, *Appl. Polym. Symp.* **28**:637.

Mueller, S. C., and Maclachlan, G. A., 1980, Plasma membranes only utilize UDP-glc supplied to the protoplasmic surface, *Plant Physiol. Suppl.* **65**:106.

Pont-Lezica, R., Romero, P. A., and Hopp, H. E., 1978, Glucosylation of membrane-bound proteins by lipid-linked glucose, *Planta* **140**:177.

Potter, J. L., and Weisman, R. A., 1976, Cellulose synthesis by extracts of *Acanthamoeba* during encystment, *Biochim. Biophys. Acta* **428**:240.

Raymond, Y., Fincher, G. B., and Maclachlan, G. A., 1978, Tissue slice and particulate β-glucan synthetase activities from *Pisum* epicotyls, *Plant Physiol.* **61**:938.

Robyt, J. F., 1979, Mechanisms involved in the biosynthesis of polysaccharides, *Trends Biochem. Sci.* **4**:47.

Shematek, E. M., Braatz, J. A., and Cabib, E., 1980, Biosynthesis of the yeast cell wall. 1. Preparation and properties of 1,3-β-glucan synthetase, *J. Biol. Chem.* **255**:888.

Shockman, G. D., 1965, Unbalanced cell-wall synthesis: Autolysis and cell-wall thickening, *Bacteriol Rev.* **29**: 345.

Shore, G., and Maclachlan, G. A., 1973, Indolacetic acid stimulates cellulose deposition and selectively enhances certain β-glucan synthetase activities, *Biochim. Biophys. Acta* **329**:271.

Shore, G., and Maclachlan, G. A., 1975, The site of cellulose synthesis, *J. Cell Biol.* **64**:557.

Shore, G., Raymond, Y., and Maclachlan, G. A., 1975, Cell surface and intracellular cellulose synthetase activities in relation to the stage and direction of cell growth, *Plant Physiol.* **56**:34.

Spencer, F. S, and Maclachlan, G. A., 1972, Changes in molecular weight of cellulose in the pea epicotyl during growth, *Plant Physiol.* **49**:58.

Spencer, F., Shore, G., Ziola, B., and Maclachlan, G. A., 1971, Particulate glucan synthetase activity: Generation and inactivation after treatments with indoleacetic acid and cycloheximide, *Arch. Biochem. Biophys.* **152**:311.

Tandecarz, J. S., and Cardini, C. E., 1978, A two-step enzymatic formation of a glucoprotein in potato tuber, *Biochim. Biophys. Acta* **543**:423.

Thomas, D., and Stanley, R. G., 1968, Glycerol activation of glucan synthesis in cell-free preparations, *Biochem. Biophys. Res. Commun.* **30**:292.

Thomas, D., Smith, J. E., and Stanley, R. G., 1969, Stereospecific regulation of plant glucan synthetase *in vitro, Can. J. Bot.* **47**:489.

Thompson, J. S., 1971, Role of autolytic enzymes in growth and morphogenesis of bacterial cell walls, *J. Theor. Biol.* **33**:63.

Torossian, K., and Maclachlan, G. A., 1980, Glucose-labelled saccharide lipids from *Pisum sativum*, *Plant Physiol. Suppl.* **65**:32.

Valent, B., and Albersheim, P., 1974, On the binding of xyloglucan to cellulose fibers, *Plant Physiol.* **54**:105.

Whelan, W. H., 1976, On the origin of primer for glycogen synthesis, *Trends Biochem. Sci.* **1**:13.

Wong, Y.-S., Fincher, G. B., and Maclachlan, G. A., 1977, Cellulase can enhance β-glucan synthesis, *Science* **195**:679.

17

Intermediates of Cellulose Synthesis in *Acetobacter*

YEHOSHUA ALONI AND MOSHE BENZIMAN

1. Introduction

For many years now, the bacterium *Acetobacter xylinum* has served as a model organism for studying cellulose biogenesis, namely, the biosynthesis of linear β1,4-glucan chains and their crystallization into cellulose fibrils. In this chapter, we shall review developments in the study of the metabolic sequence leading from glucose to cellulose in this organism. We have not, however, attempted a complete survey of the literature; for a more detailed review of the field, the reader is referred to the review by Colvin (1972).

For a better understanding of the topic reviewed, we felt it advantageous first to summarize briefly some general features of cellulose formation as they relate to general metabolic activities in *A. xylinum*.

Acetobacter xylinum is a gram-negative obligate aerobe. Its growth in static cultures is characterized by formation of a floating cellulosic pellicle that presumably insures an abundant supply of oxygen to the cells embedded in the pellicles (Schramm and Hestrin, 1954). Fibril-free suspensions of washed nonproliferating cells maintain a high capacity for cellulose synthesis (1.5×10^8 glucose residues polymerized per hour per cell). Synthesis by resting cells can occur under controlled conditions from defined exogenous substrates in a system that is essentially free from net synthesis of any other polysaccharide (Hestrin and Schramm, 1954). The range of utilizable substrates for cellulose production by resting cells includes hexoses; hexanoates; three-carbon compounds such as pyruvate, glycerol, and dihydroxyacetone; and four-carbon dicarboxylic acids of

YEHOSHUA ALONI AND MOSHE BENZIMAN • Department of Biological Chemistry, Institute of Life Sciences, The Hebrew University of Jerusalem, 91904 Jerusalem, Israel.

the citrate cycle (Schramm *et al.*, 1957a; Benziman and Burger-Rachamimov, 1962; Weinhouse and Benziman, 1974, 1976).

The polymeric product, formed in abundance (up to 45% of all added glucose), is deposited extracellularly and can be readily visualized (Hestrin and Schramm, 1954; Brown *et al.*, 1976). Synthesis proceeds without a lag period, at a rate that is linear with respect to cell concentration, and is essentially irreversible (Hestrin and Schramm, 1954; Ohad *et al.*, 1962; Colvin, 1971).

Cellulose formation is not indispensable to the metabolic utilization of carbohydrates, either for energy or for carbon. Net protein synthesis is not required for cellulose synthesis in resting cells, since the process can proceed normally in the absence of a nitrogen source (Hestrin and Schramm, 1954), and is not affected by the presence of inhibitors of protein synthesis (Webb and Colvin, 1967). Cellulose production is conditional on concurrent oxidation processes, and utilizable substrates for cellulose synthesis are all metabolically associated with the two amphibolic pathways operative in *A. xylinum:* the pentose cycle for the oxidation of carbohydrates and the citrate cycle for the oxidation of organic acids and related compounds (Gromet *et al.*, 1957; Benziman and Mazover, 1973; Swissa and Benziman, 1976).

2. Studies with Whole Cells

2.1. Primary Precursors of Cellulose

An early indirect *in vivo* approach toward the understanding of the intermediate steps in cellulose synthesis related cellulose formation to other metabolic processes occurring in *A. xylinum*. These included mainly studies on the patterns of isotopic distribution and label retention in the cellulose formed by resting cells from substrates specifically labeled with carbon-14. The results obtained indicate that cellulose arises in this organism from a metabolic pool of hexose-phosphate that is sustained by direct phosphorylation of free hexoses and by reactions of the pentose cycle, and that the conversion of hexose-phosphate to cellulose is direct in the sense that it does not necessarily include intermediary cleavage of the carbon skeleton of the hexose moiety (Schramm *et al.*, 1957a,b; Gagnaire and Travel, 1975). Condensation of triose-phosphates to the hexose-phosphate level has been indicated in similar studies as a step in gluconeogenesis and cellulose synthesis from glycerol, dihydroxyacetone, and citrate-cycle intermediates (Benziman and Burger-Rachamimov, 1962; Weinhouse and Benziman, 1976; Chumpitazi-Hermoza *et al.*, 1978; Gagnaire and Travel, 1980) and, to some extent, also from glucose (White and Wang, 1964; Everson and Colvin, 1966).

By analogy with other polysaccharide-synthesizing systems, glucose-6-phosphate (G6P), G1P, and nucleoside diphosphoglucose were the major candidates for primary precursors of cellulose. Enzyme systems that could promote formation of these compounds were demonstrable in cell-free extracts of *A.*

xylinum. These include a constitutive glucokinase for formation of G6P (Benziman and Rivetz, 1972), a phosphoglucomutase catalyzing the interconversion of G6P and G1P (Gromet *et al.*, 1957; Frei-Roitman, 1974), and a UDP-glucose pyrophosphorylase that catalyzes the synthesis of UDP-glucose from UTP and G1P. It is noteworthy that no other nucleoside diphosphoglucose pyrophosphorylase has been found in cell-free extracts of this organism (Swissa, 1978; Frei-Roitman, 1974). The individual reactions catalyzed by these enzymes *in vitro* proceed at rates greater than or in the same rough order of magnitude as the *in vivo* rate of cellulose formation from glucose. Thus, the possibility that glucose polymerization to cellulose occurs via these sequential reactions does not present difficulty on kinetic grounds.

Conclusive direct evidence that glucose-phosphates and UDPG are the primary *in vivo* precursors of cellulose remained scanty for many years. This was mainly due to the prevailing view that cellulose synthesis in *A. xylinum* occurs by an entirely extracellular process (Colvin, 1971, 1972). Evidence was therefore sought for the presence of hexose-phosphates and other sugar derivatives in the extracellular medium of cellulose-synthesizing cells. The first report on the concentration behavior with time of such components during *in vivo* synthesis was by Cooper and Manley (1975a). In experiments that utilized limiting amounts of glucose, they found that formation of G1P and UDP-glucose (determined in ethanol extracts) is very rapid and parallels that of cellulose. The levels of these compounds remain constant during cellulose formation and decrease when the rate of cellulose synthesis decreases. In contrast to the behavior of G1P, G6P concentrations (detected at lower levels and only in extracts of ultrasonically disrupted cells) did not change significantly during cellulose formation. These data were interpreted as indicating the possible involvement of UDP-glucose or G1P or both in cellulose synthesis. However, as pointed out by the authors, it may be argued that the formation of these compounds, as described, may be related not only to cellulose formation but also to the synthesis of other cell constituents.

For a number of years, our laboratory has extensively studied the regulation of general carbohydrate metabolism in *A. xylinum*. Since a substantial part of these studies were relevant to the mechanism of generation of substrates for cellulose synthesis, in recent years we have begun to focus our attention on the unsolved problems of the biogenesis of cellulose.

To assess the intermediary role of the various compounds that may be reasonably considered as potential precursors of cellulose, we initiated time-course tracer studies in which the *in vivo* flow of carbon from [^{14}C]glucose through sugar-phosphate and nucleotide-sugar pools into cellulose was quantitatively evaluated (Swissa *et al.*, 1980). Resting cells were pulse-labeled with [1-^{14}C]glucose. Pool sizes of labeled glucose-phosphate and UDP-glucose were estimated with a newly developed method by measuring the $^{14}CO_2$ released in a sequential enzymatic conversion of each of these compounds into ribulose-5-P.

The choice of [1-^{14}C]glucose as the substrate in these experiments was based on the following reasoning: previous studies have indicated that glucose catabolism in *A. xylinum* occurs mainly via a pentose cycle, that glucose is rapidly converted to gluconate, and that although glucose sustains a much more rapid initial rate of cellulose synthesis than does gluconate, the bulk of glucose carbon is assimilated to cellulose via a pentose cycle operating through the intermediate gluconate (Schramm *et al.*, 1957a,b; Weinhouse and Benziman, 1974). Since cycling in the pentose-phosphate pathway extensively dilutes the radioactivity in carbon-1 of glucose and gluconate, the incorporation of ^{14}C from [1-^{14}C]glucose into cellulose and its potential precursors would represent label transfer from noncycled hexose moieties derived directly from exogenous glucose. Consequently, labeling of cellular substances not directly related to carbohydrate metabolism is prevented. Moreover, a sequence of events similar to those that occur during a pulse-chase experiment could be expected to occur with limited amounts of this substrate. A rapid influx on the pathway to cellulose of labeled uncycled hexose moieties derived from exogenous substrate would occur from the outset and would shortly be followed by a flow of unlabeled moieties arising from pentose-cycle intermediates.

As can be seen in Fig. 1, the results obtained were in accord with these expectations and provide strong support for the concept that G6P, G1P, and UDP-glucose are primary precursors of cellulose. ^{14}C-labeled G6P, G1P, and UDP-glucose appear within 5 sec after the cells are exposed [1-^{14}C]glucose, their levels reaching a maximum after 5 min. The decrease in the level of labeled hexose-phosphates and UDP-glucose that occurs on depletion of the exogenous substrate is accounted for by the continuous incorporation of [1-^{14}C]glucose into cellulose. The close relationship between cellulose synthesis and the energy status of the cell was indicated when the intracellular level of ATP was determined in these experiments (Swissa, 1978). On addition of glucose to resting cells, an extremely rapid buildup of cellular ATP preceded that of hexose-phosphates, reaching a maximum level within 15–30 sec after the cells were exposed to the substrate.

2.2. *Possible Intermediates on the Pathway to Cellulose beyond the Level of UDP-Glucose*

We found that the conversion of [1-^{14}C]glucose into cellulose was accompanied by the incorporation of labeled glycosyl moieties into various cellular fractions: a water-soluble fraction, a lipid fraction (extractable in chloroform–methanol, 2:1, vol/vol.), and an alkali-soluble fraction (extractable with 0.5 N NaOH or 0.1% sodium dodecyl sulfate, at 100°C for 20 min) (Fig. 1) (Swissa *et al.*, 1980). Label incorporation into these fractions was observed within 1 min after the cells were exposed to [1-^{14}C]glucose, while the amount of label incorporated reached a plateau within 5 min. The kinetics of formation of these glycosylated products, when compared to that of cellulose, suggested that they may function as intermediates in cellulose synthesis. To test this possibility, we made use of a mutant blocked in cellulose synthesis. Such mutants can readily be selected when pellicle-forming *A. xylinum* cells, normally grown in static cultures, are

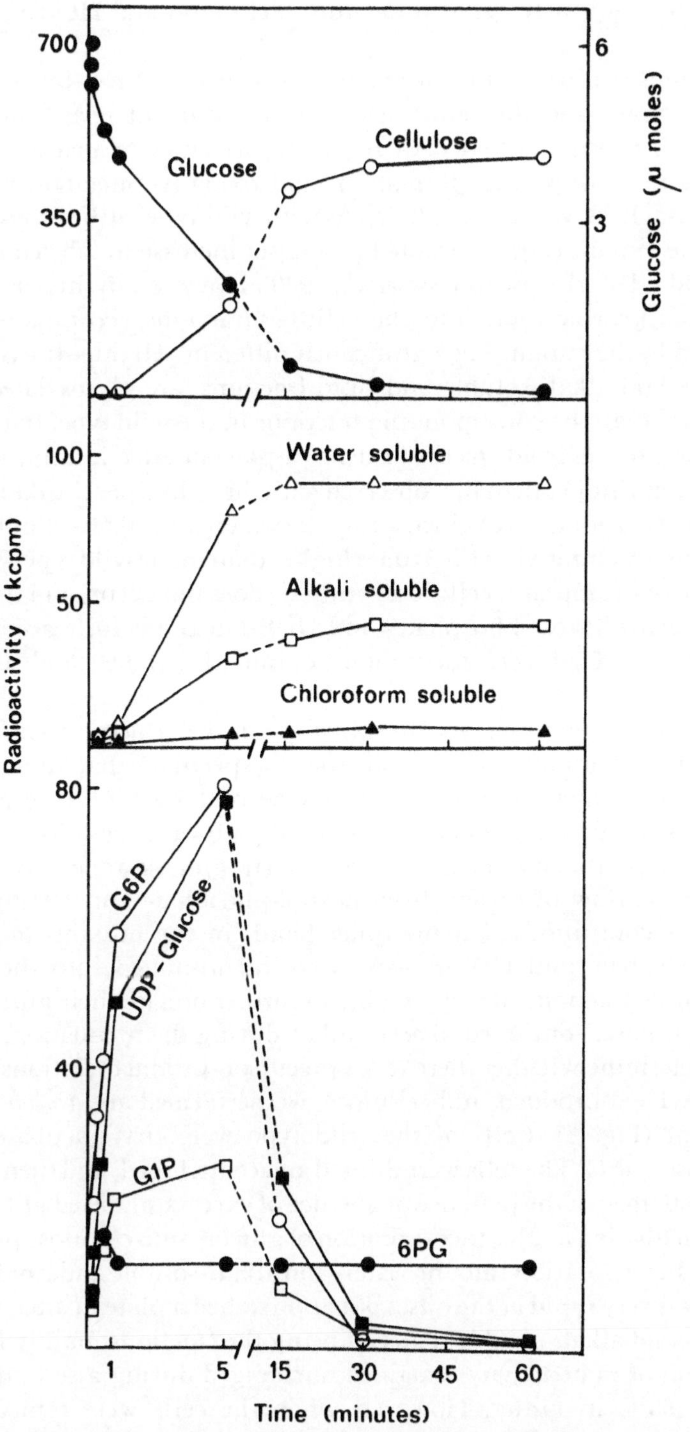

FIGURE 1. Time-course of metabolic changes during cellulose synthesis by wild-type cells of *A. xylinum*. Reaction mixtures contained 100 μmoles phosphate buffer, pH 6.0, 20 mg (dry weight) wild-type cells, and 6 μmoles [1-^{14}C]glucose (2 μCi/μmole). From Swissa *et al.* (1980).

cultured in vigorously agitated medium (Schramm and Hestrin, 1954; Forge, 1977).

Previous studies on the enzymatic constitution of cell-free extracts of the wild-type strain and the celluloseless mutant did not reveal any significant difference between the two with regard to activities related to hexose phosphorylation, UDP-glucose formation, and oxidative metabolism (Leisinger, 1966; Swissa, 1978; Weinhouse, 1977). As with wild-type, utilization of [1-^{14}C]glucose by the mutant (Fig. 2) resulted in a rapid increase in labeled hexose-phosphates and UDP-glucose (Swissa *et al.*, 1980). However, in this case, the pattern of glucose incorporation into the cellular fractions accompanying glucose utilization by the mutant was significantly different. All three fractions, namely, the water- and alkali-soluble and lipid fractions, are glycosylated to a much higher level than the corresponding fractions in the wild type. In this case, too, the labeling of the lipid fraction reaches a plateau after 15 min, but at a level about 10 times higher than that observed with the wild type. Unlike the wild-type strain, the incorporation of glucose into the water- and alkali-soluble fractions is continuous, reaching a level 4–5 times higher than in the wild type. Furthermore, in the case of the mutant, cellulose synthesis does not occur, and the disappearance of labeled hexose-phosphates and UDP-glucose is fully accounted for by concurrent [1-^{14}C]glucose incorporation into the water- and alkali-soluble fractions.

From the alterations in the kinetics of the glucose-transfer reactions observed in the mutant, it was possible to speculate that the glycosylation products may function as intermediates beyond UDP-glucose and that the mutant is blocked in subsequent stages on the pathway to cellulose. Accordingly, in the wild type, the intermediates beyond UDP-glucose appear to be in a steady state, and the flow of carbon from hexose-phosphate and UDP-glucose into cellulose is continuous. On the other hand, in the mutant, the carbons of hexose-phosphate and UDP-glucose were incorporated into the water- and alkali-soluble fractions, which, unable to further donate their glucose moieties to form cellulose, continued to accumulate during the incubation.

To determine whether there was a precursor–product relationship between the glycosylated products and cellulose, we performed an *in vivo* pulse–chase experiment (Fig. 3). Cells of the wild type were first incubated with [U-^{14}C]glucose at 0°C. The cells were diluted and centrifuged, and then reincubated at 30°C in buffer, in the presence or absence of excess unlabeled glucose. During the first incubation at 0°C, incorporation of glucose into cellulose proceeded at a slow rate. Incorporation into the water- and alkali-soluble and lipid fractions of the cells was very rapid at the onset but approached a plateau after 30 min, with the water- and alkali-soluble material being the fractions mainly labeled. The distribution of radioactivity remained unchanged during a second incubation period at 30°C in buffer. However, when the cells were reincubated with unlabeled glucose, radioactivity was chased from the water- and alkali-soluble fractions into cellulose. The rate of increase in radioactivity in the cellulose is accounted for by the concurrent loss of radioactivity, initially only from the

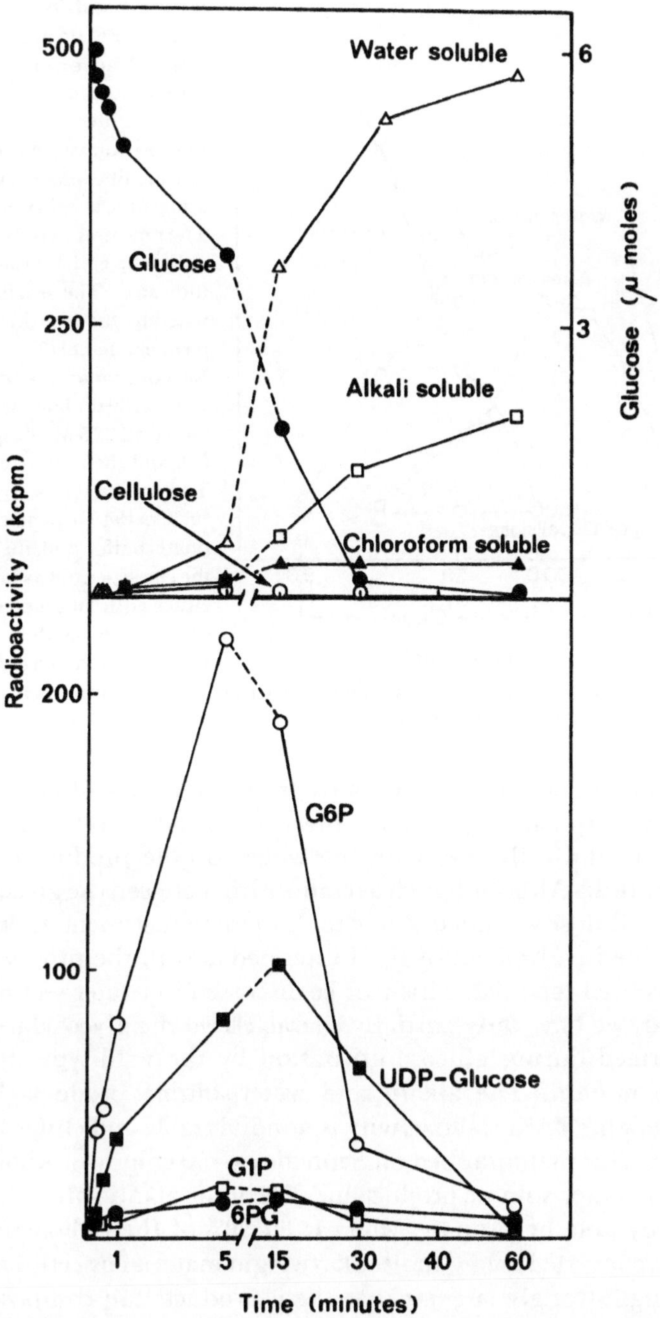

FIGURE 2. Time-course of metabolic changes during glucose utilization by a celluloseless mutant of *A. xylinum*. Reaction mixtures contained 100 μmoles phosphate buffer, pH 6.0, 20 mg (dry weight) of celluloseless mutant cells, and 6 μmoles [1-^{14}C]glucose (2 μCi/μmole). From Swissa *et al.* (1980).

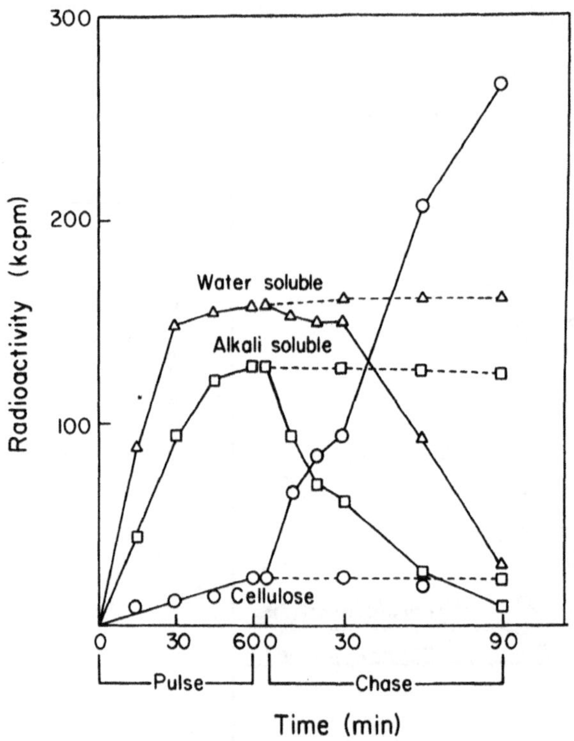

FIGURE 3. Incorporation of [^{14}C]glucose into the water- and alkali-soluble fractions and its subsequent transfer from these fractions into cellulose. *Pulse:* Reaction mixtures contained in a final volume of 0.5 ml; 1.5 μmole (U-^{14}C]glucose (5 × 10^6 cpm), 2.5 μmoles potassium acetate, 2.5 μmoles sodium phosphate buffer, pH 6.0, and 4 mg (dry weight) of wild-type cells. The mixtures were incubated at 0°C in a shaking bath for the pulse periods as indicated. The reactions were terminated by the addition of 3 ml chloroform-methanol (2:1, vol./vol.). *Chase:* Reaction mixtures pulsed for 60 min as described above were each diluted with 6 ml 0.05 M phosphate buffer, pH 6.0, and then centrifuged in the cold. The supernatants were removed and each pellet suspended in 2 ml phosphate buffer and then incubated for the chase periods as indicated, at 30°C, either with 80 μmoles unlabeled glucose (—) or with no addition (---). Reactions were terminated with 3 ml chloroform-methanol (2:1, vol./vol.). Terminated reaction mixtures were then fractionated and analyzed as described by Swissa *et al.* (1980).

alkali-soluble product and subsequently from the water-soluble product. The pattern of label transfer indicates a precursor-product relationship between these fractions and cellulose, with the water-soluble product being the one closest to cellulose. Although such a relationship between the glycosylated lipid fraction and cellulose was not demonstrable in this experiment, indirect evidence for the possible involvement of lipid intermediates in the process of cellulose synthesis has been reported. This will be discussed in a later section.

Thus far, we have only partially characterized the glycosylated polymeric products formed during glucose utilization by the wild-type strain and the celluloseless mutant. The alkali- and water-soluble products have a high molecular weight. Their radioactivity is nondialyzable and elutes with the void volume when chromatographed on Sephadex G-50 columns. About 50% of the label from the water-soluble product and 35% of the alkali-soluble product were rendered dialyzable by pronase, whereas 80–90% of the radioactivity in these fractions was converted to low-molecular-weight material by cellulase treatment. These findings strongly suggest that these products are composed of glycoproteins, containing β1,4-linked glucosyl residues that are probably heterogeneous with regard to the molecular weight of their cellodextrin moiety. The radioactivity associated with the water-soluble fraction remains nondialyzable following treatment with mild alkali (β-elimination). Thus, if a carbohydrate-peptide linkage exists, it is probably not an *O*-glycosidic linkage through serine or threonine (Kornfeld and Kornfeld, 1976).

The radioactive lipid products were shown by thin-layer chromatography to be composed 80% of polar lipids and 20% of neutral lipids. About 30% of the radioactivity associated with the lipid fraction is released as water-soluble material by mild acid treatment and an additional 20% by strong acid treatment. The radioactivity released by mild acid treatment migrates coincident with cellobiose and glucose in thin-layer chromatography. Only 10% of the radioactivity is rendered water-soluble by mild alkali treatment. On the other hand, treatment of sonicated suspensions of the lipid product with either cellulase or β-glucosidase causes 90% of the radioactivity to become water-soluble, indicating a probable β-linkage for the glucosyl moieties.

3. Studies with Cell-Free Preparations

A different approach to the study of the biosynthesis of cellulose in *A. xylinum* has been to effect the separation of the synthesizing enzyme system from the structure of the cell. In a cell-free system, the nature of the enzymes involved in the process could be studied under defined and controlled conditions, and the intermediary role of likely cellulose precursors could be assessed without the restrictions imposed by selective permeability factors existing in whole cells.

3.1. Capacity for Synthesis of a Cellulosic Product from UDP-Glucose in Vitro

Various methods for rupturing or permeabilizing *A. xylinum* cells have been examined. Treatment of cells with toluene or ether, which proved very satisfactory in permeabilizing cells for the study of various enzyme activities *in situ* (Swissa *et al.*, 1976; Kornfeld *et al.*, 1978), resulted in complete loss of their capacity for synthesizing cellulose from all the substrates tested (Swissa, 1978). This fact, in itself, indicates that an essential part of the synthetic mechanism depends on the integrity of intrinsic membrane components. Ultrasonic disruption of cells led to complete loss of activity with glucose as substrate. However, particulate fractions obtained from sonic extracts catalyzed the transfer of glucose from UDP-glucose (but not from G6P or G1P) into a cellulosic material. This was first reported by Glaser (1958) and successfully repeated in several laboratories (Ben-Hayyim and Ohad, 1965; King and Colvin, 1976; Swissa, 1978). Membranous fractions isolated by differential centrifugation from cells ruptured in a pressure cell also catalyzed glucose transfer from UDP-glucose or G1P into an alkali-insoluble product (Cooper and Manley, 1975b). Cells treated with EDTA and subsequently frozen and thawed showed no activity with glucose, but exhibited glucose transfer capability with UDP—glucose (García *et al.*, 1974). Limited hydrolysis of the cell wall by lysozyme yielded a preparation containing 90% lysed cells, which was fully active in the synthesis of cellulose from glucose but was devoid of any activity with glucose-phosphates or UDP-glucose as substrates, even when supplemented with ATP and UTP (Webb and Colvin, 1962, 1963; Dennis and Colvin, 1965).

Cell-free lysates in which all the cells have been lysed by lysozyme, as judged by the complete release into the medium of various enzyme activities, have been obtained in our laboratory. These lysates, when supplemented with exogenous magnesium ions and incubated with 40 mM glucose, maintained 50% of the cellulose-synthesizing capacity of whole cells but also catalyzed the conversion of hexose-phosphate and UDP-glucose into cellulose (Swissa, 1978).

The results obtained with the different *in vitro* systems thus far studied all link UDP-glucose to cellulose synthesis. Table I compares the cellulose-synthesizing activity of various cell-free preparations. The activity with UDP-glucose of all the *in vitro* systems appears to be very similar, irrespective of the method by which the enzyme system has been obtained. The specific activity manifested by these systems is, however, very poor. At an optimum UDP-glucose concentration of 3–4 mM [which is approximately the steady-state intracellular concentration of UDP-glucose in glucose-utilizing cells (Cooper and Manley, 1975b; Weinhouse, 1977)], the maximal *in vitro* rate of synthesis represents only a fraction of a percentage of the *in vivo* rates of synthesis from glucose exhibited by whole cells. This low synthesizing efficiency indicates that the *in vitro* systems are far from being optimized and, in view of the potential complexity of the system, may reflect the dependence of effective cellulose synthesis on the physical integrity of essential enzymatic and structural components of the system. This is probably further manifested by the nature of the final product of the *in vitro* synthesis, which, although identified as a β1,4-glucan, appears not to consist of the typical crystalline cellulose synthesized by whole cells (Ben-Hayyim and Ohad, 1965; Colvin, 1980). Another possibility to be considered, however, is that

TABLE I
"Cellulose"-Synthesizing Activity of *in Vitro* Systems[a]

Preparation	Substrates (pmoles/min per mg cells)			Reference
	UDP-glucose	Glucose-1-phosphate	Glucose-6-phosphate	
Particulate fraction	2.4	0.056	—	Glaser (1958)
	1.0	—	<0.02	Ben-Hayyim and Ohad (1965)
	4.0	—	—	Colvin (1980)
	3.1	—	—	Swissa *et al.* (1980)
	2.0	1.4	0.4	Cooper and Manley (1975b)
EDTA-treated cells	2.4	<0.03	—	Garcia *et al.* (1974)
	3.5	—	—	Cooper and Manley (1975b)
Lysozyme lysate	1.8	0.14	0.48	Swissa (1978)

[a] Data obtained from the literature were standardized for comparison by expressing cell activity as pmoles [^{14}C]glucose incorporated into alkali-insoluble product per minute per amount of enzyme derived from 1 mg dry weight of cells. The UDP-glucose concentrations in these experiments ranged from 3 to 4 mM. The activity of whole cells with glucose under comparable conditions is 1500–2000 pmoles/min per mg cells.

the synthetic system consists of a multienzyme complex, the components of which are tightly bound and react efficiently in a concerted manner. As was shown for other multienzyme complexes (Gaertner, 1978), in this case, too, it is possible that an intermediate such as UDP-glucose, normally derived from glucose, the first substrate of the sequence leading to cellulose, remains enzyme-bound and does not exchange readily with UDP-glucose when supplied in solution.

3.2. *Products Other Than Cellulose Synthesized in Vitro from UDP-Glucose*

A number of studies indicate that noncellulosic products can also be synthesized from UDP-glucose by cell-free preparations of *A. xylinum*. Some of these products may be analogous to the probable intermediates in cellulose synthesis indicated in our *in vivo* studies. García *et al.*, (1974) reported that incubation of frozen and thawed EDTA-treated cells with UDP-[^{14}C]glucose led to the formation of several labeled compounds: an alkali-insoluble substance (presumably cellulose), an alkali-soluble substance that was not characterized but was the product mainly formed, and two products soluble in organic solvents, identified as lipid-diphosphate-α-glucose and lipid-diphosphate-α-cellobiose. The nature of the lipid moieties of the latter products has not been clearly established, although their overall properties suggest the presence of a polyprenol. In the presence of [β-^{32}P]UDP-glucose, the system incorporated radioactivity only into the lipid-diphosphate derivative, which is interpreted as indicating the transfer of an α-G1P group from UDP-glucose. The reaction is substrate-specific, and neither glucose, G1P, ADP-glucose, GDP-glucose, nor TDP-glucose could substitute for UDP-glucose as glucose donor. The time-course of formation of the lipid compounds indicates that they rapidly reach a plateau that can be interpreted as a steady-state level. On the other hand, glucose incorporation into the other products is much greater and increases continuously.

Very similar results have been described by Kjosbakken and Colvin (1973), working with a particulate enzyme system isolated from sonic extracts of *A. xylinum*. This system was reported to be capable of forming acid- and alkali-labile compounds that are a combination between (a) lipid(s) and glucose, cellobiose, and perhaps higher polymers of glucose. The kinetics of their formation suggest that they are made prior to the formation of the final cellulosic product. The possibility that lipid-bound cellobiose is on the pathway of cellulose biosynthesis is intriguing, especially since for steric reasons cellobiose is an attractive unit to serve as precursor (Delmer, 1977). Unfortunately, no report exists establishing a precursor–product relationship between this sugar-lipid compound and cellulose. Furthermore, in our laboratory, we have been unable to reproduce some of the details of these reports. In addition, in a very recent communication (Couso *et al.*, 1980), it was reported that the lipid-diphosphate cellobiose formed by EDTA-treated cells is the precursor of a lipid trisaccharide identified as β-mannosyl (1 → 3) β-cellobiosyl-1-diphosphate-prenol, with UDP-glucose and GDP-mannose acting as sugar donors. They also report that formation of lipid-linked cellodextrins, other than lipid-cellobiose, could not be detected in their system. The situation becomes even more uncertain in view of the report (Sanderman, 1977; Dekker *et al.*, 1977) that the lipid-linked disaccharide formed by enzyme preparations from *A. xylinum* has been identified as maltose rather than cellobiose.

The existence of a soluble intermediate polymer of glucose that may be involved in cellulose biosynthesis has been suggested in the course of other *in vitro* studies. According to Kjosbakken and Colvin (1975), when the particulate enzyme preparation of *A. xylinum* is incubated with UDP-[^{14}C]glucose, a labeled insoluble oligomer of glucose is produced that is covalently attached to the particles. The intermediate may be removed from the particles by alkaline treatment and is then preferentially adsorbed to fragments of added cellulose. In a later communication, King and Colvin (1976) reported the formation from UDP-[^{14}C]glucose by a similar enzyme preparation of a labeled nondialyzable borate-soluble β1,4-glucan that increases in concentration in the system as the time of incubation increases. When the isolated labeled polymer was added to a standard enzyme system containing nonradioactive UDP-glucose, it became alkali-insoluble after 2 hr of incubation. The polymer gave a negative Lowry test for protein. On this and on the basis of its behavior on gel chromatography following pronase treatment, it was concluded that the polymer is not a protein or a carbohydrate attached to a protein.

At variance with these results is the report of Sanderman and Dekker (1979) that membrane preparations from *A. xylinum* utilize UDP-glucose primarily for the formation of a high-molecular-weight, alkali-soluble, noncellulosic polymer. The polymer shows unusual resistance to various degradative enzymes. On this and on the basis of the chromatographic behavior of the di- and trisaccharides liberated from the polymer by acid hydrolysis, it is characterized as a β1,2-glucan.

We have demonstrated the ability of particulate fractions of sonic extracts to promote glucose transfer from UDP-glucose into a cellulosic alkali-insoluble product, using preparations derived from both the wild-type and the celluloseless mutant of *A. xylinum* (Fig. 4) (Swissa *et al.*, 1980). The transfer by enzyme preparations of both strains was accompanied by glucosylation of lipid-, water-, and alkali-soluble fractions. As was the case when glucose was used as substrate in wild-type whole cells, glucose transfer from UDP—glucose into the various fractions of the *in vitro* system of this strain was rapid during the first 10–15 min and then declined, approaching a plateau; in this case, however, the extent of glucose incorporation into the alkali-insoluble fraction (presumably cellulose) was lower than into the water- and alkali-soluble fractions and did not increase continuously but rather leveled off after only a few minutes. Different kinetics were observed with preparations of the mutant. In this case, incorporation of radioactivity from UDP-glucose into all fractions continued to increase, and the amounts incorporated were severalfold larger than in the experiments with the

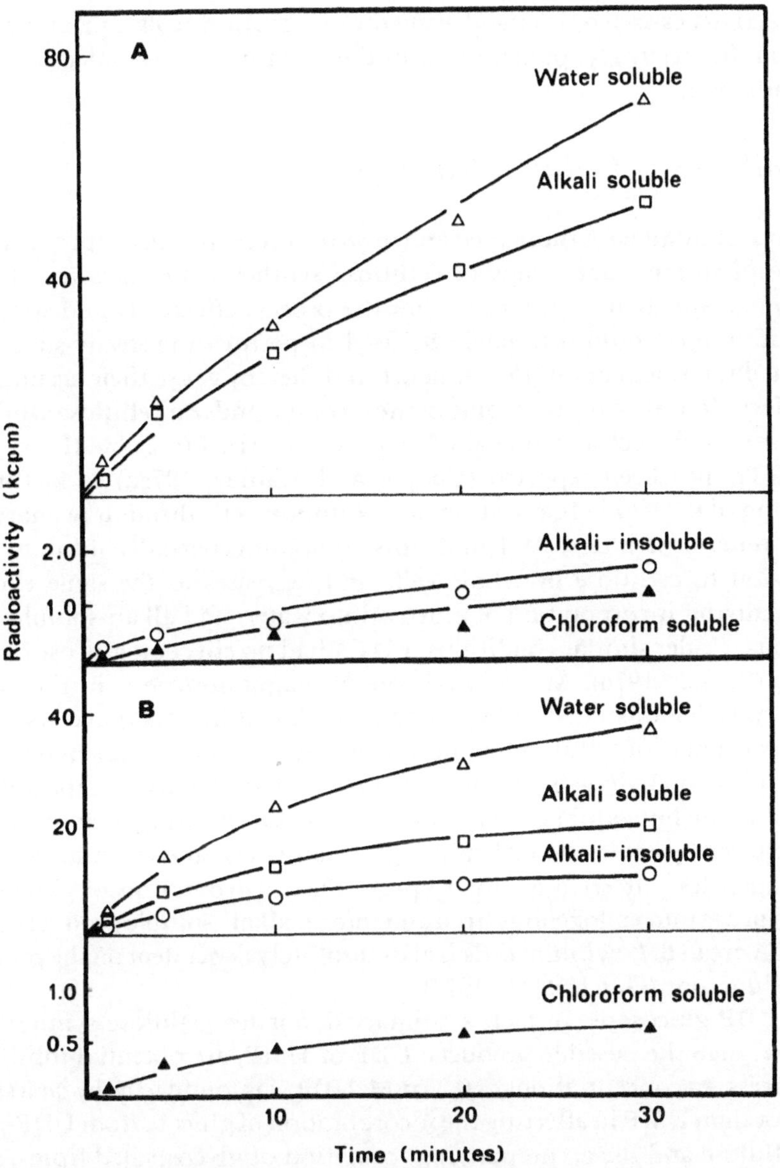

FIGURE 4. Time-course of incorporation of glucose from UDP-glucose into various fractions by a particulate system of *A. xylinum*. Reaction mixtures contained the following components in a final volume of 0.05 ml: UDP-[U-^{14}C]glucose, 1 nmole (800,000 cpm); $MgCl_2$, 0.5 μmole; Tris-HCl buffer, pH 8.0, 2.5 μmoles; and particulate enzyme preparations: (A) from the celluloseless mutant and (B) from wild-type cells, 1.5 mg of protein. From Swissa *et al.* (1980).

wild type, except for incorporation into the cellulosic product, which, in the mutant, was 5–10 times less than in the wild type.

The lability of the enzyme(s) involved in synthesis of the alkali-insoluble product in the wild-type strain could not be prevented by increasing substrate concentrations or by addition of boiled extracts. It could be due to progressive enzyme inactivation during the incubation period. This was suggested by the loss of activity caused by preincubating the preparation at 30°C prior to substrate addition. Interestingly, preparations of the mutant were less affected by such a preincubation.

4. *Inhibitors of Cellulose Synthesis*

Concomitant with the *in vivo* and *in vitro* studies that attempt to construct a scheme of intermediary steps of cellulose synthesis, we have tested various compounds for their ability to inhibit the process effectively and specifically. Such inhibitors could ultimately be used to perturb the steady-state level of suspected intermediates in the sequence and thereby cause their accumulation. The effects described by us of some of these compounds on cellulose synthesis by whole cells and a cell-free preparation are summarized in Table II.

EDTA has been reported (Cooper and Manley, 1975a) to facilitate the loosening of the rigid organization of *A. xylinum* cells through primary action on the outermost layers. At 1 mM, this compound partially inhibits glucose conversion to cellulose by whole cells and decreases to the same extent the accompanying incorporation of glucose into water- and alkali-soluble cellular functions. Under similar conditions, EDTA had no effect on glucose oxidation to CO_2 (Swissa, 1978). An essential role of magnesium ions in the synthetic process was demonstrated in experiments with cell-free lysozyme-lysates of *A. xylinum* capable of cellulose synthesis from glucose. The sediment obtained by centrifuging such lysates maintained the glucose-oxidizing capacity of the original lysate but exhibited very low cellulose-synthesizing activity. This was fully restored by addition of 2.5 mM Mg^{2+}. Manganese and calcium ions proved to be much less effective in this respect. The incorporation of glucose from UDP-glucose into endogenous lipid and into an alkali-soluble product catalyzed by EDTA-treated *A. xylinum* cells is also absolutely dependent on the presence of magnesium ions (García *et al.*, 1974).

If UDP-glucose is, in fact, a primary donor for cellulose synthesis in *A. xylinum*, then the possible products, UDP or UMP, are potential inhibitors in this process. At concentrations of 0.25 mM, UDP was found to be by far a stronger inhibitor than UMP in affecting the incorporation of glucose from UDP-glucose into cellulose and the accompanying formation of glycoslyated lipid-, alkali-, and water-soluble materials. Moreover, the inhibitory effect of a combination of the two nucleotides did not exceed that of UDP alone. These results are consistent with the notion that UDP is a product of the transfer reaction and that a nonphosphorylated glucose moiety if transferred from UDP-glucose to the appropriate acceptors in the reaction sequence leading to cellulose. In contrast to our results, both UDP and UMP were reported to inhibit the *in vitro* incorporation of glucose from UDP-glucose into endogenous lipid material (Kjosbakken and Colvin, 1973; García *et al.*, 1974). The concentrations of the nucleotides relative to that of the substrate in these experiments were, however, much greater than in ours.

The detergent Triton X-100 was found to inhibit the *in vivo* incorporation of glucose into cellulose and the glycosylation of cellular material, probably by

TABLE II
Inhibitors of *in Vivo* and *in Vitro* Cellulose Synthesis[a]

Inhibitor	Glucose incorporation (% of control)							
	In vivo				*In vitro*			
	Lipid-soluble	Alkali-soluble	Water-soluble	Cellulose	Lipid-soluble	Alkali-soluble	Water-soluble	Cellulose
EDTA (1 mM)	—	67	65	80	—	—	—	—
UMP (0.25 mM)	—	—	—	—	96	73	100	78
UDP (0.25 mM)	—	—	—	—	50	40	64	29
UMP (0.25 mM) +UDP (0.25 mM)	—	—	—	—	50	42	69	29
Triton X-100								
0.1%	44	91	29	69	218	158	124	128
0.5%	49	65	19	51	508	148	120	102
Coumarin (4 mM)	10	28	53	0	39	73	37	21
2,6-Dichlorobenzonitrile								
20 μM	—	—	—	—	59	16	28	7
1.2 mM	40	19	15	36	—	—	—	—
Bacitracin (5 mM)	103	84	79	102	117	91	97	105

[a] The *in vivo* systems contained: 50 mM phosphate buffer, pH 6.0, 40 mM [U-^{14}C]glucose, and resting cells (5 mg dry weight). Incubation: 30 min at 30°C. The *in vitro* systems contained: 50 mM Tris-HCl buffer, pH 8.0, 10 mM $MgCl_2$, 2 mM EDTA, 20 μM UDP-[U-^{14}C]glucose, and 1.5 mg protein of a particulate preparation derived from a sonic extract. Incubation: 10 min at 30°C. Incorporation of glucose into the various fractions was determined as described by Swissa *et al.* (1980).

perturbing and partially solubilizing essential membranous structures of the cells. On the other hand, the corresponding reactions with UDP-glucose in the *in vitro* system were markedly stimulated by the detergent, with the greatest effect being on glucosyl incorporation into the lipid fraction. These results are at variance with the report of García *et al.* (1974) that stated that the incorporation *in vitro* of glucose from UDP-glucose into endogenous lipid in a crude enzymatic preparation of *A. xylinum* is completely inhibited by a similar concentration of Triton X-100.

Coumarin, which has been reported to inhibit cellulose synthesis in plant systems (Hara *et al.*, 1973; Delmer, 1977; Hopp *et al.*, 1978b), strongly inhibited cellulose formation by *A. xylinum* both *in vivo* (Sato *et al.*, 1976) and *in vitro*. In both systems, it also lowered to a great extent the level of other glycosylated products formed in the course of cellulose synthesis. Coumarin was equally effective in inhibiting the glycosylation of endogenous acceptors in the mutant. Its effect on cellulose synthesis in *A. xylinum* appears to be specific. It does not affect enzyme activities related to UDP-glucose formation, the cellular level of ATP, or the oxidative capacities of the cells toward various substrates. The inhibitory effect on cellulose synthesis is reversed by washing out the coumarin (Swissa, 1978).

Dichlorobenzonitrile, another very specific and potent inhibitor of cellulose synthesis in plant systems (Montezinos and Delmer, 1980), also strongly inhibited cellulose formation by whole cells and cell-free preparations of *A. xylinum* and affected the glycosylation of endogenous acceptors in both systems in a similar manner.

The antibiotic bacitracin, which is an inhibitor of lipid-linked glycosyl-transfer reactions (Siewert and Strominger, 1967), had no effect, even at high concentrations, on the *in vivo* and *in vitro* cellulose-synthesizing system of *A. xylinum*. This does not necessarily exclude the possible involvement of a glycosylated lipid intermediate in cellulose synthesis. In bacterial systems, the inhibitory effect of bacitracin is attributed to its binding with polyprenol pyrophosphates, thereby preventing cleavage of the pyrophosphate and reutilization of the lipid-phosphate for a second cycle (Stone and Strominger, 1971). Such an inhibitory mechanism would not apply if the formation of the inferred glucolipid intermediate of cellulose synthesis involves the transfer of an unphosphorylated glucose moiety from UDP-glucose to an endogenous lipid-phosphate. Such a transfer mechanism would be compatible with the observed inhibitory effect of UDP mentioned before.

The effects and mechanism of action of the stilbene derivative Calcofluor White ST, which separates the processes of polymerization and crystallization of cellulose and greatly increases the rate of glucose polymerization to cellulose in *A. xylinum*, are covered in Chapter 14.

5. *General Assessment*

The preceding outline of the various approaches to the study of cellulose synthesis in *A. xylinum* indicates that strong evidence exists to suggest the conclusion that the initial stages of the process involve the following sequence: glucose → G6P → G1P → UDP-glucose →→ cellulose.

The probable existence of intermediary stages beyond UDP-glucose on the pathway to cellulose has also been indicated. The synthesis *in vivo* of fibrillar cellulose and *in vitro* of β1,4-glucan is accompanied by formation of glycolipids that appear to contain in their carbohydrate portion either glucose or β1,4-linked glucose moieties. Although the kinetics of formation of these glucolipids and their sensitivity to inhibitors of cellulose synthesis suggest their involvement as intermediates in cellulose synthesis, no precursor–product relationship has thus far been established between them and cellulose. On the other hand, such a relationship with cellulose was shown for the glycosylated water-soluble and alkali- or sodium-dodecyl-sulfate-extractable materials formed during cellulose synthesis *in vivo*. Furthermore, the kinetics of glucose transfer from these substances to cellulose suggest the water-soluble intermediate as being the one closest to cellulose. These substances have been tentatively identified as glucoproteins containing β1,4-linked glucose and might be similar to the protein-linked intermediates of cellulose synthesis proposed by Hopp *et al.* (1978a) for the alga *Prototheca zopfii* and functionally analogous to the proposed glucoprotein precursors of glycogen and starch (Lavintman *et al.*, 1974; Krisman and Barengo, 1975; Whelan, 1976).

In constructing a scheme for the sequence of reactions leading to the synthesis of cellulose, it is necessary to make provision for recent cytological findings related to the process. Until a few years ago, it was generally believed that the synthesis of crystalline cellulose microfibrils by *A. xylinum* is an extracellular process (Colvin, 1971, 1972). According to these ideas, cellulose fibrils are randomly deposited in the surrounding media and are not physical appendages of the bacterial cell (Ohad *et al.*, 1962; Ben-Hayyim and Ohad, 1965). However, several recent reports (Brown *et al.*, 1976; Forge and Preston, 1977; Zaar, 1977, 1979) strongly support the concept that each cell synthesizes a single ribbon of cellulose composed of crystalline microfibrils that elongate continuously, in association with multiple synthesizing sites organized in a row along the longitudinal axis of the cell that, in turn, are associated with corresponding extrusion sites in the lipopolysaccharide (LPS) layer of the bacteria. Accordingly, the glucan chains comprising the microfibrils are synthesized in immediate contact with the cell envelope and must be physically attached to the bacterial surface (Colvin and Leppard, 1977; Colvin *et al.*, 1977). Moreover, it is at this end of the growing glucan chain, which is most probably the reducing end, where the incorporation of new glucosyl residues occurs (Brown, 1979). It seems compatible with this concept that peripheral hydrophobic proteins are the anchoring sites for the growing glucan chains. In this role and cytological location, we can hypothesize the existence of an "anchor protein" that may tentatively be identified with the easily extractable water-soluble glucoprotein implicated as a possible intermediate of the synthetic process. We might further hypothesize that the alkali- or sodium-dodecyl-sulfate protein-linked intermediate is a hydrophobic integral protein and may be visualized to function as a "carrier protein" that, once glycosylated, either directly from UDP-glucose or indirectly via a sugar-lipid, subsequently donates cellodextrin moieties to the growing chain at the "anchor-protein" site.

The involvement of a lipid intermediate is conceivable, considering that UDP-glucose is probably synthesized in the cytoplasm or in the periplasm (Cooper and Manly, 1975c). A glycosylated lipid compound can mediate, as a carrier across the hydrophobic membranous region, between the level of UDP-glucose and that of the growing glucan chain at the cell-wall phase. Such a compound may be similar to the lipid intermediates of cellulose synthesis proposed for the alga *Prototheca zopfii* (Hopp *et al.*, 1978a) and functionally analogous to the lipid-bound sugar intermediates involved in the synthesis of bacterial cell-wall components (Lennarz and Scher, 1973).

The formation of a β1,4-polyglucan chain would thus require the cooperative action of distinct enzyme systems that promote generation of UDP-glucose, transport of active glucosyl units to phospholipid carriers, transfer of glucosyl or cellodextrin to hydrophobic glucoproteins, and transfer of cellodextrins to "anchor protein" in the outer cell wall. It has been proposed (Brown and Willison, 1977) that the enzyme systems that catalyze the multiple sequence of reactions leading to cellulose are arranged as a multienzyme complex, traversing the plasma membrane, the periplasmic space, and the outer LPS layer. Since a large number of glucan chains constitute the microfibril that is supposed to have an independent terminus in the bacterial envelope, a multisubunit enzyme complex would be expected to participate in the simultaneous synthesis of a number of glucan chains. The glucose-polymerizing capacity within each subunit will, in turn, depend on the availability of substrates and on the physical integrity and proper spatial arrangement of its diverse enzymatic and carrier components. The cooperative action of the multitude of synthesizing units thus constitutes the overall cellulose-synthesizing capacity of the cell. As to the mechanism of crystallization, evidence presented recently indicates that polymerization and crystallization of cellulose in *A. xylinum* are cell-directed coupled processes and that this coupling may well be essential for the *in vivo* assembly of the crystalline cellulose microfibril (Haigler *et al.*, 1980; Benziman *et al.*, 1980).

Although the *in vitro* synthesis of cellulose from UDP-glucose obtained with *A. xylinum* preparations is usually more convincing than that reported with other systems, the incomplete, inconsistent, and conflicting data so far obtained from the *in vitro* studies are hardly helpful in formulating a complete biochemical model of the biosynthetic pathway. Evidently, the methods employed in these studies were not sufficiently optimized to allow efficient isolation of the multienzyme synthesizing assembly from the structure of the cell as a fully active catalytic system. Furthermore, the extremely low activity of the *in vitro* systems necessitates the use of carrier-free radioactive substrates, which increases the possibility that relevant results may be overshadowed by enzyme activities not related to cellulose synthesis.

It is clear that to define, characterize, and finally reassemble the component parts of the cellulose-synthesizing system *in vitro*, more effort should be exerted to find ways to stabilize the complete system in cell-free preparations.

ACKNOWLEDGMENT

The research described in this chapter was supported by the Israel Commission for Basic Research, The Israel Academy of Sciences and Humanities.

References

Ben-Hayyim, G., and Ohad, I., 1965, Synthesis of cellulose by *Acetobacter xylinum*. VIII. On the formation and orientation of bacterial cellulose fibrils in the presence of acidic polysaccharides, *J. Cell Biol.* **25**:191.

Benziman, M., and Burger-Rachamimov, H., 1962, Synthesis of cellulose from pyruvate by succinate-grown cells of *Acetobacter xylinum*, *J. Bacteriol.* **84**:625.

Benziman, M., and Mazover, A., 1973, Nicotinamide adenine dinucleotide- and nicotinamide adenine dinucleotide phosphate-specific glucose-6-phosphate dehydrogenases of *Acetobacter xylinum* and their role in the regulation of the pentose cycle, *J. Biol. Chem.* **248**:1603.

Benziman, M., and Rivetz, B., 1972, Factors affecting hexose phosphorylation in *Acetobacter xylinum*, *J. Bacteriol.* **111**:325.

Benziman, M., Haigler, C. H., Brown, R. M., Jr., White, A. R., and Cooper, K. M., 1980, Cellulose biogenesis: Polymerization and crystallization are coupled processes in *Acetobacter xylinum*, *Proc. Natl. Acad. Sci. U.S.A.* **77**(11):6678.

Brown, R. M., Jr., 1979, Biogenesis of natural polymer systems with special reference to cellulose assembly and deposition, in: *Structure and Biochemistry of Natural Biological Systems* (W. M. Walk, ed.), pp. 51–123, Philip Morris, New York.

Brown, R. M., Jr., and Willison, J. H. M., 1977, Golgi apparatus and plasma membrane involvement in secretion and cell surface deposition, with special emphasis on cellulose biogenesis, in: *International Cell Biology 1976–1977* (B. R. Brinkley and K. P. Porter, eds.), pp. 267–285, The Rockefeller University Press, New York.

Brown, R. M., Jr., Willison, J. H. M., and Richardson, C. L., 1976, Cellulose biosynthesis in *Acetobacter xylinum:* Visualization of the site of synthesis and direct measurement of the *in vivo* process, *Proc. Natl. Acad. Sci. U.S.A.* **73**:4565.

Chumpitazi-Hermoza, B. F., Gagnaire, D., and Travel, F. R., 1978, Biosynthèse de cellulose bactérienée à partir de glycerol sélectivement deutérié en position 1, 2, ou 3: Etude par résonance magnétique nucléaire, *Biopolymers* **17**:2361.

Colvin, J. R., 1971, Structure and formation of the cellulose microfibirl, in: *High Polymers*, Vol. 5 (N. H. Bikales and L. Segal, eds.), pp. 695–718, Wiley-Interscience, New York.

Colvin, J. R., 1972, The structure and biosynthesis of cellulose, *CRC Crit. Rev. Macromol. Sci.* **1**:47.

Colvin, J. R., 1980, The mechanism of formation of cellulose-like microfibrils in a cell-free system from *Acetobacter xylinum*, *Planta* **149**:97.

Colvin, J. R., and Leppard, G. G., 1977, The biosynthesis of cellulose by *Acetobacter xylinum* and *Acetobacter acetigenus*, *Can. J. Microbiol.* **23**:701.

Colvin, J. R., Sowden, L. C., and Leppard, G. G., 1977, The structure of cellulose producing bacteria, *Acetobacter xylinum* and *Acetobacter acetigenus*, *Can. J. Microbiol.* **23**:790.

Cooper, D., and Manley, R. St. J., 1975a, Cellulose synthesis by *Acetobacter xylinum*. I. Low molecular weight compounds present in the region of synthesis, *Biochim. Biophys. Acta* **381**:78.

Cooper, D., and Manley, R. St. J., 1975b, Cellulose synthesis by *Acetobacter xylinum*. II. Investigation into the relation between cellulose synthesis and cell envelope components, *Biochim. Biophys. Acta* **381**:97.

Cooper, D., and Manley, R. St. J., 1975c, Cellulose synthesis by *Acetobacter xylinum*. III. Matrix, primer and lipid requirements and heat stability of the cellulose forming enzymes, *Biochim. Biophys. Acta* **381**:109.

Couso, R. O., Ielpi, L., García, R. C., and Dankert, M. A., 1980, Synthesis of mannosyl cellobiose diphosphate prenol in *Acetobacter xylinum*, *Arch. Biochem. Biophys.* **204**:434.

Dekker, R. F. H., Rietschel, E. T., and Sanderman, J., Jr., 1977, Isolation of α-glucan and lipopolysaccharide fractions from *Acetobacter xylinum*, *Arch. Microbiol.* **115**:353.

Delmer, D. P., 1977, The biosynthesis of cellulose and other plant cell wall polysaccharides, in: *Recent Advances in Phytochemistry*, Vol. 11 (F. A. Loewus and V. C. Runeckles, eds.), pp. 45–77, Plenum Press, New York.

Dennis, D. T., and Colvin, J. R., 1965, The relation between cellulose biosynthesis and the structure of the cell envelope in *Acetobacter xylinum*, in: *Cellular Ultrastructure of Woody Plants* (W. A. Côté, Jr., ed.), pp. 199–212, Syracuse University Press, Syracuse.

Everson, R. G., and Colvin, J. R., 1966, Rearrangement of glucose molecules during synthesis of bacterial cellulose, *Can J. Biochem.* **44**:1567.

Forge, A., 1977, Electron microscopy of a non-pellicle-forming strain of *Acetobacter xylinum*, *Ann. Bot.* 41:455.

Forge, A., and Preston, R. D., 1977, An electron microscope examination of *Acetobacter xylinum* showing the ultrastructure of the cells and the association of cellulose microfibrils, *Ann. Bot.* 41:437.

Frei-Roitman, J., 1974, Factors affecting the activity of phosphoglucomutase and UDP-glucose pyrophosphorylase of *Acetobacter xylinum*, M.Sc. thesis, The Hebrew University of Jerusalem.

Gaertner, F. H., 1978, Unique catalytic properties of enzyme clusters, *TIBS* **3**:63.

Gagnaire, D. Y., and Travel, F. R., 1975, Biosynthèse de cellulose bactérienne à partir de glucose sélectivement deuterié en position 6: Etude par RMN, *FEBS Lett* **60**:317.

Gagnaire, D. Y., and Travel, F. R., 1980, Biosynthèse de cellulose bactérienne à partir de D-glucose uniformément enrichi en ^{13}C, *Eur. J. Biochem.* **103**:133.

García, R. C., Recondo, E., and Dankert, M., 1974, Polysaccharide biosynthesis in *Acetobacter xylinum:* Enzymatic synthesis of lipid diphosphate and monophosphate sugars, *Eur. J. Biochem.* **43**:93.

Glaser, L., 1958, The synthesis of cellulose in cell-free extracts of *Acetobacter xylinum*, *J. Biol. Chem.* **232**:627.

Gromet, Z., Schramm, M., and Hestrin, S., 1957, Synthesis of cellulose by *Acetobacter xylinum*. IV. Enzyme systems present in a crude extract of glucose-grown cells, *Biochem. J.* **67**:679.

Haigler, C. H., Brown, R. M., Jr., and Benziman, M. 1980, Calcofluor White alters the *in vivo* assembly of cellulose microfibrils, *Science* **210**:903.

Hara, M., Umetsu, N., Miyamoto, C., and Tamari, K., 1973, Inhibition of the biosynthesis of plant cell wall materials, especially cellulose biosynthesis by coumarin, *Plant Cell Physiol.* **14**:11.

Hestrin, S., and Schramm, M., 1954, Synthesis of cellulose by *Acetobacter xylinum*. II. Preparation of freeze-dried cells capable of polymerizing glucose to cellulose, *Biochem. J.* **58**:345.

Hopp, E. H., Romero, P. A., Daleo, G. R., and Pont-Lezica, R., 1978a, Synthesis of cellulose precursors: The involvement of lipid-linked sugars, *Eur. J. Biochem.* **84**:561.

Hopp, E. H., Romero, P. A., and Pont-Lezica, R., 1978b, On the inhibition of cellulose synthesis by coumarin, *FEBS Lett.* **86**:259.

King, G. G. S., and Colvin, J. R., 1976, Intermediate polymers of cellulose biosynthesis, *Appl. Polym. Symp.* **28**:623.

Kjosbakken, J., and Colvin, J. R., 1973, Biosynthesis of cellulose by a particulate enzyme system from *Acetobacter xylinum*, in: *Biogenesis of Plant Cell Wall Polysaccharides* (F. Loewus, ed.), pp. 361–371, Academic Press, New York and London.

Kjosbakken, J., and Colvin, J. R., 1975, New evidence for an intermediate polymer of glucose in cellulose biosynthesis by *Acetobacter xylinum*, *Can. J. Microbiol.* **21**:111.

Kornfeld, R., and Kornfeld, S., 1976, Comparative aspects of glycoprotein structure, *Annu. Rev. Biochem.* **45**:217.

Kornfeld, S., Benziman, M., and Milner, Y., 1978, Regulatory properties of the α-ketoglutarate dehydrogenase complex of *Acetobacter xylinum: In situ* studies and the localization of the allosteric response in the E_1 component, *J. Biol. Chem.* **253**:5678.

Krisman, C. R., and Barengo, R., 1975, A precursor of glycogen biosynthesis: α-1,4-Glucan protein, *Eur. J. Biochem.* **52**:117.

Lavintman, N., Tandercarz, J., Carceller, N., Mendiara, S., and Cardini, C. E., 1974, Role of uridine diphosphate glucose in the biosynthesis of starch, *Eur. J. Biochem.* **50:**145.

Leisinger, T., 1966, Enzymatische Untersuchungen an Wildtyp and an einer cellulosefreien Mutante von *Acetobacter xylinus, Pathol. Microbiol.* **29:**756.

Lennarz, W. J., and Scher, M. G., 1973, The role of lipid-linked activated sugars in glycosylation reactions, in: *Membrane Structure and Mechanisms of Biological Energy Transduction* (J. Avery, ed.), pp. 441–453, Plenum Press, New York.

Montezinos, D., and Delmer, D. P., 1980, Characterization of inhibitors of cellulose synthesis in cotton fibers, *Planta* **148:**303.

Ohad, I., Danon, M. D., and Hestrin, S., 1962, Synthesis of cellulose by *Acetobacter xylinum*. V. Ultrastructure of polymer, *J. Cell Biol.* **12:**31.

Sanderman, H., Jr., 1977, Formation of an acid-labile maltosyl-lipid by enzyme preparations from *Acetobacter xylinum, FEBS Lett.* **81:**294.

Sanderman, H., Jr., and Dekker, R. F. H., 1979, β-1,2-Glucosyl transfer by membrane preparations from *Acetobacter xylinum, FEBS Lett.* **107:**237.

Sato, S., Matsuda, K., and Tamari, K., 1976, β-1,4-Glucan occurring in homogenate of *Phaseolus aureus* seedling: Possible nascent stage of cellulose biosynthesis *in vivo, Plant Cell Physiol.* **17:**1243.

Schramm, M., and Hestrin, S., 1954, Factors affecting production of cellulose at the air/liquid interface of a culture of *Acetobacter xylinum, J. Gen. Microbiol.* **11:**123.

Schramm, M., Gromet, Z., and Hestrin, S., 1957a, Synthesis of cellulose by *Acetobacter xylinum*. III. Substrates and inhibitors, *Biochem. J.* **67:**669.

Schramm, M., Gromet, Z., and Hestrin, S., 1957b, Role of hexose phosphate in synthesis of cellulose by *Acetobacter xylinum, Nature (London)* **179:**28.

Siewert, G., and Strominger, J. L., 1967, Bacitracin: An inhibitor of the dephosphorylation of lipid pyrophosphate, an intermediate in biosynthesis of the peptidoglycan of bacterial cell walls, *Proc. Natl. Acad. Sci. U.S.A.* **57:**767.

Stone, K. J., and Strominger, J. L., 1971, Mechanism of action of bacitracin: Complexation with metal ion and C_{55}-isoprenyl pyrophosphate, *Proc. Natl. Acad. Sci. U.S.A.* **68:**3223.

Swissa, M., 1978, Biosynthesis of cellulose in *Acetobacter xylinum*, Ph.D. thesis, The Hebrew University of Jerusalem.

Swissa, M., and Benziman, M., 1976, Factors affecting the activity of citrate synthase of *Acetobacter xylinum* and its possible regulatory role, *Biochem. J.* **153:**173.

Swissa, M., Weinhouse, H., and Benziman, M., 1976, Activities of citrate synthase and other enzymes of *Acetobacter xylinum in situ* and *in vitro, Biochem. J.* **153:**499.

Swissa, M., Aloni, Y., Weinhouse, H., and Benziman, M., 1980, Intermediary steps in cellulose synthesis in *Acetobacter xylinum:* Studies with whole cells and cell-free preparations of the wild type and a celluloseless mutant, *J. Bacteriol.* **143:**1142.

Webb, T. E., and Colvin, R. C., 1962, The lysis of *Acetobacter xylinum, Can. J. Microbiol.* **8:**841.

Webb, T. E., and Colvin, R. C., 1963, The effect of bacterial cell lysis and of plant extracts on cellulose production by *Acetobacter xylinum, Can. J. Biochem. Physiol.* **41:**1691.

Webb, T. E., and Colvin, J. R., 1967, The extracellular proteins of *Acetobacter xylinum* and their relationship to cellulose synthesis, *Can. J. Biochem.* **45:**465.

Weinhouse, H., 1977, Regulation of carbohydrate metabolism in *Acetobacter xylinum*, Ph.D. thesis, The Hebrew University of Jerusalem.

Weinhouse, H., and Benziman, M., 1974, Regulation of hexose phosphate metabolism in *Acetobacter xylinum, Biochem. J.* **138:**537.

Weinhouse, H., and Benziman, M., 1976, Phosphorylation of glycerol and dihydroxyacetone in *Acetobacter xylinum* and its possible regulatory role, *J. Bacteriol.* **127:**747.

Whelan, W. J., 1976, On the origin of primer for glycogen xynthesis, *TIBS* **1:**13.

White, G. A., and Wang, C. H., 1964, The dissimilation of glucose and gluconate by *Acetobacter xylinum*. I. The origin and the fate of triose phosphate, *Biochem. J.* **90:**408.

Zaar, K., 1977, The biogenesis of cellulose by *Acetobacter xylinum, Cytobiologie* **16:**1.

Zaar, K., 1979, Visualization of pores (export sites) correlated with cellulose production in the envelope of the gram-negative bacterium *Acetobacter xylinum, J. Cell Biol.* **80:**773.

18

Protein Glycosylation in Higher Plants: Recent Developments

H. Maelor Davies and Deborah P. Delmer

1. Introduction

Within the last few years, very rapid progress has been made in the investigation of the biosynthesis of the most common type of glycoproteins, i.e., those having an *N*-glycosidic link between the oligosaccharide(s) and the amide nitrogen of asparagine residues on the protein. Pathways that involve the use of glycolipid intermediates have been delineated, and new, unexpected, phenomena such as oligosaccharide processing have been discovered. In this chapter, we shall review developments in this area with respect to higher plants, concentrating on the transfer of sugars to protein rather than on details of the synthesis of glycolipid precursors. An excellent review on the formation of lipid-linked sugars in plants and their functions in glycoprotein synthesis was recently published by Elbein (1980). The reader interested in the synthesis of *O*-glycosidically linked oligosaccharides of plant glycoproteins is referred to papers by Karr (1972) on the glycosylation of extensin and Soliday and Kolattukudy (1979) on the glycosylation of fungal cutinase as examples. We shall also exclude from this chapter the protein-linked carbohydrates proposed as primers or intermediates in polysaccharide synthesis. The existence, structure, and functioning of these are still controversial issues, and evidence is limited [for example, their consideration with respect to cellulose synthesis will be found in the paper by Hopp *et al.* (1978) and for starch synthesis in the paper by Lavintman *et al.*, (1974)].

As in animals, the plant glycoproteins with *N*-glycosidic linkages fulfill many roles (for a review of functions and structure, see Lamport, 1980; *e.g.* they include lectins, enzymes, and legume seed reserve proteins. While the importance

H. Maelor Davies* and Deborah P. Delmer • MSU-DOE Plant Research Laboratory, Michigan State University, East Lansing, Michigan 18824.
*Present address: Calgene Inc., 1910 Fifth St., Suite F, Davis, CA 95616.

of such glycoproteins is obvious, the exact function of the carbohydrate group(s) is not known. Suggestions include an influence on the solubility and conformation of the protein and the provision of specific recognition sites, especially at the cell surface. Also, perhaps, there is processing of the oligosaccharide, a phenomenon that may be involved in routing the newly synthesized glycoprotein from the site of its synthesis to its final location in the cell.

2. *In Vitro Glycosylation Systems from Higher Plants*

The mechanism of protein glycosylation has been examined using *in vitro* systems from several different higher plant tissues: *Phaseolus aureus* (mung bean) roots (Alam and Hemming, 1973; Ericson *et al.*, 1978; Lehle *et al.*, 1978; Roberts and Pollard, 1975), *Pisum sativum* (pea) cotyledons (Beevers and Mense, 1977), pea stems (Durr *et al.*, 1979; Pont Lezica *et al.*, 1978), *Phaseolus vulgaris* (kidney bean) cotyledons (Ericson and Delmer, 1977), *Gossypium hirsutum* (cotton) fibers (Ericson *et al.*, 1978), *Ricinus communis* (castor bean) endosperm (Marriott and Tanner, 1979; Mellor and Lord, 1978, 1979), and *Armoracia rusticana* (horseradish) roots (Sevier and Shannon, 1977). The preparations from developing legume cotyledons are of special interest, since certain of the seed reserve proteins of legumes are glycosylated. The study of glycosylation in these systems may reveal important details concerning the factors that govern the quality and quantity of the deposited seed proteins, in addition to supplying information relevant to glycoprotein synthesis in general. They also have the advantage that a small number of glycoproteins are being synthesized in large amounts, potentially simplifying the analysis of products.

Ericson and Delmer (1977) used a crude tissue homogenate from which cell-wall and other debris was removed by centrifugation at 600*g*. The widely adopted method for an *in vitro* preparation, however, includes a further centrifugation at 30,000–175,000*g* to provide a total membrane pellet. This is usually washed once in buffer and then resuspended in buffer for use.

Such a preparation can catalyze the reactions that are now recognized as essential steps in the glycosylation of protein asparagine residues in animal and fungal systems (for a review, see Parodi and Leloir, 1979):*

1. UDP-GlcNAc + lipid-P → lipid-P-P-GlcNAc + UMP

2. UDP-GlcNAc + lipid-P-P-GlcNAC → lipid-P-P-$(GlcNAc)_2$ + UMP

3. GDP-man + lipid-P-P$(GlcNAc)_2$ → lipid-P-P$(GlcNAc)_2$man + GDP
   Repetition of step 3 with mannose donation from GDPman gives lipid-P-P$(GlcNac)_2$-man_5

*Abbreviations used in the reactions and in the following text are: (GlcNAc) *N*-acetyl-D-glucosamine; (—P—) phosphoryl; (man) D-mannose.

4. GDP-man + lipid-P → lipid-P-man + GDP

5. Lipid-P-P-$(GlcNAc)_2man_5$ + lipid-P-man → lipid-P-P-$(GlcNac)_2man_6$ + lipid-P

This sequence of reactions, with repetiton of step 5, results in the assembly of an oligosaccharide of GlcNAc and mannose residues on a carrier lipid, which in animals is a polyisoprenoid of the dolichol type. Finally, the oligosaccharide is transferred from the carrier lipid to the protein asparagine residue. [Recent studies by Kornfeld *et al.* (1980) have shown that the first five mannose residues of the growing oligosaccharide may be donated from GDP-man as in step 3, but that the sixth and subsequent ones must come from lipid-P-man.]

The *in vitro* membrane systems from higher plants incorporate radioactive GlcNAc and mannose from radiolabeled UDP-GlcNAc and GDP-man, respectively, into lipid-P-monosaccharide (extractable in $CHCl_3$-CH_3OH, 1:1 by volume). Comparisons of the properties of these products with authentic polyprenyl phosphate derivatives of GlcNAc and mannose have suggested their polyprenyl nature. Delmer *et al.* (1978) isolated the lipid-P-man made by extracts of kidney bean cotyledons; analysis of the structure of the lipid by mass spectrometry showed it to be of the dolichol type. The involvement of this type of lipid is also suggested by the ability of some extracts to use exogenously supplied dolichyl-P-man in the reaction sequence above (e.g., see Alam and Hemming, 1973) and the stimulation of the reactions by provision of dolichyl phosphate (e.g., see Mellor and Lord, 1979). There is some evidence from mung bean hypocotyl and cotton fiber preparations to suggest that the lipid acceptor for GlcNAc, although also a polyprenol, may be of a different type from that utilized for mannose (Ericson *et al.*, 1978), but it has yet to be characterized.

The plant membranes also incorporate these sugars into lipid-linked oligosaccharides (extractable in $CHCl_3$-CH_3OH-H_2O, 1:1:0.3 by volume) (steps 2, 3, and 5 in the reaction sequence above) and into a lipid-free residue fraction insoluble in chloroform–methanol mixtures. All the radiolabeled GlcNAc incorporated into this "insoluble residue" fraction can be resolubilized into aqueous buffer by digestion with protease, and this suggests that the GlcNAc-labeled products in this material are all glycoproteins. Only a portion of the mannose-labeled products can be digested with protease, the exact proportion depending on the source of the membranes. Mannose-labeled products resistant to protease are probably oligomannans, synthesized from GDP-man without the involvement of lipid carriers (Elbein and Hassid, 1966; Elbein, 1969; Smith *et al.*, 1976). Their synthesis by plant extracts is well known, but their purpose is not understood.

Note that only nucleotide sugars and divalent cation (Mg^{2+} or Mn^{2+}) need be added to the membrane preparations; lipid carriers and acceptor proteins are endogenous components of the system. To study the reactions in more detail, exogenous carriers and acceptors can be added. Thus, lipid was added, as mentioned above, to confirm the involvement of polyisoprenoids and to stimulate relevant steps. Sevier and Shannon (1977) supplied horseradish root preparations with variously treated horseradish peroxidase samples to establish the best glycoprotein acceptor, and much valuable work has been done with exogenously supplied deglycosylated glycoproteins in animal systems to establish the amino acid sequence at the glycosylated sites.

Given that plant-membrane preparations have the requisite carrier lipids, acceptor proteins, and glycosyl transferases to undertake the reactions in the scheme above, it should be possible to demonstrate the operation of the whole pathway. This would comprise steps 1 through 5, followed by reiteration of step 5 and transfer of the final oligosaccharide product to a protein asparagine residue. In considering the evidence that such a pathway does take place and that the incorporations observed *in vitro* are not the result of isolated, uncoupled reactions, it is important to keep in mind that this scheme is not the "whole story" of protein glycosylation in animals. The oligosaccharide can also receive glucose residues from lipid-linked glucose, at least in some systems. Transfer of the oligosaccharide from lipid to protein, termed "core" glycosylation, is followed by the removal of the glucose residues and some mannose residues in events termed "oligosaccharide processing" (reviewed by Parodi and Leloir, 1979). The resulting oligosaccharide is of the high mannose type composed of GlcNAc and mannose as generalized below (the numbers of branches and mannose residues are variable but always small; the two GlcNAc residues adjacent to the protein are invariant):

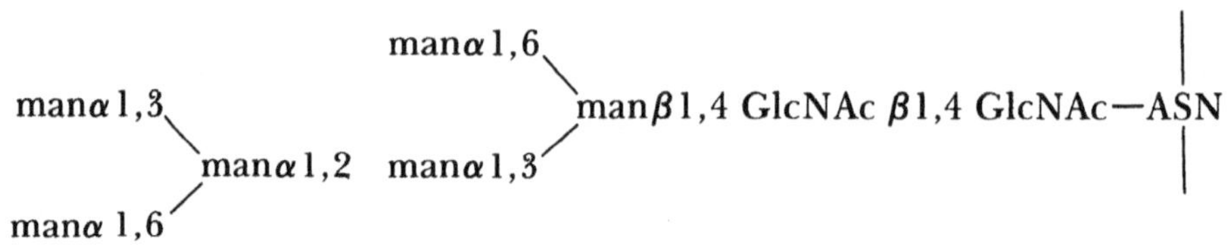

Analyses of many of the plant lectins and legume seed proteins have shown the presence of GlcNac and mannose as the predominant sugars (Delmer and Lamport, 1977; Sharon, 1974, and recently Lis and Sharon (1978) have analyzed the structure of the soybean lectin in detail and shown its similarity to the preceding generalized structure.

In some glycoproteins, the processed oligosaccharide may be converted to the "complex" type by the addition of other sugars (e.g., fucose, galactose, and sialic acid) directly from their nucleotide derivatives (Turco and Robbins, 1979). It is of some interest to note that sialic acid has never been detected as a constituent of a plant glycoprotein.

3. *The Evidence for Core Glycosylation in Vitro*

What evidence exists for the operation of the lipid-linked core glycosylation pathway in higher plant cell-free systems? Perhaps the most direct demonstration is achieved using "pulse-chase" experiments in which labeled GlcNAc or mannose is followed through successive derivatives in all or part of the reaction sequence and ultimately into glycoprotein products. For example, Forsee and Elbein (1975) showed that radioactive mannose incorporated into small lipid-linked oligosaccharides (e.g., trisaccharides) in preparations from cotton fibers could be "chased" into larger lipid-oligosaccharides (8–10 glycose units) by further incubation with unlabeled GDP-man. In the absence of exogenously supplied GDP-man, UDP-[^{3}H]GlcNAc labeled only lipid-P-P-(GlcNac)$_2$, but in the presence of GDP-man, most of the radioactivity was found in larger lipid-linked oligosaccharides of the same size as those labeled with mannose. Thus, mannose and GlcNAc were probably incorporated into the same lipid-oligosaccharide. Similar experiments were reported by Beevers and Mense (1977) with a pea cotyledon extract and by Ericson and Delmer (1978) with a kidney bean extract. Lehle *et al.* (1976) isolated GlcNAc-labeled dolichyl pyrophosphoryl di-*N*-acetyl chitobiose from their mung bean hypocotyl system and showed that incubation of the preparation with this component and unlabeled GDP-man gave rise to lipid-linked mannosyl-di-*N*-acetyl chitobiose. Also, isolated lipid-linked radiolabeled mannose or GlcNAc, when added back to the preparation, labeled some endogenous glycoproteins. More recently, Marriott and Tanner (1979) undertook similar studies with membranes from the endosperm of germinating castor bean seeds. The membranes were preincubated with GDP-man and UDP-GlcNAc, and radioactivity from a subsequent incubation with dolichyl phosphoryl [^{14}C]mannose was incorporated into macromolecular products.

Ericson and Delmer (1978), working with a preparation from developing cotyledons of kidney bean, described an interdependence of exogenously supplied GDP-man and UDP-GlcNAc for the incorporation of either mannose or GlcNAc into the insoluble-residue fraction. This interdependence is excellent evidence for the lipid-linked pathway, but, unfortunately, we have experienced difficulties in reproducing these results. We suggest two reasons for this. One is the possible existence of endogenous pools of nucleotide sugars or lipid-linked intermediates, or both, the sizes of which may vary with the preparation. The existence of large pools would preclude, for example, exposure of the extract to UDP-GlcNAc in the complete absence of GDP-man. Second, in the course of our investigations into the nature of the labeled insoluble-residue fraction, we found a different kind of glycosylation with GlcNAc (described below). It seems that in this system only a very small fraction of the GlcNAc that is incorporated into glycoproteins is transferred via core glycosylation.

We were interested to know whether the seed reserve glycoprotein vicilin [also known as "7 S globulin," "G1" (McLeester *et al.*, 1973), and "Glycoprotein II" (Pusztai and Watt, 1970), and "Phaseolin" (Osborne, 1894)] was among the macromolecular products of *in vitro* glycosylation in the kidney bean cotyledon system. Both vicilin and the minor kidney bean storage protein phytohemagglutinin (PHA) have been shown to contain typical asparagine-linked oligosaccharides comprising GlcNAc and mannose as sole constituent sugars (Ericson and Delmer, 1978; Leavitt *et al.*, 1977). We extracted GlcNAc- or mannose-

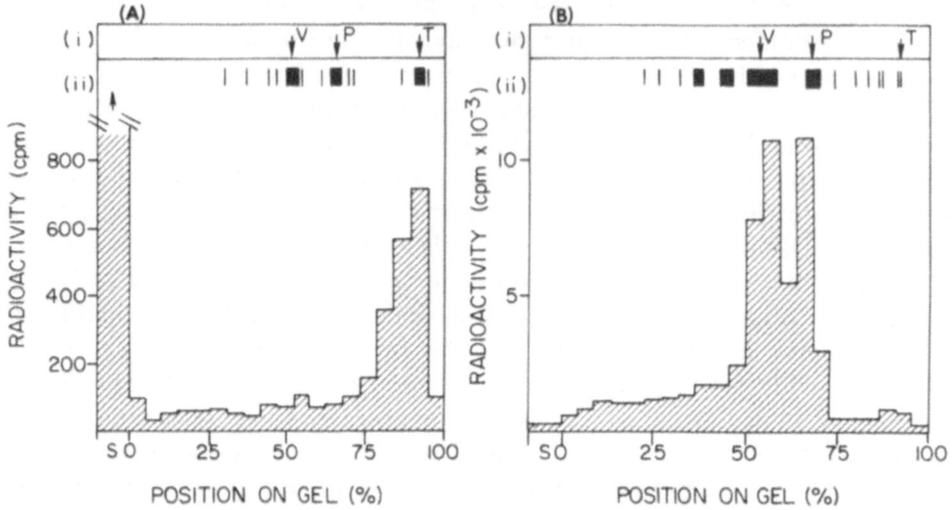

FIGURE 1. Analysis of the glycoprotein products of an *in vitro* glycosylation system from developing kidney bean (*Phaseolus vulgaris*) cotyledons. Insoluble residue components labeled *in vitro* with (A) [U-^{14}C]man (from GDP-[^{14}C]man) and (B) [6-^{3}H]GlcNAc (from UDP-[6-^{3}H]GlcNAc) were extracted with buffered SDS and subjected to SDS-polyacrylamide gel electrophoresis. The positions of vicilin (V), PHA (erythroagglutinin and leucoagglutinin) (P), and tracking dye (T) were identified (i) by inspection of a central strip from the slab gel after staining in Coomassie blue (ii). Radioactivity was determined in horizontal 3-mm slices of the slab after elution into scintillation fluid. (S) Stacking gel, which contained 33,000 cpm [U-^{14}C]man in (A). From Davies and Delmer (1979); reprinted with the kind permission of Springer-Verlag.

labeled insoluble residue in hot concentrated sodium dodecyl sulfate (SDS) and subjected the extracts to SDS–polyacrylamide gel electrophoresis (Fig. 1) (Davies and Delmer, 1979). The SDS extracted about 40% of the mannose radioactivity (the same proportion could be solubilized with protease), but except for some small peptides migrating with the tracking dye, mannose-labeled proteins were not found in the gel. In contrast, GlcNAc label was located in bands having the same mobilities as vicilin and PHA. This was confirmed using other polyacrylamide gel electrophoresis methods. Also, the attachment of labeled GlcNAc to vicilin was shown by immunochemical assay using specific antivicilin antiserum; mannose-labeled products showed only a very slight reaction.

Resolution of SDS-solubilized mannose-labeled products was possible on a gel-filtration column, and the attachment of mannose to proteins was confirmed by the markedly reduced size of the material after attack with protease (Fig. 2) (Davies and Delmer, 1981). Note the size heterogeneity of the mannose-labeled proteins. Three broad peaks were recognized in this elution profile, as indicated in Fig. 2. Material in peak I did not enter polyacrylamide gels, whereas that from peak III migrated with the tracking dye in the manner of small peptides. The pooled and concentrated fractions from peak II, however, gave a diffuse distribution of radioactivity in an SDS gel, as shown in Fig. 3. (It was often difficult to detect these glycoproteins on gels when, as in Fig. 1, the whole SDS extract was applied directly, without a preceding gel filtration. This could be

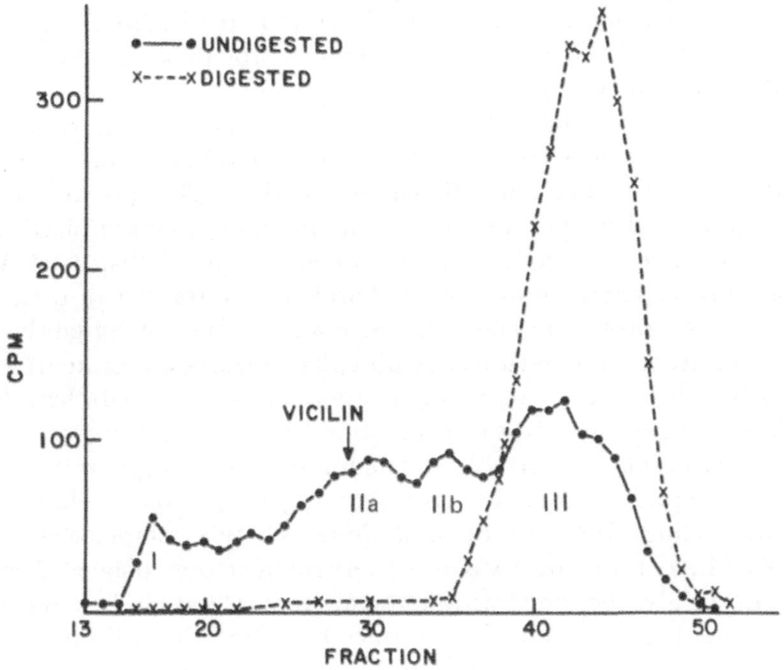

FIGURE 2. Analysis of the products of *in vitro* mannosylation in a preparation from developing cotyledons of kidney bean, by gel-filtration chromatography. Insoluble residue components labeled *in vitro* with [^{14}C]man (from GDP-[^{14}C]man) were either digested with protease K (X----X) or left untreated (●—●) before solubilization with buffered SDS and application to a column of LKB "Ultrogel" AcA 34. (↓) Position of vicilin extracted from mature seeds. Protease K converted the size of material of peaks I, IIa, and IIb to the same small size as that of peak III, indicating that the labeled mannose of I, IIa, and IIb was incorporated into glycoproteins.

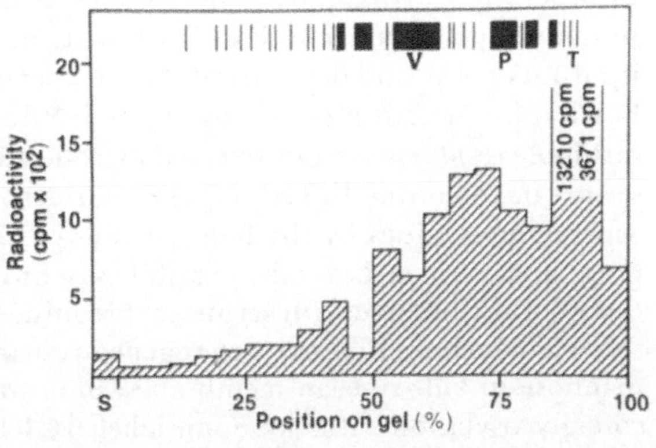

FIGURE 3. Resolution of *in vitro* mannosylated glycoproteins from developing kidney bean cotyledon membranes. The products of *in vitro* mannosylation from GDP-[U-^{14}C]man were extracted with SDS and analyzed on a gel-filtration column as described in the Fig. 2 caption. Fractions corresponding to peaks IIa and IIb (Fig. 2) were pooled, concentrated, and subjected to SDS-polyacrylamide gel electrophoresis. A central strip of the gel was stained with Coomassie blue to show the protein bands, and the positions of vicilin (V), PHA (P), and tracking dye (T) were noted. The remainder of the gel was cut into 3-mm slices for determination of the distribution of radioactivity.

accounted for by our finding that the preparations often yielded much larger amounts of peak I material than of peaks II and III.) Radioactivity was not obviously associated with any of the well-defined protein bands, and much of it migrated faster than vicilin and PHA.

Our finding that the bulk of the GlcNAc incorporated into glycoproteins *in vitro* was linked to proteins recognizable as vicilin and PHA, while much of the mannose was incorporated into different, unidentified, glycoproteins, led us to suspect that most of the GlcNAc incorporation was not taking place via core glycosylation. The core-glycosylation pathway requires that GlcNAc and mannose be incorporated into oligosaccharide before transfer of the latter to protein, so that a protein receiving GlcNAc would also concimitantly receive mannose and vice versa. Therefore, it is difficult to imagine one protein receiving GlcNAc and a different one receiving mannose (as we observed), both by core glycosylation. We examined the incorporations in more detail by comparing the linkages of the radioactive GlcNAc and mannose in lipid-oligosaccharide with those in glycopeptides prepared by proteolytic digestion of the insoluble residue (Davies and Delmer, 1981). In the lipid-oligosaccharide, radiolabeled GlcNAc was mostly 4-linked, and there was only a very small amount of labeled terminal GlcNAc. In the glycopeptide fraction, the reverse was found; there was mostly labeled terminal GlcNAc and only a trace of radioactive 4-linked GlcNAc. Because the linkage of radiolabeled GlcNAc was different, most of it could not have been incorporated into glycoproteins by transfer of core oligosaccharide from lipid. Instead, GlcNAc was probably attached terminally to oligosaccharides already linked to vicilin and PHA. This explanation received strong support from our finding that tunicamycin, an inhibitor of the synthesis of lipid-oligosaccharide in this system [and an inhibitor of lipid—P-GlcNAc synthesis in a great many systems (Elbein *et al.*, 1979)], had only a very small inhibitory effect on the incorporation of GlcNAc into insoluble residue (Fig. 4). The small amount of labeled GlcNAc in 4-linkage in the glycoprotein products suggested that a small amount of core glycosylation occurred. This was borne out by the linkage analysis for labeled mannose, which showed the same pattern in lipid-oligosaccharide as in glycopeptides. Mannose was incorporated into terminal, 2-, and 3,6- (much less) linkages in both fractions, this linkage pattern being typical for asparagine-linked core oligosaccharides. Furthermore, the incorporation of mannose into lipid-oligosaccharide and glycoproteins was inhibited by tunicamycin (Fig. 4), showing its dependence on the availability of lipid-P-GlcNAc and demonstrating the operation of the lipid-linked pathway. Finally, it is worth mentioning that GlcNAc and mannose incorporated into insoluble residue were not released by conditions suitable for β-elimination of serine- or threonine-linked oligosaccharides, thus suggesting their linkage to asparagine residues by the familiar *N*-glycosidic bond. This is an important point because it excludes the possibility of direct incorporation of GlcNAc into *O*-glycosidic linkage with serine or threonine residues *in vitro*.

We conclude that the heterogeneous array of proteins glycosylated with mannose in kidney bean membranes, *in vitro*, receive mannose by the classic core-glycosylation pathway. Some labeled GlcNAc must also be incorporated by

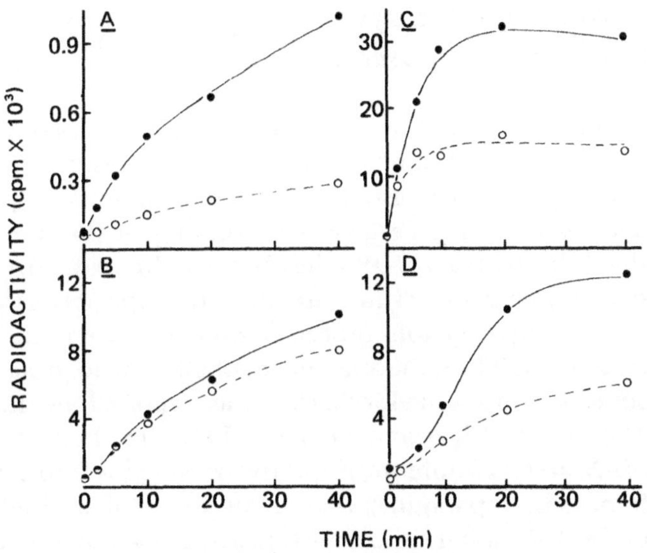

FIGURE 4. Effect of tunicamycin (0.1 mg/ml) on the synthesis of lipid-oligosaccharide and glycoproteins in a preparation from developing cotyledons of kidney bean. The membrane preparation was incubated with UDP-[6-^{3}H]GlcNAc (A, B) or GDP-[U-^{14}C]man (C, D) with (O- - -O) or without (●—●) tunicamycin, and the incorporation of the radiolabeled sugar into lipid-oligosaccharide (A, C) and protease-digestible residue (glycopeptide from glycoprotein products) (B, D) was followed.

this route, but the bulk of that supplied as UDP-[^{3}H]GlcNAc goes directly to the nonreducing termini of oligosaccharides already linked to vicilin and PHA, without the involvement of a lipid carrier.

Pont Lezica *et al.* (1978) presented some evidence for the incorporation of radiolabeled glucose from UDP-glucose into lipid-oligosaccharide and into glycoproteins in a pea epicotyl preparation. Dolichyl phosphoryl glucose would also donate glucose to these fractions, and, curiously, their results suggested glucose to be the only sugar type in the lipid-oligosaccharide, seven or eight residues being present per molecule. They did not show very conclusive evidence for the actual flow of glucose from UDP-glucose through dolichyl phosphoryl glucose, lipid-linked oligosaccharides, and into glycoprotein, however, and their result must be regarded as preliminary at this time. Other workers have found only a very small transfer of glucose from UDP-glucose to lipid-oligosaccharide in several plant systems (Durr *et al.*, 1979; D. P. Delmer, unpublished observations; A. D. Elbein, personal communication).

The discovery of glucose residues at the nonreducing ends of lipid-linked core oligosaccharides and the phenomenon of oligosaccharide processing in animal systems led us to begin an examination of lipid-linked glucosylation in our kidney bean cotyledon preparation. We have shown the incorporation of radiolabeled glucose (from radioactive UDP-glucose) into a heterogeneous array of glycoproteins giving exactly the same distribution of radioactivity on an SDS-polyacrylamide gel as mannose-labeled products (e.g., as shown in Fig. 3). (H. M. Davies and D. P. Delmer, unpublished results). It is possible that the same proteins that received mannose received glucose.

4. Glycosylation by Pathways Other Than Oligosaccharide Transfer from Lipid-Oligosaccharide

According to current thinking, core glycosylation proceeds by oligosaccharide transfer from lipid-oligosaccharide, and additions of single sugar residues can take place later, especially in the synthesis of protein-linked oligosaccharides of the complex type. Our finding of the transfer of single GlcNAc residues directly from UDP-GlcNAc to vicilin and PHA in *Phaseolus vulgaris* cotyledon preparations illustrates the latter type of reaction. There are other examples of glycosylaton processes different from the classic core-glycosylation pathway. Thus, the attachment of GlcNAc to protein without its prior incorporation into oligosaccharide was reported by Khalkhali and Marshall (1976) and by Chen and Lennarz (1977). In these animal systems, GlcNAc and di-*N*-acetyl chitobiose, respectively, were incorporated into direct linkage with protein asparagine; the involvement of a lipid carrier was established for the di-*N*-acetyl chitobiose transfer, but was not considered in the single GlcNAc attachment. A closer similarity to our result with the kidney bean system is to be found in the studies by Struck and Lennarz (1977), Kean *et al.* (1980), and Sevier and Shannon (1977). The first two groups demonstrated the tunicamycin-insensitive incorporation of GlcNAc from UDP-GlcNAc into terminal linkage on the oligosaccharide of chicken ovalbumin and bovine rhodopsin, respectively. As in our system, the incorporated GlcNAc could be released using an exoglycosidase, β-*N*-acetyl hexosaminidase. Sevier and Shannon (1977) worked with a particulate preparation from horseradish root and supplied the acceptor protein exogenously in the form of partially deglycosylated horseradish peroxidase. The transferase system used the acceptor most efficiently when only 12% of the GlcNAc was removed from it, corresponding to an average removal of one GlcNAc molecule per oligosaccharide chain. They concluded that the preparation added GlcNAc residues from UDP-GlcNAc to the nonreducing termini of oligosaccharide groups on the acceptor protein.

The kidney bean cotyledon system differs from all these examples in a significant way. The systems discussed above are involved in the synthesis of glycoproteins that have terminal GlcNAc residues, but vicilin, although a recipient of GlcNAc into terminal linkage *in vitro*, does not possess any terminal GlcNAc when isolated from mature seeds (Davies and Delmer, 1981). It is possible that the attachment of GlcNAc residues to vicilin *in vitro* is an artifact of the preparation. Alternatively, terminal GlcNAc residues may be subsequently removed in a processing event, analogous to the oligosaccharide processing described in animal systems. It is worth noting in passing that a β-*N*-acetyl glucosaminidase has been identified among several enzymes associated with protein bodies in mung bean cotyledons (Van der Wilden *et al.*, 1980). Such an activity could conceivably remove terminal GlcNAc from storage glycoproteins such as vicilin during their deposition in the developing protein body.

Currently, most of the known glycoproteins from plants are of the "high mannose" type. One can perhaps anticipate eventual description of more of the "complex" type and the demonstration of glycosylation events responsible for the addition of sugars other than mannose, glucose, and GlcNAc.

5. *Glycoprotein Products of the in Vitro Systems*

Comparatively few studies have been concerned with the identity of the endogenous proteins glycosylated by plant membrane preparatins. A preparation from the endosperm of geminating castor bean seeds glycosylates several proteins using exogenously supplied GDP-[^{14}C]man, as shown by SDS-polyacrylamide gel electrophoresis (Marriott and Tanner, 1979). The radioactive mannose was not released under conditions suitable for β-elimination, and so its attachment to protein involves an *N*-glycosidic linkage. No further characterization of the separated glycoprotein products was undertaken, but the gel results indicated that the seed lectin glycoprotein ricin was not among them. Roberts and Pollard (1975) attempted to resolve glycoprotein products from a particulate preparation of mung bean hypocotyls incubated with [^{14}C]glucosamine *in vitro*, but very low incorporations were obtained.

In their studies on glucosylation in pea epicotyls, mentioned earlier, Pont Lezica *et al.* (1978) analyzed the macromolecular products by SDS-gel electrophoresis. Several proteins were labeled when UDP-[^{14}C]glucose was supplied, but apparently only one, of approximate molecular weight 13,000, when dolichyl phosphoryl [^{14}C]glucose was provided. Similarity of this molecular weight to that of a pea seed lectin subunit, and the demonstration that a portion of the radioactive products would bind to Sephadex and could be eluted using glucose, led the authors to suggest that glucose-specific lectins were among the products. Such a conclusion requires more evidence, however, and the presence of lectins in the epicotyl, analogous to the seed lectins, needs to be substantiated.

As mentioned earlier, we obtained several lines of evidence to indicate that the kidney bean seed reserve glycoproteins vicilin and PHA are glycosylated with GlcNAc in membranes from developing cotyledons (Davies and Delmer, 1979). The incorporation of mannose can be taken as indicative of core glycosylation, and a hetergeneous array of proteins is so glycosylated (See Figs. 2 and 3) (Davies and Delmer, 1981). A small quantity of vicilin may be among these, since we did obtain a slight reaction with antivicilin antiserum, and the range of molecular weights does encompass those of vicilin and PHA subunits. At this time, however, the identities of these diverse glycoprotein products are unknown. An obvious possibility was that they represented nascent polypeptide chains of vicilin, and perhaps also of PHA and other proteins, because the occurrence of glycosylation during translation has been shown in animal and viral systems [e.g., see Hanover and Lennarz (1980) and references cited therein]. We were, however, unable to obtain any binding of the mannose-labeled glycoproteins to anion exchange resins using methods developed for binding peptidyl-transfer RNA (tRNA). The sizes of the products, as shown by gel-filtration chromatography, were not affected by incubation with ribonuclease, again suggesting the absence of any attached tRNA. Also, the glycosylation from GDP-man *in vitro* was not reproducibly affected by the presence of puromycin, an antibiotic that detaches nascent polypeptide chains from ribosomes and microsomal membranes. Having established that translation did not occur in the membrane preparation (presumably because of the removal of soluble factors and substrates), we attempted to stimulate mannosylation of proteins by increasing the supply of nascent chains (H. M. Davies and D. P. Delmer, unpublished preliminary studies). We were able to restore protein synthesis in the cotyledon membranes by adding a wheat germ translation system lacking messenger RNA, as a source of soluble factors, but there was no change in the rate of incorporation of [1-^{3}H]mannose from GDP-[1-^{3}H]man into macromolecular products. Suitable though the nascent-chain hypothesis seemed, we have therefore not yet been able to find any support for it.

Intriguingly, the results of Hanover and Lennarz (1980) with an *in vitro* glycosylation system from hen oviduct are almost identical to ours. Although hen oviduct is heavily engaged in the synthesis of ovalbumin, and glycosylation of nascent ovalbumin chains is readily demonstrated *in vivo*, nascent chains of this protein are not among the products of *in vitro* glycosylation. The proteins that received mannose *in vitro* did not bind to antiovalbumin antibodies, did not copurify with nascent chains labeled with [^{35}S]methionine *in vivo*, and were not isolated with polyribosomes prepared from the *in vitro* membrane system. Mannosylation *in vitro* was not inhibited by puromycin, and the mannosylated proteins were not bound by an anion-exchange column under conditions suitable for binding peptidyl-tRNA (J. A. Hanover, personal communication). In both kidney bean cotyledon and chicken oviduct preparations, therefore, it appears that the heterogeneous arrays of proteins receiving core oligosaccharide are not nascent polypeptide chains and are not significantly related to the principal protein synthesized by each tissue. The identity of these acceptors is a complete mystery at present. Perhaps they comprise a multiplicity of membrane proteins that are glycosylated post-translationally *in vivo*. If this is so, they are present in very small amounts, because when we stained SDS gels containing them with Coomassie blue, only very weak and diffuse staining corresponded with the location of much of the mannose radioactivity (see Fig. 3). None of the sharp bands was labeled with mannose to a significant extent relative to the diffusely stained regions. Another possibility is that they are normally unglycosylated proteins *in vivo*, which somehow gain access to the glycosylation system as an artifactual occurrence during disruption of the tissue. It will be interesting to discover to what extent the glycosylation of similar proteins can be demonstrated *in vivo*.

6. *Subcellular Localization*

Some progress has been achieved to date with the subcellular localization of elements of the glycosylation system. Beevers's group (Nagahashi and Beevers,

1978; Nagahashi *et al.*, 1978) found that the most active isolated fraction from pea cotyledons for the incorporation of GlcNAc and mannose into lipid-linked monosaccharide, lipid-linked oligosaccharide, and protease-susceptible polymer was the rough endoplasmic reticulum (RER). In the endosperm of germinating castor beans, Marriott and Tanner (1979) also showed that the transfer of GlcNAc and mannose to lipid derivatives and glycopolmyer was chiefly located in the RER, little activity being found in mitochondria and glyoxysomes. It is also worth mentioning that the glycosylation of legume seed storage proteins does not occur during their entry into the developing protein bodies; Nagahashi and Beevers (1968) found no mannose of GlcNAc transferase activities associated with these organelles in peas.

It has not yet been established whether core glycosylation can occur during the synthesis of the acceptor protein in plants, but since it can in animals, an RER location for the glycosyl transferases is understandable. There may also be glycosylation activity in the Golgi apparatus in plants, and the significance of this is not clear. Beevers and colleagues (references citied above) found that Golgi membranes from pea cotyledons did incorporate GlcNAc and mannose into lipid-oligosaccharide and trichloroacetic acid (TCA)-precipitable material, but only the GlcNAc could be released from the TCA precipitate by protease. This suggests that the Golgi apparatus is the site of direct transfer of sugars from their nucleoside diphosphates to polymers, i.e., of mannan synthesis and such reactions as the single GlcNAc transfer to vicilin and PHA that we found in kidney bean cotyledons. Lehle *et al.* (1978), using fractionated membranes from mung bean hypocotyles, showed that *in vitro* incorporation of mannose into mannolipid was highest in the RER fraction, whereas its incorporation into uncharacterized glycopolymer (possibly including mannan) was highest in Golgi membranes. Durr *et al.*, (1979), working with a pea stem preparation, reported maximal synthesis of mannolipid in the RER and the highest incorporation of mannnose into lipid-oligosaccharide and glycoproteins in the Golgi membranes.

As an alternative to the studies discussed above in which nucleotide sugars were supplied to isolated subcellular fractions, radiolabeled mannose or glucosamine can be supplied to the intact tissue. After a suitable incubation period, the tissue can be fractionated and the fractions analyzed for the incorporation and location of the labeled sugar. On first thought, this might seem the better approach, because the relative activities of fractions *in vitro* probably reflect differences in their contents of lipids and acceptor proteins as well as differences in enzyme activities. One must keep in mind, however, that newly labeled intermediates and end products are likely to move through the endomembrane system and between organelles *in vivo* so that their locations might indicate sites of accumulation rather than sites of synthesis. Nagahashi *et al.* (1980) recently used this approach with pea cotyledons. The results support their data from *in vitro* incubations described above; there was little incorporation of mannose into TCA-precipitable material in the Golgi membranes, but there was a high incorporation of glucosamine. The RER showed the highest incorporations into lipid-linked derivatives and TCA-precipitable material. Taking the results cited above, collectively, it is obvious that there is some disparity concerning the activity of the Golgi. Perhaps it is involved to an extent that depends on the type of tissue examined, but it is not clear why the Golgi should be concerned with core glycosylation, and more studies are obviously required.

An *in vivo* labeling of castor bean endosperm tissue with [^{14}C]galactose allowed Mellor and Lord (1978) to conclude that glyoxysomal glycoproteins containing galactose received the sugar in the endoplasmic reticulum, before their passage into the glyoxysome. No distinction was made between rough and smooth membranes.

7. *Other Aspects*

Ultimately, for a complete understanding of protein glycosylation, it will be necessary to dissect the various glycosyl transferases from their membrane environments. Certain glycosyl transferases have been released from membranes of animal cells with retention of activity (e.g., see Spencer and Elbein, 1980), showing that solubilization is possible, given the right conditions. Not much has yet been reported for plant systems, however. Digitonin treatment solubilized only 2.3% of the activity responsible for mannolipid synthesis from GDP-man in mung bean shoot membranes, and Triton X-100 was even less effective (Alam and Hemming, 1973). More studies concerned with transferase isolation are to be encouraged, especially the establishment of optimum detergent treatments.

It is possible to learn something of the different properties of various transferases without resorting to their isolation. Nagahashi *et al.* (1978) have initiated this kind of study by examining the divalent-cation requirements of various processes in particulate fractions from developing pea cotyledons. The incorporation of radiolabeled mannose from GDP-[^{14}C]man into lipid derivatives was stimulated more by Mn^{2+} than by Mg^{2+}, while the reverse was true for the incorporation of GlcNAc from UDP-[^{14}C]GlcNAc. A difference of cation requirement for transfer of these two sugars to glycoproteins should not have been obtained, since one enzyme system must transfer both sugars as preformed oligosaccharide. They found, however, that mannose incorporation into insoluble residue was preferentially stimulated by Mg^{2+} over Mn^{2+} and GlcNAc incorporation by Mn^{2+} over Mg^{2+}. Presumably, the results were complicated by the differential stimulation of synthesis of the lipid intermediates, affecting their concentrations in the preparation, and by the synthesis of insoluble residue mannans directly from GDP-man. Perhaps studies such as these will eventually indicate sets of incubation conditions that will allow the operation of selected individual glycosyl-transfer reactions *in situ* in the membrane, thus facilitating the study of intermediates and regulation.

An interesting question concerns whether there is any regulatory link between translation and core glycosylation. Schwaiger and Tanner (1979) inhibited glycosylation in membranes from barley aleurone tissue, to see whether the secretion of α-amylase would thereby be prevented, and also to ascertain whether the attachment of carbohydrate to the protein constituted the "signal" for its secretion from the cell. Tunicamycin did inhibit the secretion of amylase substantially, but the enzyme did not accumulate in the cell, and therefore synthesis must have been inhibited. Since tunicamycin had a much smaller effect on the overall rate of protein synthesis, this suggests that the specific synthesis of amylase stopped when its glycosylation was prevented, i.e., that glycosylation during translation is necessary for completion of translation. There are other explanations, however. There was some indication that tunicamycin may have interfered with the action of gibberellic acid in activating α-amylase synthesis in this tissue, and the authors did not rule out the possibility that lack of glycosylation stimulates degradation of the enzyme. Nevertheless, the results are of interest and represent the first example wherein an inhibitor of glycosylation suppresses the appearance of a specific glycoprotein in a higher plant system.

In concluding this chapter, we should like to draw attention to the considerable similarity in the glycosylation mechanisms of plants and animals. It would seem that the same fundamental mechanisms of protein glycosylation operate in both kingdoms, and it should be possible to transfer hypotheses and principles between them. The pathways of core glycosylation, using carrier lipids, are almost identical, and the subcellular localization of this process is probably RER in both cases. In both animal and plant systems, sugars can be added after core glycosylation by routes not requiring carrier lipids. Thus far, there has been no report of oligosaccharide processing in plants, but studies on this aspect are beginning, and it is probable that it will be found. The continuing studies with higher plants not only will contribute to our appreciation of the details of synthesis of many important plant glycoproteins but also should advance our understanding of the mechanism and perhaps also the function of glycoproteins in general.

Note added in proof: Since the preparation of this chapter the term "phaseolin" has replaced "vicilin" for the 7 S globulin of *Phaseolus vulgaris* seeds.

References

Alam, S. S., and Hemming, F. W. 1973, Polyprenol phosphates and mannosyl transferases in *Phaseolus aureus*, *Phytochemistry* **12:**1641.

Beevers, L., and Mense, R. M., 1977, Glycoprotein biosynthesis in cotyledons of *Pisum sativum* L: Involvement of lipid-linked intermediates, *Plant Physiol.* **60:**703.

Chen, W. W., and Lennarz, W. J., 1977, Metabolism of lipid-linked *N*-acetylglucosamine intermediates, *J. Biol. Chem.* **252:**3473.

Davies, H. M., and Delmer, D. P., 1979, Seed reserve protein glycosylation in an *in vitro* preparation from developing cotyledons of *Phaseolus vulgaris*, *Planta* **146:**513.

Davies, H. M., and Delmer, D. P., 1981, Glycosylation of endogenous proteins in a kidney bean membrane preparation: Core glycosylation and a lipid-independant attachment of *N*-acetyl glucosamine, *Plant Physiol.* **68**:284.

Delmer, D. P., and Lamport, D. T. A. L., 1977, The origin and significance of plant glycoproteins, in: *Cell Wall Biochemistry* (B. Solheim and J. Raa, eds.), pp. 85–104, University of Tromso, Norway.

Delmer, D. P., Kulow, C., and Ericson, M. C., 1978, Glycoprotein synthesis in plants. II. Structure of the mannolipid intermediate, *Plant Physiol.* **61**:25.

Durr, M., Bailey, D. S., and Maclachlan, G., 1979, Subcellular distribution of membrane-bound glycosyltransferases from pea stems, *Eur. J. Biochem.* **97**:445.

Elbein, A. D., 1969, Biosynthesis of a cell wall glucomannan in mung bean seedlings, *J. Biol. Chem.* **244**:1608.

Elbein, A. D., 1980, The role of lipid-linked saccharides in the biosynthesis of complex carbohydrates, *Annu. Rev. Plant Physiol.* **30**:239.

Elbein, A. D., and Hassid, W. Z., 1966, The enzymatic synthesis of a glucomannan, *Biochem. Biophys Res. Commun.* **23**:311.

Elbein, A. D., Gafford, J., and Kang, M. S., 1979, Inhibition of lipid-linked saccharide synthesis: Comparison of tunicamycin, streptovirudin, and antibiotic 24010, *Arch. Biochem. Biophys.* **196**:311.

Ericson, M. C., and Delmer, D. P., 1977, Glycoprotein synthesis in plants. I. Role of lipid intermediates, *Plant Physiol.* **59**:341.

Ericson, M. C., and Delmer, D. P., 1978, Glycoprotein synthesis in plants. III. Interaction between UDP-*N*-acetylglucosamine and GDP-mannose as substrates, *Plant Physiol.* **61**:819.

Ericson, M. C., Gafford, J. T., and Elbein, A. D., 1978, Evidence that the lipid carrier for *N*-acetylglucosamine is different from that for mannose in mung beans and cotton fibers, *Plant Physiol.* **61**: 274.

Forsee, W. T., and Elbein, A. D., 1975, Glycoprotein biosynthesis in plants: Demonstration of lipid-linked oligosaccharides of manose and *N*-acetylglucosamine, *J. Biol. Chem.* **250**:9283.

Hanover, J. A., and Lennarz, W. J., 1980, *N*-Linked glycoprotein assembly: Evidence that oligosaccharide attachment occurs within the lumen of the endoplasmic reticulum, *J. Biol. Chem.* **255**:3600.

Hopp, H. E., Romero, P. A., Daleo, G. R., and Pont Lezica, R., 1978, Synthesis of cellulose precursors: The involvement of lipid-linked sugars, *Eur. J. Biochem.* **84**:561.

Karr, A. L., 1972, Isolation of an enzyme system which will catalyse the glycosylation of extensin, *Plant Physiol.* **50**:275.

Kean, E. L., Plantner, J. J., and Poncz, L., 1980, Inhibition by tunicamycin of the incorporation of mannose and glucosamine into rhodopsin, *Fed. Porc. Fed. Am. Soc. Exp. Biol.* **39**:1899.

Khalkhali, Z., and Marshall, R. D., 1976, UDP-*N*-acetyl-D-glucosamine-asparagine sequon *N*-acetyl-β-D-glucosaminyl-transferase—activity in human serum, *Carbohydr. Res.* **49**:455.

Kornfeld, S., Chapman, A., and Fujimoto, K., 1980, Studies of the synthesis of lipid-linked oligosaccharides, *Fed. Proc. Fed. Am. Soc. Exp. Biol.* **39**:1671.

Lamport, D. T. A. L., 1980, Structure and function of plant glycoproteins, in *The Biochemistry of Plants,* Vol. 3 (P. K. Strumpf, ed.), pp. 501–541, Academic Press, New York.

Lavintman, N., Tandecarz, J., Carceller, M., Mendiara, S., and Cardini, C. E., 1974, Role of uridine diphosphate glucose in the biosynthesis of starch, *Eur. J. Biochem.* **50** :145.

Leavitt, R. D., Felsted, R. L., and Bachur, N. R. 1977, Biological and biochemical properties of *Phaseolus vulgaris* isolectins, *J. Biol. Chem.* **252**:2961.

Lehle, L., Fastaczek, F., Tanner, W., and Kauss, H., 1976, Formation of polyprenol-linked mono- and oligosaccharides in *Phaseolus aureus, Arch. Biochem. Biophys.* **175**:419.

Lehle, L., Bowles, D. J., and Tanner, W., 1978, Subcellular site of mannosyl transfer to dolichyl phosphate in *Phaseolus aureus, Plant Sci. Lett.* **11**:27.

Lis, H., and Sharon, N., 1978, Soybean agglutinin—a plant glycoprotein: Structure of the carbohydrate unit, *J. Biol. Chem.* **253**, 3468.

Marriott, K. M., and Tanner, W., 1979, Dolichylphosphate-dependent glycose transfer reactions in the endoplasmic reticulum of castor bean endosperm, *Plant Physiol.* **64:**445.

McLeester, R. C., Hall, T. C., Sun, S. M., and Bliss, F. A., 1973, Comparison of globulin proteins from *Phaseolus vulgaris* with those from *Vicia faba, Phytochemistry* **2:**85.

Mellor, R. B., and Lord, J. M., 1978, Incorporation of D-[^{14}C]galactose into organelle glycoprotein in castor bean endosperm, *Planta* **141:**329.

Mellor, R. B., and Lord, J. M., 1979, Formation of lipid-linked mono- and oligosaccharides from GDP-mannose by caster bean endosperm homogenates, *Planta* **146:**91.

Nagahashi, J., and Beevers, L., 1968, Subcellular localization of glycosyl transferases involved in glycoprotein biosynthesis in the cotyledons of *pisum sativum* L., *Plant Physiol.* **61:**451.

Nagahashi, J., Mense, R. M., and Beevers, L., 1978, Membrane-associated glycosyl transferases in cotyledons of *Pisum sativum:* Differential effects of magnesium and manganese ions, *Plant Physiol.* **62:**766.

Nagahashi, J., Browder, S. K., and Beevers, L., 1980, Glycosylation of pea cotyledon membranes, *Plant Physiol.* **65:**648.

Osborne, T. B., 1894, The proteids of the kidney bean, *J. Am. Chem. Soc.* **16:**764.

Parodi, A. J., and Leloir, L. F., 1979, The role of lipid intermediates in the glycosylation of proteins in the eucaryotic cell, *Biochim. Biophys. Acta* **559:**1.

Pont Lezica, R., Romero, P. A., and Hopp, H. E., 1978, Glucosylation of membrane-bound proteins by lipid-linked glucose, *Planta* **140:**177.

Pusztai, A., and Watt, W. B., 1970, The isolation and characterization of a major antigenic and non-hemagglutinating glycoprotein from *Phaseolus vulgaris, Biochim. Biphys. Acta* **207:**413.

Roberts, R. M., and Pollard, W. E., 1975, The incorporation of D-glucosamine into glycolipids and glycoproteins of membrane preparations from *Phaseolus aureus* hypocotyls, *Plant Physiol.* **55:**431.

Schwaiger, H., and Tanner, W., 1979, Effects of gibberellic acid and of tunicamycin on glycosyltransferase activities and on α-amylase secretion in barley, *Eur. J. Biochem.* **102:**375.

Sevier, E. D., and Shannon, L. M., 1977, Plant glycoprotein biosynthesis: Uridine diphosphate *N*-acetylglucosaminyl transferase from horseradish root, *Biochim. Biophys. Acta* **497:**578.

Sharon, N., 1974, Glycoproteins of higher plants, in: *Plant Carbohydrate Biochemistry* (J. B. Pridham, ed.), pp. 235–252, Academic Press, New York.

Smith, M. M., Axelos, M., and Péaud-Lenoël, C., 1976, Biosynthesis of mannan and mannolipids from GDP-man by membrane fractions of sycamore cell cultures, *Biochimie* **58:**1195.

Soliday, C. L., and Kolattukudy, P. E., 1979, Introduction of O-glycosidically linked mannose into proteins via mannosyl phosphoryl dolichol by microsomes from *Fusarium solani F. pisi, Arch. Biochem. Biophys.* **197:**367.

Spencer, J. P., and Elbein, A. D., 1980, Transfer of mannose from GDP-mannose to lipid-linked oligosaccharide by soluble mannosyl transferase, *Proc. Natl. Acad. Sci. U.S.A.* **77:**2524.

Struck, D. K., and Lennarz, W. J., 1977, Evidence for the participation of saccharide-lipids in the synthesis of the oligosaccharide chain of ovalbumin, *J. Biol. Chem.* **252:**1007.

Turco, S. J., and Robbins, P. W., 1979, The initial stages of processing of protein-bound oligosaccharides *in vitro, J. Biol. Chem.* **254**:4560.

Van der Wilden, W., Herman, E. M., and Chrispeels, M. J., 1980, Protein bodies of mung bean cotyledons as autophagic organelles, *Proc. Natl. Acad. Sci. U.S.A.* **77**:428.

II
STRUCTURE

19
The Structure of Cellulose Microfibrils in *Valonia*

HIROSHI HARADA AND TOSHIYUKI GOTO

1. Introduction

It is well known that cellulose is the main consitutent of plant-cell walls and that it exists as thin threads of indefinite length called microfibrils. The cellulose microfibril is an important factor influencing the chemical, physical, and mechanical properties of the cell walls, and it will give a clue to an understanding of how cellulose crystallizes and orientates *in vivo*. We chose the green alga *Valonia macrophysa* as a model, not only because its cell wall consists of a typical cellulose, but also because its wall is large enough to handle with ease.

2. Dimensions of Cellulose Microfibrils

Cellulose crystallites (micelles) have been studied since the first X-ray crystallographic measurements by Hengstenberg and Mark (1928). It has been suggested that the crystallite sizes in cotton and ramie are different from those in wood (Hengstenberg and Mark, 1928; Nagasawa, 1937; Wardrop, 1954). A little later, by electron-microscopic observations, the microfibril width was reported to be different in various kinds of cellulosic materials (Balashov and Preston, 1955).

The disparity in the size of crystalline regions of cellulose, as estimated by X-ray diffraction, and the width of the microfibrils seen in the electron microscope, led to differing concepts as to the molecular organization of

HIROSHI HARADA • Department of Wood Science and Technology, Faculty of Agriculture, Kyoto University, Kyoto 606, Japan. TOSHIYUKI GOTO • Department of Microbiology, Osaka Medical College, Takatsuki, Osaka 569, Japan.

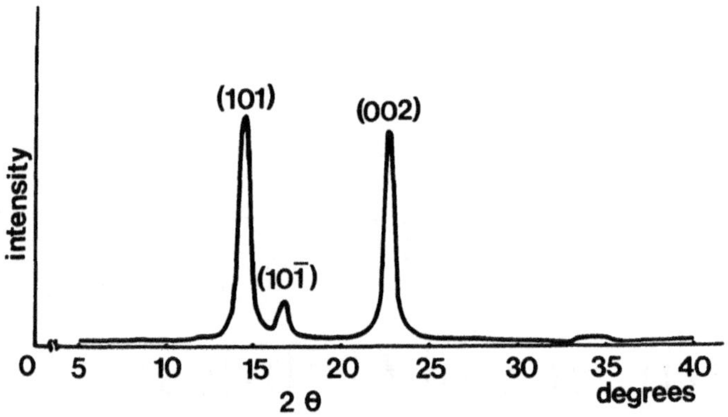

FIGURE 1. X-Ray diffractogram of *Valonia*, corrected for air scattering.

microfibrils (Wardrop, 1969). Thus, Frey-Wyssling (1954) regarded the microfibril as made up of a number of crystallites [which were later more precisely termed "elementary fibrils" by Frey-Wyssling and Mühlethaler (1963)] separated by a paracrystalline phase of cellulose, whereas Preston and Cronshaw (1958) considered that there was a central core of crystalline cellulose surrounded by a paracrystalline phase.

Cellulose materials from various sources of plant-cell walls such as *Valonia, Pinus densiflora,* and *Populus euramericana* were examined by both X-ray diffraction and electron microscopy. Each pulverized material was pressed into a tablet that was used to obtain an X-ray diffractogram. The block of each material was also examined. When the laminated cell walls of *Valonia* were examined by X-ray diffraction (Fig. 1) reflections* (002) and (101) were strong compared with reflection (10$\bar{1}$), which may indicate uniplanar orientation of cellulose crystallites in the cell wall, as suggested by Sponsler (1931) and Preston (1952) and others (see also Section 3).

The crystallite sizes were calculated at the reflections (002) and (101) by using line-broadening with the so-called Scherrer equation after subtracting the instrumental broadening (Scherrer, 1918; Klug and Alexander, 1954) (Table I). The crystallite widths of *Valonia* were 14.3 nm at reflection (002) and 11.9 nm at reflection (101).

As shown in Table I, the crystallite sizes are different depending on the sources of cellulose materials. The difference in crystallite size in this study may originate from the presence of noncellulosic cell-wall components that may restrict crystallite-size increase (Hanna and Côté, 1974), because the gelatinous layer and *Valonia* cell wall, which are unlignified and contain relatively little hemicelluloses (Norberg and Meier, 1966; Preston and Cronshaw, 1958), have larger crystallites than *Pinus* and poplar compression wood. On the other hand, cellulose unit-cell parameters have been reported to be different for various cellulosic materials (Wellard, 1954; Honjo and Watanabe, 1958; Nieduszynski and Atkins, 1970; Preston, 1974). There is also the possibility that the difference in crystallite sizes may be due essentially to the crystalline types.

*The reflections are indexed here according to the Meyer and Misch (1937) unit cell.

TABLE I
Crystallite Size and Microfibril Width

Samples	Crystallite size[a] (nm)	Microfibril width[b] (nm)
Pinus densiflora		
Untreated	2.02 (002)	—
Holocellulose	2.76 (002)	2–4 (≈ 2.5)
Populus euramericana		
Gelatinous layer	4.1 (002)	2–5 (≈ 3.0)
Compression wood	2.2 (002)	—
Valonia	14.3 (002)	15–30 (≈ 20.0)
	11.9 (101)	

[a] Reflection examined. [b] Mode width.

The same samples were also examined by electron microscopy after disintergration or ultrathin sectioning (Goto *et al.*, 1978). The microfibrils are assumed to rest on the supporting film at the most stable surface, whatever their cross-section is (Goto *et al.*, 1972). The microfibril widths were measured directly from the micrographs in the area where each microfibril could be seen separately (Figs. 2, 3, and 4). The range of widths and the most frequent (mode) width (indicated in parentheses) of various cellulose materials are listed in Table I.

On the basis of results from both X-ray line-broadening and electron microscopy, in *Pinus* cellulose, poplar gelatinous layer, and *Valonia*, all of which were different in crystallite size, the different microfibrillar widths are also confirmed. In electron-microscopic observations, uranyl acetate has been used as a negative stain. Heyn (1966) suggested that is can penetrate into whatever regions water can penetrate; i.e., the stain can penetrate all the regions except the crystalline region. This principle is based on the fact that uranyl acetate does not stain crystalline cellulose and that the amorphous cellulose can take up only water and aqueous solutions (Heyn, 1966). On this principle, the translucent parts seen in the electron micrographs may indicate crystalline regions of cellulose. Thus, the difference in microfibril widths obtained might be interpreted as reflecting the size of crystalline regions of microfibrils. It is quite interesting to note that the values from X-ray line-broadening are similar to those from the electron micrographs, although crystallite sizes were measured with the (002) reflection; however, the values obtained are not always equal to the "3.5 nm structure unit of elementary fibrils" proposed by Frey-Wyssling and Mühlethaler (1963).

3. Cross-Sectional View of Cellulose Microfibrils

Many investigations of the fine structure of cellulose microfibrils have been undertaken using electron microscopy. Frey-Wyssling (1954), Frey-Wyssling and Mühlethaler (1963), and Preston and Cronshaw (1958) proposed models to explain the cross-sectional shape of cellulose microfibrils. Frey-Wyssling and Mühlethaler (1963) assumed that each microfibril consists of so-called elementary

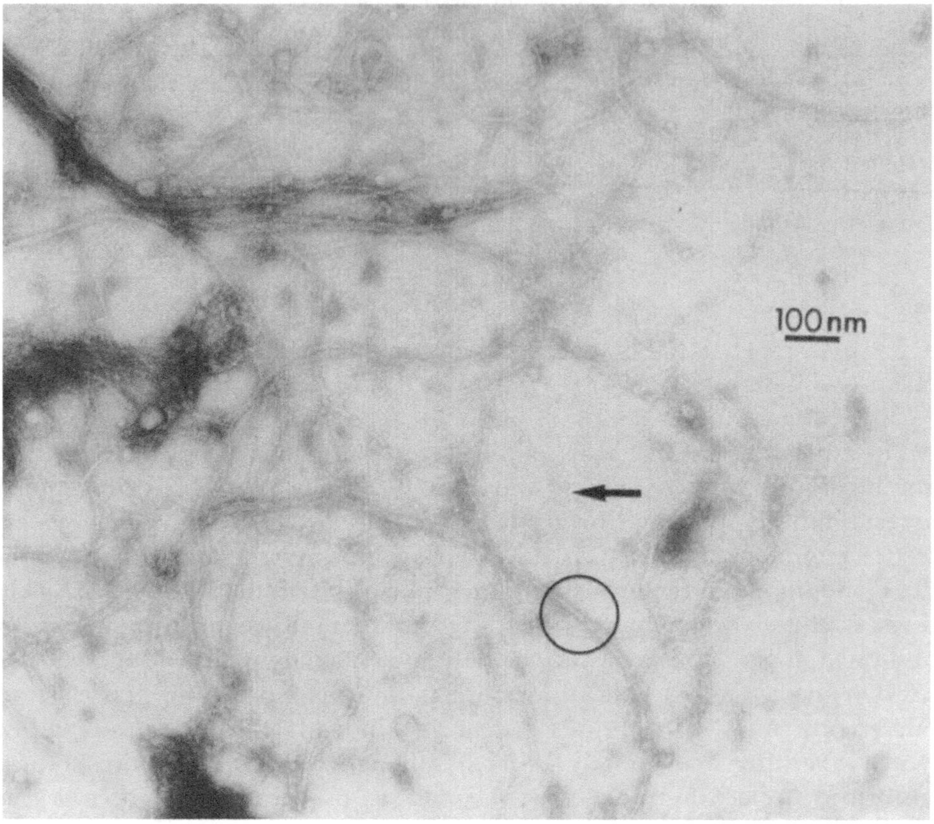

FIGURE 2. Microfibrils of *Pinus* holocellulose negatively stained with uranyl acetate. The circled area and the arrow show the aggregation of microfibrils and a kink, respectively. The holocellulose was prepared according to the method of Wise *et al.* (1946).

fibrils with a diameter of approximately 3.5 nm. Preston and Cronshaw (1958), however, considered the microfibrils of *Valonia* cellulose to be very highly crystalline, consisting of a rectangular crystalline core that may be embedded in a paracrystalline region.

The cross-sectional features of cellulose microfibrils or elementary fibrils were observed with the electron microscope by Côté and Day (1962), Heyn (1965, 1966, 1969), and Yoshimi and Harada (1969). Heyn studied the fine structure of cotton, ramie, and wood fiber through electron-microscopic examination of negatively prestained ultrathin sections. He reported that in cotton and ramie, both circular and rectangular cross-sectional shapes for elementary fibrils were observed. However, the actual cross-sectional shape is still questionable, since its appearance depends not only on the plane of sectioning but also on the variable orientation of the microfibrils in the fiber (Boylston and Rollins, 1971).

The *Valonia* cell wall is composed of a so-called crossed fibrillar structure of cellulose. This is visualized in Fig. 5, which represents the intermediate layer of the cell wall by a direct carbon replica according to the method of Imamura and

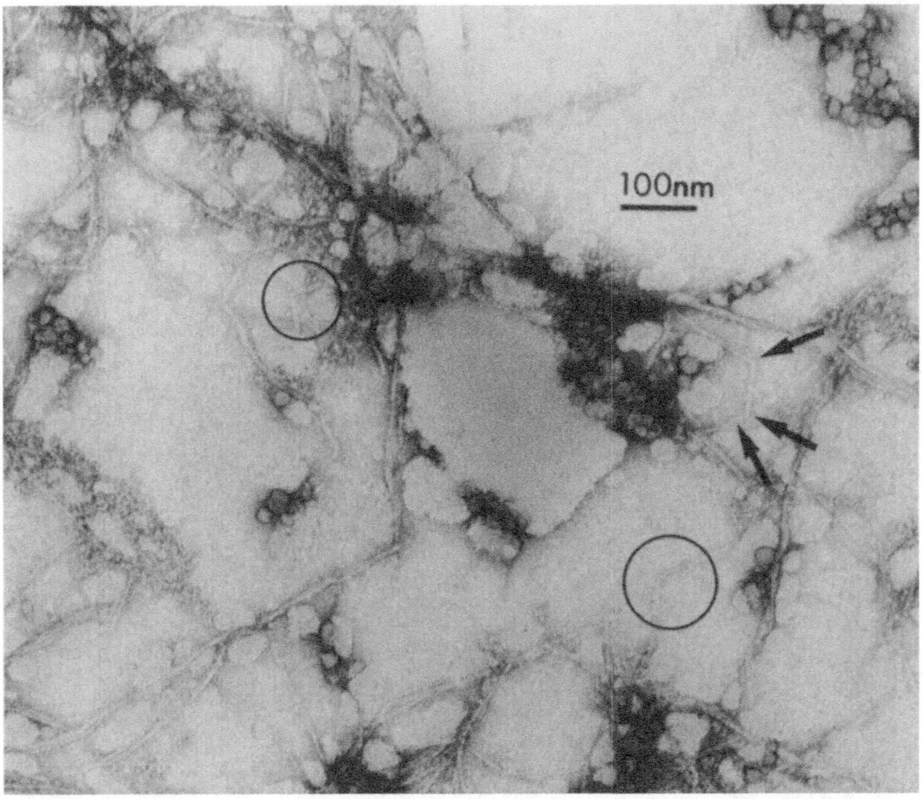

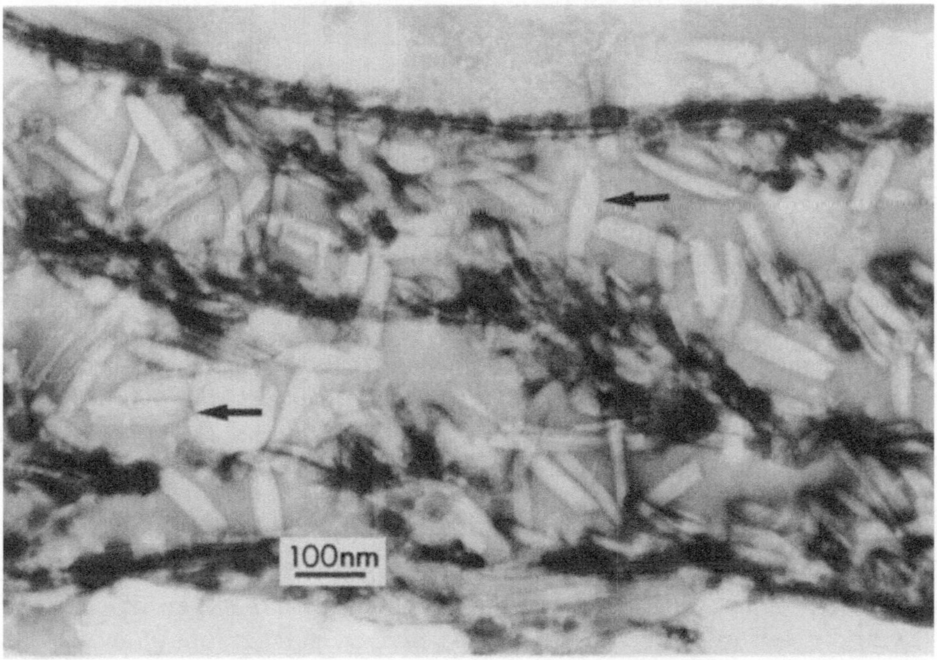

FIGURE 3. Microfibrils of poplar (*Populus euramericana*) gelatinous layer showing kinks (circled areas and arrows). Negatively stained with uranyl acetate.

FIGURE 4. *Valonia* microfibrils negatively stained with uranyl acetate. (←) Uninjured microfibrils.

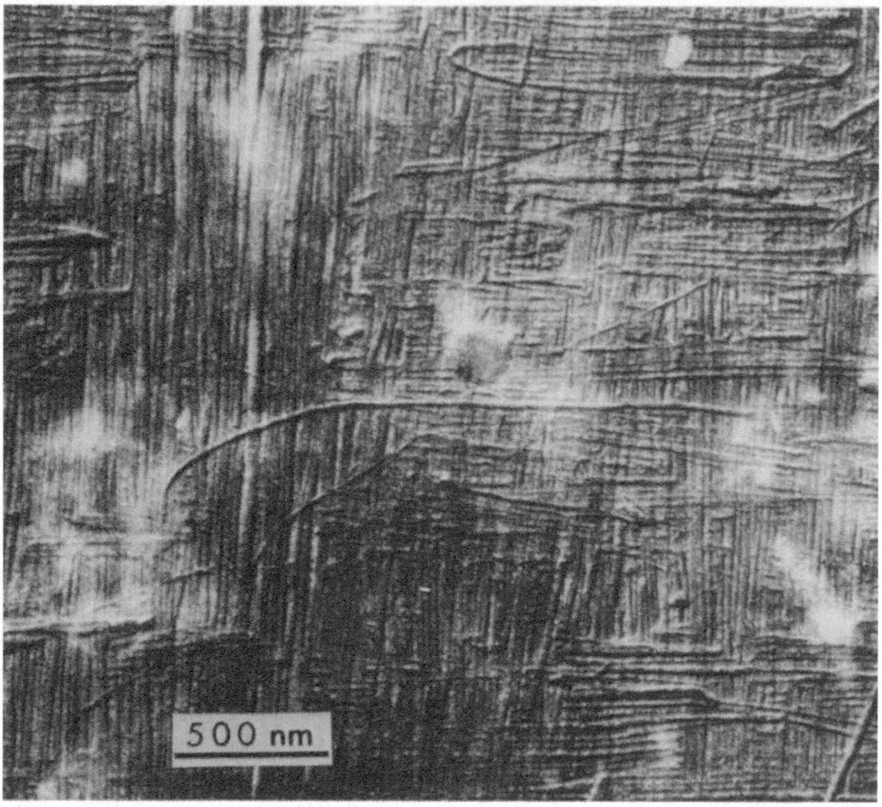

FIGURE 5. Direct carbon replica of *Valonia macrophysa* showing crossed microfibrillar structure in the intermediate layer. Negative print.

Harada (1973). The transverse section of this cell wall is observed using the negative-staining technique (Fig. 6). In the micrograph, white lines, which run transversely, indicate the longitudinal sections of the microfibrils (L), and the white dots dispersed among the white lines are the cross sections of microfibrils (C).

To demonstrate that the microfibrils (C) are oriented vertically, the transverse section in the same series as Fig. 6 is illustrated after removal of the resin followed by shadow-casting with platinum–palladium (Fig. 7). The white dots in Fig. 6 produced by the negative-staining technique are shown here as protrusions (arrows).

The negatively stained sections were scrutinized further with the electron microscope and a specimen-tilting apparatus. Some of the micrographs obtained are shown in Fig. 8. The actual microfibril cross section that is encircled in Fig. 8(1) can be seen in sharp contour. These electron micrographs show that the shape of *Valonia* cellulose microfibrils is rectangular, with the longer dimension 15–20 nm and the shorter one 7–10 nm. The longer side lies parallel to the cell-wall surface.

When the ultrathin sections are observed using the negative-staining

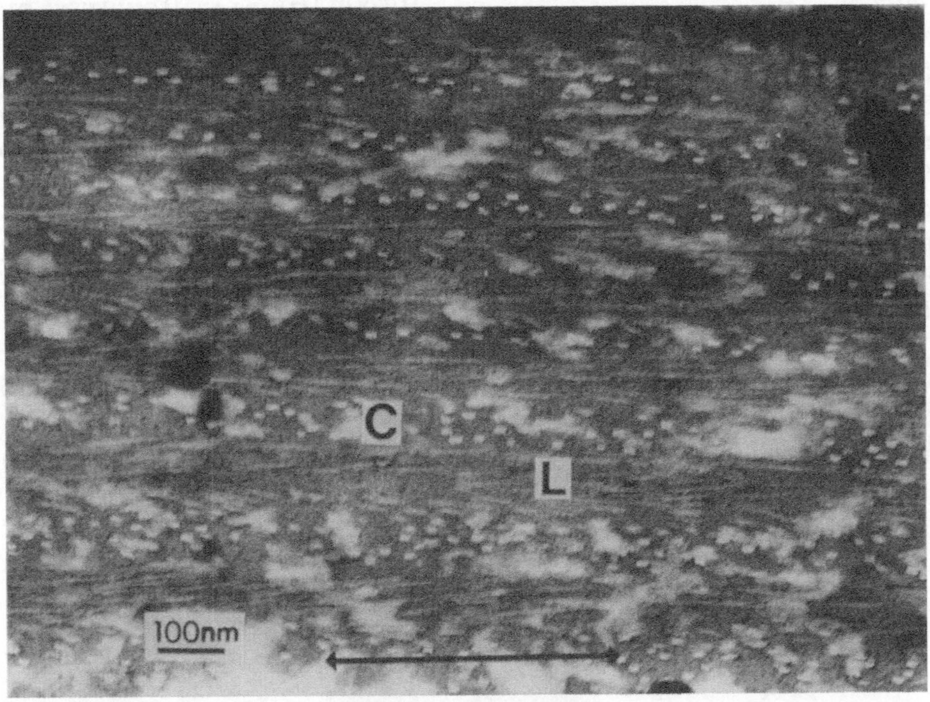

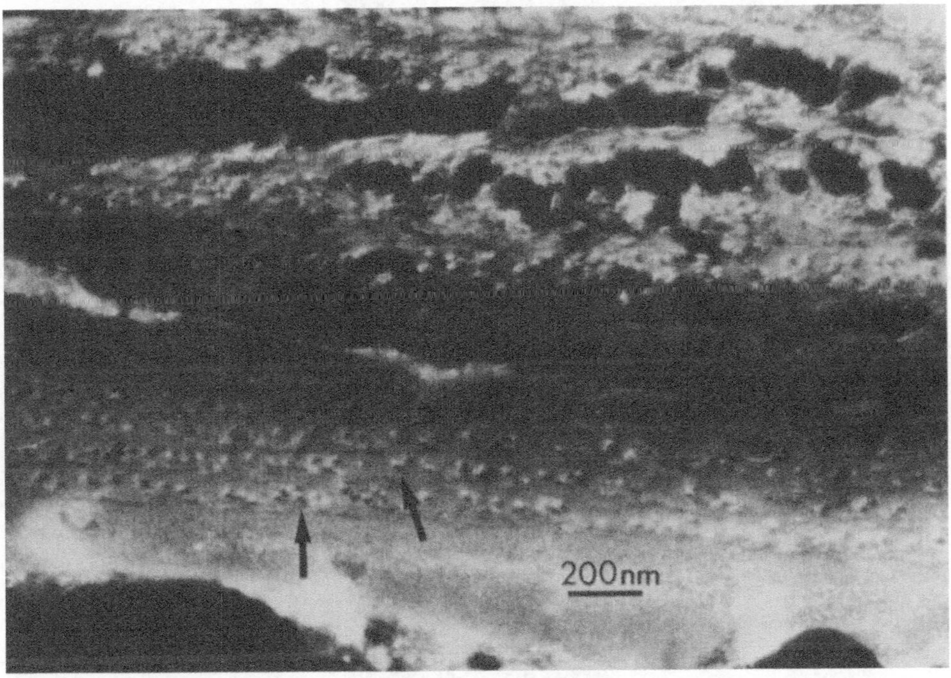

FIGURE 6. Transverse section of *Valonia* cell wall negatively stained with uranyl acetate. (↔) Direction parallel to cell-wall surface. Microfibrils are shown in cross section (C) and longitudinal section (L).

FIGURE 7. Transverse section of *Valonia* cell wall as in Fig. 6, with methacrylate removed and Pt–Pd shadow-casting. Cross sections (C) in Fig. 6 are seen here as protrusions (↑). Negative print.

technique, the thickness of the section should be taken into consideration. The thickness of the section is one of the important factors in interpreting the electron micrographs of the negatively stained sample. The thicker the sections are, the less valid the explanation can be, based on observations of the shape of the microfibrils embedded in the stain (Hanna, 1971). When the microfibrils are parallel and close-packed, the possibility of misreading the image is significantly increased by the thickness effect of sections and the plane inclination (Preston, 1971).

Now, suppose that the microfibrils are dispersed under conditions of negligible interaction among microfibrils. The resultant images of microfibrils for various tilt angles are imagined as shown in Fig. 9, in which (a–c) indicate the side views of the microfibrils embedded in the negative stain and (a′–c′) the resultant images of (a–c), respectively. When a microfibril stands vertically as in

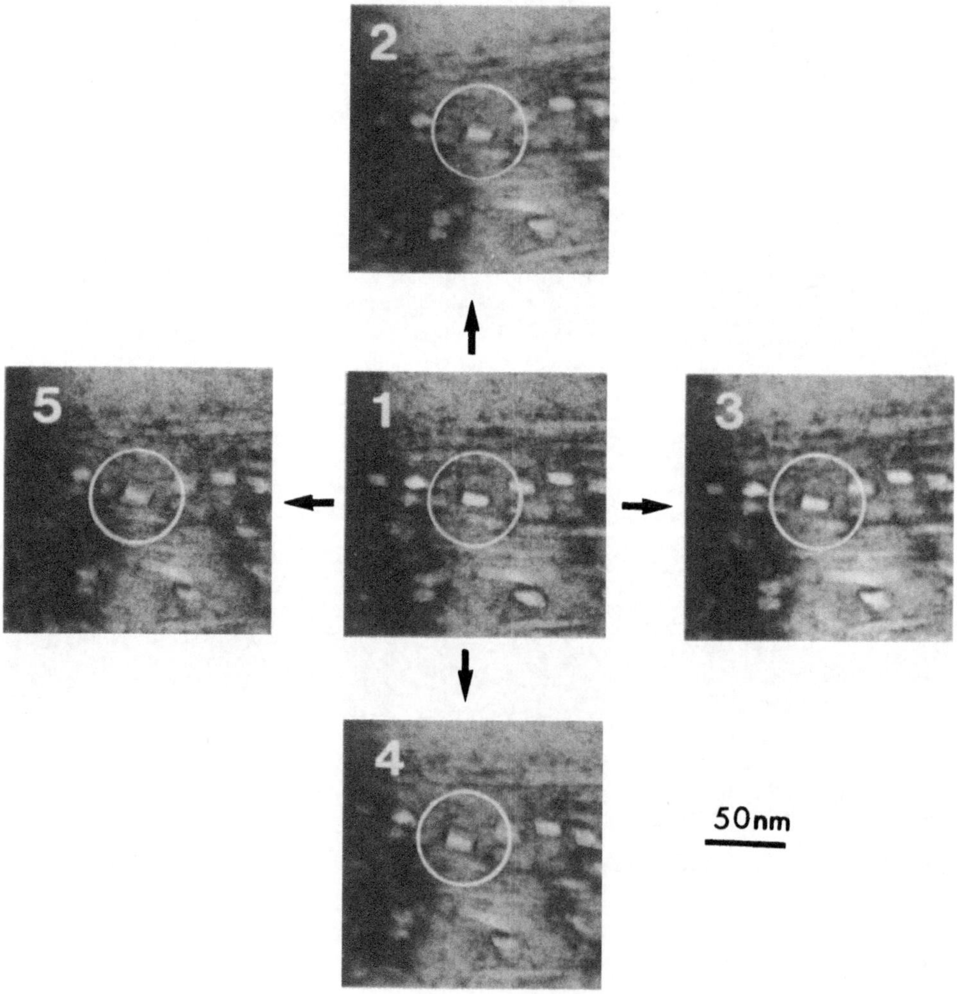

FIGURE 8. Electron micrographs of transverse section of *Valonia* cell wall with the aid of a specimen-tilting apparatus. (1) Specimen showing actual microfibrillar cross section; (2–5) specimen tilted by 5° degrees in the directions indicated by the arrows (see Fig. 9).

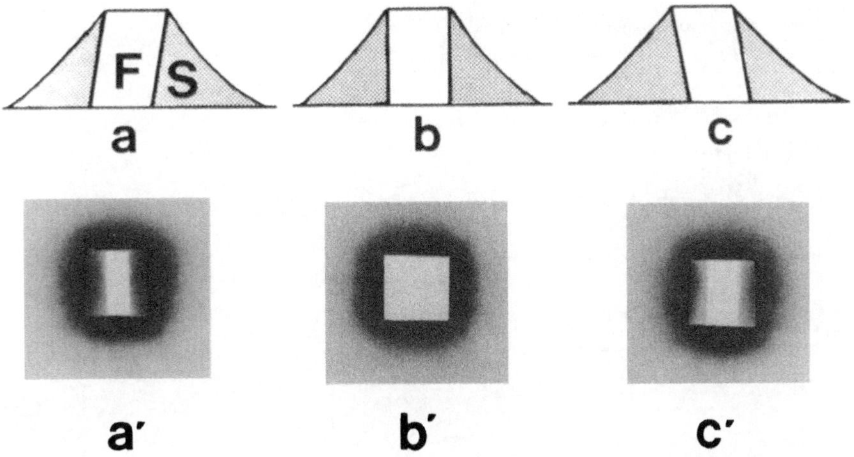

FIGURE 9. Diagrammatic representation of the relationship between the direction of tilt and the resultant image of a microfibril by using the negative-staining technique. (a–c) Side views of the microfibrils (F) embedded in negative stain (S): (a′–c′) resultant images on the photographic plate of (a–c), respectively.

Fig. 9b, the corresponding resultant image (b′) is clear, with distinct contours as in the micrograph in Fig. 8(1). However, when the microfibril is somewhat tilted as in (a) or (c), the resultant image becomes obscure as in (a′) or (c′), respectively; i.e., the boundary between the microfibril and the stain becomes partly obscure [see Fig. 8, (2–5)].

In practice, the ultrathin sections tend not to be cut precisely transversely; thus, in the micrographs, the contour between the bright portion indicating microfibrils and the dark portion of the stain surrounding the bright portion may not be clear. If the ultrathin sections are tilted by an appropriate angle, the cross section of some microfibrils will become clear where the four sides of the microfibril become distinct. As illustrated in Fig. 9, the microfibrils with distinct sides in the electron micrograph will be interpreted as showing the actual cross section. Therefore, by observing clear images of cross-sectional views of the microfibrils using the negative-staining technique, the shape of the cross section of *Valonia* microfibrils is found to be rectangular with the longer dimension 15–20 nm and the shorter one 7–10 nm.

Other cellulose materials, such as ramie, cotton, and wood, were observed by Heyn (1966, 1969), and the shapes of their microfibrils were circular, ellipsoid, square, or rectangular. But it is doubtful whether the actual cross section of the microfibrils was observed. For a conceivable explanation of this question, let us consider another model consisting of several layers of wood sticks (in place of microfibrils) adhered by varying the angles by about 9°. Figure 10 shows a photograph of a transverse section of the model. In this model, the sticks are circular in cross section, but their shapes will vary, depending on how the sticks are inclined, from the so-called beadlike through ellipsoidal to completely longitudinal sections. This explains why the interpretation will be mistaken unless the section is cut perpendicular to the long axis of the microfibril

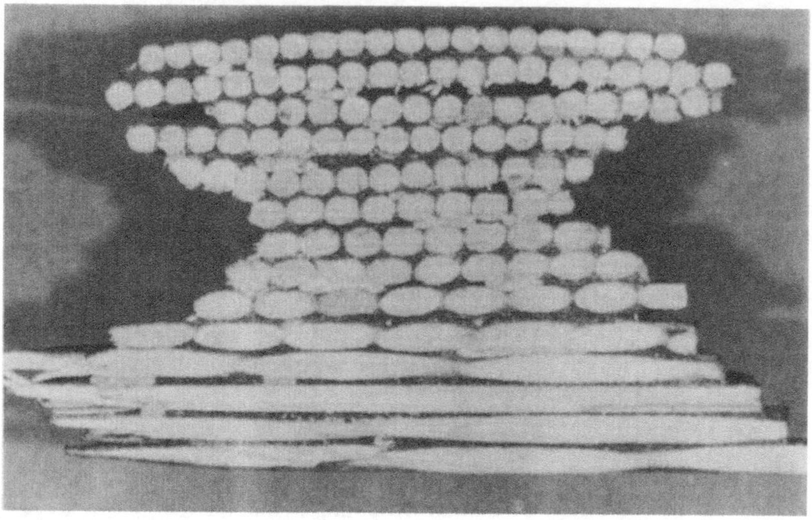

FIGURE 10. Photograph of a model of microfibrillar organization using circular wood sticks to represent microfibrils. The tilt angle increases from 0 to nearly 90°.

(Preston, 1971). Furthermore, when the thickness effect of the ultrathin section is added to the section inclination, it is clear that the shape will become more complex. Another complexity may be introduced in the observation of various cross-sectional shapes (Heyn, 1966, 1969): there may exist differences between the cellulose microfibrils in a green alga and in a higher plant such as ramie or wood (Beall and Murphey, 1970).

The electron micrographs of negatively stained microfibrils from *Valonia* do not show the less-ordered portion in a microfibril, or the portion of microfibril penetrated by negative stain, indicating that *Valonia* microfibrils are highly crystalline, which is described in detail below.

The results described above regarding the shape and size of *Valonia* microfibrils support the validity of the microfibril cross-sectional model proposed by Preston and Cronshaw (1958); i.e., the microfibril consists of a central crystalline core of about 10×20 nm surrounded by a less-ordered region.

The number of microfibrils in a lamella surface was calculated with micrographs of carbon replicas (see Fig. 5) and of the negative-staining technique (see Fig. 6). In the replica, the number of microfibrils existing in a lamella is 38–48 per micron. However, in negatively stained sections, the number of microfibrils is equal to or two or even three times that in the replica. This suggests that a lamella will consist of one, two or three rows of microfibrils. But in the *Valonia* cell wall, not only are there two directions (crossed structure) for microfibrils, but also there still remains a possibility of their having another direction ("third lamella") (Cronshaw and Preston, 1958).

In the micrographs (Fig. 6), the microfibrils in cross-sectional view between two rows of transversely oriented microfibrils (white lines in Fig. 6) are not packed in a regular pattern; rather, they are dispersed. This feature suggests that the cellulose microfibrils are arranged in the swollen state, i.e., in the "green" state or in a state wherein the wall is forced to expand in the processes of embedding, removal of resins, or negative staining. If the arrangement in the micrographs of the negatively stained section shows the "green" state, the regular arrangement of microfibrils in the replica may be interpreted as an artifact due to drying during preparation.

Even if the arrangement of microfibrils is rather dispersed in a lamella, as described above, the fact that the longer dimension of the microfibril in cross section lies perferentially parallel to the cell-wall surface is in agreement with the results of X-ray diffraction studies (Sponsler, 1931; Preston, 1952, 1974; Tanaka and Okamura, 1977). Figures 11 and 12 show the X-ray diagrams of *Valonia* cell wall where the X-ray beam was radiated normal and parallel to the cell-wall surface, respectively (Goto, 1976). Figure 11 shows two sets of intense arcs at 0.54 nm ($10\bar{1}$) and 0.39 nm (002), but the arc of 0.61 nm (101) is missing or very weak, which suggests that the (101) planes tend to lie parallel to the wall surface. In Fig. 12, the (101) plane is present as a dense arc on the equator, while the (101) is present as an arc on the meridian. This is the evidence for uniplanar orientation. In recent years, Tanaka and Okamura (1977) examined precisely the orientation of *Valonia* cellulose crystallites using pole figure analysis and confirmed this uniplanar orientation. From the results of X-ray diffraction, the longer dimension of the microfibril in cross section may correspond to the (101) plane of cellulose crystal lattice, which suggests that the (101) plane is more hydrophilic than the ($10\bar{1}$) plane (Frey-Wyssling and Mühlethaler, 1951).

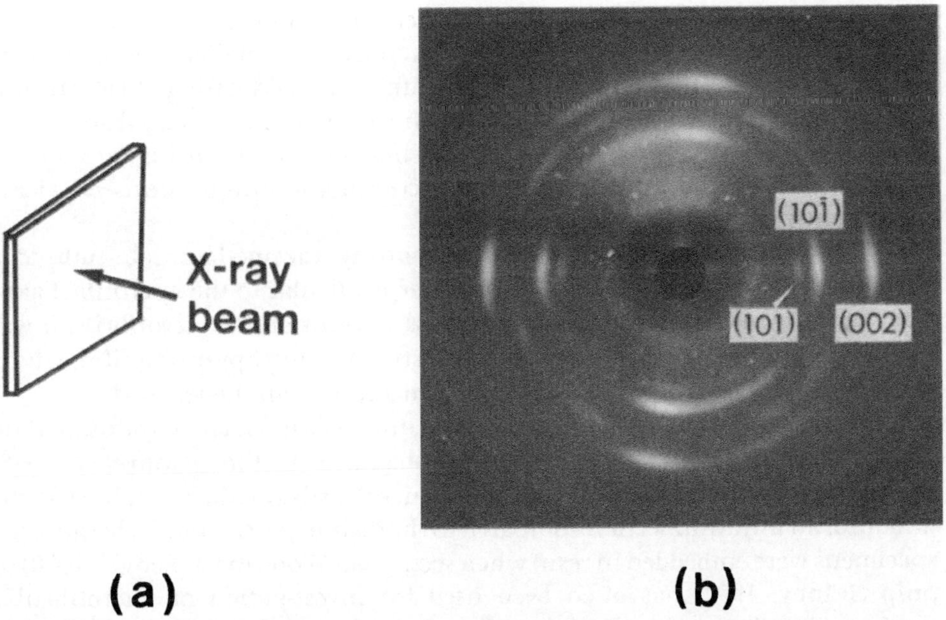

FIGURE 11. Diagrammatic representation (a) and X-ray diffractogram (b) of a single wall of *Valonia macrophysa* in which the beam was radiated normal to the wall surface.

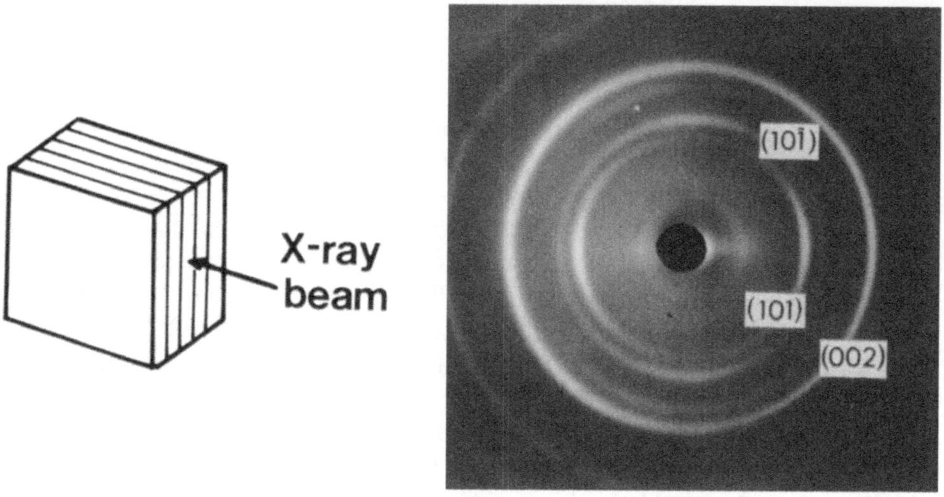

(a) **(b)**

FIGURE 12. Diagrammatic representation (a) and X-ray diffractogram (b) of a block of laminated walls of *Valonia macrophysa* in which the beam was radiated parallel to the wall surface.

4. Internal Structure of Cellulose Microfibrils

It has been a moot question whether cellulose microfibrils have a definite periodicity in the longitudinal direction, for example, whether the microfibrils contain folded molecular chains (Manley, 1964, 1971) or whether they are relatively uniformly crystalline (Mühlethaler, 1969). Moreover, cellulose microfibrils have been reported to be divided laterally into smaller fragments (e.g., elementary fibrils) that may be structural units (Frey-Wyssling, 1954). In this section, the fine structure of cellulose microfibrils in the longitudinal and lateral directions is elucidated with additional information by using a standard disintegration method and an ultrathin-sectioning technique, devised to leave the specimen uninjured (Goto *et al.*, 1978).

This sectioning technique involves cutting the methacrylate-embedded samples containing isolated microfibrils perpendicular to the microfibril axis. Although the intention was to observe cross sections of the microfibrils, it was often observed that the microfibrils were resting on the supporting film, which could be caused by removal of embedding medium from the sections.

Fortunately, it can be assumed that the microfibrils isolated by this method are equivalent to those obtained by the disintegration method (Goto *et al.*, 1975) and that the mechanical influence on the microfibrils is reduced, at least in the middle of an ultrathin section compared to the disintegrated sample, because the specimens were embedded in resin when sectioned. Wood meal (Fengel, 1970) or pulp (Rånby, 1952) has often been used for investigation of microfibrillar structure, but these microfibrils may have been damaged by hydrolysis or other treatments.

Small pieces of *Valonia* or wood were used for hydrolysis treatment, and the inner parts of the specimens that were less injured mechanically were used for ultrathin sections and electron-microscopic observation (see Figs. 16 and 18). This ultrathin-section method is, however, confined to observation of microfibrils shorter than the ultrathin sections (75–100 nm) (Goto *et al*, 1978).

4.1. Lateral Structure of Cellulose Microfibrils

For examination of the paracrystalline region surrounding the microfibrillar core, the microfibrils were stained with various stains. When the microfibrils in *Valonia* and the gelatinous layer of poplar were observed after staining with uranyl acetate and phosphotungstic acid as typical negative stains, no significant difference in the widths of the stained microfibrils was recognized (Figs. 13 and 14). In *Valonia*, there may exist wider microfibrils stained with uranyl acetate, but the reason for this is not obvious at present.

The microfibrils were also observed by using ruthenium red, which is known as a stain specific for pectin. *Valonia* microfibrils stained with ruthenium red were not found to be different in width from those stained with the other stains (Fig. 13), but a slight difference was noted with the gelatinous layer from poplar (Fig. 14). It has been reported that this layer is relatively rich in pectin (Furuya *et al*., 1970; Harada *et al*., 1971). The difference noted above may suggest that some pectin might surround the microfibrils, although it must be remembered that electron-microscopic staining is fundamentally different from optical-microscopic staining, which depends on colors. In electron microscopy, the existence of heavy metals, such as ruthenium, produces high electron contrasts to reveal a distinct dark area compared with the light area produced by cellulose on the photographic plate. The existence of pectin as indicated by use of ruthenium red is still an open question, but we would rather suppose that ruthenium red might act as a usual negative stain, such as uranyl acetate and phosphotungstic acid. It is quite possible that the difference in microfibrillar widths will depend on the ionic sizes of the three stains. It is known that the diameter of the uranyl ion, UO_2^{++}, is 0.3–0.5 nm and that of the phosphotungstate ion $PW_{12}O_{40}^{---}$ 0.8–0.9 nm [Levin (1963), quoted by Höglund (1968)], and that the size of the ruthenium red complex is 1.1 nm (Luft, 1971). These figures suggest

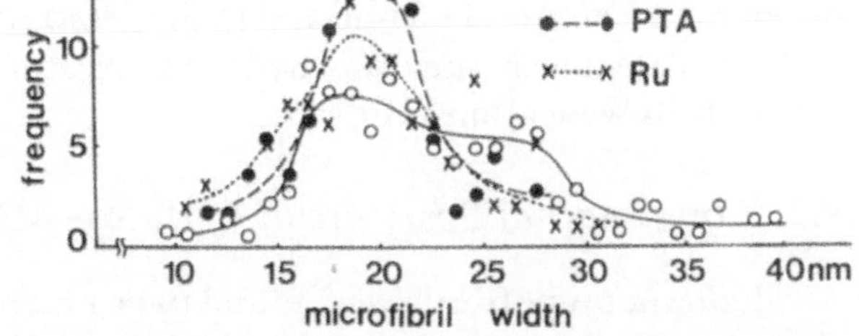

FIGURE 13. Width distribution of *Valonia* microfibrils stained with uranyl acetate (UAc), phosphotungstic acid (PTA), and ruthenium red (Ru).

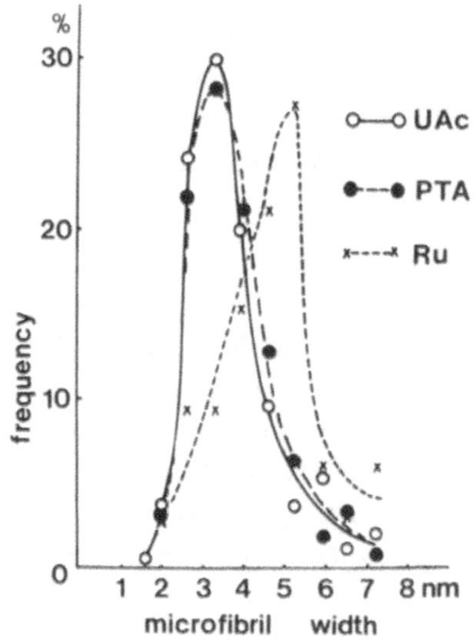

FIGURE 14. Width distribution of microfibrils from poplar gelatinous layer stained with uranyl acetate (UAc), phosphotungstic acid (PTA), and ruthenium red (Ru).

that the region around the microfibrils in the gelatinous layer is impregnated by uranyl acetate and phosphotungstic acid, but not by ruthenium red (Fig. 14), and that the region may accordingly be paracrystalline (Goto *et al.*, 1978). Negative-staining images relative to the size of the stain were also found in gamma-irradiated microfibrils (Goto *et al.*, 1974).

By mechanical disintegration and hydrolytic treatments of *Valonia*, the part where a stain such as uranyl acetate could not penetrate (the white part in Fig. 4, which is probably the crystalline region) was found to be divided into substructures (Figs. 15 and 16), which were also observed by Frey-Wyssling *et al.* (1966). The width distribution of these substructures in *Valonia* microfibrils is shown in Fig. 17, in which 3 nm is the most frequently observed width. This distribution suggests that 3 nm may be a stable size in mechanical and chemical properties, but that there is no tendency for the microfibrils to aggregate into multiples of 3 nm. Fengel (1974) and Hanna and Côté (1974) have suggested the existence of cellulose microfibrils 1–2 nm in width. The foregoing evidence throws considerable doubt on the existence of elementary fibrils, which are considered as the elemental structural unit of the plant-cell wall. It is interesting to note that the widths of the microfibrils from *Valonia* and the gelatinous layer of poplar are not of the same order in the untreated state and that the crystallite sizes that were calculated from reflection (002) by using the Scherrer equation are different in *Valonia* (14.3 nm) and in the gelatinous layer of poplar (4.1 nm) (see Table I). After mechanical and hydrolytic treatments, however, the widths of the microfibrils were almost the same.

4.2. *Longitudinal Structure of Cellulose Microfibrils*

Valonia microfibrils were found to be kinked, and at many kinking points the microfibrils were observed to be broken into substructures (see Fig. 15).

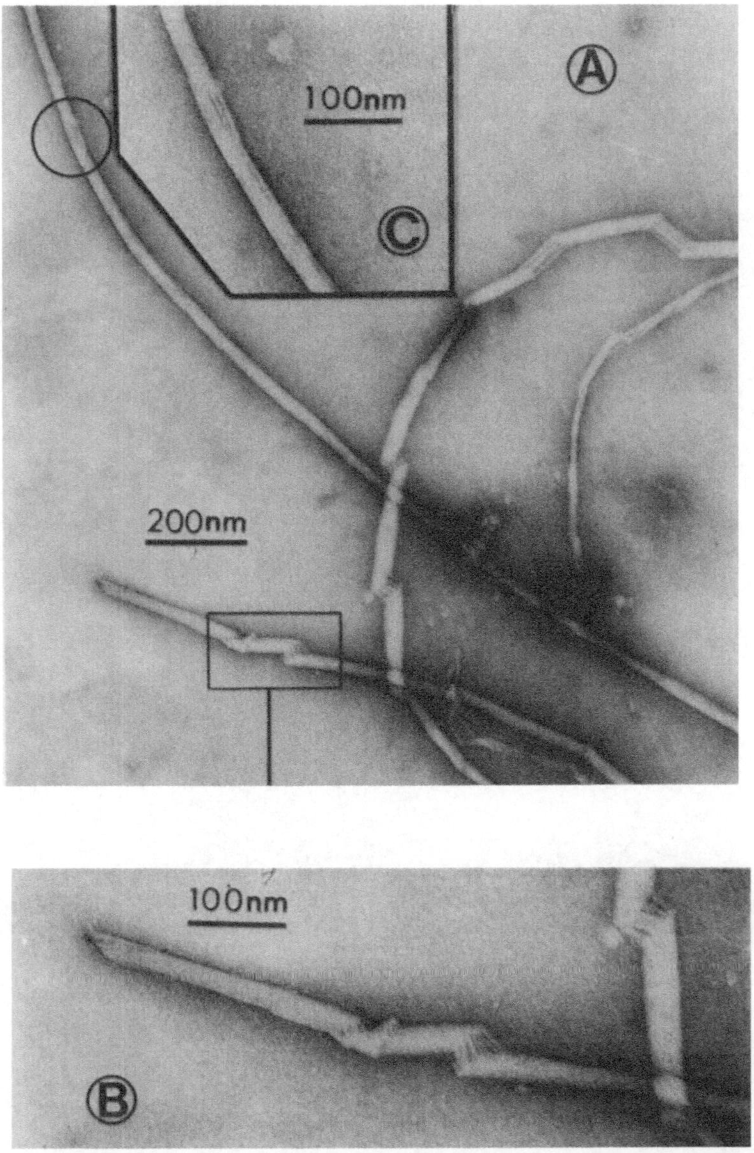

FIGURE 15. Microfibrils of *Valonia* fractured during disintegration with a homogenizer. Negatively stained with uranyl acetate. (B, C) Higher magnifications of the indicated sections of (A).

Apparently, *Valonia* microfibrils tend to split along the longitudinal axis; i.e., the bond in the lateral direction of the microfibrils is weaker than that in the longitudinal direction. In Fig. 15C can be seen stained striations in the microfibrils that may represent an initial stage of breakdown, such as that shown in Fig. 15B. These observations indicate that the cellulose-chain molecules are orientated parallel to the longitudinal axis of the microfibrils.

The microfibrils of the gelatinous layer in the disintegrated samples (see Fig. 3) were also found to have many kinks, suggesting that the microfibril is

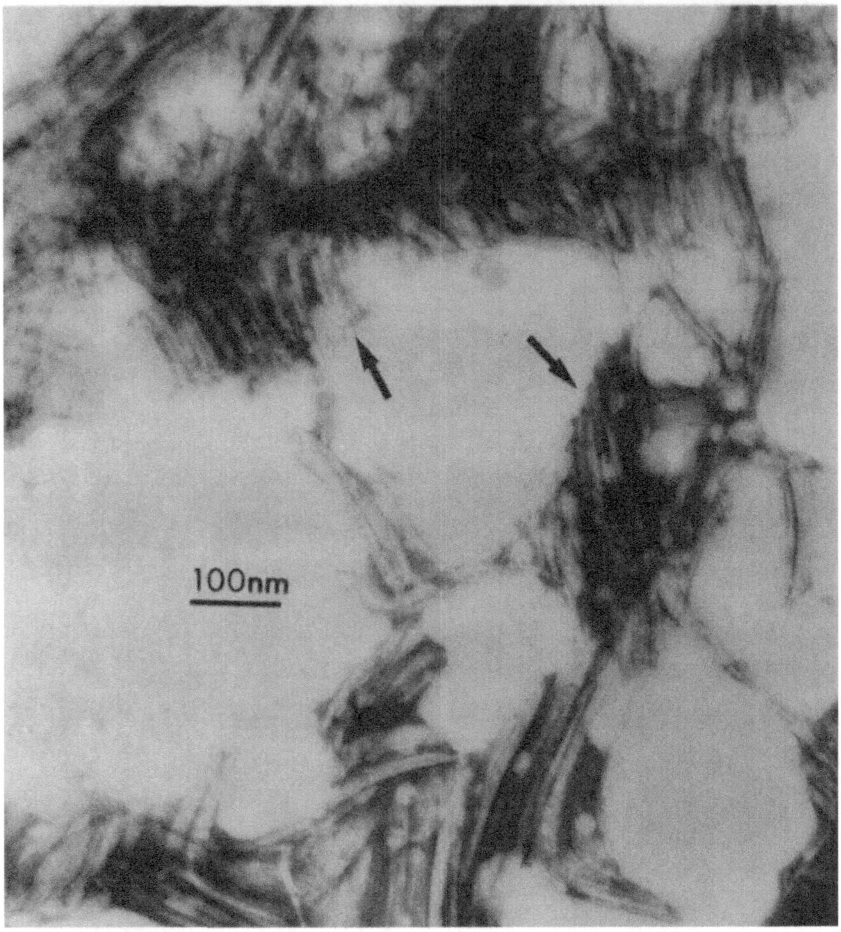

FIGURE 16. Microfibrils of *Valonia* hydrolyzed with 5 N hydrochloric acid for 300 min. The arrows show split microfibrils. Negatively stained with uranyl acetate.

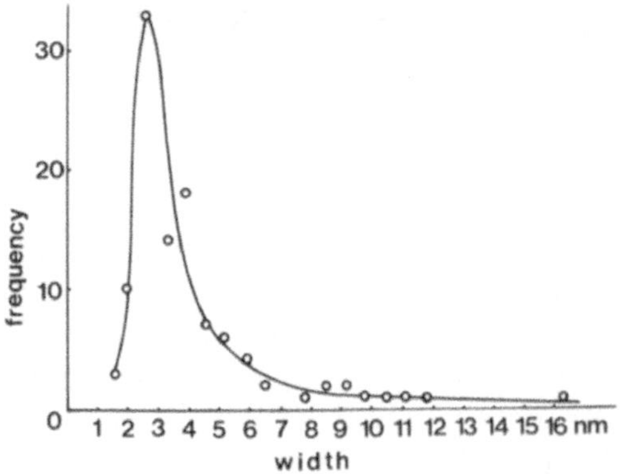

FIGURE 17. Width distribution of substructures separated from *Valonia* microfibrils during a disintegration such as that shown in Fig. 15.

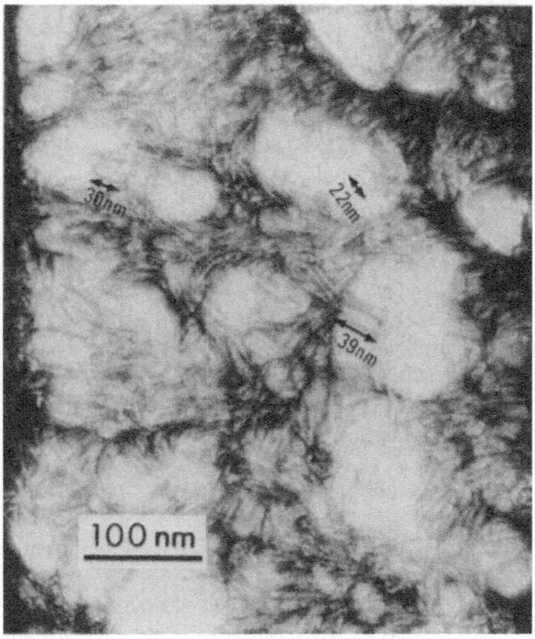

FIGURE 18. Microfibrils of poplar gelatinous layer hydrolyzed with 2.5 N hydrochloric acid for 60 min. (↔) Shortened microfibrils. The section had a thickness of 100 nm. Negatively stained with uranyl acetate.

highly crystalline as observed by Mühlethaler (1960). On hydrolysis, shorter microfibrils were observed in the gelatinous layer (Fig. 18). Their length was more than 20 nm with a maximum of 40–50 nm, which corresponds to the distance between the kinking points in the disintegrated samples (Fig. 3). In *Valonia* microfibrils, however, points of weakness along the longitudinal axis of the microfibrils could not be observed when ultrathin sections were examined (see Fig. 4).

It was never observed that the microfibrils were stained at any definite intervals in the untreated cellulose samples. Therefore, the weaker points described above are interpreted not as amorphous regions, but as locations where some crystalline dislocations [e.g., a dislocation caused by chain ends (Mühlethaler, 1969)] occurred. Such dislocations have been observed using dark-field electron microscopy (Peterlin and Ingram, 1970; Lazaro and Chiaverina, 1973).

References

Balashov, V., and Preston, R. D., 1955, Fine structure of cellulose and other microfibrillar substances, *Nature (London)* **176**:64.

Beall, F. C., and Murphey, W. K., 1970, Speculation on cellulose microfibril structure, *Wood Fiber* **2**:282.

Boylston, E. K., and Rollins, M. L., 1971, The staining of textile fibers for electron microscopy: A survey, *Microscope* **19**:255.

Côté, W. A., Jr., and Day, A. C., 1962, The G layer in gelatinous fibers—Electron microscopic studies, *Forest Prod. J*, **12**:333.

Cronshaw, J., and Preston, R. D., 1958, A re-examination of the fine structure of the walls of vesicles of the green alga *Valonia, Proc. R. Soc. B London Ser.* **148**:137.

Fengel, D., 1970, Ultrastructural behavior of cell wall polysaccharides, *Tappi* **53**:497.

Fengel, D., 1974, 10-Å-Fibrillen in Cellulose, *Naturwissenschaften* **61**:31.

Frey-Wyssling, A., 1954, The fine structure of cellulose microfibrils, *Science* **119**:80.

Frey-Wyssling, A., and Mühlethaler, K., 1951, The fine structure of cellulose, *Fortschr. Chem. Org. Naturst.* **8**:1.

Frey-Wyssling, A., and Mühlethaler, K., 1963, Die Elementarfibrillen der Cellulose, *Makromol. Chem.* **62**:25.

Frey-Wyssling, A., Mühlethaler, K., and Muggli, R., 1966, Elementarfibrillen als Grundbausteine der native Cellulose, *Holz Roh- Werkstoff* **24**:443.

Furuya, N., Takahashi, S., and Miyazaki, M., 1970, The chemical compositions of gelatinous layer from the tension wood of *Populus euro-americana, Mokuzai Gakkaishi,* **16**:26.

Goto, T., 1976, Studies on the structure of cellulose microfibrils in wood, Ph.D. thesis, Kyoto University.

Goto, T., Harada, H., and Saiki, H., 1972, Microfibrillar structure of standard cellulose powder and wood (*Pinus densiflora* Sieb. et Zucc.), *Bull. Kyoto Univ. For.* **44**:176.

Goto, T., Harada, H., and Saiki, H., 1974, Fine structure of γ-irradiated tracheid wall in *Picea abies, Bull. Kyoto Univ. For.* **46**:153.

Goto, T., Harada, H., and Saiki, H., 1975, Cross-sectional view of microfibrils in gelatinous layer of poplar tension wood (*Populus euramericana*), *Mokuzai Gakkaishi* **21**:537.

Goto, T., Harada, H., and Saiki, H., 1978, Fine structure of cellulose microfibrils in poplar gelatinous layer and *Valonia, Wood Sci. Technol.* **12**:223.

Hanna, R. B., 1971, The interpretation of high resolution electron micrographs of the cellulose elementary fibril, *J. Polym. Sci.* **C36**:409.

Hanna, R. B., and Coté, W. A., Jr., 1974, The sub-elementary fibril of plant cell wall cellulose, *Cytobiologie* **10**:102.

Harada, H., Taniguchi, T., and Kishi, A., 1971, Ultrastructure of the gelatinous layer in tension wood fibers from *Populus euramericana, Bull. Kyoto. Univ. For.* **42**:221.

Hengstenberg, J., and Mark, H ., 1928, XVIII. Über Form and Grösse der Mizelle von Zellulose und Kautschuk, *Z. Kristallogr.* **69**:271.

Heyn, A. N. J., 1965, Observations on the size and shape of the cellulose microcrystallite in cotton fiber by electron staining, *J. Appl. Phys.* **36**:2088.

Heyn, A. N. J., 1966, The microcrystalline structure of cellulose in cell walls of cotton, ramie and jute fibers as revealed by negative staining of sections, *J. Cell. Biol.* **29**:181.

Heyn, A. N. J., 1969, The elementary fibril and supermolecular structure of cellulose in soft wood fiber, *J. Ultrastruct. Res.* **26**:52

Höglund, S., 1968, Electron microscopic studies on some virus and immunoglobulin, *Arkiu. Kemi* **28**:505.

Honjo, G., and Watanabe, M., 1958, Examination of cellulose fibre by the low-temperature specimen method of electron diffraction and electron microscopy, *Nature* (*London*) **181**:326.

Imamura, Y., and Harada, H., 1973, Electron microscopic study on the development of the bordered pit in coniferous tracheids, *Wood Sci. Technol.* **7**:189.

Klug, H. P., and Alexander, L. E., 1954, *X-Ray Diffraction Procedures,* John Wiley, New York, pp. 491–538: 2nd ed. pp. 618–708 (1974).

Lazaro, R., and Chiaverina, J., 1973, Étude de la structure microfibrillaire cristalline de la cellulose de Valonia ventricosa: Progres et perspectives, *Cellul. Chem. Technol.* **7**:269.

Levin, Ö., 1963, Electron microscopic studies on protein molecules with special reference to the high molecular weight respiratory protein, thesis, quoted by Höglund, S., 1968, *Arkiu. Kemi* **28**:505.

Luft, J. H., 1971, Ruthenium red and violet. I. Chemistry, purification, methods of use for electron microscopy and mechanism of action, *Anat. Rec.* **171**:347.

Manley, R. St. J., 1964, Fine structure of native cellulose microfibril, *Nature* (*London*) **204**:1155.

Manley, R. St. J., 1971, Molecular morphology of cellulose, *J. Polym. Sci. Part A* **2**(9):1025.

Meyer, K. H., and Misch, L., 1937, Sur la constitution de la partis crystalliseé de la cellulose. VI. Positions des atomes dans le nou veau modèle spatial de la cellulose, *Helv. Chim. Acta* **20**:232.

Mühlethaler, K., 1960, Die Feinstruktur der Zellulosemikrofibrillen, *Beih. Z. Schweiz. Forstverein.* **30**:55.

Mühlethaler, K., 1969, Fine structure of natural polysaccharide systems, *J. Polym. Sci.* **C28**:305.

Nagasawa, T., 1937, Size of "micelle" in wood fibers, *J. Jpn. For. Soc.* **19**:260.

Nieduszynski, I. A., and Atkins, E. D. T., 1970, Preliminary investigation of algal cellulose. I. X-ray intensity data, *Biochim. Biophys. Acta* **222**:109.

Norberg, P. H., and Meier, H., 1966, Physical and chemical properties of the gelatinous layer in tension wood fibres of aspen (*Populus tremula* L.), *Holzforschung* **20**:174.

Peterlin, A., and Ingram, P., 1970, Morphology of secondary wall fibrils in cotton, *Text. Res. J.* **40**:345.

Preston, R. D., 1952, *The Molecular Architecture of Plant Cell Walls*, pp. 43 and 96, Chapman and Hall, London.

Preston, R. D., 1971, Negative staining and cellulose microfibril size, *J. Microsc.* **93**:7.

Preston, R. D., 1974, *The Physical Biology of Plant Cell Walls*, p. 145, Chapman and Hall, London.

Preston, R. D., and Cronshaw, J., 1958, Constitution of the fibrillar and non-fibrillar components of the walls of *Valonia ventricosa*, *Nature (London)* **181**:248.

Rånby, B. G., 1952, The cellulose micelles, *Tappi* **35**:53.

Scherrer, P., 1918, Bestimmung der Grösse und der inneren Struktur von Kolloidteilchen mittels Röntgenstrahlen, *Goettingen Nachr.* **2**:98.

Sponsler, O. L., 1931, Orientation of cellulose space lattice in the cell wall: Additional X-ray data from *Valonia* cell-wall, *Protoplasma* **12**:241.

Tanaka, F., and Okamura, K., 1977, Measurement of pole figures and orientation functions for *Valonia* cellulose, *J. Polym. Sci. Polym. Phys. Ed.* **15**:897.

Wardrop, A. B., 1954, The fine structure of the conifer tracheid, *Holzforschung* **8**:12.

Wardrop, A. B., 1969, Fiber morphology and papermaking, *Tappi* **52**:396.

Wellard, H. J., 1954, Variation in the lattice spacing of cellulose, *J. Polym. Sci.* **13**:471.

Wise, L. E., Murphy, M., and D'Addieco, A. A., 1946, Chlorite holocellulose, its fractionation and bearing on summative wood analysis and on studies on the hemicelluloses, *Paper Trade J.* **122**:35.

Yoshimi, T., and Harada, H., 1969, The structure of microfibrils in wood—the shape and size of the cross-section, Abstract paper presented at the 19th annual meeting of the Japan Wood Research Society, Sapporo, Japan, p. 56.

20

The Macromolecular Organization of Cellulose and Chitin

JOHN BLACKWELL

1. Introduction

In this chapter, we will review what is currently known about the physical structures of cellulose and chitin and discuss the implications of these structures for studies of the biosynthesis of these two polymers. Cellulose and chitin are both fiber-forming polymers, and as fibers they serve as the load-bearing components of major groups of natural composite tissues—cellulose in many plant-cell walls and chitin in the skeletal materials of many lower animals. Their chemical structures are very similar: cellulose is poly-$\beta(1\rightarrow4)$-D-glucose and chitin is its 2-acetamido derivative (see Fig. 1); not surprisingly, this leads to similar structures and morphologies. The $\beta(1\rightarrow4)$ glucan chain is inherently stiff and extended and must be judged ideal for the formation of fibrous structures. Both polymers are naturally crystalline, and their crystal structures have been the subject of X-ray crystallographic studies for over 60 years. From early studies, it was known that both cellulose and chitin could adopt a number of different polymorphic structures. These have different unit cells, but in all cases the chains have extended ribbonlike conformations, with two pyranose residues repeating in approximately 10.3 Å. The actual packing of these chains and their intermolecular hydrogen bonding proved more difficult to determine, and it has only been in the last decade that these problems have yielded to X-ray investigations. The central question for cellulose was the polarity of adjacent chains: the unit cells contain sections of two chains that could have the same (parallel) or opposite (antiparallel) sense. Native cellulose fibrils appear to

JOHN BLACKWELL • Department of Macromolecular Science, Case Western Reserve University, Cleveland, Ohio 44106.

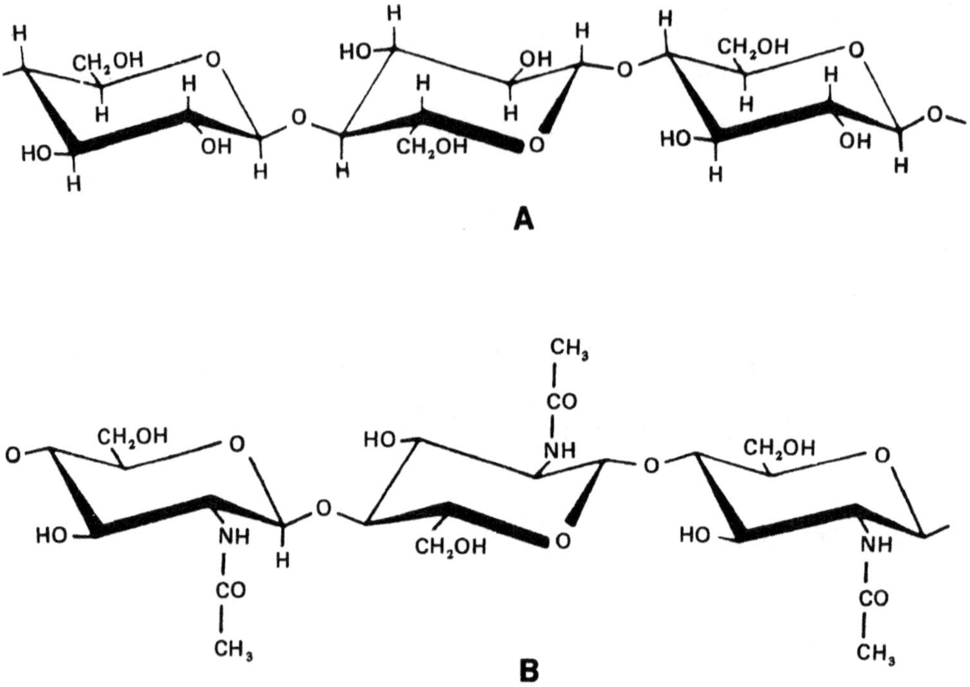

FIGURE 1. Chemical structures of cellulose (A) and chitin (B).

crystallize almost immediately after synthesis, and chain polarity is probably determined by the biosynthesis mechanism. Industrial processing of cellulose frequently involves regeneration from solution or modification of native fibers by swelling agents, which result in adoption of different crystal structures. Determination of the crystal structures leads to an understanding of the mechanism involved in these processes. In the case of chitin, our knowledge of the polarities of adjacent chains is of longer standing, although the hydrogen bonding between the chains has been determined only recently. An important conclusion is that adjcent chains are antiparallel in the most common form of chitin (α-chitin), whereas they are parallel in native cellulose, which implies major differences in their biosynthesis mechanisms.

Most native cellulosic and chitinous tissues are composites in which the fibers are suspended in a lower modulus matrix. In plant-cell walls, the cellulose fibrils are surrounded by a matrix of *amorphous* hydrated polysaccharides (e.g., xylans, mannans) known as hemicelluloses. In some systems, especially wood, the amorphous cross-linked polymer lignin is also present. In contrast, the matrix in chitinous tissues is usually protein and may also be extensively calcified (e.g., crustacean shells). Many groups of these tissues, especially the insect cuticles, are unique structural complexes of chitin and protein in which both phases are *ordered.* We have recently developed models for some of these complexes, which give insight into the interaction between the fibers and matrix in these composites.

2. *X-Ray Diffraction of Fibrous Polymers*

Before we review the structures determined for cellulose and chitin, it will be useful to summarize the application of X-ray crystallography to fibrous polymers for the benefit of readers not familiar with the technique. The X-ray diffraction pattern of native ramie fibers (cellulose I) is shown in Fig. 2. This pattern was obtained by placing a bundle of ramie fibers perpendicular to a narrow beam of monochromatic X-rays and recording the diffraction pattern on a photographic film placed a short distance behind the specimen. The pattern consists of a series of spots or reflections of different relative intensities. From the *positions* of the reflections, it is possible to determine the dimensions of the unit cell, the basic building block of the crystal. The reflections can be thought of as arising from sets of equally spaced planes within the structure; *d* is the perpendicular spacing between the planes in a particular set, and their

FIGURE 2. X-ray fiber diagram of native ramie cellulose (cellulose I). In this and all other X-ray patterns, the vertical axis corresponds to the fiber axis.

orientation with respect to the unit cell axes is defined by the Miller indices h,k,l. The *intensities* of the reflections depend on the positions of the atoms within the unit cell, and hence can be used to determine the arrangement of adjacent polymer chains. The intensity of a reflection $I(hkl)$ is proportional to $F^2(hkl)$ where $F(hkl)$ is the scattering amplitude or structure factor for that hkl plane. $F(hkl)$ can be calculated from the coordinates of the atoms:

$$F(hkl) = \sum_{j=1}^{n} f_j \exp 2\pi i(hx_j + ky_j + lz_j)$$

where x_j, y_j, z_j are the coordinates of the jth atom, expressed as fractions of the unit cell sides, and f_j is the atomic scattering factor of the jth atom, which takes account of the differences in the scattering of X-rays by different types of atoms (and can be found in standard tables). The summation is over all N atoms in the unit cell.

Thus, given the structure of the polymer, it is possible to predict its X-ray diffraction pattern. Unfortunately, the reverse procedure, determination of the structure of the polymer from the X-ray pattern, is rather more difficult. This can be achieved in a routine way for molecules that form large crystals. Such crystals typically give rise to more than a thousand reflections, and statistical methods are well developed to determine their structures. Fibrous polymers, however, are generally not completely crystalline, and the crystalline domains are relatively small. As a result, the X-ray pattern of an oriented polymer generally contains 20–50 reflections, which is too few to allow application of the methods devised for single crystals. As a result, for many years the approach was to determine polymer structures by trial and error. A model is postulated, and its calculated intensities are compared to the observed values. This model is then varied until the best agreement is obtained between the observed and calculated intensities, as judged by the crystallographic R values:

$$R = \frac{\Sigma||F_o| - |F_c||}{\Sigma|F_o|} \qquad R' = \frac{\Sigma w||F_o| - |F_c||}{\Sigma w|F_o|} \qquad R'' = \frac{\Sigma w[|F_o| - |F_c|]^2}{\Sigma w F_o^2}$$

where F_o and F_c are the observed and calculated structure factors (proportional to the square roots of the intensities), w is a weight assigned to each observation, and the summations are over all the intensity data. In general the lower the R value, the better the agreement. Typically, a "correct" or "acceptable" polymer structure can be expected to have an R value in the range of 0.15–0.30, depending on the complexity of the structure and other factors such as the accuracy of the intensity measurement.

As will be seen below, this method was applied a number of times to determine the structures of cellulose, without success. The basic unit cells for cellulose contain disaccharide units of two chains, and it was immediately apparent that successive glucose residues in each chain are related by at least an approximate twofold screw axis (i.e., by a rotation of 180° about the chain axis). Thereafter, for native cellulose, for example, models could be built in which the two chains have either the same or opposite sense. These appeared to be stereochemically acceptable and gave approximately the same agreement between observed and calculated intensities. Given this impasse, it was not possible to make much progress with regard to the hydrogen bonding and other details.

This difficulty in distinguishing parallel and antiparallel models stems from the fact that "up" and "down" chains are not greatly different, with a result that reversal of the polarity of one chain has only a secondary effect on the X-ray pattern. At this point, it should be stressed that this problem is common to many polymer structures. For most structures in which the polarity of the adjacent chains is known, this is determined from the space-group symmetry rather than from a comparison of the entire intensity data. In cellulose, both parallel and antiparallel models are compatible with the space-group symmetry, and we are presented with a more difficult problem.

In the last 10–15 years, our methods of determining polymer structures have advanced considerably (for a review, see French and Gardner, 1980). First, our knowledge of stereochemistry has been advanced by the determination of the structures of a large number of low-molecular-weight structures that serve as model compounds. For polysaccharides, it has been found that pyranose rings have very similar stereochemistry in a large number of mono- and disaccharides (Arnott and Scott, 1972), with the result that a much better model for the cellulose chain can be constructed than was available before the mid-1960s. This has led in turn to much better agreement between observed and calculated R values for all trial models, with the result that the small predicted differences between the parallel and antiparallel models become more significant. Second, development of better models made possible the application of least-squares methods to refine polymer structures. Given the limited X-ray data, it is not possible to attempt a least-square refinement of the atomic coordinates. However, when we can be confident about the model for the monomer, the helix symmetry, and the helix repeat, it is possible to refine the packing parameters that define the arrangement of chains within the unit cell. Application of this method to fibrous polymer structures is primarily due to Arnott and co-workers (see Smith and Arnott, 1978; Arnott and Wonacott, 1966), who have developed the linked-atom least-squares (LALS) programs, which have been applied to a number of crystalline polymers, including cellulose and chitin in this laboratory. In the cellulose structure, once the chain conformation is defined, determination of the chain packing requires refinement of 7 variables against 36 observed and 40 unobserved intensity data points (see below), which is easily accomplished by least-squares methods. The procedure adopted was to set up parallel and antiparallel chain models and to refine each to obtain the best agreement between the observed and calculated intensities. Selection of the best model is then on the basis of the lowest R value (R'' is used in the least-squares method). The significance of improvement in the R'' value is a standard statistical problem and can be determined from the tests devised by Hamilton (1965), based on the number of observations and refinable parameters. Our best parallel and antiparallel models for native cellulose had R'' values of 0.215 and 0.270, and on this basis the antiparallel model can be rejected at the 0.5% level; i.e., the parallel model is favored by better than 200:1. The same methods can be used to select between other less extreme differences, such as different hydrogen-bonding options.

Later versions of the LALS program have made more extensive use of stereochemical data, and simultaneously optimize the X-ray agreement and the packing by including modified potential functions, which is in fact a combination of X-ray methods and conformational analysis. This is particularly useful where even fewer X-ray data are available than for cellulose and for polymers with large complex monomer repeats.

3. *Polymorphic Structures of Cellulose*

At least four different crystalline forms of cellulose (celluloses I-IV) are recognized on the basis of their X-ray diffraction patterns and infrared spectra (Marchessault and Sarko, 1967; Blackwell and Marchessault, 1971). In each structure, the chains have approximately the same backbone conformation, with two glucose residues repeating in approximately 10.3 Å. The structures differ in terms of the packing of adjacent chains (Wellard, 1954). Cellulose I (native cellulose) is the naturally occurring form in plant-cell walls. Regeneration of cellulose from solution, e.g., in the manufacture of rayon fibers, leads to cellulose II. This structure can also be produced by the process known as mercerization, which involves swelling in caustic soda followed by removal of the swelling agent. This process is used to modify the properties of cotton fibers. The cellulose I→II transition is irreversible, which implies that cellulose II is the stable form and that native cellulose I is a metastable structure. Cellulose III can be produced by treatment of either cellulose I or cellulose II with liquid ammonia. However, the structures obtained depend on the parent material and are distinguished as celluloses III_I and III_{II}. The unit cells of these structures are approximately the same, but the infrared spectra show that they have different hydrogen-bonding networks (Marrinan and Mann, 1956). Washing with water restores the parent cellulose I or II (Segal *et al.*, 1954), indicating that forms III_I and III_{II} retain a memory of their original structure. Likewise, cellulose IV_I and IV_{II} can be prepared from celluloses I and II by treatment in hot glycerol (e.g., at 270° C); these structures can also be restored to their original forms. Treatment of cellulose I or II with concentrated acids has been reported to produce a fifth form (Ellefsen and Norman, 1962), designated cellulose X, which is similar but not identical to cellulose IV.

4. *X-Ray Studies of the Structures of Cellulose*

Investigations of the crystal structures of cellulose have a long history. The first reported studies are by Nishikawa and Ono (1913), who examined the X-ray patterns of flax and hemp, using the methods developed shortly before by the Braggs. This work was done more than ten years before the chemical structure of cellulose was established, and the X-ray studies by Meyer and Mark (1928) and Sponsler and Dore (1930) and others during that period contributed significantly to establishing that cellulose is a polymer of glucose. By the late 1920s, the unit cells of celluloses I and II had been determined, and it was known that each cell contained disaccharide sections of two cellulose chains, with an approximate twofold screw axis parallel to the chains. The first detailed trial-and-error determination of the structure of cellulose I is due to Meyer and Misch (1937), who showed that the two chains have a relative stagger of approximately $c/4$ along the fiber axis ($c \approx 10.3$ Å, the fiber repeat). It was not possible to determine the polarities of the chains, but Meyer and Misch arranged the chains *antiparallel*, arguing that it was likely that there would be a 50/50 mixture of both polarities. This was a reversal of the previous view from the same laboratory, since Meyer and Mark (1928) assumed parallel polarities.

The work of Meyer and Misch preceded the development of ideas on stereochemistry. Their model was a straight chain with successive glycosidic oxygens lying on the helix axis. The first space-filling models of cellulose were built by Hermans *et al.* (1943) and showed that the extended cellulose chain adopts a bent conformation such that the 03-H group forms an intramolecular hydrogen bond to the 05 atom of the next residue. Models for the cellulose structures were modified to use the Hermans model, and there was extensive discussion of possible intra- and intermolecular hydrogen bonding, based on the data from polarized infrared spectra (Liang and Marchessault, 1959; Marchessault and Liang, 1960). Preference for the antiparallel models for celluloses I and II was maintained, especially in view of the proposals for adjacent reentry chain folding in single crystals of many synthetic polymers. The actual crystal structures for celluloses I and II were determined in the 1970s by application of least-squares refinement and stereochemical constraints. The structures as determined in this laboratory are described below (Gardner and Blackwell, 1974; Kolpak and Blackwell, 1976). Independent work by Sarko and co-workers in Syracuse (Sarko and Muggli, 1974; Stipanovic and Sarko, 1976) has led them to similar conclusions as regards the basic structures, although there are some minor differences.

4.1. Cellulose I

Intensity data for cellulose I were obtained (Gardner and Blackwell, 1974) from the diffraction patterns of fibers drawn from the cell walls of the sea alga *Valonia ventricosa*, which is the most crystalline native cellulose. This diffraction pattern is shown in Fig. 3. Solution of the structure is complicated by the fact that 3 of the 41 observed reflections cannot be indexed by the two-chain (Meyer) and Misch) unit cell, but require a larger unit cell containing sections of eight chains. Nevertheless, the weakness of these three reflections, plus the absence of many other reflections predicted for the eight-chain (but not the two-chain) unit cell, means that the differences between the four two-chain units

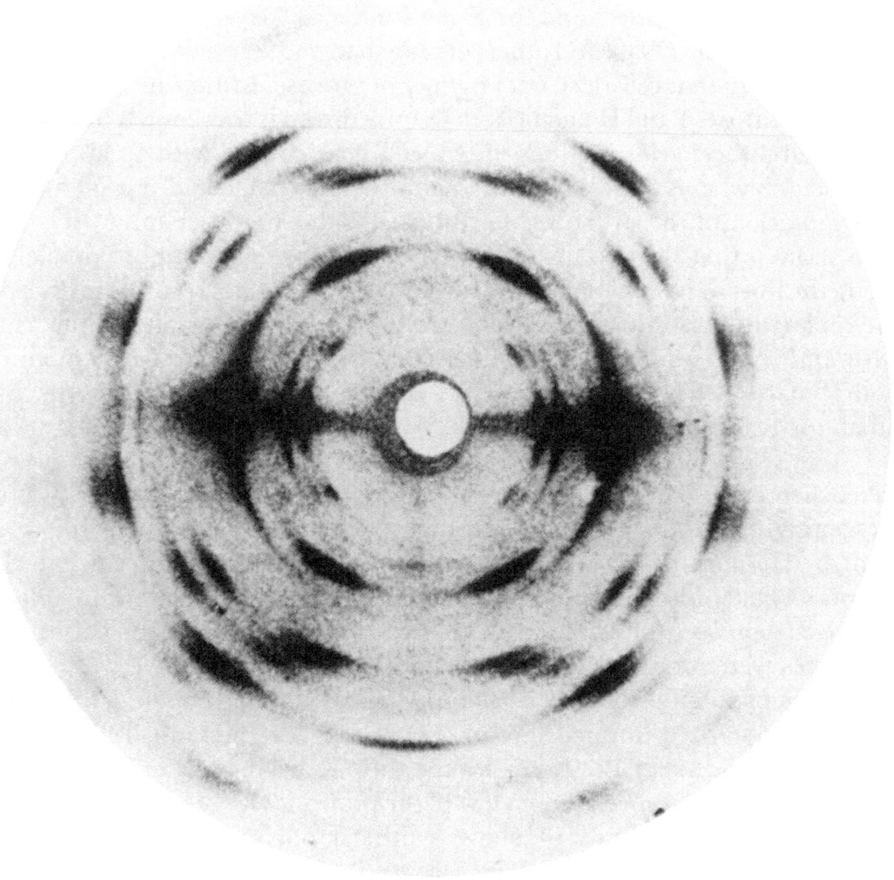

FIGURE 3. X-ray fiber diagram of *Valonia* cellulose (cellulose I). The specimen was a bundle of parallel fibers drawn from the purified cell wall. From Gardner and Blackwell (1974); reproduced with the permission of Wiley-Interscience Inc.

making up the eight-chain unit cell must be very small. Thus, the two-chain unit cell can be taken as an adequate approximation to the structure. The unit cell is monoclinic with dimensions $a = 8.17$ Å, $b = 7.86$ Å, $c = 10.38$ Å, and $\gamma = 97.0°$. For *Valonia*, odd-order $(00l)$ reflections are absent, indicating a 2_1 helical axis parallel to the chains, and thus the space group is $P2_1$. For less crystalline celluloses, a weak (001) reflection is often seen, as is the case, for example, for ramie cellulose I (see Fig. 1), which means that the 2_1 symmetry is not perfect throughout the structure. The diffraction geometry greatly enhances the intensity of meridional reflections as compared to those elsewhere on the pattern, and only a very small deviation from the perfect 2_1 symmetry would be necessary to produce a weak (001) reflection. Spurious meridional reflections of this type are a common problem for many polymer structures.

Our procedure was to construct both parallel and antiparallel packing models and to refine each using the LALS programs to see which would give the best agreement with the observed intensity data. This reduced to consideration of four basic models: (p_1) parallel "up" chains with the center chain staggered by $+c/4$; (p_2) parallel "down" chains with the center chain staggered by $+c/4$; (a_1) antiparallel chains with an "up" chain at the corner and a "down" chain at the center staggered by $-c/4$; (a_2) antiparallel chains as in a_1 but with the center chain staggered by $+c/4$. The stagger is defined by the separation of the glycosidic oxygens along the helix axis; an "up" chain has z(O5) > z(C5). The atomic coordinates for the cellulose chain were derived from standard bond lengths and angles averaged for a series of mono- and disaccharides. The chain was constrained to form a 2_1 helix with a repeat of c = 10.3 Å and to form an intramolecular 03-H · · · 05′ hydrogen bond as shown in Fig. 4. This chain is rigid except for rotational freedom at the $-CH_2OH$ side chain. The four models were refined in terms of the following parameters: ϕ_1 and ϕ_2, the orientation of the corner and center chains about their axes; S, the chain stagger; χ_1 and χ_2, the orientation of the $-CH_2OH$ groups on the corner and center chains; K, a scale factor; and B, an isotropic temperature factor. Initially χ_1 was set equal to χ_2; later work showed that independent refinement of χ_1 and χ_2 did not lead to significantly different values; i.e., the $-CH_2OH$ groups have the same orientation on the two chains.

The four models were refined against the observed intensity data, which consisted of 36 reflections (the two meridional reflections are omitted because of difficulty in making geometric corrections). The final R values for the four models were $R_{p1} = 0.179$, $R_{p2} = 0.202$, $R_{a1} = 0.207$, and $R_{a2} = 0.249$. The Hamilton test allows rejection of the a_2 model, and the p_2 model is not fully hydrogen bonded, and can also be rejected. The remaining p_1 and a_1 models were refined against the full data. (In addition to the 36 observed reflections, a further 40 are predicted within the observed range, but are too weak to be detected. The correct model should predict weak intensity for these reflections, which are incorporated using a weighting scheme $w = 1$ for observed and $w = \frac{1}{2}$ for unobserved.) This led to R' values of $R'_{p1} = 0.233$ and $R'_{a1} = 0.299$. As indicated above, the best antiparallel model can be rejected at the 0.5% confidence level, corresponding to a preference better than 200:1 for the best parallel-chain model.

The refined parallel-chain model for cellulose I is shown in Fig. 4. The final values of ϕ_1 and ϕ_2 are not significantly different; i.e., the corner and center chains have the same orientation about their long axes. The chains are staggered by 0.266c [cf. 0.25c proposed by Meyer and Misch (1937)]. The hydrogen-bonding network is shown in Fig. 4C. The orientation of the $-CH_2OH$ groups allows the formation of a second intramolecular hydrogen bond: 02′-H · · · 06. Thus, there are intramolecular hydrogen bonds on *both* sides of the glycosidic linkage, which should enhance the stiffness of the extended conformation. The 06-H hydroxyl is hydrogen bonded to 03 of the next chain along the a axis. No hydrogen bonding occurs along the b axis or the unit-cell diagonals, and hence native cellulose is seen as a series of hydrogen-bonded sheets of chains parallel to the a axis that are held together by van der Waals forces between the hydrophobic "faces" of the sugar rings.

In addition to the work of Sarko, the parallel-chain model has been

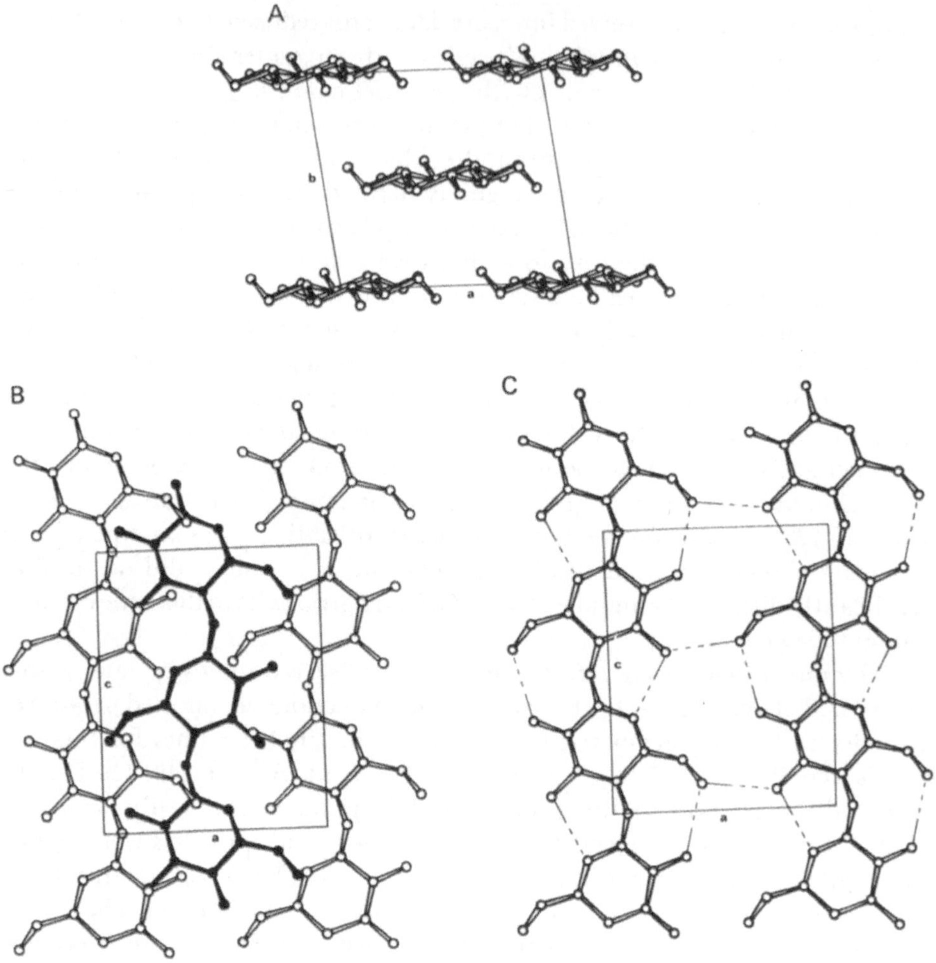

FIGURE 4. Structure of cellulose I. (A) *ab* projection (looking along the chain axes); (B) *ac* projection; (C) hydrogen-bonding network in the sheet parallel to the *ac* plane. From Gardner and Blackwell (1974); reproduced with the permission of Wiley-Interscience Inc.

confirmed by electron-diffraction work (Claffey and Blackwell, 1976). French and Murphy (1977) have repeated the refinement using our X-ray intensity data but with different models for the cellulose chain (including non-2_1 conformations) and in all cases show a significant preference for parallel chains. Finally, an independent analysis by H. Tadokoro (personal communication) also favors the parallel-chain structure.

4.2. Cellulose II

The most detailed cellulose II patterns are those obtained for Fortisan rayon, which is prepared by deacetylating highly oriented cellulose triacetate fibers. The X-ray pattern of Fortisan is shown in Fig. 5. Fortisan is not as crystalline as

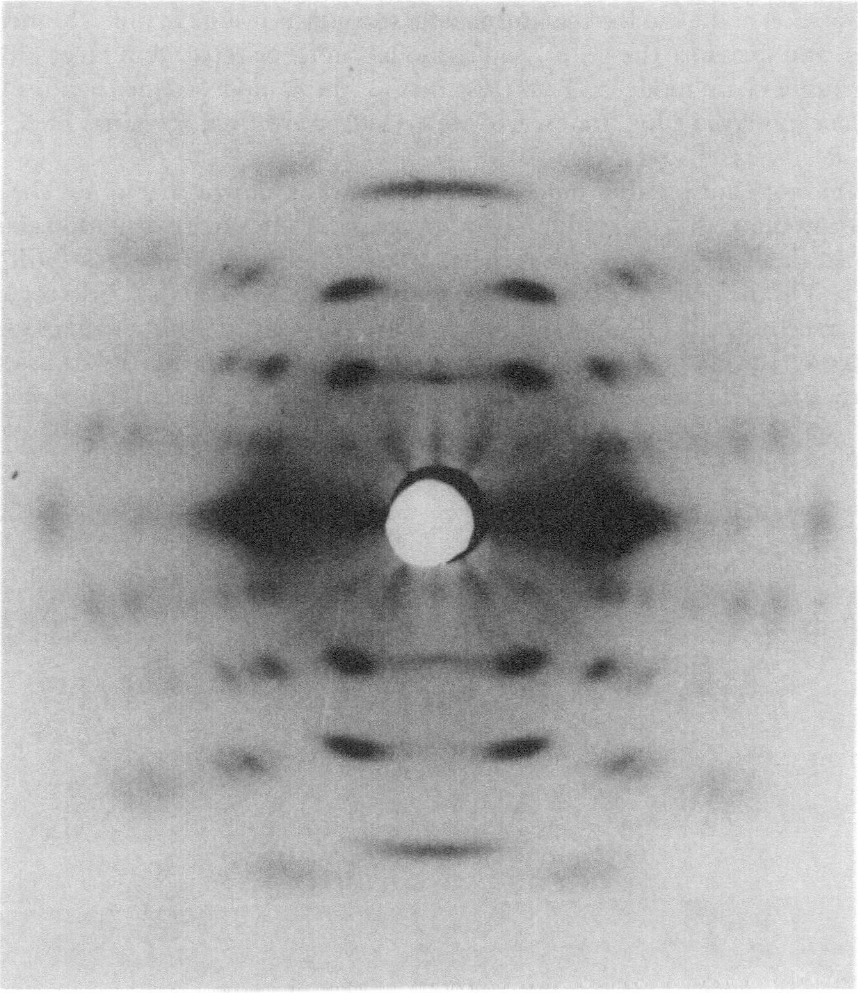

FIGURE 5. X-ray fiber diagram of Fortisan rayon fibers (cellulose I). From Kolpak and Blackwell (1976).

Valonia, but nevertheless yields some 44 observed nonmeridional reflections; in addition, 41 unobserved reflections are predicted in the region of the observed data (Kolpak and Blackwell, 1976). The unit cell is monoclinic with dimensions $a = 8.01$ Å, $b = 9.04$ Å, $c = 10.36$ Å, and $\gamma = 117.1°$. Odd-order ($00l$) reflections are absent except for a very weak (001), and a $P2_1$ space group was assumed. The structure was refined in terms of the same four models defined for cellulose I. Refinement against the observed data led to the following R values: $R_{p1} = 0.254$, $R_{p2} = 0.188$, $R_{a1} = 0.195$, and $R_{a2} = 0.171$. The p_1 model can be rejected, but it is not possible to distinguish among the other three on the basis of X-ray criteria alone. However, models p_1, p_2, and a_1 are all stereochemically unacceptable. For example, model p_2 has an oxygen–oxygen interatomic distance of 2.05 Å between adjacent molecules, whereas the absolute minimum allowable is 2.60 Å.

Modification of the models to eliminate these contacts results in much higher R values, and therefore the p_1, p_2, and a_1 models must be rejected in favor of the antiparallel-chain model a_2. Final R values for the a_2 model refined against the full data (observed plus unobserved reflections) gave final R values of $R_{a2} = 0.213$, $R'_{a2} = 0.200$, and $R''_{a2} = 0.155$.

The final antiparallel model for cellulose II is shown in Fig. 6. The *ab* projection shows that the chains have approximately the same orientation about their axes and are stacked along the short *ab* diagonal. These stacks will be stabilized by the hydrophobic (van der Waals) forces, more so than between the sheets in cellulose I. The relative stagger of the chains is $+0.216c$, again close to the quarter-stagger position. The refinement has led to different conformations of the $-CH_2OH$ groups on the center and corner chains. Adjacent center chains

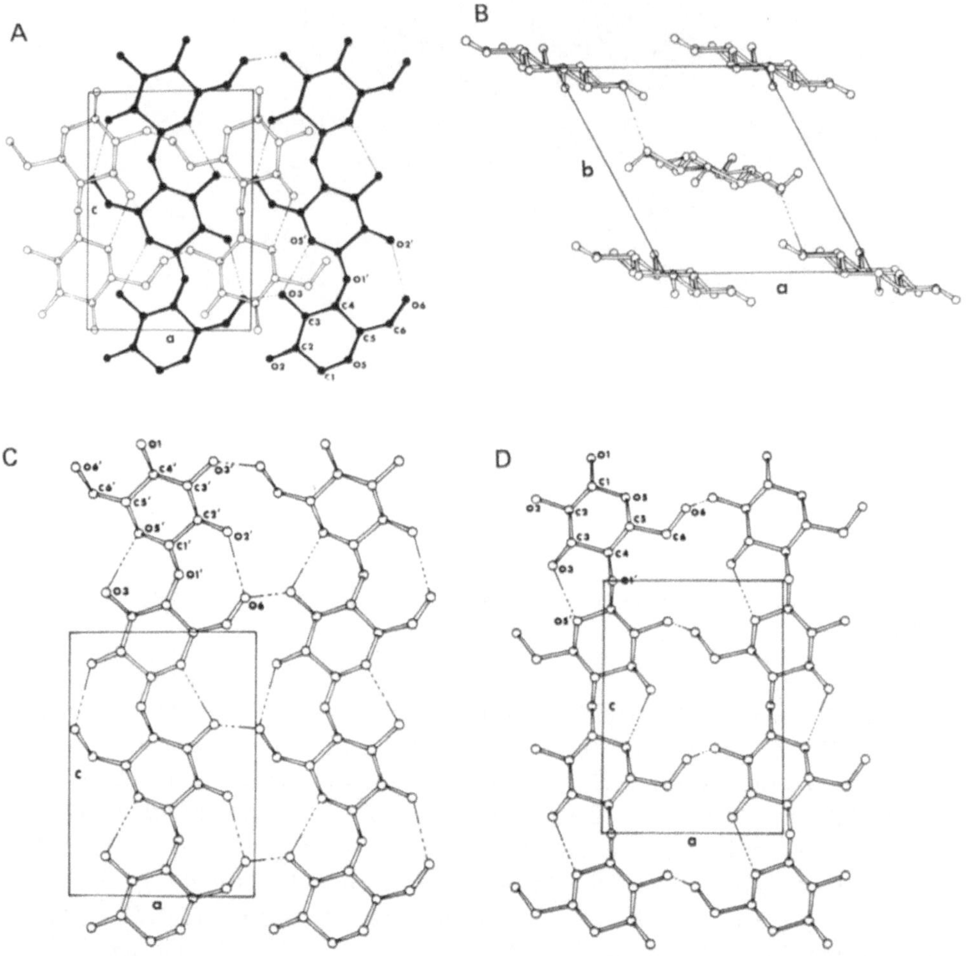

FIGURE 6. Structure of cellulose II. (A) *ac* projection; (B) *ab* projection [(---) hydrogen bonds between center and corner chains]; (C) hydrogen bonding of the center chains; (D) hydrogen bonding of the corner chains; (E) hydrogen bonding between center and corner chains. From Kolpak and Blackwell (1976).

along the *a* axis are shown in Fig. 6C. The $-CH_2OH$ groups are oriented like those in cellulose I, so as to allow formation of a second intramolecular bond 02′—H · · · 06 and an intermolecular bond 06—H · · · 03 along the *a* axis. The sheet of corner chains is shown in Fig. 6D. Here, the $-CH_2OH$ group is swung round so that it forms an 06—H · · · 02 intermolecular bond. Intramolecular bonding for 02—H is now not possible, and this group forms another intermolecular bond 02—H · · · 02′ to the next chain along the long *ab* diagonal. This bond is shown in Fig. 6E, and is indicated by the dashed lines in Fig. 6A. This extra intermolecular bonding is a major difference between cellulose II and cellulose I and probably goes a long way to explain the higher stability of the regenerated form.

Cellulose II can also be produced from cellulose I by swelling in the mercerization process. We have repeated the refinement work described above using intensity data collected for mercerized ramie cellulose and derive the same antiparallel-chain model for this structure (Kolpak *et al.*, 1978).

4.3. Implications of the Solid-State Structures of Cellulose

The structure determined for celluloses I and II give us insight into the process of biosynthesis of cellulose fibrils. Also, they lead to the interesting question of the mechanism of conversion of cellulose I to cellulose II in the solid

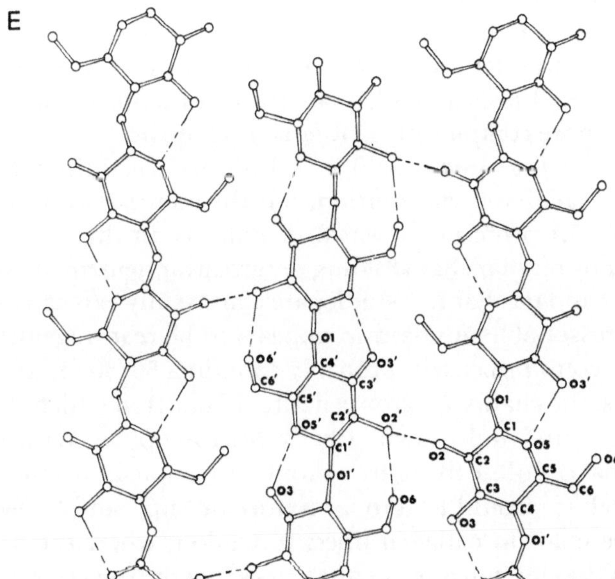

FIGURE 6. (*continued*)

(albeit swollen) state, since this involves a change from parallel to antiparallel polarity. These two processes will be considered in turn below.

The immediate conclusion from a parallel-chain model for native cellulose is that it rules out a folded-chain structure, which would necessarily contain antiparallel chains within the crystallites. Biosynthesis therefore results in microfibrils that are fully extended chain polymer single crystals, which will have the maximum attainable strength and modulus. The picture developed by Brown *et al.* (1976) is that cellulose microfibrils are synthesized from rows of enzyme complexes on the surface of the cell membrane. One can envisage each group of enzymes producing a stiff extended chain that extends outward into the aqueous medium with those from neighboring complexes. Cellulose will be insoluble under these conditions and will crystallize with its neighbors to form the microfibril. The size (diameter) of the fibril will be determined by the number of chains synthesized at that location.

Given the higher stability of cellulose II, it is interesting that cellulose I is synthesized instead. Protein fibers such as collagen are synthesized by first producing the protein molecules and then transporting them outside the cell where they self-assemble. If cellulose were to be synthesized in such a two-step process, then it seems inevitable that cellulose II would be formed. It seems likely, then, that cellulose I is required by the specifics of the biosynthesis mechanism. It would certainly seem to be simpler to synthesize a parallel-chain structure, since different enzyme complexes would be necessary for "up" and "down" chains. Incidentally, if alternating "up" and "down" chains were to be synthesized, it is difficult to see why these would not adopt the cellulose II lattice. Thus, cellulose I probably occurs in nature simply because parallel chains are synthesized and these crystallize into extended-chain fibrils. Cellulose II may be the more stable form, but cellulose I is clearly adequate for its functions, and thus there is no incentive for a more complex biosynthesis mechanism.

With regard to the conversion from cellulose I to cellulose II, it is easy to see how this can be achieved via solution, but the mercerization process presents more difficulty. In principle, reversal of polarity in the solid state could be achieved either by regular chain folding or by rearrangement of extended chains. Morphological and mechanical studies are against any extensive chain folding, and the most reasonable mechanism appears to be rearrangement of extended chains. A bulk cotton fiber will probably contain a 50/50 mixture of "up" and "down" chains: the chains are grouped into domains in which they all have the same sense, i.e., "up" and "down" fibrils. Microscopy by Willison and Brown (1978) shows that cellulose fibers are wound continuously around the cell during biosynthesis, which would lead to a mixture of "up" and "down" fibrils. An analogy can be made to collagen fibers in tendon, where the banded staining pattern seen in the electron microscope shows that the fibers consist of a random arrangement of "up" and "down" fibrils (see Parry and Craig, 1977). In mercerization, the individual chains are separated and on removal of the swelling agents will be more likely to unite with antiparallel mates, since the cellulose II structure is thermodynamically more stable.

It should be noted that the mercerization change to cellulose II usually requires several swellings and deswellings to achieve maximum conversion, and there is always some residue of unconverted cellulose I. It is difficult to see why the change should be so difficult unless there is a major structural rearrangement. Conversion to cellulose II is not achieved with swelling agents such as hydrazine and ethylene diamine, which probably do not separate the chains so extensively.

5. *Chitin*

The chitin fibrils in the skeletal tissues of lower animals serve a function similar to that of cellulose in plant-cell walls and collagen in mammalian connective tissue. It is useful to review their structures here in view of the obvious chemical and physical analogies to cellulose and also the occurrence of chitin fibrils in certain fungal-cell walls. At least two crystalline forms of chitin occur in nature, known as α- and β-chitin (Rudall, 1963), which will be seen to be analgous to cellulose II and I, respectively. Almost all chitin, including fungal chitin, is in the α-form. The β-form is very rare and has been identified in only four sources: squid pen, *Aphrodite* chaetae, pogonophore tubes, and the spines of certain diatoms (see Picken and Lotmar, 1950; Blackwell *et al.*, 1965, 1967) (see also Chapter 10). It is significant that all these are aquatic systems. With the exception of the diatom spines, all chitin is found in a composite with proteins. Our knowledge of these proteins is limited at present (for a review, see Neville, 1975), probably because study of chitinous systems lacks the impetus of medical or industrial relevance. However, the proteins are often extensively cross-linked, and the matrix is frequently calcified. X-ray studies have shown that in some tissues the protein is highly ordered (Rudall, 1963), and investigation of these systems may lead to detailed models for the fiber–matrix polysaccharide–protein interactions.

5.1. *α-Chitin*

The most oriented and crystalline α-chitin specimens are obtained from lobster tendons and are deproteinized by boiling in 5% potassium hydroxide for 24 hr. (This process has little or no effect on the chitin structure.) The X-ray pattern of deproteinized lobster tendon is shown in Fig. 7. The unit cell is orthorhombic and has dimensions a = 4.74 Å, b = 18.86 Å, and c = 10.32 Å (Minke and Blackwell, 1978). The fiber repeat (c) and the overall X-ray pattern are very similar to those for cellulose. The unit cell contains two chains, and the space group is at least approximately $P2_12_12_1$, which requires that the chains are *antiparallel*.

There have been several attempts to solve the structure in the past, notably by Meyer and Pankow (1935) and Carlstrom (1957). The major features of the structure were determined by Carlstrom, who arranged the chains in stacks along

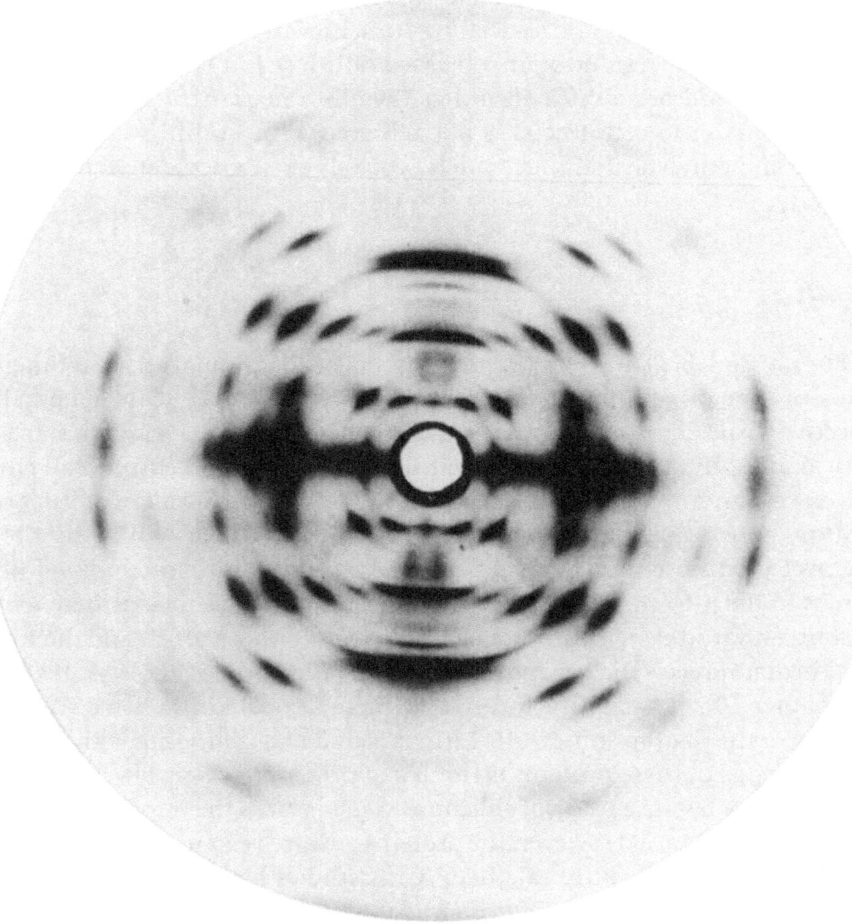

FIGURE 7. X-ray fiber diagram of deproteinized lobster tendon. From Minke and Blackwell (1978); reproduced with the permission of Academic Press Inc.

the *a* axis, linked by N—H · · · O=C hydrogen groups between the acetamido sidechains. The arrangement of the $-CH_2OH$ groups and the remaining intermolecular hydrogen bonding were determined recently in this laboratory. We refined both parallel and antiparallel models: the parallel model gave much worse agreement with the X-ray data and was totally unacceptable stereochemically. Good agreement was obtained for the antiparallel model, which had no bad contacts. The refined structure of α-chitin is shown in Fig. 8. The $-CH_2OH$ groups form a 50/50 mixture of inter- and intramolecular hydrogen bonds. The $-CH_2OH$ group on the center chain may form a donor intramolecular bond to the carbonyl on the next residue and serve as the acceptor for an intermolecular hydrogen bond for the $-CH_2OH$ group of the corner chain. Alternatively, the bonding pattern can be reversed, i.e., intra- and intermolecular bonds for the corner and center chains, respectively. The calculations suggest a random

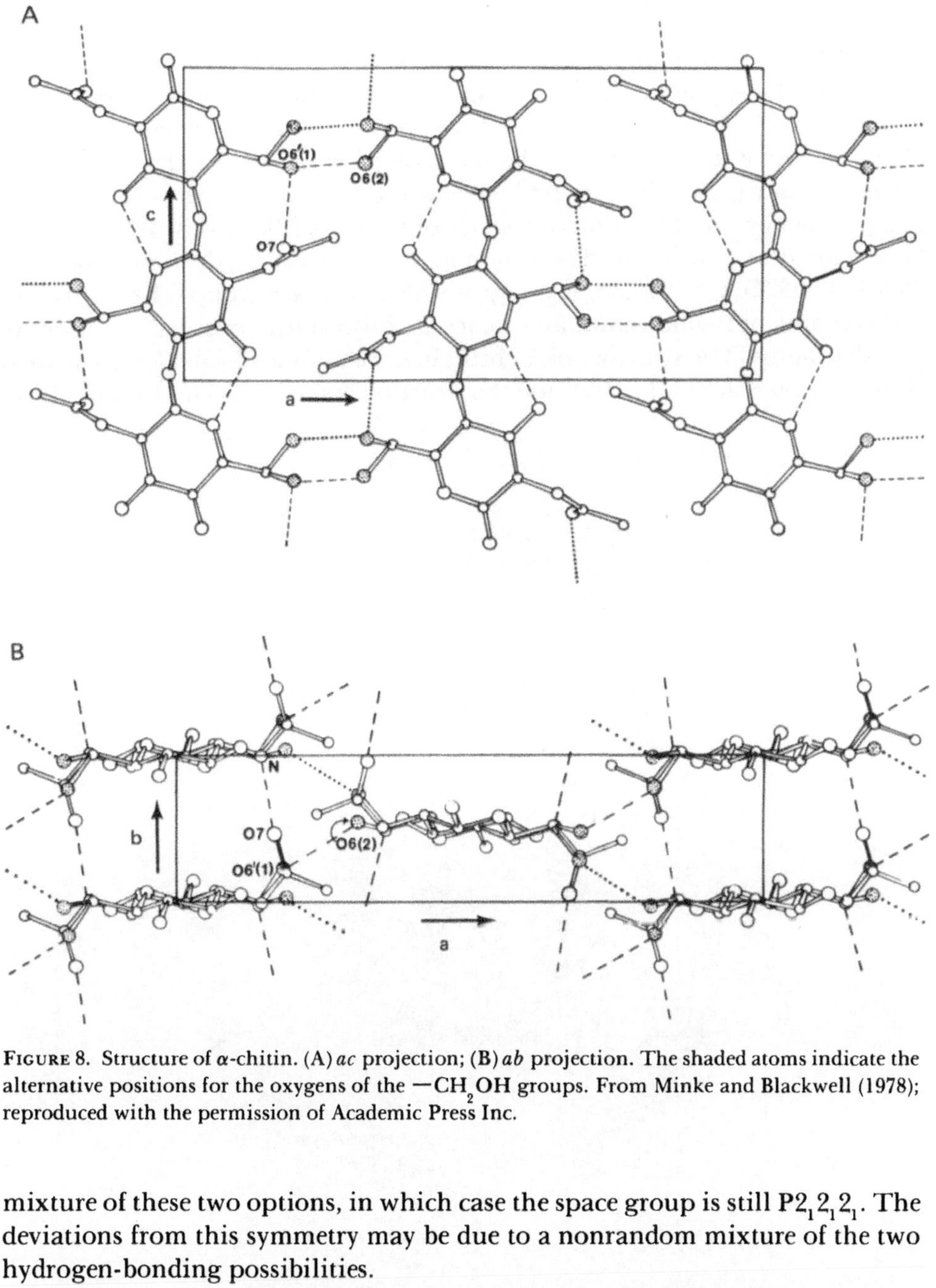

FIGURE 8. Structure of α-chitin. (A) *ac* projection; (B) *ab* projection. The shaded atoms indicate the alternative positions for the oxygens of the $-CH_2OH$ groups. From Minke and Blackwell (1978); reproduced with the permission of Academic Press Inc.

mixture of these two options, in which case the space group is still $P2_12_12_1$. The deviations from this symmetry may be due to a nonrandom mixture of the two hydrogen-bonding possibilities.

The intermolecular hydrogen bonding of the $-CH_2OH$ groups is probably responsible for the inability of chitin to swell in water. The hydrophobic forces maintain the stack of chitin chains along the *a* axis, and this is further stabilized by the hydrogen bonding between the acetamido groups. In the absence of intermolecular hydrogen bonding along the *b* axis, the structure should swell by penetration of water molecules between the stacks of chains.

5.2. β-Chitin

β-Chitin is much rarer than α-chitin; only four sources have been identified, but two of these, pogonophore tube and diatom spines, are highly crystalline, analogous to *Valonia* cellulose. The X-ray pattern of oriented diatom spines is shown in Fig. 9. The spines have been oriented by suspending them in a fibrin (protein) matrix and then stretching. Studies of the structure of β-chitin are complicated by the formation of a series of hydrates (Blackwell, 1969), but the basic anhydrous structure has a monoclinic unit cell with dimensions $a = 4.85$ Å, $b = 9.25$ Å, $c = 10.38$, Å, and $\gamma = 97.5°$; the space group is $P2_1$. This unit cell contains only one chain, and hence β-chitin must consist of an array of parallel chains. The structure of β-chitin has been refined by the LALS method (Gardner and Blackwell, 1975), and the structure is shown in Fig. 10. The chitin

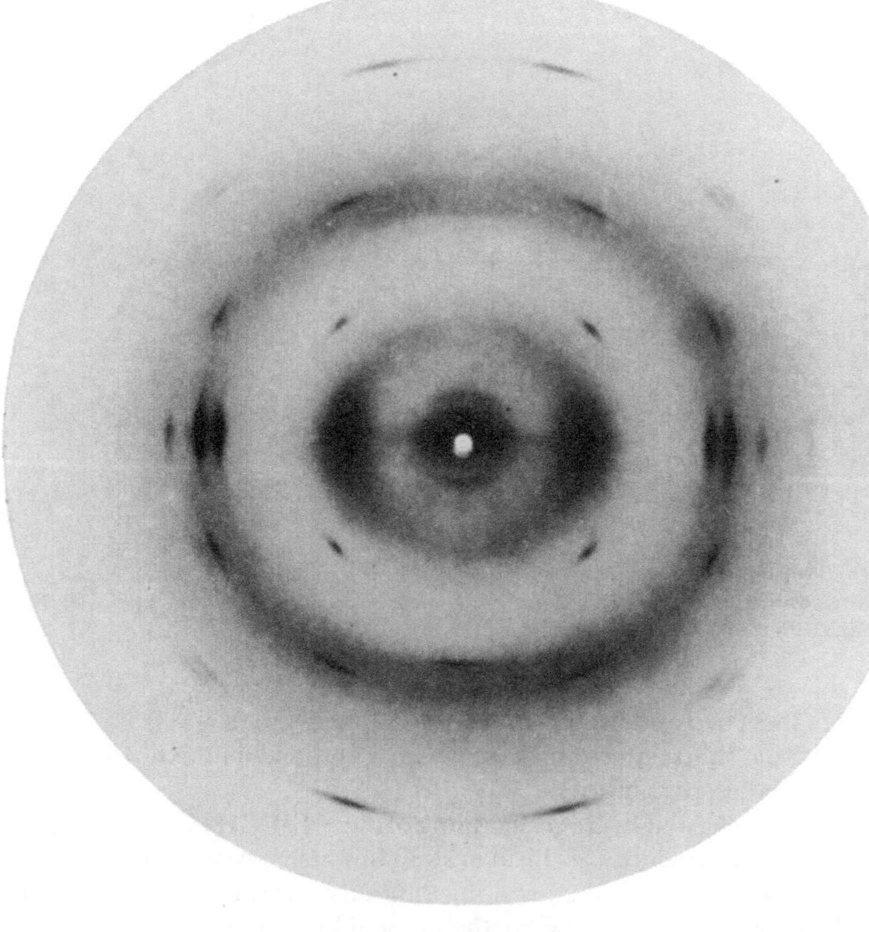

FIGURE 9. X-ray fiber diagram of diatom spines (β-chitin). The spines are suspended in a matrix of fibrin (clotted fibrinogen), which is then stretched to produce orientation. From Blackwell (1969); reproduced with the permission of Wiley-Interscience Inc.

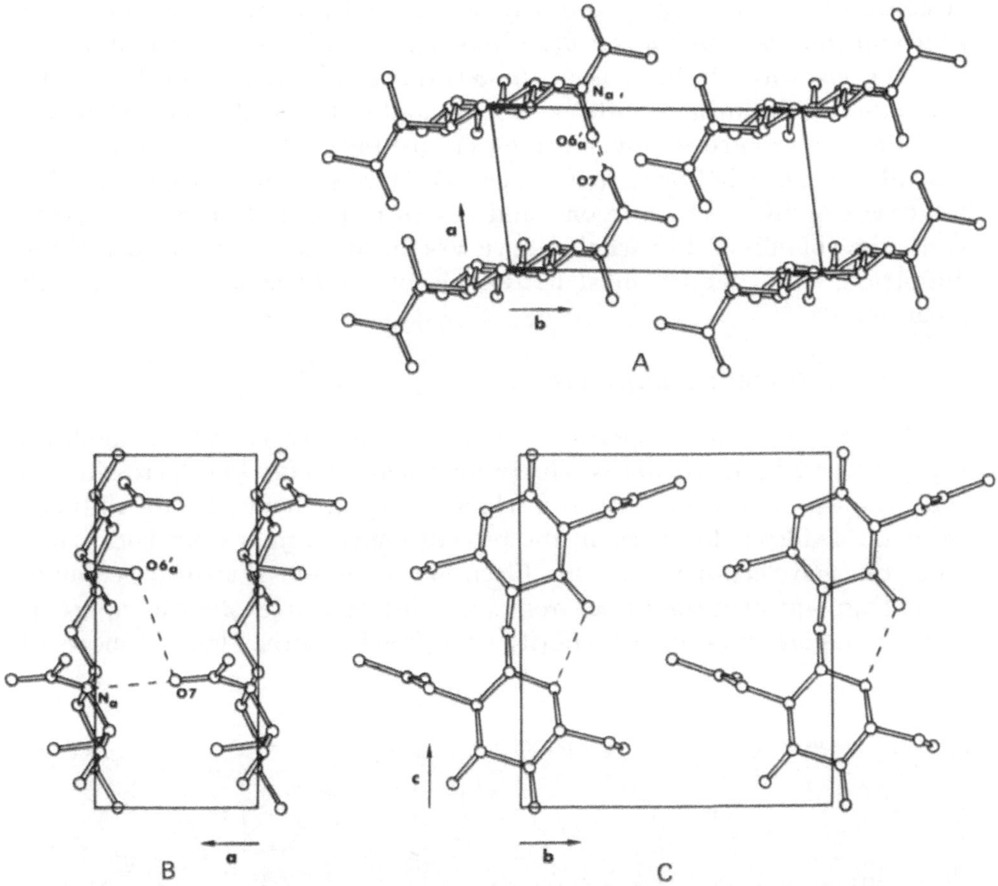

FIGURE 10. Structure of β-chitin. (A) *ab* projection; (B) *ac* projection; (C) *bc* projection. From Gardner and Blackwell (1975); reproduced with the permission of Wiley-Interscience Inc.

chains are stacked along the *a* axis and linked by N—H · · · O=C hydrogen bonds, as seen in α-chitin. The —Ch_2OH groups form intermolecular hydrogen bonds to the carbonyl of the next chain along the *a* axis. There is no intermolecular hydrogen bonding along the *b* axis, and consequently the structure is easily swollen in water.

5.3. *Analogy to Cellulose*

β-Chitin and α-chitin are shown to be parallel and antiparallel arrays of chains, analogous to cellulose I and II. It is interesting that most chitin synthesized in nature is in the antiparallel form, unlike cellulose, which is in the parallel form. β-Chitin swells extensively in water, and this may make it unsatisfactory for its structural role. Cellulose I, however, has sufficient intermolecular hydrogen bonding to prevent swelling. All four types of β-chitin are from aquatic sources, and it may be that a swellable material is not unsatisfactory in these cases; compare, for example, many aquatic plants in which the fibrous component is $\beta(1\rightarrow3)$-xylan, which is also swollen in water.

The majority of chitin is in the α-form, and thus it is likely that its biosynthesis mechanism is different from that of cellulose. Bartnicki-Garcia *et al.* (1978) have isolated an enzyme complex from fungal cells that has been shown to synthesize α-chitin fibrils (see Chapter 11). This large multienzyme complex has been described as a chitosome and is seen in the electron microscope to synthesize a bundle of chitin chains. In view of the structure determined by X-ray diffraction, this complex must be capable of synthesizing chains of both polarities.

5.4. *Chitin–Protein Complexes*

In the intact chitin-protein tissues, the protein serves as a lower modulus matrix between the chitin fibrils. This protein matrix is often highly cross-linked and is thought to serve as a defense mechanism to prevent attack of the chitin by chitinase. Evidence for order in the protein matrix comes from both X-ray diffraction and electron microscopy. Electron micrographs of stained sections of intact chitin–protein complexes from a number of sources show an approximately hexagonal array of chitin fibrils in the protein matrix. Figure 11 shows an

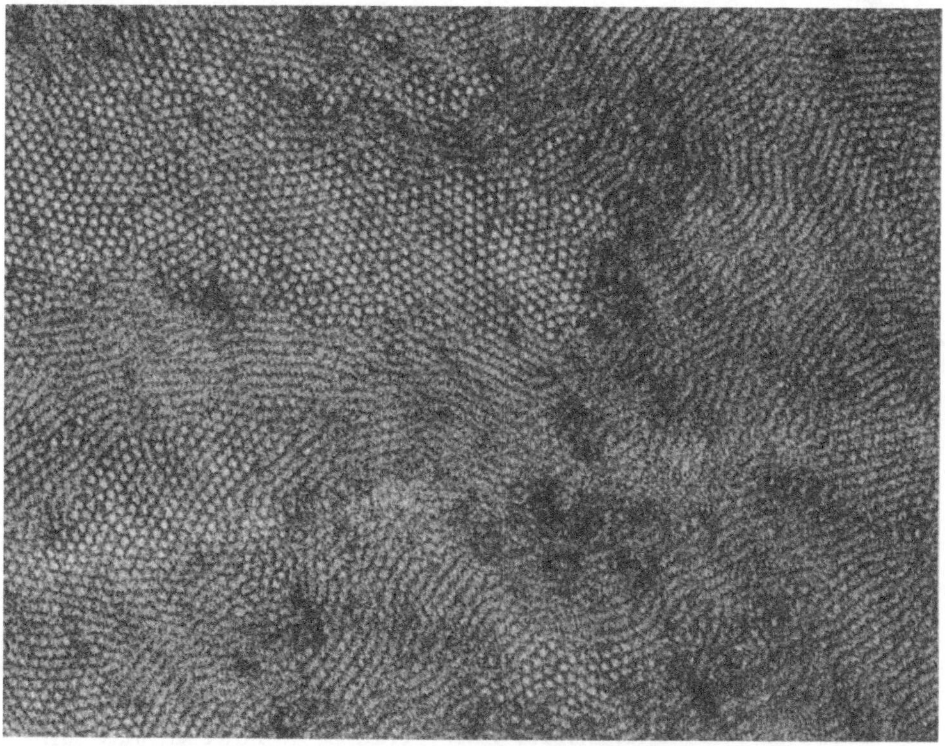

FIGURE 11. Electron micrograph of a transverse section of *Megarhyssa* ovipositor stained with lead citrate and uranyl acetate. From Germinario *et al.* (1982).

electron micrograph obtained in this laboratory (Germinario *et al.*, 1982) of a cross section of the ovipositor of the ichneumon fly *Megarhyssa*. The approximately hexagonal arrangement of the unstained fibrils is apparent against the background of the protein, which is stained with lead citrate and uranyl acetate; the interfibrillar spacing is $a \approx 75$ Å. Neville (1975) has surveyed a large number of arthropod systems and reports values of a in the range 51–83 Å.

X-ray diffraction patterns show that both the protein and the chitin are ordered. Figure 12 shows the fiber diagram of intact and deproteinized *Megarhyssa* ovipositor (Blackwell and Weih, 1980). The latter pattern is very similar to that for lobster tendon chitin (see Fig. 7) and identifies the chitin as the α-form. The features of α-chitin are recognizable in the pattern of the intact ovipositor. Specifically, layer lines 3, 6, and 9 for the complex correspond to layer lines 1, 2, and 3 for the deproteinized specimen. The additional layer lines must therefore be due entirely to an ordered protein structure. For *Megarhyssa*, the protein repeat is 30.6 Å, approximately 3 times the chitin repeat of 10.3 Å. Previous work by Rudall (1963, 1976) showed that repeats of approximately 31 Å are common in insect cuticles and other systems, including the β-chitin-containing tissues, squid pen and *Aphrodite* chaetae. However, repeats of 41 and 62 Å (i.e., 4 and 6 times the chitin repeat) are seen in some cases, e.g., locust and hornet apodemes, respectively. It is stressed, however, that in some systems, such as lobster tendons, no protein order is detected.

Relatively little is known of the chemical and physical structure of the protein, although there are significant differences in the amino acid analyses for proteins from different sources (Neville, 1975). Efforts to isolate the proteins are complicated by extensive cross-linking and by degradation, and in many cases the results suggest a number of protein moieties. Rudall and Kenchington (1973) showed that for newly emerged *Megarhyssa* ovipositor, the features of the X-ray pattern associated with the protein disappear when the specimen is treated with 7 M urea, and there is no evidence for regular protein–polysaccharide covalent links. Rudall (1976) also reports that the protein isolated from honeybee cuticle (a 31 Å system like *Megarhyssa*) shows three electrophoretic bands at molecular weights of 56,000, 27,000, and 18,000. He argues that the 56,000 band is probably a dimer of that at 27,000, which is the basic subunit. The band at 18,000 is weak and is thought to be a degradation product.

5.5. *Structure of the Megarhyssa Complex*

The additional features of the diffraction pattern of intact ovipositor not given by the purified chitin can be used to derive the structure of the ordered protein (Blackwell and Weih, 1980). These data consist primarily of the additional layer lines plus some small-angle equatorials that relate to the packing of the fibrils.

Small-angle maxima are observed on the equator at d = 62.9, 36.2, and 31.4 Å, and are indexed as the (100), (110), and (200) reflections of a hexagonal network with dimensions a = 72.5 Å, which is close to the value of approximately 75 Å seen in the electron microscope. Comparison of the intensities with the Fourier transforms of solid cylinders shows that the structure consists of close-packed cylinders with diameter 72.5 Å. We envisage that the chitin–protein complex consists of a chitin core surrounded by a sheath of protein that is close-packed as shown in Fig. 13A. The diameter of the chitin core is uncertain. Chemical analysis indicates approximately 15% protein, which corresponds to a diameter of 28 Å, as reported by Neville (1975), but some diffraction evidence suggests a diameter of 38 Å, and further studies are necessary to resolve this question.

The intensity distribution on the protein layer lines suggests that the protein forms a sixfold helix with a repeat of 30.6 Å. Layer lines 1 and 5 are stronger and the intensity is closer to the meridian than layer lines 2 and 4. The protein component of layer line 3 is very weak—the intensity is mostly due to the chitin. The maxima on layer lines 1 and 5 are off-meridional; i.e., they are split in

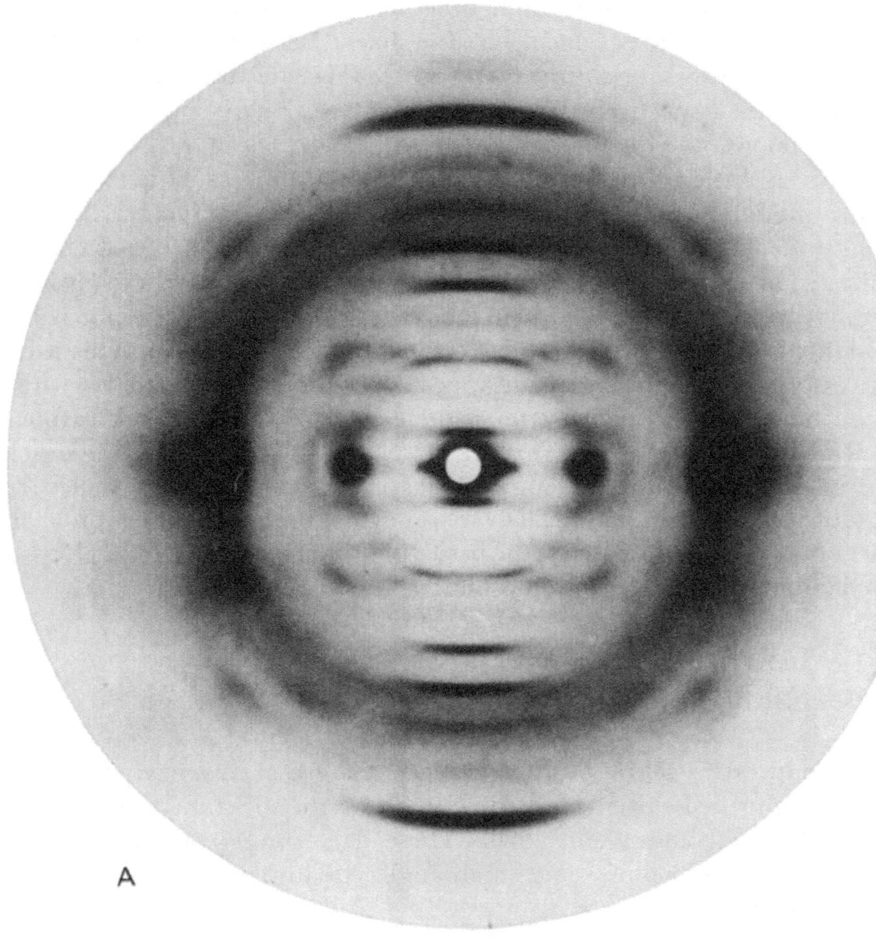

FIGURE 12. X-ray fiber diagram of (A) intact and (B) deproteinized *Megarhyssa* ovipositor. From Blackwell and Weih (1980); reproduced by permission of Academic Press Inc.

the patterns of tilted specimens and the first meridional intensity is on the 6th layer line. These are the diffraction characteristics of a helix with six subunits per turn. The intensity distribution suggests a radius of 24.2 Å for the protein helix, which is close to the center of the protein sheath in Fig. 13A.

Thus, we propose a model for the protein core consisting of a 6_1 helix of protein subunits arranged around the chitin core as shown in Fig. 13B. The proteins are shown as globular units, and as such the sheath structure is analogous to those formed by rodlike viruses. A globular protein with a molecular weight of 27,000 would have approximately the dimensions shown in Fig. 13B. Nevertheless, we know very little about the conformation of the protein, and other models that maintain the 6_1 helical symmetry are possible. X-ray methods simply indicate the arrangement of the scattering units. These units need not be separate molecules, but could be sections of several staggered chains, perhaps with some fibrous character.

Finally, close examination of the layer lines shows that they are split and

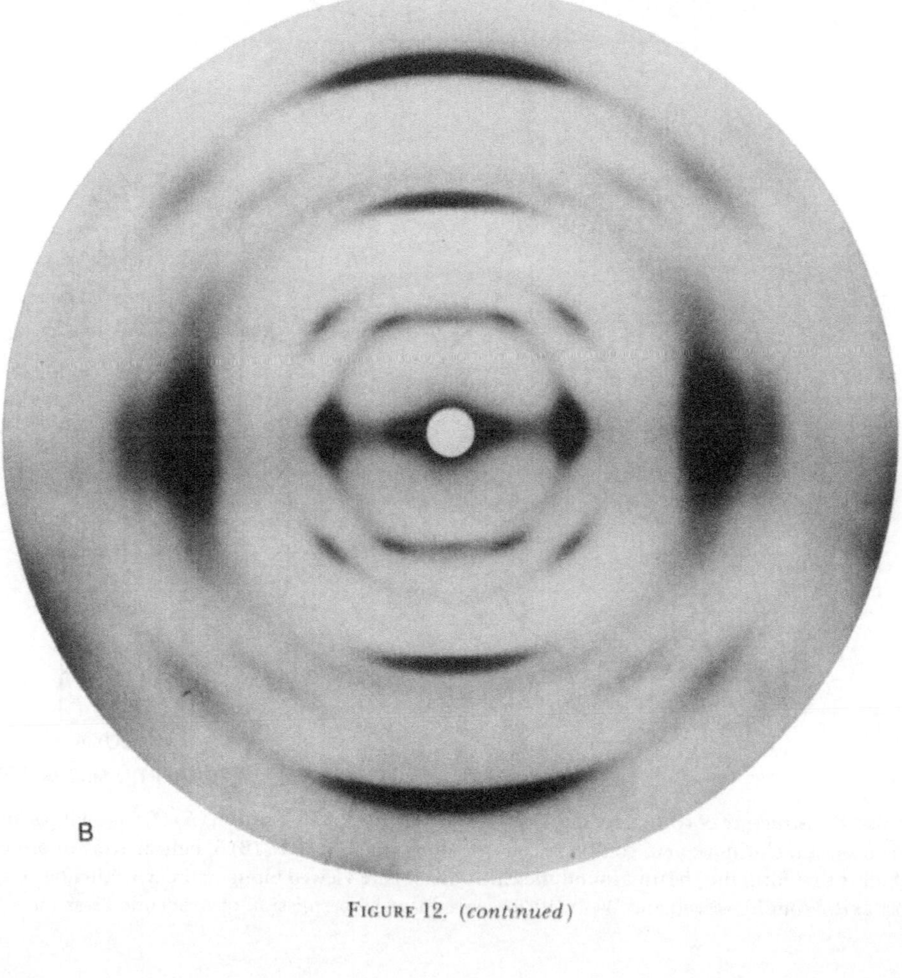

B

FIGURE 12. *(continued)*

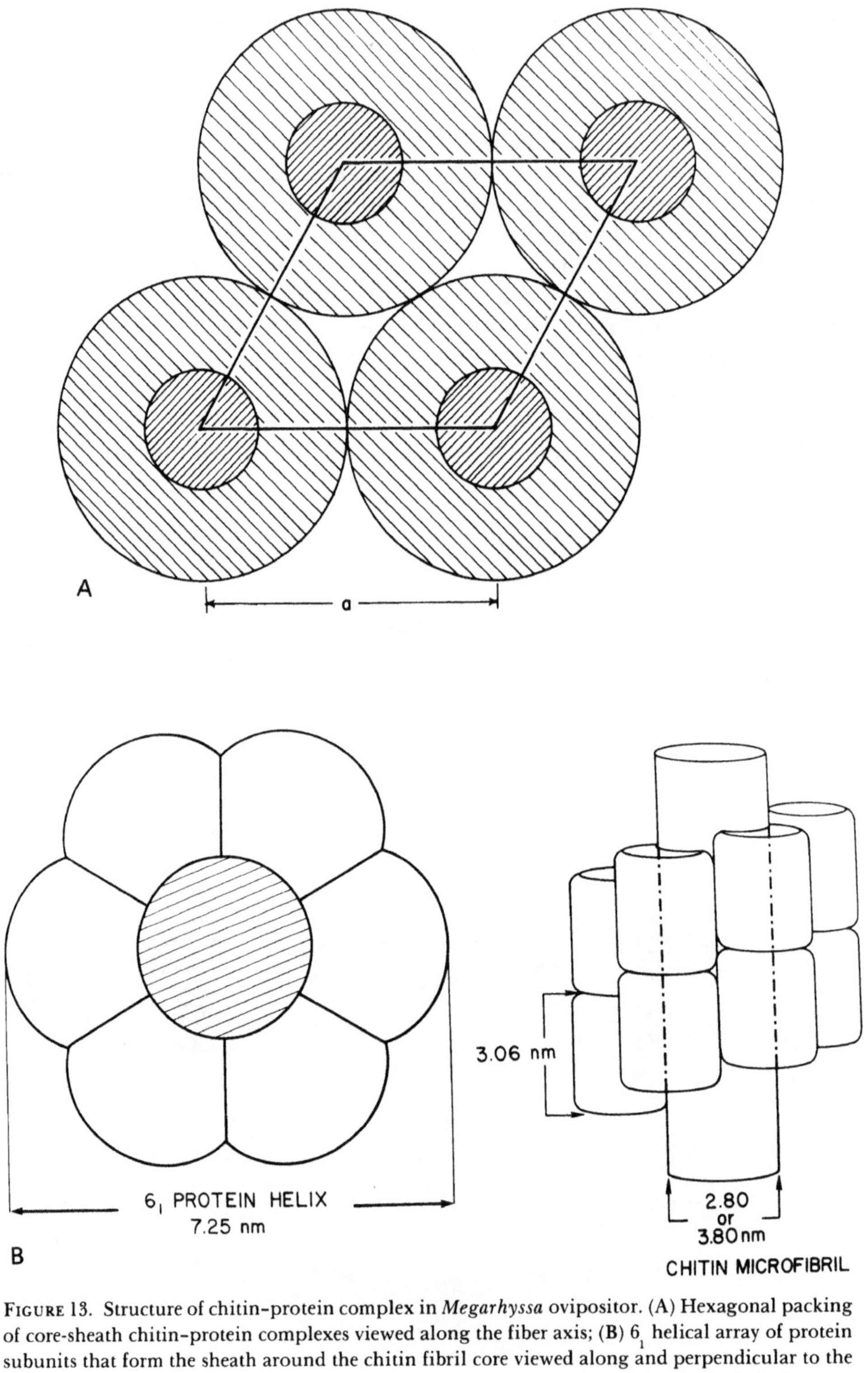

FIGURE 13. Structure of chitin–protein complex in *Megarhyssa* ovipositor. (A) Hexagonal packing of core-sheath chitin–protein complexes viewed along the fiber axis; (B) 6_1 helical array of protein subunits that form the sheath around the chitin fibril core viewed along and perpendicular to the fiber axis. From Blackwell and Weih (1980); reproduced by permission of Academic Press Inc.

that the true repeat is some multiple of 30.6 Å. Our best estimate of the repeat for *Megarhyssa* is 153 Å. Small-angle meridionals are seen for *Loligo* pen and *Aphrodite* chaetae β-chitin–protein complexes, indicating that the axial repeats are 180 and 290 Å. It is possible that there are small deviations from the 6_1 symmetry, perhaps due to a nonintegral helix or formation of a coiled coil structure.

The structure described above shows the type of interactions that occur between the fiber and matrix components of chitinous tissues. It is clear that much further work is necessary to determine the details of these complex structures. Nevertheless, this is the first biological composite for which we have a detailed model for the fiber–matrix interactions, and it remains to be seen what relationship this may have to other tissues, including plant-cell walls.

Acknowledgments

The research in this laboratory described in this chapter was done in association with my colleagues, K. H. Gardner, F. J. Kolapk, R. Minke, W. A. Weih, and L. T. Germinario, and was supported by NSF, most recently through Grant Number DMR81-07130 from the Polymer Program.

References

Arnott, S., and Scott, W. E., 1972, Accurate X-ray diffraction analysis of fibrous polysaccharides containing pyranose rings, *J. Chem. Soc. Perkins Trans.* **2**:324.

Arnott, S., and Wonacott, A. J., 1966, The refinement of the crystal and molecular structures of polymers using X-ray data and stereochemical constraints, *Polymer* **7**:157.

Bartnicki-García, S., Bracker, C. E., Reyes, E., and Ruiz-Herrera, J., 1978, Isolation of chitosomes from taxonomically diverse fungi and synthesis of chitin microfibrils *in vitro*, *Proceedings of the 1st International Conference on Chitin/Chitosan* (R. A. A. Muzzerelli and E. R. Pariser, eds.), p. 450, MIT Sea Grant, Cambridge, Massachusetts.

Blackwell, J., 1969, The structure of β-chitin, or parallel chain systems of poly *N*-acetyl-D-glucosamine, *Biopolymers* **7**:281.

Blackwell, J., and Marchessault, R.H., 1971, Infrared specroscopy of cellulose, in: *Cellulose and Cellulose Derivatives* (N. Bikales and L. E. Segal, eds.), p. 1, Wiley-Interscience, New York.

Blackwell, J., and Weih, M. A., 1980, The structures of chitin–protein complexes: The ovipositor of the ichneumon fly *Megarhyssa*, *J. Mol. Biol.* **137**:49.

Blackwell, J., Parker, K. D., and Rudall, K. M., 1965, Chitin in pogonophore tubes, *J. Mar. Biol. Soc. U.K.* **45**:659.

Blackwell, J., Parker, K. D., and Rudall, K. M., 1967, The chitin of the diatom spines, *J. Mol. Biol.* **28**:383.

Brown, R. M., Willison, J. H. M., and Richardson, C. L., 1976, Cellulose biosynthesis in *Acetobacter xylinum:* Visualization of the site of synthesis and direct measurement of the *in vivo* process, *Proc. Natl. Acad. Sci. U.S.A.* **73**:4565.

Carlstrom, D., 1957, The structure of α-chitin, *J. biochem. Biophys. Cytol.* **3**:669.

Claffey, W. B., and Blackwell, J., 1976, Electron diffraction of *Valonia* cellulose, *Biopolymers* **15**:1903.

Ellefsen, O., and Norman, N., 1962, Formation and characteristics of the crystalline cellulose X modification of cellulose, *J. Polym. Sci.* **58**:769.

French, A. D., and Gardner, K. H., 1980, Fiber diffraction methods, *Am. Chem. Soc. Symp. Ser.* **141**.
French, A. D., and Murphy, V. C., 1977, A virtual bond modeling study of cellulose I, *Am. Chem. Soc. Symp. Ser.* **48:**12.
Gardner, K. H., and Blackwell, J., 1974, The structure of native cellulose, *Biopolymers* **13:**1975.
Gardner, K. H., and Blackwell, J., 1975, Refinement of the structure of β-chitin, *Biopolymers* **14:**1575.
Germinario, L. T., Blackwell, J., and Frank, I., 1982, Reconstruction of chitin–protein complexes using correlation averaging methods (in prep.).
Hamilton, W. C., 1965, Significance test on the crystallographic R factor, *Acta Crystallogr.* **18:**502.
Hermans, P. H., de Booys, J., and Maan, C., 1943, Form and mobility of cellulose molecules, *Kolloid-Z.* **102:**169.
Kolpak, F. J., and Blackwell, J., 1976, Determination of the structure of cellulose II, *Macromolecules* **9:**273.
Kolpak, F. J., Weih, M. A., and Blackwell, J., 1978, Mercerization of cellulose. I. Determination of the structure of mercerized cotton, *Polymer* **19:**123.
Liang, C. J., and Marchessault, R. H., 1959, Infrared spectra of crystalline polysaccharides. I. Hydrogen bonds in native celluloses, *J. Polym. Sci.* **37:**385.
Marchessault, R. H., and Liang, C. Y., 1960, Infrared spectra of crystalline polysaccharides. III. Mercerized cellulose, *J. Polym. Sci.* **43:**71.
Marchessault, R. H., and Sarko, A., 1967, X-ray structure of polysaccharides, *Adv. Carbohydr. Chem.* **22:**421.
Marrinan, H. J., and Mann, J., 1956, Infrared spectra of the crystalline modifications of cellulose, *J. Polym. Sci.* **21:**301.
Meyer, K. H., and Mark, H., 1928, The structure of the crystalline components of cellulose, *Ber. Deutsch Chem. Ges.* **61B:**593.
Meyer, K. H., and Misch, L., 1937, The constitution of the crystalline part of cellulose. VI. The positions of the atoms in the new spacial model of cellulose, *Helv. Chim. Acta* **20:**232.
Meyer, K. H., and Pankow, G. W., 1935, Constitution and structure of chitin, *Helv. Chim. Acta* **18;**589.
Minke, R., and Blackwell, J., 1978, The structure of α-chitin, *J. Mol. Biol.* **120:**167.
Neville, A. C., 1975, *Biology of the Arthropod Cuticle*, Springer-Verlag, Berlin.
Nishikawa, S., and Ono, S., 1913, *Proc. Math. Phys. Soc. (Tokyo)* **7:**131.
Parry, D. A. D. and Craig, A. S., 1977, Quantitative electron microscope observations of the collagen fibrils in rat-tail tendon, *Biopolymers* **16:**1015.
Picken, L. E. R., and Lotmar, W., 1950, Oriented protein in chitinous structures, *Experientia* **6:**58.
Rudall, K. M., 1963, The chitin/protein complexes of insect cuticles, *Adv. Insect Physiol.* **1:**257.
Rudall, K. M., 1976, Molecular structure in arthropod cuticles, in: *The Insect Integument* (H. R. Hepburn, ed.), p. 271, Elsevier, New York.
Rudall, K. M., and Kenchington, W., 1973, The chitin system, *Biol. Rev.* **49:**597.
Sarko, A., and Muggli, R., 1974, Packing analysis of carbohydrates and polysaccharides. IV. *Valonia* cellulose and cellulose II, *Macromolecules* **7:**486.
Segal, L. E., Loeb, L., and Creely, J. J., 1954, An X-ray study of the decomposition product of the ethylamine–cellulose complex, *J. Polym. Sci.* **13:**193.
Smith, P. J., and Arnott, S., 1978, LALS: A linked-atom least-squares reciprocal-space refinement system incorporating stereochemical restraints to supplement sparse diffraction data, *Acta Crystallogr.* **A34:**3.
Sponsler, O. L., and Dore, W. H., 1930, Structure of ramie cellulose as deduced from X-rays, *Cellulosechemie* **11:**186.
Stipanovic, A. J., and Sarko, A., 1976, Packing analysis of carbohydrates and polysaccharides. VI. Molecular and crystal structure of regenerated cellulose II, *Macromolecules* **9:**851.
Wellard, H. J., 1954, Variation in the lattice spacing of cellulose, *J. Polym. Sci.* **13:**471.
Willison, J. H. M., and Brown, R. M., 1978, Cell wall structure and deposition in *Glaucocystis*, *J. Cell Biol.* **77:**103.

21

Comparisons between Synthetic and Natural Microfiber Systems

L. H. SAWYER AND W. GEORGE

1. Introduction: Fibers, Fibrils, and the Ultimate Fibril

The fibril, as a structural entity, enjoys the properties of thermodynamic stability, structural simplicity, and end-use versatility and is comprised of systems of macromolecules or other linearly associated materials. For these materials, the most thermodynamically stable solid form is the "ideal" extended-chain crystal (Wünderlich, 1973). The fibril may be considered as an extended-chain crystal of variable perfection with thermodynamic stability proportional to the degree of perfection. Stability increases due to intramolecular, low-energy configurations characteristic of extended molecules and from the highly viscous, strained, amorphous domains that often surround fibrils (Zachmann, 1974).

On a local scale, the linear macromolecule is a one-dimensional structure; the length dimension dominates the other two in extent and property bias. The fibril's simplicity arises from the maintenance of this basic one-dimensional local character on larger structural scales. Versatility of the fibrillar structure is inferred from its prevalence in natural and synthetic materials. In addition, the fibril appears to be a basic building block for a complex range of macrostructures.

The simplest and most basic definition of a fibril is that it is a structural entity with material properties that are biased predominantly along a linear dimension or symmetry axis; necessarily, one such property is length. Additional material properties may also be biased along the fibril axis. Fibrils may exist on various scales of size, and a given fibril may be composed of fibrillar structures of a lesser scale. An extensive colinear portion of a polymer molecule will be taken as the zeroth-order fibril. A fiber will be considered as a macroscopic composite fibril, and filament will connote an infinitely long fiber.

L. H. SAWYER AND W. GEORGE • North Carolina State University, Raleigh, North Carolina 27650.

Historically, it is interesting to point out that when the first investigators of microstructure turned their attention to biological materials, they observed fibrillar structures in great profusion. These early pioneers were limited to the use of hand lenses and the newly developed compound microscope, which often had poorer resolving power than a good lens. These instruments revealed a hierarchy of fibrils originating with macroscopically visible hairs and fibers that appeared to be made up of smaller and smaller fiberlike structures. The smallest of these fibrils was assumed to be much smaller than the smallest visible to the investigators. Grew (1682) reported such observations in the following manner:

> . . . The *Cane*, I say, is as the *Lymphaduct;* and the *Straws* are as the *Fibres* whereof it is composed. By which also appears, the admirable smallest of these *Fibres*. For there are some *Lymphaducts*, which may be reconed fifty times smaller than a *Horse-Hair*. Allowing therefore but twenty of the aforesaid *Fibres* to make a thred so big as one *Lymphaduct;* then one of the said *Fibres*, must be a thousand times smaller than a *Horse-Hair*. That these *Fibres*, whereof the *Lymphaducts* are made, are themselves made up of other *Fibres*, is not altogether improbable.

Extending this inference of smaller and smaller fibrils making up larger fibrillar structures, Nägeli (1852) anticipated, and to a greater extent Astbury and Woods (1932) recognized, the colinear polymer molecule as the inception of fibrillar structure. While differences may exist among fibrillar structures due to the chemical nature of the polymer and how far the structure has evolved toward thermodynamic equilibrium, the basic structural nature of the fibril is universal.

All fibrils may be viewed as local regions of low-entropy material meeting the previously defined requirement. Low entropy may be considered synonymous with high molecular order.* A fibril of any degree or rank within the hierarchy of fibrils may have any amount of molecular order, but only those structures with sufficiently high molecular order will be differentiable from the nonfibrillar matrix.

Basic differences exist between natural and synthetic processes of conversion of polymer into fibrils. The differences appear to reflect the relative initial entropy states of the polymer prior to incorporation into fibrils. Biological systems employ enzymatic loci for the "controlled" conversion of monomer to colinear polymer. The newly synthesized polymer molecule associates with other similar colinear molecules a short distance from the synthesis site. The associated polymer is incorporated into the low-entropy fibrillar structure before it can acquire a high configurational entropy. The same is true in those synthetic† processes that utilize certain coordination catalysts, particularly of the Ziegler–Natta class. These systems form fibrils directly. The situation is quite different, however, in synthetic and natural systems that produce filaments. In these cases, completely synthesized polymer in the relatively high-configurational-entropy, random-coil state undergoes shaping into filament. The fibril exists as a partial structural component of the filament. These systems produce fibrillar structure

*Not to be confused with the rank or order of a fibril.

†Hybrid filament-forming processes exist that are of commercial importance. They utilize natural polymers, and their filamentary products are distinguished in many circles as synthetic, a term that we shall use as inclusive of man-made.

less efficiently than those previously discussed, since zeroth-order fibrils must first be formed from a high-entropy polymer matrix and then assembled into the fibril. This requires extensive cooperative entropy reduction, which is slow in highly viscous polymer fluids. Kinetic factors enter into the thermodynamic considerations of the approach of such a system to equilibrium.

A process that produces a low-entropy material from high-entropy precursor without benefit of a true first-order phase transition and resulting free-energy driving force is denied in classic equilibrium thermodynamics. Such a process is more conveniently considered through irreversible, nonlinear thermodynamics in the sense of Prigogine (1967). Such considerations allow entropy reduction on a local scale along specific axes if entropy is produced and transported on a larger global scale. Systems capable of generating fibrillar structures will generally operate far from equilibrium, in a steady-state condition.

A fibril or fiber-forming system may have many available steady states, and it will search for the most stable. The most stable steady state represents an irreversible, open system's thermodynamically preferred state. Irreversible thermodynamics presumes thermodynamic equilibrium on the local scale. It is possible for a system, operating away from equilibrium, to gain in global stability if local regions of the system are allowed to approach thermodynamic equilibrium. The formation of fibrils in a fiber-forming process is an example of local equilibrium structures in a steady-state process. Global requirements on the system necessitate the existence of dissipative processes acting in concert with the local equilibrium structure. The inability of a system to permit the development of dissipative structures will limit the extent to which the system can achieve local equilibrium.

The preceding statements in their present form may be taken by a conservative reader as merely conjectures of the authors. We submit, however, that they are attractively consistent with the views of the Prigogine school of irreversible thermodynamics as found in Glansdorff and Prigogine (1971).

2. *Synthetic-Fiber-Forming Technologies: Common Characteristics*

Practical synthetic fiber-forming processes produce macroscopic, continuous filaments. After forming, the filaments may be chopped, drawn, or subjected to any of a large variety of additional treatments depending on the desired end use. The microfiber or fibrillar structure of the (macro) filament is determined by polymer properties and the totality of local environmental conditions that the polymer experiences throughout processing. The art of forming synthetic fiber has only recently recognized means for incorporating specific microstructure into the filament for desired properties. Previously, the fiber producer was primarily concerned with identifying and avoiding microstructures that reflected undesired process or product qualities. The practice of fiber "spinning" developed from the early work of Audemars (1855), fulfilling the earlier conjectures of Hooke (1664) and Réaumur (1734) on the conversion of nonfibrous materials into fibrous form.

The earliest interest in synthetic fibers developed from the desire to find a cheaper and more abundant source of silklike fibers. Audemars' cellulose nitrate solution was prepared from mulberry twigs utilizing ancient Chinese methods for collecting silkworm secretions in recovering the filament. This process was not suitable for commercial fiber formation, but new methods and processes were developed rapidly in the last part of the 19th and early part of the 20th century that enabled the synthetic-fiber industry to come into its own. At this time, fiber formation was a "high art," and many of the evolving technological advances were empirical in nature.

The recent awareness of the role of science in the art of fiber formation has shown that many of the older empirical advances rested on sound physical foundations. Often, the fiber producer's optimization of his spinning process led to the development of a particular microstructure within the polymer fluid or product or both, a feature first noted by Carothers and Hill (1932) for melt extrusion. The ability of the fiber producer to physically understand his empirical process has resulted in the recognition that the microstructure of a given polymeric material can be varied extensively through processing conditions. In fact, the presence and degree of fibrillar structure will depend almost completely on the process environment in which the filaments of semiflexible polymers are formed. Understanding the fiber-forming process is therefore of prime importance in examining fibrillar structures in synthetic systems.

Synthetic filaments are generally formed by one or a combination of three general processes. These are melt, dry, and wet "spinning." All three processes involve the steady-state conversion of a polymeric fluid into a solid, often glassy, filament. The primary differences among the processes are the types of fluid and the method of effecting the fluid-to-solid transformation. In the case of melt spinning, a molten, bulk polymer is cooled after being shaped to yield a filament. Both dry and wet spinning employ solutions of polymer. The dry-spun filament is solidified by evaporation of solvent in a gaseous environment. The wet-spinning process utilizes a liquid bath that solidifies the filament as a result of mass transport into and out of the fiber. The factors that dictate the use of a particular process include the chemical and physical nature of the polymer as well as the desired properties of the filament product. Combined processes such as plasticized melt spinning and air-gap wet spinning enable the fiber producer to spin additional materials and obtain special properties.

The synthetic filament-forming process can be schematically depicted as a sequence of interconnected unit operations (Fig. 1). The first step in the scheme is the formation of the spinning fluid. In closed, continuous systems, monomers are fed into a reactor where polymerization occurs to produce the necessary high-molecular-weight material. Solvent or heat or both are added to form the desired fluid as polymerization continues. Some polymerization processes are capable of directly producing fibrils that survive later operations to be incorporated into the macrofilament. Portions of the fluid are continuously drawn off to feed the downstream process units. Batch and semicontinuous systems introduce previously formed polymer, usually as a solid, into the fluidizing region where heat or solvent or both are combined with the polymer to produce the appropriate "spinning" fluid. Batch systems operate from a single

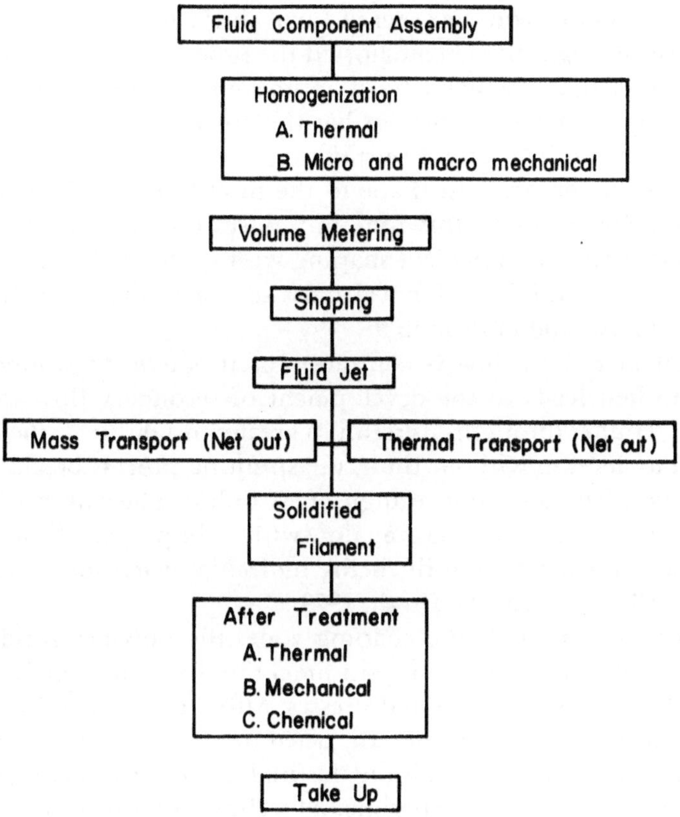

FIGURE 1. General schematic for the components of a steady-state synthetic fiber-forming process.

charge, while the semicontinuous systems accommodate multiple input sequences.

Once the fluid components are assembled in the simplest systems, they must be completely combined and homogenized so that the material properties of the fluid are continuous throughout. Thermal and mechanical mixing on the micro- as well as the macroscale are necessary for thorough homogenization. The mixing process is often combined with transport of the bulk fluid through a type of channel flow designed to assist homogenization. The high-polymer molecules of the mixing fluid have characteristically long relaxation times. This results in a "memory" or entropy inertia in the polymer fluid. Relatively small mechanical perturbations are capable of influencing the entropy inertia for a long time period. As noted earlier, experimental and theoretical evidence exists suggesting that fibrillar domains may be established in the polymer fluid and remain through the processes of filament formation.

Useful synthetic filaments should be highly uniform in shape and in properties along their axial dimension. This requires a highly controlled mass throughput. Most synthetic fiber-forming processes accomplish this control through volume-metering pumps that utilize the positive-displacement action of gear pumps. In modern production facilities, pump speeds are often controlled to tolerances on the order of 1 part in 100,000. Such accurate control over the mass throughput is mandatory if the process is to approach the stable steady-state operation characteristic of optimum fiber-producing conditions.

The shaping zone of the process has the potential for greatly influencing filament properties. The broad purpose of this zone is to impart a general cross-sectional dimension and shape to the final filament. Concurrently, the path of fluid flow through this zone introduces the potential for a range of filament microstructure. Filament shaping with specially designed spinnerette holes offers the possibility of novel cross-sectional shapes such as square, triangular, hollow, and multilobular.

The polymer fluid flow is contracted as it enters the spinnerette. This contraction often leads to the development of secondary flows and general instabilities. Higher throughputs result in greater instability. If the instabilities are allowed to develop without limit, the spinning process breaks down. The spinnerette may have an entrance angle of 90° or less. The entrance flow is such that it generates a sharper converging flow within the polymer fluid. Such flows have been shown capable of influencing molecular extension and orientation (McHugh, 1975; Silebi and McHugh, 1979).

After passing through the shaping zone, the polymer fluid leaves the physical restraints of the extrusion equipment in the form of an unconstrained fluid jet with surfaces free of normal stresses. Although the jet is free of physical constraints such as channel walls, it is acted on by a number of forces. These forces play a major role in determining the final shape, size, and properties of the filament. The accessibility to experimental technique with appropriate mathematical approximation has resulted in considerable analytical attention being directed toward this zone of the spinning process. At present, however, all schemes of analysis are incomplete primarily because of difficulties arising from associated transport processes in real systems (Slatterly, 1966; Denn and Petrie, 1975; Fisher and Denn, 1976; Han, 1976).

These difficulties in analyzing the fluid jet are represented by the next three blocks on the general spinning schematic (Fig. 1). The transport phenomena that lead to the fluid-to-solid transition must be included in any analysis. The precise nature of transport phenomena present in a given process is not well understood. Depending on the specifics of a spinning process, mass may enter or leave the system concurrent with thermal energy. The net mass and energy flows are away from the jet, causing the fluid to become progressively more viscous until it is practically solid. (Gradients in the radial direction may well be introduced.) The transition from fluid to solid often results in a heterogeneous, filamentary material. The crystalline components are usually anisotropic in shape and vary locally in character.

Transport processes assist the formation of a microstructure within the jet. Appropriate couplings of the transport flows with the axial fluid flow can be expected to enhance the fibrillar content of the solidified filament. The generally high thermodynamic stability favors fibrillar structures during solidification. Kinetic factors, such as slow nucleation and slow growth, limit the development of well-formed fibrils in many cases. These factors result in the kinetically favored folded-chain morphology contributing substantially to the total crystalline content of the final filament (Peterlin, 1979).

Once the filament is formed, it may be taken up and subjected to further processing as a second stage, or it may undergo further processing prior to take-up. The purpose of the additional processing is to improve and tailor filament properties for specific end uses. These additional treatments include thermal, mechanical, and chemical modification of the filament. All are capable of significantly altering the filament microstructure.

The final zone of the general spinning scheme, the take-up zone, can have a major effect on the fibrous character of the filament. The take-up force (associated with speeds of windup) acts on the spinning way (jet) with modifications due to aerodynamic effects and body-force effects. If the speeds of take-up are quite high, inertial effects develop within the jet and along the free surfaces (drag forces). In some cases, aerodynamic shock waves are generated along the surfaces, making it necessary for operators in production to wear ear protection. The stresses transmitted from the take-up zone back through the jet into the more liquid regions may well encourage the development of radial heterogeneity across the filament.

3. *Specific Fiber-Forming Processes*

Figure 2 is a composite schematic of the three most widely used fiber-spinning processes. While it is obvious that all three follow the general spinning scheme, the nature of the spinning fluid and the method of solidification result in considerable individuality among the processes as well as fiber properties available through a given process.

The spinning of molten glass into filament dates back to antiquity, but the application of this process to organic materials was delayed until the accidental discovery of the fiber-forming properties from melts of thermally stable polycondensate polymers. It is often recounted that in examining the stability of a newly synthesized polymer in the melt, Carothers inserted a probe into the fluid and on withdrawing it noted a thin trailing filament that he elongated by walking around the laboratory. Glass-fiber production from melts was highly developed in the 19th century (Tscheuschner, 1885). A bicomponent filament was produced in the Black Forest of Germany that was widely used in the late Victorian period for curly doll's hair and a Christmas tree decoration known as "angel hair."

Polymer may be introduced into the melt stage of the process directly from the polymerization reactor or fed continuously in chip form into an extruding device (Fig. 2) that melts and transports the polymer to the metering section. Some small-scale and experimental operations utilize batch processes where the polymer is melted in a static pool prior to transport. Polymer melts are usually highly viscous fluids and require high pressures for transport. The high

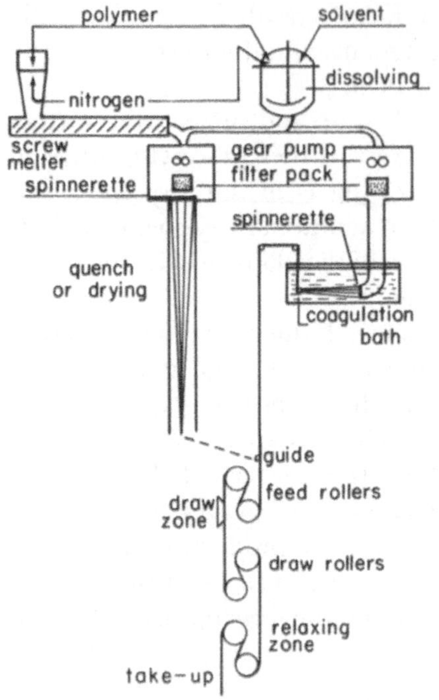

FIGURE 2. Composite filament-forming system. After Walczak (1977) by permission.

pressures and high mass flow rates characteristic in melt spinning, coupled with the shear, converging, and extensional flows inherent in a spinning operation, all act to introduce potential for fibrillar structure. This potential can be optimized by extrusion conditions and fibrils recovered in the product by appropriate quenching conditions.

The first filament-forming processes of organic materials involved solutions of natural polymers. Audemars (1855) obtained a patent covering the treatment of the bark from mulberry trees to recover a nitrocellulose derivative that was in turn drawn into filaments from an ether–alcohol solution and subsequently air dried. This advance culminated in the development of nitro rayon as a silk substitute in 1885 (Urquhart, 1952). These commercially successful dry-spun filaments were superseded in the early 20th century by wet-spun viscose rayon, which remains a viable filament process. This filament in fiber form is a replacement for natural cotton fiber. The process and product are traced to the inventions by Chardonnet (1885) in France and Cross *et al.* (1892) in England.

The minor component of dry-spinning solutions is often the polymer. Optimum concentrations for solutions of cellulosic derivatives range between 20 and 30% by weight of polymer. The availability of better solvents for synthetic polymers, such as acrylonitrile, enable spinning solutions with 40–60% by weight of polymer to be employed.

The high percentage of volatile solvent places process limitations on production. Present commercial processes utilize spinnerettes containing up to the order of 100 holes. A greater density of holes results in filament bundles that are difficult to dry with current technology. In addition, elaborate solvent-recovery systems must be coupled with the spinning process, since most of the solvent has to be recovered for reasons of safety and economy.

The large solvent content also results in dry-spun filament cross sections being exaggerated or irregular. This is primarily due to the rapid coagulation of the fluid jet surface, which collapses onto the more fluid central portions. The highly crenulated cross sections of cellulose acetate and the dog-bone shape of Orlon® result from this effect.

Wet-spinning processes were developed along with the dry processes in the late 19th century utilizing the same polymer solutions. The simplicity and relative cheapness of dry spinning made it the method of choice where either could be used (Loasby, 1952). The discovery and commercialization of viscose rayon changed this situation. Viscose could only be wet spun, and the final regenerated cellulose filament had many superior properties. From its inception in 1898, the viscose process grew rapidly to dominate the synthetic-fiber industry until the recent expansion of polyester melt filament production.

Wet spinning involves a solution in which the solvent component is usually nonvolatile. To effect solidification of a filament, the solvent must be removed or altered in a way to make it less "good." This is accomplished in a "coagulation" bath. This liquid bath contains components that either enhance the diffusion of solvent away from the polymer jet stream or diffuse into the fluid filament, changing the quality of polymer–solvent interaction. Often, both types of components are present in the bath. The net result is coagulation and solidification of the filament.

The rate of solidification is primarily dependent on the rate of transport in or out of the filament. Since coagulation is a chemical process, the surface that is the first layer exposed to the diffusion often solidifies first, forming a skin. The interior regions then solidify, usually at a slower rate because of a difference in the chemical activity of the components in the coagulation bath on the interior compared to the surface. This leads to the skin-and-core phenomenon, in which the outer layers of a filament have different structural and physical properties than the interior or core. The bath can be adjusted with the proper additives to give an all-skin filament that may resemble a hollow tubule or an all-core material with greater uniformity across the filament cross section.

There are several wet-spinning processes involving rigid polymer molecules which readily form anisotropic solutions that can be spun into highly fibrillar filaments. For instance, a high-temperature mesamorphic pitch is spun into filaments, dried, and pyrolyzed to give high-modulus carbon fibers. Also, a number of the *p*-benzamides that exhibit liquid-crystalline character in solution form filaments, using a combined dry–wet process. These are highly fibrillar and possess superior physical properties (tensile strength and modulus). The experimental and theoretical correlation between anisotropic fluid and fibrillar structure is most evident in these rigid, rodlike polymer systems. However, a similar correlation should also hold for other fluid systems if external influences can interact with the molecular components of the fluid.

While melt-, dry-, and wet-spinning processes are generally employed in "pure" form, there is often a special polymer or unusual property that necessitates a combination of these spinning processes. As noted above, air-gap spinning, a dry–wet process often used with the *p*-benzamides, relies on a region of flow below the spinnerette along the jet where cooling is achieved by evaporation prior to the introduction of the jet into the bath to yield optimum physical properties. Mist spinning is also a dry–wet process, but the quench medium, in this case, is gaseous with fine liquid droplets. Plasticized melt or gel spinning is a melt–dry process that uses a small amount of solvent to assist the formation of a high-temperature spinning fluid. The solvent is removed by evaporation, either rapidly resulting in surface texture or slowly allowing extended interaction with the solidifying microstructure. Melt–wet processes are possible in which the primary purpose of liquid quench media is to improve the speed and efficiency of thermal transport.

Several additional synthetic fiber-forming processes deserve mention because of their novel nature and their potential for introducing fibrillar structure. These are small-scale processes for the production of specialized fibers. In general, the fibrillar character of the product is the attractive feature of the process.

Bicomponent fibers (Ziabicki, 1976) involve the extrusion of two polymeric components to give a composite fiber structure. The fluid polymer blend usually segregates during the fiber-forming process, resulting in intimately associated domains of nonmiscible polymers on solidification. The components generally segregate in the cross section to form skin-core, side-by-side, or matrix-fibril structures in which a pure polymer component is present in each domain of the fiber.

The segregation of the polymeric components is intensified by the fluid flows. Each component possesses a characteristic flow response, and the interaction of these responses leads to a given fiber morphology. Bicomponent morphologies can be used to optimize the properties of each component, such as modulus or dyeability, or to impart new properties, such as self-crimp, to the composite structure.

Spray spinning involves the production of a high-velocity particle dispersion of fiber-forming fluid. The particles of fluid develop filamentary character due to aerodynamic forces that cause them to elongate. Solidification occurs after sufficient thermal (cooling) and mass transport (drying), resulting in a fibrous or fibrillar product. The fluid dispersion may be formed by spraying from a nozzle, blowing high-pressure gas across the surface of a polymer fluid, or impinging a high-velocity jet onto a dispersing surface.

Reaction spinning (Pohl, 1958) involves forming a prepolymer into a fluid jet within which polymerization occurs. The fluid prepolymer is activated as the jet is formed. Both radical and condensation processes can be utilized. The fluid jet in which the molecules grow experiences a viscous drag as it moves through a quenching zone. The jet movement is capable of orienting the fluid. The fluid flows can be assisted by external forces to form a highly oriented fibrous product.

Polymerization results in the filament solidification and may continue, even after take-up, to yield certain fiber properties.

Suspension spinning (Ziabicki, 1976) is used to convert high-temperature, nonsoluble fusable polymers into fibrous form. The nonprocessible component is suspended as a colloidal particle in a carrier fluid that, in turn, undergoes processing into fiber. The nonprocessible component is carried along by a matrix of the suspending component. The carrier matrix is then destroyed, under tension, fusing the suspension of nonprocessible polymer in a fibrous form.

4. Direct Fibril-Forming Processes

Several classes of processes involving synthetic polymer systems form fibrillar structures directly from the polymeric component. Conditions of fibril formation are usually complex (varied) in these processes, and deviation from optimum conditions results in a decrease in the fibril population. Broadly, the conditions of fibril formation that affect the presence of fibrillar structures are similar to those available in filament formation. A later section will discuss fluid requirement for fibril formation and the effect that process conditions may have on these requirements.

Fibrils in the form of shish kebabs (Fig. 3) have been recovered from a

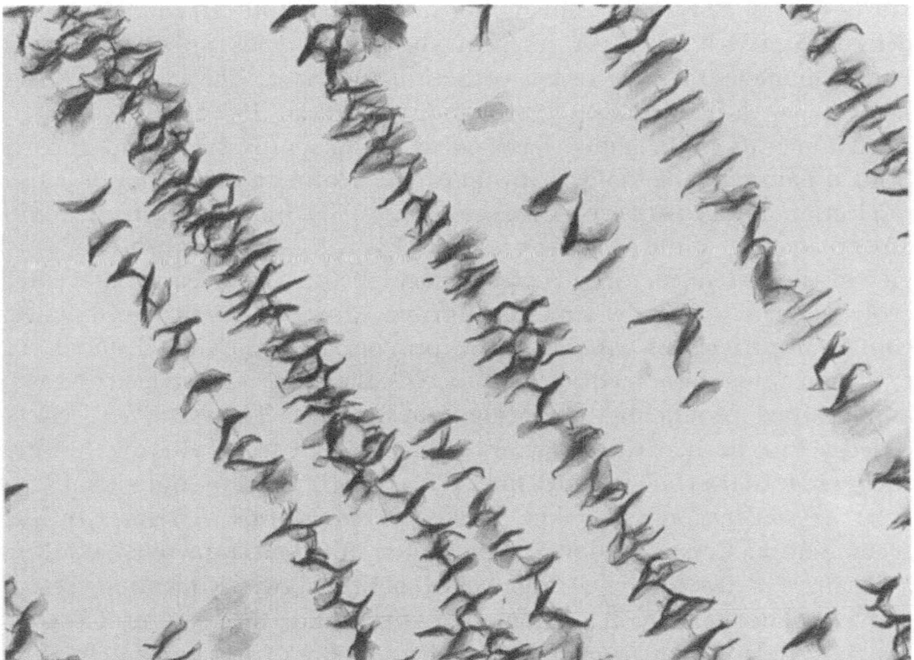

FIGURE 3. Transmission electron micrograph of polyethylene shish-kebabs produced in an initially simple shear flow. Courtesy of P. Tucker.

number of dilute polymer solutions subjected to unstable, mechanically induced shear flows. The appearance of shish kebabs follows the onset of secondary or higher-order flow patterns that represent a new steady state in the fluid motion. These added motions almost always contain zones of extensional flow. The solubilized polymer, as a component of the fluid, participates in this state. One consequence of the flow-induced state is to remove polymer from solution. The polymer comprising the shish kebabs is no longer an intimate component of the fluid, since it experiences a phase transition prior to or shortly after attainment of the new fluid-flow state.

Shish kebabs appear to have a nonequilibrium origin, but their recovered structure often appears to reflect an approach to equilibrium. Shish kebabs have been recovered with a broad range of shish diameters and kebab spacings (see Fig. 3). Recent experiments have demonstrated the potential of shifting the kebab position and relative spacing with annealing temperature (Pennings *et al.*, 1977), but the fundamental structural relationship of the kebab to the shish is unresolved.

In general, shish kebabs appear to be highly crystalline. (Some formed at high temperatures in unstable couette flows reflect a high-temperature crystal structure of the polymer involved.) The thinner shish sections, which appear continuous, are often thought to contain an array of extended molecules. Particularly clean shish kebabs of polyethylene produced by George and Tucker (1976) suggest a crystalline shish of orthorhombic unit cell containing certain defects. Further, the shish segments between kebabs appeared to have a length/diameter ratio consistent with estimated crystalline surface energies for polyethylene utilizing Wulff's theorem (Mandelkern, 1964).

Numerous investigators have noted the generally small and apparently limited nature of the diameter of the shish. Hoffman (1979) has developed a nucleation-and-growth theory that suggests such a limitation and, in addition, an explanation of the periodic presence of kebabs along a shish. The shish is viewed as an extended, concatenated molecular structure that associates and then evolves toward thermodynamic equilibrium. However, the extended molecules contain imperfections possibly in the form of configurational disorder. As the evolution along the axial direction proceeds, the imperfections are concentrated and localized, producing the concatenated structure. The assembled defects are reflected in a cumulative volume strain that has the effect of altering the internal energy state of the evolving structure. When the cumulative strain is sufficiently great, crystallization stops and a defect region forms to relieve the strain accumulation. Crystallization reinitiates after sufficient strain relief, resulting in a localized defect domain sandwiched between two sequential crystalline domains. The site of the defect domain corresponds to the position of a kebab. In his theory, Hoffman provided an additive term to the general free-energy expression for nucleation and growth of a polymer crystal that reflects the role of the cumulative strain. With this modification, he was able to calculate independently shish dimensions, the ratio of which corresponds to that predicted by the classic Wulff's theorem relating crystal dimensions inversely to facet free energies (Burton *et al.*, 1951).

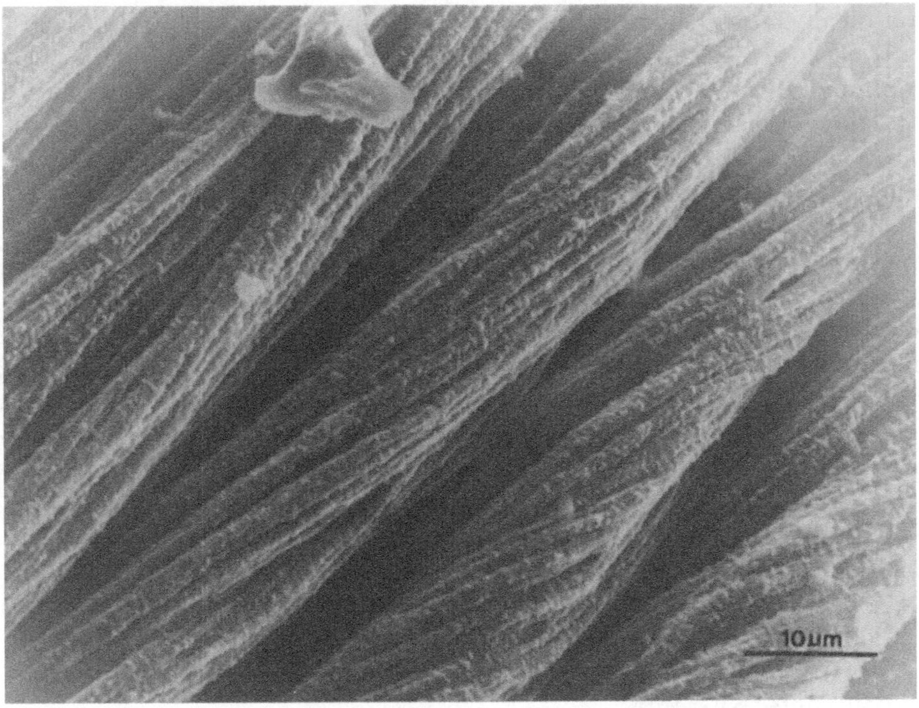

FIGURE 4. Scanning electron micrograph of poly(ethylene terephthalate) tactoids recovered from solution. Courtesy of H. Sugg.

Fibril-like structures that do not possess the shish-kebab morphology have been derived from solutions. Microscopic fields of these structures resemble stacks of rough-bark logs (Fig. 4). On the grossest scale, the structures are tactoidal (Sawyer and George, 1978). The fine structure of these directly produced fibrils varies from oriented amorphous to highly crystalline, depending on the polymer and formation method. One example of the structure (Smook *et al.*, 1980) is considered to consist of a collection of highly perfect shish structures with the rough texture possibly kebab or other overgrowth polymer. A speculation would be that all such structures possess an underlying structure resembling a shish at some degree of evolution.

Fibrils have been produced in bulk polymer through a more nearly equilibrium method than the solution processes discussed above. The bulk polymer is heated significantly above its melting temperature and subjected to extreme pressures (Wünderlich, 1964). The temperature and pressure conditions are maintained for extended periods of time, after which the polymer is cooled under pressure below its melting point. This process produces extended-chain lamellar crystals that have been shown to be composed of lateral stacks of fibrils (Wünderlich, 1973). The elevated pressure utilized in this process acts to compress the polymer, removing free volume. The driving force directs the polymer toward the most efficient volume-filling configuration possible. For linear polymers, this configuration leads to space filling with stacked, extended molecules, since bends and turns along a molecule require extra volume. The high-temperature environment provides sufficient thermal motion for the polymer to achieve a favored configuration. The acquisition of such a favored configuration is apparently cooperative, leading to the extensive lamellar structures. Besides rearrangement, the possibility of molecular cleavage and recombination in a more colinear configuration is possible at the high temperature–pressure levels present.

Several specialized polymerization processes generate fibrils as a consequence of polymerization. Stereoregular polymerization with coordination-type catalysts for simple polyolefins may form "nascent" fibrils (Wilkjord and Manley, 1968). The catalyst appears to provide a site for the stepwise, coordinated insertion of monomer units. Polymer flows from the site in an ordered, regular configuration. Association of adjacent ordered molecules close to the site of polymerization appears to provide an aggregate of zeroth-order fibrils capable of participation in higher-order, more complex fibrillar structures. Similar "nascent" fibrils have been reported from biological systems where an intimate coupling between polymerization and crystallization appears to dominate the processes for fibril formation.

5. *Fibrils in Natural Fibers—Comparison and Integration with Synthetic Systems*

Fibrils, appearing in nature in many contexts, result from a number of formation mechanisms. Several chapters in this volume, as well as a wealth of sometimes conflicting literature, deal with specific examples of natural fibrils and possible mechanisms for their formation. The present authors consider the fibril as a universal structure of linear organic polymers and other similarly configured materials. Indeed, the fibril is viewed as a consequence—a preferred consequence—of the evolution of the polymer toward thermodynamic equilibrium. This evolution is often assisted by the environment, biological or technological, in which the fibril forms.

Fibrils are observed in almost all biological systems, where they are generally incorporated as an integral part of a system or as a metabolic product of the organism apparently generated to promote its survival. Fibrillar structures are seldom, if ever, produced as a by-product; rather, it seems their presence is intimately involved with existence of living entities. Fibrils have also been discovered in nature within organic sludges and within inorganic materials of geologic origin such as asbestos, silicates, and metals. In these examples, the fibrillar structure is a rarity and appears to reflect local, exceptional events. The filaments of *Araneida* and *Bombyx mori* contain low levels of fibrils in a continuous fibrillar structure.

Similarities between specific mechanisms for the formation of fibrous structures in natural and synthetic systems are not readily recognized. On the other hand, strong structural similarities exist between fibril and fiber produced naturally and technologically. In general, the mechanisms for the formation of fibrous structures possess a common characteristic in that they operate far from the state of thermodynamic equilibrium. While the mechanisms involve nonequilibrium processes, they require that the processes produce a stable stationary state that is conducive to the evolution of a final fibrous structure approximating an ideal fibrous crystal. Limitations in structural stability may be considered to result from kinetic and other restrictive considerations imposed on the fiber-forming system through a specific mechanistic pathway. Thus, natural and synthetic fiber-forming systems exhibit similarities in the underlying principles of formation and in the final structure, but not necessarily in terms of the detailed mechanisms by which they are formed.

The processes of formation of fibrous elements in nature are highly directed to supply a specific need in a biological system. This highly directed formation has been shown to be associated with a localized enzyme complex for the case of the production of several cellulose fibrils (Romanovicz and Brown, 1976; Brown and Montezinos, 1976). An enzyme complex or similar biological control appears to be necessary in natural systems to provide the uniformity and reproducibility necessary and characteristic of biological systems. The function of a biological control site is to assemble monomers and link them in the proper sequence and steric inclination for subsequent incorporation into the fibrous structure. Incorporation or crystallization may accompany the linkage process (Stockmann, 1972), but nonequilibrium thermodynamic considerations developed from synthetic systems suggest that the fibrillar structure initially assumes a tactoidal character with initial crystallization occurring several monomer units from the polymerization sites (Sawyer and George, 1978). Evolution of the tactoidal structure into a crystalline fibril may continue far from the polymerization site in both time and space much as it does in synthetic systems.

The closest analogue in synthetic systems to a biological control mechanism is the coordination catalysts used for the synthesis of linear, stereoregular polyolefins (Wünderlich, 1973). This class of catalysts provides a site for the stepwise insertion of monomer, in steric register, into a growing polymer chain. Under appropriate conditions, chains growing from adjacent sites have been observed to aggregate into the fibrillar (shish-kebab) structures directly above and some distances beyond the catalyst surface. These shish-kebab structures generally comprise only a small portion of the synthesized polymer, and direct conversion of these microfibrils into films or fibers has not been demonstrated. The fibrillar component possesses a thermal stability that imparts properties to a polymeric material that are lost if the fibrils are destroyed. Recent work (Whitley, 1979) suggests that a large portion of fibrillar structures, once formed, may survive synthetic fiber- and film-forming processes such as melt extrusion. If such a process can be perfected, the efficiency and precision evident in biological systems may begin to be reflected in a synthetic filament-forming system.

From the practical point of view, it is the efficiency and precision in the formation of fibrous structures that really differentiate natural from synthetic systems. Natural systems have been designed by very complex evolutionary pathways to produce fibrillar structures of high uniformity and perfection as economically as possible. In contrast, synthetic systems are designed through limited experiences, incomplete understanding, and insufficient knowledge to produce fibers and films of some utility. The technology of synthetic-fiber formation is unable to assert a determinate microscopic control on the initiation of fibril formation. Out of necessity, control over the microscale is at present only weakly accomplished through macroscopic process variables, and herein one finds the major source for the low precision and efficiency in synthetic fiber-forming systems.

There appear to be two routes to improving synthetic fiber-forming systems. The first involves a better understanding of the coupling between micro- and macroconsiderations and will be discussed later. The second route involves integrating natural processes into a "man-made" fibrous product. Since the ultimate structures of natural and synthetic fibers are similar, such an integration should be possible and beneficial.

Cotton, wool, flax, and other natural macroscopic fibers have been converted into "man-made" fibrous structures since prehistoric times. These conversions preserved the structure of a complete natural fiber-forming process entirely in forming the "man-made" fiber. Wood and other cellulosic materials have been converted into rayon and other regenerated cellulosic fibers for over 100 years. These processes completely destroy the natural fibrous structure, often with molecular degradation. "Man-made" fiber structures produced from these processes usually possess very limited microscopic fibrillar character. While integration of natural and synthetic fiber-forming systems is not entirely new, an integrated process that makes use of *renewable natural polymers in the form of microfibrils* that can be engineered into larger structures via synthetic manipulation remains to be developed by technologists.

Similarity of structure between natural and synthetic fibrous materials has its origin in their basic, composite nature. A universal fiber structure may be envisioned as a two-component composite with fibrils existing in a matrix of a heterogeneous material that serves to "glue" the fibrils together, forming a continuous structural entity. The heterogeneity of the matrix may be structural or chemical, or both, such as the amorphous polysaccharide matrices associated with most cellulose fibrils or the sometimes ordered proteinaceous matrices associated with chitin (see Chapter 20). The fibrils serve as a structural load-bearing framework, while the matrix protects and modifies the properties of the final fibrous structure.

Natural systems produce composite fibers that usually contain a high portion of the fibrillar component. Synthetic systems generally form fibers low in fibrillar content. As a result, the natural fibril would be the structure of interest in an integrated natural–synthetic process. Cellulose appears to be a strong choice as a source of these natural fibrils because of its availability, physical properties, and tendency toward fibrillar structures.

As noted in Chapter 20, cellulose is an ideal material for the formation of fibrous structures. Crystalline, fibrillar cellulose exhibits two morphologies, cellulose I and cellulose II. Naturally occurring cellulose is cellulose I, while regenerated cellulose is usually cellulose II. Cellulose I contains cellulose molecules oriented parallel, whereas cellulose II contains the cellulose molecules oriented in an antiparallel fashion (Chapter 20). Cellulose II is the more thermodynamically stable structure in an absolute sense. However, cellulose is synthesized so that adjacent chains are all parallel, preventing instantaneous formation of cellulose II. The rapid association and crystallization of these parallel chains along with the significant stability of the cellulose I structure result in an activation energy for the cellulose I → II transition large enough to prevent the transition in the natural system.

Cellulose II is formed when this activation-energy requirement is mediated by an external influence such as swelling or solvation. The cellulose I structure may return on deswelling if it is not sufficiently weakened or destroyed. However, once cellulose I is destroyed, the system will seek the most thermodynamically stable structure, cellulose II, and cellulose I will never re-form. Blackwell suggests in Chapter 20 a mechanism analogous to an interdigitation to explain the cellulose I → II transition. Such a mechanism would be attractive from both thermodynamic and kinetic considerations provided that cellulose chains of opposite orientation occur sufficiently near each other.

To satisfy this "nearness" requirement in forming a crystal of antiparallel cellulose molecules from a material that possesses parallel structure within the crystalline domains, the collection of domains must be considered as an intimate admixture of positive and negative orientations. This may be accomplished during synthesis if the enzyme sites, each of which yields small bundles of parallel cellulose molecules that evolve into elemental fibrils, synthesize bundles with either a positive or a negative spatial orientation. A statistically "random" or a deterministically "paired" arrangement of plus and minus synthesis sites will result in a collection of cellulose I fibrils that will readily allow interdigitation on swelling.

The existence of enzyme sites that synthesize molecules of opposite spatial orientation is evidenced in chitin (Chapter 20). The predominance of antiparallel chitin fibrils may be due to a more intimate arrangement of plus and minus enzyme sites and to a slower association of chitin molecules, allowing the thermodynamically preferred antiparallel arrangement. Since the cellulose molecules associate rapidly, parallel orientation is maintained within each elemental fibril. The elemental fibrils may be assembled into larger fibrous structures or participate in membranes, but the crystalline domains will be continuous only in the volume of the elemental fibril, and these will all be cellulose I.

Intense swelling or solutioning of the cellulose will partially or completely destroy the cellulose I structure by breaking intermolecular associations. Either process will cause an expansion of the fibrillar structure so that adjacent fibrils may partially or completely merge. If two elemental fibrils of differing spatial orientation are adjacent, this merging may result in placing two molecules with antiparallel orientation in close proximity to each other. When solvent or swelling agent is removed, the individual cellulose molecules will reestablish intermolecular associations. Since associations between antiparallel molecules are the more stable, this association will be preferred, where possible, resulting in interdigitation. The resulting structure will be a mixture of cellulose II with some cellulose I depending on how much of the original cellulose I was destroyed and how completely the merging of oppositely oriented fibrils was accomplished.

Recent work involving solutions of cellulose has successfully produced cellulose fibrils directly from mechanically deformed solutions (Chanzy and Peguy, 1980; Kalb and Manley, 1980). The crystal structure of these fibrils is cellulose II and appears shish-kebab-like in character. Such structures have been aggregated into larger fibrous aggregates from synthetic polymer solutions, and the same can be assumed for the cellulose fibrils.

Solution approaches to forming novel cellulose fibrils have two shortcomings. The most obvious is the lack of efficiency and precision inherent in a process that first dissolves fibrils of natural origin to produce fibrils for incorporation into another form. The second involves the crystalline form of the cellulose within the fibrils. Cellulose I is superior in performance to cellulose II in a number of applications, and optimally cellulose fibrils should retain cellulose I identity. A novel system, therefore, requires that cellulose I structures be retained throughout the latter stages of synthetic filament formation.

The work of Haigler *et al.* (1980) (see also Chapter 14) provides a basis for suggesting a process integrating cellulose I fibrils into a novel synthetic fiber. *Acetobacter xylinum* naturally produces cellulose in the form of highly fibrillar ribbons. Undoubtedly, methods can be devised to spin these ribbons directly into stable filaments. Another alternative would be to use the ribbons as a cellulose source for solution spinning. However, in the first instance, the natural fibrous product is more or less complete, aggregated far beyond the basic fibrillar stage. In the second possibility, the natural product again suffers destruction, leading at best to fibrils of cellulose II. While these considerations are attractive, a more attractive, integrated process for developing a novel cellulose fiber would incorporate natural polymer in the early stages of its aggregation, thus allowing a greater amount of synthetic engineering.

Calcofluor White ST has been shown by Haigler and co-workers to alter the formation of cellulose fibrils in *Acetobacter xylinum* at the elemental stage. Calcofluor appears to associate strongly with the new tactoidal cellulose elemental fibrils immediately after synthesis and before crystallization and aggregation. The Calcofluor probably satisfies the intermolecular bonding requirements of the cellulose chain on the fibril surfaces, thus preventing crystallization and forming a barrier to further association of the elemental tactoids, which are still strongly associated and constrained so that drying of the cellulose results in cellulose I structure.

Production of cellulose in the form of swollen, never-collapsed tactoidal fibrils should allow an efficient integration of natural and synthetic fiber-forming systems. To achieve this practically, methods must be devised to maintain the cellulose tactoidal structure throughout processing and then evolve cellulose I from it as the final composite fibrillar structure is perfected. The form and composition of the matrix component must also be specified. However, in principle, it should be possible to combine cellulose fibrils with other natural or synthetic polymeric components to form fibrous structures more efficiently and precisely than presently possible in completely synthetic systems. For a totally synthetic system to approach the efficiency and precision of a natural process, the underlying principles of formation of fibrils in nonbiological systems must be understood.

6. *Fibril Formation: Anisotropic Fluids and Solidification*

We now turn our concern to the origin and nature of fibrils in synthetic filaments and thus emphasize possible routes to fibrillar structure. Most useful fibers are obtained from fluids that appear more or less isotropic under quiescent conditions. The process of converting bulk fluid to filament is a (macroscopic) symmetry-reducing operation. Highly symmetric fluids, three-dimensional in character, are transformed into predominately one-dimensional, axially symmetric, filamentous structures.

The reduction in macroscopic symmetry imparts an anisotropic character to the spinning fluid, reflecting the formation of ordered microscopic states. Macroscopic process flows, coupled to the molecular level, are in part responsible for the induced anisotropy in the fluid as evidenced in a simpler system by streaming birefringence. Indeed, some level of preferred axial orientation is a characteristic of all commercial as-spun filaments, and there is a growing list of specialized fibers evidencing well-defined fibrillar domains.

A bulk fluid can be conceptually approximated by placing its component molecules on a lattice. Such is the basis of the Flory–Huggins theory of polymer solutions (Flory, 1953; Morawetz, 1975). In its simplest form, the random coil of a typical polymer is positioned on the lattice without directional bias. A more refined treatment is called for in cases where the coil is expanded or otherwise systematically departs from the (statistical) random-coil configuration.

Flory (1956a,b) generalizes the earlier treatment of polymer solutions for cases where the expansion of the (coiled) configuration of the polymer arises strictly from asymmetric intramolecular interactions. Such polymers may be termed semirigid. The semirigid polymer has a preferred configuration due to an intramolecular internal energy difference, ϵ, between one (local) molecular configuration (lowest energy) and all other possible configurations. The preferred configuration is reflected on the lattice as a directional bias. The magnitude of ϵ determines the relative "rigidity" of any particular intramolecular bond. Assuming that all bonds are energetically similar, the inverse rigidity or "flexibility," f_0, is given by Flory as

$$f_0 = \frac{(z - 2)\exp\dfrac{-\epsilon}{kT}}{1 + (z-2)\exp\dfrac{-\epsilon}{kT}} \tag{1}$$

Here, z is the lattice coordination number, T the absolute temperature, and k Boltzmann's constant. From equation (1), f_0 is found to be bounded by zero and one. Values near zero correspond to large ϵ, indicating a rigid polymer. Values of f_0 approaching one are characteristic of flexible polymers; ϵ is small.

The inherent flexibility of polymer in solution plays an important role in the nature and thermodynamic stability of the solution. The change in free energy, ΔG_m, obtained on forming a solution from pure solvent and pure, ordered polymer is the sum of contributions due to the polymer disorientation, ΔG_d, and the mixing of disoriented polymer with solvent, ΔG_m^*. Thus, ΔG_m equals

$$\Delta G_m = \Delta G_m^* + \Delta G_d \tag{2a}$$

or

$$\Delta G_m = \left\{RT[n_1 \ln V_1 + n_2 \ln V_2 + \chi_1 x n_2 V_1] - n_2 RT\left[\ln x + \ln\left(\frac{z}{2e}\right) + (x-2)\ln\frac{1}{(1-f_0)e}\right]\right\} \tag{2b}$$

The second term on the right-hand side of equation (2a), the free energy of disorientation, contains the contributions of intramolecular associations for semirigid polymers. Subscripts (1) and (2) represent solvent and solute, respectively. The number of molecules of a component is represented by n and its volume fraction by V. The size of a polymer molecule is indicated by the number of segments, x. Solvent–solute interaction is denoted by the parameter χ_1.

The total free energy of solutioning, ΔG_m, must be negative for a stable solution. As pointed out by Flory, there is a critical flexibility, f_c, in the limit of large x that drives ΔG_m positive for a particular polymer concentration. The value of f_c increases with V_2 to the limit of 0.63 in the case of bulk polymer ($V_2 = 1$). The flexibility of a molecule affects ΔG_m only through the ΔG_d term. Therefore, for polymers with flexibilities less than 0.63, an increased volume fraction may be placed into solution if a portion of the polymer molecules remains ordered in solution. This ordered polymer portion, V_2', makes a contribution to ΔG_m due only to ΔG_m^*. Such a solution contains two coexisting polymer phases, an anisotropic and an isotropic phase. The composition of each phase can be determined by equating the chemical potentials of each of the components in the two phases.

In this treatment, the anisotropic molecules are considered to be rodlike, $f_0 = 0$, and the isotropic molecules are characterized by $f_0 > f_c$. This places the rigorous requirement of perfect colinearity on molecules entering the anisotropic phase. Recently, Krigbaum and Salaris (1978) have considered an anisotropic phase characterized by $f_0 > 0$, if noncolinear segments of adjacent molecules are cooperatively arranged. This enables one to describe liquid crystalline phases in semirigid polymers.

The existence of an anisotropic phase in a polymer solution is of importance in the formation of fibrils. As pointed out earlier, an anisotropic material inherently possesses the low-entropy form preferred for fibrillar structure, and such structures are known to form from anisotropic fluids. Therefore, it seems logical to believe that the fibrils formed in polymer solutions with both isotropic and anisotropic phases grow from the anisotropic phase. In addition, if the anisotropic character of a polymer solution can be spontaneously increased, a greater potential for fibrillar structure may be an expected consequence.

A further extension and application of Flory's 1956 treatment for semi-flexible molecules provides a theoretical basis for a process that may be responsible for the formation of fibrillar structure in synthetic fiber-forming systems. Following suggestions by Frenkel (1974), we argue that a spontaneous shift in the effective flexibility of a polymer molecule may be effected by environmental parameters that are coupled to the polymer state. From equation (1), these parameters are ϵ and T for a given lattice. Frenkel assumes that ϵ can be perturbed by an amount $\Delta\epsilon$ through imposed mechanical fields and changes in the solvent-solute interaction. Frenkel incorporates these perturbation effects into polymer solution theory by writing a new flexibility parameter, f:

$$f = \frac{(z-2)\exp-\left(\dfrac{\epsilon+\Delta\epsilon}{kT}\right)}{1+(z-2)\exp-\left(\dfrac{\epsilon+\Delta\epsilon}{kT}\right)} = f_0\left[\frac{\exp-\left(\dfrac{\Delta\epsilon}{kT}\right)}{1-f_0+f_0\exp-\left(\dfrac{\Delta\epsilon}{kT}\right)}\right] \tag{3}$$

where the unperturbed flexibility of the system is denoted by f_0. Perturbation is reflected in f through $\Delta\epsilon$, which in turn reflects changes not only in ϵ but also in χ_1.

This form of the flexibility parameter indicates that a reduction in f is possible due to environmental effects that increase ϵ, decrease solvation, and reduce temperature.

Figure 5 (Flory, 1956a) depicts the volume fraction of polymer coexisting in isotropic and anisotropic phases for two values of f. This plot actually represents two polymers of differing flexibility. However, the figure also depicts the consequences of spontaneously changing the effective f of a single polymer. The end result of spontaneously decreasing the flexibility of polymer in solution will depend on how the reduction is effected as well as the relative magnitude and direction of the change. In any event, the stability of the solution will always be affected, often resulting in increasing the potential for new structure.

A hypothetical example designed to demonstrate the possible structure consequences of dynamically reducing the f of a polymer fluid follows: A

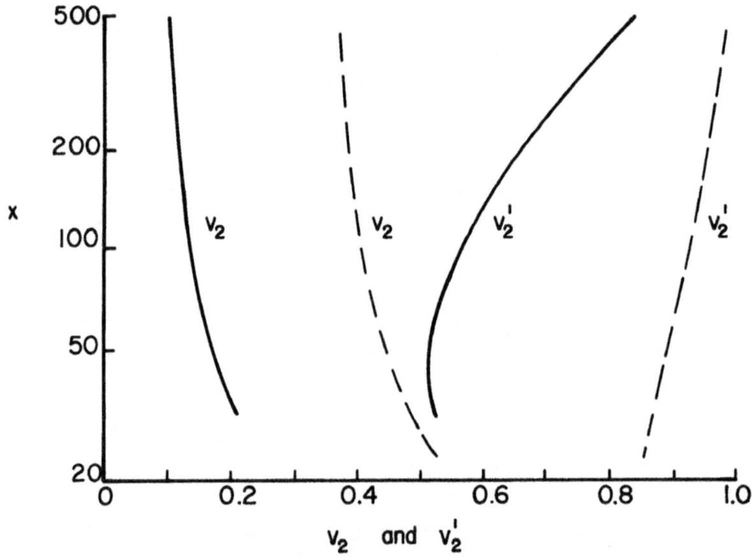

FIGURE 5. Plots of compositions of disordered (V_2) and ordered (V_2') phases in equilibrium as functions of x (logarithmic scale). $z = 12$ and $f = 0.10$ (——) and $f = 0.30$ (---). After Flory (1956a); used by permission.

semiflexible polymer is combined with a diluent to produce a saturated polymeric fluid at an elevated temperature. The polymer has a flexibility less than the critical flexibility so that isotropic and anisotropic phases are initially present in the fluid. However, the fluid is thermodynamically stable. The fluid is then subjected to a macroscopic flow field that is capable of altering the flexibility of the polymeric component. Concurrently, solvent may be removed or an antisolvent may be added to the moving fluid. The polymeric component experiences a spontaneous reduction in its effective flexibility as predicted by equation (3), and the fluid becomes thermodynamically unstable. Under the aforementioned influences, a larger proportion of the polymeric component competes to exist in the anisotropic phase, but as shown in Fig. 5, the stable volume fraction of anisotropic polymer is reduced for reduced flexibilities. *The fluid must exist as an unstable, partially ordered, supersaturated fluid until it can evolve to a more thermodynamically stable state*. If the system is closed to solvent flows, this evolution will result in polymer leaving the fluid state.

The phase-separated polymer will achieve fibrillar morphology if the temperature of the system is sufficiently high to allow nucleation and growth of only the more thermodynamically stable, solid polymer morphologies. The presence of fibrils will become significant in reasonable time if the fluid is allowed to become mechanically unstable and develop secondary flow patterns. Such flows have been shown to selectively transport microstructures with large axial ratios and collect this structure within small regions (Silebi and McHugh, 1979; Nicolis, 1975). Once a number of colinear molecules has aggregated, nucleation and growth of fibrils proceeds relatively free from the limitations of population and diffusion, according to mechanisms such as described by Hoffman (1979) in his theory on fibril formation.

Hoffman's nucleation-and-growth theory for fibrils is predicated on the assumption that flow produces statistically elongated polymer molecules. An extended molecule forms "convenient" nucleation sites for fibrillar growth. The growing fibrillar nucleus is approximated as a square parallelepiped of volume, $V = a^2 l$, where a is the side dimension of the end surface and l is the length of the structure. The structure possesses surfaces with corresponding surface free energies of σ for the lateral surface and σ_e for the bundle-end surface. This theory deviates from lamellar-crystal nucleation-and-growth theory in the nature of the bundle-end surface and the effective contributions of σ_e to the total free-energy state of the growing crystal. Hoffman suggests that growth of fibrillar crystals results in an increase in the surface energy of the bundle-end. This arises from a cumulative buildup of surface stress with associated strain at the front of crystal growth. Growth will continue until the cumulative effect of the strain becomes large enough to inhibit additional consolidation.

Considering the effect of cumulative strain, the free-energy change of the growing fibrous crystal, $\Delta\phi$, is written by Hoffman (1979) as

$$\Delta\phi = 2a^2\sigma_{eo} + 4al\sigma - a^2l(\Delta f) + a^2l\,\Delta f\,\alpha\,(V/V_c)^{2/3} \tag{4}$$

where Δf represents the volume free energy of fusion and α is a volume strain parameter.

The last term in equation (4) reflects the presence of cumulative strain. As the crystal dimensions increase, this term begins to dominate $\Delta\phi$ and eventually causes $\Delta\phi$ to become positive. Prior to this, $\Delta\phi$ reaches a minimum for given a and l, which represent the stable fibrous crystal size. Random, simultaneous nucleation at various sites along the preexisting extended molecule results in crystal growth fronts impinging on each other and localizing strain regions. These localized strain regions may evolve in time toward equilibrium, but the strain is locked in by the crystal domains on each side. The high-energy, strained regions represent sites for further growth of folded-chain lamella or kebab structures. The magnitudes of the length and diameter of the fibrillar crystal are independently limited by minimizing the free energy function with respect to each.

7. *Discussion: Nonequilibrium Thermodynamics*

In this work, we attempt to identify the fundamental nature of the fibril, particularly the nature of its synthesis. The material order existing within the fibril is a thermodynamic order in the sense of Prigogine and Lefever (1975) and Nicolis (1975). They hold that the route to such order is to be understood as a consequence of irreversible dissipative processes occurring near to or beyond regions of instability in nonlinear systems. On entering a region of instability, either wholly or through fluctuations, a process gains the capacity for generating symmetry-breaking phenomena. One natural consequence of instability may be the generation of order along a single spatial dimension, which for our purposes

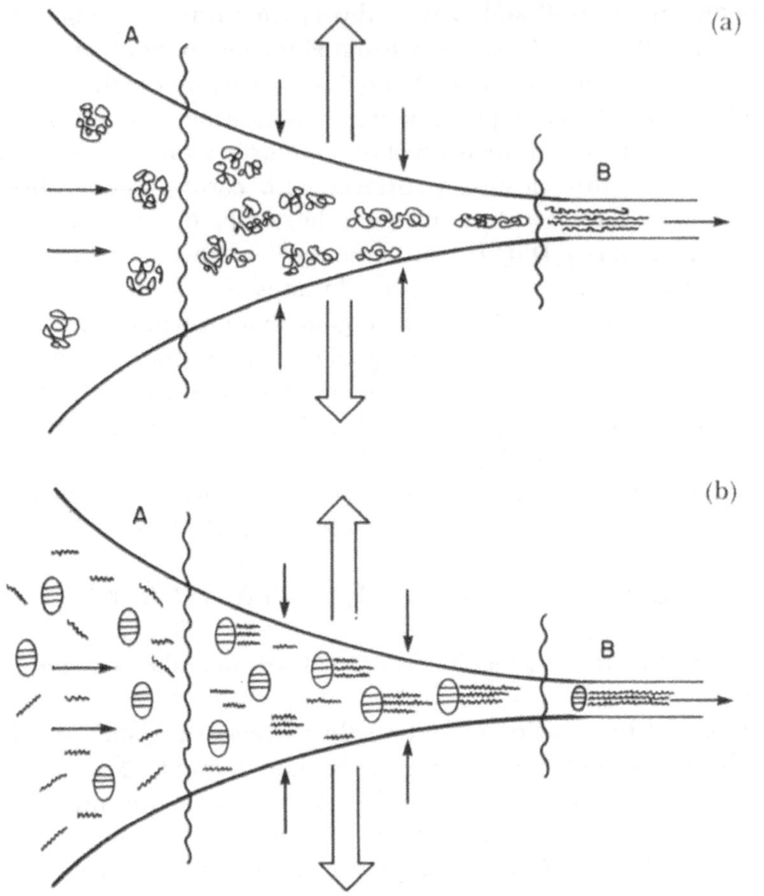

FIGURE 6. Fibril-forming process. (a) Moving from Region A to Region B, Random-coil polymer molecules, subjected to environmental pertubations, assume a more colinear character as thermal (and other entropic) flows are transported away; (b) Monomer (〰) is converted to polymer on a catalytic site (⊜) as thermal (and other entropic) flows are transported away. The process begins in Region A with addition of monomer to the catalyst leading to adjacent long chains capable of crystallizing in fibrillar form.

represents the axis of the fibril. The formation of order along a particular direction requires coordinated dissipation in directions more or less perpendicular to the dimension of order (Fig. 6). The ability to dissipate doubtless limits the degree and arrangement of order within a volume element destined to become the fibril. Such limitations on the formation of "structure" are consistent with the complex composite nature of fibrils and the processes in which they arise.

The underlying features of the processes depicted in Fig. 6 are due to George and Tucker (1976). They present a treatment that in essence is a special case of Prigogine's evolution theorem as it relates to the formation of fibrils within a filament (Fig. 6a), and we now extend it to the case of nascent fibril formation (Fig. 6b). Mass, m, enters a region at A with entropy content S_A and leaves the region at B with entropy content S_B, $S_A > S_B$. Heat (and other entropic) flows leave through the truncated tubular surface, Ω, compensating for the change in configurational entropy accompanying fibril formation. The steady-state requirement on the global entropy of $dS/dt = 0$ gives for the local entropy production

$$\frac{d_i S}{dt} = \dot{m}\,(S_B - S_A) - \int_\Omega \frac{\vec{q}}{T} \cdot \vec{n}\, d\Omega \geqslant 0 \tag{5}$$

where q is the total net flux across the surface Ω.

Positive internal entropy production is considered to be the basis for the quasi-regular structure found in fibrils, since $\dot{m}$ is limited for a given $\vec{q}$ ($\vec{q}$ is taken as negative for flow out of the region).

In the case of Fig. 6b, the stationary flow of mass, $\dot{m}$, entering the fibril-forming system consists of monomer (~) and catalyst (⊜). As the system's entropy is reduced through conversion of monomer into polymer, a stabilizing flux, indicated by the outgoing arrows between A and B, is presumed to be such that equation (5) is satisfied. In this way, a stationary rate of mass flow into and through the system is maintained, being measurable by the rate of appearance of fibrils in the efflux. Lindenmeyer (1980) has considered fiber formation in more generality, starting with Prigogine's evolution theorem. In particularizing the flow and transport process globally, he has devised an argument related to diffusive transport that limits the size of the fibril or fiber. Further, he suggests that the fibril formation arising from the dissipative processes characteristic of the nonequilibrium processes of filament formation is basically inhomogeneous. He remarks that the development of inhomogeneity is favored as a process for accommodating the dissipation. This deduction of Lindenmeyer's reinforces the conception of a basic composite nature of fibrils.

It should be possible to describe all systems, natural or synthetic, capable of forming fibrous structures through the formalisms of nonequilibrium thermodynamics. The authors believe that such descriptions will reveal fundamental features of fibrous systems that will allow in-depth comparisons between the natural and the synthetic. At present, limited comparison and contrast between the two systems can be used to improve knowledge of one system through the other. This approach has provided limited success by revealing the importance of the surface of elemental fibrils and by allowing the concepts of tactoidal and composite structure to be applied to both natural and synthetic systems. Continued exploration drawing on data from natural and synthetic systems should permit the development of a more universal description of microfiber systems.

References

Astbury, W., and Woods, H., 1932, 4. The molecular structures of textile fibers, *J. Text. Inst. Trans.* **23**:T17.

Audemars, G., 1855, British patent No. 283.

Blackwell, J., and Kolpak, F., 1975, Cellulose microfibrils as disordered arrays of elementary fibrils, *Appl. Polym. Symp.* **28**:751.

Brown, M., and Montezinos, D., 1976, Cellulose microfibrils: Visualization of biosynthetic and orienting complexes in association with the plasma membrane, *Proc. Natl. Acad. Sci. U.S.A.* **73**:143.

Burton, W., Cabera, N., and Frank, F., 1951, The growth of crystals and the equilibrium structure of their surfaces, *Philos. Trans. R. Soc. London Ser. A* **243**:299.

Carothers, W., and Hill, J., 1932, Polymerization and ring formation. XV. Artificial fibers from synthetic linear condensation superpolymers, *J. Am. Chem. Soc.* **54**:1579.

Chanzy, H., and Peguy, A., 1980, Oriented cellulose films and fibers from a mesophase system, *J. Polym. Sci. Part A-2* **18**:1137.

Chardonnet, H., 1885, French patent No. 165,349.

Cross, C., Bevan, F., and Beadle (no initial), 1892, British patent No. 8,700.

Denn, M., and Petrie, C., 1975, Mechanics of steady spinning of a viscoelastic liquid, *Am. Inst. Chem. Engs. J.* **21**:791.

Fisher, R., and Denn, M., 1976, A theory of isothermal melt spinning and draw resonance, *Am. Inst. Chem. Engs. J.* **22**:236.

Flory, P., 1953, *Principles of Polymer Chemistry*, Cornell University Press, Ithaca, New York.

Flory, P., 1956a, Statistical thermodynamics of semi-flexible chain molecules, *Proc. R. Soc. London Ser. A* **234**:60.

Flory, P., 1956b, Phase equilibria in solutions of rod-like particles, *Proc. R. Soc. London Ser. A* **234**:73.

Frenkel, S., 1974, Thermokinetics of formation of ordered structures in polymer solutions and gels, *Pure Appl. Chem.* **38**:117.

George, W., and Tucker, P., 1976, The morphology of fibrous polymeric crystals formed isothermally in stationary shear fields, *Polym. Eng. Sci.* **16**:212.

Glansdorff, P., and Prigogine, I., 1971, *Thermodynamic Theory of Structure, Stability, and Fluctuation*, Wiley-Interscience, New York.

Grew, N., 1682, *The Anatomy of Plants*, Johnson Reprint (1965), New York.

Haigler, C., Brown, M., and Benziman, M., 1980a and b, Calcofluor White ST alters the *in vivo* assembly of cellulose microfibrils, *Science* **210**:903.

Han, C., 1976, *Rheology in Polymer Processing*, Academic Press, New York.

Hoffman, J., 1979, On the formation of polymer fibrils by flow-induced crystallization, *Polymer* **20**:1071.

Hooke, R., 1664, *Micrographia*.

Kalb, B., and Manley, R., 1980, Hydrodynamically induced formation of cellulose fibers. I. Observations on the formation of cellulose fibers from stirred solutions, *J. Polym. Sci. Part A-2* **18**:707.

Krigbaum, W., and Salaris, F., 1978, Thermodynamics of nematic phases formed from semiflexible chain polymers, *J. Polym. Sci. Part A-2* **16**:883.

Lindenmeyer, P., 1980, The fundamental basis for fiber formation and fiber structure, *Text. Res. J.* **50**:395.

Loasby, G., 1952, The development of the synthetic fibers, in: *The Development of Some Man-Made Fibers*, pp. 50–79, The Textile Institute, Manchester.

Mandelkern, L., 1964, *Crystallization of Polymers*, McGraw-Hill, New York.

McHugh, A., 1975, Flow-induced crystallization from solution: The relative effects of extension and shearing flow fields, *J. Appl. Polym. Sci.* **19**:125.

Morawetz, H., 1975, *Macromolecules in Solution*, John Wiley, New York.

Nägeli, C., 1852, *Micellartheorie*, Ostwalds Klassiker No. 227, 1928, Leipzig.

Nicolis, G., 1975, Dissipative instabilities, structure, and evolution, in: *Membranes, Dissipative Structures, and Evolution* (G. Nicolis and R. Lefever, eds.), pp. 29–47, Interscience, New York.

Pennings, A. J., Lageveen, R., and deVries, R. S., 1977, Hydrodynamically induced crystallization of polymers from solution. VII. On the origin of persistent lamellar overgrowth of shish-kebab fibers, *Colloid Polym. Sci.* **255**:532.

Peterlin, A., 1979, Crystallization phenomena, *Int. J. Polym. Mater.* **7**:1.

Pohl, H., 1958, Reaction spinning of fibers, *Text. Res. J.* **28**:473.

Prigogine, I., 1967, *Thermodynamics of Irreversible Processes,* Interscience, New York.

Prigogine, I., and Lefever, R., 1975, Stability and self-organization in open systems, in: *Membranes, Dissipative Structures, and Evolution* (G. Nicolis and R. Lefever, eds.), pp. 1-28, Interscience, New York.

Réaumur, 1734, *Mémoire pour Servir à l'Histoire des Insectes.*

Romanovicz, D., and Brown, M., 1976, Biogenesis and structure of Golgi-derived cellulose scales in *Pleurochrysis.* II. Scale composition and supramolecular structure, *Appl. Polym. Symp.* **28**:587.

Sawyer, L., and George, W., 1978, Tactoidal precursors of fibrous morphologies, *Polym. Sci. Part B* **16**:193.

Silebi, C., and McHugh, A., 1979, Effect of nonhomogeneous shear rates on the flow-induced crystallization of flexible macromolecules, *J. Polym. Sci. Part A-2* **17**:1469.

Slatterly, J., 1966, Spinning of a Noll simple fluid, *Am. Inst. Chem. Engs. J.* **12**:456.

Smook, J., Torfs, J. C., Van Hutten, P. F., and Pennings, A. J., 1980, Ultra-high strength polyethylene by hot drawing of surface growth fibers, *Polym. Bull.* **2**:293.

Stockmann, V. E., 1972, Developing a hypothesis: Native cellulose elementary fibrils are formed with metastable structure, *Biopolymers* **11**:251.

Tscheuschner, E., 1885, *Handbuch der Glasfabrikation nach allen ihren Haupt- und Nebenzweigen,* 5th ed., Bernh. Friedr. Voigt, Weimar.

Urquhart, A., 1952, Cellulose-derivative rayons, past, present and future, in: *The Development of Some Man-Made Fibres,* pp. 7-24, The Textile Institute, Manchester.

Walczak, Z., 1977, *Formation of Synthetic Fibers,* Gordon and Breach, New York.

Whitley, C., 1979, Unpublished master's thesis, North Carolina State University, Raleigh.

Wilkjord, A., and Manley, R. St. J., 1968, The morphology of fibrillar polyethylene crystals, *J. Macromol. Sci. Phys.* **2**:501.

Wünderlich, B., 1964, Polyethylene crystallized from the melt under elevated pressure, *J. Polym. Sci. Part A* **2**:3697.

Wünderlich, B., 1973, *Macromolecular Physics,* Vols. I and II, Academic Press, New York.

Zachman, H., 1974, Theory of nucleation and crystal growth in concentrated solutions, *Pure Appl. Chem.* **38**:79.

Ziabicki, A., 1976, *Fundamentals of Fibre Formation,* John Wiley, London.

III
DEGRADATION

22

β-Glucanases in Higher Plants: Localization, Potential Functions, and Regulation

DESH PAL S. VERMA, VISHNU KUMAR, AND GORDON A. MACLACHLAN

1. Introduction

Polysaccharides with D-glucopyranose units joined in glucosidic linkages are the most abundant organic substances on earth. The major β-linked glucan polysaccharides are β1,4-glucan (cellulose) and β1,3-glucans (e.g., curdlan, callose). Other β-glucans have a backbone of β-1,3-linked glucose units associated with variable and low proportions of internal β1,4- or β1,6-linkages (e.g., pachyman, laminarin) or associated in a regular pattern with relatively high proportions of β1,4-linkages (mixed-linkage glucans).

For catabolizing these products, there are present in nature a number of hydrolases distributed among various organisms. In higher plants, they are involved in a number of functions, from cell growth to the maturation of fruits, and thus are of fundamental importance for plant growth and development. Depending on their primary mode of attack on the β-glucan molecules, these enzymes have been divided broadly into endo- and exo- forms. Further classification of the glucanases is based on their substrate specificity. It is important to note that the exoglucanases do not readily hydrolyze β-glucans that have been oxidized at the non-reducing ends (Barras and Stone, 1969a). A general outline and classification of different types of endoglucanases, with their

DESH PAL S. VERMA, VISHNU KUMAR, AND GORDON A. MACLACHLAN • Department of Biology, McGill University, Montreal, Quebec, H3A 1B1 Canada.

characteristic substrates and cleavage specificity for different linkages, is presented in Table I.

In general, many details are known about the substrate specificity and enzymatic characteristics of endoglucanases, but comparatively little information is available on the location and biological function of these enzymes in higher plants. Since most of the exoglucanases and endoglucanases (types "B" and "C") are primarily restricted to lower organisms, the detailed treatment of these enzymes is beyond the scope of this chapter, and only types "A" and "D" glucanases are considered. Among the endoglucanases, cellulase (type "D", EC 3.2.1.4) has been studied most extensively because of its potential function in cell-wall expansion and dissolution during various phases of plant growth.

Polysaccharide decomposition is also brought about by a number of microorganisms by producing specific enzymes when they encounter cell walls of plants. This process may permit them to survive saprophytically on plant debris, or to penetrate and parasitize living plant cells, or to do both. The degradation products thus formed may be available to the microorganism for its nourishment. In higher plants, however, β-glucanase activity does not appear to be substrate-induced but, rather, to be controlled by various plant-growth regulators. There are occasions in the life of most plant cells when it is necessary

TABLE I
Outline of Exoglucanases and Endoglucanases

Exoglucanases	Endoglucanases
Remove glucose or cellobiose residues from nonreducing end of oligo- and polysaccharides; inactive on cellobiose and laminaribiose (Reese and Mandels, 1959; Barras and Stone, 1969a; Farkas *et al.*, 1973).	*Type A* (EC 3.2.1.39), e.g., in *Nicotiana glutinosa*. Specifically hydrolyzes β-1,3-glucosidic linkages in homogeneous polymers, provided there is a laminaribiosyl residue adjacent to the cleavage site, e.g., G3G3G3G3G (Moore and Stone, 1972b). *Type B* (EC 3.2.1.6), e.g., in *Rhizopus arrhizus*. Cleaves either β-1,3- or β-1,4-glucosidic linkages provided the substrate contains a laminaribiosyl unit at the nonreducing side of the hydrolyzed linkage, i.e., −G3G3G or −G3G4G (Parrish *et al.*, 1960; Reese and Mandels, 1959). *Type C* (mixed-linkage β-glucanase, EC 3.2.1.73), e.g., in *Bacillus subtilis*. Cleaves β-1,4-glucosidic linkages of mixed-linkage glucans, provided there is a 4-*O*-laminaribiosyl-D-glucose residue on the nonreducing side, i.e., G4G3G4G (Reese and Perlin, 1963). *Type D* (endocellulase, EC 3.2.1.4), e.g., in *Streptomyces*, *Basidiomycete* QM 806, *Pisum sativum*. Hydrolyzes β-1,4-linkages and does not require the presence of 1,3 linkages, i.e., G4G4G4G (Parrish *et al.*, 1960; Wong *et al.*, 1977a).

to digest already deposited polysaccharides, for example, in specialized differentiating tissues such as abscission zones, perforating cell plates, fusing tracheids, and senescing and softening tissues of fruits. Most important, cellulase activity may be needed in growing cells, where partial hydrolysis of cellulose microfibrils in the primary cell walls may be necessary to "loosen" the more or less rigid wall framework to allow for cell expansion.

2. *β1,3-Glucanases*

These enzymes show a wide range of variability as regards their minimum glucosyl requirement, the type of linkages hydrolyzed, the substrate specificity, and the ultimate products formed due to enzymatic reactions (Parrish *et al.*, 1960; Reese and Mandels, 1959, 1963; Reese and Perlin, 1963; Moore and Stone, 1972a, b; Wong, 1979; Wong and Maclachlan, 1979). Members of various classes widely distributed in bacteria, algae, fungi, higher plants, and invertebrates have been shown to possess these enzyme activities.

2.1. Lower Organisms

One of the roles of β1,3-glucanases in Saprophytic fungi and bacteria is clearly in nutrition, since they hydrolyze β1,3-glucans to generate a mixture of utilizable saccharides (Norkrans, 1967). In algae, seasonal fluctuations are observed in concentrations of storage β-glucans of *Laminaria* (Meeuse, 1962), and of paramylon (β1,3-glucan granules) in Euglenophyceae when grown in the dark (Barras and Stone, 1969a, b). The decrease in laminarin levels observed during cell-wall formation in zygotes of *Fucus* (Quatrano and Stevens, 1976) also appears to be due to β1,3-glucanases. Some pathogenic fungal species secrete the enzyme when they come in contact with the cell wall of the host (Albersheim and Anderson-Prouty, 1975; Bateman and Basham, 1976; Pegg, 1976). This may help penetration and even degrade β1,3-glucans elaborated as a result of host-parasite interaction (Albersheim, 1976; Woodward *et al.*, 1979). In some other cases, the β1,3-glucanases appear to be responsible for hyphal-cell elongation, or branching through softening of the wall.

Whether the primary role of these enzymes in pathogens is nutritive or for infection processes is not well established and is a field of active research today. In *Basidiomycete* QM 806, the synthesis and secretion of an exoglucanase is reported to be from an induction signal in the form of starvation. Thus, the enzyme potentially has a nutritive function. In *Claviceps* species, β1,3-glucanases appear to be involved in the reclamation of extracellular β-glucans secreted previously by the fungus itself (Reese and Mandels, 1959). Likewise, mycolaminarins of *Oomycetes* are unquestionably utilized during cellular differentiation or carbon starvation or both (see Bartnicki-García, 1973); for this degradation, particulate (membrane-bound) β1,3-glucanases are responsible. Such enzymes are also known to degrade glucans in higher fungi, e.g., in *Sclerotinia sclerotiorum* during germination of sclerotia.

In *Saprolegnia monoica* and *Neurospora crassa*, lytic vesicles that contain β1,3-glucanases appear to be responsible for transporting the hydrolases to the sites of wall growth at apices, conidia, and hyphal branches. The discharge of enzyme(s) into the cell walls in these regions results in softening and initiation of growth centers. Leighton and Stock (1970) reported that during germination of conidia of *Microsporium gypseum*, the lysosomal vesicles move to the spore wall and discharge enzymes for initiation of germination. Likewise, during the process of budding in yeast, which is conceived as a modified type of tip growth in hyphal mycelia, β1,3-glucanases are accumulated in lysosomal vesicles derived by proliferation of endoplasmic reticulum and are then released at the site of bud formation (Villa *et al.*, 1976). This process is of particular interest in that the hydrolase activity is directed to a correct locus at a suitable time and thus avoids general lysis of the whole cell wall. It has been suggested that the enzymes are released to the site and are converted from a latent form, although the possibility of *de novo* synthesis of hydrolases upon a specific signal at an appropriate place or time or both cannot be ruled out (Farkas *et al.*, 1973).

2.2. *Higher Plants*

There is no known storage product in the form of β1,3-glucan in the cytoplasm of higher plants. It is not likely, therefore, that higher plant β1,3-glucanases play a role in plant nutrition or mobilization of polysaccharides. In endosperm and pollen mother-cell walls and in stem tissues of some monocots, however, there is evidence that endogenous hydrolases are important in regulation and turnover of mixed-linkage substrates. The paucity of information about the role of these enzymes in plants is due to the fact that only a few have been purified to the degree of homogeneity needed to establish their modes of action and substrate affinities (Barras and Stone, 1969a; Abeles *et al.*, 1970; Moore and Stone, 1972a, b; Kato *et al.*, 1972; Manners and Wilson, 1974; Ballance and Manners, 1978). Their subcellular distribution in different tissues is also uncertain. Depending on evidence that callose is deposited extracellularly, these enzymes are considered to be functional only after being secreted (Fulcher *et al.*, 1976; Morrison and O'Brien, 1976). It was suggested (Matile *et al.*, 1971) that these enzymes are sequestered in vacuoles and are mobilized by unknown mechanisms to the site of action, e.g., to combat invasive organisms (Pegg, 1976) or to degrade wall glucans during development (Masuda, 1968) and to help in recovery from wound response by dissolving deposits of wound glucans (Currier, 1957).

2.2.1. *Cell Growth and Differentiation*

During this process, a delicate balance appears to exist between wall synthesis (by synthetases) and wall lysis (by glucanases). This balance can be shifted in favor of lysis through a number of stimuli resulting in a complete disintegration of the wall. Changes that result from auxin treatment inside the cell wall appear to promote the release of xyloglucans (Labavitch and Ray, 1974), which may help to bring about plasticity and allow the wall to grow. It has also been suggested that this release of xyloglucans results from the action of wall hydrolases (Jacobs and Ray, 1975). Kivilaan *et al.* (1971) showed that during autolysis of cell walls, free glucose and glucans are released into the incubation medium, which implies the action of glucanase(s). These enzymes have been reported in *Avena* coleoptiles (Heyn, 1969; Wada *et al.*, 1968) and in *Pisum* epicotyls (Datko and Maclachlan, 1968). Moreover, it has also been shown that fungal and plant preparations containing β-1,3-glucanase activities could mimic the effect of auxin on elongation when added to tissue segments of oat and barley coleoptiles (Masuda and Yamamoto, 1970; Masuda *et al.*, 1973; Nevins, 1975a, b). But similar results were not obtained when the same preparations were added to Jerusalem artichoke tuber slices. An enzyme preparation extracted from bean hypocotyls containing both β1,3-glucanase and β-glucosidase also modified the extensibility of cell walls of *Avena* coleoptiles to an extent similar to that observed when coleoptiles were treated with indoleacetic acid (IAA) (Nevins, 1975b). The effect was inhibited by nojirimycin an inhibitor of many β-glucosidases and exoglucanases (Reese and Parrish, 1971). These results are suggestive of the possibility that auxins act via effects on β-glucanase, but the proposition remains unproved, in part because the enzyme preparations that were employed in the experiments cited above may have contained other enzyme activities or components besides β-glucanase and β-glucosidase.

The cell wall of cereal endosperms also contains, among other components, 1,3-/1,4 mixed-linkage β-glucans (MacLeod and McCorquodale, 1958). During seed germination, the dissolution of endosperm walls occurs, and this may facilitate access of various hydrolases generated in aleurone layers (Taiz and Jones, 1970). In germinating rice (Pamiano and Juliano, 1973), rye (Ballance and Manners, 1978), and wheat (Clarke and Stone, 1962), there are progressive increases in β1,3-glucanase activities. In barley seeds, three different glucanases, namely, (1) strictly endo-β-1,3-glucanase (type A), (2) nonspecific endo-β-1,3-glucanase (type C), and (3) a cellulase (type D), have been reported (Manners and Marshall, 1969). Among these, types "C" glucanases are of interest, since they are responsible for hydrolysis of β-1,3-/1,4 mixed-linkage glucans. Fulcher *et al.* (1977) reported the presence of callose deposits on the endosperm cell wall that are depolymerized by type A glucanases.

2.2.2. *Wounded Tissues and Invasion by Foreign Organisms and Callose Deposition*

In higher plants, glucanase activity is also considered to play a role in their defensive mechanisms. During this process, these enzymes are secreted and attack the cell walls of invading fungal organisms without harming the host. Abeles *et al.* (1971) have shown that if bean plants are exposed to ethylene gas before exposure to rust (*Uromyces phaseoli*), which elicits β1,3-glucanase activity, it slows the onset of disease on attempted infection. Likewise, in tobacco plants, the level of glucanase activity increases with the development of tobacco mosaic virus and tomato spotted-wilt-virus lesion (Nevins *et al.*, 1977). This increase in enzyme activity may be related to the wounding of plant tissues. A temporary callose deposition in sieve elements is often a result of infection by fungus or virus. On the other hand the β1,3-glucanase activities that concurrently develop with infection may also depolymerize the host-elaborated callose, thus allowing translocation to continue and facilitate the spread of the infecting agent.

It has been observed that during the infection of leguminous plants by *Rhizobium*, callose is deposited at the site of infection-thread initiation, but not elsewhere in the infection route (Kumarshinghe and Nutman, 1977), and this has been thought to be a factor responsible for root-hair curling. It has also been suggested that *Rhizobium* causes the hydrolysis of callose in the young infection thread and thus may facilitate the changes in wall structure associated with thread initiation. Whether callose formation is a response to infection, as a defensive mechanism of the host or a helping factor for entrance of rhizobia, remains to be shown.

The relationship between β1,3-glucanase activity and callose in higher plants has been the subject of many investigations. For example, in *Lilium* anthers, the microspores are released after the callose-like special wall of the tetrads has been dissolved by the hydrolytic action of β1,3-glucanases, which are known to be secreted by parental tapetal cells. This activity increases dramatically during spore dispersal (Stieglitz and Stern, 1973). Callose is often closely associated with particular regions of the cell wall, e.g., plasmodesmata (Smith and McCully, 1977), cell plates (Fulcher *et al.*, 1976), at the junction of aleurone and subaleurone layers (Fulcher *et al.*, 1977). It is also present in developing cotton hairs, in walls of cultured cells (Mollard and Barnoud, 1976), in cell walls of gymnosperms, and in pollen, pollen tubes, and embryo sacs (Mascarenhas. 1975; Cresti and Van Went, 1976). It is suggested that callose plays an important role in newly formed cells (Currier, 1957), where it may act as an initial 'gel' that separates daughter cells and traps other wall components into an increasingly firm wall barrier (Fulcher *et al.*, 1976). Callose generally disappears as the cell wall thickens and strengthens, as a result of continued deposition of cellulose and other components, and it is here that β1,3-glucanases may be playing a role. In pollen grains, callose appears to protect the male gametophyte from wounding and infection (Mascarenhas, 1975). In the endothelial cells of the embryo sac, callose acts as a barrier between zygote and the maternal tissue, and it may act as a hygroscopic water reserve. In sieve elements, callose is deposited after wounding and may serve to restrict transport. After the removal of the wound stimulus, a slow resorption of callose deposits occurs, presumably due to activity of glucanases secreted into the cell wall. It is possible that callose provides readily usable glucose and energy for neighboring tissues, following its dissolution by specific endogenous β-glucanases, provided these are secreted into the cell wall without causing destruction of other wall components.

3. β1,4-Glucanase (Cellulase)

3.1. Cell Expansion and Cell-Wall Maturation

The cell wall of higher plants is a multicomponent structure containing semicrystalline cellulosic microfibrils, that provide the architectural framework of the wall, and noncellulosic matrix including other polysaccharides and structural protein (Preston, 1974). The constraints placed on the plant cell by these components, singly or by cross-linking with one another, have been considered as a limiting factor for cell enlargement. During the process of cell expansion, synthesis and deposition of cellulose and other polysaccharides is unquestionably essential for regulating continued and irreversible growth and morphogenesis in plants.

Despite the abundance of data available on cellulose structure and fiber modification in relation to industrial applications, little is known about the fundamental mechanisms of cellulose microfibril assembly and "disassembly" at the cellular level. Electron micrographs of growing primary walls do not reveal discontinuities in the microfibrillar framework at any stage of growth (Maclachlan, 1977b), yet radioautographs show clearly that new cellulose is inserted into the inner surface of the wall at multiple loci during growth (Ray, 1962; Baker and Ray, 1965a, b; Mueller and Maclachlan, 1980). This raises questions, not only about how the expanding cells loosen or relax the circumscribing network of partially crystalline polymers, but also about how new cellulose molecules are inserted into the framework. It may be that one or more cellulases are required for weakening the framework and also for generating new primer ends in the cellulose matrix (see below). Some aspects of these problems are perhaps now answerable, since presumptive cellulose-synthesizing complexes have been visualized (R. M. Brown, Jr., and Willison, 1977). These complexes exist in association with plasma membranes in the form of unusually ordered structures at the end of growing microfibrils in algae (Brown and Montezinos, 1976; Montezinos and Brown, 1976; Willison and Brown, 1978). Such terminal complexes have also been observed in higher plants (Mueller *et al.*, 1976; Willison and Brown, 1977; Willison and Grout, 1978). Recent observations (Mueller and Brown, 1980) suggest that the microfibril-synthesizing complex may comprise a transmembrane component. Further studies are needed to explore whether cellulases are associated with such synthetase complexes (Maclachlan, 1977a) as an essential component of the mechanism whereby precursors are inserted into the wall framework.

To understand how wall extension is controlled, it is necessary to identify the key reactions and molecules required for microfibril extension and reorientation. If there exists a coordinated multienzyme complex of corresponding hydrolases and synthetases, and perhaps even repairing reactions at cell surface, factors controlling the activities of these enzymes could therefore be directly or indirectly in control of cell growth. Hydrolases may play an essential role in wall extension especially in cells undergoing tip elongation or multinet growth (see Roland and Vian, 1979). Cellulase activity has been shown to be associated with cell enlargement and loosening activity has been demonstrated in pistils and pollen tubes of the day-lily plant (Konar and Stanley, 1969).

Multinet growth in plant cells usually has two components, elongation along the cell axis and lateral cell expansion. Various theories have been put forward for explanation of the former type of growth (Preston, 1961; Cleland, 1971; P. J. Davies, 1973; Evans, 1974). Several observations have led to a belief that plant hormones stimulate elongation by acting at the level of transcription to cause synthesis of a highly labile protein that can act as a wall-loosening agent. This could allow cell extension to be driven by cell turgor pressure. The polysaccharidases are prime candidates for the wall-loosening factors and cellulase has been thought to be one such protein.

Wall extensibility as measured with the Instron analyser is increased by exogenous cellulase (Olson *et al.*, 1965; Ruesink, 1969). This enzyme activity has been found in wall fractions of growing regions of the plant (Heyn, 1969; Maclachlan and Perrault, 1964; Fan and Maclachlan, 1967a, b) and has been localized *in situ* in the walls of growing roots (Bal, 1973). In pea epicotyls, the activity is correlated with the regions of most active wall breakdown (Maclachlan and Young, 1962; Matchett and Nance, 1962). In decapitated pea epicotyls, there is a correlation between the auxin-induced growth and cellulase activity (Fan and Maclachlan, 1966). It has been shown that auxin treatment in pea epicotyls generates two distinct endocellulases that have marked differences in their amino acid composition and physical and immunological properties (Byrne *et al.*, 1975). These hydrolyze in a random manner higher-DP cellodextrins as well as carboxymethylcellulose (CMC) and some mixed-linkage glucans containing β1,4-linkages (Wong *et al.*, 1977a). Auxin treatment of the epicotyls evokes a preferential increase in the levels of both cellulases (Datko and Maclachlan, 1968, 1970). Increase in the activity of these enzymes occurs *de novo* (see below). At least one of the enzymes is secreted and becomes associated with wall microfibrils (Bal *et al.*, 1976). Thus, cellulase has all the characteristics attributed to the wall-loosening factor, i.e., it is a wall-associated protein that is generated in response to auxin during cell-elongation growth.

Evidence has been presented against the role for cellulase in certain forms of cell expansion. In *Avena* coleoptile, for example, cellulase treatment does not appear to affect wall extension (Wada *et al.*, 1968). Exogenously applied auxin does not cause any major increase in cellulase activity for some hours, though cell extension begins much sooner (Rayle *et al.*, 1970; Nissl and Zenk, 1969). There is no increase in cellulase activity in gibberellin-treated barley seedlings, although other cell-wall-hydrolyzing enzymes respond positively. This has led to the suggestion of possible involvement of β1,3-glucanase, since this enzyme not only increases the wall extensibility of coleoptiles (Masuda *et al.*, 1967, 1970; Masuda, 1968) but also appears to provoke limited cell elongation when added exogenously (Masuda *et al.*, 1967; Masuda, 1968; Tanimoto and Masuda, 1968; Wada *et al.*, 1968).

Arguments to suggest that modest levels of cellulase activity may paradoxically play a constructive role in cellulose deposition during growth have been presented and discussed by Maclachlan (1976, 1977a, b). Direct evidence for this was obtained by studies on the incorporation of [^{14}C]sucrose into alkali-insoluble glucan by excised pea segments, which were enhanced by treatment with small amounts of fungal cellulase in the medium. Similar experiments have been conducted using tissue slices pretreated with buffer-soluble or -insoluble cellulase, followed by incubation in labeled UDP-glucose (Wong *et al.*, 1977b). Both cellulases markedly enhance the rate of glycosyl transfer to alkali-insoluble glucan when used at moderate concentration. If high levels of fungal or pea cellulases are used, they reduce the yield of newly synthesized glucan. These results imply that a limiting factor for cellulose deposition in peas is the availability of endogenous β1,4-glucan chain ends that may act as "acceptor" molecules for transglycosylation. Partial endohydrolysis by cellulase could provide a means to overcome this limitation (see Maclachlan, 1976).

With respect to growth due to lateral cell expansion, wall loosening is again a primary requirement. Cellulase activity has been associated with enlargement of cells in several tissues; pea epicotyls (Fan and Maclachlan, 1966, 1967a, b; Ridge and Osborne, 1969) as well in bean stems and petioles (Linkins *et al.*, 1973; Lewis *et al.*, 1972, 1975; Lewis and Koehler, 1979), tobacco leaves (Tracey, 1950, 1959), and lily pistils (Konar and Stanley, 1969). Application of exogenous IAA causes massive swelling in tissues of the growing region of epicotyl, which also shows vascular differentiation and root-primordia development. These events are accompanied by a dramatic increase in cellulase activity. However, it is simplistic to conclude that all treatments that cause tissue swelling are mediated by cellulase. Ethylene, for example, induces swelling in etiolated pea epicotyls, but causes only a slight increase in cellulase activity (Ridge and Osborne, 1969; Maclachlan, 1977a and b).

During development and differentation, the walls of certain cells are weakened or dissolved. Examples include the removal of end walls of cells during formation of vessels and articulated laticifers (see Esau, 1967), formation of sieve-plate pores (Northcote and Wooding, 1966), and pore formation between guard cells of stomata (W.V. Brown and Johnson 1962). Wall dissolution has also been observed during seed germination, in lacunae development in plants with hollow stems, and in the cortex in advance of developing adventitious roots (Fan and Maclachlan, 1967a; Lewis *et al.*, 1975). A massive degradation of the cell wall occurs (degradation of cellulose and hemicellulose) during corolla senescence in *Ipomea tricolor* (Wiemken-Gehrig *et al.*, 1974), and dissolution of entire cell walls, leading to masses of free-floating nuclei, can be observed in quiescent onion-root meristems (Bal and Payne 1972). Cellulase activity levels have been correlated with many of these above processes. An active cellulase has been extracted from developing vessels of barley roots (Sassen, 1965) and from xylem and phloem elements of *Acer* (Sheldrake, 1970). It has been observed that xylem and phloem tissues have in general higher cellulase activities than cambium, while callus without differentiation showed a very low activity.

Sheldrake (1970) thus concluded that cellulase is directly involved in the process of vascular differention. In a recent study, Kawase (1979) showed that development of aerenchyma in waterlogged *Helianthus annuus* was accompanied by an increase in cellulase activity in response to ethylene production triggered by an anaerobic environment.

A morphogenetic role for cellulase has also been shown in *Achlya ambisexualis*. It has been observed (Thomas and Mullins, 1967) that sexual hormone antheridiol (inducing antheridial branching in male strains of *Achlya*) elicits the rise in cellulase and only those strains that respond to hormones show a concomitant increase in cellulase. Such a response is inhibited by protein-synthesis inhibitors. Also during vegetative branching, cellulase activity increases, and thus this enzyme may play a role in wall softening at the site of branch initiation.

3.1.1. Intracellular Localization of Cellulases

The subcellular localization of cellulases has been studied by Bal *et al.* (1976) using electron microscopy and cytological techniques. In tissues treated with auxin, wall degradation is evident, and cuprous oxide precipitation [a measure of cellulase activity (Bal, 1972, 1973)] was found to be particularly intense at these loci (Fig. 1). Immunocytochemical studies using specific antibodies showed that

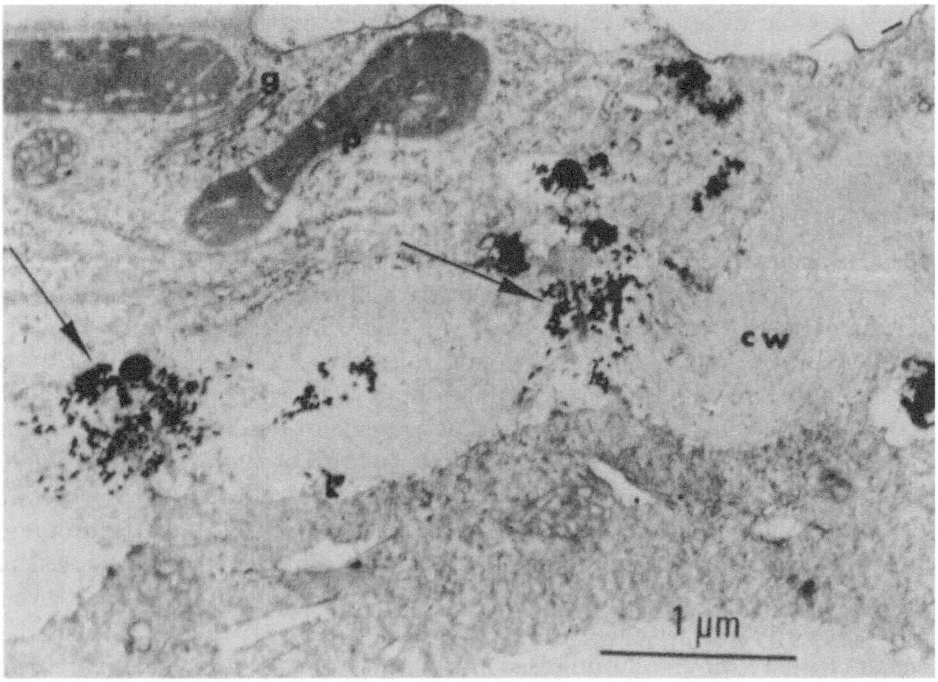

FIGURE 1. Localization of cellulolytic activity in auxin-treated pea epicotyls. (→) Cuprous oxide precipitate at the site of reaction using carboxymethyl cellulose as a substrate. (cw) Cell wall; (g) Golgi; (p) proplastid. From Bal *et al.* (1976).

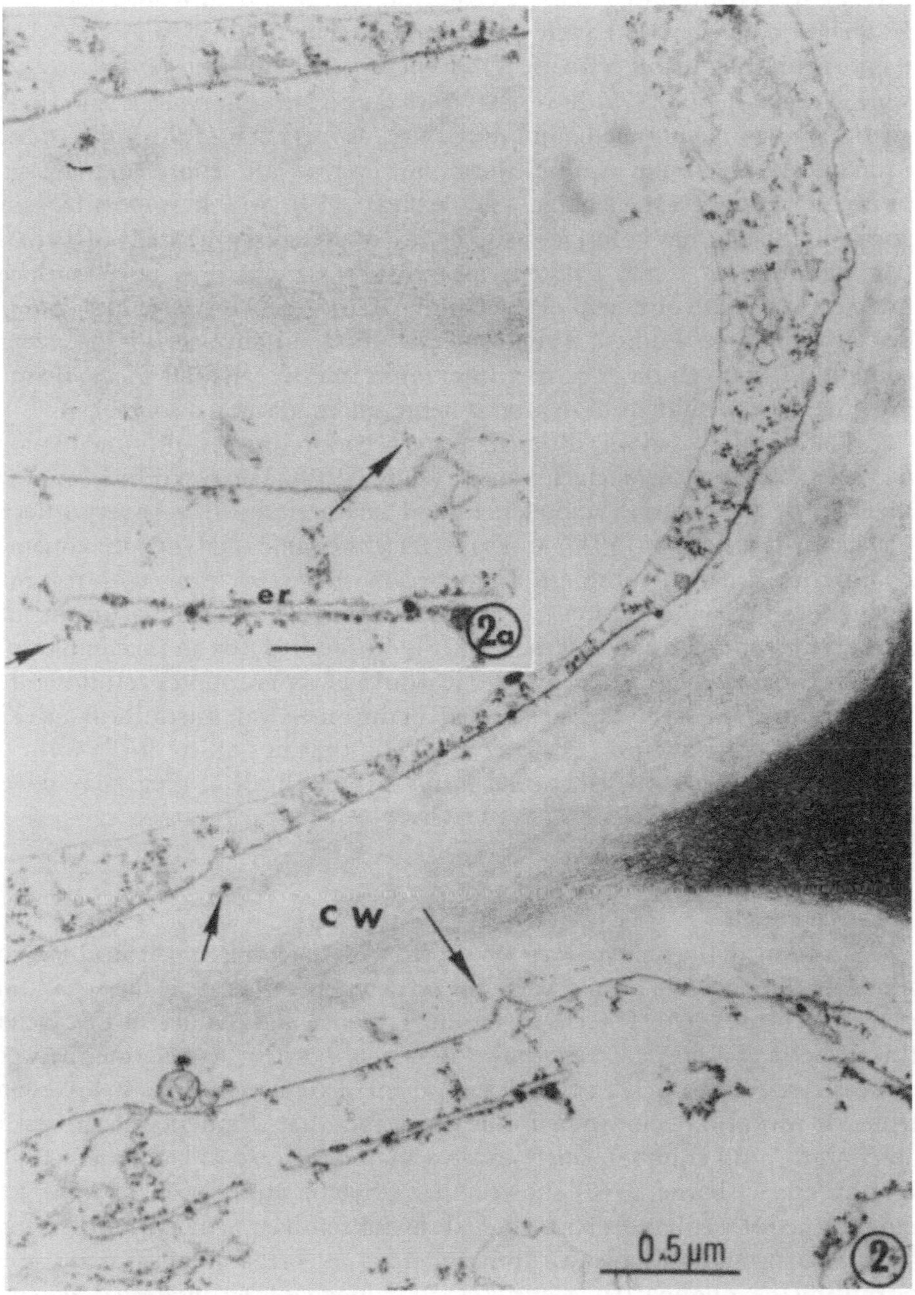

FIGURE 2. Immunological localization of one of the cellulases (buffer-insoluble) in pea epicotyls, using ferritin-conjugated antibodies. (→) Ferritin particles in association with cell-wall microfibrils and the endoplasmic reticulum (er). (cw) Cell wall. From Bal *et al.* (1976).

buffer-insoluble cellulase is localized on microfibrils at the inner surface of the cell wall with a small amount associated with the endoplasmic reticulum (ER) (Fig. 2). Buffer-soluble cellulase, in contrast, is present in distended parts of the ER. These regions of ER have few ribosomes attached to them (smooth ER), and it is possible that the ER cellulase is nascent protein being transported to the cell wall. Verma *et al.* (1975) have shown that the latter enzyme is specifically synthesized by membrane-bound polysomes. Conversion of the buffer-soluble form of cellulase to the insoluble form, however, would require an increase in molecular weight (Maclachlan and Wong, 1979), which is possible since covalent linkages may be formed between two discretely synthesized polypeptides (Ciechanover *et al.*, 1980). How enzyme transport takes place to the cell surface is not clear, as no involvement of the Golgi complexes or microtubules could be detected in these studies. These observations lend support to the general contention that cellulases have a functional role in cell-wall metabolism in growing cells, with their biosynthesis being under hormonal control.

The question as to why different forms of endocellulases are found in plants has been discussed by Maclachlan and Wong (1979). The two cellulases of pea that have been purified to homogeneity and have been shown to be very different physically (Byrne *et al.*, 1975) appear to have the same catalytic mechanism in hydrolyzing different substrates. Comparison of the two forms with regard to their subcellular distribution between cell organelles and compartments after auxin treatment indicates that the former is synthesized first and accumulates in vesicles (microsome fraction), where it would never encounter cellulose. The buffer-insoluble enzyme is concentrated on the inner-wall microfibrils (see Fig. 2). It has been suggested (Lewis *et al.*, 1972) that not all multiple forms of cellulase are necessarily functional forms, although all of them may have a capacity to hydrolyze 1,4-linkages *in vitro*.

3.2. Abscission Zone

Shedding of the plant parts is associated with the formation of an abscission zone where cell walls break apart and cell contents appear in the separation cavity (Webster, 1968). There is increasing evidence that cellulase is involved in this process. Several workers have shown a positive correlation between increased cellulase levels and the development of abscission zones in leaves and fruits. It was also demonstrated that the agents that stimulate the process of abscission could simultaneously cause a major increase in cellulase activity. Horton and Osborne (1967) showed that ethylene application in *Phaseolus vulgaris* petioles enhances both abscission and cellulase activity. Furthermore, an inverse correlation between cellulase activity and break strength (as measured by pulling the zone apart) in bean abscission zones has been observed (Lewis and Varner, 1970). Two forms of cellulase were also found to exist in the abscission zone. One of these is synthesized *de novo* in response to ethylene treatment and could be solubilized with a high-salt-fortified buffer. This induction is suppressed by auxin treatment. These enzymes have been isolated and partially characterized from abscising beans and citrus tissues (Lewis *et al.*, 1972; Reid *et al.*, 1974; Huberman *et al.*, 1975; Huberman and Goren, 1979).

It has been suggested that ethylene stimulates release of cellulases into the cell wall. This hypothesis is based on results obtained from vacuum infiltration of tissues segments with buffers, followed by low-speed centrifugation to recover the buffer from "the walls and cell-wall spaces" (Abeles and Leather, 1971). The type of cellulase extracted by this technique was shown to be almost exclusively (greater than 95%) exocellular cellulase with an isoelectric point (pI) at pH 9.5. This form of cellulase is not present in freshly harvested nonabscising tissues and can be detected only during abscission or ethylene treatment (Reid *et al.*, 1974). The other form of cellulase, cellulase pI 4.5, can be extracted at zero time and has the same isoelectric point as the form obtained by grinding the ethylene-treated tissue. Whether the proteins in both cases are the same has not been established. Furthermore, the specificity of this method is doubtful, since the filtration medium was fortified with 1 M NaCl, which dissolve both cellulases (see Byrne, 1974).

The most important question that emerges is the relationship between the form of cellulase and its function(s) during abscission. The possibility that the pI 4.5 form may be altered and transported out of the cell to become the 9.5 form was tested (Koehler *et al.*, 1976) by analyzing the membrane-bound cellulase activity. It was concluded that 5–10% of the total cellulase enzyme is plasma-membrane-bound. The presence of this form of cellulase was suggested to indicate that enzyme movement occurs from cytoplasm to its site of action in the cell wall. Huberman and Goren (1979) separated enzymes that were located at the site of action (exocellular enzymes) from enzymes that were still in the cytoplasm (see Rasmussen, 1973). In further studies of the activity of isoenzymes and their kinetics in abscission zones, the results indicated that some isoenzymes were not involved in the abcission process, although they were present in the tissue. They found marked similarity between the enzymatic properties of exo- and endocellular isoenzymes from nonabscising zones. The analysis of cellulase activity in different fractions led them to propose that practically all the enzymes synthesized in the cell eventually reach the cell wall. Whenever an increase occurs in cellulase activity that is not followed by abscission, it may be that the enzyme does not reach the substrate.

Ratner *et al.* (1969) observed that in citrus leaf explants, cycloheximide treatment at a low concentration retards abscission and inhibits development of cellulase activity. However, there is evidence that cycloheximide concentrations that inhibit protein synthesis nevertheless may promote ethylene production and increase cellulase activity in fruit abscission (Rasmussen, 1973, 1975). An increase in cellulase activity in the peduncle abscission zone of nonabscising mature fruits has been observed (Goren *et al.*, 1973; Greenberg *et al.*, 1975). This increase is enhanced by ethylene and delayed by 2,4-dichlorophenoxyacetic acid (2,4-D) treatments. However, these changes were always independent of abscission, since the mature fruits lose their ability to abscise at this zone a very short time after setting. Instead, the abscission occurs at the calyx area. This phenomenon was explained on the basis of the existence of several isoenzymes of cellulases, some of which may not be directly related to the separation process (Lewis *et al.*, 1972; Goren and Huberman, 1976). The relative activity of these isozymes varied with application of ethylene and 2,4-D. Thus, the role of cellulases in abscission appears to be a tissue-specific phenomenon. For some tissues, e.g., begonia flower, it has even been suggested (Hanish-Ten-Cate *et al.*, 1975) that cellulase activity is not the cause but the result of abscission.

3.3. Role of Cellulases in Fruit Ripening

A characteristic of the ripening process that is common to most fruits is a decrease in the firmness of the fruit. The loss in firmness is usually associated with the cell-wall-degrading enzymes (Dilley, 1970; Pilnik and Voragen, 1970). Their activity presumably leads to observed chemical and structural alterations in the cell walls. In pear, Sterling (1954) reported that whereas lignified walls and the stone cells did not show much change during fruit formation, in overripe fruits the parenchymatous cell walls became extremely thin and tenuous. In avocado fruits, structural differences in the cell walls from hard and soft fruits were demonstrated by Pesis *et al.* (1978). In hard fruits, the middle lamella is obvious, and cellulose microfibrils are packed tightly on both sides of it. In soft fruits, however, no middle lamella is observed, and many microfibrils are reported to be missing or diagonally rearranged. Ben-Arie *et al.*, (1979) followed structural changes in cell walls during ripening of apple and pear fruits and showed predominant dissolution of middle lamella with some architectural changes in the cellulose microfibrils, including disintegration. In fully ripe fruits, almost all the fibrillar arrangement in the cell wall is lost and empty regions are formed. This is similar to a response shown by tissue disks kept in polygalacturonase and cellulase enzyme solutions. In apple, polygalacturonase alone is sufficient to dissolve the middle lamella. Polygalacturonase and polymethylesterase have been shown to be active during fruit ripening in avocado (Barmore and Rouse, 1976).

Cellulase plays a specific role in the maturation and softening of the fruits (Hall, 1964; Dickinson and McCollum, 1964; Hasegawa and Smolensky, 1971; Pharr and Dickinson, 1973; Sobotka and Stelzig, 1974; Pesis *et al.*, 1978; Ben-Arie *et al.*, 1979). In tomato fruits the cellulase activity is high at the small-green-fruit stage, gradually falling off during fruit swelling and rising again to high levels with incipient ripeness (Hobson, 1968). Sobotka and Watada (1971) showed that cellulase in extracts of the Manhattan variety of tomato increased to a maximum 2 days after the "breakers-stage" (incipient color) of ripeness and then decreased sharply later. Cellulase from this system was purified (Sobotka and Stelzig, 1974) and shown to consist of two components that hydrolyze (in the presence of 0.1 M D-glucono-δ-lactone) cellodextrins into detectable amounts of glucose. In other varieties of tomato fruits, C_1 activity is claimed to be absent whereas cellobiase activity has been shown to be present (Pharr and Dickinson, 1973).

The cellulase activity in softening avocado fruits during the natural and ethylene-induced ripening process has been studied by Pesis *et al.* (1978). It is directly correlated with aspects of the ripening process such as the climacteric rise in respiration, ethylene evolution, and softening. Exogenous ethylene treatment induces the activity depending on the stage of maturity of fruit; the more mature the fruit, the higher the response. In hard avocado fruits, only a little cellulase activity could be detected which declined progressively from the distal to the proximal end of the fruit. Some changes in hemicellulose-degrading enzymes in Japanese pear fruits in connection with fruit development, softening, and ripening have also been observed (Yamaki and Kakiuchi, 1979). These enzyme activities are low during cell enlargement but increase with ripening. Exocellulase activities have been suggested to be associated with softening, while endocellulase seems to function in cellular breakdown during overripening.

3.4. Fungal and Bacterial Invasion

Some bacteria and fungi are excellent cellulose decomposers in more or less anaerobic environments, e.g., in rumen of cattle, while others are aerobic and found in soils and sea (Kadota, 1959). Depending on the relationship with the host, these microbes may be pathogenic (parasitic) or symbiotic.

3.4.1. Pathogenesis

In the case of fungi, the success of pathogenicity depends partly on hyphal organization and partly on the continuous release of cell-wall-hydrolyzing enzymes, which give the organisms a penetrating capacity. A cellulase complex has been shown to be one of the major enzyme activities that is synthesized during the process of infection. The literature on cellulose-decomposing bacteria and fungi has been reviewed earlier by Norkrans (1967). In some pathogenic bacteria, e.g., *Pseudomonas solanacearum*, and a variety of pathogenic fungi such as *Fusarium oxysporum* (Husain and Diamond, 1960; Deese and Stahmann, 1962), a positive correlation between cellulase activity and pathogenicity has been demonstrated. Representatives of white and brown rot (Danylyak *et al.*, 1977) and dark micromycetes (Pavlenko and Lizak, 1978) are known to have cellulase activity. There are on the list many parasitic fungi that show cellulase activity in culture filtrates, e.g., *Botryodiplodia theobromae* (Ali and Saxena, 1978), *Coriolus versicolour* (Mel'Ny Chuck *et al.*, 1978), *Alternaria* species (Hasija *et al.*, 1979), *Rhizoctonia bataticola* (Ramasami and Shanmugam, 1976), several fruit rot fungi (Laxminarayana and Reddy, 1978), and some species of *Fusarium*, specifically *Fusarium oxysporum* (Roeb *et al.*, 1977; Olutiola, 1978).

Many cellulolytic organisms can solubilize more than 80% of cotton fibers, although culture filtrates from the same organisms often degrade only a small fraction of such substrates. These observations led Reese and co-workers to postulate the existence of a nonhydrolytic enzyme C_1, responsible for the initial attack on native cellulose, in addition to a hydrolase "C_x" that completes the degradation. Mandels and Reese (1964) further elaborated the concept of a "cellulase complex." Separation of a component with classic C_1 characteristics was first clearly demonstrated by Selby and Maitland (1967) from culture filtrates of *Trichoderma viridae.* Among two other fractions isolated, one (C_x) showed high activity toward CMC and low activity toward cellobiose, while the third component had high activity toward cellobiose and low activity toward CMC. All three components were required to degrade cotton-fiber cellulose. Similar C_1 activity was shown in *Trichoderma koningii* and *Fusarium solani* (Wood, 1968, 1969) and culture filtrates of *T. viridae* (Berghem and Petterson, 1974). Eriksson (1969) considers that the C_1 enzyme from *Chrysosporium lignorum* acts endwise, splitting off cellobiose units. Evidence for the existence of exocellulase (specifically attacking at one end to remove cellobiose or glucose units) has been provided in several fungi (see Reese, 1977). Most of the cellulases reported are of endotype, and only a few microorganisms (King, 1963) and higher plants (Sobotka and Stelzig, 1974) have been shown to possess these enzymes. Cooper *et al.* (1978) studied the macerating activity of several purified enzymes on isolated cell walls and on intact tissues of tomato cultivars. They observed that the treatment of walls with pectic enzymes and hemicellulases did not increase the final rate of degradation by cellulase; the rate was higher at first, suggesting that initial penetration of matrix by the enzyme is a limiting factor.

From the recent studies reviewed at a symposium of the American Chemical Society (R. D. Brown and Jurasek, 1979), it is clear that synergism exists between components of fungal cellulase complexes, but the nature and sequence of reaction remained a matter of debate. Wood and McCrae (1979) provided evidence for primary attack by a C_x component followed by C_1, which is considered to be a cellobiohydrolase (exocellulase). Eriksson *et al.*, (1976) have shown that the cellulase complex of certain fungi operates only aerobically, and they have identified a component that catalyzes cellulose oxidation as an intermediate step in breakdown. It is probable that there exist different strategies in various enzyme complexes for degrading natural products like cellulose, which provide the best adaptive mode for growth of an organism during pathogenesis.

3.4.2. *Symbiosis*

One of the most important symbiotic relationships is established by bacteria (*Rhizobium* species) with leguminous plants where they are involved with nitrogen fixation. During the infection process, the bacterium is housed in an invagination tube that has continuity with the host cell wall. The extension and deposition of the new cell-wall material is apparent at the site of infection thread initiation thus suggesting the involvement of hydrolyzing enzymes in this process. *Rhizobium* does not appear to completely destroy the wall of the root hair at the point of attachment. Instead, the cell wall of the host is invaginated due to plasticity at that point. This invagination may not be due to cellulase generated by *Rhizobium*, since the free-living form does not produce significant amounts of this enzyme (Mosse, 1964). Recently, however, Martinez-Molina *et al.* (1979) have observed some cellulase activity in *Rhizobium* grown in culture. It has been proposed that a polygalacturonase might be required for invagination (Ljunggren, 1969). The involvement of cell-wall-hydrolyzing enzymes in the process of infection remained controversial (see Dart, 1974, 1977; Verma, 1982) because infection is a localized event and the enzyme may not be detectable when total enzyme activity in the whole root is measured (Hunter and Elkan, 1975).

Verma *et al.* (1978a, b) localized the cell-wall-hydrolyzing enzymes in the process of infection. They showed that the activities of both cellulase and pectinase increase during infection and that these activities were compartmentalized (Fig. 3). Cellulase appears to be generated by the host rather than the bacterium. In contrast, pectinase activity is higher in *Rhizobium.* The latter occurs in small amounts in free-living rhizobia (Hubbell *et al.*, 1978; Verma *et al.*, 1978a) and may be sufficient for causing pliability in the cell wall at the point of infection. Callaham (1979) has suggested on the basis of ultrastructural observations that *Rhizobium* enzymes are responsible for the degradative penetration of the host cell wall. Eventual dissolution of the infection-thread

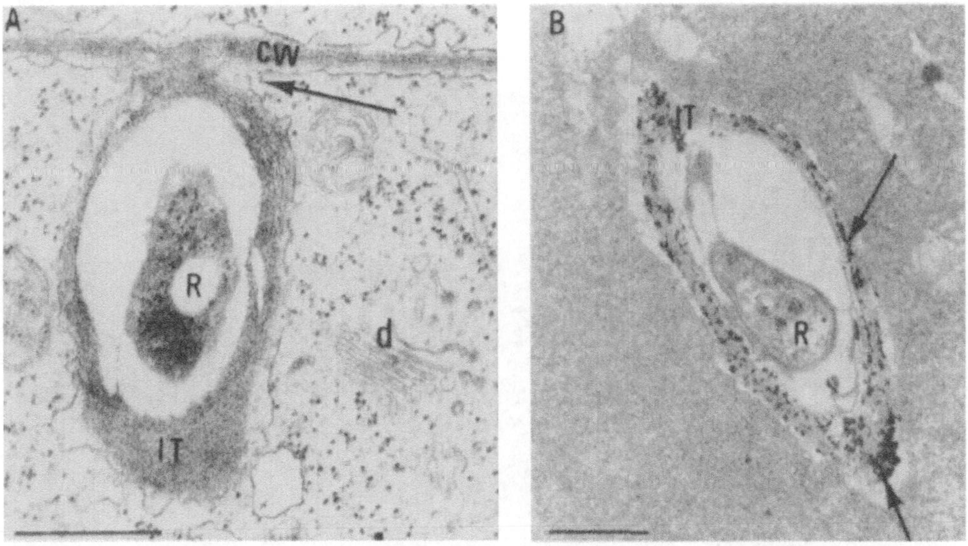

FIGURE 3. Ultrastructure of an infection thread (A) and localization of cellulolytic activity (B) in soybean root nodules. The arrow in (A) indicates the continuity of the host plasma membrane and cell wall and in (B) the reaction product of cellulase. (R) *Rhizobium*; (CW) cell wall; (d) dictyosomes; (IT) infection thread. From Verma *et al.* (1978b).

wall, however, takes place with the help of cellulase, and this activity has been localized in the cell-wall matrix of the infection thread (Fig. 4). Since the production of this enzyme in plants is under the control of phytohormones (Fan and Maclachlan, 1967a, b; Verma *et al.*, 1975) and *Rhizobium* can produce these hormones, a cooperative action of both partners is contemplated in infection and development of root nodules (see Verma, 1982). Hubbell *et al.* (1978) have suggested a general role for cell-wall-hydrolyzing enzymes in the infection process. In soybeans, most of the infection thread is dissolved by cellulase(s), and any residual matrix that remains is sequestered into membrane vesicles through endocytosis (Bassett *et al.*, 1977). Although most of the infection-thread wall

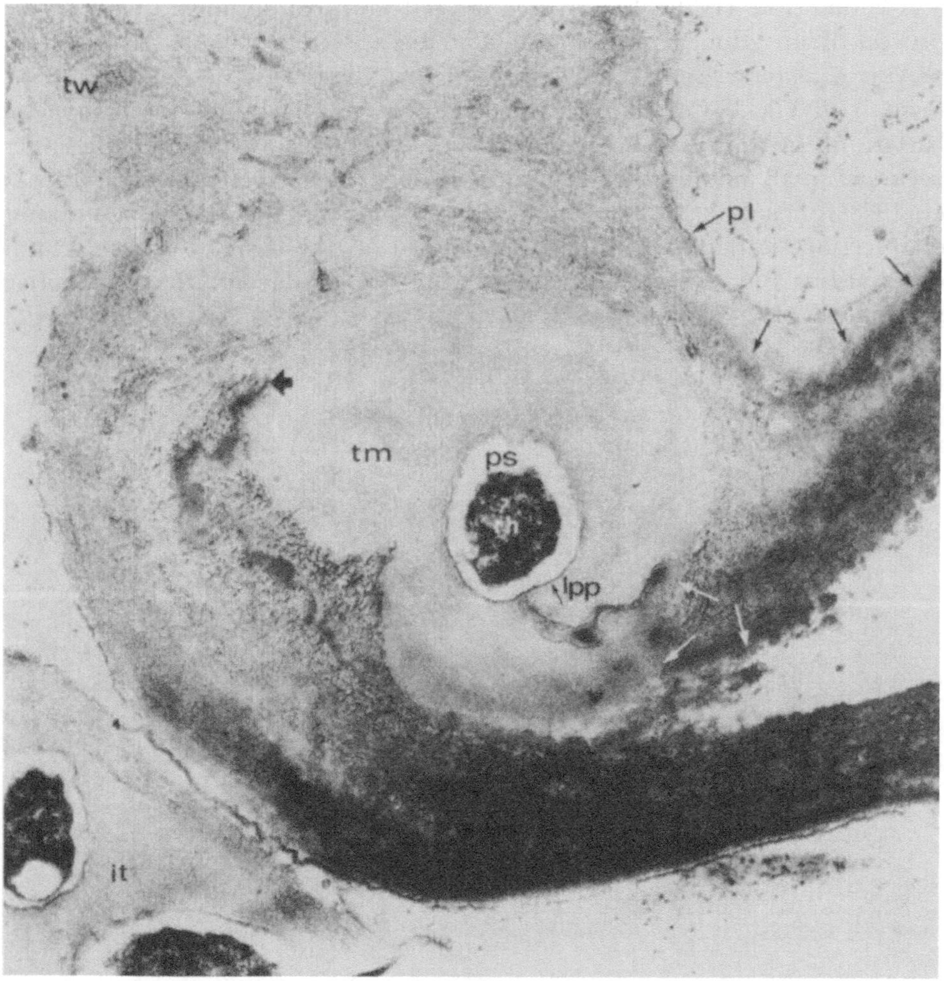

FIGURE 4. Electron micrograph of the infection site in *Trifolium* showing wall loosening at the site of contact with *Rhizobium*. Note the loosening of the cell-wall microfibrils at the site of infection. (tm) Thread matrix; (tw) thread wall; (pl) plasmalemma; (ps) periplasmic space; (lpp) lipopolysaccharide; (rh) *Rhizobium*; (it) infection thread; (tw) thread wall. From Callaham (1979).

material disintegrates, vacuoles containing fibrillar material have often been observed (Bassett *et al.*, 1977; Bal *et al.*, 1980). Cellulase activity in cultured soybean cells (Ozawa and Yamaguchi, 1980) in association with *Rhizobium* has been shown to increase in the first hour of contact.

4. Regulation of β-Glucanases

4.1. Substrate Induction

In lower organisms, cellulase is generally considered as an "adaptive" or "inducible" enzyme, and cellulose was thought to be the "inducer." Consideration of how an insoluble substrate like cellulose could act as an inducer across the fungal membrane led Reese and Mandels (see Reese, 1963) to the idea that cellulase enzymes are produced in small amounts even in the absence of inducer. This low enzyme level is sufficient to cause some initial hydrolysis of the extracellular insoluble substrates, releasing soluble products that may then enter the cell and act as an inducer. According to this hypothesis, cellobiose or soluble cellodextrins and not cellulose may be considered as inducer for this enzyme. It was observed by Reese and Mandels (1963) that cellobiose was indeed an effective inducer of cellulase, provided very low levels of the sugar were used. In some fungi, soluble derivatives of cellulose such as CMC and hydroxymethyl cellulose also can elicit cellulase production, perhaps by the same mechanism. Apart from cellulose derivatives, certain other disaccharides (e.g., sophorose, lactose, which are not β1,4-linked) can act as inducers, e.g., in *Trichoderma viridae* (Nisizawa *et al.*, 1971). Cellobiose and glucose did not induce cellulase in these species. Glucose appears to be a repressor in some plant pathogens, e.g., *in Pyranochaeta terrestris* (J. C. Horton and Keen, 1966), probably through catabolite repression. In *Myrothecium verrucaria* and some wood-rotting fungi, it has been observed that glucose supplied as a carbon source in the medium resulted in the production of a limited amount of cellulase but at about the time that glucose was exhausted, the production increased manyfold (Hulme and Stranks, 1970). Similar results were obtained by replacing glucose by glycerol or glycine. These observations led to the conclusion that neither a sugar nor cellulose is required *per se* for production of cellulase but, rather, that the enzyme is generated as a general consequence of limited metabolism.

This hypothesis also explains to some extent the observation that substances like sophorose induce cellulase in *T. viridae*, where there is an extremely slow catabolism of this dimer (Hulme and Stranks, 1970). Cellobiose likewise is a more effective inducer of cellulase when its consumption is lowered, e.g., by derivatization, lowering the temperature, or adding toxic materials (Reese and Mandels, 1963) in the presence of surfactants or in stationary cultures (Norkrans, 1963a; Johanson, 1966).

Cellulase is produced constitutively by a number of cellulolytic organisms (see Reese, 1977) including several pathogenic forms, e.g., *Diplodia zeae* and

Fomes marginatus (see Norkrans, 1963b), even when high levels of glucose are present as the sole source of carbon. It has been suggested that glucose in *D. zeae* controls the release and not the synthesis of cellulase (BeMiller *et al.*, 1969). In some cases, there is evidence of other polysaccharidases being generated along with cellulase, e.g., xylanase in *M. verrucaria* (Hulme and Stranks, 1970); mannanase in *Chrysosporium lignorium* (Eriksson and Rzedowski, 1969), mannanase and xylanase in *Polyporus adustrus* (Eriksson and Goodell, 1974), and exopolygalacturonase, endo-β-1,4-galactanase, two arabinanases, and endo-1,4-xylanase in *Fusarium roseum* (Mullen and Bateman, 1975). Cellulase production is accompanied by polygalacturonase, L-galactosidase, β-glucosidase, and β-xylosidase in *Pyrenochaeta lycopersici* (Goodenough and Kempton, 1976). An interesting observation in these studies is the appearance of enzymes in a definite order. In most cases, cellulase appears after the activities of other enzymes have diminished (Jones *et al.*, 1972; Mullen and Bateman, 1975).

In *Achlya*, where cellulase activity is reported to be elicited by antheridiol hormone (Mullins and Ellis, 1974; Thomas and Mullins, 1967), the phenomenon can be mimicked by amino acids (Mullins and Ellis, 1974) or even by shake cultures. Its secretion has been shown to be selectively inhibited by cytochalasin β, a drug that binds to microfilaments (Thomas *et al.*, 1974).

4.2. Role of Phytohormones

In higher plants, the regulation of cellulase activity appears to be dependent on multiple factors. It has not so far been established what controls the observed appearance and disappearance of this enzyme activity in fruits (Sobotka and Stelzig, 1974; Pharr and Dickinson, 1973; Hasagawa and Smolensky, 1971; Yamaki and Kakiuchi, 1979) and floral parts, but in other tissues, there is evidence that the activity is controlled by phytohormones such as auxin and ethylene. It has also been claimed that ethylene controls excretion of cellulase in abscission zones, as well as in nonabscising zones (see Huberman and Goren, 1979). Growth regulators (ethylene and 2,4-D) affect the activity of almost all isoenzymes of cellulase irrespective of their location. Gibberellins on the other hand, do not appear to have any effect on cellulase level in these tissues (E. Davies and Maclachlan, 1968; Bonn, 1970).

A direct correlation has been shown to exist between wall-bound β1,3-glucanase activity and IAA treatment in segments of *Avena* coleoptiles. The response is inhibited by cycloheximide (Masuda and Yamamoto, 1970). Similar hormonally evoked activities have also been observed when IAA is applied to etiolated barley coleoptiles (Tanimoto and Masuda, 1968). The β1,3-glucanase activity in pea epicotyls, which is augmented by auxin treatment (Fan and Maclachlan, 1966; Datko and Maclachlan, 1968), is due to the response of only one of two forms of the enzyme (Wong and Maclachlan, 1980). There may also be changes in the intracellular distribution of such enzymes (E. Davies and Maclachlan, 1969). Masuda and colleagues (Masuda, 1968; Masuda *et al.*, 1970) reported that exogenous addition of fungal β1,3-glucanase provoked the elongation of barley-coleoptile segments. This observation strengthens the notion that such enzymes have a physiological role in initiating cell expansion. However, Nevins *et al.* (1977) did not observe such an elongation response when *Avena* sections were treated with highly purified β1,3-glucanase. It has also been observed that enzyme inhibitors like nojirimycin suppresses IAA-induced growth of excised *Avena* coleoptile and pea epicotyl segments (Nevins, 1975a). What other enzyme derivatives this may inhibit has not been documented to the extent required to attribute its effects solely to β-glucosidase action.

During seed germination, β1,3-glucanases have potential functions in degradation of wall barriers in seed reserves. In these instances, gibberellic acid (GA) appears to be a regulatory hormone. GA induces *de novo* synthesis and secretion of amylase in isolated barley aleurone layers (Jacobsen and Varner, 1967) and it also enhances the release of β1,3-glucanase from aleurone layers. However, this hormone does not have any significant effect on β-glucanase synthesis (Taiz and Jones, 1970). Other authors (Manners *et al.*, 1971) have confirmed that there is only a slight increase in the total activity of type A β1,3-glucanase in extracts when whole barley grains are germinated in the presence of GA. However, there is evidence that the hormone regulates its intracellular distribution. When the time-course of β1,3-glucanase release from aleurone layers is compared to that of amylase synthesis and release, the former is reported to be totally secreted by a time only 20% of the latter is released (Taiz and Jones, 1970). GA has also been shown to stimulate the activity of enzymes capable of hydrolyzing 1,3-/1,4-mixed linkage β-glucans (i.e., Type C enzyme) in whole grain. Such enzymatic activity appears in barley-grain before the rise in amylase and protease activity, and therefore it may well be important in the physiology of seed germination.

Abeles and Forrence (1970) presented some evidence that β1,3-glucanases in excised and aging petiolar tissue of *Phaseolus vulgaris* are subject to temporal and hormonal regulation. The hydrolytic activity in aging tissues increases 6 hr after excision and can be reversed by application of IAA, GA, and cytokinins individually. On the other hand, ethylene, an activator of aging and abscission, and a wound-elicited product of metabolism, enhances β1,3-glucanase activity. Cycloheximide prevents this enhancement (Abeles, 1969). In tobacco (Moore and Stone, 1972c, d), the activity of endo-β1,3-glucanase increases markedly during the course of senescence and is unaffected by removal of inhibitor or presence of activator. In senescing tobacco lead disks, where a comparatively high level of β1,3-glucanase activity is present, IAA or GA did not change the normal pattern of loss of proteins of chlorophyll in relation to the level of hydrolase. However, cytokinin delays the loss of protein and chlorophyll and also delays and reduces the rise in hydrolase activity (Moore and Stone, 1972c). On the other hand, abscissic acid, which increases the rate of loss of protein and chlorophyll, also causes a decrease in the rate and extent of the rise in hydrolases.

There is sufficient evidence to suggest that the levels and distribution of the β-glucanases are generally subject to regulation by growth regulators and physiological conditions. Since substrates for these enzymes in higher plants are generally extracellular and localized in specific tissues, it is not necessarily the amounts of enzyme that are critical, but the regulation of their subcellular distribution that controls their activity *in vivo*. Combined studies are needed that use techniques of both cell biology and biochemistry in order to elucidate the complex interactions between regulatory elements and responses of the plant that are mediated by β-glucanases.

References

Abeles, F. B., 1969, Abscission: Role of cellulases, *Plant Physiol.* **44:**447.

Abeles, F. B., and Forrence, L. E., 1970, Temporal and hormonal control of β-3-glucanase in *Phaseolus vulgaris, Plant Physiol.* **45:**395.

Abeles, F. B., and Leather, G. R., 1971, Abscission: Control of cellulase secretion by ethylene, *Planta* **97:**87.

Abeles, F. B., Bosshart, R. P., Forrence, L. E., and Habig, W. H., 1970, Preparation and purification of glucanase and chitinase from bean leaves, *Plant Physiol.* **47:**129.

Abeles, F. B., Leather, G. R., Forrence, L. E., and Cracker, L. E., 1971, Abscission: Regulation of senescence, protein synthesis and enzyme secretion by ethylene, *Hortic. Sci.* **6:**371.

Albersheim, p., 1976, The primary cell wall, in: *Plant Biochemistry* (J. Bonner and J. E. Varner, eds.), pp. 226–274, Academic Press, New York.

Albersheim, P., and Anderson-Prouty, A. J., 1975, Carbohydrates, proteins, cell surfaces and the biochemistry of pathogenesis, *Annu. Rev. Plant Physiol.* **26:**31.

Ali, S. S., and Saxena, S. B., 1978, Production of cellulase by *Botryodiplodia theobromae, Acta. Bot. Indica* **6**(Suppl):132.

Baker, D. B., and Ray, P. M., 1965a, Direct and indirect effects of auxin on cell-wall synthesis in oat coleoptile tissue, *Plant Physiol.* **40:**345.

Baker, D. B., and Ray, P. M., 1965b, The effect of auxin on synthesis of oat coleoptile cell-wall constituents, *Plant Physiol.* **40:**353.

Bal, A. K., 1972, Localization of cellulase in plant cells, Proceedings of the 4th International Congress on Histochemistry and Cytochemistry, (T. Takevchi, K. Ogawa and S. Fujita, eds.), pp. 301–302.

Bal, A. K., 1973, Cellulase in: *Electron Microscopy of Enzymes*, Vol. 3 (M. A. Hyat, ed.), pp. 68–76, Van Nostrand, New York.

Bal, A. K., and Payne, J. F., 1972, Endoplasmic reticulum and cell wall breakdown in quiescent root meristem of *Allium cepa* L., *Z. Pflanzenphysiol.* **66S:**265.

Bal, A. K., Verma, D. P. S., Byrne, H., and Maclachlan, G. A., 1976, Subcellular localization of cellulases in auxin-treated pea, *J. Cell Biol.* **69:**97.

Bal, A. K., Shantharam, S., and Verma, D. P. S., 1980, Changes in the outer cell wall of *Rhizobium* during development of root nodule symbiosis in soybean, *Can. J. Microbiol.* **26:**1096.

Ballance, G. M., and Manners, D. J., 1978, Partial purification and properties of an endo-1,3-β-D-glucanase from germinated rye, *Phytochemistry* **17:**1539.

Barmore, C. R., and Rouse, A. H., 1976, Pectin esterase activity in controlled atmosphere stored avocado, *J. Am. Soc. Hortic. Sci.* **101:**294.

Barras, D. R., and Stone, B. A., 1969a, β-1,3-Glucan hydrolase from *Euglena gracilis*. II. Purification and properties of β-1,3-glucan exo-hydrolase, *Biochim. Biophys. Acta* **191:**329.

Barras, D. R., and Stone, B. A., 1969b, Carbohydrate composition and metabolism in *Euglena*, in: *The Biology of Euglena*, Vol. II (D. E. Buetow, ed.), pp. 149–191, Academic Press, New York and London.

Bartnicki-García, S., 1973, Fundamental aspects of hyphal differentiation, in: *Microbial Differentiation* (J. M. Ashworth and J. E. Smith, eds.), pp. 245–267, Cambridge University Press, Oxford.

Bassett, B., Goodman, R. N., and Novacky, A., 1977, Ultrastructure of soybean nodules. I. Release of *Rhizobia* from the infection thread, *Can. J. Microbiol.* **23**:573.

Bateman, D. F., and Bashman, H. G., 1976, Degradation of plant cell walls and membranes by microbial enzymes, in: *Physiological Plant Pathology, Encyclopedia of Plant Physiology New Series*, Vol. 4 (R. Heitfuss and P. H. Williams, eds.), pp. 316–355, Springer-Verlag, Berlin.

BeMiller, J. N., Diane, O., Tegtmeier, D. O., and Pappelis, A. J., 1969, Constitutive cellulolytic enzymes of *Diplodia zeae*, in: *Cellulases and Their Applications* (R. F. Gould, ed.), pp. 188–196, *Advances in Chemistry Series*, No. 95, American Chemical Society, Washington, D.C.

Ben-Arie, R., Kislev, N., and Frenkel, C., 1979, Ultra structural changes in the cell-walls of ripening apple and pear fruit, *Plant Physiol.* **64**:197.

Berghem, L. E. R., and Petterson, L. G., 1974, The mechanism of enzymatic cellulose degradation: Isolation and some properties of a β-glucosidase from *Trichoderma viridae, Eur. J. Biochem.* **46**:295.

Bonn, B. D., 1970, Hormonal specificity for regulation of cellulase activity and growth in pea epicotyls, M.Sc. thesis., Biology Department, McGill University, Montreal, Canada.

Brown, R. D., and Jurasek, L. (eds.), 1979, *Hydrolysis of Cellulose: Mechanism of Enzymatic and Acid Catalysis, Advances in Chemistry Series*, No. 181, American Chemical Society, Washington, D.C.

Brown, R. M., Jr., and Montezinos, D., 1976, Cellulose microfibrils: Visualization of biosynthetic and orienting complexes in association with the plasma-membrane, *Proc. Natl. Acad. Sci. U.S.A.* **73**:143.

Brown, R. M., and Willison, J. H. M., 1977, Golgi apparatus and plasma membrane involvement in secretion and cell surface deposition, with special emphasis on cellulose, in: *International Cell Biology 1976–77* (B. R. Brinkley and K. R. Porter, eds.), pp. 267–283, The Rockefeller University Press, New York.

Brown, W. V., and Johnson, C., Sr., 1962, The fine structure of the grass guard cells, *Am. J. Bot.* **49**:110.

Byrne, H., 1974, Auxin-regulated cellulases from *Pisum sativum*: Purification, characterization and development, Ph.D. thesis, McGill University, Montreal, Canada.

Byrne, H., Christou, N. V., Verma, D. P. S., and Maclachlan, G. A., 1975, Purification and characterization of two cellulases from auxin treated pea epicotyls, *J. Biol. Chem.* **250**:1012.

Callaham, D. A., 1979, A structural basis for infection of root hairs of *Trifolium ripens* by *Rhizobium trifolii*, M.Sc. thesis, University of Massachusetts, Amherst.

Ciechanover, A., Heller, H., Elias, S., Haas, A. L., and Hershko, A., 1980. ATP dependent conjugation of reticulocyte protein with the polypeptide required for protein degradation, *Proc. Natl. Acad. Sci. U.S.A.* **77**:1365.

Clarke, A. E., and Stone, B. A., 1962, β-1,3-Glucanase hydrolysis from the grape wine (*Vitis vinifera*) and other plants, *Phytochemistry* **1**:175.

Cleland, R., 1971, Cell-wall extension, *Annu. Rev. Plant Physiol.* **22**:197.

Cooper, R. M., Rankin, B., and Wood, R. K. S., 1978, Cell-wall-degrading enzymes of vascular wilt fungi. II. Properties and modes of action of polysaccharidases of *Verticillium albo-atrum* and *Fusarium oxysporum* f.sp. *lycopersici, Physiol. Plant Pathol.* **13**:101.

Cresti, M., and Van Went, J. L., 1976, Callose deposition and plug formation in *petunia* pollen tubes *in situ, Planta* **133**:35.

Currier, H. P., 1957, Callose substances in plant cells, *Am. J. Bot.* **44**:678.

Danylyak, M. I., Mel'NyChuk, H. H., and Baben Ko Ye, I., 1977, Comparative study of C-2 cellulase and C-x endo glucanase activites in representatives of white and brown rot, *Ukr. Bot. Zh.* **34**:348.

Dart, P. J., 1974, The infection process, in: *The Biology of Nitrogen Fixation* (A. Quispel, ed.), pp. 381–429, North-Holland, Amsterdam.

Dart, P. J., 1977, Infection and development of leguminous nodules, in: *A Treatise on Dinitrogen Fixation*, Section III, *Biology* (R. W. F. Hardy and W. S. Silver, eds.), pp. 367–472, John Wiley, New York.

Datko, A. H., and Maclachlan, G. A., 1968, Indoleacetic acid and the synthesis of glucanases and pectic enzymes, *Plant Physiol.* **43**:735.

Datko, A. H., and Maclachlan, G. A., 1970, Patterns of development of glycosidase activity in pea epicotyls, *Can. J. Bot.* **48:**1165.

Davies, E., and Maclachlan, G. A., 1968, Effects of indoleacetic acid on intra cellular distribution of β-glucanase activities in the pea epicotyl, *Arch. Biochem. Biophys.* **128:**595.

Davies, E., and Maclachlan, G. A., 1969, Generation of cellulase activity during protein synthesis by pea microsomes *in vitro*, *Arch. Biochem. Biophys.* **129:**581.

Davies, P. J., 1973, Current theories on mode of action of auxin, *Bot. Rev.* **39:**139.

Deese, D. C., and Stahmann, M. A., 1962, Pectic enzymes and cellulase formation by *Fusarium oxysporum* f. sp. *Cubense* on stem tissues from resistant and susceptible banana plants, *Phytopathology* **52:**247.

Dickinson, D. B., and McCollum, D. P., 1964, Cellulase in tomato fruits, *Nature (London)* **203:**525.

Dilley, D. R., 1970, Enzymes, in: *The Biochemistry of Fruits and Their Products*, Vol. I (A. C. Hulme, ed.), p. 184, Academic Press, New York.

Eriksson, K. E., 1969, New methods for the investigation of cellulases, in: *Cellulases and Their Applications* (R. F. Gould, ed.), pp. 83–104, American Chemical Society, Washington, D.C.

Eriksson, K. E., and Goodell, E. W., 1974, Pleotropic mutants of the wood rotting fungus *Polyporus adustus* lacking cellulase, mannase and xylanase, *Can. J. Microbiol.* **20:**371.

Eriksson, K. E., and Razedowski, W., 1969, Extra-cellular enzyme system utilized by the fungus *Chrysosporium lignorum* for the breakdown of cellulose. I. Studies on enzyme production, *Arch. Biochem. Biophys.* **129:**683.

Eriksson, K. E., Petterson, B., and Westermark, U., 1976, Oxidation: An important enzyme reaction in fungal degradation of cellulose, *FEBS Lett.* **49:**282.

Esau, K., 1967, *Plant Anatomy*, 2nd ed., John Wiley, New York.

Evans, M. L., 1974, Rapid responses to plant hormones, *Annu. Rev. Plant Physiol.* **25:**195.

Fan, D. F., and Maclachlan, G. A., 1966, Control of cellulase activity by indoleacetic acid, *Can. J. Bot.* **44:**1025.

Fan, D. F., and Maclachlan, G. A., 1967a, Massive synthesis of ribonucleic acid and cellulase in pea epicotyl in response to indoleacetic acid with and without current cell division, *Plant Physiol.* **42:**1114.

Fan, D. F., and Maclachlan, G. A., 1967b, Studies on regulation of cellulase activity and growth in excised pea epicotyl sections, *Can. J. Bot.* **45:**1837.

Farkas, V., Biely, P., and Baue, S., 1973, Extra cellular β-glucanases of the yeast *Saccharomyces cerevisae*, *Biochim. Biophys. Acta* **321:**246.

Fulcher, R. G., McCully, M. E., Satterfield, G., and Southerland, J., 1976, β-1,3-Glucans may be associated with cell plate formation during cytokinesis, *Can. J. Biol.* **54:**539.

Fulcher, R. G., Satterfield, G., McCully, M. E., and Wood, P. J., 1977, Observations on the aleuron layer. II. Fluorescence microscopy of the aleurone–subaleurone junction with emphasis on possible β-1,3-glucan deposits in barley, *Aust. J. Plant Physiol.* **4:**917.

Goodenough, P. W., and Kempton, R. J., 1976, The activity of cell wall degrading enzymes in tomato roots infected with *Pyrenochaeta lycopersici* and effect of sugar concentrations in these roots on disease development, *Physiol. Plant Pathol.* **9:**313.

Goren, R., and Huberman, M., 1976, Effects of ethylene and 2,4-D on the activity of cellulase isoenzymes in abscission zones of the developing orange fruit, *Physiol. Plant.* **37:**123.

Goren, R., Teitelbaum, G., and Ratner, A., 1973, The role of cellulase in the abscission of citrus leaves and fruits in relation to exogenous treatments with growth regulators, *Acta Hortic.* **34:**359.

Greenburg, J., Goren, R., and Riov, J., 1975, The role of cellulase and polygalacturonase in abscission of young and mature shamouti orange fruits, *Physiol. Plant.* **34:**1.

Hall, C. B., 1964, Cellulase activity in tomato fruits according to portion and maturity, *Bot. Gaz.* **125:**156.

Hänish-Ten-Cate, C. H., Van Netter, J., Dortland, J. F., and Bruinsma, J., 1975, Cell wall solubilization in pedicel abscission of begonia hybrid flower buds, *Physiol. Plant.* **33:**276.

Hasegawa, S., and Smolensky, D. C., 1971, Cellulase in dates and its role in fruit softening, *J. Food Sci.* **36:**966.

Hasija, S. K., Gulyas, F., and Szegi, J., 1979, Cellulose decomposition by phytopathogenic *Alternaria* spp., *Acta Phytopathol. Acad. Sci. Hung.* **14:**13.

Heyn, A. N. J., 1969, Glucanase activity in coleoptiles of *Avena*, *Arch. Biochem. Biophys.* **132:**442.

Hobson, E. G., 1968, Cellulase activity during maturation and ripening of tomato fruit, *J. Food Sci.* **33:**588.

Horton, J. C., and Keen, N. T., 1966, Regulation of induced cellulase synthesis in *Pyrenochaeta terrestris* Gorez *et al.* by utilizable carbon compounds, *Can. J. Microbiol.* **12:**209.

Horton, R. F., and Osborne, D. J., 1967, Senescence, abscission and cellulase in *Phaseolus vulgaris*, *Nature (London)* **214:**1086.

Hubbell, D. M., Norales, V. M., and Umali-García, M., 1978, Pectolytic enzymes in *Rhizobium*, *Appl. Env. Microbiol.* **35:**210.

Huberman, M., and Goren, R., 1979, Exo- and endocellular cellulase and polygalacturonase in abscission zones of developing orange fruits, *Physiol. Plant* **45:**189.

Humberman, M., Goren, R., and Birk, Y., 1975, The effects of pH ionic strength and ethylene on the extraction of cellulase from abscission zones of citrus leaf explants, *Plant Physiol.* **55:**941.

Hulme, M. A., and Stranks, D. W., 1970, Induction and regulation of production of cellulase by fungi, *Nature (London)*, **226:**469.

Hunter, W. T., and Elkan, G. H., 1975, Role of pectic and cellulotytic enzymes in the invasion of the soybean by *Rhizobium japonicum*, *Can. J. Microbiol.* **21:**1254.

Husain, A., and Diamond, A. E., 1960, Role of cellulolytic enzymes in pathogenesis by *Fusarium oxysporum* f. lycopersici, *Phytopathology* **50:**329.

Jacobs, M., and Ray, P. M., 1975, Promotion of xyloglucan metabolism by acid pH, *Plant Physiol.* **56:**373.

Jacobsen, J. V., and Varner, J. E., 1967, Gibberellic acid induced synthesis of protease by isolated aleurone payer of barley, *Plant Physiol.* **42:**1569.

Johanson, M., 1966, A comparison between the cellulolytic activity of white and brown rot fungi, I. The activity of insoluble cellulose, *Physiol. Plant.* **19:**709.

Jones, T. M., Anderson, A. J., and Albersheim, P., 1972, Host pathogen interactions. IV. Studies on the polysaccharide degrading enzymes secreted by *Fusarium oxysporum* f. sp. *Lycopersici*, *Physiol. Plant Pathol.* **2:**153.

Kadota, H., 1959, Cellulose decomposing bacteria in the sea, in: *Symposium on Marine Boring and Fouling Organisms* (D. L. Ray, ed.), pp. 332–340, University of Washington Press, Seattle.

Kato, K., Yamada, A., and Noguchi, M., 1972, Purification and some properties of β-1,3-glucanase of suspension cultured tobacco cells, *Agric. Biol. Chem.* **37:**1269.

Kawase, M., 1979, Role of cellulase in aerenchyma development in sunflower, *Am. J. Bot.* **66:**183.

King, K. W., 1963, Endwise degradation of cellulose, in: *Symposium on Advances in Enzymic Hydrolysis of Cellulose and Related Matter* (E. T. Reese, ed.), pp. 159–170, Pergamon Press, New York.

Kivilaan, A., Bandursky, R. S., and Schulze, A., 1971, A partial characterization of an autolytic solubilized cell-wall glucan, *Plant Physiol.* **48:**389.

Koehler, D. E., Leonard, R. T., Vanderwoude, W. J., Linkins, A. E., and Lewis, L. N., 1976, Association of latent cellulase activity with plasma membranes from kidney bean abscission zones, *Plant Physiol.* **58:**324.

Konar, R. N., and Stanley, R. G., 1969, Wall softening enzymes in the gynoecium and pollen of *Hemarocallis fulva*, *Planta* **84:**304.

Kumarshinghe, R. M. K., and Nutman, P. S., 1977, Rhizobium stimulated callose formation in clover root hairs and its relation to infection, *J. Exp. Bot.* **28:**961.

Labavitch, J. M., and Ray, P. M., 1974, Relationship between promotion of xyloglucan metabolism and induction of elongation by indole-acetic acid, *Plant Physiol.* **54:**499.

Laxminarayana, P., and Reddy, S. M., 1978, Enzymatic studies on certain fruit rot fungi. I. Production of cellulase *in vitro* and *in vivo*, *Proc. Indian Acad. Sci. Sect. B* **87:**135.

Leighton, T. J., and Stock, J. J., 1970, Isolation and preliminary characterization of developmental mutants from *Microsporium gypseum*, *J. Bacteriol.* **104:**834.

Lewis, L. N., and Koehler, D. E., 1979, Cellulase in the kidney bean seedling, *Planta* **146:**1.

Lewis, L. N., and Varner, J. E., 1970, Synthesis of cellulase during abscission of *Phaseolus vulgaris* leaf explants, *Plant Physiol.* **46**:194.

Lewis, L. N., Lew, F. T., Reid, P. O., and Barnes, J. E., 1972, Isoenzymes of cellulase in the abscission zone of *Phaseolus vulgaris*, in: *Plant Growth Substances* (D. J. Carr *et al.*, eds.), pp. 234–239, Springer-Verlag, Berlin.

Lewis, L. N., Linkins, A. E., O'Sullivan, S., and Reid, P. D., 1975, Two forms of cellulase in bean plants, in: *Plant Growth Substances* (S. Kikuchi, ed.), pp. 708–718. Hirokawa, Tokyo.

Linkins, A. E., Lewis, L. N., and Palmer, R. L., 1973, Hormonally induced changes in the stem and petiole anatomy and cellulase enzyme patterns in *Phaseolus vulgaris*, *Plant Physiol.* **52**:554.

Ljunggren, H., 1969, Mechanism and pattern of *Rhizobium* invasion into leguminous root hairs, *Physiol. Plant Suppl.* **V**:1.

Maclachlan, G. A., 1976, A potential role for endo-cellulase in cellulose biosynthesis, *Appl. Polym. Symp.* **28**:645.

Maclachlan, G. A., 1977a, Cellulose metabolism and cell growth, in: *Plant Growth regulation*, 9th International Conference on Plant Growth Substances (P. E. Pilet, ed.), pp. 13–20, Springer-Verlag, Berlin and Heidelberg.

Maclachlan, G. A., 1977b, Cellulose metabolism in growing cells, *Trends Biochem. Sci.* **1977**:226–228.

Maclachlan, G. A., and Perrault, J., 1964, Cellulase from pea epicotyls, *Nature (London)* **204**:81.

Maclachlan, G. A., and Wong, Y. S., 1979, Two pea cellulases display the same catalytic mechanism despite major differences in physical properties, in: *Hydrolysis of Cellulose: Mechanisms of Enzymatic and Acid Catalysis* (R. D. Brown, Jr., and L. Jurasek, eds.), *Advances in Chemistry Series* No. 181, American Chemical Society, Washington, D.C.

Maclachlan, G. A., and Young, M., 1962, Breakdown and synthesis of cell-walls during growth, *Nature (London)* **195**:1319.

MacLeod, A. M., and McCorquodale, H., 1958, Water soluble carbohydrates of seeds of graminae, *New Phytol.* **57**:168.

Mandels, M., and Reese, E. T., 1964, Fungal cellulases and the microbial decomposition of cellulosic fabrics, *Dev. Ind. Microbiol.* **5**:5.

Manners, D. J., and Marshall, J. J., 1969, Studies on carbohydrate metabolising enzymes. XXII. The β-glucanase system of malted barley, *J. Inst. Brew. (London)* **75**:550.

Manners, D. J., and Wilson, G., 1974, Purification and properties of an endo-(1-3)-β-D-glucanese from malted barley, *Carbohydr. Res.* **37**:9.

Manners, D. J., Palmer, G. H., Wilson, G., and Yellowees, D., 1971, Effect of gibberellic acid on the development of some cereal carbohydrases, *Biochem. J.* **125**:30.

Martinez-Molina, E., Morales, V. M., and Hubbell, D. H., 1979, Hydrolytic enzyme production of *Rhizobium*, *Appl. Env. Microbiol.* **38**:9186.

Mascarenhas, J. P., 1975, The biochemistry of angiosperm pollen development, *Bot. Rev.* **41**:260.

Masuda, Y., 1968, Role of cell-wall degrading enzymes in cell-wall loosening in oat coleoptiles, *Planta* **83**:171.

Masuda, Y., and Yamamoto, R., 1970, Effect of auxin on β-1,3-glucanase activity in *Avena* coleoptile, *Dev. Growth Differ.* **11**:287.

Masuda, Y., Tanimoto, E., and Wada, S., 1967, Effect of β-1,3-glucanase on the elongation growth of oat coleoptile, *Bot. Mag. Tokyo* **80**:100.

Masuda, Y., Oi, S., and Satamura, Y., 1970, Further studies on the role of cell-wall degrading enzymes in cell-wall loosening of oat coleoptiles, *Plant Cell Physiol.* **11**:631.

Matchett, W. M, and Nance, J. F., 1962, Cell wall breakdown and growth in pea seedling systems, *Am. J. Bot.* **49**:311.

Matile, P., Cortat, M., Wiemken, A., and Frey-Wyssling, A., 1971, Isolation of glucanase containing particles from budding *Saccharomyces cerevisiae*, *Proc. Natl. Acad. Sci. U.S.A.* **68**:636.

Meeuse, B. J. D., 1962, Storage products, in: *Physiology and Biochemistry of Algae* (R. A. Levin, ed.), pp. 289–313, Academic Press, London.

Mel'Ny Chuck, H. H., Kolesnyeva, H. V., and Danylyak, M. I., 1978, Activity dynamics of cellulase β-1,4-glucan-4-glucano hydrolases of different strains of *Coriolus versicolor* and *Coniophora cerebella*, *Ukr. Bot. Zh.* **35**:629.

Mollard, A., and Barnoud, F., 1976, Une glucane hemicellulosique β-1,3- dans les parois des cellules de Rosier cultivées *in vitro*, *Physiol. Veg.* **14:**233.

Montezinos, D., and Brown, R. M., Jr., 1976, Surface architecture of the plant cell: Biogenesis of the cell wall with special emphasis on the role of plasma membrane in cellulose biosynthesis, *J. Supramol. Struct.* **5:**277.

Moore, A. E., and Stone, B. A., 1972a, A β-1,3-glucan hydrolase from *Nicotiana glutinosa* leaves, I. Extraction, purification and physical properties, *Biochim. Biophys. Acta* **258:**238.

Moore, A. E., and Stone, B. A., 1972b, A β-1,3-glucan hydrolase from *Nicotiana glutinosa* leaves, II. Specificity action pattern and inhibitor studies, *Biochim. Biophys. Acta* **258:**248.

Moore, A. E., and Stone, B. A., 1972c, Effect of senescence and hormone treatment on the activity of β-1,3-glucan hydrolase in *Nicotiana glutinosa* leaves, *Planta* **104:**93.

Moore, A. E., and Stone, B. A., 1972d, Effect of infection with TMV and other viruses on the level of β-1,3-glucan hydrolase in leaves of *Nicotiana glutinosa*, *Virology* **50:**791.

Morrison, I. N., and O'Brien, T. P., 1976, Cytokinesis in the developing wheat grain: Division with and without phragmoplast, *Planta* **130:**57.

Mosse, B., 1964, Electron microscopic studies of nodule development in some clover species, *J. Gen. Microbiol.* **36:**49.

Mueller, S. C., and Brown, R. M., Jr., 1980, Evidence for an intramembrane component associated with a cellulose microfibril synthesizing complex in higher plants, *J. Cell Biol.* **84:**315.

Mueller, S. C., and Maclachlan, G. A., 1980, Plant membranes only utilize UDP-glucose supplied to the protoplasmic surface, *Plant Physiol. Suppl.* **65:**106.

Mueller, S. C., Brown, R. M., Jr., and Scott, T. K., 1976, Cellulosic microfibrils: Nascent stages of synthesis in a higher plant cell, *Science* **194:**949.

Mullen, J. J., and Bateman, D. F., 1975, Polysaccharide degrading enzymes produced by *Fusarium roseum* "Avenaceum" in culture and during pathogenesis, *Physiol. Plant Pathol.* **6:**233.

Mullins, J. T., and Ellis, E. A., 1974, Sexual morphogenesis in *Achlya*: Ultra structural basis for the hormonal induction of antheridial hyphae, *Proc. Natl. Acad. Sci. U.S.A.* **71:**1347.

Nevins, D. J., 1975a, The effect of nojirimycin on plant growth and its applications concerning role for exo-β-glucanases in auxin-induced cell expansion, *Plant Cell Physiol.* **16:**347.

Nevins, D. J., 1975b, The *in vitro* stimulation of IAA-induced modification of *Avena* cell wall polysaccharides by an exoglucanase, *Plant Cell Physiol.* **16:**495.

Nevins, D. J., Huber, D. J., Yamamoto, R., and Loescher, W. H., 1977, β-D-Glucan of *Avena* coleoptile cell walls, *Plant Physiol.* **60:**617.

Nisizawa, T., Suzuki, H., Nakayama, M., and Nisizawa, K., 1971, *De novo* synthesis of cellulase induced by sophorose in *Trichoderma viridae* cells, *J. Biochem.* **70:**375.

Nissl, D., and Zenk, M. H., 1969, Evidence against induction of protein synthesis during auxin induced initial elongation of *Avena* coleoptiles, *Planta* **89:**32.

Norkrans, B., 1963a, Influence of some culture conditions on fungal cellulase production, *Physiol. Plant.* **16:**11.

Norkrans, B., 1963b, Degradation of cellulose, *Annu. Rev. Phytopathol.* **1:**325.

Norkrans, B., 1967, Cellulose and cellulosis, *Adv. Appl. Microbiol.* **9:**91.

Northcote, D. H., and Wooding, F. B. P., 1966, Development of sieve tubes in *Acer pseudoplatanus*, *Proc. R. Soc. London Ser. B* **163:**524.

Olson, A. C., Bonner, J., and Moore, D. J., 1965, Force extension analysis of *Avena* coleoptile cell elongation, *Planta* **66:**126.

Olutiola, P. O., 1978, Growth sporulation and production of pectic and cellulolytic enzymes in *Fusarium oxysporum*, *Trans. Br. Mycol. Soc.* **70:**109.

Ozawa, T., and Yamaguchi, M., 1980, Increase in cellulase activity in cultured soybean cells caused by *Rhizobium japonicum*, *Plant Cell Physiol.* **21:**331.

Pamiano, E. P., and Juliano, B. O., 1973, Changes in the activity of some hydrolases, peroxidases and cellulases in the rice seed during germination, *Plant Physiol.* **52:**274.

Parrish, F. W., Perlin, A. S., and Reese, E. T., 1960, Selective enzymolysis of poly-β-glucans and the structure of the polymers, *Can. J. Chem.* **38:**2094.

Pavlenko, V. F., and Lizak, Y. U. V., 1978, Study of the cellulolytic characteristics of dark micromycetes, *Mikol. Fitopatol.* **12:**296.

Pegg, G. F., 1976, The response of ethylene treated tomato plants to infection by *Verticillium-alboatrum*, *Physiol. Plant Pathol.* **9:**215.

Pesis, E., Fuchs, Y., and Zauberman, G., 1978, Cellulase activity and fruit softening in avocado, *Plant Physiol.* **61:**416.

Pharr, D. M., and Dickinson, D. B., 1973, Partial characterization of C_x cellulase and cellobiase from ripening tomato fruits, *Plant Physiol.* **51:**577.

Pilnik, W., and Voragen, A. G. J., 1970, Pectic substances and other uronides, in: *The Biochemistry of Fruits and Their Products* (A. C. Hulme, ed.), Academic Press, London.

Preston, R. D., 1961, The growth of the cell wall, *Encycl. Plant. Physiol.* **14:**311.

Preston, R. D., 1974, *The Physical Biology of Plant Cell Walls*, Chapman and Hall, London.

Quatrano, R. S., and Stevens, P. T., 1976, Cell wall assembly in *Fucus* zygotes. I. Characterization of the polysaccharide components, *Plant Physiol.* **50:**224.

Ramasami, R., and Shanmugam, N., 1976, Studies on pectolytic and cellulolytic enzymes of *Rhizoctonia-bataticola in vitro* and *in vivo*, *Indian Phytopathol.* **29:**305.

Rasmussen, G. K., 1973, Changes in cellulase and pectinase activities in fruit tissues and separation zones of citrus treated with cycloheximide, *Plant Physiol.* **51:**626.

Rasmussen, G. K., 1975, Cellulase activity, endogenous abscissic-acid and ethylene in 4 citrus cultivars during maturation, *Plant Physiol.* **55:**765.

Ratner, A., Goren, R., and Monselise, S. P., 1969, Activity of pectin esterase and cellulase in the abscission zone of citrus leaf explants, *Plant Physiol.* **44:**1717.

Ray, P. M., 1962, Cell wall synthesis and cell elongation in oat coleoptile tissue, *Am. J. Bot.* **49:**928.

Rayle, D. L., Haughton, P. M., and Cleland, R., 1970, An *in vitro* system that stimulates plant cell extension growth, *Proc. Natl. Acad. Sci. U.S.A.* **67:**1814.

Reese, E. T. (ed.), 1963, *Advances in Enzymic Hydrolysis of Cellulose and related Materials*, Pergamon Press, New York.

Reese, E. T., 1977, Degradation of polymeric carbohydrates by microbial enzymes, *Recent Adv. Phytochem.* **11:**311.

Reese, E. T., and Mandels, M., 1959, β-D-1,3-Glucanases in fungi, *Can. J. Microbiol.* **5:**173.

Reese, E. T., and Mandels, M., 1963, Enzymatic hydrolysis of β-glucans, in: *Enzymic Hydrolysis of Cellulose and Related Materials* (E. T. Reese, ed.), pp. 197–234, Pergamon Press, New York.

Reese, E. T., and Parrish, F. W., 1971, Nojirimycin and D-glucono-1,5-lactone as inhibitors of carbohydrases, *Carbohydr. Res.* **18:**381.

Reese, E. T., and Perlin, A. S., 1963, Enzymic preparation of 3-*O*-cellobiosyl D-glucose, *Biochem. Biophys. Res. Commun.* **12:**194.

Reid, P. D., Strong, H. G., Lew, F., and Lewis, L. N., 1974, Cellulase and abscission in red kidney bean (*Phaseolus vulgaris*), *Plant Physiol.* **53:**732.

Ridge, I., and Osborne, D. J., 1969, Cell growth and cellulases, regulation by ethylene and indole-3-acetic acid in shoots of *Pisum sativum*, *Nature (London)* **223:**310.

Roeb, L., Stegemann, H., and Langerfield, E., 1977, Variation in the activity of polysaccharide and protein degrading enzymes and potato rotting pathogens of the genus *Fusarium*, *Potato Res.* **20:**23.

Roland, J. C., and Vian, B., 1979, The wall of the growing plant cell: Its three dimensional organization, *Int. Rev. Cytol.* **61:**129.

Ruesink, A. W., 1969, Polysaccharidases and the control of cell-wall elongation, *Planta* **89:**95.

Sassen, M. M. A., 1965, Breakdown of the plant cell wall during the cell fusion process, *Acta Bot. Neer.* **14:**165.

Selby, K., and Maitland, C. C., 1967, The cellulase of *Trichoderma viridae:* Separation of its components involved in the solubilization of cotton, *Biochem. J.* **104:**716.

Sheldrake, A. R., 1970, Cellulase and cell differentiation in *Acer pseudoplatanus*, *Planta* **95:**167.

Smith, M. M., and McCully, M. E., 1977, Mild temperature stress and callose synthesis, *Planta* **136:**65.

Sobotka, F. E., and Stelzig, D. A., 1974, An apparent cellulase complex in tomato (*Lycopersicon esculatum* L.) fruits, *Plant Physiol.* **53:**759.

Sobotka, F. E., and Watada, A. E., 1971, Cellulase in high pigment and crimson tomato fruit, *J. Am. Soc. Hortic.* **96:**705.

Sterling, M., 1954, Sclereid development and the texture of Bartlet pears, *J. Food Res.* **19:**433.

Stieglitz, H., and Stern, H., 1973, Regulation of β-1,3-glucanase activity in developing anthers of lilium, *Dev. Biol.* **34:**169.

Taiz, L., and Jones, R. L., 1970, Gibberellic acid β-1,3-glucanase and the cell-walls of barley aleurone layers, *Planta* **92:**73.

Tanimoto, E., and Masuda, Y., 1968, Effect of auxin on cell-wall degrading enzymes, *Physiol. Plant.* **21:**820.

Thomas, D. des S., and Mullins, J. T., 1967, Role of enzymatic wall softening in plant morphogenesis: Hormonal induction in *Achlya, Science* **156:**84.

Thomas, D. des S., Lutzac, M., and Manavathu, E., 1974, Cytochalasin selectivity inhibits synthesis of a secretory protein cellulase in *Achlya, Nature (London)* **249:**140.

Tracey, M. V., 1950, Cellulase from leaves and roots of tobacco, *Biochem. J.* **47:**431.

Tracey, M. V., 1959, Role of cellulase in nature, in: *Marine Boring and Fouling Organisms* (D. L. Ray, ed.), pp. 253–264, University of Washington Press, Seattle.

Verma, D. P. S., 1982, Plant-*Rhizobium* interactions in symbiotic nitrogen fixation, in: *Molecular Biology of Plant Development* (H. Smith and D. Grierson, eds.), pp. 437–466, Blackwell, Oxford.

Verma, D. P. S., Maclachlan, G. A., Byrne, H., and Evings, D., 1975, Regulation and *in vitro* translation of messenger ribonucleic acid for cellulase from auxin treated pea epicotyls, *J. Biol. Chem.* **250:**1019.

Verma, D. P. S., Zogbi, V., and Bal, A. K., 1978a, A cooperative action of plant and *Rhizobium* in dissolving host cell-wall during development of symbiosis, *Plant. Sci. Lett.* **13:**137.

Verma, D. P. S., Kazazian, V., Zogbi, V., and Bal, A. K., 1978b, Isolation and characterization of the membrane envelope enclosing the bacterioids in soybean root nodules, *J. Cell Biol.* **78:**919.

Villa, T. G., Notario, V., Benitex, T., and Villanueva, J. R., 1976, Purification of an exo-1,3-β-glucanase from *Candida utilis, Can. J. Biochem.* **54:**927.

Wada, S., Tanimoto, E., and Masuda, Y., 1968, Cell elongation and metabolic turnover of the cell-wall as affected by auxin and cell-wall degrading enzymes, *Plant Cell Physiol.* **9:**123.

Webster, B. D., 1968, Anatomical aspects of abscission, *Plant Physiol.* **43:**1512.

Wiemken-Gehrig, V., Wiemken, A., and Matile, P., 1974, Mobilisation von Zellwardstoffen in der welkenden Blute von *Ipomoea tricolor* cav, *Planta* **115:**297.

Willison, J. H. M., and Brown, R. M., Jr., 1977, An examination of the developing cotton fibre: Wall and plasmalemma, *Protoplasma* **92:**21.

Willison, J. H. M., and Brown, R. M., Jr., 1978, Cell wall structure and deposition in *Glaucoystis, J. Cell Biol.* **72:**103.

Willison, J. H. M., and Grout, B. W. W., 1978, Further observations on cell wall formation around isolated protoplasts of tobacco and tomato, *Planta* **140:**57.

Wong, Y. S., 1979, Studies on endo-1,3-β-D-glucanases from *Pisum sativum:* Purification, development and enzymic properties, Ph.D. thesis, McGill University, Montreal, Canada.

Wong, Y. S., and Maclachlan, G. A., 1979, 1,3-β-D-Glucanases from *Pisum sativum* seedlings. II. Substrate specificities and enzymic action patterns, *Biochim. Biophys. Acta* **571:**256.

Wong, Y. S., and Maclachlan, G. A., 1980, 1,3-β-D Glucanases from *Pisum sativum* seedlings. III. Development and distribution of endogenous substrates, *Plant Physiol.* **65:**222.

Wong, Y. S., Fincher, G. B., and Maclachan, G. A., 1977a, Kinetic properties and substrate specificities of two cellulases from auxin treated pea epicotyls. *J. Biol. Chem.* **252:**1402.

Wong, Y. S., Fincher, G. B., and Maclachlan, G. A., 1977b, Cellulase can enhance β-glucan synthesis, *Science* **195:**679.

Wood, T. M., 1968, Cellulolytic enzyme system of *Trichoderma koningii:* Separation of components attacking native cotton, *Biochem. J.* **109:**217.

Wood, T. M., 1969, The cellulase of *Fusarium solani:* Resolution of the enzyme complex, *Biochem. J.* **115:**457.

Wood, T. M., and McCrae, S. I., 1979, Synergism between enzymes involved in the solubilisation of native cellulose, in: *Hydrolysis of Cellulose: Mechanisms of Enzymatic and Acid Catalysis* (R. D. Brown and L. Jurasek, eds.), pp. 181–209, American Chemical Society, Washington, D.C.

Woodward, J. R., Keane, P. J., and Stone, B. A., 1979, β-Glucans and β-glucan hydrolases in plant pathogenesis with special reference to wilt inducing toxins from *Phytophthora* species, in: *Fungal Polysaccharides* (P. A. Standford and K. Masuda, eds.), Advances in Chemistry Series, American Chemical Society, Washington, D.C.

Yamaki, S., and Kakiuchi, N., 1979, Changes in hemicellulose degrading enzymes during development and ripening of Japanese pear fruit, *Plant Cell Physiol.* **20:**301.

23

Visualization of Cellulases and Cellulose Degradation

ALAN R. WHITE

1. Introduction

1.1. Significance of Cellulases

Cellulases have biological and economic significance in their destructive roles in plant pathogenesis (W. Brown, 1965; Albersheim, 1969) and wood decay (Scheffer and Cowling, 1966; Liese, 1970); in their possible beneficial use in converting waste cellulose into glucose, which may then be converted into the valuable chemical ethanol (Wilkie, 1975); and in their role in cell-wall loosening in higher plants to allow cell elongation and growth (Ridge and Osborne, 1969). The cellulase enzyme system of *Trichoderma reesei* QM9414 was chosen for this study for two reasons. First, *Trichoderma* is famous for its ability to degrade highly crystalline cellulose and is one of the most efficient of the cellulolytic fungi (Mandels, 1975). Second, because *Trichoderma* is such a good source of cellulase, its enzyme system has been studied more extensively than that of any other cellulolytic organism. Consequently, the biochemical characterizations and activities of the various *Trichoderma* cellulase enzymes are well known.

In most enzyme–substrate relationships, the substrate is soluble and the enzyme is either soluble or bound to some cell-related structure. The enzyme and substrate must meet by chance in solution, or the substrate must diffuse to the bound enzyme before an enzyme–substrate complex can be formed. The relationship of the cellulase system to its substrate is quite different. Cellulose is an insoluble polymer, and the soluble cellulase enzymes must move to this insoluble surface before any binding and catalytic activity is possible. Most

ALAN R. WHITE • Department of Botany, University of North Carolina at Chapel Hill, Chapel Hill, North Carolina 27514.

assays for cellulase activity measure soluble reducing sugars released during the process of cellulose degradation, but only cellooligosaccharides composed of fewer than seven glucose monomers have significant solubility in aqueous solvents (Gum and Brown, 1977b). Thus, biochemical techniques cannot measure the degradation process from the time the enzyme binds to the insoluble cellulose substrate until soluble reducing sugars are released. Reese (1975b) includes the characterization of insoluble degradation products among the topics of cellulase research that require further careful study. High-resolution electron-microscopic techniques offer a unique opportunity to study the morphological changes in cellulose structure that occur during the intermediate stages of cellulose degradation.

1.2. Biochemical Characterization of the Cellulase Enzyme System

The ability of a variety of organisms to degrade cellulose is a long-noted phenomenon (Ward, 1888, 1898; Seilliere, 1905). Pringsheim (1912) set forth the "classic" theory of cellulase action by suggesting that two enzymes were involved in cellulose hydrolysis. He showed that cellulolytic bacteria produced two products during cellulose degradation, glucose and cellobiose, and postulated that one enzyme (cellulase) cleaved cellobiose from the chains of glucose in cellulose, while a second enzyme (cellobiase) split the cellobiose into two glucose molecules. By 1940, enzymatic cellulose decomposition was seen as a hydrolytic process that required at least two enzymes. It was suspected that one of these enzymes, the classic cellulase, had an affinity for long chains of glucose, while the other, cellobiase, had an affinity for short chains of glucose (Grassmann *et al.*, 1933).

In the early 1950s, Reese and co-workers (Reese *et al.*, 1950; Reese and Levinson, 1952; Reese, 1956) noticed that large numbers of organisms are capable of hydrolyzing the soluble cellulose derivative carboxymethylcellulose, but relatively few organisms are capable of efficiently hydrolyzing native crystalline cellulose. They concluded that truly cellulolytic organisms possess enzymes that the others lack. The enzymes were termed C_1 and C_x. C_1 was postulated to attack crystalline cellulose, either producing short chains or decrystallizing the glucan chains so that the cellulose was then susceptible to attack by the hydrolytic C_x enzyme. Organisms that are capable of hydrolyzing only noncrystalline cellulose derivatives possess only the C_x enzyme, while truly cellulolytic organisms possess both the C_1 and C_x enzymes.

Throughout the 1950s and 1960s, investigators worked to purify and characterize the C_1 and C_x enzymes (Li *et al.*, 1965; Selby and Maitland, 1967; Wood, 1968; Eriksson and Rzedowski, 1969). Wood and McCrae (1972) and Halliwell and Griffin (1973) reported that their purified C_1 enzyme was actually an enzyme known as cellobiohydrolase. Brown and co-workers (Emert *et al.*, 1974) have suggested that the old C_1-C_x terminology be abandoned in favor of more specific names for the enzymes. The Commission on Enzymes of the International Union of Biochemistry has assigned systematic names for the various cellulases, and there is now some agreement that the cellulase enzyme system contains three major enzymes the combined actions of which lead to the efficient degradation of cellulose.

Figure 1 illustrates a currently proposed scheme of cellulase action (Wood and McCrae, 1972; Reese, 1975a; Gritzali and Brown, 1979) in which endo-1,4-β-D-glucanase (EC 3.2.1.4) randomly cleaves internal glucosidic bonds within an unbroken glucan chain. The newly created nonreducing chain ends then become the substrate for 1,4-β-D-glucan cellobiohydrolase (EC 3.2.1.91), which cleaves cellobiose dimers from the glucan chain and releases them into solution. The hydrolysis of cellulose into the glucose end product is completed by β-glucosidase (EC 3.2.1.21), which splits cellobiose into glucose monomers. The creation by endoglucanase (EG) of nonreducing glucan chain ends that are sites of catalytic action for cellobiohydrolase (CBH) leads to a synergism in the overall rate of cellulose degradation. The synergistic relationship between the actions of EG and CBH is crucial for efficient cellulose degradation (Wood and McCrae, 1978).

The cellulases of the imperfect fungus *Trichoderma reesei* QM9414 have been well characterized (Gum and Brown, 1976, 1977a; Shoemaker and Brown, 1978a,b; Gritzali and Brown, 1979). There are at least four endoglucanase enzymes, referred to as endoglucanases I, II, III, and IV. There are two known cellobiohydrolases: cellobiohydrolases I and II. Cellobiohydrolase I has four known forms designated A, B, C, and D. Gritzali and Brown (1979) have published an excellent review of the purification procedures and characteristics of each of the various enzymes and forms.

1.3. High-Resolution Electron Microscopy

To successfully attain the goal of observing details of structures measuring less than 5.0 nm, specialized high-resolution techniques must be used. In normal electron microscopy, films of polymers 40–60 nm thick are used as support for the specimen. When the object of interest measures less than 5 nm, a film of such thickness presents problems. The image must be formed through a supporting structure that is 10 times as thick as the specimen, and resolution is lost. Ultrathin films of carbon that are less than 2.5 nm thick may be used as a specimen-support matrix, eliminating much of the noise arising from thicker polymer support films and permitting better resolution. Thin carbon films may be produced by indirectly evaporating carbon onto the surface of clean, freshly cleaved mica (Whiting and Ottensmeyer, 1972; Ottensmeyer *et al.*, 1975). Direct deposition of evaporated carbon is blocked by a slab of glass positioned directly above the mica. Glass microscope slides placed at an angle on each side of the mica act as baffles to deflect carbon onto the mica surface. Indirect carbon deposition produces very thin and evenly distributed carbon films. These carbon films may then be floated onto a water surface and picked up on electron-microscope grids covered with Parlodion micronets, which give added support

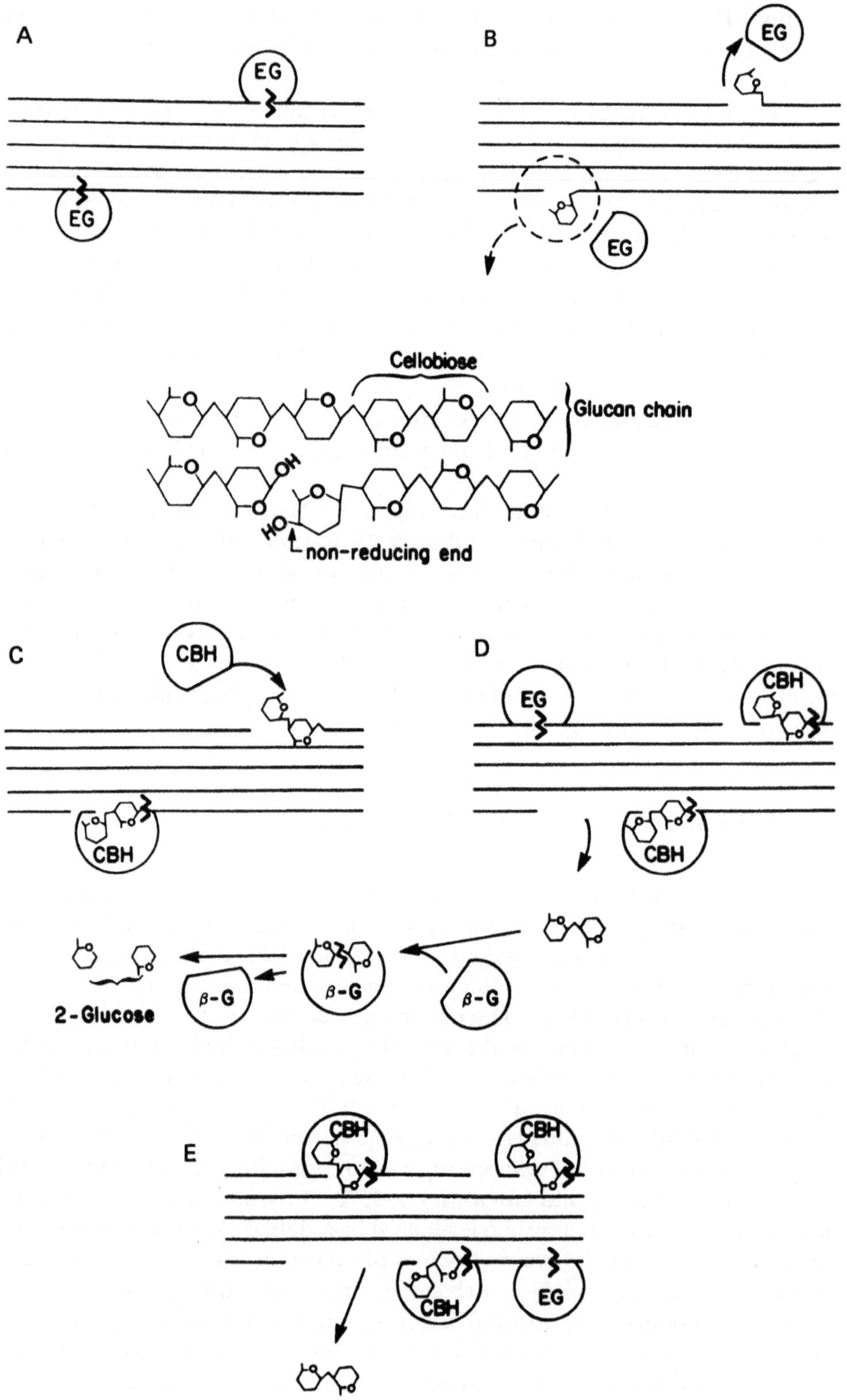

A
EG
EG
B
EG
EG
Cellobiose
Glucan chain
OH
HO
non-reducing end
C
CBH
CBH
D
EG
CBH
CBH
β-G
β-G
β-G
2-Glucose
E
CBH
CBH
CBH
EG

to the fragile films. Micronets are easily and reproducibly manufactured by the methods described by Pease (1975). Micronet-thin carbon grids may then, with care, be used for negative staining in the same manner as grids coated with thicker polymer films.

For high-resolution electron microscopy of cellulose and cellulose degradation, *Acetobacter xylinum* strain ATCC 23769 from the American Type Culture Collection was grown in 100 ml glucose medium (Hestrin and Schramm, 1954) at 30° C for 24 hr. Cellulose pellicles were removed from the air–liquid interface of the medium and soaked in three 20-min changes of cold 50 mM phosphate buffer, pH 7.0 (150 ml/pellicle). Cells and buffer were released from several cubic centimeters of the washed pellicles by twisting a pellicle around a wooden applicator stick. Cell suspensions were kept at 0° C until needed and used without dilution. Cells were transferred to micronet-coated electron-microscopy grids supporting thin carbon films by touching the surface of a grid to a drop of the washed cell suspension. Cellulose was generated by floating a grid with adhering cells face down on a drop of glucose incubation medium in a spot plate for 7–10 min at room temperature.

Micronet-thin carbon grids with adhering cells and cellulose were incubated in separate drops of cellulase solutions within a moist chamber at room temperature. The cellulase solutions used in degradation experiments were as follows: (1) complete cellulase enzyme system, 1.0 mg/ml in 50 mM citrate buffer, pH 4.8; (2) purified endo-1,4-β-D-glucanase IV, 0.15 or 0.5 mg/ml in 50 mM citrate buffer, pH 4.8; and (3) purified 1,4-β-D-glucan cellobiohydrolase I (D), 0.5 mg/ml in 50 mM citrate buffer, pH 4.8. Purified samples of CBH I (D), EG IV, and the complete cellulase enzyme system of *Trichoderma reesei* QM9414 were gifts from Ross D. Brown, Jr., Department of Food Sciences and Human Nutrition, University of Florida, Gainesville, Florida. Enzymes were purified as described by Gum and Brown (1977a), Shoemaker and Brown (1978b), and Gritzali and Brown (1979).

After cellulase incubation, grids were washed with several drops of 50 mM citrate buffer and negatively stained with 1% uranyl acetate in 50 mM citrate buffer, pH 4.8. Bacitracin was used as a spreading agent in the negative stain (Gregory and Pirie, 1973). All preparations were examined with a Hitachi HU-11E electron microscope operating at 75 KV. For calibration purposes, a 52,000-line/inch grating replica and cationized ferritin molecules were used.

FIGURE 1. A schematic representation of cellulase action. (A) The *Trichoderma* cellulase enzyme system consists of three major enzymes: endoglucanase (EG), cellobiohydrolase (CBH), and β-glucosidase (β-G). EG binds randomly to the surface of the cellulose microfibril. The catalytic action of EG breaks a glucosyl bond within a glucan chain. (B) EG leaves the microfibril surface. The nick in the glucan chain exposes a reducing and a nonreducing end. (C) CBH can act only on the free nonreducing end of a glucan chain. The catalytic action of CBH cleaves a cellobiose unit from the nonreducing chain end. (D) Cellobiose is released into solution, where it is split into glucose monomers by β-G. CBH moves to the newly created free nonreducing end and continues to cleave cellobiose units from the glucan chain. (E) EG nicks continuous glucan chains, but releases little soluble reducing sugar. CBH is catalytically active only at glucan chain ends. Thus, EG creates sites at which CBH may act. The result is synergistic degradation of cellulose.

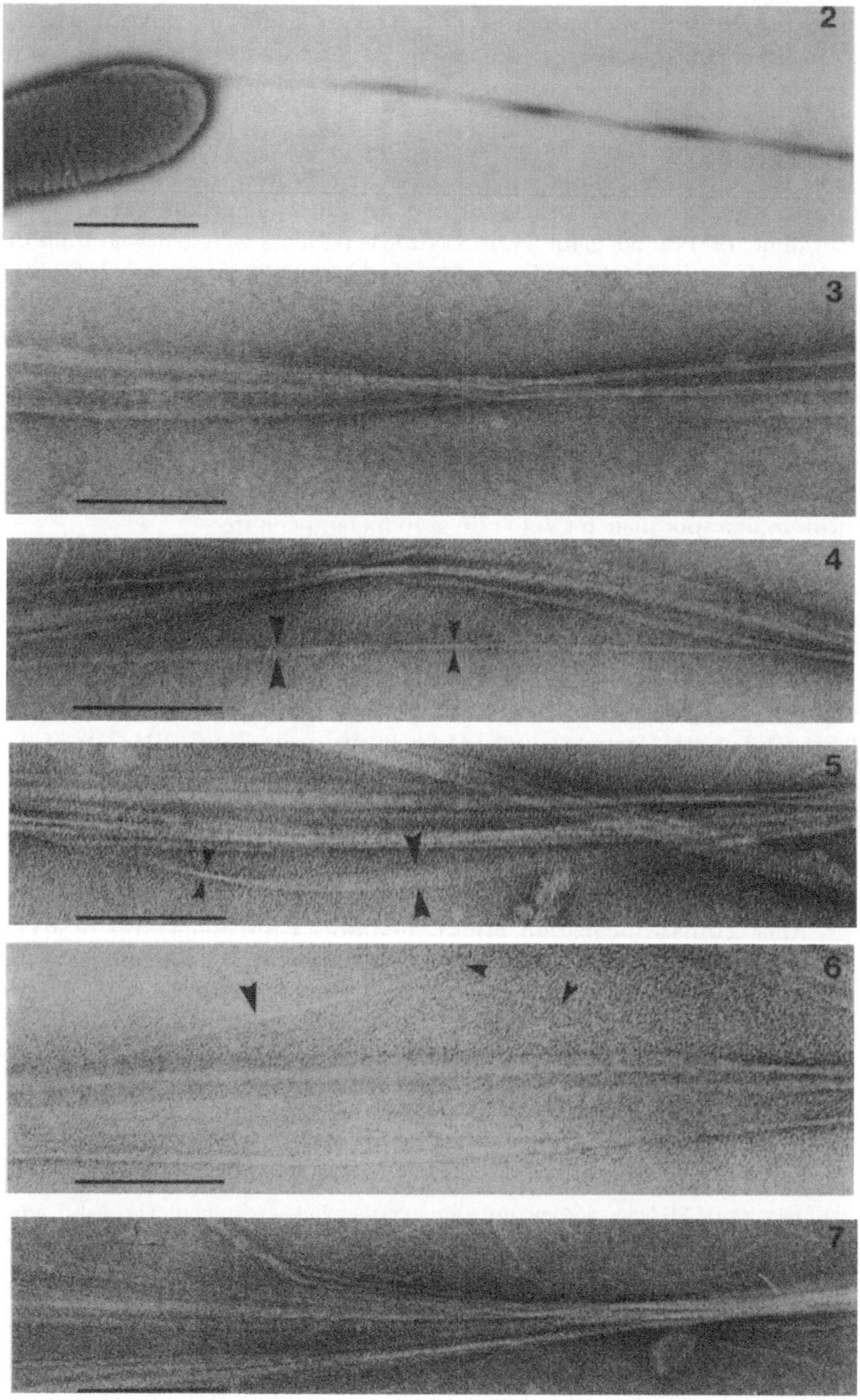
2
3
4
5
6
7

1.4. The Cellulose Substrate

There are many sources of cellulose; however, most natural cellulose is associated with chemically complex cell walls. Plant-cell walls consist of cellulosic microfibrils embedded in a matrix of hemicelluloses, various glycans, pectins, lignin, and cell-wall proteins (Keegstra *et al.*, 1973). Algae generally have less complicated cell-wall constructions, although cellulose in these cell walls is also usually embedded in a noncellulosic matrix (Montezinos and Brown, 1976; Romanovicz and Brown, 1976; Willison and Brown, 1978). The noncellulosic matrix prohibits cellulase attack by denying the enzymes access to the cellulose surface. Cellulose must either be cleaned of matrix components or found in a naked natural state before efficient enzymatic degradation is possible. The cellulose produced by the gram-negative bacterium *A. xylinum* provides a nearly ideal model substrate for degradation studies. *Acetobacter xylinum* cellulose lacks a noncellulosic matrix; it is highly crystalline; it is not consolidated into a thick cell wall, yet the microfibrils are organized into the simple higher structure of a ribbon; the dimensions of the ribbon are small, approaching the size of the cellulase enzymes themselves; and the cellulose may be easily generated in thin films directly on electron-microscope grids at the beginning of each experiment (Colvin, 1972; Gardner and Blackwell, 1974; R. M. Brown, Jr., *et al.*, 1976; White and Brown, 1981).

Acetobacter xylinum cellulose consists of ribbons of microfibrils generated at the surface of the bacterial cell (R. M. Brown, Jr., *et al.*, 1976). Each ribbon is between 40 and 60 nm wide and is composed of 50–80 microfibrils, each 3.0–3.5 nm wide. The ribbons are twisted with a periodicity of approximately 1 μm (Figs. 2 and 3). In a currently proposed mechanism of ribbon formation (R. M. Brown, Jr., *et al.*, 1976; Zaar, 1979; Blackwell and Kolpak, 1976), microfibrils are generated by multienzyme complexes and extruded through pores in the bacterial cell wall. These pores are arranged in rows within the lipopolysaccharide layer. Several microfibrils may coalesce into discrete bundles that subsequently aggregate into a full ribbon. The ribbon is thus composed of

FIGURE 2. An *A. xylinum* cell with attached cellulose ribbon. Note the periodic twisting of the ribbon. ×25,000. Scale bar: 1 μm.

FIGURE 3. A composite ribbon showing a twist and surface striations arising from irregular coalescence of bundles of microfibrils. ×200,000. Scale bar: 100 nm.

FIGURE 4. A bundle of microfibrils separated from the composite ribbon. The maximum width of the bundle is 6 nm (large arrows) and the minimum width is 3 nm (small arrows). ×200,000. Scale bar: 100 nm.

FIGURE 5. A ribbon and a twisting, separated microfibrillar bundle that has a maximum width of 12.5 nm (large arrows) and a minimum width of 3 nm (small arrows). ×200,000. Scale bar: 100 nm.

FIGURE 6. A ribbon with a separated 6-nm microfibrillar bundle (large arrow) that has split into two 3-nm microfibrils (small arrows). ×200,000. Scale bar: 100 nm.

FIGURE 7. A composite ribbon that has splayed into microfibrillar bundles of various sizes. ×200,000. Scale bar: 100 nm.

8

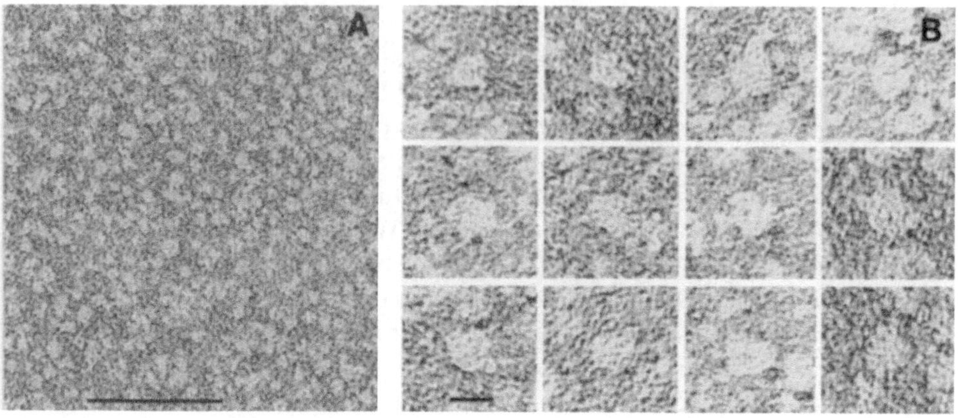

9

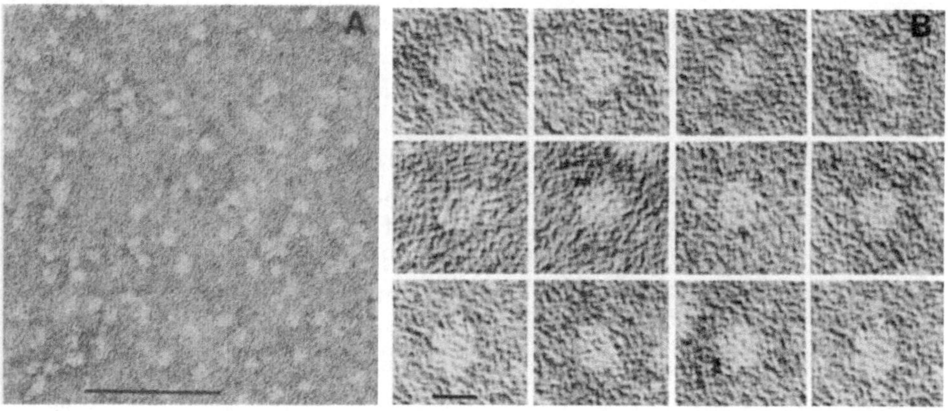

10

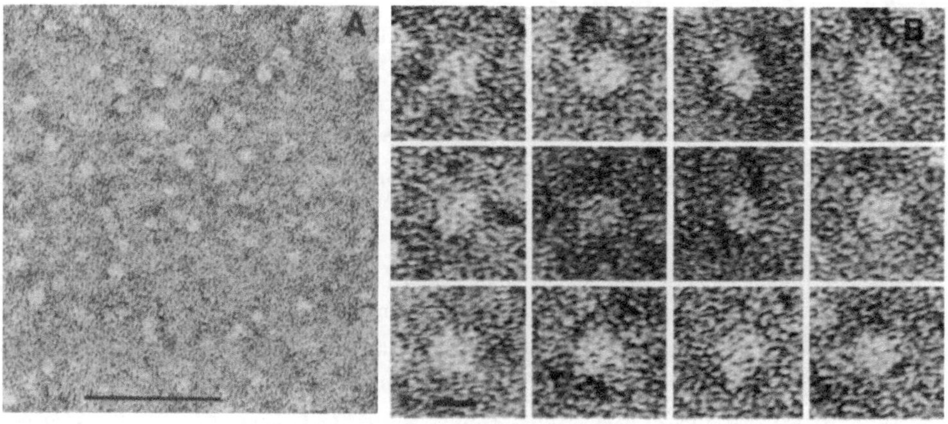

several levels of hydrogen-bonded structures. Glucan chains crystallize into microfibrils, intermicrofibrillar hydrogen bonding holds individual microfibrils into bundles, and these bundles are hydrogen bonded along their surfaces to form the ribbon. Bundles of microfibrils appear in negative-staining micrographs as striations on the ribbon surface (Fig. 3). Native cellulose from *A. xylinum* has a cellulose I crystallite size of 6.5 and 7.5 nm in its wet and dry states, respectively (Gardner and Blackwell, 1974).

Bundles of microfibrils are occasionally observed separated from the main *A. xylinum* cellulose ribbon (Figs. 4, 5, 6, and 7). These bundles are often twisted and typically exhibit minimum and maximum widths of 3 and 6 nm, respectively (Fig. 4). The maximum width of a microfibril bundle may be much larger, but is usually a multiple of 3 nm. Figure 5 shows such a twisting bundle that has a minimum thickness of 3 nm and a width of 12.5 nm. This could represent the coalescence of two 3 × 6 nm bundles into a wide and flat bundle. The 3 × 6 nm bundles may be made up of smaller discrete units. In Fig. 6, a 6-nm bundle splits into two 3-nm units. Haigler *et al.* (1980) and Benziman *et al.* (1980) propose that the 3-nm structures are in turn formed by fasciation of still smaller 1.5-nm units. There is controversy over which unit actually represents the basic microfibril size in *A. xylinum* cellulose (see Chapter 14). Figure 7 shows a ribbon that is splayed into bundles of various sizes.

2. *High-Resolution Electron Microscopy of Cellulase Components*

High-resolution negative staining has been used to study the interrelationships between viral coat proteins (Horne, 1978), the subunit structure of enzymes (Josephs, 1971; Green *et al.*, 1972), and the structure of enzymes themselves (Massover, 1978). It is possible to visualize structural features within individual molecules measuring 1 nm or less with these methods. In this study, samples of the complete *Trichoderma* cellulase enzyme system, purified CBH I

FIGURE 8. Cellulase enzyme system. (A) Complete cellulase enzyme system of *Trichoderma* negatively stained on a thin carbon support. Two of the three enzymes, EG and CBH, have reported molecular weights between 37,000 and 53,000 (Gritzali and Brown, 1979). β-Glucosidase has a somewhat smaller reported molecular weight of 26,000 (Iwasaki *et al.*, 1964). Negative staining of the complete enzyme system reveals particles of heterogeneous sizes. ×335,000. Scale bar: 50 nm. (B) Selected particles from the complete enzyme system highly magnified. These large particles have diameters between 4 and 7 nm. ×900,000. Scale bar: 5 nm.

FIGURE 9. Purified cellobiohydrolase. (A) Purified cellobiohydrolase I (C) negatively stained on a thin carbon support. Particles are more nearly homogeneous in size than in the complete enzyme system. ×335,000. Scale bar: 50 nm. (B) Selected particles from the cellobiohydrolase I (C) preparations highly magnified. Particles have diameters between 5 and 7 nm. ×900,000. Scale bar: 5 nm.

FIGURE 10. Purified endoglucanase. (A) Purified endoglucanase IV negatively stained on a thin carbon support. These particles are also homogeneous in size. ×335,000. Scale bar: 50 nm. (B) Selected particles from the endoglucanase IV preparations highly magnified. Particles have diameters between 5 and 7 nm. ×900,000. Scale bar: 5 nm.

(C), and purified EG IV were spread onto micronet-thin carbon-coated grids at protein concentrations of approximately 0.02 mg/ml. Grids were negatively stained and examined as described in Section 1.3. For calibration purposes, crystallized catalase and cationized ferritin molecules were used.

Negative staining of samples of the complete cellulase enzyme system reveals particles ranging from 4 to 7 nm in diameter (Fig. 8A). The mean diameter of the particles is 5.4 ± 0.1 nm. In negatively stained samples of purified CBH I (C) and EG IV, particles are more homogenous in size (Figs. 9 and 10). CBH I (C) diameters range from 5 to 7 nm with a mean diameter of 5.8 ± 0.3 nm. EG IV diameters range from 5 to 7 nm with a mean diameter of 6.1 ± 0.3 nm. In general, the purified component enzymes are composed of particles with more uniform sizes than those of the complete enzyme system. This is to be expected, since the complete system is comprised of enzymes that have reported molecular weights of 26,000 for β-glucosidase, 37,000–52,000 for EG, and 53,000 for CBH (Iwasaki *et al.*, 1964; Shoemaker and Brown, 1978b; Gum and Brown, 1977a). Comparison of images of EG IV and CBH I (C) reveals particles of approximately the same size and shape. Although it is difficult to assign exact three-dimensional shapes to the images of enzymes in a micrograph, it appears that these particles are approximately spherical with no distinguishing morphological surface features immediately obvious. It should be noted that cellulase enzymes do not seem to form complexes in either purified or mixed solutions. In all cases, observed particles are solitary and distributed randomly on the carbon support film.

3. *Cellulose Degradation by the Complete Cellulase Enzyme System*

3.1. *Initial Binding of Enzymes to the Substrate*

When cellulose ribbons from *A. xylinum* are exposed to 1 mg/ml cellulase solutions, immediately washed with buffer, and negatively stained with uranyl acetate, the surface of the ribbon is covered with particles that are not washed from the ribbon. Figures 11A and B show a comparison of ribbons before and after exposure to cellulases. The surface striations visible on control ribbons are obscured by bound particles in Fig. 11B. These particles have diameters ranging from 3 to 7 nm with a mean diameter of 5.4 ± 0.2 nm. The size range of bound particles overlaps well with the size range of particles observed in solutions of cellulase components reported in the previous section. In higher-magnification micrographs, individual particles are visible on the surface of cellulose ribbons (Fig. 12). No physical complexes of enzyme components are evident at the substrate surface. In some areas, particles appear to be arranged linearly, probably because they are bound to the linear arrangement of the microfibril bundles that make up the ribbon.

3.2. Progressive Degradation of Cellulose Ribbons

Acetobacter xylinum cellulose incubated with cellulase enzymes is completely degraded within 30 min into products not visible in the electron microscope. The first morphological changes in ribbon structure become apparent within 5–10 min when ribbons splay into smaller bundles of microfibrils (Fig. 13). The microfibril bundles and remaining composite ribbon are both coated with particles. Initially, microfibril bundles are straight, as though rigid. Splaying of bundles becomes more extensive after 10 min, and the bundles begin to show signs of further degradation. Microfibril bundles are often observed with sharp curves or bends (Fig. 14) that are rarely seen after shorter incubations. This may indicate that the integrity of microfibril structure is being

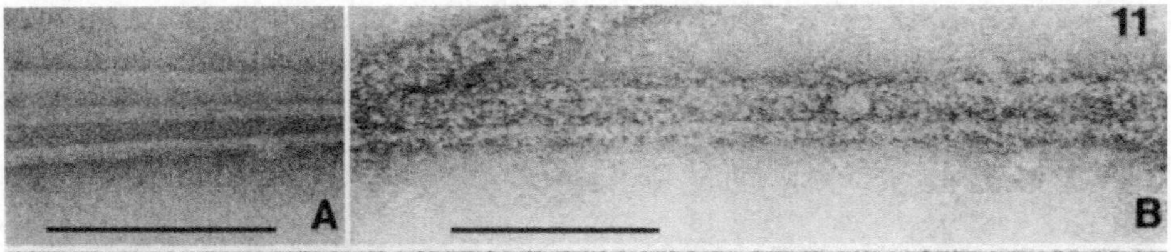

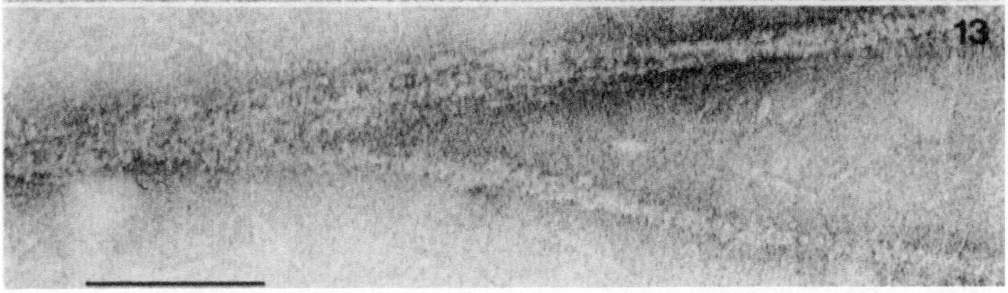

FIGURE 11. (A) Normal *A. xylinum* cellulose ribbon not exposed to cellulases. Note the sharply defined surface striations. ×200,000. Scale bar: 100 nm. (B) *Acetobacter xylinum* cellulose exposed to cellulases for a few seconds and then washed with buffer. Note that particles bound to the ribbon obscure the surface striations. ×200,000. Scale bar: 100 nm.

FIGURE 12. Cellulose ribbon exposed to cellulases for a short time. The ribbon surface is coated with particles. Surface striations (arrows) are faintly visible in this high-magnification micrograph. ×385,000. Scale bar: 50 nm.

FIGURE 13. A ribbon exhibiting bundle-splaying following 5 min of cellulase incubation. ×200,000. Scale bar: 100 nm.

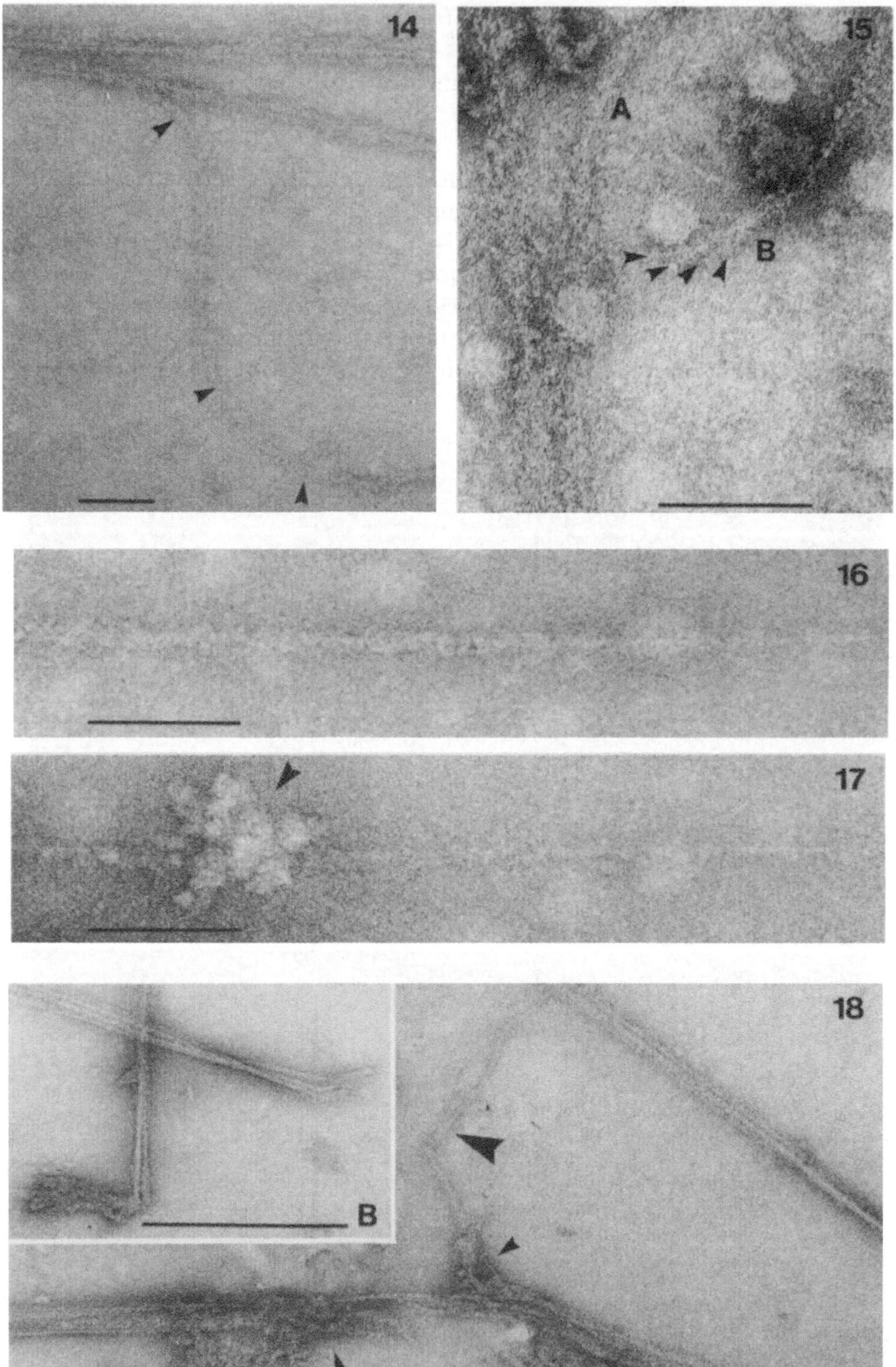
14
15
A
B
16
17
18
B
A

destroyed, leading to more flexibility within the bundles. Figure 15 shows a ribbon that has splayed into three bundles, one of which (arrows) appears to be further degraded and is almost separated from the ribbon. The region of the bundle that is most degraded has a frayed appearance, and this bundle may have split into individual microfibrils. In most cases, the bundles do not separate into smaller bundles or individual microfibrils. After 20 min of incubation, few whole ribbons remain. Bundles often display tapered ends (Figs. 16 and 17) and may be associated with amorphous fragments (Fig. 17). The microfibril bundles are progressively thinned until, after 30 min of incubation, no visible products remain.

The characteristic splaying of ribbons into microfibril bundles may be the result of the structure within the cellulose ribbons. The ribbons are composed of discrete groups of hydrogen-bonded microfibrils that coalesce into the composite ribbon. The surface glucan chains of one bundle hydrogen bond to the surface glucan chains of adjacent bundles. The least extensive hydrogen bonding may be between the bundles (Blackwell and Kolpak, 1976), and these are the first bonds to be broken during the degradation process. When cellulases begin to attack the ribbon, EG breaks glucosidic bonds within glucan chains on the surface of bundles. The free nonreducing chain ends become available sites for CBH action, which removes the nicked surface layer of the bundle. As bundle surfaces are eroded, the hydrogen-bonded glucan chains that are holding adjacent bundles together are weakened or removed, and the bundles slip apart. As degradation progresses, combined EG and CBH action on the cellulose microfibrils continues to erode the surface layers of glucan chains. Tighter hydrogen bonding between microfibrils within each bundle could prevent splaying into smaller groups of microfibrils. Cellulase action is possible only on exposed surfaces, which are gradually worn away, thinning the bundles. The amorphous fragments that appear (Fig. 17) during degradation may represent masses of

FIGURE 14. Bundle-splaying following 10 min of cellulose incubation. The sharp bends in the small bundle (arrows) may be the result of damage to the structural integrity of the microfibrils. ×100,000. Scale bar: 100 nm.

FIGURE 15. Bundle-splaying following 10 min of cellulase incubation. Bundle (A) remains integrated with the ribbon. Bundle (B) appears to be separated from the ribbon and degraded further to small strands approximately 2.5 nm wide (arrows). These may represent individual microfibrils. ×200,000. Scale bar: 100 nm.

FIGURE 16. A microfibrillar bundle remaining after 20 min of cellulase incubation. Note the low contrast and tapering of the bundle toward the right. ×200,000. Scale bar: 100 nm.

FIGURE 17. A bundle and associated amorphous fragments (arrow) following 20 min of cellulase incubation. This bundle tapers to the right in a manner similar to the bundle in Fig. 16. ×200,000. Scale bar: 100 nm.

FIGURE 18. (A) A ribbon that is fragmented instead of exhibiting bundle-splaying following 15 min of cellulase incubation. Note the bent microfibrillar bundle (large arrow) and associated amorphous fragments (small arrows). ×80,000. Scale bar: 500 nm. (B) Two severed ribbon ends following 15 min of cellulase incubation. Note amorphous masses associated with the ends of the ribbons. ×80,000. Scale bar: 500 nm.

glucan chains or chain fragments that have been decrystallized and partially degraded, but not fully solubilized.

3.3. Ribbon Fragmentation

Occasionally during the intermediate stages of degradation, discrete fragments of ribbons are observed. In Fig. 18A, a ribbon is almost severed in one area, while adjacent regions of the ribbon are still structurally intact. A fragmented bundle of microfibrils lies nearby (large arrow) and may still be connected to the ribbon by an amorphous region of cellulose. Other amorphous fragments are associated with the intact ribbon (small arrows). Figure 18B shows two severed ribbon ends that also have associated amorphous fragments. The cause of this localized degradation is unknown; however, it could arise from locally elevated concentrations of EG that act on one area of a ribbon preferentially over other regions. This phenomenon could also be the result of structurally weak or poorly crystallized regions within the ribbon that are degraded more easily than surrounding regions. If the ribbon was bent at one spot, the microfibrils may have been mechanically spread apart, increasing exposed microfibril surface area, and allowing the enzymes greater access to cellulose surfaces (Rowland and Roberts, 1972; Rowland, 1975; Fan *et al.*, 1980). Degradation could proceed faster in such a region, eventually eroding through the whole ribbon and producing discrete, fragmented ribbon ends. The amorphous fragments associated with the ribbon ends may, again, represent masses of partially degraded and decrystallized glucan chains that have not been hydrolyzed completely into soluble sugars or oligosaccharides.

4. Cellulose Degradation by Purified and Reconstituted Cellulase Components

4.1. Purified Cellobiohydrolase I (D) and Endoglucanase IV

Cellulose incubated separately with purified samples of cellobiohydrolase I (D) and endoglucanse IV is not degraded in the same manner as cellulose incubated with the complete cellulase enzyme system. The cellulase system is estimated to contain about 60% CBH I (D) and 15% EG (Gritzali and Brown, 1979). To approximate the action of each of the enzymes in the mixture, purified CBH I (D) and EG IV were prepared at protein concentrations of 0.5 and 0.15 mg/ml, respectively. Section 3.2 showed that no visible cellulose remains after 30 min of incubation with the cellulase system at a protein concentration of 1 mg/ml. CBH I (D) produced few morphological changes in *A. xylinum* cellulose after 30 min of incubation even though particles are bound to ribbon surfaces (Fig. 19). At 0.15 mg/ml, EG IV also produced no observable alteration of ribbon structure within 30 min. EG IV does begin to produce ribbon-splaying after 30 min at a higher protein concentration of 0.5 mg/ml (Fig. 20). In both cases, numerous recognizable ribbons are observed even after 60 min of incubation. When EG IV and CBH I (D) are present simultaneously at protein concentrations of 0.15 and 0.5 mg/ml, respectively, in an incubation admixture, ribbons are completely degraded within 30 min. *β*-Glucosidase was not present in this admixture, but this enzyme is estimated to account for only about 1% of the total protein, and its role in the solubilization of crystalline cellulose does not appear to be essential (Gritzali and Brown, 1979).

The known specificities of the EG and CBH enzymes (EG cleaves unbroken glucan chains, while CBH requires a free, nonreducing glucan chain end for substantial catalytic activity) suggest that in the initial stages of degradation, EG should be the most active enzyme because few nonreducing ends are available for CBH to act on. EG must provide the primary degradative action by hydrolyzing glucosidic bonds on the microfibril surfaces, thereby disrupting the organization of glucan chains required to maintain hydrogen bonding between bundles. When CBH is also present, it begins to cleave cellobiose from newly created chain ends, eventually eroding away the original crystalline surface, completely disrupting the forces that hold microfibril bundles into the ribbon configuration, and exposing new glucan chain surfaces for EG to act on. CBH thus accelerates the breakdown of ribbons into bundles. When EG alone is present, it may break enough glucosidic bonds to weaken the microfibril surfaces that are holding the bundles into the ribbon configuration and, by itself, lead to ribbon-splaying. This is consistent with the findings of Wood and McCrae (1978) that EG increases the capacity for uptake of alkali and reduces the tensile strength of cellulose. However, it should be noted that although EG produces visible changes in ribbon structure, the degradative action of this enzyme when alone is slow compared to the synergistic degradation that occurs when EG and CBH are simultaneously present in an admixture.

4.2. Sequential Incubation with Endoglucanase and Cellobiohydrolase

Incubation of *A. xylinum* cellulose in 0.15 mg/ml EG IV, followed by a distilled-water wash and incubation in 0.5 mg/ml CBH I (D), produces results similar to those of incubations with the purified enzymes alone. After 5 min of EG IV incubation followed by a short exposure of CBH I (D), ribbon surfaces are coated with particles (Fig. 21). During such short incubations, the particles may appear to line up along microfibrils and microfibril bundles (Fig. 22) in a manner similar to short incubations with the complete enzyme system. EG IV incubation for 30 min followed by CBH I (D) incubation for 30 min may produce some signs of cellulose degradation. In Fig. 23, a short section of a microfibril bundle is separated from the composite ribbon. Both the ribbon and the bundle are bending in a manner uncharacteristic of undegraded *A. xylinum* cellulose, suggesting that the structural integrity of the ribbon has been disrupted. In longer sequential incubations, numerous ribbons or bundles of microfibrils are still visible. Figure 24 shows a ribbon incubated for 60 min in EG IV followed by 15 min in CBH I (D). The ribbon is thinner than normal, suggesting that either

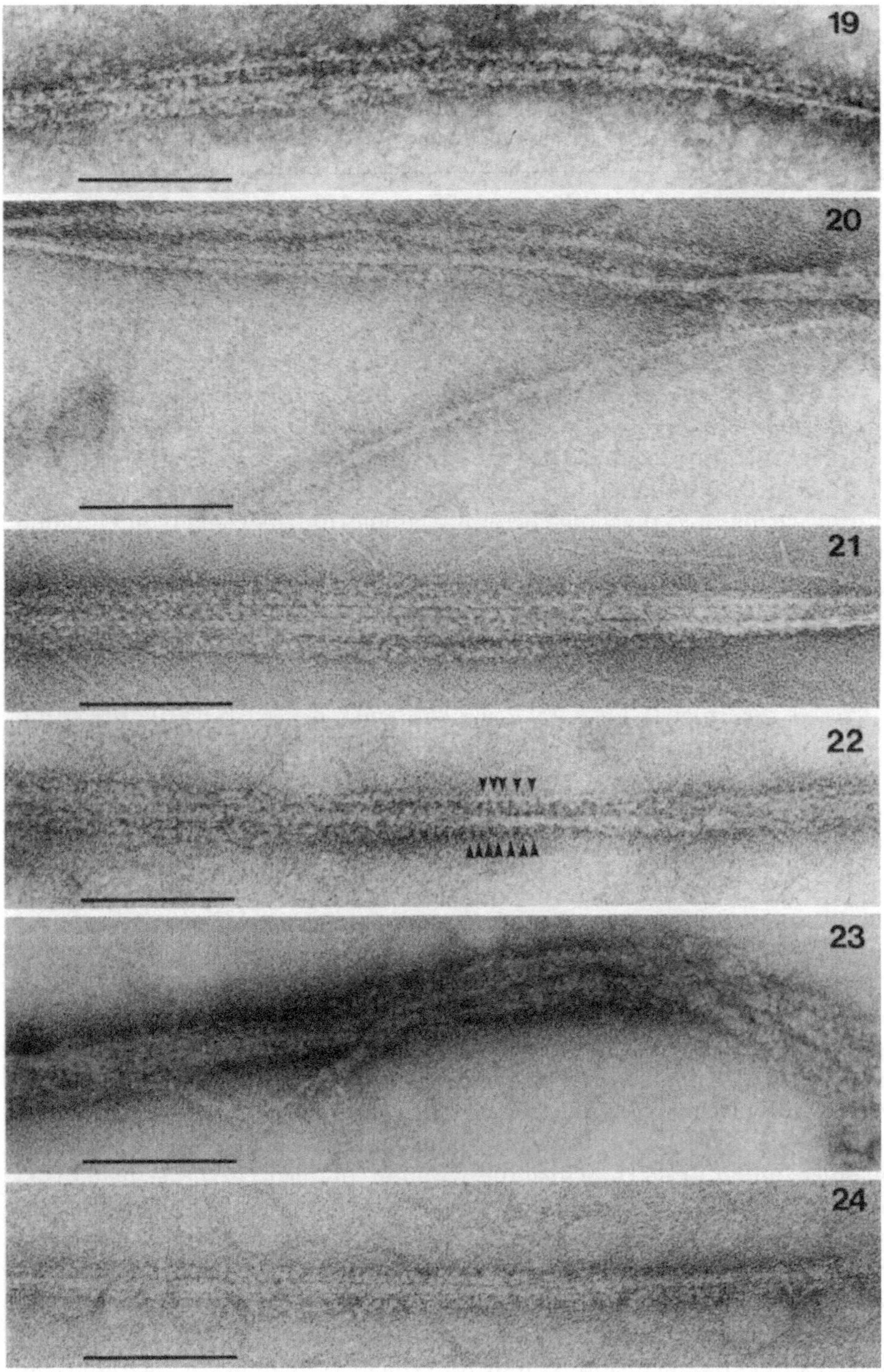
19
20
21
22
23
24

bundles have splayed off or the surface has been eroded significantly. Note that even after a total of 75 min of sequential incubation, the ribbon is not completely degraded.

Sequential incubation illustrates the need for simultaneous presence of both EG and CBH before synergistic action is possible. In the case of sequential incubation, EG randomly cleaves glucan chains on the surface of cellulose microfibrils and is capable of continued action, but it can act only on exposed glucan chains. Its random cleaving action, essentially cleaving only one bond at a time, is not efficient at removing surface layers, and fresh glucan chains are not exposed. Degradation is limited by the rate at which surface chains are removed. The EG enzymes are then washed away, and CBH is allowed to cleave cellobiose from the exposed nonreducing chain ends. However, when all nicked chains have been removed, CBH no longer has available sites for enzymatic action, and degradation is again sharply reduced. At that point CBH is faced with a cellulose surface very similar to the fresh ribbon that it is unable to efficiently degrade. When EG and CBH are in synergistic action, each creates sites of action for the other. EG nicks continuous surface chains, while CBH removes those surface chains, exposing new, unbroken chains.

5. *Conclusions and Future Perspectives*

It has been possible to establish the visual patterns of the process of cellulose degradation utilizing high-resolution electron-microscopic techniques. Particles that are presumed to be cellulase enzymes are initially observed bound to the cellulose ribbons. Within 10 min, the ribbon is split along its long axis into bundles of microfibrils that are subsequently thinned until they are completely

FIGURE 19. An *A. xylinum* cellulose ribbon following 30 min of incubation in purified 1,4-β-D-glucan cellobiohydrolase I (D). Particles are bound to the ribbon surface, but general ribbon morphology has not changed. ×200,000. Scale bar: 100 nm.

FIGURE 20. A ribbon incubated for 30 min in purified endo-1,4-β-D-glucanase IV. This ribbon exhibits some bundle-splaying; however, whole ribbons are still present. ×200,000. Scale bar: 100 nm.

FIGURE 21. A cellulose ribbon after 5 min of incubation in purified EG IV followed by a short exposure to purified CBH I (D). Note that the ribbon surface is coated with particles. ×200,000. Scale bar: 100 nm.

FIGURE 22. A ribbon after short sequential incubations in EG IV (5 min) and CBH I (D) (5 min). In certain regions, particles appear as linear arrays along microfibrils and microfibrillar bundles on the ribbon surface (arrows). ×200,000. Scale bar: 100 nm.

FIGURE 23. A ribbon after longer sequential incubations in EG IV (30 min) and CBH I (D) (30 min). This ribbon shows signs of degradation. A short section of a ribbon bundle is separated from the ribbon, and both the ribbon and the bundle are bent. ×200,000. Scale bar: 100 nm.

FIGURE 24. A ribbon after sequential incubations in EG IV (60 min) and CBH I (D) (15 min). This ribbon is thinner than normal, indicating that microfibrillar bundles may have split off or the ribbon surface may have been significantly eroded. ×200,000. Scale bar: 100 nm.

dissolved. Incubations with purified components of the cellulase system produce less dramatic changes in ribbon morphology. Purified cellobiohydrolase I (D) produced no visible change in cellulose structure. Purified endoglucanase IV did produce some splaying of ribbons into microfibril bundles, but in both cases whole ribbons were present even after 60 min of incubation. Sequential incubations in purified endoglucanase IV and cellobiohydrolase I (D) illustrated the necessity of simultaneous presence of these two enzymes before synergism is possible.

High-resolution electron microscopy has been successfully used to examine aspects of cellulose degradation that have been difficult to study with biochemical methods. These results suggest a promising future for applications of electron microscopy in cellulase research. Perhaps most important is the positive identification of the particles observed bound to the cellulose surfaces as cellulase enzymes. This could be accomplished by demonstrating that purified antibodies against the individual cellulase enzymes form antibody-antigen complexes with bound particles. Similar techniques have been used to localize specific proteins within ribosomes (Lake, 1979), and these methods should be applicable to the cellulase problem. If antibodies against the enzyme components of the cellulase system are labeled with different marker molecules, it may be possible to determine whether the component enzymes have characteristic binding patterns on the cellulose surface.

High-resolution dark-field electron microscopy may reveal structural details of the enzymes themselves. Ottensmeyer *et al.* (1975, 1977) have succeeded in imaging unstained, unshadowed protein and cyclic peptide antibiotic molecules. Their images have revealed astonishing details of macromolecular fine structure at resolutions below 1 nm. The scanning transmission electron microscope has also recently been used for dark-field imaging of unstained biological macromolecules with resolution of fine structure as small as 0.25 nm (Ohtsuki *et al.*, 1979). Dark-field electron microscopy could be used to study the macromolecular fine structure of the various cellulase enzymes as well. It may even be possible to image cellooligosaccharides bound to the active sites of individual enzymes. The success of high-resolution techniques in imaging biological macromolecules and their interactions underscores the increasing mergence of biochemistry and electron microscopy. Hopefully, the combination of these methods will produce a synergistic effect as beneficial to our understanding of enzyme action as the cooperation between cellulases is to the efficient degradation of cellulose.

References

Albersheim, P., 1969, Biochemistry of the cell wall in relation to infective processes, *Annu. Rev. Phytopathol* **7**:171.

Benziman, M., Haigler, C. H., Brown, R. M., Jr., White, A. R., and Cooper, K. M., 1980, Cellulose biogenesis: Polymerization and crystallization are coupled processes in *Acetobacter xylinum*, *Proc. Natl. Acad. Sci. U.S.A.* **77**:6678.

Blackwell, J., and Kolpak, F. J., 1976, Cellulose microfibrils as disordered arrays of elementary fibrils, *Appl. Polym. Symp.* **28**:751.

Brown, R. M., Jr., Willison, J. H. M., and Richardson, C. L., 1976, Cellulose biosynthesis in *Acetobacter xylinum*: Visualization of the site of synthesis and direct measurement of the *in vivo* process, *Proc. Natl. Acad. Sci. U.S.A.* **73**:4565.

Brown, W., 1965, Toxins and cell wall dissolving enzymes in relation to plant disease, *Annu. Rev. Phytopathol.* **3**:1.

Colvin, J. R, 1972, The structure and biosynthesis of cellulose, *CRC Crit. Rev. Macromol. Sci.* **1**:47.

Emert, G. H., Gum, E. K., Jr., Lang, J. A., Lin, T. H., and Brown, R. D., Jr., 1974, Cellulases, in: *Food Related Enzymes* (J. Whitaker, ed.), pp. 79–100, *Advances in Chemistry Series* No. 136, American Chemical Society, Washington, D.C.

Eriksson, K.-E., and Rzedowski, W., 1969, Extracellular enzyme system utilized by the fungus *Chrysosporium lignorum* for the breakdown of cellulose. I. Studies on the enzyme production, *Arch. Biochem. Biophys.* **129**:683.

Fan, L. T., Lee, H. Y., and Beardmore, D. H., 1980, Mechanism of the enzymatic hydrolysis of cellulose: Effects of major structural features of cellulose on enzymatic hydrolysis, *Biotechnol. Bioeng.* **22**:177.

Gardner, K. H., and Blackwell, J., 1974, The structure of native cellulose, *Biopolymers* **13**:1975.

Grassmann, W., Zechmeister, L., Toth, G., and Stadler, R., 1933, Uber den enzymatischen Abban der Cellulose und ihrer Spaltproduckte, *Justus Liebigs Ann. Chem.* **503**:167.

Green, N. M., Valentine, R. C., Wrigley, N. C., Ahmad, F., Jacobson, B., and Wood, H. G., 1972, Transcarbamylase. XI. Electron microscopy and subunit structure, *J. Biol. Chem.* **247**:6284.

Gregory, D. W., and Pirie, B. J. S., 1973, Wetting agents for biological electron microscopy. I. General considerations and negative staining, *J. Microsc. (Oxford)* **99**:251.

Gritzali, M., and Brown, R. D., Jr., 1979, The cellulase system of *Trichoderma*: Relationships between purified extracellular enzymes from induced or cellulose-grown cells, in: *Hydrolysis of Cellulose: Mechanisms of Enzymatic and Acid Catalysis* (R. D. Brown, Jr., and L. Jurasek, eds.), pp. 237–260, *Advances in Chemistry Series*, No. 181, American Chemical Society, Washington, D.C.

Gum, E. K., Jr., and Brown, R. D., Jr., 1976, Structural characterization of a glycoprotein cellulase, 1,4-β-D-glucan cellobiohydrolase C from *Trichoderma viride*, *Biochim. Biophys. Acta* **446**:371.

Gum, E. K., Jr., and Brown, R. D., Jr., 1977a, Comparison of four purified extracellular 1,4-β-D-glucan cellobiohydrolase enzymes from *Trichoderma viride*, *Biochim. Biophys. Acta* **492**:225.

Gum, E. K., Jr., and Brown, R. D., Jr., 1977b, Two alternative HPLC separation methods for reduced and normal cellooligosaccharides, *Anal. Biochem.* **82**:372.

Haigler, C. H., Brown, R. M., Jr., and Benziman, M., 1980, Calcofluor White ST alters the *in vivo* assembly of cellulose microfibrils, *Science* **210**:903.

Halliwell, G., and Griffin, M., 1973, The nature and mode of action of the cellulolytic component C_1 of *Trichoderma koningii* on native cellulose, *Biochem. J.* **135**:587.

Hestrin, S., and Schramm, M., 1954, Synthesis of cellulose by *Acetobacter xylinum*. 2. Preparation of freeze-dried cells capable of polymerizing glucose to cellulose, *Biochem. J.* **58**:345.

Horne, R. W., 1978, Special specimen preparation methods for image processing in transmission electron microscopy: A review, *J. Microsc.* **113**:241.

Iwasaki, T., Hayashi, K., and Funatsu, M., 1964, Purification and characterization of two types of cellulase from *Trichoderma koningii*, *J. Biochem. (Tokyo)* **55**:209.

Josephs, R., 1971, Electron microscopic studies on glutamic dehydrogenase: Subunit structure of individual molecules and linear associates, *J. Mol. Biol.* **55**:147.

Keegstra, K., Talmadge. K. W., Bauer, W. D., and Albersheim. P., 1973, The structure of plant cell walls. III. A model of the walls of suspension-cultured sycamore cells based on the interconnections of the macromolecular components, *Plant Physiol.* **51**:188.

Lake, J. A., 1979, Practical aspects of immune electron microscopy, *Methods Enzymol.* **61**:250.

Li, L. H., Flora, R. M., and King, K. W., 1965, Individual roles of cellulase components derived from *Trichoderma viride*, *Arch. Biochem. Biophys.* **111**:439.

Liese, W., 1970, Ultrastructural aspects of woody tissue degradation, *Annu. Rev. Phytopathol.* **8**:231.

Mandels, M., 1975, Microbial sources of cellulase, *Biotechnol. Bioeng. Symp.* **5**:81.
Massover, W. H., 1978, The ultrastructure of ferritin macromolecules. III. Mineralized iron in ferritin is attached to the protein shell, *J. Mol. Biol.* **123**:721.
Montezinos, D., and Brown, R. M., Jr., 1976, Surface architecture of the plant cell: Biogenesis of the cell wall, with special emphasis on the role of the plasma membrane in cellulose biosynthesis, *J. Supramol. Struct.* **5**:277 (229).
Ohtsuki, M., Isaacson, M. S., and Crewe, A. V., 1979, Dark field imaging of biological macromolecules with the scanning transmission electron microscope, *Proc. Natl. Acad. Sci. U.S.A.* **76**:1228.
Ottensmeyer, F. P., Whiting, R. F., Schmidt, E. E., and Clemens, R. S., 1975, Electron microtephroscopy of proteins, A close look at the ashes of myokinase and protamine, *J. Ultrastruct. Res.* **52**:193.
Ottensmeyer, F. P., Andrew J. W., Bazett-Jones, D. P., Chau, A. S. K., and Hewitt, J., 1977, Signal to noise enhancement in dark field electron micrographs of vasopressin: Filtering of arrays of images in reciprocal space, *J. Microsc. (Oxford)* **109**:259.
Pease, D. C., 1975, Micronets for electron microscopy, *Micron* **6**:85.
Pringsheim, H., 1912, Uber den fermentativen Abban der Zellulose, *Z. Physiol. Chem.* **78**:266.
Reese, E. T., 1956, A microbiological process report: Enzymatic hydrolysis of cellulose, *Appl. Microbiol.* **4**:39.
Reese, E. T., 1975a, Polysaccharases and the hydrolysis of insoluble substrates, in: *Biological Transformation of Wood* (W. Liese, ed.), pp. 165–181, Springer-Verlag, Berlin.
Reese, E. T., 1975b, Summary statement on the enzyme system, *Biotechnol. Bioeng. Symp.* **5**:77.
Reese, E. T., and Levinson, H. S., 1952, A comparative study of the breakdown of cellulose by microorganisms, *Physiol. Plant.* **5**:345.
Reese, E. T., Siu, R. G., and Levinson, H. S., 1950, The biological degradation of soluble cellulose derivatives and its relationship to the mechanism of cellulose hydrolysis, *J. Bacteriol.* **59**:485.
Ridge, I., and Osborne, D. J., 1969, Cell growth and cellulases: Regulation by ethylene and indole-3-acetic acid in shoots of *Pisum sativum*, *Nature (London)* **223**:318.
Romanovicz, D. K., and Brown, R. M., Jr., 1976, Biogenesis and structure of Golgi-derived cellulosic scales in *Pleurochrysis*. II. Scale composition and supramolecular structure, *Appl. Polym. Symp.* **28**:587.
Rowland, S. P., 1975, Selected aspects of the structure and accessibility of cellulose as they relate to hydrolysis, *Biotechnol. Bioeng. Symp.* **5**:183.
Rowland, S. P., and Roberts, E. J., 1972, The nature of accessible surfaces in the microstructure of cotton cellulose, *J. Polym. Sci. Polym. Chem. Ed.* **10**:2447.
Scheffer, T. C., and Cowling, E. B., 1966, Natural resistance of wood to microbial deterioration, *Annu. Rev. Phytopathol.* **4**:147.
Seilliere, G., 1905, Sur la digestion de cellulose, *C. R. Seances Soc. Biol. Ses Fil.* **58**:409.
Selby, K ., and Maitland, C. C., 1967, The cellulase of *Trichoderma viride:* Separation of the components involved in the solubilization of cotton, *Biochem. J.*, **104**:716.
Shoemaker, S. P., and Brown, R. D., Jr., 1978a, Enzymic activities of endo-1,4-β-D-glucanases purified from *Trichoderma viride*, *Biochim. Biophys. Acta* **523**:133.
Shoemaker, S. P., and Brown, R. D. Jr., 1978b, Characterization of endo-1,4-β-D-glucanases purified from *Trichoderma viride*, *Biochim. Biophys. Acta* **523**:147.
Ward, H. M., 1888, A lily-disease, *Ann. Bot. (London)* **2**:319.
Ward, H. M., 1898, *Penicillium* as a wood-destroying fungus, *Ann. Bot. (London)* **12**:565.
White, A. R., and Brown, R. M., Jr., 1981, Enzymatic hydrolysis of cellulose: Visual characterization of the process, *Proc. Natl. Acad. Sci. U.S.A.* **78**:1047.
Whiting, R. F., and Ottensmeyer, F. P., 1972, Heavy atoms in model compounds and nucleic acids imaged by darkfield transmission electron microscopy, *J. Mol. Biol.* **67**:173.
Wilkie, C. R. (ed.), 1975, *Cellulose as a Chemical and Energy Source*, John Wiley, New York.
Willison, J. H. M., and Brown, R. M., Jr., 1978, Cell wall structure and deposition in *Glaucocystis*, *J. Cell Biol.* **77**:103.

Wood, T. M., 1968, Cellulolytic enzyme system of *Trichoderma koningii:* Separation of components attacking native cotton, *Biochem. J.* **109**:217.

Wood, T. M., and McCrae, S. I., 1972, The purification and properties of the C_1 component of *Trichoderma koningii* cellulase, *Biochem. J.* **128**:1183.

Wood, T. M., and McCrae, S. I., 1978, Purification and properties of some endoglucanase components with special reference to their action on cellulose when acting along and in synergism with the cellobiohydrolase, *Biochem. J.* **171**:61.

Zaar, K., 1979, Visualization of pores (export sites) correlated with cellulose production in the envelope of the gram negative bacterium *A. xylinum, J. Cell Biol.* **80**:773.

Index

Abscission zone, 470
Acanthamoeba, 332
Acceptor, for chitin chain growth, 218
Acceptor molecules, 327
  in glucan synthesis, 336
Acetobacter xylinum, 106, 114, 120, 128, 226, 260, 273, 300, 342, 446, 493
  as a model for cellulase action, 494
  as a model system in cellulose, 275
Achlya, 478
Achyla ambisexualis, 468
Acid phosphatase, in Golgi apparatus, 135
Agaricus bisporus, 211
Aggregation time, in cellulose synthesis, 283
Aleurone tissue, from barley, 377
Algae, as model systems in cellulose, 4
Alginic acid, in *Fucus*, 47
Amylase, 377
Analogues, of synthetic and natural systems, 443
Anchoring sites, for glucan chain polymerization, 357
Aniline blue
  fluorescence in root hairs, 173
  specificity questioned, 177
  reaction, for callose, 177
Anisotropic fluids, 447
Antibodies, to cellulases, 506
Antiparallel chains, of cellulose, 409
Aphrodite, 417, 423
Araneida, 442
Armorarcia rusticana, 364
Arylsulfatase, in Golgi apparatus, 135
Asparagine, 364
Aspergillus flavus, 209
Aspergillus nidulans, 208
Asymmetric intramolecular interactions, 447
ATPase, lack of, in chitosomes, 215
Autogenous inhibitors, of chitin synthetase, 211
Auxin
  role in cell wall elongation, 463
  role in protoplast division, 155
Avena, 463, 479
Avena sativa, 217

Bacillus subtilis, 460
Bacterial growth, 303
Bacterial invasion, role of cellulase in, 473
Bacterial utilization of alginate, fucoidin, or cellulose, 57
Basidiomycete QM 806, 460
Bicomponent fibers, 438
Blastocladiella emersonii, 209
Bombyx mori, 442
Botryodiplodia theobromae, 473
Breakage loci, within Golgi cellulose, 137
Budding division, in tobacco protoplasts, 154
Buffers, effects on cellulose synthesis, 232
Bundle fasciation, in *Acetobacter*, 291
Bundle splaying, of microfibrils in *Acetobacter*, 286, 501

Calcium
  role in release of wall polymerization, 54
  role in wall structure, 38
  stimulation of cellulose synthesis, 232
Calcofluor, 274, 276, 446
  binding to glucans, 275
  cell wall formation in *Fucus*, 48
  chitin microfibril inhibition, 199
  isomerization by light, 278
  staining of wall protoplast, 153
Calcofluor-altered cellulose, reversibility of, 280
Calcofluor-band cellulose, in *Acetobacter*, 280
Callose, 167, 238, 463

Callose deposits,
composition of, 117
induced by UDP-glucose, 175
Carbohydrate metabolism, of *Acetobacter*, 343
Carboxymethylcellulose (CMC), 274, 332
effects on cellulose synthesis, 285
Carrier lipids, in core glycosylation, 377
Cell expansion, 465
Cell plates, callose in, 181
Cell shape, maintenance of, 162
Cell wall
of higher plants, reviews, 24
nuclear division, 151
periodic acid stain, *Oocystis*, 11
structure, *Oocystis*, 5
Cell wall assembly, *in vitro*, 39
Cell wall birefringence, time-course of appearance of, 49
Cell wall cellulose, 253
Cell wall formation, influence of cell division on, 149
Cell wall maturation, 465
Cell wall microfibrils, organized patterns, 5
Cell wall regeneration, by protoplasts, 61
Cell-directed, self-assembly, of cellulose, 273
Cell-free cellulose synthesis, 349
criteria for, 227
Cell-free glucan synthesis, 329
Cellobiohydrolase, 491
Cellobiose, 334
Cellodextrans, 331
as primers, 334
Cellohexaose, 335
Cellotetraose, 332
Cellulase, 461, 489
biochemical characterization of, 490
blocks mitosis in protoplast, 161
intracellular localization of, 468
purified, 496
in symbiotic relationships, 474
visualization of, 489
Cellulose
from cotton, 253
criteria for determination, 66
crystal structure of, 404
crystallinity in presence of Calcofluor, 280
determination, new procedure, 66
folded chain structure of, 416
in *Fucus*, 47, 54
macromolecular organization of, 403
microfibril formation, 302
polymorphic structures of, 408
secondary structure, 67
Cellulose (*cont.*)
solid-state structure of, 415
solution of, 446
transition, 408
unit cell, 385
Cellulose degradation, by purified cellulases, 502
Cellulose I, 273, 408, 409, 445
Cellulose II, 274, 408, 412, 445
Cellulose microfibrils
crossed polylamellate, 107
cross-sectional dimensions of, 383
internal structure of, 394
lateral structure of, 395
longitudinal structure of, 396
orientation
by granule bands, 13
Oocystis, 13
Cellulose nitrate, 247
Cellulose synthesis
in vitro, 225
kinetics of, 68
by protoplasts, 64
site of, 228
Cellulose tricarbanilates, 309
Celluloseless mutant, of *Acetobacter*, 346
Centric diatoms, chitin formation in, 186, 193
Chaetomorpha, 3
Chain elongation, of cellulose, 337
Charales, scale formation in, 143
Chitin, 237, 417
α-chitin, 207, 220, 417
structure of, 419
analogy to cellulose, 421
β-chitin, 420
appendages, in algae, 193
formation, 186
structure of, 421
cytological site of synthesis, 185
formation, in algae, 185
macromolecular organization of, 403
microfibril synthesis *in vitro*, 207
x-ray diffraction of, 186
Chitin chain formation, mechanisms of, 218
Chitin fibril, synthesizing apparatus, 193
Chitin microfibrils, biosynthesis of, 212
Chitin synthesis
rate of, 193
regulation of, 220
Chitin synthetase, 207
activation of, 210
dissociated forms of, 217
general properties, 208
kinetics of, 208

Chitin synthetase (*cont.*)
purification, 219
solubilization, 213
structure of, 214
Chitin-protein complexes, 422
structure of, 426
Chitodextrins, role in chitin synthesis, 209
Chitosome, 204
and chitin biosynthesis, 213
lipid composition of, 215
shell of, 215
subunits, 217
Chlamydomonas, 39, 57
Chlorococcalean algae, spike formation in, 204
Chrysochromulina, 129, 131, 133, 135, 141, 142
Chrysosporium lignorum, 474, 478
Citrate cycle, 342
Cladophora, 3
Cladosporium fluvum, 179, 180
Clavicaps, 461
Cochliobolus miyabeanus, 208
Colchicine, 157, 268
treatment of *Zea mays*, 95
Colinear molecules, aggregation of, 450
Collagen, 416
Comparative structure, of natural and synthetic fiber, 444
Complete cellulase system, 498
Composite fibers, 444
Composite filament, 436
Composite ribbon, in *Acetobacter*, 275
Congo Red, inhibition of chitin in algae, 199
Consecutive processes of polymerization and crystallization, 274
Controlled proteolysis, in chitin synthetase activation, 211
Conversion kinetics
of bacterial cellulose synthesis, 260
of cell wall cellulose, 254
Coprinus cinereus, 209
Core glycosylation, 366, 370, 377
in vitro, 366
Coriolus versicolour, 473
Cortical microtubules, in chitin formation, 194
Cotton fibers, detached, 226
Coumarin, 67, 157
inhibitor of cellulose, *Oocystis*, 11
inhibitor of cellulose synthesis, 356
Coupled program of development, 162
Crystalline core, of cellulose microfibril, 392
Crystallinity of cellulose, in wet state, *Acetobacter*, 280
Crystallite size, 383
Crystallite width, of *Valonia* cellulose, 385
Crystallization
of microfibrils, 204
of synthetic fibers, 440
Culture medium, influence on protoplast development, 151
Cunninghamella elegans, 208
Curving configurations of microfibrils, 93
Cyclotella, 186
Cyclotella cryptica, 186
Cyclotella meneghiniana, 186
Cytochalasin B, 157
Cytokinesis
in *Oocystis*, 4
in protoplasts, 153
Cytokinin, protoplast division, 155
Cytological studies, of cell wall regeneration, 64
Cytoplasmic coat, in chitin synthesis, 191

Darkfield electron microscopy, 506
Daucus carota, 76, 77
Degree of polymerization (DP)
of cellulose, 243
of cellulose from protoplast, 67
weight-average of, 313
α-Dextran, synthesis, 328
Dextran sucrase, 328
Dichlobenil, 159
Dichlorobenzonitrile (DCB), 157, 356
Digitonin, 376
dissociation of chitin synthetase, 208
role of dissociation of chitosomes, 217
Dihydroxyacetone, 341
Dimethylsulfoxide, 309
Diplodia zeae, 477
Dipolar interactions, of Calcofluor to cellulose, 275
Direct fibril forming processes, 439
Dispersed wall, 24
Division process in higher plant cells, 150
DMSO-paraformaldehyde, solvent for cellulose, 310
Dolichol lipid, 365
Dry spinning, 432

E-fracture face, in *Zea mays*, 90
EDTA, 354
effects on microfibril assembly, *Oocystis*, 10
Elemental fibril, 445
Elementary fibrils, 386
Elementary microfibril, 273
β-Elimination, 333, 373
Endocellulases, 470

Endogenous acceptor, 332
Endoglucanase, 459, 491
Endomembrane system, 143
Endoplasmic reticulum, 470
  changes induced by UDP-glucose, 173
  role in microfibril orientation, 101
  rough, site of chitosome synthesis, 215
Endopolygalacturonase, 30, 39
Endosperm, of castor bean, 376
Energy sources, for cell wall construction, 56
Ethylene, treatment of *Zea mays*, 101
Ethylene terephthalate, 441
Euphorbiaceae, 161
Exo-β-1, 3-glucanase, from *Fucus*, 56
Exocytes, 34
Exocytosis rate, *Pleurochrysis*, Golgi, 139
Exoglucanase, 460
Exoglycosidase, 372
Expanding walls, causes of controversy concerning, 25
Extrusion pores, in *Acetobacter*, 275

Fertilization, in *Fucus*, 54
Fibers, 429
Fibril formation, theory of, 450
Fibril forming processes, 452
Fibrils, 429
Fibroid coils, of chitin, 215
Filaments, synthetic, 432
First-order kinetics, of bacterial cellulose synthesis, 322
First-order rate constant, alpha and beta, 312
Fixation of radish root hairs, difficulties, 169
Flagella, vestigial, in *Glaucocystis*, 114
Fluid jet, 434
Fluorescein diacetate, 235
Fluorescence microscopy, 275
Fomes marginatus, 478
Free energy, 448
Freeze etch, *Oocystis*, 6, 7
Freeze fracture, of cotton fibers, 117
Freezing procedure, *Zea mays*, 95
Frequency of division, protoplasts, 163
Fructan sucrase, 328
Fruit ripening, role of cellulase in, 472
Fucan, in *Fucus*, 47
L-Fucose in Golgi scales, 131
Fucus, 39, 47, 461
Fucus distichus, 53, 54
Fucus vesiculosus, 49, 54
Fungal cell wall growth, 207
Fungal invasion, role of cellulase in, 473
Fusarium oxysporum, 473
Fusarium roseum, 478
Fusarium solani, 474
Fuzz fibers, of cotton, 116

Galactofucan, in *Fucus*, 47
Galactouronan, 39
GDP-glucose, 329
GDP-mannose, 367
Gel permeation chromatography, 300
Genetic control, of secretion, 140
Gerl complex, *Pleurochrysis*, Golgi, 140
Glaucocystis, 3, 4, 90, 106, 107
Globules, involvement in cellulose synthesis, 120
Glucan
  β-1,3 glucan, 226
    in *Fucus* eggs, 56
  β-1,4 glucan, in Golgi apparatus, 128
Glucan chains, polarity of, 403
Glucan synthesis, from grass tissues, 226
Glucanases
  β-glucanases, 459
    in higher plants, 459
    regulation of, 477
  β-1,3 glucanases, in higher plants, 462
  β-1,4 glucanases, 465
  role in cell growth and differentiation, 462
Glucokinase, 343
Gluconate, 344
Glucoproteins, 357
Glucosamine, 376
Glucose pyrophosphorylase, 343
Glucose-6-phosphate, 342
β-Glucosidase, 492
GluNac (*N*-acetylglucosamine), 372
Glutaraldehyde, pretreatment on freeze-fracture, 91
Glycerol, 341
Glycogen, 327
Glycolipids, 337
  of chitosomes, 215
Glycoprotein II, 367
Glycoproteins, 348
  as acceptors, 336
  plant, 363
Glycosyl transfer, 327
Glycosyl transferases, 366
Glycosylation, Golgi apparatus, site of, 127
Glycocalyx, 23
Glyoxysomal glycoproteins, 376
Golgi apparatus, 127
  periplasm production, *Oocystis*, 9
  scale biogenesis, 134
Golgi complex, role in β-glucan synthesis, 226

Golgi pathway, of non-celluloses, 128
Golgi vesicles, in radish root hairs, 181
Gossypium hirsutum, 116, 364
Gossypium herbaceum, 258
Granule bands, *Oocystis*, 14
Growth theory, for fibril formation, 451
α-D-Guluronic acid, in *Fucus*, 47

Helianthus annuus, 468
Hemicellulose, 404
Hexanoates, 341
Hexoses, 341
Hierarchical assembly of cellulose, 287
High resolution electron microscopy, 491
Hormonal control of mitosis in protoplasts, 155
Host–parasite relations, 163
Hydrogen bonding
  calcofluor to cellulose, 275
  intramolecular, 403
Hydrolases, destruction of synthetases, 329
Hydrophobic enzymes, 337
Hydrophobic forces, 414
Hymenomonas carterae, 137

Immunological localization of cellulases, 471
In vitro cellulose synthesis, 349
In vitro chitin synthesis, future developments, 219
In vitro glycoprotein products, 373
In vitro glycosylation, 368, 374
  higher plants, 364
In vitro mannosylation, 369
Infrared absorption spectra, Golgi scales, 134
Inhibitors
  of cell wall formation, 158
  of cellulose synthesis, 354
Intermediates in cellulose synthesis, 341
Intrinsic viscosity of cellulose, 246
Impomea tricolor, 467
Irreversible thermodynamics, 431

Kinetic force in microfibril movement, 203
Kinks of microfibrils, 397

Lacunae in *Glaucocystis*, 111
LALS program for analysis of polymer structure, 407
Lamellation of elongating walls, 27
Laminaria, 461
Laminaridextrans, 333
Lateral cell expansion, 467
Lectins, 363
Life cycle, *Oocystis*, 4
Ligase, 328
Lilium, 464
Lint fibers of cotton, 116
Lipid carriers
  in chitin assembly, 218
  in chitin synthesis, 218
Lipid intermediates, 238
  in cellulose synthesis, 358
Lipid-diphosphate cellobiose, 352
Lipid-linked acceptors, 337
Lipopolysaccharide envelope, 357
Loligo, 427
Lorica formation, 194
  inhibition by Calcofluor, 200
Low entropy, 430
Lycopersicum peruvianum, 180, 181
Lysozyme, 349

Magnesium, stimulation of cellulose synthesis, 232
Man-made fibers, 444
Mannans, 404
  β-1,4 mannans, 66
Mannose, 366
β-D-Mannuronic acid, in *Fucus*, 47
Mass accumulation, 303
  rate of, 305
Matrix polysaccharides, role in ordering of cell walls, 37
Megarhyssa complex, structure of, 423
Melt spinning, 432
Membrane flow, 127, 141
  role in microfibril orientation, 102
Membrane potential in cellulose synthesis, 237
Membrane-associated particles from protoplasts, 77
Mercerization, 417
Metabolic sequence in cellulose synthesis, 341
Metastable cellulose I, 274
Micrasterias, 36, 90, 142, 294
Micrasterias denticulata, 293
Microfiber systems
  natural, 429
  synthetic, 429
Microfibers, 429
Microfibril
  assembly, 3
  elongation, of chitin, 194
  shape, 391
  swollen state of, 393
  tear, freeze etch, 89

Microfibril bundles, in *Acetobacter*, 275
Microfibril deposition
  alteration of, 99
  kinetics of cytology, 69
  model, in *Glaucocystis*, 109
  unidirectional, in *Zea mays*, 91
Microfibril formation, study of, with deep etch technique, 71
Microfibril orientation, 141
  cytoplasmic control of, 121
  loss of Calcofluor, 203
  role of microtubules, *Oocystis*, 16
Microfibril synthesis in primary walls, 75
Microfibril tears, 93
Microfibrillar bands, 89
Microfilaments, visualized by freeze fracture, 96
Micronet procedure, for high resolution electron microscopy, 493
Microscopic techniques, in study of cell wall regeneration, 70
Microsporium gypseum, 462
*Mic*rotubule-associated proteins, 204
Microtubule–membrane interactions, in algae, 194
Microtubules
  role in cellulose assembly, 267
  roles in determination of microfibril orientation, 76
  role in microfibril orientation, 122
  visualized by freeze fracture, 97
Models
  of Calcofluor-disrupted cellulose assembly, in *Acetobacter*, 279
  of cellulose assembly in *Acetobacter*, 279
  of cellulose assembly in *Oocystis*, 16
  of cellulose structure, 411
  of chitin fibril synthesis, 203
  pattern of microfibril deposition, 112
  systems for cellulose biogenesis, 3, 275
Molecular weight, 243
  determination techniques, 246
  monodisperse, 265
Molecular weight average, 308
  of bacterial cellulose, 301
Molecular weight distribution, 243, 249
Molecules, visualization of, 506
Monoclinic unit cell, 410
Mortierella vinacea, 208
Mougeotia, 159
Mucor rouxii, 208, 237
Multienzyme complex, 358
  in cellulose synthesis, 351
Multinet growth, 119, 466
Multivesicular bodies, 217
Mung bean, hypocotyl, 32
Mutants
  blocked in cellulose synthesis, 344
  lacking proteases, 211
Myrothecium verrucaria, 477

N-acetylglucosamine, 185
β-*N*-acetyl glucosaminidase, 373
β-*N*-acetyl hexosaminidase, 372
Native band cellulose in *Acetobacter*, 292
Natural fibers, comparison with synthetic fibers, 442
Network formation in cell walls, 38
Neurospora crassa, 208, 209, 462
Neutral red, 235
Nicotiana glutinosa, 460
Non-cellulosic constituents of regenerating walls, 78
Non-reducing end of glucan chain, 492
Nonstatistical step polymerization, 249
Nucleation theory of fiber formation, 451
Nucleosides, effects on chitin synthesis, 209
Nucleotide sugars, role in chitin synthesis, 209
Number average, molecular weight and degree of polymerization, 244

Oligomeric polymerization, 317
Oligosaccharide processing, 366, 377
Oocystis, 3, 5–7, 9, 10, 13, 28, 89, 90, 106, 142, 294
  life cycle, culture, 4
  microtubules, cellulose, 15
Oocystis apiculata, 4
Oomycetes, 461
Optimal conditions, for *in vitro* cellulose synthesis, 238
Ordered granule hypothesis
  for cellulose synthesis, 274
  of Preston, 204
Ordered wall, 24
Oriented microfibrils in *Zea mays*, 91
Osmotic potential, 230
Osmotica, high molecular weight, influence on cellulose synthesis, 230
Ovipositor
  deproteinized, 424
  source of chitin, 422
Ovules of cotton, 229

P-fracture face, *Zea mays*, 89
Paired complexes, cellulose synthesis, *Oocystis*, 12
PAT–AG cytochemical test for cell walls, 27

Pathogenesis, role of cellulase in, 473
Pathways, polysaccharide incorporation, 51
Patterns, of microfibril orientation, bow-shaped, 31
Pea, membranes, 333
Pectin, 37, 395
Pectinase, in protoplast production, 62
Pellicle formation, in *Acetobacter*, 341
Pellicle balloons, of *Acetobacter*, 301
Pelvetia, 36
Penicillium verruculosum, 227
Pentose cycle, 342
Periplasm, *Oocystis*, 10
Periplasmic space, 358
Petunia, 77, 181
Phaseolus aureus, 217, 364
Phaseolus vulgaris, 364, 368, 470, 479
Phosphoglucomutase, 343
Phragmoplast formation, inhibition of, 162
Phycomyces blakesleeanus, 210, 216
Phytohemagglutinin, 367
Phytohormones, 478
Pinus, 385
Pinus densiflora, 384
Pisum sativum, 95, 364, 460
Pit fields
  influence on microfibril orientation, 94
  in *Zea mays*, 93
Plant cell wall assembly, comparison with *Oocystis*, 19
Plasma membrane
  freeze etch, *Oocystis*, 6
  invagination in chitin synthesis, 193
  linear particles, cellulose, 7
  particles, cellulose, 6
  role of Golgi in cellulose synthesis, 142
  site of cellulose synthesis, *Oocystis*, 17
  site for β-1,3-glucan synthesis, 181
  structure, *Oocystis*, 19
Plasma membrane pathway, of cellulose assembly, 128
Plasmalemma, control of mitosis, 162
Pleurochrysis, 4, 54, 129, 228
Poisson distribution, of molecular weight, 251
Poisson function, as a predictor of chain elongation, 304
Poisson polydispersity, 317
Poisson polymerization function, 301
Poisson polymerization process, 316
Polar organization of microfibrils, 197
Pollen tubes, callose formation, 180
Polydispersity of bacterial cellulose, 319
Polyethylene, 440
Polyethylene glycol 4000, protection of cellulose synthesis, 228
Polyisoprenoids, 365
Polymer flexibility, 448
Polymer fluid, 431, 434
Polymer introduction in *Acetobacter*, 305
Polymer melt, 435
Polymer solutions, theory of, 447
Polymeric precursor
  of bacterial cellulose, 320
  of cellulose, 301
Polymerization, of microfibrils, 204
Polymerization kinetics
  of bacterial cellulose synthesis, 299
  of cellulose synthesis, 250
Polymerization rate increase of cellulose, by Calcofluor, 282
Polymolecular mixtures, 244
Polyoxins, as inhibitors of chitin synthesis, 210
Polyporus adustrus, 478
Polyribosomes, associated with Golgi, 140
Polysaccharide
  decomposition 460
  as hormones, 23
  spontaneous assembly, 41
Populus euramericana, 384
Potassium, inhibition of cellulose synthesis, 237
Poteriochromonas stipitata, chitin lorica formation, 185, 194
Pressure potential gradient, of plasma membrane, 228
Prigogine theorem, 452
Primary cell wall, constituents, 65
Primary precursors, of cellulose, 342
Primary wall, 24
  expansion, 24
Primer, protein, 336
Primer requirements for β glucan synthesis, 327
Process, steady-state of fiber formation, 433
Protease, role in activation of chitin, 210
Protein core, 425
Protein glycosylation, 376
  in higher plants, 363
Proteoglycans, of protoplasts, 81
Protoplast budding, 82
Protoplast rotation, in *Glaucocystis*, 114
Protoplasts, 333
  advantages as systems for wall biogenesis, 62
  of tomato, 75
Prototheca zopfii, 357, 358
Pseudomonas solanacearum, 473
Pyrenochaeta lycopersici, 478
Pyrenochaeta terrestris, 477
Pyruvate, 341

Quantum leap hypothesis, 73
Quince slime, small fibrils of cellulose, 280

Radial microfibrils, of Golgi scales, 129
Radial precursor pool, *Pleurochrysis*, Golgi, 136
Ramie, 386
Raphanus sativus, 169
Rate of synthesis of cellulose, 113
Rate limitations, of cell-directed cellulose synthesis, 293
Reaction spinning, 438
Regenerating walls, cytology of, 82
Renewable natural polymers, 444
Resealing, of cut cotton fibers, 234
Reserve proteins, of legume seeds, 363
Resting cells, of *Acetobacter*, 341
Reversal, in direction of ribbon synthesis, 292
Reviews, cell wall structure and composition, 64
Rhamnogalacturonan II, 34
Rhizobium, 175, 179, 464, 474, 475
Rhizoctonia bataticola, 473
Rhizoids, of *Fucus*, 46
Rhizopus, 227
Rhizopus arrhizus, 460
Ribbon fragmentation of cellulose microfibrils, 502
Ribbons, of microfibrils, 495
Ricinus communis, 364
Roelofson, site of microfibril synthesis, 18
Root hairs, damaged, 171, 173
Rose cells, cultured microfibrils determined by x-ray diffraction, 67
Rosettes
  found in diverse systems, 90
  particle in cellulose synthesis, 89
  in pea stem, 92
Rotation centers, in the poles of *Glaucocystis*, 107
Rough endoplasmic reticulum, 375

Saccharomyces, 332
Saccharomyces carlsbergensis, 209
Saline medium, influence on division, 153
Saprolegnia monoica, 462
Scale exocytosis, *Pleurochrysis*, Golgi, 139
Scale fraction A, 129
Scale subcomponents, *Pleurochrysis*, 133
Scales, cellulosic, 128
Schizophyllum commune, 210
Sclerotinia sclerotiorum, 462
Secondary wall formation, 24
Self-assembly of cell walls, 39
Sequential incubation with cellulases, 503
Shaping zone of synthetic fiber formation, 434
Shish kebab, 439, 440
Simultaneous process of polymerization and crystallization, 274
Skimmia, 36
Skimmia japonica, 77
Solanum tuberosum, 77
  protoplasts, 77
Soluble cellulose precursors, 301
Spiral microfibrils in Golgi scales, 131
Splaying of microfibrils, 499
Spray spinning, 438
Starch, biosynthesis, 327
Statistical step polymerization, 248
Stereoregular polymerization, 442
Streptomyces, 227, 460
  cellulase, 66
Streptomyces cacaoi var. *asoensis*, 210
Streptomyces tendae, 210
Structure, of cellulose microfibrils, 383
Structure-controlled mechanism of cellulose, 257
Subcellular localization, of glycosylation, 375
Subelementary fibrils, 75, 273
Substrates, for cellulose synthesis, 341
Sulfated polysaccharides, in *Fucus*, 51
Sulfation
  of radial microfibrils, 135
  role of Golgi, 127
Sulfotransferase, in *Fucus*, 53
Summative helices, 121
Symbiosis, 474
Synthetic fibers, 432

Tactoids, 441
Tannic acid, terminal complex fixative, *Oocystis*, 14
Technologies, for synthetic fibers, 431
Terminal complexes
  cellulose, *Oocystis*, 10
  cellulose synthesis, comparison of, 18
  factors affecting, 90
  linear, *Oocystis*, 8
  in *Zea mays* mesocotyl, 88
Tetrahydrofuran, 310
Thalassiosira, 186
Thalassiosira fluviatilis, 192
Theoretical implications, of cell-directed self-assembly, 294
Thermodynamic stability, of cellulose, 416
Threshold concentration, of Calcofluor, on cellulose synthesis, 284
Tip growth, of microfibrils, 105
Tobacco mesophyll protoplasts, 151
Transglycosylation, 330

Transmembrane complex, in cellulose synthesis, 89, 329
Trichoderma koningi, 474
Trichoderma reesei, 489
Trichoderma verde, 227
Trichoderma viridae, 474, 477
Trifluoroacetic acid, 129
Triton X-100, 376
  inhibitor of cellulose synthesis, 354
  preparation of cytoskeleton, in algae, 194
Tunicamycin, 370, 377
Turgor, 162
  role in cellulose synthesis, 228, 234

UDP-glucose, 169, 225, 329, 344
  influence on root hairs, 175
UDP-glucose utilization, radish root hairs, 171
Ultimate fibrils, 429
Ultracryotomy, on ultrathin section, 28
Ultrastructural cytochemistry for study of walls, 27
Uniplanar orientation, of cellulose microfibrils, 393
Unit cell for cellulose, 406
Umprimed synthesis, 328
Uromyces phaseoli, 463
UV degradation, of Calcofluor cellulose, 278

Valonia, 3, 253, 257, 259, 383, 420
Valonia macrophysa, 383
Venturia inaequalis, 209
Vicia faba, 77
Vicilin, 367
Vinca rosea, 80
Viscometry, in molecular weight determination of cellulose, 246
Vitis vinifera, 177

Wall architecture, in cotton, 118
Wall bodies, in cellulose synthesis, 76
Wall extensibility, 466
Wall models, plywood, 30
Wall organization, three dimensional study of, 24
Wall subunits, positioning and ordering, 32
Walls, multilayered, 29
Wet spinning, 432
  processes, 437
Winding layer, of *Gossypium*, 116
Woven scales, *Chrysochromulina*, 142

X-ray crystallography, of cellulose, 227
X-ray diffraction, 383
  of fibrous polymers, 405
  of Golgi-cellulose, 132
X-ray spacings, Golgi cellulose, 133
Xylans, 404
  β-1,4 xylans, 66
Xylofucan, in *Fucus*, 47
Xyloglucans, 334
  in higher plants, 47
Xylose, in Golgi scales, 131

Z-stage, of scale unfolding, *Pleurochrysis*, 137
Zea mays, 87
Zeroth moment, of number distribution, 311
Zygotes, of *Fucus*, 45
Zymogenic state of chitin synthetase, 210

GPSR Compliance

The European Union's (EU) General Product Safety Regulation (GPSR) is a set of rules that requires consumer products to be safe and our obligations to ensure this.

If you have any concerns about our products, you can contact us on ProductSafety@springernature.com

In case Publisher is established outside the EU, the EU authorized representative is:

Springer Nature Customer Service Center GmbH
Europaplatz 3
69115 Heidelberg, Germany

Batch number: 10838292

Printed by Printforce, the Netherlands